D1759055

Loudspeaker and
Headphone Handbook

Loudspeaker and Headphone Handbook

Third Edition

Edited by

John Borwick

With specialist contributors

Focal Press

OXFORD AUCKLAND BOSTON JOHANNESBURG MELBOURNE NEW DELHI

Focal Press
An imprint of Butterworth-Heinemann
Linacre House, Jordan Hill, Oxford OX2 8DP
225 Wildwood Avenue, Woburn, MA 01801–2041
A division of Reed Educational and Professional Publishing Ltd

 A member of the Reed Elsevier plc group

First published 1988
Second edition 1994
Reprinted 1997, 1998
Third edition 2001

British Library Cataloguing in Publication Data
Loudspeaker and headphone handbook. – 3rd ed.
  1. Headphones – Handbooks, manuals, etc.   2. Loudspeakers –
  Handbooks, manuals, etc.
  I. Borwick, John
  621.38284

Library of Congress Cataloguing in Publication Data
Loudspeaker and headphone handbook/edited by John Borwick, with
specialist contributors – 3rd ed.
    p. cm
Includes bibliographical references and index.
ISBN 0-240-51578-1 (alk. paper)
1. Headphones – Handbooks, manuals etc.   2. Loudspeakers –
Handbooks, manuals, etc.   I. Borwick, John.

TK5983.5 .L68
621.382′84–dc21   00–069177

ISBN   0 240 51578 1

Typeset by Florence Production Ltd, Stoodleigh, Devon
Printed and bound in Great Britain

FOR EVERY TITLE THAT WE PUBLISH, BUTTERWORTH-HEINEMANN
WILL PAY FOR BTCV TO PLANT AND CARE FOR A TREE.

Contents

Preface to the third edition

There have been significant advances in technology during the six years since the publication of the second edition of this book. These have been driven by increasingly more sophisticated computer-aided design systems and digital audio processing, and have affected research procedures, the loudspeakers themselves and the range of user applications.

My starting point for preparing this new edition was to approach the original authors and invite them to revise their texts to bring them fully up to date, and this is mainly what has happened. However, there are exceptions. For example, the authors of Chapters 1 and 2 (Ford and Kelly) were planning to retire and asked to be excused. These chapters have therefore been rewritten from scratch by two new authors (Holland and Watkinson). As well as updating the coverage of their respective topics, they have introduced new ideas currently of high interest – mutual coupling, for example, in Chapter 1 and specialized types of drive units to be used in active designs for very small enclosures in Chapter 2.

The death of Peter Baxandall has necessitated a different editorial approach for Chapter 3 but everyone I spoke to described his comprehensive coverage of electrostatic loudspeakers as beyond improvement. Accordingly this chapter is virtually unaltered except for some suggestions kindly supplied by Peter Walker of Quad.

By contrast, Chapter 4 is an entirely new contribution by Graham Bank, Director of Research, outlining the theory, construction and wide range of potential applications of the most interesting innovation of recent years, the Distributed Mode Loudspeaker.

Another death, that of Glyn Adams, affected the decision about Chapter 7 on The Room Environment. Once again, the existing text covers the basic theory so comprehensively that I have left it almost unchanged but introduced an add-on Chapter 8 to cover recent developments in the application of acoustic theory to listening room design – contributed by Philip Newell. My own Chapter 12 on Measurements, a field in which dramatic changes have occurred, has been thoroughly revised and updated by Julian Wright. Carl Poldy has reworked his seminal Chapter 14 on Headphones and added new material on computer simulation techniques.

This edition has inevitably grown in size but new text, illustrations and references add up to the most comprehensive handbook on loudspeakers and headphones currently available.

John Borwick

Contributors

Graham Bank earned a degree in Applied Physics from the University of Bradford in 1969 and, after a period as a Research Assistant, gained an MSc in 1973. On completing a research programme in 1997, he was awarded a PhD in Electrical Systems Engineering at the University of Essex. He is currently the Director of Research at NXT, researching flat panel loudspeaker technology for both the consumer and professional markets.

John Borwick graduated from Edinburgh University with a BSc (Physics) degree and, after 4 years as Signal Officer in the RAF and 11 years at the BBC as studio manager and instructor, helped to set up and run for 10 years the Tonmeister degree course at the University of Surrey. He was Secretary of the Association of Professional Recording Services for many years and is now an Honorary Member. He is also a Fellow of the AES where he served terms as Chairman (British Section) and Vice President (Europe).

Martin Colloms graduated from the Regent Street Polytechnic in 1971 and is a Chartered Engineer and a Member of the Institution of Electrical Engineers. He co-founded Monitor Audio in 1973 and is now a freelance writer and audio consultant.

Laurie Fincham received a BSc degree in Electrical Engineering from Bristol University and served a 2-year apprenticeship with Rediffusion before working for Goodmans Loudspeakers and Celestion. In 1968 he joined KEF Electronics as Technical Director where, during his 25-year stay, he pioneered Fast Fourier Transform techniques. He emigrated to California in 1993, first as Senior VP of Engineering at Infinity Systems and since 1997 as Director of Engineering for the THX Division of Lucasfilm Ltd. He is a Fellow of the AES and active on IEC standards groups, while retaining a keen interest in music, playing acoustic bass with a jazz group.

Mark R. Gander, BS, MSEE; Fellow AES; Member ASA, IEEE, SMPTE, has been with JBL since 1976 holding positions as Transducer Engineer, Applications Engineer, Product Manager, Vice-President Marketing and Vice-President Engineering, and is currently Vice-President Strategic Development for JBL Professional. Many of his papers have been published in the *AES Journal*. He has served as a Governor of the Society and edited Volumes 3 and 4 of the AES Anthology *Loudspeakers*.

Keith Holland is a lecturer at the Institute of Sound and Vibration Research, University of Southampton, where he was awarded a PhD for a thesis on horn loudspeakers. Dr Holland is currently involved in research into many aspects of noise control and audio.

Peter Mapp holds degrees in Applied Physics and Acoustics and has a particular interest in speech intelligibility measurement and prediction. He has been a visiting lecturer at Salford University since 1996 and is Principal of Peter Mapp Associates, where he has been responsible for the design of over 380 sound systems, both in the UK and internationally.

Philip Newell has 30 years' experience in the recording industry, having been involved in the design of over 200 studios including the Manor and Townhouse Studios. He was Technical and Special Projects Director of the Virgin Records recording division and designed the world's first purpose-built 24-track mobile recording vehicle. He is a member of both the Institute of Acoustics (UK) and the Audio Engineering Society.

Sean Olive is Manager of Subjective Evaluation for Harman International where he oversees all listening tests and psychoacoustic research on Harman products. He has a Master's Degree in Sound Recording from McGill University and is a Fellow and past Governor of the Audio Engineering Society.

Carl Poldy graduated with a BSc degree in physics from the University of Nottingham and gained a PhD in solid state physics from the University of Durham. After five years at the Technical University of Vienna, lecturing and researching into the magnetism of rare earth intermetallic alloys, he spent 15 years with AKG Acoustics, Vienna before joining PSS Philips Speaker Systems in 1993.

Floyd E. Toole studied electrical engineering at the University of New Brunswick, and at the Imperial College of Science and Technology, University of London where he received a PhD. He spent 25 years at Canada's National Research Council, investigating the psychoacoustics of loudspeakers and rooms. He has received two Audio Engineering Society Publications Awards, an AES Fellowship and the AES Silver Medal. He is a Member of the ASA and a past President of the AES. Since 1991 he has been Corporate Vice-President Acoustical Engineering for Harman International.

John Watkinson is a fellow of the AES and holds a Masters degree in Sound and Vibration. He is a consultant and has written twenty textbooks.

John Woodgate is an electronics consultant based in Essex, specializing in standards and how to comply with them, particularly in the field of audio.

Julian Wright is Head of Research at Celestion International Limited and Director of Information Technology at Wordwright Associates Limited. He received a BSc degree in Electroacoustics in 1984 and an MSc in Applied Acoustics in 1992 from Salford University, and is a fellow of the Institute of Acoustics. Currently his main professional interests are in Finite Element Modelling and software design. In his leisure time he is an active musician and genealogist.

1 Principles of sound radiation

Keith R. Holland

1.1 Introduction

Loudspeakers are transducers that generate sound in response to an electrical input signal. The mechanism behind this conversion varies from loudspeaker to loudspeaker, but in most cases involves some form of motor assembly attached to a diaphragm. The alternating force generated by the motor assembly, in response to the electrical signal, causes the diaphragm to vibrate. This in turn moves the air in contact with the diaphragm and gives rise to the radiation of sound. This chapter is concerned with the acoustic part of that transduction mechanism; that is, the radiation of sound by the vibrating diaphragm.

The radiation of sound by vibrating surfaces is a common part of everyday life. The sound of footsteps on a wooden floor or the transmission of sound through a closed window are typical examples. However, despite the apparent physical simplicity of sound radiation, for all but the most basic cases, its analysis is far from straightforward[1]. The typical loudspeaker, consisting of one or more vibrating diaphragms on one side of a rectangular cabinet, represents a physical system of sufficient complexity that, to this day, accurate and reliable predictions of sound radiation are rare, if not non-existent. One need not be deterred, however, as much may be learned by studying simpler systems that possess some of the more important physical characteristics of loudspeakers.

To begin to understand the mechanism of sound radiation, it is necessary to establish the means by which a sound 'signal' is transported from a source through the air to our ears. To this end, Section 1.2 begins with a description of sound propagation, within which many of the concepts and terms found in the remainder of the chapter are defined. Those readers already familiar with the mechanisms of sound propagation may wish to skip over this section. The sound radiated by idealistic, simple sources such as the point monopole source is then described in Section 1.3, where it is shown how a number of these simple sources may be combined to form more complex sources such as idealized loudspeaker diaphragms. These concepts are then further developed in Section 1.4 to include the radiation of sound from multiple sources, such as multi-driver and multi-channel loudspeaker systems. In Section 1.5, the limitations of the idealized loudspeaker model are discussed along with the effects of finite cabinet size and the presence of walls on sound radiation. Section 1.6 deals with the radiation of sound by horn loudspeakers, and Section 1.7 ends the chapter with a brief discussion of non-linear sound propagation.

Many of the concepts described may be applied to the radiation of sound in media other than air; indeed, many acoustics textbooks treat sound propagation in air as a special case, preferring the use of the term 'fluid', which includes both gases and liquids. For the sake of clarity, in the analysis that follows, the propagating medium will be assumed to be air; this is a book about loudspeakers after all!

1.2 Acoustic wave propagation

Acoustic waves are essentially small local changes in the physical properties of the air which propagate through it at a finite speed. The mechanisms involved in the propagation of acoustic waves can be described in a number of different ways depending upon the particular cause, or source of the sound. With conventional loud-speakers that source is the movement of a diaphragm, so it is appropriate here to begin with a description of sound propagation away from a simple moving diaphragm.

The process of sound propagation is illustrated in Fig. 1.1. For simplicity, the figure depicts a diaphragm mounted in the end of a uniform pipe, the walls of which constrain the acoustic waves to propagate in one dimension only. Before the diaphragm moves (Fig. 1.1(a)), the pressure in the pipe is the same everywhere and

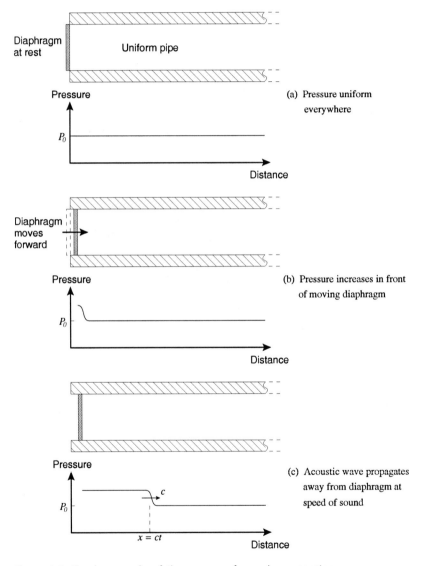

Figure 1.1. Simple example of the process of sound propagation.

equal to the static (atmospheric) pressure P_0. As the diaphragm moves forwards (Fig. 1.1(b)), it causes the air in contact with it to move, compressing the air adjacent to it and bringing about an increase in the local air pressure and density. The difference between the pressure in the disturbed air and that of the still air in the rest of the pipe gives rise to a force which causes the air to move from the region of high pressure towards the region of low pressure. This process then continues forwards and the disturbance is seen to propagate away from the source in the form of an acoustic wave. Because air has mass, and hence inertia, it takes a finite time for the disturbance to propagate through the air; a disturbance 'leaves' a source and 'arrives' at another point in space some time later (Fig. 1.1(c)). The rate at which disturbances propagate through the air is known as the *speed of sound* which has the symbol 'c'; and after a time of t seconds, the wave has propagated a distance of $x = ct$ metres. For most purposes, the speed of sound can be considered to be constant and independent of the particular nature of the disturbance (though it does vary with temperature).

A one-dimensional wave, such as that shown in Fig. 1.1, is known as a *plane wave*. A wave propagating in one direction only (e.g. left-to-right) is known as a *progressive wave*.

1.2.1 Frequency, wavenumber and wavelength

Most sound consists of alternate positive and negative pressures (above and below the static pressure) so, in common with other fields of study where alternating signals are involved, it is useful to think of sounds in terms of their frequency content. Fourier analysis tells us that any signal can be constructed from a number of single-frequency signals or sine-waves[2]. Therefore, if we know how an acoustic wave behaves over a range of frequencies, we can predict how it would behave for any signal. Usefully, using the sine-wave as a signal greatly simplifies the mathematics involved so most acoustic analysis is carried out frequency-by-frequency (see Appendix).

An acoustic wave can be defined as a function of time and space. At a fixed point in space, through which an acoustic wave propagates, the acoustic pressure would be observed to change with time; it is this temporal variation in pressure that is detected by a microphone and by our ears. However, the acoustic wave is also propagating through the air so, if we were able to freeze time, we would observe a pressure which changed with position. It is interesting to note that if we were to 'ride' on an acoustic wave at the speed of sound (rather like a surfer) we could, under certain conditions such as the one-dimensional wave illustrated in Fig. 1.1, observe no variation in pressure at all.

For a single-frequency acoustic wave, both the temporal and the spatial variations in pressure take the form of sine-waves. Whereas the use of the term *frequency* to quantify the number of alternate positive and negative 'cycles' that occur in a given time is widespread, the spatial equivalent, *wavenumber*, is less common. The wavenumber is defined as the number of alternate positive and negative cycles that occur in a given distance; it has the units of radians per metre and usually has the symbol k. The direct temporal counterpart to wavenumber is *radial frequency* which has units of radians per second and the symbol ω, As there are 2π radians in a cycle, the relationship between frequency (f) in Hz (Hertz, or cycles-per-second as it used to be called) and radial frequency is a simple one: $\omega = 2\pi f$.

The temporal and spatial frequencies are linked by the speed of sound. If a cycle of an acoustic wave takes t seconds to complete, the wave would travel ct metres in that time so the relationship between radial frequency and wavenumber is simply

$$k = \omega/c \qquad (1.1)$$

Acoustic *wavelength* is the distance occupied by one cycle of a single-frequency acoustic wave (the spatial equivalent of time period). It has the units of metres and usually has the symbol λ. The relationships between frequency, wavenumber and wavelength are

$$c = f\lambda \tag{1.2}$$

and

$$k = 2\pi/\lambda \tag{1.3}$$

1.2.2 Mathematical description of an acoustic wave

The study of the radiation and propagation of sound is greatly simplified through the use of mathematical tools such as complex numbers. Although most of this chapter (and others in this book) can be read and understood with little knowledge of mathematics, it is perhaps unfortunate that in order to get the most out of the text, the reader should at least be aware of these mathematical tools and their uses. The Appendix at the end of this chapter contains a brief explanation of some of the mathematics that is to follow; readers who are unfamiliar with complex numbers may find this useful.

The pressure in an acoustic wave varies as a function of space and time. For a one-dimensional (plane) wave at a single frequency, the pressure at any point in time t and at any position x may be written

$$\hat{p}(x, f, t) = \hat{P}e^{j(\omega t - kx)} \tag{1.4}$$

where $\hat{P}$ is the amplitude of the sine wave, $(\omega t - kx)$ is the phase and the 'f' is to remind us that the equation refers to a single frequency. The $^\wedge$ symbol over the variables indicates that they are complex, i.e. they possess both amplitude and phase. The complex exponential representation is a useful mathematical tool for manipulating periodic functions such as sine-waves (see Appendix); it should be borne in mind, though, that the actual pressure at time t and position x is purely real. It is often useful in acoustics to separate the time-dependent part from the spatially dependent part of the phase term; indeed, very often the time-dependent part is assumed known and is left out of the equations until it is required. For the sake of clarity, a single-frequency plane progressive wave may then be described as

$$\hat{p}(x, f) = \hat{P}e^{-jkx} \tag{1.5}$$

Equation (1.5) gives us a compact mathematical representation of a plane acoustic wave such as occurs at low frequencies in pipes and ducts, and at all frequencies at large distances away from sources in the free-field.

1.2.3 Wave interference and the standing wave

The description so far has been limited to waves propagating in one direction only. These waves only occur in reality under free-field conditions and where there is only one sound source. When reflective objects are present, or there are multiple sources, more than one acoustic wave may propagate through a given point at the same time. If the waves have the same frequency, such as is the case for reflections, an interference field is set up. An important point to note about the interference between waves is that a wave remains unchanged when another wave interferes with it (a sort of 'non-interference'). This means that the pressures in each of the waves can simply be summed to yield the total sound field; a process known as linear superposition. By way of example, the acoustic pressure within a uniform pipe due to a forward-propagating wave and its reflection from the end of the pipe takes the form

$$\hat{p}(x, f) = \hat{P}e^{-jkx} + \hat{Q}e^{jkx} \tag{1.6}$$

where $\hat{P}$ and $\hat{Q}$ are the amplitudes of the forward and backward propagating waves respectively (the − and + signs in the exponents indicate the direction of propagation). The sound field described in equation (1.6) represents what is known as a *standing-wave field*, a pattern of alternate areas of high- and low-pressure amplitude which is fixed in space. The difference in acoustic pressure between the areas of high pressure and areas of low pressure depends upon the relative magnitudes of $\hat{P}$ and $\hat{Q}$, and is known as the standing wave ratio.

In the special case when $\hat{P} = \hat{Q}$, equation (1.6) can be simplified to yield

$$p(x,f) = 2\hat{P} \cos{(kx)} \tag{1.7}$$

which is known as a pure standing wave. The standing-wave ratio is infinite for a pure standing wave, as the acoustic pressure is zero at positions where $kx = \pi/2$, $3\pi/2$, etc.

Standing-wave fields do not only exist when two waves are travelling in opposite directions. Figure 1.8 in Section 1.4 shows a two-dimensional representation of the sound field radiated by two sources radiating the same frequency; the pattern of light and dark regions is fixed in space and is a standing-wave field. A standing-wave field cannot exist if the interfering waves have different frequencies.

1.2.4 Particle velocity

The description of sound propagation in Section 1.2 mentioned the motion of the air in response to local pressure differences. This localized motion is often described in terms of acoustic particle velocity, where the term 'particle' here refers to a small quantity of air that is assumed to move as a whole. Although we tend to think of a sound field as a distribution of pressure fluctuations, any sound field may be equally well described in terms of a distibution of particle velocity. It should be borne in mind, however, that acoustic particle velocity is a vector quantity, possessing both a magnitude and a direction; acoustic pressure, on the other hand, is a scalar quantity and has no direction.

A particle of air will move in response to a difference in pressure either side of it. The relationship between acoustic pressure and acoustic particle velocity can therefore be written in terms of Newton's law of motion: force is equal to mass times acceleration. The force in this case comes from the rate of change of pressure with distance (the pressure gradient), the mass is represented by the local static air density and the acceleration is the rate of change of particle velocity with time:

Force = mass × acceleration

therefore

$$\frac{\partial p}{\partial x} = -\rho\frac{\partial u}{\partial t} \tag{1.8}$$

where p is acoustic pressure, x is the direction of propagation, ρ is the density of air and u is the particle velocity (the minus sign is merely a convention: an increase in pressure with increasing distance causes the air to accelerate backwards). For a single-frequency wave, the particle velocity varies with time as $u \propto e^{j\omega t}$, so $\partial u/\partial t = j\omega u$, and equation (1.6) may be simplified thus:

$$\hat{u}(x,f) = \frac{-1}{j\omega\rho}\frac{\partial\hat{p}}{\partial x} \tag{1.9}$$

For a plane, progressive sound wave, the pressure varies with distance as $\hat{p} \propto e^{-jkx}$, so $\partial\hat{p}/\partial x = -jk\hat{p}$ and equation (1.9) becomes

$$\hat{u}(x,f) = \frac{\hat{p}(x,f)}{\rho c} \tag{1.10}$$

The ratio $\hat{p}/u = \rho c$ for a plane progressive wave is known as the characteristic impedance of air (see Section 1.2.6).

The concept of a particle velocity proves very useful when considering sound radiation; indeed, the particle velocity of the air immediately adjacent to a vibrating
surface is the same as the velocity of the surface.

Particle velocity is usually very small and bears no relationship to the speed of
sound. For example, at normal speech levels, the particle velocity of the air at our
ears is of the order of 0.1 mm/s and that close to a jet aircraft is nearer 50 m/s, yet
the sound waves propagate at the same speed of approximately 340 m/s in both cases
(see Section 1.7). Note that, for air disturbances small enough to be called sound,
the particle velocity is always small compared to the speed of sound.

1.2.5 Sound intensity and sound power

Sound waves have the capability of transporting energy from one place to another,
even though the air itself does not move far from its static, equilibrium position. A
source of sound may generate acoustic power which may then be transported via
sound waves to a receiver which is caused to vibrate in response (the use of the word
'may' is deliberate; it is possible for sound waves to exist without the transportation
of energy). The energy in a sound wave takes two forms: kinetic energy, which is
associated with particle velocity, and potential energy, which is associated with pressure. For a single-frequency sound wave, the sound intensity is the product of the
pressure and that proportion of particle velocity that is in phase with the pressure,
averaged in time over one cycle.

For a plane progressive wave, the pressure and particle velocity are wholly in phase
(see equation (1.10)), and the sound intensity (I) is then

$$I = \left\langle \hat{p} \times \frac{\hat{p}}{\rho c} \right\rangle = \frac{|\hat{p}|^2}{2\rho c} \tag{1.11}$$

where $\langle \rangle$ denotes a time average and $|\hat{p}|$ denotes the amplitude or modulus of the
pressure. It is worth noting here that $|\hat{p}|^2/2$ is the mean-squared pressure which is
the square of the rms pressure.

Sound intensity may also be defined as the amount of acoustic power carried by
a sound wave per unit area. The relationship between sound intensity and sound
power is then

Sound power = sound intensity × area

or

$$W = \int_S I \, dS \tag{1.12}$$

The total sound power radiated by a source then results from integrating the sound
intensity over any imagined surface surrounding the source.

Interested readers are referred to reference 3 for more detailed information
concerning sound intensity and its applications.

1.2.6 Acoustic impedance

Equation (1.8) shows that there is a relationship between the distribution of pressure in a sound field and the distribution of particle velocity. The ratio of pressure
to particle velocity at any point (and direction) in a sound field is known as acoustic
impedance and is very important when considering the sound power radiated by a
source. Acoustic impedance usually has the symbol Z and is defined as follows:

$$\hat{Z}(f) = \frac{\hat{p}(f)}{\hat{u}(f)} \tag{1.13}$$

Note that acoustic impedance is defined at a particular frequency; no time-domain equivalent exists; the relationship between the instantaneous values of pressure and particle velocity is not generally easy to define except for single-frequency waves.

Acoustic impedance, like other forms of impedance such as electrical and mechanical, is a quantity that expresses how difficult the air is to move; a low value of impedance tells us that the air moves easily in response to an applied pressure (low pressure, high velocity), and a high value, that it is hard to move (high pressure, low velocity).

1.2.7 Radiation impedance

Both acoustic intensity and acoustic impedance can be expressed in terms of pressure and particle velocity. By combining the definitions of intensity and impedance, it is possible to define acoustic intensity in terms of acoustic impedance and particle velocity

$$I_S = \frac{|\hat{u}|^2}{2} \, \text{Re}\{\hat{Z}\} \tag{1.14}$$

where $\text{Re}\{\hat{Z}\}$ denotes just the real part of the impedance. Given that the particle velocity next to a vibrating surface is equal to the velocity of the surface, and that acoustic power is the integral of intensity over an area, it is possible to apply equations (1.12) and (1.14) to calculate the acoustic power radiated by a vibrating surface. Acoustic impedance evaluated on a vibrating surface is known as radiation impedance; for a given vibration velocity, the higher the real part of the radiation impedance, the higher the power output. The imaginary part of the radiation impedance serves only to add reactive loading to a vibrating surface; this loading takes the form of added mass or stiffness.

The real part of the radiation impedance, when divided by the characteristic impedance of air (ρc), is often termed the *radiation efficiency* as it determines how much acoustic power is radiated per unit velocity. If the radiation impedance is purely imaginary (the real part is zero), the sound source can radiate no acoustic power, although it may still cause acoustic pressure. An example of this type of situation is a loudspeaker diaphragm mounted between two rigid, sealed cabinets. The air in the cabinets acts like a spring (at low frequencies at least) so the radiation impedance on the surfaces of the diaphragm is a negative reactance and, although the motion of the diaphragm gives rise to pressure changes within the cabinets, no acoustic power is radiated.

1.2.8 Sound pressure level and the decibel

Many observable physical phenomena cover a truly enormous dynamic range, and sound is no exception. The changes in pressure in the air due to the quietest of audible sounds are of the order of $20 \, \mu\text{Pa}$ (20 micro-Pascals), that is 0.00002 Pa, whereas those that are due to sounds on the threshold of ear-pain are of the order of 20 Pa, a ratio of one to one million. When the very loudest sounds, such as those generated by jet engines and rockets, are considered, this ratio becomes nearer to one to 1000 million! Clearly, the usual, linear number system is inefficient for an everyday description of such a wide dynamic range, so the concept of the Bel was introduced to compress wide dynamic ranges into manageable numbers. The Bel is simply the *logarithm of the ratio of two powers*; the decibel is one tenth of a Bel.

Acoustic pressure is measured in Pascals (Newtons per square metre), which do not have the units of power. In order to express acoustic pressure in decibels, it is therefore necessary to square the pressure and divide it by a squared reference pressure. For convenience, the squaring of the two pressures is usually taken outside the logarithm (a useful property of logarithms); the formula for converting from acoustic pressure to decibels can then be written

$$\text{decibels} = 10 \times \log_{10}\left\{\frac{p^2}{p_0^2}\right\} = 20 \times \log_{10}\left\{\frac{p}{p_0}\right\} \tag{1.15}$$

where p is the acoustic pressure of interest and p_0 is the reference pressure. When 20 μPa is used as the reference pressure, along with the rms value of p, sound pressure expressed in decibels is referred to as sound pressure level (SPL).

The acoustic dynamic range mentioned above can be expressed in decibels as sound pressure levels of 0 dB for the quietest sounds, 120 dB for the threshold of pain and 180 dB for the loudest sounds.

Decibels are also used to express electrical quantities such as voltages and currents, in which case the reference quantity will depend upon the application. When dealing with quantities that already have the units of power, such as sound power, the squaring inside the logarithm is unnecessary. Sound power level (SWL) is defined as acoustic power expressed in decibels relative to 1 pW (pico-Watts, or 1×10^{-12} Watts).

1.3 Sources of sound

Sound may be produced by any of a number of different mechanisms, including turbulent flow (e.g. wind noise), fluctuating forces (e.g. vibrating strings) and volume injection (e.g. most loudspeakers). The analysis of loudspeakers is concerned primarily with volume injection sources, although other source types can be important in some designs.

1.3.1 The point monopole

The simplest form of volume injection source is the point monopole. This idealized source consists of an infinitesimal point at which air is introduced and removed. Such a source can be thought of as a pulsating sphere of zero radius, so can never be realized in practice, but it is a useful theoretical tool nevertheless.

The sound radiated by a point monopole is the same in all directions (omnidirectional) and, under ideal free-field acoustic conditions, consists of spherical waves propagating away from the source. As the radiated wave propagates outwards, the spherical symmetry of the source, and hence the radiated sound field, dictates that the acoustic pressure must reduce as the acoustic energy becomes 'spread' over a larger area. This decrease in pressure with increasing radius is known as the 'inverse square law' (because the sound intensity, which is proportional to the square of pressure, reduces as the square of the distance from the source).

The spherical sound field radiated by a point monopole at a single radial frequency ω, takes the form

$$\hat{p}(r,f) \propto \frac{e^{-jkr}}{r} \tag{1.16}$$

where r is the distance from the source and k is the wavenumber. One should note that according to equation (1.16), the acoustic pressure tends to infinity as the radius tends to zero – an impossible situation which further relegates the concept of the point monopole to the realms of theory.

The constant of proportionality in equation (1.16) is a function of the strength of the monopole, which can be quantified in terms of the 'rate of injection of air' or volume velocity, which has units of cubic metres per second. The relationship between the radiated pressure field and the volume velocity of a point monopole can be shown[4] to be

$$\hat{p}(r,f) = \frac{j\rho ck\hat{q}\,e^{-jkr}}{4\pi r} \tag{1.17}$$

where $\hat{q}$ is the volume velocity.

The particle velocity in a spherically expanding wave field is in the radial direction and can be calculated from equations (1.8) and (1.16) with the substitution of r for x:

$$\hat{u}(r,f) = \left(1 - \frac{j}{kr}\right)\frac{\hat{p}(r,f)}{\rho c} \tag{1.18}$$

The part of the particle velocity that is in phase with the pressure is represented by the left term in the brackets only, so the sound intensity at any radius r is then

$$I_S(r) = \left\langle (p(r,f) \times \frac{p(r,f)}{\rho c}\right\rangle = \frac{|\hat{p}|^2}{2\rho c} \tag{1.19}$$

which is identical to equation (1.11) for a plane progressive wave. Since the pressure falls as $1/r$, the sound intensity is seen to fall as $1/r^2$, and although the 'in-phase' part of the particle velocity also falls as $1/r$, the part in quadrature (90°) falls as $1/r^2$.

The sound power output of the monopole can be deduced by integrating the sound intensity over the surface of a sphere (of any radius) surrounding the source thus,

$$W_S = \int_S \frac{|\hat{p}|^2}{2\rho c}\,dS = \frac{\rho ck^2|\hat{q}|^2}{8\pi} \tag{1.20}$$

1.3.2 The monopole on a surface

Equation (1.16) shows that the sound field radiated by a point monopole under ideal, free-field conditions consists of an outward-propagating spherical wave. Under conditions other than free-field, though, estimating the sound field radiated by a point monopole is more difficult. However, an equally simple sound field, of special importance to loudspeaker analysis, is radiated by a point monopole mounted on a rigid, plane surface. Under these conditions, all of the volume velocity of the monopole is constrained to radiate sound into a hemisphere, instead of a sphere. The monopole thus radiates twice the sound pressure into half of the space. The sound field radiated by a point monopole mounted on a rigid, plane surface is therefore

$$\hat{p}(r,f) = \frac{j\rho ck\hat{q}\,e^{-jkr}}{2\pi r} \tag{1.21}$$

but it exists on only one side of the plane. It is interesting to substitute this value of pressure into the first half of equation (1.20). Doing this we find that, although the surface area over which the integration is made is halved, the value of $|\hat{p}|^2$ is four times greater; the sound power output of the monopole is now double its value under free-field conditions (see Section 1.4).

1.3.3 The point dipole

If two monopoles of equal volume velocity but opposite phase are brought close to each other, the result is what is known as a dipole. Figure 1.2 illustrates the geometry of the dipole. On a plane between the two monopoles (y-axis in Fig. 1.2), on which all points are equidistant from both sources, the sound field radiated by one

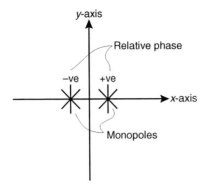

Figure 1.2. The dipole realized as two monopoles of opposite phase.

monopole completely cancels that from the other and zero pressure results. Along a line through the monopoles (*x*-axis), the nearest monopole to any point will radiate a slightly stronger sound field than the furthest one, due to the 1/*r* dependence of the monopole sound field, and a small (compared to that of a single monopole) sound pressure remains. The sound field along the *x*-axis has a different polarity on one side of the dipole to the other. Assuming that the distance between the monopoles is small compared to the distance to the observation point, the sound field takes the form

$$\hat{p}(r, \theta) = \frac{\hat{D} \cos(\theta)\, e^{-jkr}}{r} \tag{1.22}$$

where $\hat{D}$ is a function of frequency and the spacing between the monopoles and is known as the dipole strength. The sound field in equation (1.22) is often described as having a 'figure-of-eight' directivity pattern. As one might expect with two mono-poles nearly cancelling one another out, the sound power radiated by a dipole is considerably lower than that of one monopole in isolation.

A practical example of a source with near-dipole type radiation (at low frequencies at least) is a diaphragm exposed on both sides, such as found in most electrostatic loudpeakers.

1.3.4 Near- and far-fields

A point monopole source radiates a spherically symmetric sound field in which the acoustic pressure falls as the inverse of the distance from the source ($p \propto 1/r$). On the other hand, equation (1.18) tells us that the particle velocity does not obey this inverse law, but instead falls approximately as 1/*r* for $kr > 1$ and $1/r^2$ for $kr < 1$. The region beyond $kr = 1$, where both the pressure and particle velocity fall as 1/*r*, is known as the hydrodynamic far-field. In the far-field, the propagation of sound away from the source is little different from that of a plane progressive wave. The region close to the source, where $kr < 1$ and the particle velocity falls as $1/r^2$, is known as the hydrodynamic near-field. In this region, the propagation of sound is hampered by the curvature of the wave, and large particle velocities are required to generate small pressures. It is important to note that the extent of the hydrodynamic near-field is frequency dependent.

The behaviour of the sound field in the hydrodynamic near-field can be explained as follows. An outward movement of the particles of air, due to the action of a source, is accompanied by an increase in area occupied by the particles as the wave expands. Therefore, as well as the increase in pressure in front of the particles that gives rise to sound propagation, there exists a reduction in pressure due to the particles moving further apart. The 'propagating pressure' is in-phase with and proportional to the particle velocity and the 'stretching pressure' is in-phase with and proportional to the particle displacement. As velocity is the rate of change of displacement with time, the displacement at high frequencies is less than at low frequencies for the same velocity, so the relative magnitudes of the propagating and stretching pressures are dependent upon both frequency and radius.

The situation is more complex when finite-sized sources are considered. A second definition of the near-field, which is completely different from the hydrodynamic near-field described above, is the geometric near-field. The geometric near-field is a region close to a finite-sized source in which the sound field is dependent upon the dimensions of the source, and does not, in general, follow the inverse-square law. The extent of the geometric near field is defined as being the distance from the source within which the pressure does not follow the inverse-square law.

1.3.5 The loudspeaker as a point monopole

In practice, the point monopole serves as a useful approximation to a real sound source providing the real source satisfies two conditions:

(a) the source is physically small compared to a wavelength of the sound being radiated,
(b) all the radiating parts of the source operate with the same phase.

If l is a typical dimension of the source (e.g. length of a loudspeaker cabinet side), then the first condition may be written $kl < 1$ which, for a typical loudspeaker having a cabinet with a maximum dimension of 400 mm, is true for frequencies below $f < c/2\pi l \approx 140$ Hz. The second condition is satisfied by a single, rigid loudspeaker diaphragm mounted in a sealed cabinet; it is not satisfied when passive radiating elements such as bass reflex ports are present, or if the diaphragm is operated without a cabinet. In the former case, the relative phases of the diaphragm and the port are frequency-dependent, and in the latter case, the rear of the diaphragm vibrates in phase-opposition to the front, and the resulting radiation is that of a dipole (see Section 1.3.3) rather than a monopole. It follows, therefore, that equations (1.17) and (1.20) may usefully be applied to (some) loudspeakers radiating into free-field at low frequencies, in which case, the volume velocity is taken to be the velocity of the radiating diaphragm multiplied by its radiating area:

$$\hat{q}_d = \hat{u}_d S_d \tag{1.23}$$

where $\hat{u}_d$ is the velocity and S_d is the area of the diaphragm.

1.3.6 Sound radiation from a loudspeaker diaphragm

The point monopole serves as a simple, yet useful, model of a loudspeaker at low frequencies. As frequency is raised, however, the sound field radiated by a loudspeaker becomes dependent upon the size and shape of the diaphragm and cabinet, and on the details of the vibration of the diaphragm(s); a simple point monopole model will no longer suffice.

A useful technique for the analysis of the radiation of sound from non-simple sources, such as loudspeaker diaphragms, is to replace the vibrating parts with a distribution of equivalent point monopole sources. Given that we know the sound field radiated by a single monopole under ideal conditions (equation (1.17)), it should be possible to estimate the sound field radiated by a vibrating surface by summing up the contribution of all of the equivalent monopoles. The problem with this technique is that we do not, in general, know the sound field radiated by any one of the monopoles in the presence of the rest of the source. There is one specific set of conditions under which this technique can be used, however: the special case of a flat vibrating diaphragm mounted flush in an otherwise infinite, rigid, plane surface.

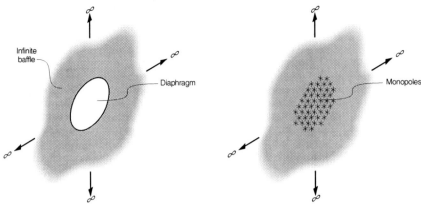

Figure 1.3. Representation of a flat, circular diaphragm as a distribution of monopoles.

This is the *baffled piston* model that serves as the basic starting point for nearly all studies into the radiation of sound from loudspeakers.

Figure 1.3 shows a flat, circular diaphragm mounted in a rigid, plane surface and its representation as a distribution of monopoles. As the entire diaphragm is surrounded by a plane surface, each equivalent monopole source can be considered in isolation. The sound field radiated by a single monopole is then that given in equation (1.21) for a monopole on a surface. Calculation of the sound field radiated by the baffled piston simply involves summing up (with due regard for phase) the contributions of all of the equivalent monopoles

$$\hat{p}(R,f) = \sum_N \hat{p}_n(R,f) = \frac{j\rho ck}{2\pi} \sum_N \left(\hat{q}_n \frac{e^{-jkr_n}}{r_n} \right) \tag{1.24}$$

where R is the observation point, N is the total number of monopoles, $\hat{q}_n$ is the volume velocity of monopole n, and r_n is the distance from that monopole to R. One should note that, in general, all the q_n could be different and that in all but one special case (see below), all the r_n, which represent the path lengths travelled by the sound from each monopole to R, will be different.

If the vibration of the diaphragm is assumed to be uniform over its surface (a *perfect piston*), the $\hat{q}_n$ can be taken out of the summation and replaced by $\hat{q}_d$, the total volume velocity of the diaphragm, as defined in equation (1.23). A further simplification results from assuming that the point R is in the far-field (see Section 1.3.4), i.e. it is sufficiently far from the diaphragm ($R \gg a$ where a is the radius of the diaphragm) that the r_n in the denominator (but not in the phase term) can be taken out of the summation. Having made the perfect piston and far-field simplifications, equation (1.24) can be further simplified by dividing the diaphragm into more and more monopoles until, in the limit of an infinite number of monopoles, the summation becomes an integral

$$\hat{p}(R,f) = \frac{j\rho ck\hat{q}_d}{2\pi R} \int_S e^{-jkr(S)} \, dS \tag{1.25}$$

The expression for the phase term, and hence the result of the integral, is dependent upon the geometry of the diaphragm. For a circular piston, and other simple geometries, the integral may be evaluated analytically; for more complex shapes, numerical integration is required, in which case, one may as well revert to the use of equation (1.24). The details of the integration are beyond the scope of this book; interested readers may refer to reference 4.

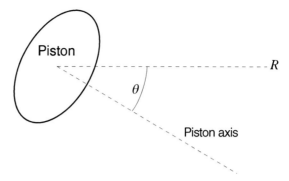

Figure 1.4. Geometry for piston directivity.

The sound field radiated by a perfect, circular piston mounted in an infinite baffle to a point in the far-field is then, with the inclusion of the result of the integration,

$$\hat{p}(R, f, \theta) = \frac{j\rho ck\hat{q}_d\, e^{-jkR}}{2\pi R} \left\{ \frac{2\mathbf{J}_1\,[ka\sin(\theta)]}{ka\sin(\theta)} \right\} \tag{1.26}$$

where $\mathbf{J}_1[\ ]$ is a Bessel function, the value of which can be looked up in tables or computed, a is the radius of the diaphragm, and θ is the angle between a line joining the point R to the centre of the piston and the piston axis as shown in Fig. 1.4. The bracketed term, containing the Bessel function, is known as the *directivity function* for the piston. Figure 1.5 shows values of the circular piston directivity function as polar plots for values of ka ranging from 0.5 to 20 (approximately 250 Hz to 10 kHz for a 200 mm diameter diaphragm). A more simple measure of the directivity of a loudspeaker is the *coverage angle*, defined as the angle over which the response remains within one half (-6 dB) of the response on the axis at $\theta = 0$. By setting the directivity function in equation (1.26) to equal 0.5, the coverage angle for a piston in a baffle is found to be when $ka\sin(\theta) \approx 2.2$. Coverage angle is usually specified as the inclusive angle, 2θ.

Setting the value of θ to zero in equation (1.26) gives the pressure radiated along the axis of the piston (in the far-field) which is known as the *on-axis frequency response*

$$\hat{p}_0(R, f) = \frac{j\rho ck\hat{q}_d\, e^{-jkR}}{2\pi R} \tag{1.27}$$

There is a very marked similarity between this expression and that quoted in equation (1.21) for the sound field radiated by a point monopole on a surface. This is no fluke; to an observer at R, in the far-field along the piston axis, the piston looks identical to a monopole with the same volume velocity because the path lengths from all parts of the piston to R are all virtually the same.

Assuming perfect piston vibration, infinite plane baffle mounting and far-field observation, equation (1.26) provides us with a useful model of the sound field radiated by a loudspeaker diaphragm. If the far-field on-axis frequency response is all that is required, equation (1.27) tells us that the loudspeaker diaphragm can be treated as if it were a simple point monopole on a surface.

It is useful at this point to look at the on-axis frequency response of a typical loudspeaker drive-unit. It is shown in Chapter 2 that, to a first approximation, the diaphragm velocity decreases as the inverse of frequency above the fundamental resonance frequency of the drive-unit. Also, equation (1.27) shows that, for a given diaphragm velocity, the on-axis frequency response increases with frequency. The net

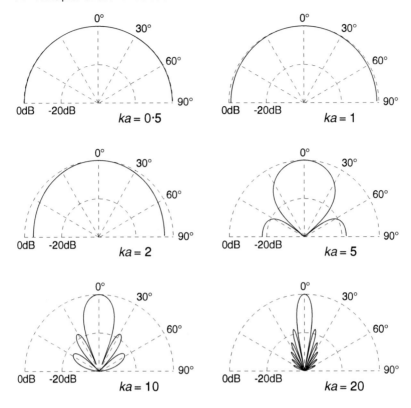

Figure 1.5. Circular piston directivity function for various values of *ka*.

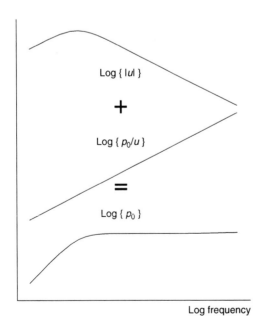

Figure 1.6. Graphical demonstration of the reason for the flat on-axis frequency response of an idealized loudspeaker diaphragm in an infinite baffle. The roll-off in diaphragm velocity above the fundamental resonance frequency is offset by the rising relationship between diaphragm velocity and the on-axis pressure.

result is that the on-axis frequency response of an idealized loudspeaker drive-unit is independent of frequency above the resonance frequency. Figure 1.6 shows a graphical demonstration of where this 'flat' response comes from. In practice, this seemingly perfect on-axis response is compromised by non-piston diaphragm behaviour and voice-coil inductance etc. (see Chapter 2). Even with a perfect, rigid, piston for a diaphragm and a zero-inductance voice-coil, the flat on-axis response has limited use; the piston directivity function in equation (1.26) and Fig. 1.5 tells us that as frequency increases, so the radiation becomes more directional and is effectively 'beamed' tighter along the axis.

1.3.7 Power output of a loudspeaker diaphragm

At low frequencies, the free-field power output of a loudspeaker diaphragm is the same as that of a point monopole of equivalent volume velocity and is given by equation (1.20); for a loudspeaker mounted in an infinite baffle, the power output is that of a monopole on a surface – double that of the free-field monopole. At higher frequencies, however, a more general approach is required. In Section 1.2.5 it was shown that the sound power output of a vibrating surface, such as a loudspeaker diaphragm, results from integrating the sound intensity on a sphere surrounding the source. For a perfect circular piston mounted in an infinite baffle, the sound pressure at a radius r and an angle to the axis of θ is given by equation (1.26). The sound intensity also varies with angle, therefore, and determination of the sound power output requires integration of the mean-square pressure over the entire hemisphere of radiation. The details of the necessary integration can be found in many acoustics textbooks (e.g. reference 4). Also, in Section 1.2.7, it was shown that the same sound power output can be calculated from the radiation impedance; the resultant radiation impedance is a complicated function of frequency and is therefore only shown graphically in Fig. 1.7. So, is a radiation impedance function that is too complex to include here of any use in the analysis of loudspeakers? Yes. At high and low frequencies, the radiation impedance function can be closely approximated by much simpler expressions. At low frequencies, (values of $ka < 1$, where a is the diaphragm radius),

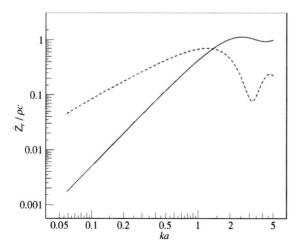

Figure 1.7. Radiation impedance of a rigid, circular piston in an infinite baffle, —— real part, - - - imaginary part; ρc is the characteristic impedance of air.

$$\hat{Z}_r \approx \rho c \left\{ \frac{(ka)^2}{2} + j \frac{8ka}{3\pi} \right\} \qquad (1.28)$$

and at high frequencies ($ka > 2$),

$$\hat{Z}_r \approx \rho c \qquad (1.29)$$

The sound power output at low frequencies is then

$$W_S \approx S_d \frac{|\hat{u}_d|^2}{2} \rho c \frac{(ka)^2}{2} \approx \frac{\rho c k^2 |\hat{q}_d|^2}{4\pi} \qquad (1.30)$$

which is identical to that for a monopole on a surface.

In practice, the high-frequency approximation is seldom used, but the low-frequency approximation serves us well. This is partly because most loudspeaker diaphragms are pretty good approximations to perfect pistons at low frequencies, but not at higher frequencies, partly because it is the imaginary part of the radiation impedance that most affects diaphragm motion (and this vanishes at high frequencies) and partly because the total power output is not usually of much interest at frequencies where the diaphragm has a complex directivity pattern.

It is interesting to note that the radiation impedance curve shown in Fig. 1.7 does not continue to rise at frequencies where $ka > 1$, because of the interference between the radiation from different parts of the diaphragm; this is exactly the same phenomenon that gives rise to the narrowing of the directivity pattern at high frequencies in Fig. 1.5. In fact, a glance at the expression for the on-axis frequency response (equation (1.27)) shows that there is no 'flattening-out' of the response as seen in the radiation impedance. The narrowing of the directivity pattern at high frequencies exactly offsets the flattening-out of the radiation impedance curve when considering the on-axis response.

1.4 Multiple sources and mutual coupling

There are many situations involving loudspeakers where more than one diaphragm radiates sound at the same time. If the diaphragms are all receiving different (uncorrelated) signals, then the combined sound power output is simply the sum of the power outputs of the individual diaphragms. If, however, two or more diaphragms receive the same signal, the situation becomes more complicated due to mutual coupling, the term given to the interaction between two or more sources of sound radiating the same signal.

1.4.1 Sound field radiated by two sources

A single, simple source of sound, such as a point monopole or a loudspeaker diaphragm at low frequencies, radiates a spherically symmetric sound field which is omnidirectional. When a second simple source is introduced, the two sound fields interfere and a standing-wave pattern is set up, consisting of areas of high and low pressure which are fixed in space. Figure 1.8 shows a typical standing-wave field set up by two identical monopole sources spaced three wavelengths apart. The dark areas indicate regions of high sound pressure and the light areas regions of low pressure. The standing-wave pattern exists because the path length from one source to any point is different from the path length from the other (except for points on a plane between the sources). The areas of high pressure occur where the path lengths are either the same, or are multiples of a wavelength different, so that the two sound fields sum in phase. When the path lengths differ by odd multiples of half a wavelength, the two sound fields are in phase opposition, and tend to cancel each other. The standing-wave pattern extends out into the far-field ($r \gg d$ where d is the

Figure 1.8. Standing-wave field set up by two identical monopoles spaced three wavelengths apart; dark areas represent regions of high acoustic pressure and light areas regions of low pressure.

distance between the sources), where a complicated directivity pattern is observed. Figure 1.9 shows a polar plot of the far-field directivity of the two sources in Fig. 1.8. Even though each source is omnidirectional in isolation, the combined sound radiation of the two sources is very directional. Figure 1.10 shows a polar plot of the directivity of the same two sources at a lower frequency where they are one-eighth of a wavelength apart; the field is seen to be near omnidirectional and close to double the pressure radiated by a single source. Clearly, if the two sources are close together compared to the wavelength of the sound being radiated, then the path length differences at all angles represent only small phase shifts and the two sound fields sum almost exactly.

1.4.2 Power output of two sources – directivity considerations

The sound fields and directivity depicted in Figs 1.9 and 1.10 are shown in two dimensions only. Figures 1.11 and 1.12 are representations of the directivity shown in Figs 1.9 and 1.10 extended to three dimensions. In Section 1.2.5 it was shown that the power output of a source can be found by integrating the sound intensity over a surface surrounding the source. In Figs 1.11 and 1.12, the sound intensity in any direction is proportional to the square of the distance from the centre to the surface of the plot at that angle; the total power output then results from integrating this squared 'radius' over all angles. Assuming that each source in isolation radiates the same power at all frequencies, we can apply the above argument to Figs 1.11 and 1.12. By studying the two figures, it can be seen that the combined power output of the two sources is higher at low frequencies than it is at high frequencies, This is because, as stated in Section 1.4.1, the sound fields radiated by the two sources sum everywhere almost in phase at low frequencies; there is very little cancellation. The low frequency polar plot (Fig. 1.12) is almost spherical with a radius of 2. Integrating the square of this radius over all angles leads to a power output four times greater than that of a single source (a sphere with a radius of 1). Clearly, we have a case of

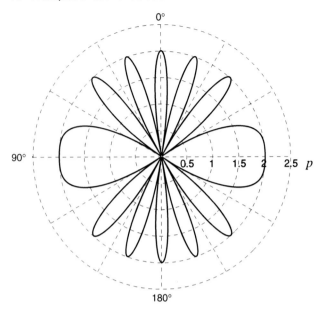

Figure 1.9. Far-field directivity of two identical monopoles spaced three wavelengths apart.

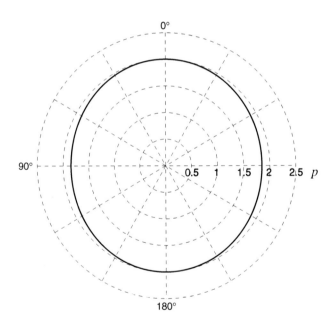

Figure 1.10. Far-field directivity of two identical monopoles spaced one-eighth of a wavelength apart.

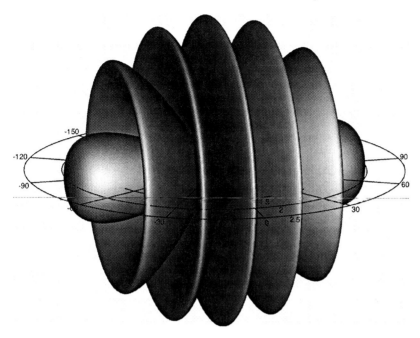

Figure 1.11. Three-dimensional representation of the far-field directivity of two identical monopoles spaced three wavelengths apart.

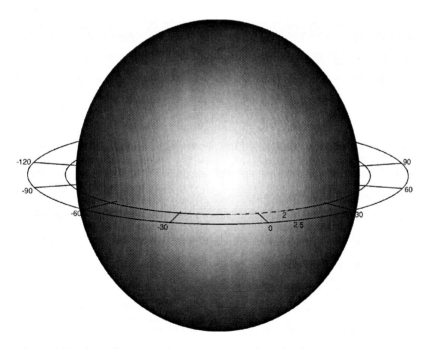

Figure 1.12. Three-dimensional representation of the far-field directivity of two identical monopoles spaced one-eighth of a wavelength apart.

$1 + 1 = 4!$ Bringing a second source close to a first effectively doubles the power output of both sources.

At high frequencies (Fig. 1.11), the two sound fields sum in phase for approximately half of the angles where the radius is 2; for the other half they tend to cancel each other and the radius is near zero. The net result is approximately double the power output compared to a single source and we are back to $1 + 1 = 2$.

At low frequencies then, the sound field radiated by two sources is near omnidirectional, and the power output of each source is doubled. At high frequencies, the sound field radiated by two sources is a complicated function of angle and the power output of each source is unchanged.

1.4.3 Power output of two sources – radiation impedance considerations

In Section 1.2.7, it was shown that the power output of a source can be calculated by considering the radiation impedance. The radiation impedance for an idealized loudspeaker diaphragm (perfect piston) mounted in an infinite baffle is given in equation (1.28), and is defined as the ratio of the pressure on the surface of the diaphragm to the velocity of that diaphragm. Introducing a second diaphragm will modify this radiation impedance as the sound field radiated by the second will contribute to the pressure on the surface of the first, and vice-versa. The total pressure on the surface of one diaphragm therefore has two components, one due to its own velocity, and another due to the velocity of the other diaphragm. In general, because of the piston directivity function (see equation (1.26)) the contribution of one diaphragm to the pressure on the other is a complicated function of frequency and the angle between the two diaphragm axes; also, the pressure contribution may vary across the diaphragm surface. At low frequencies, however, where the wavelength is large compared to the radius of the diaphragms (but not, necessarily, compared to the distance between them) the directivity function is equal to one at all angles, the radiated field of the second source becomes that of a monopole on a surface and the pressure contribution can be assumed to be uniform over the surface of the first source, thus:

$$\hat{p}_2 = \hat{Z}_{r1}\hat{u}_d + \frac{j\rho ck\hat{q}_d\, e^{-jkR}}{2\pi R} \tag{1.31}$$

therefore $$\hat{Z}_{r2} = \frac{\hat{p}_2}{\hat{u}_d} = \hat{Z}_{r1} + \frac{j\rho cka^2\, e^{-jkR}}{2R} \tag{1.32}$$

where $\hat{p}_2$ is the total pressure on the surface of one diaphragm, $\hat{Z}_{r1}$ is the radiation impedance of one diaphragm in isolation, $\hat{Z}_{r2}$ is the radiation impedance of one diaphragm in the presence of another, $\hat{u}_d$ and $\hat{q}_d$ are the velocity and volume velocity of one of the diaphragms, R is the distance between the diaphragms and a is the diaphragm radius. Equation (1.14) tells us that the power output of a source is directly proportional to the real part of the radiation impedance. The ratio of the power output of one source in the presence of a second (W_2) to that of the same source in isolation (W_1) is therefore the real part of $\hat{Z}_{r2}$ divided by the real part of $\hat{Z}_{r1}$,

$$\frac{W_2}{W_1} = \frac{\mathrm{Re}\{\hat{Z}_{r1}\} + \dfrac{\rho cka^2 \sin{(kR)}}{2R}}{\mathrm{Re}\{\hat{Z}_{r1}\}} \tag{1.33}$$

Substituting the low-frequency approximation for $\hat{Z}_{r1}$ (equation (1.28)), and re-arranging yields

$$\frac{W_2}{W_1} = 1 + \frac{\sin{(kR)}}{kR} \tag{1.34}$$

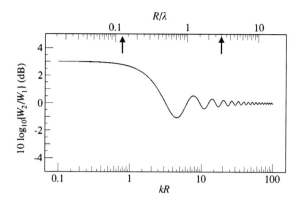

Figure 1.13. Mutual coupling between a pair of loudspeaker diaphragms spaced a distance R apart. The two arrows represent the spacings of one-eighth and three wavelengths that correspond to the polar plots in Figs 1.11 and 1.12. W_1 is the power output of a single loudspeaker in isolation and W_2 is the power output of a single loudspeaker in the presence of another.

For small values of kR, sin $(kR)/kR$ is close to one and the power output is doubled in accordance with Section 1.4.2 from a directivity consideration. For large values of kR, sin $(kR)/kR$ vanishes and the power output is the same as it is for the diaphragm in isolation, again, as described in Section 1.4.2. Equation (1.34) is a statement of the degree of mutual coupling between two sources which is dependent upon the product of frequency (in the form of k) and the distance between the diaphragms R. Figure 1.13 shows a plot of the mutual coupling between two loudspeaker diaphragms as a function of kR. Also marked on a second horizontal scale is the ratio of distance apart to wavelength, R/λ, the two frequencies that correspond to the polar plots in Figs 1.11 and 1.12 are marked on this scale with arrows.

The fact that equation (1.33) is derived from low-frequency considerations should not worry us unduly. It is true that the closer the sources are to each other, the higher the frequency up to which mutual coupling is significant. However, even for the worst case when the two diaphragms are touching ($R = 2a$), Fig. 1.13 shows that the mutual coupling between the sources is only really significant up to frequencies where $kR \approx 3$ (or $ka \approx 1.5$). At these frequencies the radiation from each diaphragm in isolation is near omnidirectional.

A comparison between the description of mutual coupling from a directivity viewpoint and that from a radiation impedance viewpoint shows that they yield exactly the same result. This is perhaps not surprising as the total power radiated by the sources (the radiation impedance description) must equal the total power passing through a sphere surrounding the sources (the directivity description) in the absence of any energy loss mechanisms. Nevertheless, it is intriguing that the effect on the radiation impedance, and hence radiated power, of the additional pressure on one diaphragm by the action of another can be predicted entirely by studying only the interference patterns generated in the far-field.

1.4.4 Practical implications of mutual coupling – 1: radiation efficiency

The existence of mutual coupling between two loudspeaker diaphragms is a mixed blessing. In the previous two sections, it was shown that at low frequencies, where the distance between the diaphragms is less than about one quarter of a wavelength, the combined power output of the two diaphragms is four times greater ($+6\,dB$) than that of one diaphragm in isolation. This additional 'free' power output turns out to be very useful, and is the main reason why large diaphragms are used to radiate low frequencies. Equation (1.30) states that the power output of a loudspeaker diaphragm on an infinite baffle is proportional to the volume velocity of the diaphragm $\hat{q}_d$ multiplied by $(ka)^2$. For a circular diaphragm, $\hat{q} = \pi a^2 \hat{u}_d$, so for a given diaphragm velocity the power output is proportional to the diaphragm radius raised to the fourth power. It therefore follows that a doubling of diaphragm area, by introducing a second diaphragm close to the first for example, yields a fourfold increase in power output. Thus, the high radiation efficiency of large diaphragms at low frequencies can be thought of as being due to mutual coupling between all the different parts of the diaphragm. Similarly, when two (or more) diaphragms are mounted close together, perhaps sharing the same cabinet, the radiation impedance at low frequencies is similar to that of a single diaphragm with a surface area equal to the combined areas of the diaphragms.

1.4.5 Practical implications of mutual coupling – 2: The stereo pair

The downside of mutual coupling occurs when two loudspeaker diaphragms are spaced a significant distance apart and used to reproduce stereophonic signals. Under free-field conditions with the listener situated on a plane equidistant from the two loudspeakers (the axis) the sound fields radiated by the loudspeakers sum at the listener's ears in phase at all frequencies (assuming both ears are on the axis); the sound field is exactly double that radiated by one loudspeaker. If the listener moves away from the axis, the different path lengths from the two loudspeakers to the ear give rise to frequency dependent interference. Figure 1.14 shows the frequency responses of a pair of loudspeakers at two points away from the axis relative to the response of a single loudspeaker; the dashed line represents the response on the axis. The response shapes shown in Fig. 1.14, which are known as comb filtering, are a result of alternate constructive and destructive interference due to the changing relative phase of the two signals as frequency is raised. Comparing the two response plots (and any number of similar ones), it is clear that the response is similar every-

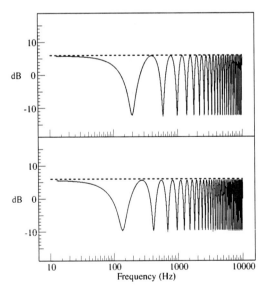

Figure 1.14. Frequency response of a pair of loudspeakers spaced 3 m apart at two points away from the plane equidistant from the loudspeakers (the stereo axis) relative to the response of a single loudspeaker. The dashed line represents the response on the axis.

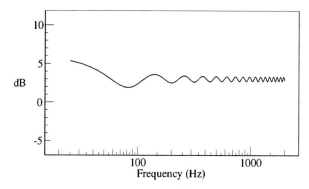

Figure 1.15. Combined power output of a pair of loudspeakers spaced 3 m apart relative to that of a single loudspeaker.

where only at low frequencies – the frequency range in which mutual coupling occurs – and that a flat response occurs only on the axis. The comb-filtered response that occurs at off-axis positions is an accepted limitation of two-channel stereo reproduction over loudspeakers and will not be considered further in this chapter. However, what is important when considering the radiation of sound by loudspeakers is the effect that mutual coupling has on stereo reproduction under normal listening conditions.

Most stereophonic reproduction is carried out under non-free-field acoustic conditions, i.e. in rooms. The sound field that exists under these conditions consists of the direct sound, which is the sound radiated by the loudspeakers in the direction of the listener (as in free-field conditions), and the reverberant sound, which is a complicated sum of all of the reflections from all of the walls (see Chapter 8). The reverberant sound field is the result of the radiation of sound by the loudspeakers in all directions, and is therefore related to the combined power output of the two loudspeakers (see Section 1.4.2). For a listener on the axis in a room, the direct sound from the stereo pair of loudspeakers will have the same response as that radiated from a single loudspeaker, but raised in level by 6 dB. In contrast, the total power output, and hence the reverberant sound field, will show a 6 dB increase at low frequencies due to mutual coupling, but only a 3 dB increase at higher frequencies compared to that due to a single loudspeaker. The result is a shift in the frequency balance of sounds as they are moved across the stereo sound stage from fully left or fully right to centre. Figure 1.15 shows the total power output of a pair of loudspeakers mounted 3 m apart; a typical spacing for a stereo pair. This response may be calculated directly from equation (1.34) (easy) or by integrating responses such as those in Fig. 1.11 over the surface of a sphere surrounding the loudspeakers (hard).

1.4.6 Practical implications of mutual coupling – 3: diaphragm loading

In the discussion of mutual coupling above it is assumed that a loudspeaker diaphragm is a pure velocity source which means that the velocity of the diaphragm is unaffected by the acoustic pressure on it. Thus a doubling of the radiation impedance gives rise to a doubling of power output. Whereas this is a fairly good first approximation to the dynamics of loudspeaker diaphragms, in practice some 'slowing down' or 'speeding up' of the diaphragm motion will result from the different forces (or loads) associated with changes in radiation impedance. It is worthwhile here to show estimates of the additional load on a loudspeaker diaphragm, brought about by mutual coupling with a second loudspeaker, to give some insight into when the perfect velocity source assumption is likely to be valid.

The radiation impedance of a baffled loudspeaker diaphragm in the presence of a second diaphragm is given by equation (1.32). Using this expression, the additional radiation impedance on the diaphragm due to the motion of a second, identical diaphragm can be calculated as a fraction of the radiation impedance of a single diaphragm. Figure 1.16 shows the result of evaluating this fraction for a pair of 250 mm diameter loudspeakers mounted on an infinite baffle at a frequency of 30 Hz. The curve is very insensitive to frequency and is in fact valid for all frequencies for which the loudspeaker diaphragm can be considered omnidirectional, and for which the low-frequency approximation for radiation impedance holds (in this case up to about 300 Hz). However, the curve is sensitive to diaphragm size but, as acoustics tends to follow geometric scaling laws very closely, it is valid for different diaphragm sizes if the distance between the sources is adjusted by an equivalent amount (it would scale exactly if frequency were changed as well). A second horizontal scale representing the ratio of the distance apart to the diaphragm radius (R/a), is included as a good approximate guide for all 'loudspeaker-sized' diaphragms. It should be noted that equation (1.34) is based on the assumption that the second diaphragm acts as a point monopole. This is clearly not the case for very small values of R/a, but in practice the errors are small, at about 0.5% for $R/a = 3$ rising to 3% for $R/a = 2$ (diaphragms touching).

It is clear from Fig. 1.16 that the additional loading on one diaphragm due to the motion of another drops to below 10% of the loading for a single diaphragm alone when the spacing between the diaphragms exceeds about six diaphragm radii. Whether these additional loads are significant or not depends upon the electro-acoustic characteristics of the loudspeakers. A loudspeaker having a lightweight diaphragm and weak magnet system, for example, will be more sensitive to acoustic loading than one having a heavy diaphragm and powerful magnet system. However,

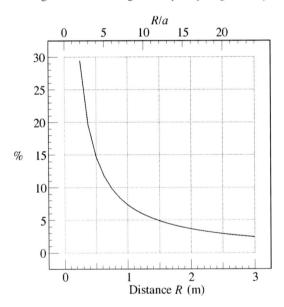

Figure 1.16. Percentage additional radiation loading on a loudspeaker diaphragm of 250 mm diameter at a frequency of 30 Hz due to the presence of a second identical diaphragm a distance R away from the first. The top horizontal scale shows approximate values for the ratio of the distance apart to the diaphragm radius for most 'loudspeaker-sized' diaphragms; this is possible as the curve is essentially frequency independent for all frequencies where the low-frequency approximation for radiation impedance holds (see text).

simulations suggest that with conventional moving-coil drive-units the change in output is of the order of 1 dB or so for a pair of very weak drive-units mounted alongside each other. For more robust drive-units, or in any case when the distance between the diaphragms is increased beyond about four diaphragm radii, the change is so small as to be insignificant.

For two diaphragms mounted close together, the doubling of the power output due to mutual coupling can be explained quite easily in terms of a doubling of acoustic pressure on each drive-unit. What is perhaps less obvious is why there is still a doubling of power output at low frequencies when the diaphragms are separated by many diaphragm radii. Figure 1.16 shows that for two 250 mm drive-units mounted 3 m apart as a typical stereo pair, the pressure on each diaphragm due to the motion of the other is only increased by about 2.5%, yet the power output is still doubled. The answer lies in the phase of the additional pressure. At low frequencies, the real part of the radiation impedance (responsible for power radiation) is very small compared to the imaginary part, so most of the pressure on a diaphragm due to its own motion acts as an additional mass and does not contribute directly to sound radiation. The pressure radiated by a second, distant diaphragm, however, reaches the first with just the right phase relative to its velocity to increase the real part of the radiation impedance and hence power output.

For most conventional moving-coil drive-units then, the additional loading due to the presence of a second diaphragm has little effect on the motion of the diaphragm with the result that a doubling of power output results whether the diaphragms are close together or far apart (provided they are still within a quarter wavelength of each other). This is not the case for loudspeakers with very light diaphragms and weak motor systems; electrostatic loudspeakers, for example. Loudspeakers of this type can be approximated by pressure sources (the pressure on the diaphragm is independent of the load) and are affected by mutual coupling in a different way: a doubling of power output results when they are far apart, but when they are close together, the increase in loading reduces the diaphragm motion and the increase in power output is negated. Most electrostatic loudspeakers are in fact dipole radiators (see Section 1.3.3) which, when mounted with both loudspeakers of a pair facing the same direction, cannot mutually couple, as one loudspeaker is on the null axis of the other, and thus cannot affect the pressure on the diaphragm of the other.

1.4.7 Multiple diaphragms

When multiple diaphragms receive the same signal, mutual coupling occurs between each diaphragm and each of the others. Equation (1.34) can therefore be extended to include multiple diaphragms, thus:

$$\frac{W_m}{W_1} = 1 + \sum_{n=1}^{N} \frac{\sin{(kR_n)}}{kR_n} \tag{1.36}$$

where W_m is the power output of one diaphragm in the presence of N other diaphragms at distances R_n. When estimating the degree of additional loading on a diaphragm, due regard must be paid to the relative phases of the contributions from the different additional diaphragms. The total radiation impedance is then

$$\hat{Z}_{rm} = \hat{Z}_{r1} + \sum_{n=1}^{N} \frac{j\rho cka^2\,e^{-jkR_n}}{2R_n} \tag{1.37}$$

1.5 Limitations of the infinite baffle loudspeaker model

So far, our idealized loudspeaker has consisted of one or more perfect, circular pistons mounted in an otherwise infinite, rigid baffle. The models developed thus far go a long way towards estimating the sound field radiated by a real loudspeaker mounted flush in a wall at frequencies where its diaphragm(s) can be assumed to behave as a piston (see Chapter 2). However, many real loudspeakers have diaphragms mounted on the sides of finite-sized cabinets which can be poor approximations to infinite baffles. Thus it is useful to establish the differences between the sound field radiated in this case and that by our ideal, infinitely baffled diaphragm. To make matters worse, these loudspeaker cabinets are often placed with their backs against, or some distance away from, a rigid back wall. Again, estimates of the likely combined effect of the cabinet and the back wall should prove useful. Ultimately, one must consider the sound field radiated by loudspeaker diaphragms in cabinets, in the presence of multiple walls. The answers to the latter problem takes us into the realms of room acoustics, a highly complex subject which is covered in more detail in Chapter 8.

1.5.1 Finite-sized cabinets and edge diffraction

In Section 1.3.5 it was stated that the sound field radiated by a loudspeaker diaphragm mounted in a wall of a finite-sized cabinet can be approximated by that of a monopole in free-space (equation (1.17)), provided the wavelength is large compared to any of the cabinet dimensions. At low frequencies, the sound is therefore radiated into a full sphere. At higher frequencies, where the wavelengths are small compared to the front wall of the cabinet, the cabinet may be approximated by an infinite baffle. In this case, the sound is constrained, by the cabinet, to be radiated into a hemisphere. At still higher frequencies, the wavelength is small compared to the dimensions of the diaphragm and the sound is beamed along the axis regardless of whether there is a baffle or not. Clearly, as frequency is raised, there is a transition from free-field monopole behaviour to that of a monopole on a surface and on to a directional piston. If we limit our interest to the on-axis frequency response, we need only consider the first two frequency ranges as the sound field radiated by a piston along its axis is the same as that of the equivalent monopole on a surface. A comparison between equation (1.17) for a free-field monopole and equation (1.21) for a monopole on a surface shows that there is a factor of 2 (or 6 dB) difference in the radiated sound fields. Whereas the on-axis frequency response of an idealized loudspeaker drive-unit in an infinite baffle is uniform (see Section 1.3.6), that of the same drive-unit mounted in a finite-sized cabinet is not. The degree of response non-uniformity and the frequency range in which it occurs depends upon the cabinet size and shape.

The mechanisms behind the effective 'unbaffling' of the diaphragm at low frequencies, and the resultant non-uniform frequency response, associated with finite-sized cabinets can best be understood by considering the diffraction of sound around the edges of the cabinet. The theory of the diffraction of sound waves is very involved and beyond the scope of this book. Interested readers are referred to reference 4 for more details. Instead, the phenomenon will be dealt with here in a conceptual manner.

As a sound wave radiates away from a source on a finite-sized cabinet wall, it spreads out as it propagates in the manner of half of a spherical wave. When the wave reaches the edge of the wall, it suddenly has to expand more rapidly to fill the space where there is no wall (see Fig. 1.17). There are two consequences of this sudden expansion. First, some of the sound effectively 'turns' the corner around the edge and carries on propagating into the region behind the plane of the source. Second, the sudden increase in expansion rate of the wave creates a lower sound pressure in front of the wall, near the edge, than would exist if the edge were not there. This drop in pressure then propagates away from the edge into the region in front of the plane of the source. The sound wave that propagates behind the plane of the source is in phase with the wave that is incident on the edge and the one that propagates to the front is in phase opposition. These two 'secondary' sound waves are known as diffracted waves and they 'appear' to emanate from the edge; the total

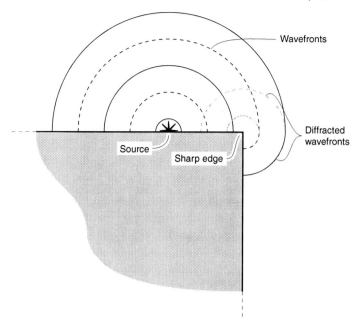

Figure 1.17. Graphical representation of the sudden increase in the rate of expansion of a wavefront at a sharp edge; the diffracted wave in the shadow region behind the source plane has the same phase as the wave incident on the edge, the diffracted wave in front of the source plane is phase-reversed.

sound field may then be thought of as being the sum of the direct wave from the source (as if it were on an infinite baffle) and the diffracted waves. The direct wave exists only in front of the baffle; the region behind is known as the 'shadow' region where only the diffracted wave exists.

At low frequencies, the diffracted waves from all of the edges of the finite-sized cabinet sum to yield a sound field with almost exactly one half of the pressure radiated by the source on an infinite baffle ($\hat{p}_b$). Thus behind the cabinet there is a pressure of $\hat{p}_b/2$ (diffracted wave only) and in front of the cabinet the direct wave ($\hat{p}_b$) plus the negative-phased diffracted wave ($-\hat{p}_b/2$) giving a total pressure of $\hat{p}_b/2$ everywhere – exactly as expected for a monopole in free-space. Assuming that the edge is infinitely sharp (has no radius of curvature), there can be no difference between the strength of the diffracted wave at low frequencies and that at high frequencies (the edge remains sharp regardless of scale). The only difference, therefore, between the diffracted waves at low frequencies and those at higher frequencies is the effect that the path length differences between the source and different parts of the edge has on the radiated field. The diffracted waves from those parts of the edge further away from the source will be delayed relative to those from the nearer parts, giving rise to significant phase differences at high frequencies but not at low frequencies. The net result is a strong diffracted sound field at low frequencies and a weak diffracted sound field at high frequencies.

Figure 1.18 shows the results of a computer simulation of the typical effect that a finite-sized cabinet has on the frequency response of a loudspeaker. Figure 1.18(a) is the on-axis frequency response of an idealized loudspeaker drive-unit mounted in an infinite baffle. The response is seen to be uniform over a wide range of frequencies. Figure 1.18(b) is the frequency response of the same drive-unit mounted on the front of a cabinet of dimensions 400 mm high by 300 mm wide by 250 mm deep. The

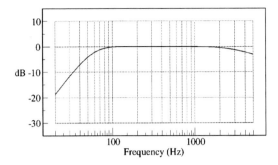

Figure 1.18(a). On-axis frequency response of an idealized loudspeaker diaphragm mounted in an infinite baffle; the response is uniform over a wide range of frequencies.

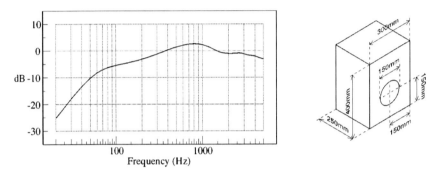

Figure 1.18(b). On-axis frequency response of the same loudspeaker diaphragm mounted on the front face of a finite-sized cabinet (the rear enclosure size is assumed to be the same in both cases). The response has reduced by 6 dB at low frequencies and is uneven at higher frequencies. The differences between this and Fig. 1.18(a) are due to diffraction from the edges of the cabinet.

6 dB decrease in response at low frequencies, due to the change in radiation from baffled to unbaffled, is evident from a comparison between Figs 1.18(a) and (b). Also evident is an unevenness in the response in the mid-range of frequencies. These response irregularities are due to path length differences from the diaphragm to the different parts of the diffracting edges and on to the on-axis observation point; unlike the low-frequency behaviour, these are dependent upon the detailed geometry of the driver and cabinet and the position of the observation point.

1.5.2 Loudspeakers near walls

When a source of sound is operated in the presence of a rigid wall, the waves that propagate towards the wall are reflected back in the same way that light reflects from a mirror. Indeed, taking the light analogy further, one can think of the reflected waves as having emanated from an identical 'image' source beyond the wall. The source and its image behave in exactly the same way as two identical sources spaced apart by twice the distance from the source to the wall, including all the effects of interference and mutual coupling described in Section 1.4. The sound field radiated by a loudspeaker in the presence of a reflective wall is therefore a complicated function of distance, frequency and observation position.

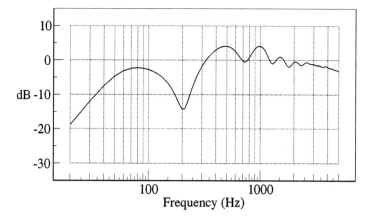

Figure 1.19(a). On-axis frequency response of the same diaphragm and cabinet as in Fig. 1.18(b), but with the rear of the cabinet against a rigid wall. Interference between the direct sound from the loudspeaker and that reflected from the wall produces a comb-filtered response, but the response at low frequencies is restored to that for the infinite baffle case (Fig. 1.18(a)).

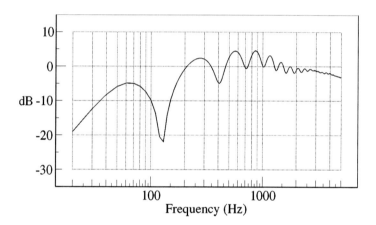

Figure 1.19(b). As Fig. 1.19(a) but with the rear of the cabinet 0.25 m from the rigid wall.

Figure 1.19(a) shows the on-axis frequency response of the driver described in Section 1.5.1 in the same cabinet but mounted with the cabinet back against a rigid wall. The waves that diffract around the front edges of the cabinet are now reflected from the wall and propagated forward to interfere with both the direct sound from the driver and the front diffracted wave. The result is alternating constructive and destructive interference as frequency is raised. Figure 1.19(b) is as Fig. 1.19(a), but with the rear of the cabinet moved 0.25 m away from the wall. In both cases, the low-frequency response is raised by 6 dB back to the level of the response of the driver in the infinite baffle. Clearly, 'sinking' the cabinet into the wall until the front of the cabinet is flush removes both the wall reflections and the cabinet edge diffraction and we return to the uniform response of the infinite baffle shown in Fig. 1.18(a).

1.6 Horns

Despite the useful effects of mutual coupling, the radiation efficiency of even large loudspeaker diaphragms is small at low frequencies. For example, a diaphragm with a diameter of 250 mm has a radiation efficiency (proportional to the real part of the radiation impedance – see Section 1.2.7) of just 0.7% at 50 Hz when mounted in an infinite baffle, and half that when mounted in a cabinet. Sound power output is proportional to the product of the mean-squared velocity and the radiation efficiency, so a low radiation efficiency means that a high diaphragm velocity is required to radiate a given sound power. The only way in which the radiation efficiency can be increased is to increase the size of the radiating area, but larger diaphragms have more mass (if rigidity is to be maintained) which means that greater input forces are required to generate the necessary diaphragm velocity (see Chapter 2).

Electroacoustic efficiency is defined as the sound power output radiated by a loudspeaker per unit electrical power input. Because of the relatively high mass and small radiating area electro-acoustic efficiencies for typical loudspeaker drive-units in baffles or cabinets are of the order of only 1–5%. Horn loudspeakers combine the high radiation efficiency of a large diaphragm with the low mass of a small diaphragm in a single unit. This is achieved by coupling a small diaphragm to a large radiating area via a gradually tapering flare. This arrangement can result in electro-acoustic efficiencies of 10–50%, or ten times the power output of the direct-radiating loudspeaker for the same electrical input. Additionally, horns can be employed to control the directivity of a loudspeaker and this, along with the high sound power output capability, is why they are used extensively in public address loudspeaker systems.

The following sections describe, in a conceptual rather than mathematical way, how horns increase the radiation efficiency of loudspeakers, how they control directivity, and why there is often the need to compromise one aspect of the performance of a horn to enhance another.

1.6.1 The horn as a transformer

The discussion of near- and far-fields in Section 1.3.4 showed that, in the hydrodynamic near-field, the change in area of an acoustic wave as it propagates gives rise to a 'stretching pressure' which is additional to the pressure required for sound propagation. The stretching pressure does not contribute to sound propagation as it is in phase quadrature (90°) with the particle velocity, so the acoustic impedance in the near-field is dominated by reactance (see Section 1.2.7). As a consequence, large particle velocities are required to generate small sound pressures when the rate of change of area with distance of the acoustic wave is significant. It is this stretching phenomenon that is responsible for the low radiation efficiency of direct-radiating loudspeakers at low frequencies. Physically, one can imagine the air moving sideways out of the way, in response to the motion of the loudspeaker diaphragm, instead of moving backwards and forwards. In the hydrodynamic far-field, the stretching pressure is minimal, the acoustic impedance is dominated by resistance, and efficient sound propagation takes place. The only difference between the sound fields in the near- and far-fields is the rate of change of area with distance of the acoustic wave; the flare of a horn is a device for controlling this rate of change of area with distance, and hence the efficiency of sound propagation.

Horns are waveguides that have a cross-sectional area which increases, steadily or otherwise, from a small throat at one end to a large mouth at the other. An acoustic wave within a horn therefore has to expand as it propagates from throat to mouth. The manner in which acoustic waves propagate along a horn is so dependent upon the exact nature of this expansion that the acoustic performance of a horn can be radically changed by quite small changes in flare-shape. It is usually assumed in acoustics that changes in geometry that are small compared to the wavelength of the sound of interest do not have a large effect on the behaviour of the sound waves,

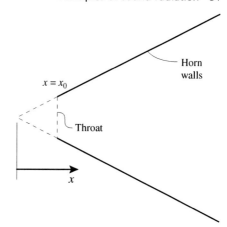

Figure 1.20. Geometry of a conical horn. The origin for the axial coordinate is usually taken as the imagined apex of the cone.

so why should horns be any different? The answer lies in the stretching pressure argument above. The concept of a stretching pressure can be applied to horns by considering *flare-rate*. Flare-rate is defined as the rate of change of area with distance divided by the area, and usually has the symbol m:

$$m(x) = \frac{1}{S(x)} \frac{dS(x)}{dx} \qquad (1.39)$$

where $S(x)$ is the cross-sectional area at axial position x. The simplest flare shape is the conical horn, which has straight sides in cross-section and a cross-sectional area defined by

$$S(x) = S(0) \left(\frac{x}{x_0}\right)^2 \qquad (1.40)$$

where $S(0)$ is the area of the throat (at $x = 0$) and x_0 is the distance from the apex of the horn to the throat as shown in Fig. 1.20. The sound field within a conical horn can be thought of as part of a spherical wave field, and has a flare-rate which is dependent on distance from the apex:

$$m(x) = \frac{2}{x} \qquad (1.41)$$

The flare rate in a conical horn (and in a spherical wave field) is therefore high for small x and low for large x. For a spherical wave field, the radius r at which the resistive and reactive components of the acoustic impedance are equal in magnitude is when $kr = 1$ (see Section 1.3.4), at which point the flare-rate is, with the substitution of x for r,

$$m = 2k \qquad (1.42)$$

Thus for positions within a conical horn where $kx < 1$, the acoustic impedance is dominated by reactance and the propagation is near-field-like. For positions where $kx > 1$, the impedance is resistive and the propagation is far-field-like. The radial dependence of the flare-rate in a conical horn (and a spherical wave) gives rise to a gradual transition from the reactive, near-field dominated behaviour associated with the stretching pressure, to the resistive, radiating, far-field dominated propagation as a wave propagates from throat to mouth. The transition from near- to far-field dominance is gradual with increasing frequency and/or distance from apex, so distinct 'zones' of propagation are not clearly evident.

A common flare shape for loudspeaker horns is the exponential. An *exponential horn* has a cross-sectional area defined by

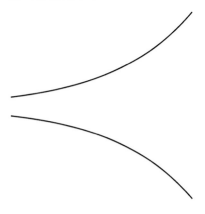

Figure 1.21. The flare shape of an exponential horn.

$$S(x) = S(0)\, e^{mx} \qquad (1.43)$$

The flare shape of the exponential horn is shown in Fig. 1.21. The flare-rate of an exponential horn is constant along the length of the horn ($m(x) = m$), giving rise to a behaviour that is quite different from the conical horn. With reference to equation (1.42), at frequencies where $k < m/2$, throughout the entire length of the horn, the reactive, near-field-type propagation dominates and, if the horn is sufficiently long, an almost totally reactive impedance exists everywhere. At frequencies where $k > m/2$, again throughout the entire length of the horn, the far-field-type propagation dominates leading to an almost totally resistive impedance everywhere. The frequency where $k = m/2$ is known as the *cut-off frequency* of an exponential horn and marks a sudden transition from inefficient sound propagation within the horn to efficient sound propagation. The cut-off frequency is then

$$k_c = \frac{m}{2}$$

Therefore

$$f_c = \frac{mc}{4\pi}\, (\text{Hz}) \qquad (1.44)$$

Physically, propagation within an exponential horn above cut-off is similar to a spherical wave of large radius, with minimal stretching pressure, and that below cut-off, similar to a spherical wave of small radius, dominated by the stretching pressure. The sharp cut-off phenomenon clearly occurs because the transition from one type of propagation to the other occurs simultaneously throughout the entire length of the horn as the frequency is raised through cut-off. The acoustic impedance at the throat of an infinite-length exponential horn is shown in Fig. 1.22, which clearly illustrates that, at frequencies below cut-off, the real part of the acoustic impedance is zero, which means that a source at the throat can generate no acoustic power (see Section 1.2.7). At frequencies above the cut-off frequency, the real part of the acoustic impedance is close to the characteristic impedance of air; a source at the throat therefore generates acoustic power with a radiation efficiency of 100%.

In practice, horns have a finite length and, unless the mouth of the horn is large compared to a wavelength, an acoustic wave propagating towards the mouth sees a sudden change in acoustic impedance from that within the horn to that outside, and some of the wave is reflected back down the horn. A standing-wave field is set up between the forward propagating wave and its reflection (see Section 1.2.3), which leads to comb-filtering in the acoustic impedance. Figure 1.23 shows the radiation efficiency at the throat of a typical finite-length exponential horn. Also shown are

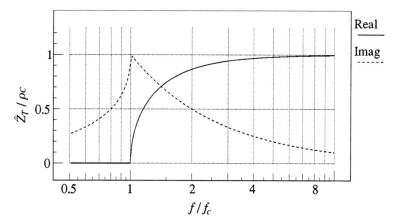

Figure 1.22. Acoustic impedance at the throat of an infinite-length exponential horn. f/f_c is the ratio of frequency to cut-off frequency and ρc is the characteristic impedance of air. No acoustic power can be radiated below the cut-off frequency as the real part of the acoustic impedance is zero.

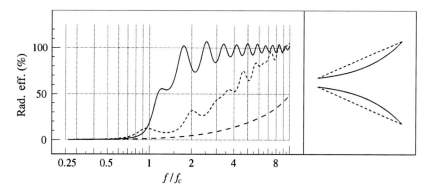

Figure 1.23. Radiation efficiency of an exponential horn (solid line) compared to that of a conical horn (short-dashed line) of the same overall size. Relatively small changes in the flare shape of a horn can have a large effect on the efficiency at low frequencies. The third curve (long-dashed line) is the radiation efficiency of a baffled piston having the same size as the throats of the horns.

the radiation efficiency of a conical horn having the same overall dimensions, and that of a piston the size of the throat mounted on an infinite baffle. The frequency scale is normalized to the cut-off frequency of the exponential horn. The comb-filtering, due to the standing wave field within the horn, can be seen, as can the improvement in radiation efficiency of the conical horn over the baffled piston, and of the exponential horn over the conical horn (at frequencies above cut-off).

The exponential horn acts as an efficient impedance matching transformer at frequencies above cut-off by giving the small throat approximately the radiation efficiency of the large mouth. The power output of a source mounted at the throat of a horn is proportional to the product of its volume velocity and the radiation efficiency at the throat; thus, a small loudspeaker diaphragm mounted at the throat of an exponential horn can radiate low frequencies with high efficiency. Below cut-off, however, the horn flare effectively does nothing, and the radiation efficiency is then similar to the diaphragm mounted on an infinite baffle. This seemingly ideal situation is marred somewhat by the sheer physical size of horn flare required for the efficient radiation of low frequencies. The cut-off frequency is proportional to the flare rate of a horn, which in turn is a function of the throat and mouth sizes and the length of the horn, thus

$$S(L) = S(0)\, e^{mL}$$

so

$$m = \frac{1}{L} \ln\left\{\frac{S(L)}{S(0)}\right\} \tag{1.45}$$

where L is the length of the horn, and $\ln\{\ \}$ denotes the natural logarithm. For a given cut-off frequency and throat size, the length of the horn is determined by the size of the mouth. To avoid gross reflections from the mouth, leading to a strong standing wave field within the horn, and consequently an uneven frequency response, the mouth has to be sufficiently large to act as an efficient radiator of the lowest frequency of interest. In practice, this will be the case if the circumference of the mouth is larger than a wavelength. For the efficient radiation of low frequencies, the mouth is then very large. Also, a low cut-off frequency requires a low flare-rate which, along with the large mouth, requires a long horn. By way of example, a horn required to radiate sound efficiently down to 50 Hz from a loudspeaker with a diaphragm diameter of 200 mm would need a mouth diameter of over 2 metres, and would need to be over 3 metres long! Compromises in the flare-rate raise the cut-off frequency, and compromises in the mouth size gives rise to an uneven frequency response. Reference 5 is a classic paper on the optimum matching of mouth size and flare-rate.

A radiation efficiency of 100% is not usually sufficient to yield the very high electroacoustic efficiencies of 10% to 50% quoted in the introduction of this section. However, unlike 'real' efficiency figures, which compare power output with power input, the radiation efficiency can be greater than 100% as the figure is relative to the radiation of acoustic power into the characteristic impedance of air, ρc. Arranging for a source to see a radiation resistance greater than ρc results in radiation efficiencies greater than 100%. A technique known as compression is used to increase the radiation efficiency of horn drivers; all that is required is for the horn to have a throat that is smaller than the diaphragm of the driver, as shown in Fig. 1.24.

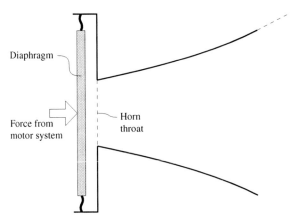

Diaphragm

Force from
motor system

Horn
throat

Figure 1.24. Representation of the principle behind the compression driver. Radiation efficiencies of greater than 100% can be achieved by making the horn throat smaller than the diaphragm.

Assuming that the cavity between the diaphragm and the throat is small compared to a wavelength, it can be shown that the acoustic impedance at the diaphragm is approximately that at the throat multiplied by the ratio of the diaphragm area to the throat area, known as the compression ratio

$$Z_d \approx Z_T \frac{S_d}{S_T} \tag{1.46}$$

where Z_d and S_d are the acoustic impedance and area at the diaphragm, and Z_T and S_T are the acoustic impedance and area at the throat. A compression ratio of 4:1 thus gives a radiation efficiency of 400% at the diaphragm. The 'trick' to achieving optimum electroacoustic efficiency is to match the acoustic impedance to the mechanical impedance (mass, damping, compliance etc.) of the driver. If the compression ratio is too high, the velocity of the diaphragm will be reduced by the additional acoustic load and the gain in efficiency is reduced. This can, however, have the benefit of 'smoothing' the frequency response irregularities brought about by insufficient mouth size, etc. Some dedicated compression drivers operate with compression ratios of 10:1 or more.

1.6.2 Directivity control

In addition to their usefulness as acoustic transformers, horns can be used to control the directivity of a loudspeaker. Equation (1.26) and Fig. 1.5 in Section 1.3.6 show that the directivity of a piston in a baffle narrows as frequency is raised. For many loudspeaker applications, this frequency-dependent directivity is undesirable. In a public address system, for example (as discussed in Chapter 10), the sound radiated from a loudspeaker may be required to 'cover' a region of an audience without too much sound being radiated in other directions where it may increase reverberation. What is required in these circumstances is a loudspeaker with a directivity pattern that can be specified and that is independent of frequency. By attaching a specifically designed horn flare to a loudspeaker driver, this goal can be achieved over a wide range of frequencies.

Consider the simple, straight-sided horn shown in Fig. 1.20. The directivity of this horn can be divided into three frequency regions as shown in Fig. 1.25. At low

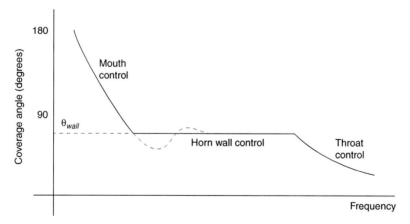

Figure 1.25. Simplified representation of the coverage angle of a straight-sided horn. At low frequencies, the coverage angle is determined by the size of the mouth, and at high frequencies by the size of the throat; the coverage angle in the frequency range between the two is fairly even with frequency and roughly equal to the angle between the horn walls (θ_{wall}). The dashed line shows a narrowing of the coverage angle at the lower end of the wall control frequency range which is often encountered in real horn designs.

frequencies, the coverage angle (see Section 1.3.6) reduces with increasing frequency in a manner determined by the size of the horn mouth, similar to a piston with the dimensions of the mouth. Above a certain frequency, the coverage angle is essentially constant with frequency and is equal to the angle of the horn walls. At high frequencies, the coverage angle again decreases with increasing frequency in a manner determined by the size of the throat, similar to a piston with dimensions of the throat. Thus the frequency range over which the coverage angle is constant is determined by the sizes of the mouth and of the throat of the horn. The coverage angle within this frequency range is determined by the angle of the horn walls. This behaviour is best understood by considering what happens as frequency is reduced. At very high frequencies, the throat beams with a coverage angle which is narrower than the horn walls as if the horn were not there. As frequency is lowered, the coverage angle (of the throat) widens to that of the horn walls and can go no wider. As frequency is further lowered, the coverage angle remains essentially the same as the horn walls until the mouth (as a source) begins to become 'compact' compared to a wavelength and the coverage angle is further increased, eventually becoming omni-directional at very low frequencies. The coverage angle shown in Fig. 1.25 is, of course, a simplification of the actual coverage angle of a horn. In practice, the mouth does not behave as a piston and there is almost always some narrowing of the directivity at the transition frequency between mouth control and horn wall control. A typical example of this is shown as a dashed line in Fig. 1.25. Different coverage angles in the vertical and horizontal planes can be achieved by setting the horn walls to different angles in the two planes.

1.6.3 Horn design compromises

Sections 1.6.1 and 1.6.2 describe two different attributes of horn loudspeakers. Ideally, a horn would be designed to take advantage of both attributes, resulting in a high-efficiency loudspeaker with a smooth frequency response and constant directivity over a wide frequency range. However, very often a horn designed to optimize one aspect of performance must compromise other aspects. For example, the straight-sided horn in Fig. 1.20 may exhibit good directivity control but, being a conical-type horn, will not have the radiation efficiency of an exponential horn of the same size. The curved walls of an exponential horn, on the other hand, do not control directivity as well as straight-sided horns. Early attempts at achieving high efficiency and directivity control in one plane led to the design of the so-called *sectoral horn* or *radial horn* shown in Fig. 1.26. In this design, the two side-walls of the horn are straight, and set to the desired horizontal coverage angle. The vertical dimensions of the horn are then adjusted to yield an overall exponential flare. Whereas the goals of high efficiency and good horizontal directivity control can be achieved with a sectoral horn, the severely compromised vertical directivity can be a problem. Given

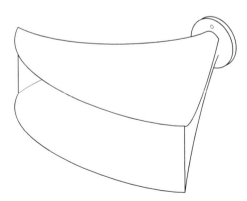

Figure 1.26. Sectoral or radial horn. The walls controlling the horizontal directivity are set to the desired coverage angle. The shape of the other two walls is adjusted to maintain an overall exponential flare resulting in less than ideal vertical directivity.

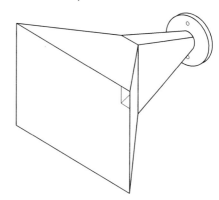

Figure 1.27. Constant directivity horn. Different horn wall angles in the two planes can be achieved using compound flares. Sharp discontinuities within the flare can set up strong standing-wave fields leading to an uneven frequency response.

that a minimum mouth dimension is required for directivity control down to a low frequency, setting the horizontal and vertical walls to different angles, for example 90° by 60°, means that different horn lengths are required in the two planes. To overcome this problem, later designs used compound flares[6] so that the exit angles of the horn walls can be different in the two planes, but the mouth dimensions and overall horn length remain the same. The so-called *constant directivity horn* (CD) is shown in Fig. 1.27. The sudden flare discontinuities introduced into the horn with these designs result in strong standing wave fields within the flare which can compromise frequency response smoothness. In fact, this is true of almost any flare discontinuity in almost any horn. Modern public address horn designs employ smooth transitions between the different flare sections and exponential throat sections to achieve a good overall compromise.

The control of directivity down to low frequencies requires a very large horn. For example, in a horn designed to communicate speech, directivity control may be desirable down to 250 Hz at a coverage angle of 60°. This can only be achieved with a horn mouth greater than 1.5 m across. The same horn may have an upper frequency limit of 8 kHz, which needs a throat no greater than 35 mm across. Maintaining 60° walls between throat and mouth then requires a horn length of about 1.3 m. Attempts to control directivity with smaller devices will almost always fail.

A large number of papers have been written on the subject of horn loudspeakers. As well as references 5 and 6 mentioned above, interested readers are referred to references 4 and 7 for a mathematical approach, 8 for an in-depth discussion on horns for high-quality applications and 9 for a very thorough list of historical references on the subject.

1.7 Non-linear acoustics

In the vast majority of studies in acoustics, and loudspeakers in particular, the acoustic pressures and particle velocities encountered are sufficiently small that the processes of sound radiation and propagation can be assumed to be linear. If a system or process is linear, then there are several rules that govern what happens to signals when they pass through the system or process. These rules include the *principle of superposition*, which states that the response to signal (A + B) is equal to the response to signal (A) + the response to signal (B). Most of the analysis tools and methods used in this chapter, and most other texts on acoustics, such as Fourier analysis and the frequency response function, rely entirely on the principle of superposition, and hence linearity. When a system or process is non-linear, the principle of superposition no longer applies, and the usual analysis methods cannot be used. In this section, the conditions under which acoustic radiation and propagation may become non-linear are discussed along with some examples of the degree of non-linear acoustic behaviour encountered in loudspeakers.

1.7.1 Finite-amplitude acoustics

The speed of sound in air is dependent upon the thermodynamic properties of the air. It may be calculated as follows:

$$c = \sqrt{\gamma RT} = \sqrt{\gamma P/\rho} \tag{1.47}$$

where γ is the ratio of the specific heats, R is the ideal gas constant, T is the absolute temperature, P is the absolute pressure and ρ is the density of air. In all but the most extreme of conditions, γ and R may be considered constant for air with values of 1.4 and 287 J/kgK respectively, but the values of T, P and ρ depend upon the local conditions. From the second half of equation (1.47), it is clear that $P = \rho RT$ (the equation of state), so if the temperature is constant, changes in P must be accompanied by corresponding changes in ρ. An acoustic wave consists of alternate positive and negative pressures above and below the static pressure and, as this is an isentropic process, the relationship between the pressure and the density is given by

$$P \propto \rho^\gamma \tag{1.48}$$

which is non-linear. In linear acoustic theory, the relationship between pressure and density is assumed to be linear, which is a good approximation if the changes in pressure are small compared to the static pressure. A linear relationship between pressure and density means that the temperature does not change, so neither does the speed of sound. However, when the changes in pressure are significant compared to the static pressure, changes in temperature and hence speed of sound cannot be ignored.

In addition, when an acoustic wave exists in flowing air, the speed of propagation is increased in the direction of the flow, and decreased in the direction against the flow; the acoustic wave is 'convected' along with the flow. Although steady air flow is not usually encountered where loudspeakers are operated, the particle velocity associated with acoustic wave propagation can be thought of as an alternating, unsteady flow. Again, if the particle velocities are small compared to the speed of sound (see Section 1.2.4), the effect can be neglected, but in situations where the particle velocities are significant compared to the speed of sound, the dependence of the speed of propagation on the particle velocity cannot be ignored.

Combining the effects of finite acoustic pressure and particle velocity, the instantaneous speed of propagation at time t is given by

$$c(t) = c_0 \left\{ \frac{P_0 + p(t)}{P_0} \right\}^{(\gamma-1)/2\gamma} + u(t) \tag{1.49}$$

where c_0 is the speed of sound at static pressure P_0, $p(t)$ is the instantaneous acoustic pressure and $u(t)$ is the instantaneous acoustic particle velocity.

The result of all of this is that the speed of propagation increases with increasing pressure and particle velocity, and decreases with decreases in pressure and particle velocity. For a plane progressive wave, positive pressures are accompanied by positive particle velocities (see Section 1.2.4), and the speed of propagation is therefore higher in the positive half-cycle of an acoustic wave than it is in the negative half-cycle. The positive half-cycle then propagates faster than the negative half cycle and the waveform distorts as it propagates. Figure 1.28 shows the distortion, known as *waveform steepening*, that occurs in the propagation of sound when the acoustic pressures are significant compared to the static pressure and/or the acoustic particle velocities are significant compared to the static speed of sound.

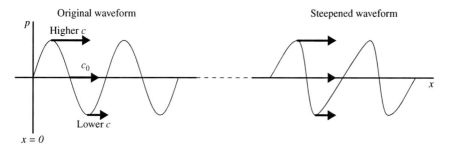

Figure 1.28. Waveform steepening due to acoustic pressures that are significant compared to the static pressure and/or acoustic particle velocities that are significant compared to the speed of sound c_0.

1.7.2 Examples of non-linear acoustics in loudspeakers

Equation (1.49) states that the instantaneous speed of sound propagation is dependent upon the instantaneous values of pressure and particle velocity. At the sound levels typically encountered when loudspeakers are operated, the effect is so small as to be negligible and the resultant linear approximation is sufficiently accurate. However, there are some situations where this is not the case. Two common examples are the high sound pressures in the throats of horn loudspeakers, and the high diaphragm velocities of long-throw low-frequency drive-units.

When horn loudspeakers equipped with compression drivers are used to generate high output levels, the pressure in the throat of the horn can exceed 160 dB SPL, with even higher levels at the diaphragm. Sound propagation is non-linear at these levels and the acoustic waveform distorts as it propagates along the horn. If the horn flares rapidly away from the throat, then these levels are maintained only over a short distance and the distortion is minimized. Horns having throat sections that flare slowly suffer greater waveform distortion (it is interesting to note that the rich harmonic content of a trombone at fortissimo is due to this phenomenon). Investigations[8] have shown that the distortion produced by high-quality horn loudspeakers only exceeds that from high-quality conventional loudspeakers when the horn system is producing output levels beyond the capability of the conventional loudspeakers.

The use of small, long-throw woofers in compact, high-power loudspeaker systems can also introduce non-linear distortion. Equation (1.30) in Section 1.3.7 shows that the power output of a loudspeaker diaphragm is proportional to the square of the volume velocity of the diaphragm. For a given sound power output, the required diaphragm velocity is therefore proportional to the inverse of the diaphragm area. Consider two loudspeakers, one with a diaphragm diameter of 260 mm, the other with a diameter of 65 mm. In order to radiate the same amount of acoustic power at low frequencies, the smaller loudspeaker requires a velocity of 16 times that of the large loudspeaker, as it has 1/16 of the area. The rms velocity of the large loudspeaker when radiating a sound pressure level of 104 dB at 1 m at a frequency of 100 Hz is approximately 0.5 m/s. The same sound pressure level from the smaller loudspeaker requires 8 m/s. Whereas 0.5 m/s may be considered insignificant compared to the speed of sound (≈ 340 m/s), 8 m/s represents peak-to-peak changes in the speed of sound of around 8%.

A secondary effect, which is a direct consequence of particle velocities that are significant compared to the speed of sound, is the so-called Doppler distortion. If, at the same time as radiating the 100 Hz signal above, the small loudspeaker were also radiating a 1 kHz signal, the cyclic approach and recession of the diaphragm due to the low-frequency signal would frequency modulate the radiation of the higher-frequency signal by approximately 70 Hz.

The arguments above tend to imply that there is a maximum amount of sound that can be radiated linearly by a loudspeaker of given dimensions, regardless of any improvements in transducer technology. Clearly, if bigger sounds are required, then bigger (or more) loudspeakers are needed. The huge stacks of loudspeakers seen at outdoor concerts are not just for show ...

References

1.   FAHY, F J, *Sound and Structural Vibration*, Academic Press, London.
2.   RANDALL, R B, *Frequency Analysis*, Bruel & Kjaer, Denmark.
3.   FAHY, F J, *Sound Intensity*, Elsevier, London.
4.   PIERCE, A D, *Acoustics*, Acoustical Society of America, New York.
5.   KEELE D B Jr, 'Optimum horn mouth size', Presented at the 46th Convention of the Audio Engineering Society, AES Preprint No. 933 (1973).
6.   KEELE, D B Jr, 'What's so sacred about exponential horns?' Presented at the 51st Convention of the Audio Engineering Society, AES Preprint No. 1038 (1975).
7.   HOLLAND, K R, FAHY, F J and MORFEY, C L, 'Prediction and measurement of the one-parameter behaviour of horns', *JAES*, **39**(5), 315–337 (1991).
8.   NEWELL, P R, *Studio Monitoring Design*, Focal Press, Oxford.
9.   EISNER, E, 'Complete solutions of the "Webster" horn equation', *Journal of the Acoustical Society of America*, **41**(4) Part 2, 1126–1146 (1967).

Appendix: Complex numbers and the complex exponential

A1.1 Complex numbers

A complex number can be thought of as a point on a two-dimensional map, known as the complex plane. As with any map, a complex number can be represented by its coordinates measured from a central point or origin as shown in Fig. A1.1. All conventional or real numbers lie along the horizontal axis of the complex plane, which is known as the real axis, with positive numbers lying to the right and negative numbers to the left of the origin. The value or size of a real number is then represented by its distance from the origin (the number '0'). All conventional arithmetic – addition, subtraction, multiplication etc. – takes place along this line.

If a real number is multiplied by -1, the line or vector joining the number to the origin is rotated through 180° so that it points in the opposite direction; the number 3 becomes -3, for example. A rotation of the vector through 90°, such that the number now lies along the vertical axis, can similarly be represented by multiplication by something, this time by the operator 'j' (or sometimes 'i'). multiplication by 'j' twice has the same result as multiplication by -1 so it follows that

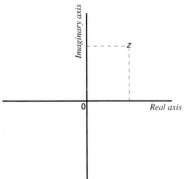

Figure A1.1. The representation of a complex number by its coordinates on a two-dimensional map known as the complex plane.

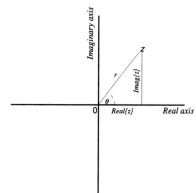

Figure A1.2. The relationships between the real part, imaginary part, magnitude and phase of a complex number.

$$j \times j = -1$$

Therefore

$$j = \sqrt{-1} \qquad (A.1.1)$$

The square root of -1 does not exist in real mathematics, so numbers that lie along the vertical axis of the complex plane are known as imaginary numbers. A complex number can lie anywhere on the complex plane and therefore has both a real coordinate, known as its real part, and an imaginary coordinate or imaginary part. Complex numbers may, therefore, be written in the form

$$\hat{z} = x + jy \qquad (A1.2)$$

where x represents the real part of complex number $\hat{z}$ and y represents the imaginary part, identified by the multiplication by 'j'; the $^\wedge$ over the variable z being used to show that it is complex.

A1.2 Polar representation

As with real numbers, the size or magnitude of a complex number is determined by its distance from the origin, which, by considering the right-angled triangle made by the complex number, its real part and the origin (see Fig. A1.2), is given by Pythagoras' theorem:

$$|\hat{z}| = r = \sqrt{x^2 + y^2} \qquad (A1.3)$$

where $|\hat{z}|$ denotes the magnitude of complex number $\hat{z}$. A complex number having this magnitude can lie anywhere on a circle of radius r, so a second number, known as the phase, is required to pin-point the number on this circle. The phase of a complex number is defined as the angle between the line joining the number to the origin and the positive real axis as shown in Fig. A1.2. Considering the triangle again, the phase of a complex number can be written in terms of the real and imaginary parts using trigonometry:

$$\angle \hat{z} = \theta = \tan^{-1} \left\{ \frac{y}{x} \right\} \qquad (A1.4)$$

where $\angle \hat{z}$ denotes the phase of $\hat{z}$. Using trigonometry again, the real and imaginary parts can be rewritten in terms of the magnitude and phase:

$$x = r \cos (\theta) \quad \text{and} \quad y = r \sin (\theta) \qquad (A1.5a,b)$$

The complex number, $\hat{z} = x + jy$, can therefore be written

$$\hat{z} = r\,(\cos\,(\theta) + j\,\sin\,(\theta)) \tag{A1.6}$$

$$= r\,e^{j\theta} \tag{A1.7}$$

The right-hand-side of equation (A1.7) is known as a complex exponential (which can be written $r\exp\,(j\theta)$), and the relationship between the cos ( ) and sin ( ) in equation (A1.6) and the exponential in (A1.7) is known as Euler's theorem, which will not be proven here. The complex exponential proves to be a very useful way of representing a complex number.

Any complex number may therefore be written in polar form as a complex exponential or in Cartesian form as real and imaginary parts; equations (A1.3) to (A1.7) allowing conversion between the two forms.

A1.3 Complex arithmetic

The arithmetic manipulation of complex numbers is relatively straightforward. Addition or subtraction of complex numbers is carried out in Cartesian form by dealing with the real and imaginary parts separately, thus:

$$(a + jb) \pm (c + jd) = (a \pm c) + j(b \pm d) \tag{A1.8}$$

Multiplication and division is best carried out in polar form by dealing with the magnitudes and phases separately, thus:

$$p\,e^{jq} \times r\,e^{js} = pr\,e^{j(q+s)} \quad\text{and}\quad \frac{p\,e^{jq}}{r\,e^{js}} = \frac{p}{r}\,e^{j(q-s)} \tag{A.9a, b}$$

Multiplication and division can also be carried out in Cartesian form, as follows:

$$(a + jb) \times (c + jd) = (ac - bd) + j(bc + ad) \tag{A1.10}$$

$$\frac{(a + jb)}{(c + jd)} = \frac{(a + jb)(c - jd)}{(c + jd)(c - jd)} = \frac{(ac + bd) + j(bc - ad)}{c^2 + d^2} \tag{A1.11}$$

where it must be remembered that $j \times j = -1$. The addition and subtraction of complex numbers in polar form involves conversion to Cartesian form, application of equation (A1.8), and then conversion of the result back to polar form.

A1.4 Differentiation and integration of complex numbers

The usefulness of the complex exponential as a compact representation of a complex number becomes most apparent when carrying out differentiation and integration. Differentiation of a complex exponential takes the form

$$\frac{d}{dx}\,(e^{jnx}) = jn\,e^{jnx} \tag{A1.12}$$

Similarly, integration takes the form

$$\int e^{jnx}\,dx = \frac{e^{jnx}}{jn} \tag{A1.13}$$

Differentiation just involves multiplication and integration just involves division.

A1.5 The single-frequency signal or sine-wave

Consider a complex number having a fixed magnitude r and a phase θ which changes with time at a fixed rate. The path of the complex number describes a circle on the complex plane centred at the origin with a radius r, and the complex number moves

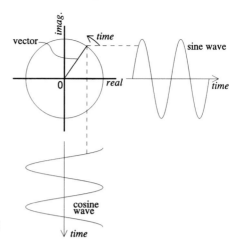

Figure A1.3. The projection of a rotating vector on the real and imaginary axes.

around the circle with an angular velocity of ω radians per second (see Fig. A1.3). The complex exponential representation of such a number is

$$\hat{z}(t) = r \, e^{j\omega t} \tag{A1.14}$$

as the magnitude is fixed and the phase varies with time. According to equation (A1.5), the real part of $\hat{z}(t)$ is $r \cos(\omega t)$ and the imaginary part is $r \sin(\omega t)$, as can be seen in Fig. A1.3; the real and imaginary parts are identical except for a shift of one-quarter of a cycle along the time axis, which represents a phase difference of 90° or multiplication by j. A single-frequency signal can therefore equally well be described by $\sin(\omega t)$ or $\cos(\omega t)$, so a more general description of a single-frequency signal is the complex exponential $e^{j\omega t}$, which, according to equations (A1.6) and (A1.7), is the sum of a cosine-wave and a sine-wave multiplied by j.

The single-frequency signal (often called the sine-wave) is very important in the analysis of linear systems, such as most audio equipment, as it is the only signal which, when used as the input to a linear system, appears at the output modified only in magnitude and phase; its shape or form remains unchanged. Thus the effect that any linear system has on a sine-wave input can be entirely described by a single complex number. In general, there is a different complex number for every different frequency; the complete set of complex numbers is then the frequency response function of the system. The inverse Fourier transform of the frequency response function is the impulse response of the system, which can be used to predict the output of the system in response to any arbitrary input signal.

2 Transducer drive mechanisms

John Watkinson

2.1 A short history

This is not a history book, and this brief section serves only to create a context. Transducer history basically began with Alexander Graham Bell's patent of 1876[1]. Bell had been involved in trying to teach the deaf to speak and wanted a way of displaying speech graphically to help with that. He needed a transducer for the purpose and ended up inventing the telephone.

The traditional telephone receiver is shown in Fig. 2.1(a). The input signal is applied to a solenoid to produce a magnetic field which is an analog of the audio waveform. The field attracts a thin soft iron diaphragm. Ordinarily the attractive force would be a rectified version of the input as in Fig. 2.1(b), but the presence of a permanent magnet biases the system so that the applied signal causes a unipolar but varying field.

The relatively massive iron diaphragm was a serious source of resonances and moving-iron receivers had to be abandoned in the search for fidelity, although countless millions were produced for telephony. The telephonic origin of the transducer led to the term 'receiver' being used initially for larger transducers. These larger devices could produce a higher sound level and, to distinguish them from the telephone earpiece, they were described as 'loudspeaking': the origin of the modern term.

The frequencies involved in an audio transducer are rather high by the standards of mechanical engineering and it is reasonably obvious that it will be much easier to move parts at high frequencies when they are very light. Subsequent developments

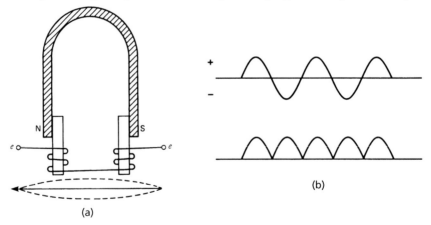

Figure 2.1. (a) Bell telephone receiver uses moving iron diaphragm. (b) Without bias magnet, input waveform (top) would be rectified (bottom).

followed this path. The moving-coil motor used in a loudspeaker was patented by Sir Oliver Lodge in 1898 but, in the absence of suitable amplification equipment, it could not enter wide use. This was remedied by the development of the vacuum tube or thermionic valve which led to wireless (to distinguish it from telephony using wires) broadcasting and a mass market for equipment. The term 'receiver' was then adopted to describe a wireless set or radio, and the term 'loudspeaker' was then firmly adopted for the transducer.

The landmark for the moving-coil loudspeaker was the work of Rice and Kellog[2] in the 1920s which essentially described the direct radiating moving-coil loudspeaker as it is still known today. Much of this chapter will be devoted to the consequences of that work.

The product which Rice and Kellog developed was known as the Radiola 104. This was an 'active' loudspeaker because the 610 mm square cabinet had an integral 10 watt Class-A amplifier to drive the coil in the 152 mm diameter drive unit. Power was also needed to energize the field in which the coil moved. This was because the permanent magnets of the day lacked the necessary field strength. The field coil had substantial inductance and did double duty as the smoothing choke in the HT supply to the amplifier.

Developments in magnet technology made it possible to replace the field coil with a suitable permanent magnet towards the end of the 1930s and there has been little change in the concept since then. Figure 2.2 shows the structure of a typical low-cost unit containing an annular ferrite magnet. The magnet produces a radial field in which the coil operates. The coil drives the centre of the diaphragm or cone which is supported by a *spider* allowing axial but not radial movement. The perimeter of the cone is supported by a flexible *surround*. The end of the coil is blanked off by a domed *dust cap* which is acoustically part of the cone. When the cone moves towards the magnet, air under the dust cap and the spider will be compressed and suitable vents must be provided to allow it to escape. If this is not done the air will force its way out at high speed causing turbulence and resulting in noise which is known as *chuffing*.

The development of 'talking pictures', as motion picture films with a soundtrack were first known, brought about a need for loudspeakers which could produce adequate sound levels in large auditoria. Given the limited power available from the amplifiers of the day, the only practical way of meeting the requirements of the cinema was to develop extremely efficient loudspeakers. The direct radiating speaker is very inefficient because the mass of air it can influence is very small compared to its own moving mass. The use of a horn makes a loudspeaker more efficient because the mass of air which can be influenced is now determined by the area of the mouth

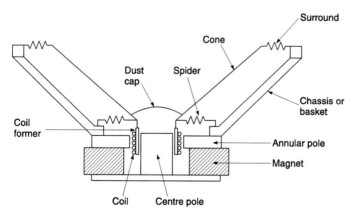

Figure 2.2. The components of a moving-coil loudspeaker.

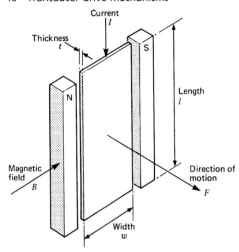

Figure 2.3. Basic dimensions of a ribbon in a magnetic field.

of the horn (see Chapter 1). A good way of considering the horn is that it is effectively an acoustic transformer providing a good match between the impedance of sound propagating in air and the rather different impedance of a practical diaphragm assembly which must be relatively massive. As will be seen later in this chapter, horns which are capable of a wide frequency range need to be very large and once powerful amplification became available at low cost the need for the horn was reduced except for special purposes.

Another application of the philosophy that the moving parts should be light was the development of the ribbon loudspeaker. Here the magnetic motor principle is retained, but the coil and the diaphragm become one and the same part. As Fig. 2.3 shows, the diaphragm becomes a current sheet which is driven all over its surface.

The electrostatic speaker (see Chapter 3) continues the theme of minimal moving mass, but it obtains its driving force in a different way. No magnets are involved.

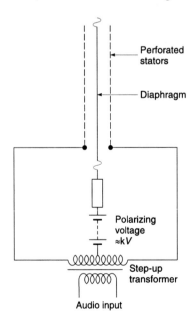

Figure 2.4. The electrostatic loudspeaker uses the force experienced by a charge in an electric field. The charge is obtained by polarizing the diaphragm.

Instead, the force is created by applying an electric field to a charge trapped in the diaphragm. Figure 2.4 shows that the electric field is created by applying a voltage between a pair of fixed electrodes. The charge in the diaphragm is provided by a high-voltage polarizing supply and the charge is trapped by making the diaphragm from a material of extremely high resistivity.

The main difficulty with the electrostatic speaker is that the sound created at the diaphragm has to pass through the stationary electrodes. The necessary requirements for acoustic transparency and electrical efficiency are contradictory and an optimum technical compromise may be difficult to manufacture economically. The great advantage of the electrostatic loudspeaker is that, when the constant charge requirements are met, the force on the charge depends only on the applied field and is independent of its position between the electrodes. In other words the fundamental transduction mechanism is utterly linear and so very low distortion can be achieved. This concept was first suggested by Carlo V. Bocciarelli and subsequently analysed in detail by Frederick V. Hunt[3]. The validity of the theory has been clearly demonstrated by the performance of the Quad electrostatic speakers developed by Peter Walker and described in detail in Chapter 3.

One of Peter Walker's significant contributions to the art was to divide the diaphragm into annular sections driven by different signals. This essentially created a phased array which could completely overcome the beaming problems of a planar transducer, allowing electrostatic speakers of considerable size and power to retain optimal directivity.

Another advantage of the electrostatic speaker is that there is no heat-dissipation mechanism in the speaker apart from a negligible dielectric loss and consequently there is no thermal stress on the components. The electrostatic speaker itself is the most efficient transducer known. However, the speaker has a high capacitance and the amplifier has to drive charge in and out of that capacitance in order to cause a voltage swing. The amplifier works with an adverse power factor and a conventional linear amplifier will therefore be inefficient. In an active electrostatic hybrid speaker, the panel amplifier could easily dissipate more than the woofer amplifier. However, amplifier topologies which remain efficient with adverse power factors, such as switching amplifiers, can be used to advantage in electrostatic speakers.

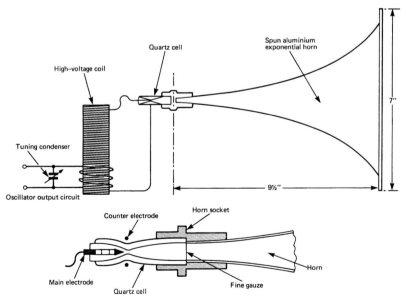

Figure 2.5. Detail of the Ionophone assembly with the output circuit of the oscillator.

The ribbon and the electrostatic speaker share the principle that the driving force is created at the point where it is to be used and so there is no requirement for mechanical vibrations to travel from one point to another as is the case with the moving-coil speaker. As the propagation speed of vibration is finite, this makes it difficult for a moving-coil transducer to have a minimum phase characteristic whereas the ribbon and the electrostatic do this naturally.

The ultimate loudspeaker might be one with no moving parts in which the air itself is persuaded to move. This has been elusive and the only commercially available device was the Ionophone, first produced by S. Klein in 1951. Figure 2.5 shows that the transduction mechanism in the Ionophone is the variation in the amplitude of an ionic discharge. The discharge is created by power from a radio frequency oscillator which is amplitude modulated by the audio signal. The displacement available is limited and adequate sound level can only be obtained by using a horn. The unit was produced by Fane in the UK for a period but was discontinued in 1968.

The loudspeaker drive unit of the 1980s hardly differed from those designed by Rice and Kellog sixty years earlier, except that it had primarily become a commoditized component designed for simple mass production rather than quality. With a few noteworthy exceptions, progress had almost been replaced by stultification. It is a matter of some concern that the loudspeaker industry is in danger of falling into disrepute. The intending purchaser of a drive unit or a complete loudspeaker cannot rely on the specification. Virtually all specifications are free of any information regarding linear or harmonic distortion and the result is a range of units with apparently identical specifications which sound completely different, both from one another and from the original sound. Almost invariably the power level at which the sound is reasonably undistorted will be a small fraction of the advertised power handling figure. The 1990s brought a brighter prospect when the falling cost and size of electronics allowed the economic development of the active loudspeaker.

The traditional passive loudspeaker is designed to be connected to a flat frequency response wideband amplifier acting as a voltage source. This causes too many compromises for accurate results. In fact all that is required is that the overall frequency, time and directional response of the loudspeaker and its associated electronics is correct. What goes on between the electronics and the transducers is actually irrelevant to the user.

The greatest problem with the progress of the active loudspeaker is that (with a few notable exceptions) passive loudspeaker manufacturers in many cases do not have the intellectual property needed to build active electronics and amplifier manufacturers lack the intellectual property needed to build transducers. Manufacturers of commoditized products regard a change of direction as an unnecessary risk and tend to resist this inevitable development, leaving the high ground of loudspeaker technology to newcomers.

2.2 The diaphragm

One of the great contradictions in loudspeaker design is that there is no optimal size for the diaphragm, thus whatever the designer does is wrong and compromise is an inherent part of the process. Figure 2.6 shows the criteria for the radiation of low frequencies. Here the diaphragm is small compared to the wavelength of the sound radiated and so all that matters is the diaphragm area and displacement. A small diaphragm will need a long travel or 'throw' and this will result in a very inefficient motor and practical difficulties in the surround and spider. As human hearing is relatively insensitive at low frequencies, a loudspeaker which can only reproduce the lowest audible frequencies at low power is pointless. Thus, in practice, radiation of significant power at low frequencies will require a large-diameter diaphragm. At some point the diaphragm size is limited by structural rigidity and further power increase will require the installation of multiple drivers.

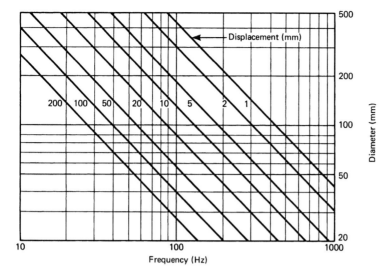

Figure 2.6. Peak amplitude of piston needed to radiate 1 W.

This approach is beneficial at low frequencies where the drive units are in one another's near field. In the case of a pair of woofers, the radiation resistance seen by each is doubled by the presence of the other, so that four times as much power can be radiated.

Unfortunately any diaphragm which is adequate for low-frequency use will be unsuitable at high audio frequencies. Chapter 1 showed that large rigid pistons have an extremely high value of *ka* at high frequencies and that this would result in *beaming* or high directivity which is undesirable. A partial solution is to use a number of drive units which each handle only part of the frequency range. Those producing low frequencies are called *woofers* and will have large diaphragms with considerable travel whereas those producing high frequencies are called *tweeters* and will have small diaphragms whose movement is seldom visible. In some systems mid-range units or *squawkers* are also used, having characteristics mid-way between the other types. A frequency-dividing system or *crossover network* is required to limit the range of signals fed to each unit (see Chapter 5).

A particular difficulty of this approach is that the integration of the multiple drive units into one coherent source in time, frequency and spatial domains is non-trivial. Most multiple drive unit loudspeakers are sub-optimal in one or more of these respects. One aspect of diaphragms which is fundamental to an understanding of loudspeakers is that the speed of propagation of vibrations through them is finite. The vibration is imparted where the coil former is attached. At low frequencies, the period of the signal is long compared to the speed of propagation and so, with suitable structural design, the entire diaphragm can move in essentially the same phase as if it were a rigid piston. As frequency rises, this will cease to be the case. The finite propagation speed will result in phase shifts between the motion of different parts of the diaphragm. The diaphragm is no longer pistonic but acts as a phased array. Where pistonic motion is a requirement, as in a sub-woofer, a cone material having a high propagation speed is required.

The finite speed of propagation of vibrations from the coil to the part of the diaphragm which is radiating causes a delay in the radiated sound waveform relative to the electrical input waveform. The result of this delay is that, acoustically, the diaphragm appears to be some distance behind its true location. The point from which the sound appears to have come from by virtue of its timing is called the *acoustic source*. In most moving-coil designs the acoustic source will be in the vicinity of the coil. Some treatments state that the acoustic source is in the centre of the coil, but this is not necessarily true as the speed of sound within the coil former will generally be higher than it is in air.

In quality speakers, the diaphragm should be designed to avoid uncontrolled breakup. One useful step is the selection of a suitable shape. This in fact is the origin of the term 'cone', which is used today almost interchangeably with diaphragm even when the shape is not a cone. Rolling a thin sheet of material into a conical shape increases the stiffness enormously.

Figure 2.7(a) shows a drive unit of moderate quality, having a paper cone which is corrugated at the edge to act as an integral surround. As can be seen in (b) the frequency response is pretty awful. This is due to uncontrolled cone breakup. Figure 2.7(c) shows the various modes involved. Modes 1, 2 and 3 are pure bell modes which occur when the circumference is an integral number of wavelengths. Modes 4, 5, 7 and 9 are concentric modes which occur when the distance from the coil to the surround and back is an integral number of wavelengths. Mode 6 is a combination of the two.

In poor-quality loudspeakers used, for example, in most transistor radios, deliberate breakup is used to increase apparent efficiency at mid and high frequencies. The finite critical bandwidth of the ear means that, with a complex spectrum, the increase in loudness due to the resonant peaks is sensed rather than any loss due to a response dip. These resonances can be very fatiguing to the listener.

In the flat panel transducers developed by NXT (see Chapter 4), the moving-coil motor creates a point drive in a relatively insubstantial diaphragm which is intended to break up. Careful design is required to make the breakup chaotic so that no irritating dips and peaks occur in the response. The advantage of this approach is simply a very thin construction which allows a loudspeaker to be located where a conventional transducer would be impossible, and it is in applications such as this where the NXT transducer is meeting most success. The development of a transparent diaphragm allows any visual display to be given an audio capability.

The transient or time response of such a transducer is questionable and stereophonic imaging is inferior to the best conventional practice [but see Chapter 4, Section 4.17: Editor]. However, the chaotic behaviour of the NXT panel may be appropriate for the rear speakers of a surround-sound system in which the rear ambience may benefit.

In drive units designed for electrical musical instruments (see Chapter 11), nonlinearity may be deliberately employed to create harmonics to produce a richer sound. This may also be true of the amplifier. In this case the quality criteria are quite different because the amplifier and speaker are part of the musical instrument.

Where higher quality is required, suitable termination at the diaphragm surround must be provided. This suppresses concentric modes because vibrations arriving from the coil are not reflected. Bell modes are also damped. This property may be inherent in the surround material or an additional damping element may be added.

Bell modes can also be discouraged by curving the diaphragm profile so that the diaphragm has compound curvature. In the case of a woofer, which needs a large yet pistonic diaphragm, one way of avoiding breakup while at the same time increasing efficiency is to use a large coil diameter. Figure 2.8(a) shows that with a traditional design all of the coil thrust is concentrated at the centre of the cone and the stress on the cone material is high adjacent to the coil. The perimeter of the cone is a long way from the coil and so the cone needs to be stiff and therefore heavy to prevent breakup. Note that (b) shows that the acoustic source of the speaker is also a long way back from the baffle.

Figure 2.8(c) shows that with a large diameter coil the stress due to drive thrust in the coil former is reduced. Around half of the area of the diaphragm is now inside the coil and the thrust of the coil is now divided between two diaphragm sections, reducing the stress on the diaphragm material by a factor of about ten. The inner

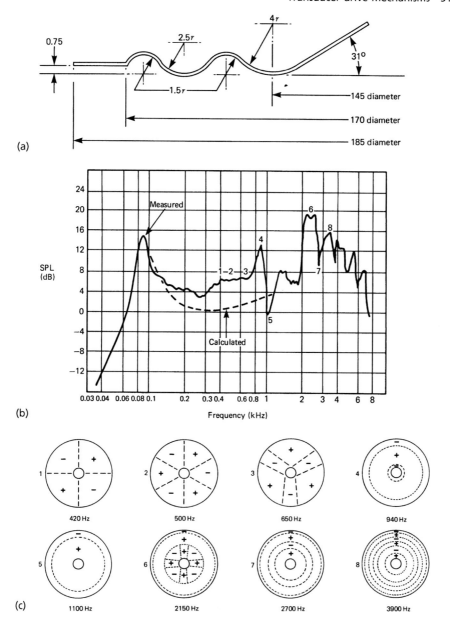

Figure 2.7. (a) Detail of the edge of a felted paper cone. (b) Power available response of a 200 mm diameter loudspeaker. (c) Nodal patterns of the cone shown in (a).

part of the diaphragm is now a dome whose diameter is constrained by the coil former. This gives a very rigid structure and so the material can be very thin and light. Also the part of the diaphragm outside the coil is now relatively narrow and adequate stiffness can be achieved with a light section.

Such a diaphragm can be very rigid while having significantly lower mass than a conventional cone. The saving in diaphragm mass can be used to adopt a heavier

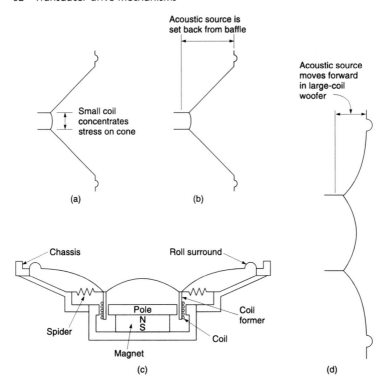

Figure 2.8. Traditional woofer (a) has small coil concentrating stress on cone apex. (b) Acoustic centre is set back from the baffle making time alignment with tweeter difficult. (c) Large coil diameter reduces stress on diaphragm. (d) Acoustic centre moves forward.

coil which will produce more driving force. The result is that the speaker can be made more efficient simply by choosing a more logical diaphragm design. A further benefit of this approach is that the acoustic source of the woofer is now closer to the baffle, making it easier to time-align the woofer and tweeter signals (d). Figure 2.9 shows a drive unit based on this principle. Turning now to high frequencies, the fact that the diaphragm becomes a phased array can be turned to advantage. Figure 2.10 shows that, if the flare angle of a cone-type moving coil unit is correct for the material, the forward component of the speed of vibrations in the cone can be made slightly less than the speed of sound in the air, so that nearly spherical wavefronts can be launched. The cone is acting as a mechanical transmission line for vibrations which start at the coil former and work outwards.

A frequency-dependent loss can be introduced into the transmission line either by using a suitably lossy material or by attaching a layer of such material to the existing cone. It is also possible to achieve this result by concentric corrugations in the cone profile[4]. When this is done, the higher the frequency, the smaller is the area of the cone which radiates. Correctly implemented, the result is a constant dispersion drive unit. As stated above, there are vibrations travelling out across the cone surface and the cone surround must act as a matched terminator so that there can be no reflections.

In the Manger transducer[5], the directivity issue is addressed by making the flat diaphragm from a material which allows it to act as a transmission line to bending waves. Figure 2.11 shows clearly the termination at the perimeter of the diaphragm. The diaphragm of the Manger transducer mechanically implements the delays needed

Figure 2.9. Production large-coil woofer based on concepts of Fig. 2.8. (Courtesy Morel Ltd.)

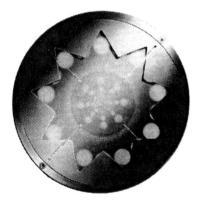

Figure 2.10. A cone built as a lossy transmission line can reduce its diameter as a function of frequency, giving smooth directivity characteristics.

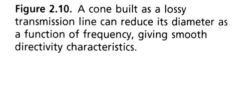

Figure 2.11. The Manger transducer uses a flat diaphragm which is a transmission line to bending waves. Note the termination damping at the perimeter of the diaphragm. (Courtesy Manger Products.)

to create a phased array. The result is a moving-coil transducer which is more akin to an electrostatic in its transient response and clarity.

2.3 Diaphragm material

The choice of diaphragm material must be the result of balancing efficiency and sound quality with material cost and ease of fabrication. In general, the higher the stiffness for a given weight, the better the material. In fact this corresponds to the criterion for a high vibration propagation speed. The other important parameter is the mechanical Q-factor, as this affects the ability of the material to damp resonances.

Where high efficiency is important, as in portable equipment, paper cones will be employed. The paper may be hard resin impregnated or filled, pressed and calendered. Sometimes a mineral filler is added to the resin. The approach is that by minimizing losses the greatest amount of resonance will occur and the efficiency will be improved. The sound quality is generally poor, particularly the transient response.

A softer type of paper will have better damping and give better transient response with some loss of efficiency. For many years, paper was the only material available but in the 1950s thermoplastics arrived. Vacuum forming of thin thermoplastic sheet gave a more repeatable result than paper. Some care is needed to ensure that the plasticizer does not migrate in service leaving a brittle cone, but otherwise such an approach is highly suited to mass production at low cost. Clearly thermoplastic cones are unsuitable for very high power applications because the heat from the coil will soften the cone material.

The aerospace industry is also concerned with materials having high stiffness-to-weight ratio, and this has led to the development of composites such as Kevlar and carbon fibre which have the advantage of maintaining their properties both with age and at higher temperatures than thermoplastics can sustain.

A sandwich construction can be used to obtain high stiffness. A pair of thin stressed skins, typically of aluminium, can be separated by expanded plastics foam. Another sandwich construction uses a thin aluminium cone which is spun or pressed to shape and then hard anodized. The anodized layers act as the stressed skins.

Aluminium and beryllium are attractive as cone materials, but the latter is difficult to work as well as being poisonous. Large aluminium cones can have very sharp resonances at the top of the working band. These can be overcome in moderately sized cones using techniques such as compound curvature.

Some success has been had with expanded polystyrene diaphragms which are solid in that the rear is conical but the front face is flat. The main difficulty is that vibrations from the coil can reflect within the body of the diaphragm, causing diffraction patterns. However, in woofers this is not an issue.

2.4 Magnetism

Most loudspeakers rely on permanent magnets and good design requires more than a superficial acquaintance with the principles. A magnetic field can be created by passing a current through a solenoid, which is no more than a coil of wire, and this is exactly what was used in early loudspeakers. When the current ceases, the magnetism disappears. However, many materials, some quite common, display a permanent magnetic field with no apparent power source.

Magnetism of this kind results from the orbiting of electrons within atoms. Different orbits can hold a different number of electrons. The distribution of electrons determines whether the element is diamagnetic (non-magnetic) or paramagnetic (magnetic characteristics are possible). Diamagnetic materials have an even number of electrons in each orbit where half of them spin in each direction cancelling any resultant magnetic moment. Fortunately the transition elements have an odd number of electrons in certain orbits and the magnetic moment due to electronic spin is not cancelled out. In ferromagnetic materials such as iron, cobalt or nickel, the resultant electron spins can be aligned and the most powerful magnetic behaviour is obtained.

It is not immediately clear how a material in which electron spins are parallel could ever exist in an unmagnetized state or how it could be partially magnetized by a relatively small external field. The theory of magnetic domains has been developed to explain it. Figure 2.12(a) shows a ferromagnetic bar which is demagnetized. It has no net magnetic moment because it is divided into domains or volumes which have equal and opposite moments. Ferromagnetic material divides into domains in order

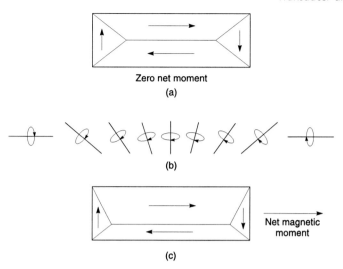

Zero net moment

(a)

(b)

Net magnetic
moment

(c)

Figure 2.12. (a) A magnetic material can have a zero net movement if it is divided into domains as shown here. Domain walls (b) are areas in which the magnetic spin gradually changes from one domain to another. The stresses which result store energy. When some domains dominate, a net magnetic moment can exist as in (c).

to reduce its magnetostatic energy. Within a domain wall, which is around 0.1 micrometres thick, the axis of spin gradually rotates from one state to another. An external field is capable of disturbing the equilibrium of the domain wall by favouring one axis of spin over the other. The result is that the domain wall moves and one domain becomes larger at the expense of another. In this way the net magnetic moment of the bar is no longer zero as shown in (c).

For small distances, the domain wall motion is linear and reversible if the change in the applied field is reversed. However, larger movements are irreversible because heat is dissipated as the wall jumps to reduce its energy. Following such a domain wall jump, the material remains magnetized after the external field is removed and an opposing external field must be applied which must do further work to bring the domain wall back again. This is a process of hysteresis where work must be done to move each way. Were it not for this non-linear mechanism, permanent magnets would not exist and this book would be a lot shorter. Note that above a certain temperature, known as the Curie temperature of the material concerned, permanent magnetism is lost. A magnetic material will take on an applied field as it cools again.

Figure 2.13 shows a hysteresis loop which is obtained by plotting the magnetization B when the external field H is swept to and fro. On the macroscopic scale, the loop appears to be a smooth curve, whereas on a small scale it is in fact composed of a large number of small jumps. These were first discovered by Barkhausen. Starting from the unmagnetized state at the origin, as an external field is applied, the response is initially linear and the slope is given by the susceptibility. As the applied field is increased, a point is reached where the magnetization ceases to increase. This is the saturation magnetization B_s. If the applied field is removed, the magnetization falls, not to zero, but to the remanent magnetization B_r which makes permanent magnets possible. The ratio of B_r to B_s is called the squareness ratio. Squareness is beneficial in magnets as it increases the remanent magnetization.

If an increasing external field is applied in the opposite direction, the curve continues to the point where the magnetization is zero. The field required to achieve this is called the intrinsic coercive force $_mH_c$. This corner of the hysteresis curve is

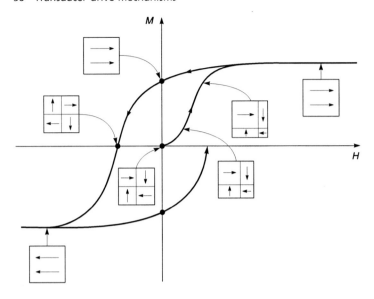

Figure 2.13. A hysteris loop which comes about because of the non-linear behaviour of magnetic materials. If this characteristic were absent, magnetic recording would not exist.

the most important area for permanent magnets and is known as the demagnetization curve. Figure 2.14 shows some demagnetization curves for various types of magnetic materials. Top right of the curve is the short circuit flux/unit area which would be available if a hypothetical (and unobtainable) zero reluctance material bridged the poles. Bottom left is the open circuit mmf/unit length which would be available if the magnet were immersed in a hypothetical magnetic insulator. There is a lot of similarity here with an electrical cell having an internal resistance.

Maximum power transfer VI_{max} is when the load and internal resistances are equal. In a magnetic circuit the greatest efficiency BH_{max} is where the external reluctance

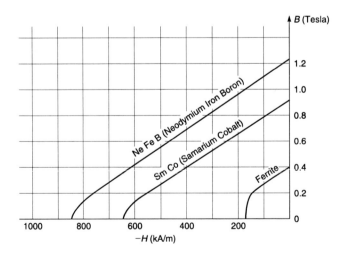

Figure 2.14. Demagnetization curves for various magnetic materials.

matches the internal reluctance. Working at some other point requires a larger and more expensive magnet. The BH_{max} parameter is used to compare the power of magnetic materials. The units are kiloJoules per cubic metre, but the older unit of MegaGauss-Oersteds will also be found.

At the turn of the twentieth century, the primary permanent magnetic material was glass-hard carbon steel which offered about 1.6 kJ/m^3. In 1920 Honda and Takei[6] discovered the cobalt steels. In 1934 Horsburgh and Tetley developed the cobalt–iron–nickel–aluminium system, later further improved with copper. This went by the name of 'Alnico' and offered 12.8 kJ/m^3. In 1938 Oliver and Shedden discovered that cooling the material from above its Curie temperature in a magnetic field dramatically increased the BH product. By 1948 BH products of 60 kJ/m^3 were available at moderate cost and were widely used in loudspeakers under the name of Alcomax. Another material popular in loudspeakers is Ticonal which contains titanium, cobalt, nickel, iron, aluminium and copper.

Around 1930 the telephone industry was looking for non-conductive magnetically soft materials to reduce eddy current losses in transformers. This led to the discovery of the ferrites. The most common of these is barium ferrite which is made by replacing the ferrous ion in ferrous ferrite with a barium ion. The BH product of barium ferrite is relatively poor at only about 30 kJ/m^3, but it is incredibly cheap. Strontium ferrite magnets are also used. In the 1970s the price of cobalt went up by a factor of twenty because of political problems in Zaire, the principal source. This basically priced magnets using cobalt out of the mass loudspeaker market, forcing commodity speaker manufacturers to adopt ferrite. The hurried conversion to ferrite resulted in some poor magnetic circuit design, a tradition which persists to this day. Ferrite has such low B_r that a large area magnet is needed. When a replacement was needed for cobalt-based magnets, most manufacturers chose to retain the same cone and coil dimensions. This meant that the ferrite magnet had to be fitted outside the coil, a suboptimal configuration creating a large leakage area. Consequently traditional ferrite loudspeakers attract anything ferrous nearby and distort the picture on CRTs. It is to be hoped that legislation regarding stray fields emitted by equipment will bring this practice to a halt in the near future. Subsequently magnet technology continued to improve, with the development of samarium cobalt magnets offering around 160 kJ/m^3 and subsequently neodymium iron boron magnets offering a remarkable 280 kJ/m^3. A magnet of this kind requires 10% of the volume of a ferrite magnet to provide the same field. The rare earth magnets are very powerful, but the highest energy types have a low Curie temperature which means they are restricted in operating temperature.

The goal of the magnet and magnetic circuit is to create a radial magnetic field in an annular gap in which the coil moves. The field in the gap has to be paid for. The gap has a finite volume due to its radial spacing and its length along the coil axis. If the gap spacing is increased, the reluctance goes up and the length of the magnet has to be increased to drive the same amount of flux through the gap. If the gap length is increased, the flux density B goes down unless a magnet of larger cross-sectional area is used. Thus the magnet volume tends to be proportional to the gap volume.

It is useful to use electrical analogs to obtain a feel for what is happening in a magnetic circuit. Magnetomotive force (mmf) is the property which tries to drive flux around a magnetic circuit and this is analogous to voltage. Reluctance is the property of materials which resists the flow of flux and this is analogous to resistance, having the same relationship with length and cross-sectional area. Magnetic flux is measured in Webers and is analogous to current. The flux density B is measured in Tesla or Webers per square metre.

Figure 2.15 shows a simple equivalent circuit. The magnet is modelled by a source of mmf with an internal reluctance. As stated above, the greatest efficiency is obtained when the load reluctance is equal to the internal reluctance of the magnet. Working at any other point simply wastes money by requiring a larger magnet. The external

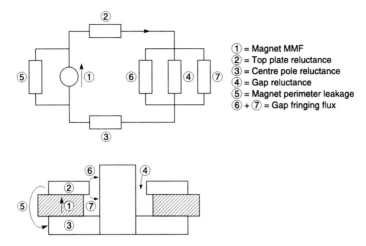

① = Magnet MMF
② = Top plate reluctance
③ = Centre pole reluctance
④ = Gap reluctance
⑤ = Magnet perimeter leakage
⑥ + ⑦ = Gap fringing flux

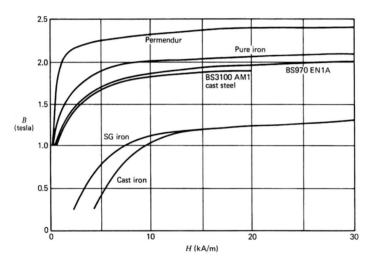

Figure 2.15. Magnetic circuits can be modelled by electrical analogs as shown here. As there is no magnetic equivalent of an insulator, the leakage paths are widely distributed.

Figure 2.16. Induction curves for soft iron.

or load reluctance is dominated by that of the air gap where the coil operates. There will be some reluctance in series with the air gap due to the pole pieces.

The reluctance of the pole pieces can be deduced from the induction curves for the pole material. Figure 2.16 shows these curves for typical materials. Note that it is not practicable to run the pole pieces close to saturation as this simply wastes too much magnet mmf and encourages leakage. For gap flux densities up to about 1.7 Tesla, free-cutting steel such as EN1A gives adequate performance at very low cost. For higher flux densities, Permendur is needed. This is difficult to machine and needs heat treatment afterwards and in most cases simply is not worth the trouble. The induction curve for cast iron is included in Fig. 2.16 to show that it is not worth considering.

In a practical magnet not all of the available flux passes through the air gap because the air in the gap does not differ from the air elsewhere around the magnet and the

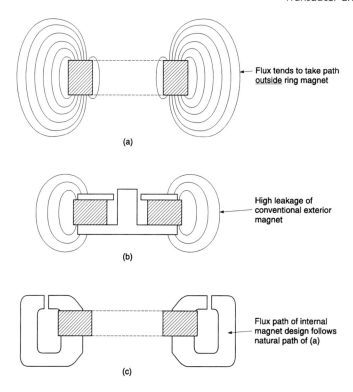

Flux tends to take path
outside ring magnet

(a)

High leakage of
conventional exterior
magnet

(b)

Flux path of internal
magnet design follows
natural path of (a)

(c)

Figure 2.17. (a) Ring magnet in space. As lines of flux mutually repel, little flux passes through the centre hole. Instead flux prefers to travel outside magnet. (b) Conventional magnet design opposes natural path of flux, causing high leakage. (c) Large coil design allows magnet to be inside coil so flux follows natural path and leakage is reduced.

flux is happy to take a shorter route home via a leakage path which is modelled as various parallel reluctances as shown. As there are no practical magnetic equivalents of insulators, the art of magnetic circuit design is to choose a configuration in which the pole pieces guide the flux where it would tend to go naturally. Designs which force the flux in unnatural directions are doomed to high leakage, needing a larger magnet and possibly also screening.

In a loudspeaker the coil is circular for practical reasons and the magnet will therefore be toroidal. Figure 2.17(a) shows a toroidal axially magnetized permanent magnet in space. Note that the flux passes around the outside of the magnet, not through the hole. This is because lines of flux are mutually repulsive and they would have to get closer together to go down the hole.

The traditional loudspeaker has a ferrite magnet outside the coil, as shown in (b) and so flux must be brought inwards to the gap; a direction in which it does not naturally wish to go. As a result a lot of flux continues to flow outwards as leakage. This can be slightly reduced by undercutting the pole pieces as shown. Figure 2.17(c) shows that, if a larger coil diameter is used, the coil is outside the magnet and so the pole pieces lead flux in a direction in which it would naturally go, making the leakage negligible.

Higher performance can be obtained with high-energy magnetic materials such as neodymium iron boron. In this case a small cross-section magnet is needed, which will fit inside all but the smallest coils. This has a shorter perimeter and less leakage. Thus, although rare earth magnets are more expensive, the cost is offset by the fact

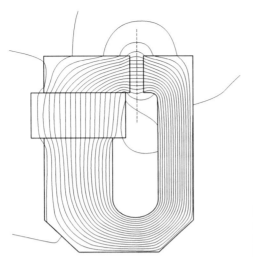

Figure 2.18. Computer simulation of magnetic circuit with Neodymium magnet inside coil. Note low leakage. (Courtesy Celtic Audio Ltd.)

that in a practical design more of the flux goes through the gap. Unless a rare earth magnet design is very bad, no screening for stray flux is needed at all. Small magnets have an acoustic advantage of causing less blockage to the sound leaving the rear of the diaphragm.

Figure 2.18 shows the Neodymium magnetic circuit developed by the author for a 200 mm woofer. Note the almost complete absence of leakage and, as a direct result, how much of the magnet flux actually passes through the gap. The absence of leakage and the higher energy of Neodymium mean that the magnet here has only one-twentieth the volume of an external ferrite magnet to do the same job.

The use of a large coil outside the magnet is difficult with ferrite magnets on small drive units, because insufficient cross-sectional area exists, but for woofers of 30 cm and above in diameter, an external coil solution is worth considering with ferrite. This has a number of advantages as was mentioned in Section 2.2.

When designing a magnetic circuit, the starting point is usually a requirement for a given Bl product. Some preliminary possibilities for a practicable coil will then result in a specification for the coil diameter and thickness, the length of the coil to be immersed in the field and the flux density needed. After allowing some clearance space inside and outside the coil to permit free movement, the dimensions of the gap are established. Clearance may range from 0.15 mm for tweeters to 0.4 mm for large woofers with a domestic speaker having a clearance of typically 0.25 mm. Where coils are expected to run at high temperature, extra clearance on the outer diameter will be needed to allow for coil expansion.

The gap dimensions allow the gap reluctance to be established. This figure is normally increased by about 10% to allow for the reluctance of the pole pieces. This will then give the load reluctance which should be the same as the internal reluctance of the magnet. If the flux density in the gap is specified, the total gap flux is then obtained from the gap area. This will then be increased by a factor to allow for leakage. In an internal design this might be 5–10%. In a large traditional external ferrite design it might be 30%. The flux density of the magnetic material selected at its BH_{max} point is then divided into the total flux to give the magnet cross-sectional area required.

The total reluctance is multiplied by the total flux to give the mmf required. The length of the magnet is found by dividing the mmf by the H value for the selected magnetic material at its BH_{max} point. The above will result in a magnet which is working at its optimum load point.

It should be noted that, in an internal design, the magnet area calculated must always be somewhat less than the area of the inner pole so that a breathing hole can be allowed in the centre of the magnet. If this is not the case, the design will have to be iterated with a larger coil or the flux requirement will have to be reduced.

2.5 The coil

The coil is the part of the loudspeaker which develops the driving force. This is its only desirable feature; everything else the coil does is a drawback. The coil designer has to maximize the desirable while minimizing the drawbacks. If the moving-coil motor is to be linear, the force produced must depend only on the current and not on the position. In the case of a tweeter, shown in Fig. 2.19(a), the travel is usually so small that the coil can be made the same length as the flux gap, as the leakage flux effectively extends the gap.

In woofers the travel is significant and, in order to obtain linearity, the coil can be made shorter (b) or longer (c) than the flux gap. Clearly (c) is less efficient than (a) as power is wasted driving current through those parts of the coil which are not in the gap. However, (b) requires a massive magnet structure and is little used because of the weight and cost. It is important that the flux distribution is symmetrical otherwise distortion will result. Figure 2.20(a) shows a typical low-cost speaker with casual pole design. Figure 2.20(b) shows a better approach. Figure 2.20(c) shows the computer simulation of the flux distribution in the gap, of a magnetic circuit designed by the author, indicating the required symmetry at the ends of the gap.

The force created in the coil is caused by the flow of current interacting with the field from the magnet. The lines of flux of a quiescent speaker motor are precisely orthogonal to the direction of motion and act as though they are in tension as shown in Fig. 2.21(a). The only way a force can be created is if the lines of force are deflected as in (b) so that a component of tension acts in the direction of coil motion. As a result, when the coil accelerates forwards, the gap flux is pushed backwards and vice versa. The motion of the flux within the magnet and the magnetic circuit can produce distortion.

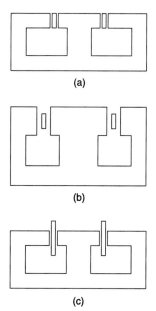

(a)

(b)

(c)

Figure 2.19. (a) Coil same length as gap. (b) Coil shorter than gap. (c) Coil longer than gap.

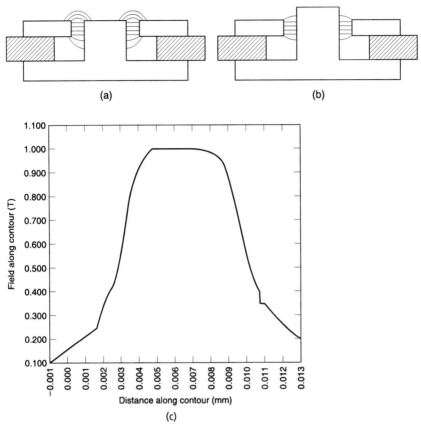

Figure 2.20. (a) Poor magnet design with assymmetrical flux. (b) Better design achieves flux symmetry. (c) Computer simulation of flux distribution in gap showing symmetry needed for low distortion. (Courtesy Celtic Audio Ltd.)

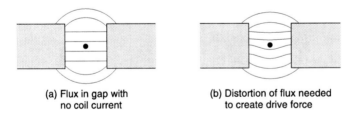

(a) Flux in gap with no coil current

(b) Distortion of flux needed to create drive force

Figure 2.21. In a quiescent speaker, lines of flux are straight as in (a). When coil produces a force, lines are distorted as at (b). This is the origin of flux modulation.

Inside the magnet, the only way that the flux can move is by the shifting of domain walls. This requires energy, is hysteretic and fundamentally non-linear. Thus the flux does not move smoothly, but in a series of jumps. In magnetic tape and disk heads the result is called Barkhausen noise. In loudspeakers the effect is that of program-modulated noise, similar but not identical to the modulation noise of an analog tape recorder. One way of stopping flux modulation is to use an electrically conductive magnet material. If the magnet is conductive, field shifts will have to generate huge eddy currents in a short circuit. Neodymium magnets are clearly superior here as they are highly conductive.

Flux modulation in steel pole pieces is resisted to some extent because steel is a conductor, but copper-plated pole pieces have been used as copper has much lower resistivity than steel. Some designers incorporate a substantial copper ring surrounding the centre pole.

It is quite instructive to compare transducer construction technologies practically. One comparison which is striking is to make a tweeter in which the magnetic circuit can be interchanged. Using the same software (moving parts) with a traditional ferrite magnet having mass-produced poles, or with a designer neodymium magnetic circuit, produces a remarkable difference in clarity.

The force on the coil is given by multiplying the current I by the length of wire, l, in the magnetic field and then by the field strength, B. The last two parameters are seldom quoted singly as only their product, known as Bl, matters. The unit of the Bl product is normally Tesla metres. As the product of B, l and I is a force in Newtons, it should be clear that an equivalent unit of Bl product is Newtons per Ampere which seems more appropriate given the job of the coil.

Unfortunately Bl is not the product of the flux density and the length of the coil. Figure 2.20(c) showed that the flux density in the gap is not constant, but falls at the edges. Fringing flux effectively extends the gap, but not all of the coil is in the field. In the case of a woofer, there may be more coil length outside the field than inside, and the part outside contributes no force. The actual Bl product is found by calculating the integral of the field strength over the length of the coil, or by actual measurement. The result of fringing is that the effective gap length is usually 0.5 to 1 mm longer than the pole dimension. Strictly the linear travel of the coil in each direction is the amount by which the coil overhangs the fringing field.

Above resonance, the moving part of the speaker is mass controlled and so, if the moving mass is known, it is not hard to find the acceleration for a given force. If the cone remains a rigid piston, knowing its effective area allows the resultant sound pressure to be calculated.

For a given cone area, increasing the Bl product increases output, whereas increasing the moving mass reduces output. Output is specified as the sensitivity of the driver. This may be power sensitivity or voltage sensitivity. Power sensitivity is independent of coil impedance whereas voltage sensitivity is not. In practice both are needed. Power sensitivity allows the necessary amplifier power to be calculated, but it does not specify whether that power is delivered as a low current at a high voltage or a high current at a low voltage. The voltage sensitivity allows that to be worked out. In fact the power sensitivity and the voltage sensitivity are related by the impedance.

The coil has finite resistance and this is undesirable as it results in heating. Long-throw woofers are especially vulnerable as the heat is developed over the whole coil whereas only that part in the field of the gap is producing a force.

The most efficiency (or the lowest magnet cost) is obtained when the gap is filled with coil material. Unfortunately this ideal can never be reached because some of the gap volume is occupied by clearances to allow motion and by insulation and, generally, a coil former. The interaction of mass and wire conductivity leads to another approach. Copper has very good conductivity, but its specific gravity is quite high: at 8.9 it is not far short of that of lead at 11.3. Thus in some applications a better result may be obtained by using aluminium whose specific gravity is only 2.7. Although the conductivity of aluminium is only 60% as good as copper, weight for weight the conductivity is twice as good.

For a long-throw woofer where the coil mass is a significant part of the moving mass, an aluminium coil can give a significant improvement in efficiency, especially if the wire is square or rectangular. The cross-sectional area of the aluminium has to be higher, and this requires a wider gap. If the same magnet is retained, this will

To find wire diameter, given coil dimensions and resistance:
let d.c. resistance = R, resistivity ω Ω cm, coil length = L_{cm}, coil diameter = D_{cm}, wire diameter = d_{cm}, turns = T, layers = l, wire length = πDT and $T = Ll/d$

$$R = \frac{4DT\omega}{d^2} = \frac{4DLl\omega}{d^3}$$

$$\therefore d^3 = \frac{4DLl\omega}{R} \quad \therefore d = \sqrt[3]{\frac{4DLl\omega}{R}}$$

To find wire mass:

Mass = length × area × ρ

$$= \frac{TD\pi d^2}{4}\rho$$

Figure 2.22. Derivation of wire diameter needed for a given coil dimension and resistance. Note result goes as the cube of the wire diameter which makes diameter tolerance important.

reduce the Bl product slightly, but the moving mass will have been reduced by a greater factor and so the efficiency rises.

The problem with aluminium wire is that it is difficult to make connections to it because of oxidization. This is solved by giving it a copper coating, resulting in wire known as CCA or copper clad aluminium. For real perfectionists, a silver coat may go on top of the copper in wire intended for tweeters. The presence of the coating changes the effective conductivity and the manufacturers typically quote it as a percentage of what it would be for solid copper of the same dimensions. CCA is also slightly denser than pure aluminium.

The coil is commonly wound on a former which is extended to carry the coil thrust to the cone. Coil formers can be made from Kraft paper, but these will not withstand temperatures above about 100°C. In high-power speakers, a Nomex (polyamide) former allows the temperature to rise to 150°C, whereas a Kapton (polyimide) former will survive 350°C, provided the coil insulation and adhesives can also withstand such treatment. Aluminium coil formers can also be used, provided a small gap is used to prevent eddy current loss. Carbon fibre is also suitable as a former material. When designing a coil, the d.c. resistance will normally be decided upon at an early stage. It is useful to be able to vary the coil diameter, length and number of layers whilst holding the resistance constant. This is complicated because, if the coil diameter changes, the resistance will change and this will require a different cross-sectional area of wire which will change the coil length and so on. Figure 2.22 shows the derivation of an equation for the coil wire diameter needed for a given coil. It is then easy to find the nearest preferred wire gauge from manufacturers' tables. The figures for this wire allow the mass of the wire to be calculated. The wire diameter also allows the number of turns in the gap to be calculated, from which the Bl product can be obtained.

It is useful to try a range of coil diameters. Even if the coil is heavier, the speaker is not necessarily less efficient as a larger coil may allow a larger magnet, a more efficient magnetic circuit or a lighter cone to be used. The sensitivity is determined by the ratio of Bl product to the total moving mass.

2.6 The case for square wire

As has been seen above, the efficiency of a moving-coil speaker is pretty miserable and so anything which offers an improvement is worth looking at. The efficiency depends upon the *Bl* product and the overall moving mass. Making the magnet more powerful is an obvious approach, but this is expensive. An alternative is to look at ways of getting more *Bl* from an existing magnet.

This is where square and rectangular wire comes into consideration. Figure 2.23(a) shows a coil made with round wire. The packing efficiency is poor because circles do not fit together at all well. The result is that the magnetic field is only partially used. Figure 2.23(b) shows that the packing efficiency is better if square wire is used. The question is, how much better. The author kept coming across the above argument, but the result was never quantified, so it was decided to work it out. It turned out to be more complex than might at first be thought.

It is important to compare like with like, so that as far as possible everything should be kept the same except for the cross-sectional shape of the wire. Thus it is necessary to consider two coils, having the same overall length, the same d.c. resistance and the same mass, differing only in the wire section.

This is not easy because, if the cross-sectional area of the wire is kept the same, the coil having the square wire will be shorter because of the better packing. This will reduce the linear travel of the cone and the overhung coil length and so we are not comparing like with like. On the other hand, if the coils are made the same length, the coil having the square wire will have more turns and hence higher resistance.

A fair approach to the problem is to allow rectangular wire so that exactly the same mass, resistance and coil length become possible. Figure 2.23(c) shows that, if the rectangular wire has the same cross-sectional area, it will have the same mass and resistance per unit length as the round wire and, if the diameter is the same as the section height, the coil will have the same length too.

Note that the *Bl* product does not change, but the thickness of the coil falls to about three-quarters of the value for round wire, allowing a smaller gap volume and consequently a smaller magnet for the same field strength. However, the magnet will

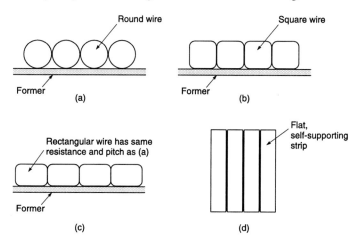

Figure 2.23. (a) Round wire does not pack well and loses efficiency. (b) Square or rectangular wire packs better. (c) Coil with rectangular wire having same dimensions and resistance as (a), needs slightly narrower gap, allowing magnet size to be reduced 10–15%. (d) Elimination of the former with self-supporting coil allows 25% improvement in motor power for given magnet.

not be three-quarters as big because, as before, much of a practical gap is used up with coil former and clearance space. Perhaps a 10% or 15% reduction in magnet size, or a similar improvement in *Bl* for the same magnet would be feasible.

Bearing in mind the phenomenal cost of square wire and the enormous difficulty in obtaining it, the likelihood is that the saving on the magnet would be eclipsed by the extra cost of the wire below a certain size of drive unit. It is only in large drive units where the magnet cost becomes an issue and small magnetic efficiency gains become worthwhile. In any case a 10% improvement in *B* can often be had by some attention to magnetic circuit design.

So the efficiency argument for square or rectangular wire is tenuous if used as a replacement for round copper wire in an existing design. It is certainly much less effective than the improvement obtained by going from copper to aluminium.

However, the most tangible advantage of square or rectangular wire is that successive turns can easily be bonded together so that the coil former can be dispensed with. This has a number of advantages. The elimination of the coil former reduces the amount of wasted volume in the gap and reduces the moving mass a little. The wasted volume is now only due to the inner and outer coil clearance spaces. A self-supporting coil has better cooling because both sides are exposed and can radiate to both poles.

Combining the elimination of the coil former with the slightly better packing of the square wire gives a tangible reduction in gap volume as shown in Fig. 2.23(d). The more powerful the speaker, the greater the ratio of coil volume to clearance volume and the more relevant the approach becomes. An improvement of 25% or 2 dB becomes possible in a large woofer. Self-supporting coils of this kind are found in all kinds of applications such as vibration table actuators and in the positioners of giant disk drives.

For small or low-powered speakers this approach does not make economic sense. If square or rectangular wire is contemplated, the speaker has to be designed from the outset to use it. If the coil is not self-supporting, most of the advantage is lost. In midrange and tweeter units, square wire may give some improvement in the velocity of propagation of vibrations through the coil because the turns abut more firmly against one another.

2.7 The suspension

The surround at the front edge of the cone, and the spider which is normally attached to the coil, together form the suspension. The combined purpose of these two is to allow axial motion of the diaphragm while preventing lateral motion or rocking. The surround is also charged with forming a seal around the edge of the diaphragm, whereas the spider has no sealing function but instead is intended to act as an axial stiffness tending to restore the diaphragm to its neutral position. In reality the surround will also have some stiffness. The combination of these two stiffnesses along with the moving mass will determine the free-air resonant frequency of the drive unit.

Surrounds can simply be a corrugated extension of the paper cone in cheap speakers, but these offer little damping and are not very linear. Better results are obtained when the surround is a purpose made unit. Highly compliant surrounds can be made from textile or fibrous material impregnated with a sealant such as Plastiflex, a polyvinylacetyl compound. Alternatively, impervious materials such as Neoprene or plasticized PVC can be used. In the case of PVC the plasticizer must be a non-migrant compound such as dibutyl sebacate. Plasticizer migration causes the characteristics of the material to change and ultimately it may fail.

An important function of the surround is to terminate vibrations propagating through the cone. Surround damping is also important to prevent resonance of the surround itself. Note that part of the surround moves with the cone and effectively adds to the moving mass. This reduces the efficiency, an effect which is tolerated in a quality unit. The increase in mass due to the spider can generally be neglected.

As the surround is hysteretic and non-linear, the rear suspension or spider is or should be the dominant stiffness. In early speakers the rear suspension was made from a thin material much of which was cut away to leave a large number of slim 'legs', hence the term spider. Today, although the name has stuck, the rear suspension is most commonly a concentrically corrugated fabric disk which has been impregnated with a phenolic resin. For a given mechanical arrangement, the stiffness can be controlled by the weave of the material and the amount of resin impregnant.

The spider and surround will be lossy and will determine the mechanical Q or damping factor of the drive unit. However, the dominant damping mechanism will usually be that due to the zero output impedance of the amplifier seen by the coil.

2.8 Motor performance

During the development of a drive unit it is useful to make some measurements of the motor performance. In the context of performance, the overall force on the moving mass is all that matters. In that case the coil and magnet are one significant force-producing mechanism and the spider/surround and displacement are the other. Both must be measured to give a clear picture.

The simulations of the magnetic circuit will give a good idea of the flux density to be expected in the gap, but it is useful to be able to measure it. There are several approaches. One of these is to use a specially designed meter having a probe operating on the Hall effect. Such probes can be made small enough to fit into the gap on practically all magnet designs. In order to make an accurate assessment of the symmetry of the gap flux distribution, it will be necessary to support the probe on some kind of fixture which allows accurate linear movement of the probe through the gap.

The gap flux can also be explored using a search coil. This has the same diameter as the coil intended for the drive unit, but the length of the coil should be the same as the thickness of the pole pieces. An apparatus similar to that of Fig. 2.24 can be used. The coil can be immersed in the gap at various displacements. If a known current is passed through a coil whose dimensions are known, the resultant force allows the flux density in the gap to be calculated.

The current remains fixed, but the coil can then be brought to different positions with respect to the pole pieces, making a note of the force at each step. It will then be possible to see how symmetrical the flux distribution is. Note that this is easier if the force gauge is of a type which is stiff; i.e. it does not deflect with an applied force. If a conventional spring balance is used, the body of the balance will have to be moved at each step to compensate for the deflection of the balance mechanism.

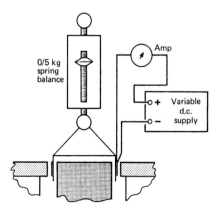

Figure 2.24. Schematic of force measurement.

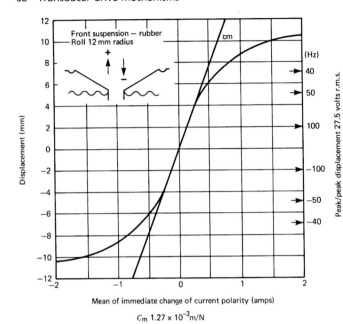

Figure 2.25.
Displacement/
coil current.

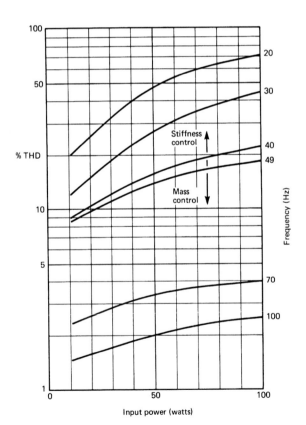

Figure 2.26.
Total harmonic
distortion/power/
frequency.

Using an actual coil, the linearity of the magnet/coil combination over its intended travel can be explored. If this is satisfactory the spider and surround can then be tested. Here the coil current is made variable in steps in both directions and the displacement of the coil, which is now controlled by the spider and surround, is measured. These may be fitted individually for investigation. The current should be varied over a complete cycle to reveal any hysteresis.

Spiders can be surprisingly non-linear. Figure 2.25 shows the displacement/force curves for a typical unit, which results in primarily (and unwelcome) third harmonic distortion. Figure 2.26 shows the result of a distortion test.

The distortion due to non-linearity in the diaphragm support and the gap/coil system will become worse as frequency goes down because the amplitude of motion increases at low frequencies. This effect was shown in Fig. 2.6. Note also that, below resonance, the cone is stiffness controlled and the spider linearity becomes important. In a passive speaker the drive unit will not be used in the stiffness control region because this is below the LF cut-off frequency but, in an active design where a high fundamental resonant frequency has been phase and amplitude equalized to extend the bass response, the spider linearity becomes important.

The majority of spiders in use today are significantly non-linear and as a result traditional drive units designed for passive speakers generally turn out to be inadequate for active designs. Drive units designed for active speakers must have more highly linear suspensions. Suspension distortion can be reduced dramatically by studying how spiders actually work and formally designing them rather than just copying a traditional design.

2.9 The chassis

The chassis or basket is a part which is often taken for granted, but it can have an effect on performance. At a basic level, the chassis provides mounting points so that the drive unit can be fixed in an enclosure, and supports the cone and the magnet assembly. A little thought will show that the chassis material and construction can affect the magnetic and thermal properties of the speaker as well as being a potential source of colouration.

Taking the example of a woofer, Fig. 2.27 shows that the cone is supported in two planes. The first is the plane of the surround which serves as a flexible pressure seal between the moving cone and the stationary chassis. The second is the plane of the spider which supports the coil or the neck of the cone. The cone is located within the spider plane so that it can only move axially, and a (hopefully) linear restoring force is provided against cone travel to keep the cone in the centre of its range of travel.

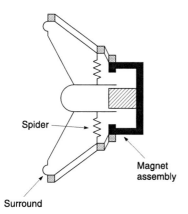

Spider

Magnet
assembly

Surround

Figure 2.27. The functions of the basket or chassis.

Figure 2.27 also shows that the magnet assembly is supported by the chassis. Magnets and steel pole pieces are inevitably heavy, and the chassis has to be rigid enough to ensure that the relationship between the spider and the pole pieces is held constant. If this is not achieved, the coil may rub on the pole pieces. In many applications it has to be accepted that the loudspeaker will be dropped or handled roughly. The chassis has to withstand very high transient forces from the magnet under these conditions.

Apart from the audible advantages of rare-earth magnets which have been advanced in this chapter, there is now a non-audible advantage which is that the lighter rare earth magnet will place less stress on the chassis during rough handling, giving a distinct reliability advantage in applications such as PA, as well as making the unit easier to transport.

When the cone is driven forwards, the magnet assembly experiences the Newtonian reaction backwards. It is often stated that the chassis of a loudspeaker has to be incredibly rigid to withstand the reaction of the magnet. This is a myth. If the relative masses of the cone and the magnet assembly are considered, it is clear that with a ratio of nearly 100:1 the magnet is not going anywhere.

Figure 2.28(a) shows what really happens when the cone is driven forwards. The pressure in the enclosure goes down, and atmospheric pressure flexes the front of the enclosure inwards, actually moving the entire drive unit. The solution is properly to engineer the enclosure for rigidity, a technique which is actively avoided in many traditional wooden box speakers. A useful improvement can be obtained by bracing the rear of the woofer magnet against the opposite wall of the enclosure with a suitable strut. Figure 2.28(b) shows that the atmospheric pressure forces cancel at the magnet.

Chassis can be made from a wide variety of materials, but the choice is far from obvious and depends upon the production volume and power handling. Candidates include cast alloys, pressed steel and injection moulding. With external ferrite magnets, a steel chassis can increase flux leakage, requiring a larger magnet for the same performance. Alnico and rare-earth designs do not have this problem. Aluminium or plastics chassis avoid flux leakage in ferrite magnets.

The high strength and ductility of steel means that chassis can be made in quantity by pressing from a relatively thin sheet with a very low material cost. For rigidity, thin sheet cannot be left flat over any significant area and so the pressing will need to be complex to ensure that every edge is flanged. The press tool also has to punch out holes to allow the pressure from the back of the cone to escape. This results in

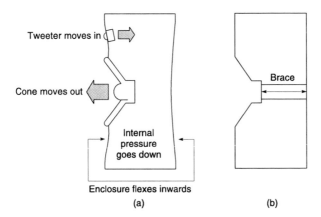

Figure 2.28. (a) Enclosure flexing due to internal pressure change. (b) Bracing enclosure panel to magnet reduces flexing.

Figure 2.29. In this high-power woofer, dissipation is aided by an aluminium chassis which is outside the enclosure. (Courtesy Volt Loudspeakers Ltd.)

a loss of strength, and many pressed steel chassis tend to err on the side of plenty of metal rather than freedom of air movement.

For high-power speakers, thermal considerations are usually uppermost. The coil dissipates a lot of heat, most of which has to be removed through the pole pieces of the magnet. Mounting the magnet assembly on a substantial cast aluminium chassis is a good way of losing that heat with minimum temperature rise. Casting allows the complex ribbed structures needed for rigidity to be replicated with ease. For volume production, die-casting is a natural technique. The die tools may cost thousands of pounds, but the unit cost is very low. In the woofer designed by David Lyth, shown in Fig. 2.29, the construction is inverted so that the cast aluminium chassis is on the outside of the cone. This allows a direct cooling path to the outside of the loudspeaker.

Cast chassis will also be found on low-volume, prototype and special-purpose drive units. These may be sand cast because the tooling costs are quite low, requiring only very basic wooden patterns which are well within the scope of the home builder. Sand casting is, however, labour intensive and becomes uneconomic as quantity rises.

Casting may advantageously be taken further than just the chassis. In production engineered active speaker designs, one casting may form the entire front of the speaker, having integral woofer and tweeter chassis, mountings for the power transformer and also acting as the heat sink for the amplifier.

Advances in engineering plastics have meant that chassis can practically be moulded. It is easy to mould integrally any required details and, as plastic is non-conducting, the connecting tags can be mounted directly in the plastic, rather than on a separate tagstrip. Developments in numerically controlled machining mean that the moulding tools can be produced rapidly and with reasonable economy. Plastics tend to be poor thermal conductors, and so their use is limited to drive units of reasonable efficiency where heat dissipation is not an issue.

2.10 Efficiency

The efficiency of a direct radiator moving-coil loudspeaker is doomed from the outset to be very low. The reason is that the density of air is low and so the acoustic impedance seen by the diaphragm is correspondingly low. To deliver power into a

$$R_m = 1.57 \frac{p}{c} \omega^2 r^4 = K\omega^2 r^4 \qquad\qquad\qquad\qquad (a)$$

where R_m is acoustic impedance, 1.57 is the baffle factor, ω^2 is frequency in rad/s, r is cone radius.

$$v = \frac{Bli}{\omega M_m} \qquad\qquad\qquad\qquad (b)$$

Here v is cone velocity,. ω is frequency, B is magnetic flux density, l is the length of motor coil inside the gap, i is current and M_m is the moving mass. W_a is acoustic power,

$$W_a = v^2 \times R_m = \left(\frac{Bli}{\omega M_m}\right)^2 \times K\omega^2 r^4 = Kr^4 \left(\frac{Bli}{M_m}\right)^2 \qquad\qquad (c)$$

so,

$$Eff = \frac{Kr^4 B^2 l^2 i^2}{M_m^2 i^2 R_E} = Kr^4 \times \frac{B^2}{M_m^2} \times \frac{l^2}{R_E} \qquad\qquad\qquad (d)$$

Figure 2.30. Derivation of transducer efficiency. At (a) the acoustic impedance is defined. (b) Here the cone velocity is obtained. (c) Using (a) and (b), it is possible to obtain the acoustic power radiated. (d) Ratio of acoustic power to input power derives efficiency.

low impedance requires a high diaphragm or cone velocity. This in itself is not a problem. The difficulty is that any practical cone and coil which has sufficient structural integrity must have some mass and the impedance presented by this mass always dwarfs the acoustic impedance in a real moving-coil direct radiator speaker. Effectively a massive cone is being moved in order to carry with it a tiny air mass. It is the equivalent of delivering pillows in an eight-wheeler: not very efficient.

The efficiency of a speaker is defined as the ratio of the acoustic power coming out to the electrical power going in. The electrical power is used in three ways. One is the resistive heating of the coil, the second is the mechanical damping of the drive unit structure and the absorbent in the enclosure, which is very small above resonance and can be neglected. The third is the acoustic power transferred. In practice, because the efficiency is so low, the acoustic power can also be neglected, leaving the input power effectively equal to the dissipation in the coil. This is a great simplification as the coil dissipation is easy to calculate, being the product of the coil resistance and the square of the current.

Next the acoustic output power must be calculated. In electricity, the power would be the product of the resistance and the square of the current. In acoustics, it is the product of the acoustic impedance and the square of the cone velocity.

Figure 2.30(a) shows the expression for the acoustic impedance which was defined in Chapter 1. This can be simplified by taking out the fixed values, such as the density of air, the speed of sound and the correction factor for the kind of baffle and replacing them with a constant K which cannot be changed by any feature of the drive unit. Note that the acoustic impedance is a function of frequency squared.

Figure 2.30(b) shows how the cone velocity is obtained. Clearly it is proportional to the motor force, Bli. Because the system is mass controlled, the velocity is inversely proportional to frequency and mass.

In Fig. 2.30(c) the acoustic power is derived by multiplying the acoustic impedance by the square of the velocity. Note that, when this is done, the frequency term cancels out. This is the principle of the mass controlled speaker: the power is independent of frequency or, in other words, there is a flat frequency response.

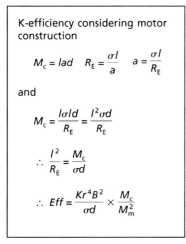

K-efficiency considering motor construction

$$M_c = lad \quad R_E = \frac{\sigma l}{a} \quad a = \frac{\sigma l}{R_E}$$

and

$$M_c = \frac{l\sigma ld}{R_E} = \frac{l^2 \sigma d}{R_E}$$

$$\therefore \frac{l^2}{R_E} = \frac{M_c}{\sigma d}$$

$$\therefore Eff = \frac{Kr^4 B^2}{\sigma d} \times \frac{M_c}{M_m^2}$$

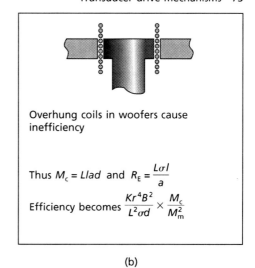

Overhung coils in woofers cause inefficiency

Thus $M_c = Llad$ and $R_E = \frac{L\sigma l}{a}$

Efficiency becomes $\dfrac{Kr^4 B^2}{L^2 \sigma d} \times \dfrac{M_c}{M_m^2}$

(a)   (b)

Figure 2.31. (a) Re-expressing the efficiency equation with the component masses helps the design process. (b) Only the coil in the gap provides thrust, the remainder adds detrimental mass.

The efficiency is derived in Fig. 2.30(d). Note that the current disappears from the expression. Thus an efficient speaker is simply a matter of using a strong magnet with plenty of wire in the gap and a large cone area, while keeping the weight and the coil resistance down. Unfortunately those requirements are contradictory because the more powerful we make the motor, the heavier the coil gets. Not so simple after all.

A better approach is to express the efficiency in a different way so that the effect of the mass of the motor is easier to interpret. This is the approach taken in Fig. 2.31. If we assume a speaker in which the coil is the same length as, or shorter than, the gap, then all of the coil is contributing to the motor force. The mass of the voice coil, M_c, is given by its volume multiplied by the density of the coil material. The resistance of the voice coil also follows from its dimensions and resistivity.

With the efficiency re-expressed, it is easier to see what to do. First, a powerful magnet helps but, as will be seen, this cannot be taken too far in a passive speaker because the result is an overdamped system which causes premature roll-off of low frequencies. It is also clear that when choosing a coil material, the product of the density and the resistivity is what matters, hence the superiority of aluminium.

It should be appreciated that M_m, the moving mass, incorporates M_c, the coil mass. The most striking result of the efficiency expression, which is counter-intuitive, is that making the coil heavier without changing the total moving mass increases the efficiency. As a result, for woofers at least, the cone should be as light as possible so that the largest proportion of the moving mass is concentrated in the coil. The epitome of this is the ribbon speaker where the diaphragm *is* the coil. The use of a large-diameter coil as described in Section 2.2 will assist in lightening the cone and increasing the proportion of the moving mass devoted to the coil.

The above expressions fail when the speaker coil and cone no longer move as a rigid body. In woofers, the cone and coil are certainly rigidly coupled over the useful frequency range, and so the approach is useful except for the assumption that all of the coil is in the gap which is generally not true for woofers. In order to have sufficient travel, woofers must use overhung or underhung coils so that the Bl product remains constant as the coil moves.

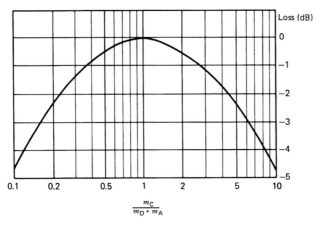

$$\frac{m_C}{m_D + m_A}$$

Figure 2.32. Efficiency loss in a direct-radiator loudspeaker.

If the coil length is described as L times the gap length, L can be called the over-hang factor. If the coil is simply extended by putting more turns above and below the gap, the coil is obviously L times as heavy and has L times as much resistance. Figure 2.31(b) shows that the efficiency expression now contains L squared in the denominator, suggesting that a large overhang factor is detrimental.

This is one of the reasons why really long-throw woofers are rare; in the overhung configuration they are simply very inefficient because so little of the coil is producing thrust and the remainder is adding to the mass and resistance. In the underhung configuration the magnet structure becomes very expensive. In practice, to obtain a given SPL it may be more efficient to put two normal woofers side by side rather than trying to engineer a very long-throw device.

In the case of a tweeter, the finite propagation speed of vibrations in the mechanism means that the coil and cone do not move in the same phase and the above arguments are not relevant. Instead the maximum power transfer between coil and cone occurs when their mass is equal as shown in Fig. 2.32.

2.11 Power handling and heat dissipation

In any inefficient device, the wasted power almost always shows up as heat and the loudspeaker is no exception. Most of the energy loss appears as heat in the coil. Most manufacturers rate their loudspeakers by the amount of power they can survive, measured in Watts, under certain conditions, without sustaining damage. This is known as the 'power-handling' figure and is absolutely meaningless. It is a physical survival figure, not a reproduction-without-distortion figure. Consequently at full rated power the loudspeaker output will invariably be distorted beyond recognition. Although the speaker will never be damaged by any useful signal it can reproduce without distortion, the user is given no idea what that useful signal is.

The power-handling figure becomes more meaningless, the more closely it is examined. First, the user of a loudspeaker is only interested in the quantity and quality of the actual sound output. In order to estimate the sound output potential, the efficiency figure will need to be consulted. However, dividing the power-handling figure by the efficiency does not yield the useful acoustic power output. Most speaker specifications are relatively silent about distortion, possibly because alongside other pieces of audio equipment the figures are generally unimpressive. Consequently the only way to proceed is to estimate the maximum acoustic power the speaker can radiate. This will be limited by the area and the available linear travel of the diaphragm at the lowest usable frequency. In the case of a woofer, generally the free air resonant frequency can be used. Using the published efficiency figure, the power of an amplifier which could drive the speaker to its maximum undistorted level can then be calculated. This might be called the useful power and it is astonishing how much smaller this figure is than the power-handling figure.

A further twist is that the power-handling figure is not the power actually dissipated. In fact it is the power that would be dissipated by a resistor having the same resistance as the nominal impedance of the speaker. As a drive unit is a complex or reactive load as well as being temperature dependent, the actual power delivered by the amplifier will be substantially less than the power delivered to an ideal resistor.

As power is dissipated, the increase in temperature increases the resistance of the coil. The temperature coefficient of resistance of copper and aluminium is about 0.4% per degree C. A temperature rise of 250°C will double the resistance of the coil. This has the effect of reducing the voltage sensitivity and thus the output of the speaker. To an extent a moving-coil speaker is self-regulating because, as the coil temperature and resistance go up, less power is drawn from the usual voltage source type amplifier. Consequently a 200 Watt rated loudspeaker may actually be drawing only 100 Watts on the same drive voltage when the coil has fully warmed up, as well as producing half as much sound power! Note that sensitivity is normally quoted with a cold coil!

Coil heating causes an effect known as compression. It is most noticeable on transients where the dynamic temperature rise of the coil actually modulates the current and so distorts the waveform. The effect of a raised coil temperature is that the electromagnetic damping of the cone by the coil is reduced so that the Q factor when hot will not be what it is when cold.

In drive units which offer high power-handling, the designer is backed into a corner by the survival requirement. The cone will have to be heavier to survive the mechanical ordeal, and so the speaker will be less efficient. This means that more heat will be developed, which in turn means that the coil temperature will rise and the coil will expand. A 250°C rise in coil temperature could cause a 50 mm voice coil to expand in diameter by 0.4 mm. This requires a greater clearance in the outer pole piece, raising the reluctance of the gap and reducing the Bl product, further reducing efficiency. Consequently high-power handling speakers are seldom very efficient and therefore need to consume that high power and suffer more from thermal compression.

In conventional voltage-driven loudspeakers there is no solution to thermal compression, but with current drive the problem is eliminated. This does, of course require a specially designed amplifier and is possible only in active speakers. It should be clear that thermal compression is absent in electrostatic speakers whose distortion is independent of level. In fact, where low distortion is a criterion, a suitably designed (and large) phased array electrostatic can be louder than a moving coil. Another approach to the elimination of thermal compression is to produce an efficient transducer so that heating is minimized. The horn transducer is at an advantage here. If the coil cannot cool itself adequately, the temperature has to rise. Eventually something will give. This could be the coil insulation, the glue holding the coil to the former, the former itself or the neck of the coil. Although recent developments in former and adhesive materials have allowed coils to run at higher temperatures, this brings only an improvement in reliability. Unfortunately, however, the improved heat resistance of a coil is often used to raise the power-handling figure even further beyond the useful power. This results only in even greater thermal compression effects. Consequently a precision loudspeaker for quality monitoring purposes does not actually need high-temperature coil technology except to survive abuse.

It follows that, for quality reasons, moving-coil drivers must be designed with effective cooling measures so that the coil temperature excursion is minimized. The coil can shed heat by radiation and by conduction. Conduction is limited because the coil former is attached only to the cone and the spider. Most spiders are thermal insulators, as are most cone materials, with the notable exception of aluminium cones,

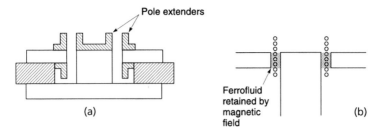

Figure 2.33. (a) Non-ferrous metal pole extenders help coil cooling. (b) Ferrofluid is retained by the flux and conducts heat to poles.

which can contribute significantly to coil cooling if the attachment of the coil former to the cone is designed as a thermal path.

If a large-diameter coil is selected, it may be possible to use a single-wound coil rather than double-wound. This will have twice the surface area and will be able to cool itself more effectively, adding to the list of arguments in favour of large-diameter coils in woofers.

Heat can be lost to the air around the coil but, if this is static, the air temperature will simply rise. Some ventilation of the coil area is necessary. The pumping action of the coil entering the magnet or of the spider or both can be used, but this must be approached with caution. With large excursions, the air flow may become turbulent resulting in 'chuffing' where the speaker produces program-modulated noise. Air passages act in a non-linear fashion and may partially rectify. At high level a poorly designed speaker may act as a unidirectional pump and displace its own neutral point. Radiation loses heat to the pole pieces and these can be advantageously blackened to improve their absorption. The coil and former can also be blackened. The pole pieces can be extended in the vicinity of the coil using aluminium plates as shown in Fig. 2.33(a). These will act as heat sinks for the overhung part of the coil without affecting the magnetic field. Ultimately the magnet assembly must shed the heat and this can be assisted with external fins, or a thermal path to a cast aluminium chassis.

The ultimate cooling method is ferrofluid, a magnetic liquid shown in Fig. 2.33(b) which is retained in the magnet gap by the magnetic field itself and conducts heat to the pole pieces. Ferrofluid became available around 1974, and consists of a colloid of magnetic particles (Fe_3O_4) of about 0.01 mm in diameter suspended in a low vapour pressure diester based on ethylexyl-azelate. Brownian motion of the carrier molecules keeps the particles permanently suspended over a practical temperature range of $-10°$ to $+100°C$. The use of ferrofluid does result in some viscous damping due to shearing in the fluid. The viscosity of the fluid is somewhat temperature dependent as shown in Fig. 2.34.

Ferrofluid is generally designed to saturate at flux levels considerably below those present in the gap so that the reluctance of the gap is not affected. It is compatible with most coil, adhesive and former materials, except for paper and rubber-based materials. Kapton absorbs some of the carrier but without ill effect. Figure 2.35 shows the cooling effect of ferrofluid in a typical tweeter.

Loudspeakers cannot use ferrofluid indiscriminately; they must be designed to use it. A speaker designed to use coil pumping for cooling cannot use ferrofluid, as the fluid would block the air path. The pumping action of the coil would force bubbles through the ferrofluid and make noises unless suitable ventilation was provided.

The speaker with massive power-handling capacity which actually produces no more SPL than an efficient unit which can handle less power is a triumph of marketing and ignorance over acoustics. When active speakers are considered, the amount of power consumed by the transducers and the power developed in the amplifiers are

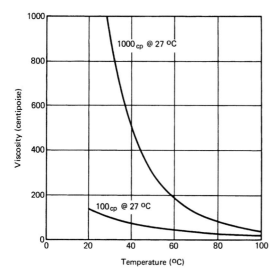

Figure 2.34. Viscosity versus temperature for 100 gauss ferrofluid.

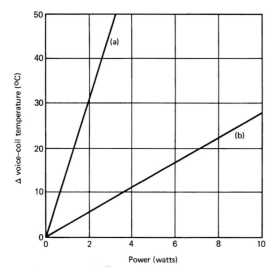

Figure 2.35. Increase of voice coil temperature for 25 mm tweeter. (a) Without ferrofluid. (b) With ferrofluid.

both the problem of the designer. The only parameter the user needs to know is the output sound power.

When the goal is to create a system producing a certain acoustic power, the system designer can then make balanced choices. For every efficiency improvement in the transducer, which might incur a cost, a smaller amplifier and power supply can be used. The saving there could well pay for the small extra cost in the transducer. The fact that the amplifier in use is under the control of the designer means that over-specified power handling is unnecessary and a further gain of efficiency can be realized. When a loudspeaker and amplifier are designed together in this way, the overall cost will actually be less than that of a passive speaker driven by a generic amplifier. Once there is an economic advantage as well as a technical advantage, active technology becomes inevitable.

2.12 The dome driver

A type of drive unit having a dome-shaped diaphragm has become popular for repro-duction of frequencies from the mid-range upwards. Generally the surround and the diaphragm are pressed as one component from a woven or fabric material, and the coil is attached to the edge of the dome. The dome section and the surround section may be impregnated or doped with a material which determines the stiffness and damping factor. The moving mass can be very low, and so high efficiency is gener-ally a characteristic of such designs. A further advantage is that in high-powered domes, the coil diameter is large and so cooling is easily arranged. In larger dome units a second suspension may be used to ensure axial movement.

The dome driver is surrounded by myth, and one of the greatest of these is illus-trated in Fig. 2.36. An ideal sound radiator would be a sphere whose diameter could somehow grow and shrink as shown at (a). This would offer excellent dispersion characteristics, radiating sound equally well over a range of directions. Superficially, the dome resembles such a spherical radiator, and many people then assume that the dome will have the same characteristics. Unfortunately this is not true. Figure 2.36(b) shows that the dome is a section of a sphere moving on a single axis and the results are quite different[7].

As was shown in Chapter 1, to establish the radiating effect of a diaphragm of finite size at some point, it is necessary to integrate the amplitude and phase of radi-ation over the whole diaphragm. This is non-trivial because the finite speed of propagation of vibrations within the dome and the lossy nature of the dome fabric will cause different parts of the diaphragm to vibrate at different phases and ampli-tudes.

Some analysis can be made by assuming the dome to be pistonic, i.e. all parts of it vibrate with the same phase and amplitude. Figure 2.37 shows the on-axis response for domes of various contours ranging from a flat disc to a hemisphere. The roll-off at high values of ka is due to the phase shift between the centre and the perimeter of the dome, which is obviously greatest in the hemispherical dome. However, none of this is of any consequence for loudspeakers because in practice domes always operate at ka values below 4 or 5. If that part of Fig. 2.37 is considered, it will be seen that practically it makes no difference whether the diaphragm is flat or domed.

Figure 2.38(a) shows the same analysis except that it is for a point 22.5° off-axis. Figure 2.38(b) is for a point 45° off-axis. In both figures it can be seen that the off-axis response has an undesirable falling characteristic due to beaming. Note that up to $ka = 4$ the results are virtually identical for all dome profiles. Thus the results both off- and on-axis show that the shape of the dome is irrelevant! In both cases the in-service result is indistiguishable from that for a flat disc. In fact it would not make any difference if the dome were concave instead of convex and some drive

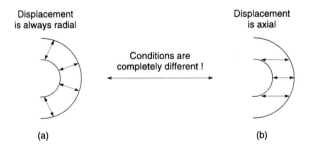

Figure 2.36. The myth of the dome. At (a) an ideal radiator would be a pulsating sphere. Dome at (b) resembles a sphere, but it does not pulsate; it moves in one axis and the results are quite different.

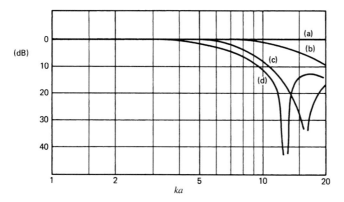

Figure 2.37. On-axis response of a rigid dome, radius *a*, of various heights *h*. (a) Flat disc: *h* = 0. (b) *h* = *a*/4. (c) *h* = *a*/2. (d) = hemisphere: *h* = *a*.

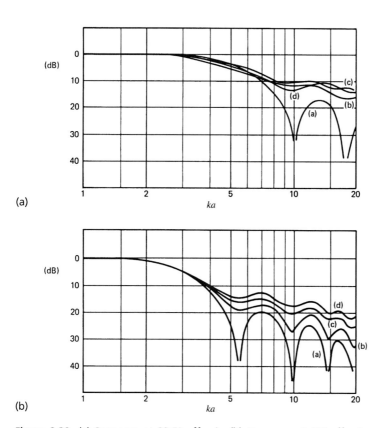

Figure 2.38. (a) Response at 22.5° off-axis. (b) Response at 45° off-axis.

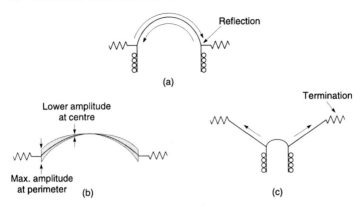

Figure 2.39. The problems of dome drivers. (a) Rigid dome cannot be terminated; vibrations travel across the dome and reflect. (b) Solution to reflections is the 'soft' dome which attenuates vibrations travelling from the coil. At high frequency, centre of dome decouples, leaving an annular radiating area (c). This causes an irregular directivity function and off-axis time smear.

units do use concave domes. The dome suffers from just as much beaming as a flat disc. Clearly the intuitive argument for a dome resembling a pulsating sphere is totally incorrect.

The assumption has been made in the above that the diaphragm is rigid, and this is not the case. Consequently, before drawing any conclusions, the effect of vibration propagation in the dome must be considered. Figure 2.39(a) shows that, in a lossless or 'hard' dome, vibrations launched from the coil travel right across the dome where they can reflect from the coil junction again. This is not desirable and is the reason for the use of a 'soft' or lossy dome material as an alternative to the hard dome. Figure 2.39(b) shows what then happens. Vibrations from the cone propagate inwards with reducing amplitude. This is beneficial on-axis, because the amplitude of the vibration of the centre of the dome is reduced, causing the cancellation effect of Fig. 2.38 to diminish.

However, when the off-axis condition is considered, the soft-dome driver exhibits exactly the opposite of what is wanted. As frequency rises, the centre of the dome decouples and only the perimeter is fully driven, producing an annular radiator of maximum size. In a particular direction the dome behaves like two displaced radiators and this will result in interference patterns being created. Consequently the polar characteristics of real soft domes are actually worse than the performance of a rigid disc. The off-axis sound is likely to have a highly variable frequency response which causes the reverberant field to be coloured.

In cone-type radiators, the vibrations from the coil propagate outwards and diminish due to damping. The surround acts as a mechanical terminator preventing reflections. As a result the diaphragm of a cone-type transducer becomes effectively smaller as frequency rises and can and does have better polar characteristics than the rigid disc (see Fig. 2.39(c)).

The majority of dome drivers use very light fabric diaphragms which are relatively transparent to sound. Figure 2.40 shows that these are often combined with magnetic circuits which are effectively solid. The result is that sound radiated rearward by the dome can reflect from the magnet and pass out through the dome with a delay. This problem can be eliminated by a suitable magnetic circuit design in which the rear radiation from the dome enters a transmission line and is absorbed.

One must conclude from physics that the dome driver is acoustically questionable. The main advantage of the dome driver is that a very large coil can be fitted which

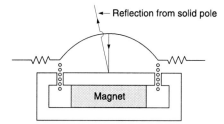

Figure 2.40. Most domes have solid magnetic circuits which act as a mirror for rear radiation.

will allow high power dissipation. This allows high SPL to be generated in acoustically treated dead rooms where only the direct sound will be heard and the poor directivity will be concealed.

2.13 The horn driver

The main advantage of the horn is efficiency (see Chapter 1) and so it becomes useful where high acoustic power is needed. A high-quality direct radiator speaker has an efficiency of typically 1% whereas horns can offer between 10% and 50%. With today's cheap amplification, high sound levels can be produced by arrays of direct radiators, but the directivity of such arrays is very irregular because of interference between the discrete drivers. The horn is a single source and avoids that difficulty. The high efficiency of the horn means that thermal compression effects in the coil can also be avoided.

Another advantage of the horn transducer is that with careful design it can produce a close approximation to an expanding spherical wave from a point source and thus has more desirable directivity characteristics than a dome driver.

The horn loudspeaker consists of the transducer proper, often called a pressure or compression driver, and the horn structure. These may be made by different manufacturers and there are some standards for the mechanical interface between the two. In public address units a coarse screw-thread rather like a pipe fitting may be used, whereas high-quality units employ a flange with axial bolts.

The horn itself may be made from any convenient material which is sufficiently rigid. In early cinema speakers, the horn was made of single curvature plywood on a wooden frame. For public address purposes, weather proofing is needed and horns may be made of glassfibre or a combination of diecast and sheet metal. For land-based foghorns, concrete is usefully employed in horns of impressive size which can be heard tens of kilometres away.

The horn acts as an acoustic transformer and presents a high acoustic impedance to the drive unit which can therefore deliver more acoustic power for a given volume velocity. Compared to a direct radiator with the same diaphragm area as the area of the horn mouth, the moving mass of the horn diaphragm is much smaller and the directivity will also be much better.

The acoustics of the horn are treated in Chapter 1 and only the salient points will be repeated here. A horn has a low-frequency cut-off below which its efficiency plummets. The cut-off frequency is determined by the rate of flare. For low-frequency reproduction this has to be very low, requiring the horn to be physically long. This effectively precludes the use of horns to reproduce low frequencies in equipment of reasonable size. At high frequencies, beaming is avoided in a horn by using a throat which must have a diameter of the same order as the shortest wavelength to be reproduced so that the entire horn mouth is illuminated.

The compression drive unit works on the moving-coil principle, but optimized for working into a narrow throat which displays a high acoustic impedance that is primarily resistive. This requires the diaphragm to move with a velocity which is

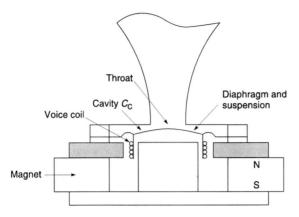

Figure 2.41. Horn driver unit.

independent of frequency, so the mass control of the direct radiator working above resonance is inappropriate. Instead resistance control is used with the highly damped resonant frequency in the middle of the band. Strictly the resonance should be at a frequency which is given by the square root of the product of the upper and lower frequencies to be reproduced. Figure 2.41 shows a simple drive unit. As with all speakers, the moving mass must be as light as practicable, but the greater air loads experienced by the pressure driver mean that diaphragm stiffness is important. An edge-driven dome is often used for rigidity. The dome is often concave and may be made of thermosetting-resin-impregnated fabric or metal such as aluminium, beryllium or titanium. The voice coil in larger units may be of rectangular aluminium wire so that it can be self-supporting.

It will be seen from Fig. 2.41 that there is an annular cavity above the suspension. This has an acoustic compliance and at high frequencies air would rather move into this cavity than down the horn, producing a filtering effect. Consequently the size of this cavity must be minimized. One approach is to use a very narrow suspension. In this case tangential or diamond pattern corrugations may be used as they offer better linearity in a suspension of restricted width. The air chambers in front of and behind the diaphragm act as stiffnesses in parallel with the diaphragm suspension stiffness and this sets the fundamental resonant frequency.

At high frequencies the wavelength can be less than the diaphragm diameter and this will result in unfavourable frequency-dependent cancellations and reinforcements in the throat of the horn leading to irregularities in the frequency response. The solution is to incorporate what is known as a phase plug between the diaphragm and the throat. This is a component containing concentric annular passages each of which has a cross-section which is small compared to the wavelength so that only plane waves can pass down them. As the speed of propagation in the diaphragm is finite, the centre of the diaphragm will vibrate with a delay with respect to the edge. If the propagation speed in the diaphragm is known, this can be corrected by using a phase plug having passages which become progressively shorter towards the centre of the diaphragm.

Essentially the acoustic distance from the coil to the horn throat is constant over the whole diaphragm. The result is that a coherent wavefront is launched into the throat due to the in-phase summation of the pressures from each of the annuli.

The phase plug acts as a second transformer in series with the horn. The impedance ratio is given by the square of the ratio of the diaphragm area to the throat area.

Possibly the greatest drawback of the horn loudspeaker is that the high pressures in the throat which allow high efficiency are also the source of non-linearity. Figure

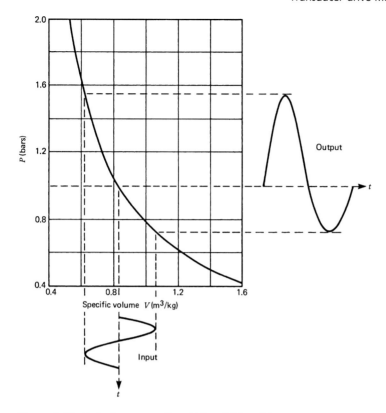

Figure 2.42. Plot of gas equation $PV^r = 1.26 \times 10^4$.

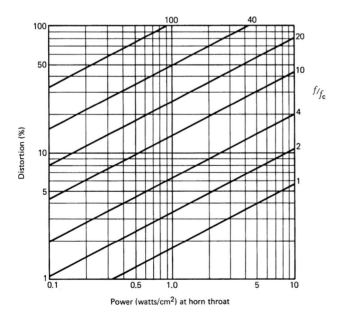

Figure 2.43. Percentage second harmonic distortion as a function of power at horn throat, with the ratio of the frequency f to the cut-off frequency f_c.

2.42 shows the relationship between pressure and volume for air at 20°C. The pressure variations which characterize normal sounds in free air are miniscule and exercise such a small region of the curve that the slope is to all intents and purposes constant, so that sound propagation is linear. However, the high pressures built up in the throat of efficient horn speakers operating at high mouth SPL will cause significant excursions along the curve of Fig. 2.42. The result of this non-linearity is harmonic distortion. Figure 2.43 shows that the second harmonic distortion is a function of the intensity at the horn throat, and the ratio of the actual frequency to the horn cut-off frequency. The length of the throat at which the high SPL is maintained also has an effect.

By reducing the throat SPL and choosing a horn flare which minimizes the length of the horn where high pressures exist, the distortion can be effectively overcome. This approach can be used in high-quality monitoring speakers[8].

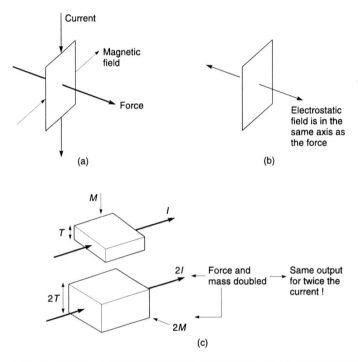

Material	p ($\times 10^3$ kg/m³)	δ ($\times 10^{-8}$ Ω/m)	$p\delta$ ($\times 10^5$)	Sensitivity (dB)
Al	2.7	2.67	7.21	0
Be	1.8	4.2	7.56	−0.4
Cu	8.96	1.6	14.34	−6.0
Ag	10.4	1.59	16.54	−7.2

(d)

Figure 2.44. Ribbon and electrostatic speakers both allow uniform diaphragm drive. At (a) the ribbon needs a transverse magnetic field, making the magnetic circuit very inefficient and limiting ribbon area. At (b) the electrostatic field is normal to the diaphragm allowing a diaphragm of arbitrary size. (c) Creating an efficient ribbon speaker dictates minimal ribbon thickness. (d) Best material has minimal density–resistivity product.

2.14 The ribbon loudspeaker

The ribbon loudspeaker works on the principle shown in Fig. 2.44(a). The diaphragm is conductive and is positioned in a transverse magnetic field. The diaphragm is supported at each end. Current flowing along the diaphragm creates a force perpendicular to it. This mechanism produces an even thrust on the diaphragm which can therefore be extremely thin and light. There are some parallels with the electrostatic speaker in which the driving force is also evenly created over the entire surface of a light diaphragm. The two systems have in common the fact that the transduction mechanism is basically linear and has a desirable minimum phase characteristic.

However, there is a great deal of difference between the two when an attempt is made to put the principle into practice. As Fig. 2.44(b) shows, the electrostatic speaker produces the drive force in the same direction as the applied (electrostatic) field so that the diaphragm size can be as large as required and the electrostatic speaker can be very efficient.

In contrast, the ribbon speaker produces the drive force transversely to the applied (magnetic) field. This means that the diaphragm is in the plane of the magnetic field lines. As the diaphragm is very thin, only a small part of the magnetic field passes through it. A further difficulty is that the width of the diaphragm is severely limited by the difficulty of producing a magnetic field in a very wide air gap. As the gap has high reluctance, leakage is a serious problem. In the Decca/Kelly 'London' ribbon speaker the gap efficiency was less than 3%.

The low gap efficiency means that the equivalent of the Bl product is very low. Fortunately the diaphragm resistance is also very low so that large currents can be used. A step-down transformer is needed to obtain such currents from a conventional audio amplifier.

These difficulties mean that a full-frequency range ribbon loudspeaker is impossible whereas it is possible with electrostatic technology. Ribbon speakers require separate low-frequency units. They may be direct radiating or horn loaded. Direct radiating ribbon speakers must be very heavy owing to the size of the magnet system and the uniform drive over the diaphragm area causes directivity problems as ka becomes too large. Horn-loaded ribbon speakers avoid this difficulty and can give very good directivity. The horn loading also brings the efficiency in line with direct radiating moving-coil transducers but, unless it is well designed, the coloration of the horn may lose some of the inherent clarity of the ribbon transduction system. For these reasons this alternative to the moving coil has not enjoyed as much success as the electrostatic speaker.

The choice of ribbon material is reflected by the need for efficiency. Figure 2.44(c) shows a hypothetical ribbon element, driven by a voltage source. If the thickness were to be doubled, twice as much current would flow, doubling the power and the force created, but the mass would also be doubled, making the amplitude the same. The efficiency would be halved. Thus the ideal diaphragm is one which is as thin as structural requirements allow. Practical diaphragms will be corrugated. This increases the transverse rigidity and reduces the resonant frequency.

The amplitude achieved goes as the ratio of current to mass, but the power required goes as the product of the current and the resistivity. Consequently the greatest efficiency is obtained with a diaphragm material having the smallest product of resistivity and density. Figure 2.44(d) shows that aluminium is the ideal material.

A more recent variation of the ribbon theme is the Heil transducer shown in Fig. 2.45. The diaphragm is a thin plastic sheet carrying a serpentine conductor of metal foil which makes the drive current flow alternately up and down the diaphragm. The diaphragm is corrugated such that alternate foil tracks face one another. The whole is immersed in a magnetic field which is perpendicular to the diaphragm. When current flows it will be in opposite directions on alternate folds of the corrugations and so these will tend to move together or apart. Air is thus displaced to produce sound.

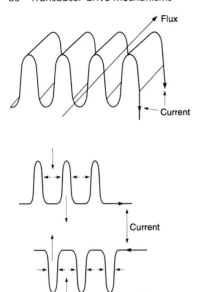

Figure 2.45. In the Heil transducer the serpentine current path causes adjacent corrugations of the diaphragm to move together or apart to produce sound.

The magnetic circuit must have a large gap to allow the insertion of the corrugated diaphragm and it must be perforated in some way to allow the sound to escape. These effects combine to produce an inefficient and hence massive magnet structure. The greatest drawback of the Heil transducer is that the uniform output over the whole diaphragm causes difficulties in achieving a suitable directivity characteristic.

2.15 Moving masses

Accurate sound reproduction depends totally on loudspeakers which contain parts moving (hopefully) in sympathy with the audio waveform. In order to build accurate loudspeakers, some knowledge of the physics of moving masses is essential.

Isaac Newton explained that a mass would remain at rest or travel at constant velocity unless some net force acted upon it. If such a force acts, the result is an acceleration where $F = ma$.

Figure 2.46 shows a mass supported on a spring. This configuration is found widely in loudspeakers because all practical diaphragms need compliant support to keep them in place and prevent air leaking from one side to the other while still allowing vibration. An ideal spring produces a restoring force which is proportional to the displacement. The constant of proportionality is called the stiffness which is the reciprocal of the compliance. When such a system is displaced and released, the displacement performs a sinusoidal function called simple harmonic motion or SHM. When more energy is put into the system, it oscillates at the same frequency but the amplitude has to increase so that the restoring force can be greater. Eventually the resonance of a mass on a spring dies away. The faster energy is taken out of the system, the greater the rate of decay. Any mechanism which removes energy from a resonant system is called damping. In a loudspeaker this could come from losses in the spider, the surround, or from electromagnetic damping in the coil. Acoustic radiation also extracts energy.

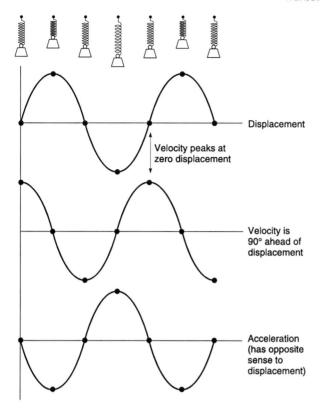

Figure 2.46. The displacement, velocity and acceleration of a body executing simple harmonic motion (SHM).

The motion of a rigid body can be completely determined by the mass, the stiffness and the damping factor. As audio signals contain a wide range of frequencies, it is important to consider what happens when resonant systems are excited by them. Figure 2.47 shows the displacement, velocity and acceleration of a mass, stiffness, damping system excited by a constant amplitude sinusoidal force acting on the mass at various frequencies.

Below resonance, the frequency of excitation is low and little force is needed to accelerate the mass. The force needed to deflect the spring is greater and so the system is said to be stiffness controlled. The amplitude is independent of frequency, described as constant amplitude operation, and so the velocity rises at 6 dB/octave towards resonance.

Above resonance, the inertia of the mass is greater than the stiffness of the spring and the system is said to be mass controlled. With a constant force there is constant acceleration yet, as frequency rises, there is less time for the acceleration to act. Thus velocity is inversely proportional to frequency which in engineering terminology is −6 dB/octave. It is shown in Chapter 1 that the radiating ability of a diaphragm is proportional to velocity, i.e. it rises at 6 dB/octave. The frequency response in this region is flat because the two slopes cancel. Mass control is universally used in the traditional moving coil loudspeaker for this reason.

It will be clear that the behaviour just noted has a direct parallel in the behaviour of an electronic damped tuned circuit consisting of an inductor, a capacitor and a resistor and the mathematics of both are one and the same. By converting mechanical

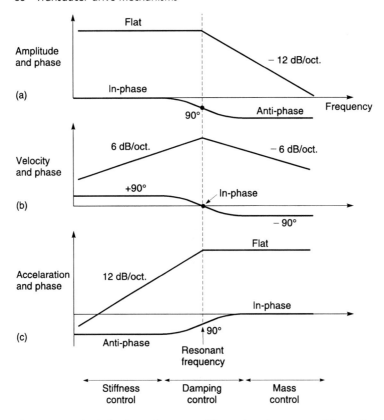

Figure 2.47. The behaviour of a mass-stiffness-damping system: (a) amplitude, (b) velocity, (c) acceleration.

parameters into electrical parameters the behaviour of a mechanism can be analysed as if it were an electronic circuit. This is particularly useful way of modelling loud-speakers.

In mass-controlled loudspeakers we are interested in the cone acceleration since this determines the acoustic waveform radiated. It will be seen from Fig. 2.47 that far above resonance the acceleration is in phase with the input signal, which is the desired result. However, as frequency falls the phase response deteriorates to 90° at resonance and a phase reversal below. The rate of phase change in the vicinity of resonance is a function of the overall damping factor.

Because of this phase characteristic the polarity of a loudspeaker is a matter of opinion. Manufacturers mark one terminal with a red spot or a + sign as an aid to wiring in the correct polarity. However, some use the convention that a positive d.c. voltage (e.g. from a battery) will cause outward motion of the cone, whereas others use the convention that the positive half cycle of an a.c. voltage at a frequency above resonance will cause outward motion. It will be seen from Fig. 2.47 that these two conventions are, of course, in phase opposition. The a.c. definition makes more sense as that is how the speaker is used, however most manufacturers use the d.c. definition.

Whatever the labelling scheme, it is important that the overall system from micro-phone to speaker shall be phased such that, within the audio band, when the microphone diaphragm moves inwards due to a pressure increase, the loudspeaker diaphragm moves outwards to create a pressure increase. A system of this kind is said to have 'absolute phase'. On certain sounds, especially percussion, greater realism

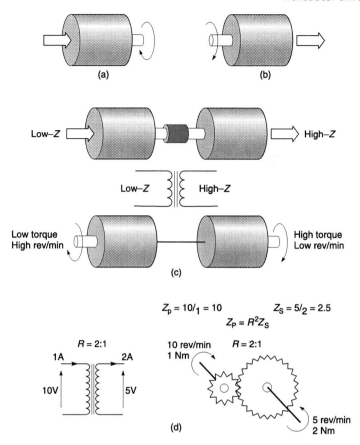

Figure 2.48. (a) Motor creates mechanical power. (b) Generator creates electrical power. Loudspeaker is both at once. (c) Motor generator allows conversion of electrical impedance like a transformer. (d) Generator-motor allows conversion of mechanical impedance like a gearbox.

will be obtained on loudspeakers of suitable precision when absolute phase is employed.

One direct consequence of the phase lag in the drive unit is that the acoustic source of the loudspeaker moves backwards as resonance is traversed. The sound source appears at a distance which is a function of frequency. Genuine sound sources do not have this attribute so the loudspeaker is creating an artifact.

2.16 Modelling the moving-coil motor

The linear moving-coil motor used in most loudspeakers is in many ways simply a linear version of the ubiquitous rotary electric motor, and it shares many characteristics. Figure 2.48(a) shows that a motor is intended to convert electricity to mechanical power, whereas a generator (b) is intended to convert mechanical power into electricity. The use of the word 'intended' is deliberate because there is actually no difference between a motor and a generator except the direction of energy flow. A motor-generator as shown at (c) may produce an output at a different voltage to the input and so it is an impedance converter having the attributes of a transformer. A generator-motor (d) can produce an output at a different rpm to the input and has the attributes of a mechanical gearbox.

Transformers and gearboxes are impedance converters. The impedance connected to the output is reflected into the input by a ratio which is the square of the turns or gear ratio. Figure 2.48 shows that a transformer and a gearbox both having a 2:1 ratio have an impedance ratio of 4:1. An electric motor, a generator and the moving-coil motor of a loudspeaker are all impedance converters, where mechanical impedance on one side is reflected to the other side as electrical impedance and vice versa.

The fact that the electrical impedance of a transducer could be influenced by an acoustic load reflected through it was first discovered by Kennelly and Pierce in 1912 and it was they who coined the term 'motional impedance'. The concept was first put on a sound mathematical footing by Hunt in 1954[3].

Fortunately, in the MKS system, mechanical and electrical units are interchangeable without the use of constants, so that a given power in Watts can result from the product of a current and a voltage or the product of a force and a velocity. Thank goodness!

There are two advantages of this approach; first, the electromechanical system can be modelled by converting all the mechanical parts to their electrical equivalent, and second, the electrical load which the power amplifier will see can be calculated.

Figure 2.49 shows that the impedance conversion of a moving-coil motor can be simulated by treating it as a transformer. On the left side the quantities are genuinely electrical; on the right side mechanical quantities are modelled electrically. The connection between the two sides is that the equations of force-creation and back-emf-creation must always hold simultaneously. The force on the coil (in Newtons) is given by the product of the flux density in the gap, B (Tesla), the length of coil wire actually in the gap l (metres) and the current (A). The back-emf (V) is the product of B (Tesla), l (metres) and the velocity (metres per second). It is clear that on the secondary side the velocity must be an analog of the voltage and the force must be an analog of the current. This is called a mobility analogy.

In both cases it is only the product of B and l which is of any consequence. This, as discussed earlier, is known as the Bl product and its units are expressed (interchangeably) in Tesla-metres or Newtons per Ampere.

Figure 2.49 also shows how the electrical impedance can be calculated as a function of the mechanical impedance. Note how the impedance ratio is the square of the Bl product, hence the use of Bl:1 to describe the turns ratio of the virtual transformer.

It is also of considerable significance that the mechanical impedance is inverted. In other words the electrical impedance seen by the amplifier is the reciprocal of the mechanical impedance seen by the coil. This is initially surprising, but in fact it does explain why the current taken by an electric motor rises when its load increases.

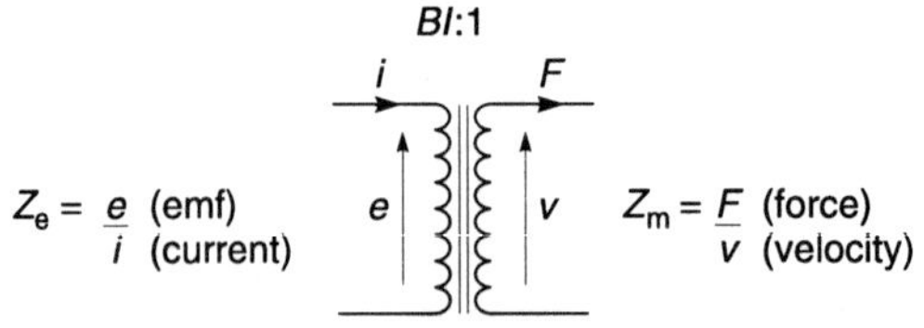

Bl:1

$Z_e = \dfrac{e}{i} \; \dfrac{(emf)}{(current)}$

e v $Z_m = \dfrac{F}{v} \; \dfrac{(force)}{(velocity)}$

$e = Blv$ (back emf creation)
$F = Bli$ (force creation)

$$Z_e = \frac{e}{i} = \frac{(Bl)^2 v}{F} = \frac{(Bl)^2}{Z_m}$$

Figure 2.49. The characteristics of a moving-coil motor can be described by two equations, one for each direction. This can be simulated by a transformer as shown here.

Figure 2.50 shows the essential parts of a moving-coil speaker. On the left is the coil having a *Bl* product and a d.c. resistance R_e. The force from the coil excites a mass-spring-damper system. The force is distributed between the mechanical imped-ances due to moving mass, the compliance and the resistance seen by the mass. In a typical speaker there is also a relatively small load due to the inefficient sound radiation which will be neglected until the next section.

As was shown earlier, a mass-controlled system has a velocity response which falls with frequency at 6 dB/octave and a compliance controlled system has a velocity response which rises at 6 dB/octave. This corresponds to the electrical behaviour of reactive devices. Figure 2.49 shows that current is analogous to force and voltage is analogous to velocity. An inductor in series with the current/force would produce a voltage/velocity proportional to frequency. This is the behaviour of a compliance.

As a result, a compliance driven by a moving-coil motor can be replaced by an inductor without altering the load seen by the amplifier. By a similar argument a mass can be replaced by a capacitor. It is easy to prove that this works in practice. A small motor gearbox from the local model shop can be connected to a resistor, an inductor or a capacitor in turn to see the result of trying to turn the shaft. With the resistor, the motor is simply harder to turn. With a capacitor (several thousand microfarads), the motor is initially harder to turn, but continues running for some time in the same direction when the shaft is released, because the system is acting like a flywheel. With a suitable inductor (one of sufficient inductance is hard to find!) the motor will reverse direction when the shaft is released, because the system is acting like a spring.

2.17 The electrical analog of a drive unit

It is extremely useful to model the behaviour of a loudspeaker by creating an elec-trical circuit which behaves in the same way. This is generally considered a good way of understanding what is happening, but this is to miss the point that a suitable choice of model makes it relatively easy to overcome many of the undesirable character-istics of the speaker by creating an opposing process in a signal processor prior to the power amplifier. The mobility model allows this to be done because it converts the acoustic and mechanical processes in a loudspeaker into a complex load seen by the power amplifier.

Figure 2.50 shows that there is a two-stage process to be modelled. Electrical energy is converted to mechanical movement, which in turn is converted to acoustic output into an air load. Understanding the process will be eased if the model is assembled gradually.

Figure 2.51(a) shows the mechanical model. Initially the loudspeaker is electrically modelled by direct replacement of the mechanical parameters with electrical ones as in (b). However, as the moving-coil motor acts like an impedance inverting trans-former, the parameters can be brought to the primary side by inverting and then impedance converting them using the square of the *Bl* product as in (c). Once this is done the transformer disappears.

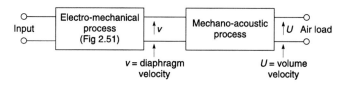

Figure 2.50. The loudspeaker is a two-stage conversion process.

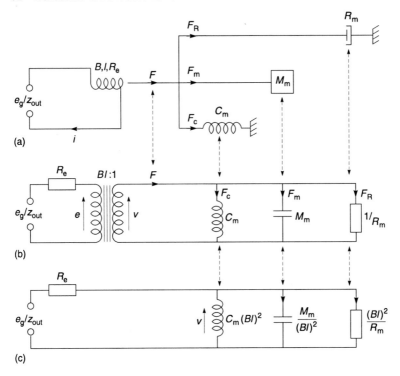

Figure 2.51. (a) First modelling of moving-coil speaker (for low frequencies) shows the sources of force on the moving mass. (b) Simple electrical modelling of (a) places components on transformer secondary. (c) Components can be moved to transformer primary in equivalent circuit.

In most cases the output impedance of the amplifier will be zero because the amplifier uses heavy negative feedback which makes it a voltage source. As a result, the coil resistance is essentially in parallel with the damping resistance. In practical moving-coil woofers the damping effect of the coil resistance is usually much greater than the mechanical damping and so the equivalent circuit can be simplified to that of Fig. 2.52.

It is the velocity of the diaphragm which is responsible for sound radiation and, as Chapter 1 shows, this will be proportional to the square of the diaphragm area. Sound radiation is modelled by applying the mechanical voltage/velocity to the primary of another virtual transformer, as shown in Fig. 2.53, although this one does not impedance invert. On the secondary side the voltage in the model is proportional to the volume velocity U and the current is proportional to the sound pressure. The air load is correctly modelled by the acoustic admittance (the reciprocal of the impedance). The turns ratio of the transformer is numerically equal to the diaphragm area.

Figure 2.53 also shows that the air load is complex and consists of a reactive term due to the mass of the air adjacent to the diaphragm and a resistive term due to the sound radiation. Considering the drive unit alone, in the absence of an adequate baffle, the air load contains a further reactive element due to the volume velocity short circuiting from one side of the diaphragm to the other at low frequencies.

The second transformer can be eliminated by bringing the secondary load to the primary side via an impedance conversion. This will be the square of the turns ratio, giving the correct result that radiation is proportional to the square of the diaphragm

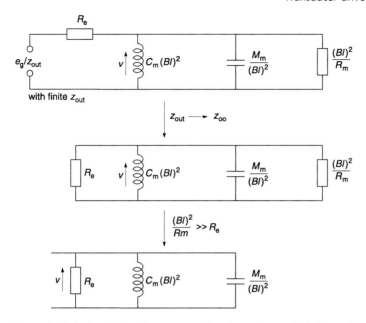

Figure 2.52. When driven by a conventional amplifier which is a voltage source, source impedance disappears.

area. The air load can be brought to the primary of the first transformer by inverting and multiplying by the square of the Bl product. Figure 2.53 further shows that the result of the inversion is that the reactive mass of the air load is seen by the amplifier as a capacitor, just as the mass of the diaphragm is. The capacitor is in series with the acoustic resistance term. The resistor is shunted by an inductance when there is no baffle.

Figure 2.54 shows the equivalent circuit derived in this way. It consists of a damped tuned circuit in parallel with the acoustic load. In practice, moving coils have finite inductance and this has been included in the model. Assuming a rigid diaphragm, the entire moving-coil drive unit can be modelled as just seven components. This is a tremendous advantage because (except at high frequencies) a few simple calculations can eliminate a lot of tedious experiment. The model can be analysed to reveal what actually happens in a loudspeaker at different frequencies. The resulting frequency response is derived in Fig. 2.55 and comprises four main regions.

Considering first the mechanical response, Fig. 2.55(a) shows that, below resonance, in zone A, the compliance dominates and as there is a series resistor R_e, the voltage across the inductor of Fig. 2.54, which represents the cone velocity, rises at 6 dB/octave. At resonance, zone B, the velocity is a function of the Q factor. Above the resonant frequency, in zone C the mass component in Fig. 2.54 dominates. It will be clear that the the effect of R_e and the impedance of the capacitor is a velocity which falls at 6 dB/octave.

But how does this frequency dependent mechanical velocity affect the sound radiated? Sound radiated by a real speaker corresponds to power dissipated in the resistor representing the resistive part of the acoustic impedance. Figure 2.55(b) shows that at low frequencies this resistor is short-circuited by an inductance which models the ease with which air can flow from front to back of the diaphragm. This is suggesting the use of a baffle or enclosure of some sort. Figure 2.55(b) also shows that in the mid-range, above the frequency where baffling is needed, the dominant mechanical component is the capacitance modelling the mass of the diaphragm and

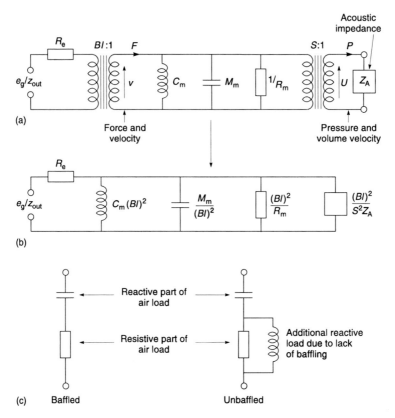

Figure 2.53. (a) Air load is reactive and can be modelled as shown. (b) In the absence of a baffle there is an additional component to model the short circuit flow around the diaphragm. (c) Air load is brought to primary side by inversion and impedance conversion.

the dominant acoustic component is the capacitor modelling the reactive component of the acoustic impedance. When this is the case, the input current will divide between the two capacitors of Fig. 2.54 in a fixed ratio and so the input current which reaches the resistive load is independent of frequency. This is the elegant result first outlined by Rice and Kellog, that a mass-controlled diaphragm can have a flat frequency response.

As frequency rises further, the ka parameter increases and, when it reaches a value of 2, the magnitudes of the reactive and resistive components become equal. In the model of Fig. 2.54, this occurs when the impedance of the capacitor modelling the acoustic reactance equals the value of the load resistance. Above this frequency, the input current no longer divides equally between the diaphragm mass and the acoustic mass. The impedance of the acoustic mass capacitance has become small and the acoustic impedance is dominated by the load resistance. The capacitor modelling the diaphragm mass shunts current away from the load and so above $ka = 2$ the output falls at 6 dB/octave as Fig. 2.55(b) also shows. In some cases the coil inductance will further increase this roll-off to 12 dB per octave.

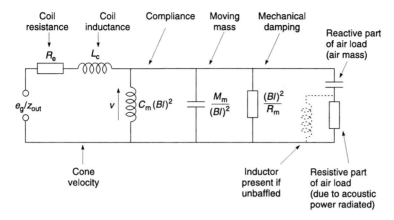

Figure 2.54. The equivalent circuit of a moving-coil drive unit which models both the operation of the transducer and the actual electrical load seen by the amplifier.

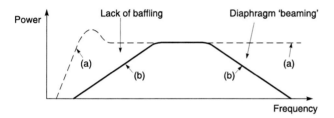

Figure 2.55. (a) The mechanical frequency response of the model of Fig. 2.54. (b) The acoustic response. Note LF loss due to lack of baffling and HF loss due to high *ka* value.

2.18 Modelling the enclosure

In the previous section it was seen that an unbaffled drive unit suffers a loss of output at low frequencies because the volume velocity is effectively short circuited by an easy path from the front to rear of the diaphragm. The simplest solution is to mount the drive unit in a sealed enclosure (see Chapter 7). When this is done, the inductance across the acoustic resistance disappears, but the enclosed air acts as a further stiffness in parallel with the stiffness of the drive unit. In other words the compliance seen by the moving cone goes down. This is modelled by Fig. 2.56.

The result is that the fundamental resonant frequency of the driver goes up when it is mounted in a practical enclosure. This can be calculated if all the parameters are known, but Fig. 2.57(a) shows how it can be measured. A signal generator and a suitable resistor provide a nearly constant current source and the impedance can be plotted by using the expression shown. Of course modern computerized speaker testers can do this with a few commands and plot the result automatically.

Figure 2.57(b) shows the result for a medium-sized woofer, the impedance peaks sharply indicating the value of the resonant frequency. This is as might be expected from a tuned circuit. Below the resonant frequency the stiffness dominates. As the frequency falls, the inductor in Fig. 2.54 progressively shunts away cone velocity so that, below resonance, it goes down at 6 dB/octave. Now the radiation characteristic compounds the effect so that the frequency response falls at 12 dB/octave.

Under mass control, the velocity experiences a 90° phase lag whereas under compliance control there is a 90° phase lead. At resonance the phase angle is zero. Around

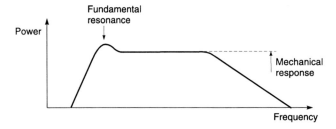

Figure 2.56. When the drive unit is enclosed, the baffling effect removes the inductor modelling the acoustic short circuit, but the stiffness of the enclosure adds to the drive unit's own stiffness.

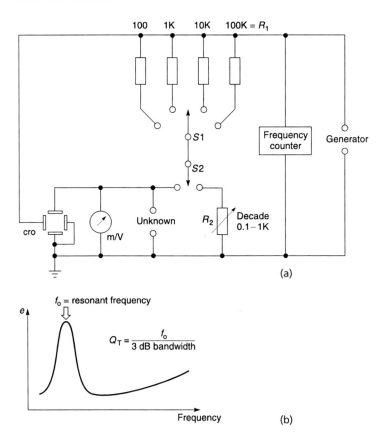

Figure 2.57. (a) Simple test rig to find resonant frequency and mechanical Q factor. X-Y oscilloscope displays straight line at 45° at resonance. (b) Impedance function of typical woofer. Initial peak is due to resonance, later rising trend is due to coil inductance.

the resonant frequency the speaker undergoes a phase reversal. The sharpness of this reversal is a function of the Q-factor of the resonance.

Clearly if this phase reversal is within the audio band it will violate the requirement for minimum phase and do nothing for the time response of percussive transients. In low-cost equipment this result is accepted but, for precision work, the

Figure 2.58. An active speaker in which the fundamental resonance of the woofer system is two octaves above the low frequency turnover point. (Courtesy Celtic Audio Ltd.)

optimum approach is to build an active speaker in which the resonant behavour of the drive unit is cancelled in amplitude and phase by an equal and opposite electronic transfer function. The low-frequency roll-off of the speaker can then be set electronically below the audio band. A useful consequence is that not only can the overall transfer function display minimum phase, but the size of the enclosure can also be dramatically reduced because the fundamental resonance of the drive unit is no longer relevant.

Figure 2.58 shows an active speaker in which the fundamental resonance of the woofers occurs two octaves above the lowest frequency reproduced.

The arrangement of Fig. 2.57 is designed to show the resonance clearly and, because a current source is used, the Q factor measured there is not the one which will result when driven by an amplifier. The electrical Q due to the damping of the coil resistance appears in parallel with the Q of Fig. 2.57. Figure 2.59(a) shows how the electrical Q factor is calculated once the resonant frequency is known.

Figure 2.59(b) shows the effect of different Q factors on the frequency response. In a traditional passive speaker the designer had to juggle the Bl product to avoid an obviously 'honky' high Q response, but also to avoid a premature roll-off of response with a low Q factor.

$$\frac{1}{Q} = \frac{1}{Q_T} + \frac{1}{Q_E}$$

In the above, $1/Q$ is total Q, $1/Q_T$ is mechanical Q and $1/Q_E$ is electrical Q.

$$Q_E = \frac{2\pi f_o M R_e}{(Bl)^2}$$

$$\therefore \frac{1}{Q} = \frac{1}{Q_T} + \frac{(Bl)^2}{2\pi f_o M R_e}$$

(a)

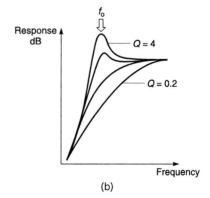

(b)

Figure 2.59. (a) Calculation of overall Q factor from mechanical and electrical components. (b) Effect of driver Q factor on frequency response of passive speaker. Active speakers eliminate this dependence.

As an aside, it is important that the resistance of the cable between the amplifier and the speaker is low enough to ensure that the effective damping resistance is not significantly changed. Despite the pseudoscience one reads, this is the only attribute a loudspeaker cable actually needs. If general-purpose electric cable of sufficient cross-section is used, the stray capacitance and inductance will always be negligible except where the separation between the amplifier and speaker is such that the term telephony is more appropriate [but see Chapter 6, Section 6.16: Editor].

A good way of reducing the size of a sealed enclosure without raising the resonant frequency is to pack the volume with a critical density of acoustic wool. The specific heat of the wool and the close coupling to the air prevents the temperature of the air rising as it is compressed and the result is that the volume can be reduced by about 15%. An excess wool density simply reduces the available volume and the resonance will rise again. Using the apparatus of Fig. 2.57 it is easy to find the optimum density of packing which is where the resonant frequency is the lowest. Unfortunately the presence of the wool also increases the damping and this may result in an overdamped condition in a passive speaker. However, in an active speaker the damping of the wool is simply taken into account in the phase and amplitude equalization and the effect is then entirely beneficial.

The efficiency of a loudspeaker goes as the square of the Bl product. Using rare earth magnets, some pretty impressive Bl products are now possible as a matter of straightforward design, resulting in highly efficient drive units. Unfortunately the designer of the passive speaker cannot use them because they have very low Q and suffer premature roll-off in a conventional enclosure. However, this is not a problem in the active speaker which simply equalizes. There is thus a case for designing drive units specifically for active speakers. In the author's experience commercially available drive units designed for passive use are seldom suitable for active applications.

In particular, active loudspeakers can operate the drive unit in the stiffness-controlled domain, where the linearity of the suspension is critical to the distortion. During the development of the Cabar speaker, it was found that no commercially available woofer could meet the linearity specification and a drive unit had to be specially designed[9].

Fortunately designing a drive unit for an active speaker is relatively easy, because the exact value of many of the parameters such as the Q factor is not particularly important. When the parameters are being electronically equalized, what matters instead is consistency from one unit to the next. By using a small enclosure, the compliance of the air spring dominates and this swamps variations in drive unit compliance. Moving mass and cone area will not change, nor will Bl or R_e and so an active equalized speaker should not go out of adjustment as it ages.

2.19 Low-frequency reproduction

The only criterion we have for the accuracy of a loudspeaker is the sensitivity of the human hearing system. If a reproduction system is more accurate than the human ear in some respect then its shortcomings in that respect will not be audible. Consequently it behoves loudspeaker designers to study psychoacoustics in order to establish suitable performance criteria. In low-frequency reproduction, directivity is seldom a problem but the frequency response and the phase linearity requirements can cause some difficulties.

In real life, transient noises produce a one-off pressure step whose source is accurately and instinctively located by the human hearing system. Figure 2.60 shows an idealized transient pressure waveform following an acoustic event. Only the initial transient pressure change is required for location. The time of arrival of the transient at the two ears will be different and will locate the source laterally within a processing delay of around a millisecond, far before any perception of timbre has taken place[10].

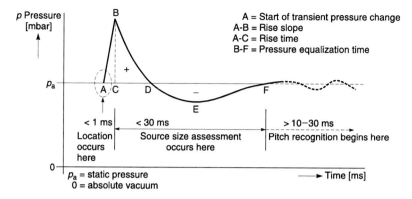

Figure 2.60. Real acoustic event produces a pressure step. Initial step is used for spatial location, equalization time signifies size of source. (Courtesy Manger Schallwandlerbau.)

Following the event which generated the transient, the air pressure equalizes. The time taken for this equalization varies and allows the listener to establish the likely size of the sound source. The larger the source, the longer the pressure-equalization time.

The above results suggest that anything in a sound reproduction system which impairs the reproduction of a transient pressure change will damage both localization and the assessment of the pressure-equalization time. Clearly in an audio system which claims to offer any degree of precision, every component must be able to reproduce transients accurately and should therefore display a minimum-phase characteristic.

An unenclosed diaphragm acts as a dipole and therefore becomes extremely inefficient at low frequencies because air simply moves from the front to the back, short-circuiting the radiation. In practice the radiation from the two sides of the diaphragm must be kept separate to create an efficient LF radiator. The concept of the 'infinite baffle' is one in which the drive unit is mounted in an aperture in an endless plate (see Chapter 7). Such a device cannot be made and in practice the infinite baffle is folded round to make an 'enclosure'. Sealed enclosures are often and erroneously called infinite baffles, but those of practical size do not have the same result. Unfortunately the enclosed air acts as a spring because inward movement of the diaphragm reduces the volume, raising the pressure.

The stiffness of this air spring acts in parallel with the stiffness of the diaphragm supports. The mass of the diaphragm and the total stiffness determines the frequency of 'fundamental resonance' of the loudspeaker. To obtain reproduction of lowest frequencies, the resonance must be kept low and in a passive speaker this implies a large enclosure to reduce the stiffness of the air spring, and a high compliance drive unit. Of course, with modern active loudspeakers a large enclosure is no longer required as was seen in Section 2.18.

Before economical and precise electronic signal processing and inexpensive amplification were available, a number of passive schemes were developed in an attempt to improve the LF response of smaller enclosures. These include the reflex cabinet shown in Fig. 2.61(a) which has a port containing an air mass. This is designed to resonate with the air spring at a frequency below that of the fundamental resonance of the driver so that, as the driver response falls off, the port output takes over. In some designs the air mass is replaced by a compliantly mounted diaphragm having no coil, known as an ABR or auxiliary bass radiator (b).

Another alternative is the transmission line speaker shown in (c) in which the rear wave from the driver is passed down a long damped labyrinth which emerges at a

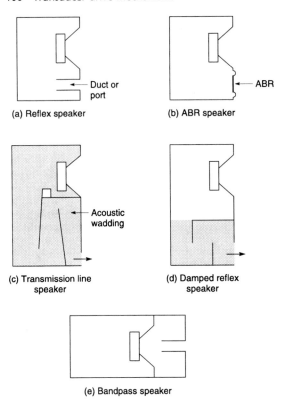

(a) Reflex speaker

(b) ABR speaker

(c) Transmission line
speaker

(d) Damped reflex
speaker

(e) Bandpass speaker

Figure 2.61. Various schemes have been devised to extend the low-frequency response of passive loudspeakers at the expense of the time response. (a) The bass reflex enclosure uses an air mass in a duct which resonates with the air spring of the enclosure. (b) The auxiliary bass radiator, or drone cone, replaces the air mass and duct with a diaphragm. (c) The rear radiation of the drive unit is taken down a transmission line which delays the rear signal by half a cycle at the frequency where it is intended to augment the front signal. (d) In some compact speakers the duct is too short to achieve a half cycle delay at the lowest reproduced frequency. Instead there is an enclosure behind the driver and the air in the duct resonates with this, damped by the filling in the duct. (e) The bandpass enclosure augments LF by using at least two overlapping resonances of the drive unit with an enclosed volume and an air mass in a duct with another volume.

port. The length is designed to introduce a delay which corresponds to a 180° phase shift at the frequency where the port output is meant to augment the driver output. A true transmission line loudspeaker is quite large in order to make the labyrinth long enough. Some smaller models are available which are advertised as working on the transmission line principle but in fact the labyrinth is far too short for the claimed frequency response and there is a chamber behind the drive unit as shown in (d) which actually makes these heavily damped reflex cabinets.

More recently the *bandpass* enclosure (e) has become popular, probably because suitable computer programs are now available to assist the otherwise difficult design calculations. The bandpass enclosure has two chambers with the drive unit between them. All radiation is via the port.

The reflex, ABR, bandpass and transmission line principles certainly do extend the frequency response of loudspeakers, or permit a reduction is size when compared to a conventional sealed enclosure and this has led to their wide adoption. However, concentrating on the frequency domain alone does not tell the whole story. Although the frequency domain performance is enhanced, the time domain performance is worsened. The LF extension is obtained only on continuous tone and not on transients. In this respect the bandpass enclosure is the worst offender whereas the true transmission line causes the least harm.

None of these traditional techniques can offer sufficient accuracy in the time domain to allow the mechanism of Fig. 2.60 to operate. Instead LF transients suffer badly from linear distortion because the leading edges of the transients are removed and reproduced after the signal has finished to give the phenomenon of *hangover*. The low-frequency content of the sound lags behind the high frequencies in an unnatural way. This can be measured as an exaggerated rearward shift in the acoustic source position in the lowest frequencies reproduced. In other words the input waveform is simply not reproduced accurately using these techniques, as is easily revealed by comparison with the original sound.

Today the shortcomings of these techniques need no longer be suffered as active technology clearly outperforms them using simply constructed sealed enclosures of compact size. Where the greatest precision is not a requirement, passive techniques will continue primarily because of tradition, and also where economy is paramount.

2.20 The compound loudspeaker

Resonant techniques such as reflex tuning allow a smaller enclosure but with the penalty of linear distortion. The compound loudspeaker is an alternative which allows a smaller enclosure while potentially retaining minimum phase. Figure 2.62 shows that the generic compound loudspeaker is one in which there are two drive units in tandem such that only the outermost one actually radiates. In general the goal is to drive the inner diaphragm in such a way that the outer diaphragm believes itself to be installed in a box of infinite internal volume. This requirement can be met if the pressure in the intermediate chamber remains constant. Tiefenbrun[11] coined the term 'isobaric' to describe a loudspeaker operating in this way.

However, the apparatus disclosed in Tiefenbrun's patent does not achieve isobaric conditions because both drive units are identical and are driven by the same signal. Unfortunately the rear speaker is in a finite enclosure which will raise its fundamental resonant frequency. What is needed is for the rear drive unit to be equalized in phase and amplitude so that it can traverse its higher resonant frequency while allowing the front drive unit to operate down to its free-air resonant frequency.

What actually happens in an isobaric speaker with identical drive signals is that the two diaphragms are effectively forced to move in step by the small sealed intermediate chamber. The result is that the system behaves very nearly like a single drive unit having twice the mass. This lowers the resonant frequency compared to what it would be with a single driver. The result is beneficial and commercially available units are audibly superior to reflex types of the same enclosure volume, because they have less linear distortion of transients and a smaller migration of the acoustic source.

Shelton and George[12] described an apparatus which actually can achieve isobaric conditions. Figure 2.63 shows that there is a pressure sensor or microphone in the

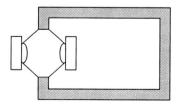

Figure 2.62. The isobaric or compound loudspeaker outperforms ported speakers because it is more faithful to the applied waveform.

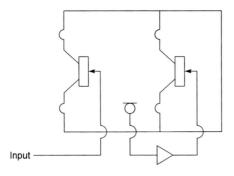

Figure 2.63. Compound speaker using pressure sensing and feedback to maintain constant pressure behind front driver. Front driver then acts as if it were in an infinite enclosure.

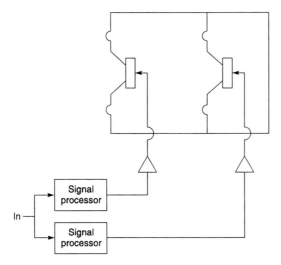

Figure 2.64. Advanced compound speaker uses feedforward signal processing on both drive units to obtain isobaric condition as well as independence of driver Q factor.

isobaric chamber and this is part of a feedback loop which drives the rear diaphragm in whatever way is needed to keep the pressure constant. The inventors correctly identify the requirement to equalize the fundamental resonance of the rear driver in its enclosure. The author is not aware of any commercially available product, which is a great shame because it would have allowed minimum phase LF reproduction in a very small enclosure.

Figure 2.64 shows an advanced compound system which was pioneered by Celtic Audio[13] and is in production. Very small enclosures are possible because the resonant behaviour of both drive units is equalized in amplitude and phase so that their actual resonant frequencies and Q factor are both irrelevant. The low-frequency response of the system is determined entirely by an active high-pass filter which is necessary to prevent excessive diaphragm excursion. Above that frequency the system displays linear phase.

2.21 Motional feedback

The motion of a loudspeaker cone is supposed to be an analog of some original sound and so the more accurately the cone moves, the more accurate will be the

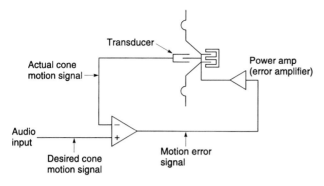

Figure 2.65. General concept of motional feedback. See text for details.

reproduction. It is possible to increase the accuracy of certain aspects of loudspeaker performance using motional feedback (see also Chapter 6, Section 6.12). The principle is shown in Fig. 2.65. The input audio signal represents the required movement of the cone and a transducer of some kind measures the actual movement. The two are compared to create an error and the error is amplified to drive the cone in such a sense as to minimize the error.

It will be clear that a linear transducer in a well-engineered feedback process results in very accurate motion of the feedback transducer. Whether this results in accurate cone motion is another matter. This is because the entire motion of the cone is represented by one signal. If the cone cannot be regarded as a rigid piston then the feedback signal does not represent it.

A ceiling on rigidity is set by the speed of sound in the cone material since this determines how fast a disturbance can propagate. At low frequencies the propagation time is negligible compared to the period of the signal and so a well-designed cone can move as a piston. At high frequencies this simply is not so. The propagation time of sound through a tweeter is a significant part of a cycle and there are significant phase shifts between the motion of the various parts. This may well be desirable as a means to optimize the directivity.

As a result, applying motional feedback to tweeters is not useful. In practice it is restricted to woofers and even then it only really works if the woofer cone is rigid. Consider what happens to a feedback system handling a transient like the sound of a drum. The cone has to accelerate rapidly and if it does not the error will simply get larger until it does.

Assuming the amplifier and power supply are up to it, the coil current and resultant force will become very large. If the drive unit does not have a good enough transient response and is not sufficiently well built, the result will be cone breakup or even disintegration. It is thus a myth that motional feedback can be used to improve a cheap drive unit. The feedback can only control one point on the cone. If the other parts do not want to move like that the whole exercise is fruitless.

It may seem obvious, but the benefits of negative feedback are only obtained when the feedback is working. This is defined as the error being negligibly small. In fact this is a general truth which applies to all uses of negative feedback including autopilots, servos and audio amplifiers. If the system ever gets into a state where the error is large, then the feedback has lost control and the system is said to be working 'open loop'.

One of the greatest myths about the use of negative feedback is that it just increases the bandwidth of a system. Figure 2.66(a) shows the full power frequency response of a real system with a finite frequency response. Figure 2.66(b) shows the same system after the application of some ideal negative feedback, which, of course, reduces the gain. Note that, although bandwidth has been increased, this has been done by

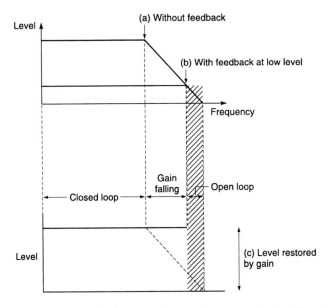

Figure 2.66. Feedback cannot increase full power bandwidth. At (a) open loop frequency response. Negative feedback (b) lowers maximum gain so bandwidth increases *for small signals only.* (c) If level is restored by external gain stage, system is driven open loop and benefit of feedback is not obtained.

reducing the power output over the whole band down to the highest level which was possible at the band edge without feedback. Thus the power bandwidth has not increased at all.

If negative feedback increases bandwidth, then it can only do so at the expense of output power. If we take the naive view that feedback simply increases bandwidth, and we expect the same power output, we simply add a gain stage to counteract the gain loss due to the feedback, and the result is in Fig. 2.66(c). In the shaded area the system goes open loop and the output is heavily distorted. This is the origin of the hi-fi pseudoscience that negative feedback is a bad thing. The truth is that the full-power frequency response of a feedback system can never be better than the open-loop response. It does not matter how accurately the error signal has been derived, or how much gain has been applied to it if the power source simply cannot slew the load fast enough.

What feedback can do, when properly applied, is to reduce distortion. However, distortion is the process of creating harmonics. If the open-loop response of a system is not good enough, the system cannot respond fast enough to cancel the distortion products.

It is also important to consider what happens under overload. A conventional passive speaker simply gets increasingly distorted under overload as the spider gets pushed to the end of its travel. A feedback speaker will try to follow the input signal until something limits. The result is a cliff-edge effect where perfect linearity gives way to gross distortion. In practice there is a danger of physical damage and so the input signal has to be limited at the signal level stage to an amplitude that the speaker system can always follow.

Consequently, to build a well-engineered motional feedback loudspeaker, it is necessary to have a basic amplifier/power supply/drive unit combination with a very wide open-loop full-power bandwidth. When the loop is closed with feedback the distortion goes down.

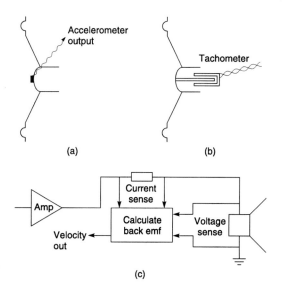

Figure 2.67. Cone motion measurement techniques. (a) Accelerometer on cone. (b) Magnetic tachometer. (c) Extraction of back emf.

To prevent distortion due to slewing rate limits, the amplifier and power supply have to be adequately rated. To withstand the merciless amplified error signal, the drive unit has to be rigidly built. Thus the practical motional feedback loudspeaker is not going to be cheap and this is its main drawback. The main advantages of motional feedback are that the size of the enclosure and the level of linear and non-linear distortion can be simultaneously reduced. With most other speaker technologies these two goals are usually mutually exclusive.

In a real LF loudspeaker the main sources of distortion will be the drive unit itself and the non-linear air spring due to the enclosure. In principle, motional feedback can reduce the effect of both. The performance of any feedback mechanism is limited by the accuracy of the feedback signal. If the feedback signal does not represent the cone motion accurately, then the cone motion cannot be controlled accurately.

Designers have found various ways of measuring the cone motion. It does not matter whether the displacement, velocity or acceleration is measured, as these parameters can be exchanged in signal processing circuitry by integration or differentiation.

One approach is to use an accelerometer. Figure 2.67(a) shows that this is simply a small inertial mass mounted via a force sensor on the cone. The force sensor is typically a piezo-electric crystal requiring a high input impedance amplifier. An alternative is to measure the cone velocity using a separate moving coil as shown in (b). This is mechanically complicated and expensive but, when well engineered, the coil voltage is directly proportional to the velocity.

Some low-frequency drive units designed for sub-woofers have dual coils so that a single unit can be driven by a traditional stereo amplifier. The two channels are simply mechanically added in the coil former. Amateur designs have appeared ingeniously using one of the dual coils as a velocity feedback coil. While this works, the efficiency is low because half of the magnetic energy in the gap is wasted. There is also a possibility of mutual inductance between the two coils which are effectively an accidental transformer.

Figure 2.67(c) shows another possibility which uses a relatively conventional drive unit. Here, the current passing through the coil is sampled by a sense resistor, and the voltage across the coil is measured. If the coil resistance is known, the voltage across the coil due to Ohmic loss can be calculated from the current. Any remaining voltage across the coil must be due to back emf which is proportional to the coil velocity.

A suitable signal processor can extract the back emf which can then be used in a feedback loop. One difficulty with this method is that, as the temperature of the coil changes, its resistance will change, causing an error in the emf calculation. A solution is to connect a length of the same type of wire used in the coil in series with the main coil so that it experiences the same heating current. The voltage across this compensating coil can be sensed to allow for temperature changes. The compensating coil may conveniently be fitted at the end of the coil former but it is important that it is screened from the magnetic circuit.

The back emf extraction approach seems attractive, but it does rely heavily on the linearity of the main magnetic circuit and coil. The integral of the flux cutting the coil must be independent of coil position, or the back emf measurement is not accurate. This technique cannot be used to linearize a cheap drive unit because these invariably have position-dependent flux problems. As a result, it works only in drive units which are already quite linear.

The attraction of the accelerometer approach should now be clear. The addition of the accelerometer to the drive unit is fairly simple, and its operational accuracy is independent of the drive unit so that a less-than-perfect driver can be linearized, provided, of course, that the diaphragm is sufficiently rigid to allow the accelerometer output to be representative of the diaphragm motion.

The degree of linearization achieved with motional feedback is a function of the open-loop gain available. The more gain that can be used, the smaller the residual error will be. The natural conclusion is that the ideal gain is infinite, like an operational amplifier. Unfortunately this cannot be achieved because real drive units have a sub-optimal phase response. Negative feedback will fail in the presence of phase shifts within the loop, because these can result in positive feedback if the loop gain is above unity when 180° of shift has occurred.

There is a change in phase response of 180° as the resonant frequency is traversed, the rapidity of the phase change being a function of the Q factor of the resonance. This is affected by the design of the drive unit and by the nature of the enclosure and its filling. Clearly, if the feedback loop contains a speaker whose phase response can reverse, an equivalent but opposite phase compensation circuit must be included in the loop so that the feedback will remain negative at all frequencies.

Once the phase reversal of the drive unit is compensated, the feedback can be used to flatten the response of the driver well below its natural resonance. The low-frequency response is now determined electronically. It cannot be set arbitrarily low, however, as the drive unit may not have enough displacement to reproduce very low frequencies. Thus a motional feedback speaker may have a small enclosure, but it will have to contain a drive unit which would be considered disproportionately large in a conventional speaker.

References

1. BELL, A G, US Patent 174,465 (1876).
2. RICE, C W and KELLOG, E W, 'Notes on the development of a new type of hornless loud-speaker'. *JAIEE*, **12**, 461–480 (1925).
3. HUNT, F V, *Electroacoustics*, Harvard University Press, Cambridge, MA (1954).
4. BALLANTINE, S, US Patent 1,876,831 (1930).
5. MANGER, J W, Private communication.
6. HONDA, K and SAITO, S, 'MS magnetic steels'. *Phys. Rev.*, **16** (1920).
7. KATES, J M, 'Radiation from a dome'. *JAES*, **24** (1976).

8. NEWELL, P R, *Studio Monitoring Design*, Ch.10, Focal Press, Oxford (1995).
9. SALTER, R J and WATKINSON, J R, 'The design of an unusual drive unit', presented at AES Conference, 'The Ins and Outs of Audio', London (1998).
10. MANGER, J W, 'Freefield pressure step function tests acoustic transmission chain', AES Preprint 2336 (E6), Montreux (1986).
11. TIEFENBRUN, I, UK Patent 1,500,711.
12. SHELTON, J and GEORGE, R, UK Patent Appl. 2,122,051.
13. SALTER, R J and WATKINSON J R, UK Patent Appl. 2,297,880.

3 Electrostatic loudspeakers

Peter J. Baxandall
revised by John Borwick

3.1 Introduction

The electrostatic principle has inherent advantages which make possible the construction of loudspeakers with lower coloration, better transient response, lower non-linearity distortion, and radiation characteristics more suitably related to room acoustics, than can be achieved using other techniques.

The diaphragm in an electrostatic loudspeaker can be so light and flexible that the electric forces developed, over most of the audio spectrum, effectively act directly on the air load to create sound waves. The effects of complex vibrational modes in the diaphragm can be almost eliminated[33] making the acoustic performance theoretically predictable with good accuracy and in a relatively simple manner.

Although a few not very successful electrostatic loudspeakers were marketed in the 1920s and 1930s, it was not until after 1950 that the virtues of constant-charge polarization became appreciated[1]. This, together with the availability of sheet plastic materials with the right characteristics for diaphragms, made satisfactory full-frequency-range electrostatic loudspeakers feasible. The British Quad loudspeaker, marketed in 1957, was the first commercial full-frequency-range design to appear, being the outcome of pioneering work by P. J. Walker and D. T. N. Williamson[2-4].

Though the theoretical principles underlying the operation of a basic electrostatic loudspeaker are essentially quite simple, there are nevertheless many surprisingly difficult technological problems to be overcome before a thoroughly reliable production design can be achieved. Very high voltages and signal-circuit impedances are involved, and the diaphragm material must have carefully controlled mechanical and electrical characteristics not required in any other application.

Another feature of present designs is that, although the loudspeakers are often regarded as inconveniently large, the maximum sound volume capability does not always fully satisfy demands. However, for normal domestic listening, the achievable levels are usually found to be quite adequate.

Manufacturing costs are also inevitably rather high, with the result that the moving-coil loudspeaker continues to supply most of the world's demands, though few owners of electrostatic loudspeakers would be happy to revert to using moving-coil ones. Also, the availability of good electrostatic loudspeakers as a standard of comparison has undoubtedly played a very significant part in the evolution of moving-coil designs of much-improved performance.

3.2 Electrostatic drive theory

The very thorough theoretical treatment by Professor Hunt[1] is often taken as being the standard work in this field, and rightly so, but the present author feels that much of the material there presented can be better expressed in simpler ways without loss of scientific soundness and with a considerable gain in lucidity.

The principles of electrostatics, though basically simple, are less familiar to most engineers than those of ordinary electronic circuits, and it is evident from the number of wrong explanations of the working of constant-charge electrostatic loudspeakers that have appeared in print that many people have found the subject difficult and confusing – as indeed the author did at one time. The above considerations are felt to justify the inclusion of a good deal of basic theory in this chapter.

3.2.1 Electrostatic forces

The earliest electrostatic loudspeakers[1] were essentially simple two-terminal parallel-plate capacitors, one plate being the flexible electrically conductive diaphragm. The electrical force acting on each plate is given by

$$F = \varepsilon_0 \frac{AV^2}{2s^2}$$

$$(3.1)$$

where
F = force (N)*
ε_0 = 8.854×10^{-12} F/m
A = area of each plate (m²)
V = voltage between plates (V)
S = distance between plates (m)

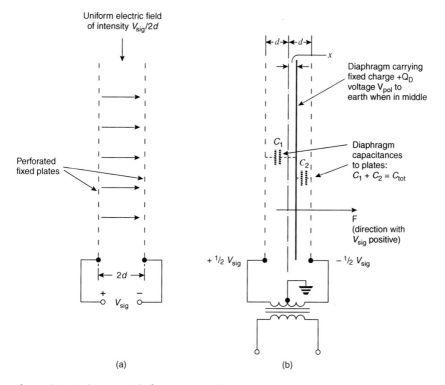

Figure 3.1. Basic essentials for constant-charge operation.

---

*The magnitude of 1 newton may be easily conceived and remembered, for it is a remarkable coincidence that it is the force with which a medium-sized apple is attracted by the Earth!

Even in the earliest days it was appreciated that the square-law force/voltage relationship of equation (3.1) was an unfortunate one when the aim was to make a linear transducer, and push–pull operation was proposed as early as 1924 as a means of reducing the resultant non-linearity distortion[1]. The construction was as in Fig. 3.1(b), the diaphragm being made of metal foil and held at a constant high polarizing voltage by connection to a d.c. supply. With such an arrangement, however, obtaining low distortion ideally requires perfect symmetry of construction and a diaphragm amplitude small compared with the plate spacing, and this confined satisfactory operation to fairly high frequencies.

The above performance limitation remained for about two decades, the stalemate being broken only when the virtues of operating the diaphragm with constant polarizing charge instead of constant polarizing voltage became appreciated.

It is possible to establish that constant-charge operation gives zero distortion via detailed analysis based on equation (3.1) – and indeed this is what is done below for obtaining the full relevant equations – but this approach hides the very great theoretical simplicity and beauty of the constant-charge idea.

It would seem that all early thinking on electrostatic loudspeakers was absolutely tied to the notion that the device was 'really a capacitor', and that equation (3.1), for the force between the plates of a capacitor, was therefore inherently and necessarily the only starting point for deducing the performance. It is now suggested, however, that if any of the early workers had stopped to ask whether there is some other relationship in electrostatics that is inherently linear and could be exploited, and had followed this up, then the virtues of the modern constant-charge scheme would have been appreciated much earlier.

Referring to Fig. 3.1(a), the application of a signal voltage V_{sig} to the two plates produces a uniform electric field of intensity $E = V_{sig}/2d$ in the space between them. If a small positive charge Q is placed anywhere in this space, it will experience a force F towards the right, when V_{sig} is positive, given by

$$F = Q \times \frac{V_{sig}}{2d} \qquad (3.2)$$

where F = force (N)
$\quad\quad\ \ Q$ = charge (C)
$\quad\quad\ \ V_{sig}$ = signal voltage (V)
$\quad\quad\ \ d$ = half-spacing (m)

(Calling the spacing $2d$ rather than d is convenient for later working.)

Thus, if Q is kept constant, the force varies linearly with signal voltage, just as desired. In practice the charge is carried on a light diaphragm, as in Fig. 3.1(b).

Though equation (3.2) gives the total force acting on the charge with good accuracy when the charge is small, it does not tell the whole story when the charge is large enough for its own field to be comparable in magnitude to that produced by the signal. It then turns out that constant-charge operation still gives a force component in accordance with equation (3.2), linearly proportional to the signal voltage and independent of the diaphragm position, but there is now a second force component independent of the signal voltage as such, but varying linearly with the position of the diaphragm. The system is therefore still perfectly linear, though the behaviour is now slightly less simple and is analysed in detail in the next section.

3.2.2 Basic analysis for constant-charge operation

With reference to Fig. 3.1(b), suppose initially that the signal voltage is zero and the diaphragm is in the middle. Now, connect the diaphragm temporarily to a positive polarizing voltage V_{pol} to charge it up, and then disconnect it. Assuming perfect insulation, the charge Q_D will be retained indefinitely. If the diaphragm material has a high electrical resistivity, then V_{pol} must be connected for long enough to ensure that a uniform voltage is reached over the whole area of the diaphragm. However, provided this condition is fulfilled, it is unimportant for the purposes of the following analysis whether the diaphragm is highly conductive or an almost perfect insulator. It is assumed, for the time being, that the diaphragm motion will be over its whole area.

With the diaphragm thus charged, the next requirement is to determine its voltage V_D when displaced a distance x in the presence of a signal voltage V_{sig}.

There are two quite independent mechanisms which determine the diaphragm voltage:

(a) When the diaphragm is displaced from the middle, the two capacitances C_1 and C_2 between it and the fixed plates become unequal. Consequently, even if the diaphragm is uncharged, the application of a signal voltage will produce a voltage on the diaphragm, because the diaphragm may be regarded as being at the output point of an unbalanced a.c. bridge circuit. It is possible, therefore, to find the diaphragm voltage by first determining the capacitances C_1 and C_2 as functions of plate area, x and d, and then calculating the bridge output voltage. But there is no need to do this once it is appreciated that the uncharged diaphragm simply takes up the potential that would exist at that position in the absence of the diaphragm. An uncharged diaphragm, assumed infinitely thin, is simply non-existent from an electrical point of view. It is therefore evident that the voltage of the diaphragm when uncharged is given by

$$V_{D_a} = -\frac{1}{2} V_{sig} \left(\frac{x}{d}\right) \tag{3.3}$$

(b) Now suppose that $V_{sig} = 0$ but that the diaphragm has been charged to the voltage V_{pol} when in the middle. If it is now moved away from the middle, its total capacitance to earth increases, so its voltage must fall, in accordance with the fundamental relationship

$$Q = CV \tag{3.4}$$

The capacitance of a parallel-plate air-dielectric capacitor is given by:

$$C = \varepsilon_0 \frac{A}{s} \tag{3.5}$$

where C = capacitance (F)
   ε_0 = 8.854 × 10^{-12} F/m
   A = area of each plate (m^2)
   s = distance between plates (m)

Applying equation (3.5) to Fig. 3.1(b) gives:

$$C_{tot} = \varepsilon_0 A[(d - x)^{-1} + (d + x)^{-1}] \tag{3.6}$$
$$= 2d\varepsilon_0 A/(d^2 - x^2)$$

If the diaphragm carries a charge Q_D, then from equations (3.4) and (3.6):

$$V_{D_b} = Q_D \frac{(d^2 - x^2)}{2d\varepsilon_0 A} \tag{3.7}$$

where V_{D_b} signifies the diaphragm voltage component due to the present mechanism (b).

Because the diaphragm was charged to V_{pol} at $x = 0$, and because, from equation (3.6), C_{tot} is then $2\varepsilon_0 A/d$, it follows from equation (3.4) that:

$$Q_D = \left(\frac{2\varepsilon_0 A}{d}\right) V_{pol} \tag{3.8}$$

and substituting this in equation (3.7) gives:

$$V_{D_b} = V_{pol}\left[1 - \left(\frac{x}{d}\right)^2\right] \tag{3.9}$$

When, as in practice, both the above mechanisms operate together, the total diaphragm voltage V_D is given by:

$$V_D = V_{D_b} + V_{D_a} \tag{3.10}$$

Hence from equations (3.9) and (3.3):

$$V_D = V_{pol}\left[1 - \left(\frac{x}{d}\right)^2\right] - \frac{1}{2}V_{sig}\left(\frac{x}{d}\right) \tag{3.11}$$

That this simple linear addition of the two effects is justified can be seen by imagining that the diaphragm is first moved a distance x by an external agency, with $V_{sig} = 0$, giving V_{D_b} as in equation (3.9), and that V_{sig} is then switched on. Since V_{sig} is being applied to a perfectly linear circuit, the principle of superposition applies and the total diaphragm voltage must therefore be as given by equation (3.10).

The total electrical force F acting from left to right on the diaphragm in Fig. 3.1(b) may be determined by applying equation (3.1) to the two 'capacitors' on either side of it, and is given by

$$F = \frac{1}{2}\varepsilon_0 A \left\{\left[\frac{V_D + 1/2V_{sig}}{d - x}\right]^2 - \left[\frac{V_D - 1/2V_{sig}}{d + x}\right]^2\right\} \tag{3.12}$$

Substituting for V_D in this from equation (3.11) leads finally to the result

$$F = \varepsilon_0 A \left[\frac{2V_{pol}^2 x}{d^3} + \frac{V_{pol}V_{sig}}{d^2}\right] \tag{3.13}$$

This is a most important and crucial equation, and it is worth pondering over its significance and how it fits in with some other notions.

The second term is simply a force proportional to the signal voltage and independent of the diaphragm position x. This is the wanted effect which was previously considered and which led to equation (3.2). That this is true may be seen by replacing Q in equation (3.2) by its $Q = CV$ equivalent as given by equation (3.8).

The first term in equation (3.13) is a force on the diaphragm directly proportional to its displacement x and in the same direction as x, so that it represents a linear negative compliance. The symbol C_{me} will be used for this, signifying mechanical compliance caused by electrical effects, so that

$$C_{me} = -\frac{d^3}{2V_{pol}^2\varepsilon_0 A} \tag{3.14}$$

Though this negative compliance is produced by electrical effects, it can be regarded from now on as being just one of the mechanical elements to be driven by the signal-dependent force F_{sig}, given by:

$$F_{sig} = \frac{\varepsilon_0 A V_{pol}V_{sig}}{d^2} \tag{3.15}$$

3.2.3 Transduction coefficients

For a more complete understanding of the behaviour of an electrostatic loudspeaker, it is desirable to evolve an electrical-impedance circuit, showing how the impedance at the loudspeaker input terminals is related to electrical and mechanical parameters in the system. To do this, knowledge of two transduction coefficients is required, one being for electrical-to-mechanical transduction and the other for mechanical-to-electrical. The convenient ones to choose are:

(a) The signal force per unit signal voltage:
(b) The current generated, *with the loudspeaker terminals short-circuited*, for unit diaphragm velocity.

Now (a) has already been determined, being given by equation (3.15), conveniently rewritten in the form

$$\frac{F_{sig}}{V_{sig}} = \frac{\varepsilon_0 A V_{pol}}{d^2} \tag{3.16}$$

Transduction coefficient (b) will now be deduced but, as a preamble to this, it is a good thing to remember that when a capacitor is said to be charged, it does not as a whole contain any more charge than when it is discharged, because the charging current feeds the same charge into one terminal that it takes out from the other. It is the plates that are charged, with equal and opposite charges, not the capacitor as a whole.

Referring again to Fig. 3.1(b), the constant diaphragm charge Q_D is shared between the two 'capacitors' C_1 and C_2. The diaphragm voltage is V_D ($= V_{pol}$ when the diaphragm is in the middle), so that both C_1 and C_2 have this voltage, and the ratio of their charges Q_1 and Q_2 is therefore equal to the capacitance ratio. Since C_1 is proportional to $1/(d + x)$ and C_2 to $1/(d - x)$, it follows that

$$\frac{Q_1}{Q_2} = \frac{d - x}{d + x} \tag{3.17}$$

(Q_1 and Q_2 are the charges on the perforated plates.) But

$$Q_1 + Q_2 = -Q_D \tag{3.18}$$

Solving equations (3.17) and (3.18) for Q_1 and Q_2 gives:

$$Q_1 = - Q_D \frac{d - x}{2d} \tag{3.19}$$

$$Q_2 = - Q_D \frac{d + x}{2d} \tag{3.20}$$

Differentiating gives

$$\frac{dQ_1}{dx} = + \frac{Q_D}{2d} \tag{3.21}$$

$$\frac{dQ_2}{dx} = - \frac{Q_D}{2d} \tag{3.22}$$

Thus a diaphragm displacement Δx gives a change in Q_1 of $+Q_D \Delta x/2d$ and a change in Q_2 of $-Q_D \Delta x/2d$, so that a quantity of charge as given by equation (3.23) moves from the right plate across through the short-circuited transformer secondary to the left plate:

$$\Delta Q = Q_D \frac{\Delta x}{2d} \tag{3.23}$$

Because the system is perfectly linear, it follows that

$$\frac{dQ}{dt} = \frac{Q_D}{2d} \times \frac{dx}{dt} \tag{3.24}$$

Here dQ/dt is the current flowing from right to left in the transformer winding, and will be called I_{mot}, the motional current. dx/dt is the diaphragm velocity U, so that

$$I_{mot} = \frac{Q_D}{2d} \times U \tag{3.25}$$

Substituting for Q_D from equation (3.8) gives

$$\frac{I_{mot}}{U} = \frac{\varepsilon_0 A V_{pol}}{d^2} \tag{3.26}$$

Comparing equations (3.26) and (3.16), it can be seen that the transduction coefficients on this basis are the same for either direction of signal transfer. Thus, calling the common transduction coefficient α, it is given by

$$\alpha = \frac{\varepsilon_0 A V_{pol}}{d^2} \tag{3.27}$$

3.2.4 Electrical-impedance circuit

From equations (3.16) and (3.26) it follows that

$$\frac{V_{sig}}{I_{mot}} = \frac{F_{sig}}{U} \times \frac{d^4}{(\varepsilon^2 A V_{pol})^2} \tag{3.28}$$

or, from equation (3.27)

$$\frac{V_{sig}}{I_{mot}} = \frac{F_{sig}}{U} \times \frac{1}{\alpha^2} \tag{3.29}$$

V_{sig}/I_{mot} is the motional electrical impedance of the loudspeaker, Z_{em}. F_{sig}/U is the total mechanical impedance loading on the diaphragm, Z_m, which, as already explained, must be taken to include a component due to the electrically produced negative compliance as given by equation (3.14). Thus equation (3.29) may be written as

$$Z_{em} = Z_m \frac{1}{\alpha^2} \tag{3.30}$$

This relationship of direct proportionality between electrical motional impedances and mechanical impedances is more straightforward than for moving-coil loudspeakers, for in the latter case it is electrical admittances that are proportional to mechanical impedances. The result is that for electrostatic loudspeakers the electrical-analogy circuit representation of the mechanical system is of exactly the same configuration as the electrical-impedance circuit representation.

If the diaphragm of an electrostatic loudspeaker were to be prevented from moving, there would be infinite motional impedance in the electrical circuit. But the total electrical impedance would still be finite – simply that of the capacitance, which will be called C_0, between the fixed plates. Hence the electrical circuit is as shown in Fig. 3.2.

C_{me} is as given by equation (3.14), and is negative. Multiplying by α^2 (see equation (3.27)) to obtain the corresponding electrical capacitance gives a value of

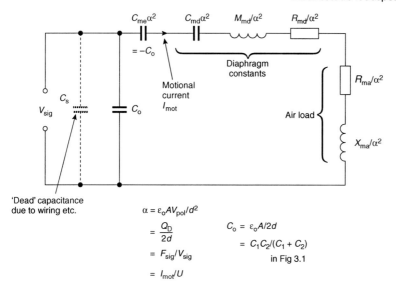

$$\alpha = \varepsilon_0 A V_{pol}/d^2$$

$$= \frac{Q_D}{2d}$$

$$= F_{sig}/V_{sig}$$

$$= I_{mot}/U$$

$$C_0 = \varepsilon_0 A/2d$$

$$= C_1 C_2/(C_1 + C_2)$$

in Fig 3.1

Figure 3.2. Electrical-impedance circuit.

$-\varepsilon_0 A/2d$. Now, using the basic capacitance equation (3.5) to determine C_0, since $s = 2d$, gives

$$C_0 = \frac{\varepsilon_0 A}{2d} \tag{3.31}$$

Thus the electrical negative capacitance in the Fig. 3.2 circuit must have a value of $-C_0$, and this simple fact should always be borne in mind.

3.2.5 Analogous electrical circuit

Figure 3.3 shows how the system, as viewed from the mechanical side, may be represented by means of an analogous electrical circuit containing mechanical impedances, force and velocity.

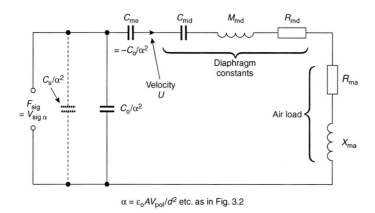

$\alpha = \varepsilon_0 A V_{pol}/d^2$ etc. as in Fig. 3.2

Figure 3.3. Analogous electrical circuit representing the mechanical system.

3.2.6 Advantages of very high resistivity diaphragm

The above theory assumes that the diaphragm will retain its charge Q_D indefinitely once it has been charged up and isolated. In practice, inevitable slight insulation leakage necessitates continuous replenishment of the charge, and the obvious way to achieve this is to connect the electrically conductive diaphragm via a very high value resistor R_0 to the polarizing-voltage supply V_{pol}, the time-constant of R_0 and the diaphragm capacitance being long compared with the periodic time at low audio frequencies. A value for R_0 of over 100 MΩ is likely to be required and, if almost the whole of V_{pol} is actually to appear on the diaphragm, the associated leakage resistance from the diaphragm to earth must be thousands of MΩ – under all conditions of humidity, moreover.

The use of a highly-conductive diaphragm, however, is unsatisfactory also for the following reasons:

(a) In practice, the diaphragm will not move equally over its whole area at all frequencies – for one thing, it is fixed to a frame at its edges. Consequently, though the total charge may be held nearly enough constant, charge will move about the surface of the diaphragm in a manner dependent on the variation in amplitude over the area[34]. Hence any particular small area is liable not to operate under the desired constant charge condition, leading to non-linearity distortion.
(b) If a spark occurs between the diaphragm and a fixed plate, the voltage of the whole diaphragm surface will suddenly change, altering the sensitivity until conditions have had time to return to normal. This effect gives rise to audible distortion. Damage may also be done by the spark. The spark problem is of major practical importance, because it is necessary to work at electric-field intensities quite close to the ultimate breakdown point of air if sufficient acoustic output power is to be made available. The problem may be mitigated to some extent by sheathing the fixed perforated metal electrodes with insulating material, which should have very slight remanent conductivity to ensure that it does not itself become charged and thus prevent the wanted voltage gradients from appearing in the air gaps. It is difficult in practice to make such sheathing completely effective everywhere, and one weak point may cause trouble if a conducting diaphragm is used[3].

The solution to both of the above problems is to make the diaphragm of insulating plastic sheet, such as Mylar, suitably treated to give it slight surface conductivity. The sheet is, in effect, mounted on a conducting frame to which the polarizing voltage, which may be several kV, is applied. Then if a spark occurs it merely discharges a small local area. Moreover, the effective capacitance feeding the spark is very small, giving a very weak and inaudible spark which does no damage.

A very high resistivity diaphragm also ensures that the charge on every cm^2 remains almost constant, giving the desired very low distortion. If, however, the surface resistivity is made too high, the loudspeaker takes too long to reach full sensitivity after switching on the polarizing supply. Preferably the surface resisitivity should be within the range 10^9 to 10^{11} Ω per square.*

3.2.7 Stability of the diaphragm

As already explained, polarizing the diaphragm with a constant charge brings into existence a linear negative compliance as given by equation (3.14). Such a negative compliance, or the corresponding negative stiffness, is inherently unstable on its own. To obtain stability, it is necessary to stretch the diaphragm sufficiently tightly to give a positive stiffness at least as great as the negative stiffness. In practice, an adequate

---

* A square piece of thin uniformly resistive material has a resistance value, measured between opposite edges, which is independent of its size. Hence surface resistivity may be expressed in Ω per square.

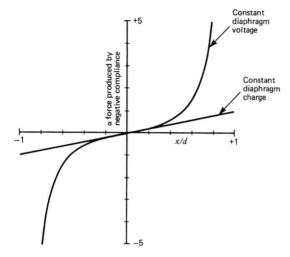

Figure 3.4. Theoretical negative-compliance characteristics. The forces plotted are the electric forces exerted on the diaphragm by the negative-compliance mechanism. For the curve, the vertical quantity is that within the curly brackets of equation (3.34).

'factor of safety' must be allowed, and the criterion adopted by Quad is that the diaphragm tension should be such that the capacitance at the fixed-plate terminals, measured at a very low frequency, e.g. 5 Hz, well below the resonance frequencies, shall increase by a factor of approximately 1.4 when the polarizing supply is switched on. From Fig. 3.2, in which all impedances to the right of $C_m d\alpha^2$ may be neglected at the low measuring frequency, it is easily deduced that this criterion corresponds to $C_m d\alpha^2 = 0.286 C_0$, the diaphragm stretching stiffness then being 3.50 times the negative stiffness.

If the diaphragm is moved away from the middle very slowly, say by a sustained strong draught, its voltage will remain equal to V_{pol}, in all positions, Q varying appropriately. The negative-compliance force on it then varies non-linearly with position, as shown in Fig. 3.4. The equation for this curve is easily derived, as follows.

Referring to Fig. 3.1(b), and bearing in mind that the diaphragm voltage is now equal to V_{pol} for all values of x, equation (3.1) can be applied to C_1 and C_2 to give the forces of attraction on the diaphragm with $V_{sig} = 0$. Thus

Force towards right $= -\varepsilon_0 A V_{pol}^2/(d - x)^2$

Force towards left $= \varepsilon_0 A V_{pol}^2/(d + x)^2$ (3.32)

Hence F, the total negative-compliance force towards the right, is given by

$$F = \tfrac{1}{2}\varepsilon_0 A V_{pol}^2[(d - x)^{-2} - (d + x)^{-2}] \tag{3.33}$$

and a little algebraic manipulation shows that this may be more conveniently expressed in the form:

$$F = \frac{2\varepsilon_0 A V_{pol}^2}{d^2}\left\{\frac{x/d}{[1 - (x/d)^2]^2}\right\} \tag{3.34}$$

This is the equation from which the Fig. 3.4 curve was plotted.

When x is very small compared with d, the denominator becomes very nearly unity and the equation then approximates to

$$F = 2\varepsilon_0 A V_{pol}^2 x/d^3 \quad \text{(for } x/d \ll 1) \tag{3.35}$$

This is the same as the first term in equation (3.13), showing that, for small diaphragm displacements, the electrically produced negative compliance is the same for constant-voltage or constant-charge polarization. The result is as expected because, even under constant-charge conditions, the diaphragm voltage V_D remains almost constant for small displacements, at $V_D = V_{pol}$, passing through a broad maximum in the middle.

It is very easy for confusion over signs to arise in the present context and some thought is necessary to avoid this. A difficulty is that there are two possible conventions which may be adopted; but whichever is chosen, a force is called positive when it is in the direction of positive displacement, i.e. towards the right in Fig. 3.1.

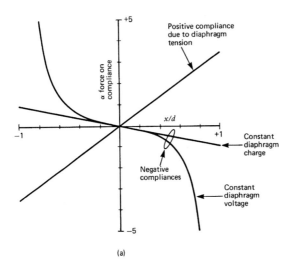

(a)

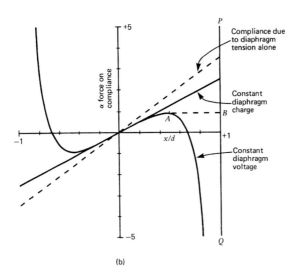

(b)

Figure 3.5. (a) Theoretical compliance characteristics. In this diagram the forces are those exerted on the compliances. The diaphragm stiffness is 3.5 times the linear (constant-charge) negative stiffness, following the criterion adopted by Quad. (b) Theoretical total-compliance characteristics corresponding to Fig. 3.5(a).

The force represented by the curve in Fig. 3.4 acts *on the diaphragm* and is produced by the electrical negative-compliance mechanism. The curve occupies the first and third quadrants because the force is in the same direction as the displacement.

According to the usual convention, however, the force/displacement graph for an ordinary positive-compliance spring is also plotted in the first and third quadrants, but the force is now that exerted *on the spring*, not the reaction force produced by the spring.

It seems natural to work on the Fig. 3.4 basis when explaining how the electrical forces acting on the diaphragm give rise to a negative compliance; but, as already mentioned, once the existence of the negative compliance has been established, it is best treated simply as one of the mechanical elements to be driven, rather than as a generator of force. From this point onwards, therefore, the negative-compliance characteristic will be plotted as in Fig. 3.5(a), which follows the normal convention for springs. The force required to drive the negative compliance is negative when the displacement is positive, because it is necessary to pull the diaphragm back to counteract the tendency for the negative compliance to make it 'run away'.

Also included in Fig. 3.5(a) is a typical line representing the positive compliance or stiffness resulting from diaphragm tension. A good straight line can indeed be obtained in practice with modern plastic diaphragm materials, for the tension remains almost constant during vibration and the amplitudes are small compared with the surface dimensions of the diaphragm. Saran has sometimes been used when widths of low-frequency radiating areas are rather small, because it has a much lower Young's modulus than Mylar, but unfortunately its long-term stability is rather poor.

The total force required to drive the combined negative compliance and stretching compliance may be obtained by adding the appropriate characteristics in Fig. 3.5(a), and the result is shown in Fig. 3.5(b). The straight full-line characteristic in Fig. 3.5(b), which applies to signal-frequency diaphragm movements, raises no complications, for it is a simple linear positive compliance of somewhat greater magnitude than that relating to the diaphragm tension alone. The 'factor of safety' applicable to this aspect of the design problem is discussed near the beginning of Section 3.2.7.

The curved characteristic, which applies for slow diaphragm movements, leads to more subtle effects, however. Imagine that an external force is applied to the diaphragm and gradually increased in magnitude in a positive direction. The diaphragm deflection x increases in a straightforward manner until the point A is reached. What happens if the force is increased slightly beyond the value required to reach point A? Clearly the compliances cannot accept such an increased force so long as the curved characteristic remains the relevant one. What occurs is that the operating point suddenly jumps from A to B, i.e. the diaphragm flops over onto the fixed plate. Once this has happened, it then becomes able, of course, to accept any applied force, the operating point now moving up and down the line PQ as the force varies.

It is usually found, in practice, that the polarizing voltage must be switched off, or at least reduced, before recovery from the 'stuck-over' state occurs. In well-designed units, however, under normal music-reproduction conditions, the dwell-time at large diaphragm displacements is never long enough for the curved characteristic in Fig. 3.5(b) to become relevant, and perfectly stable operation in accordance with the straight constant-charge characteristic is obtained right up to $x/d = 1$.

The type of negative compliance depicted in Fig. 3.5(a) has an electrical analogy – an open-circuit-stable negative capacitance. A linear negative capacitance is easily produced using the circuit of Fig. 3.6(a), and non-linearity may be introduced by adding diodes as in Fig. 3.6(b).

These circuits are open-circuit stable because, with the input on open-circuit, there is ideally 100% negative feedback, but less positive feedback, making the negative feedback dominant. On short-circuiting the input, only positive feedback remains, giving instability.

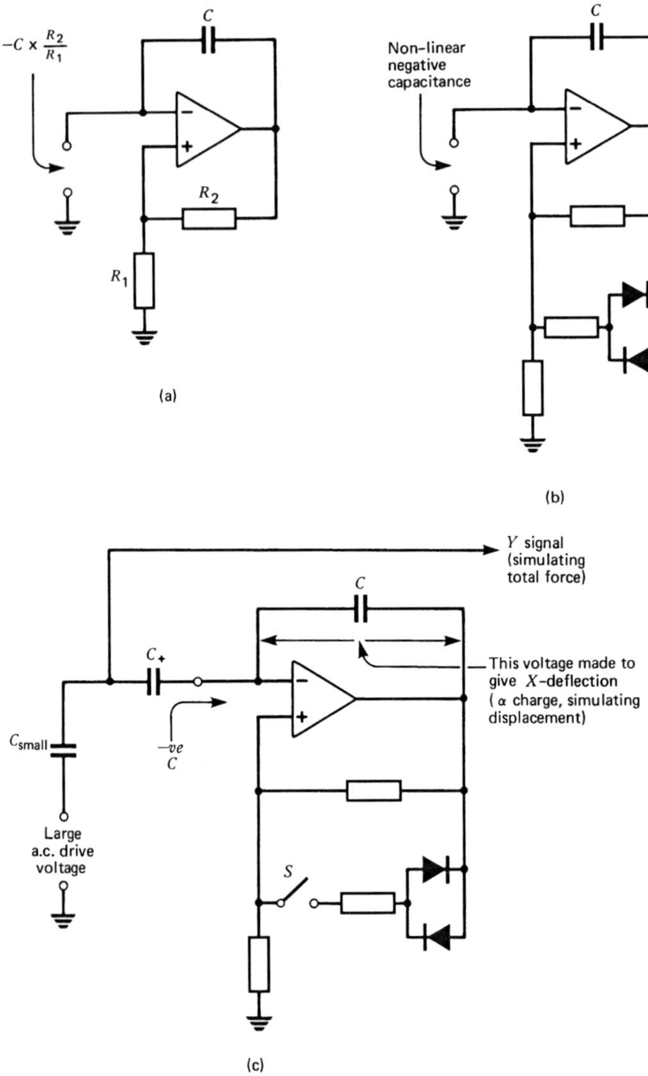

(a)

(b)

(c)

Figure 3.6. Circuits (a) and (b) simulate linear and non-linear negative compliances respectively. Circuit (c) simulates an electrostatic loudspeaker whose diaphragm is constrained by an external agency to move with a certain displacement.

To display the negative-capacitance characteristics of the Fig. 3.6 circuits on an oscilloscope, it is therefore necessary to feed them from a low-capacitance source, as in Fig. 3.6(c). By adding the capacitor marked C_+, the circuit may be made to simulate the loudspeaker-diaphragm situation, where the negative compliance is always associated with a positive compliance. The Y-voltage is the sum of the voltages across C, and the negative capacitance, just as in Fig. 3.5(b) the force plotted is the sum of the forces exerted on the negative and positive compliances.

The photograph in Fig. 3.7 was obtained with a circuit essentially as in Fig. 3.6(c), though slightly elaborated to provide proper d.c. biasing of the operational ampli-

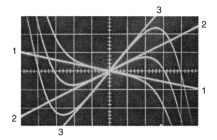

Figure 3.7. Displays obtained with the Fig. 3.6(c) circuit under three conditions:
1. C_+ omitted, giving negative capacitance only.
2. C_+ set to be 1/3.5 of negative capacitance value.
3. C_+ set to be 1/7 of negative capacitance value.
In each condition a line was obtained with S open and a curve with S closed.

fier. The values of C_+ were selected to simulate positive stiffnesses of zero, 3.5 and 7 times the negative stiffness, the first two values being as for the Fig. 3.5(a) and (b) curves respectively.

The curve shapes in the photograph are not, of course, identical to the calculated ones in Fig. 3.5, for the law of the 0A47 diodes used is not ideally suitable, but the important point to be made is that the whole of these characteristics can readily be displayed in a perfectly stable manner, whereas, in the loudspeaker situation, parts of the characteristics are said to be 'inaccessible'[1]. This is because, in the Fig. 3.6(c) simulator circuit, the drive is fed in via a small capacitor, corresponding to driving the loudspeaker diaphragm from a low-compliance source and thus constraining it to move with a certain definite amplitude, the force then being the dependent variable.

To simulate the true loudspeaker situation, the small capacitor in Fig. 3.6(c) must be removed, giving low-impedance voltage drive, equivalent to driving the diaphragm from a force source. With this done, and C_+ selected, say, to simulate a positive stiffness 3.5 times the negative stiffness, there is no difficulty in displaying the straight-line characteristic corresponding to constant-diaphragm-charge operation. However, with S in Fig. 3.6(c) closed, to simulate constant-diaphragm-voltage operation, it is found, on turning up the drive level from zero, that the beginning of the curved characteristic can be obtained, but that as soon as the input is increased to the point where the characteristic becomes horizontal (zero incremental stiffness), the circuit jumps into a state of overload, simulating diaphragm collapse onto a fixed plate. Thus only the parts of Fig. 3.7 that are shown in Fig. 3.8 can now be displayed, the other parts indeed being inaccessible.

Finally, it should perhaps be mentioned that a stability limit of $|x/d| \leqslant 1/3$ has sometimes been quoted for slow diaphragm deflection in a push–pull electrostatic loudspeaker. This, however, is quite wrong, and would appear to have arisen from insufficiently careful reading of reference 1, where this figure is given on pages 183 and 197, but only with reference to single-sided electrostatic loudspeakers. For these it is quite true that, on gradually turning up the polarizing voltage, the diaphragm slowly moves across towards the fixed plate, fall-in theoretically occurring at $|x/d| \leqslant 1/3$.

The Fig. 3.7 and Fig. 3.8 results show that with a push–pull system the higher the diaphragm tension (or the lower the polarizing voltage for a given tension), the closer to the fixed plates can the diaphragm be slowly deflected before flop-over occurs.

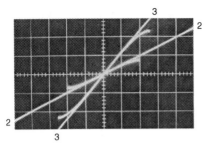

Figure 3.8. Displays obtained under the same conditions as for Fig. 3.7 but with C_{small} removed and the circuit driven by a low-impedance voltage source.

3.2.8 Behaviour with plates on open-circuit

The behaviour of a constant-charge electrostatic loudspeaker when used backwards as a microphone, with the fixed plates on open-circuit, is not normally a matter of any immediate practical concern, but it is nevertheless convenient to investigate it at this point, since a knowledge of what happens is required later on in connection with exploiting the reciprocity principle, leading to Walker's equation and the ideas behind the Quad ESL63 loudspeaker.

One approach is via the equivalent circuits of Figs 3.2 and 3.3, in which, for present purposes, the stray-plus-transformer-winding capacitance C_s will be neglected.

With the terminals in Fig. 3.2 on open-circuit, the impedance to the left of the 'diaphragm constants', consisting of $-C_0$ and C_0 in series, is zero. Similarly, with reference to Fig. 3.3, the diaphragm sees zero mechanical impedance on the left, i.e. no electrical forces whatever are developed on it.

From equations (3.26) and (3.31) it follows that

$$I_{mot} = U \times 2C_0 V_{pol}/d \tag{3.36}$$

Now the voltage V_{sig} produced by I_{mot} flowing in C_0 is given by

$$V_{sig} = \frac{1}{C_0} \int I_{mot}\, dt \tag{3.37}$$

so that, substituting for I_{mot} from (3.36)

$$V_{sig} = -2V_{pol}x/d \tag{3.38}$$

where V_{sig} represents the open-circuit output voltage developed between the fixed plates due to a diaphragm displacement x. The sign convention is as in Fig. 3.1.

It is now of interest to consider what happens to the diaphragm voltage V_D under these open-circuit conditions, and the answer is obtained by substituting equation (3.38) in equation (3.11), which gives:

$$V_D = V_{pol} \tag{3.39}$$

Implicit in doing this is the assumption that V_{sig} is developed in a balanced manner with respect to earth, which will be the case when a transformer is used as in Fig. 3.1.

Under constant-charge operating conditions there is, of course, no theoretical need for the plate circuit to be balanced, and one of the plates may be earthed if preferred. The performance is ideally quite unaffected by the choice of earthing point but, in a practical loudspeaker, push-pull feeding of the plates has the great advantage of minimizing the peak voltages developed with respect to earth and thus easing insulation breakdown problems.

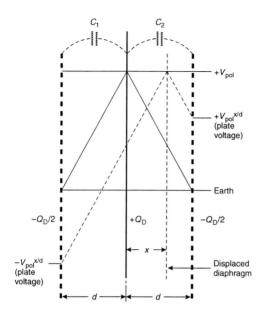

Figure 3.9. Diagram to illustrate the behaviour of a push–pull electrostatic loudspeaker with the plates on open-circuit.

Though the above is a perfectly valid way to arrive at the two results (3.38) and (3.39), and to show that with the plates on open-circuit no resultant electrical force whatever is developed on the diaphragm, a more vivid and direct approach is as follows.

Suppose the diaphragm is initially in the middle, at a voltage $+V_{pol}$, and that both plates are at earth potential, being on signal-frequency open-circuit but constrained to operate in a balanced manner with respect to earth.

The two capacitances C_1 and C_2 (see Fig. 3.9) are equally charged, the diaphragm having a total charge $+Q_D$ and each fixed plate a charge $-Q_D/2$. The voltage gradients are as represented by the slanting full lines.

With the plates on signal-frequency open-circuit, all charges must stay the same if the diaphragm is displaced for a short time. Thus, since $Q = CV$, the voltage across each capacitance must vary inversely as the capacitance value and, since the latter is inversely proportional to the plate spacing, the voltage must vary in direct proportion to the plate spacing. From this it follows that when the diaphragm is displaced a distance x the voltage distribution must become as shown in broken line, giving a voltage on the right plate of $+V_{pol}x/d$ and on the left plate of $-V_{pol}x/d$. This is in agreement with equation (3.38), and from the geometry of the diagram it is also evident that V_D must remain constant, which is in accordance with equation (3.39).

Now equation (3.1) is equivalent to saying that the force on a surface of area A is equal to $\varepsilon_0 A$ times the square of the voltage gradient at that surface. In Fig. 3.9 the voltage gradients are equal on the two surfaces of the diaphragm for all positions of the latter, so that the resultant force on it must be zero in all positions.

It will be noticed that the diaphragm voltage is necessarily equal to the plate voltage at $x = d$, the capacitance between them then being theoretically infinite.

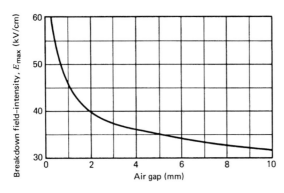

Figure 3.10. Breakdown field-intensity curve for air at atmospheric pressure, under uniform-field conditions between smooth flat plates.

3.2.9 Maximum attainable forces at high frequencies

The maximum diaphragm forces per unit area attainable in a simple air-spaced electrostatic loudspeaker arranged as in Fig. 3.1 are limited by the maximum field intensity E_{max} that air at atmospheric pressure can withstand. This varies with the gap length, a curve for the relevant uniform-field condition being given in Fig. 3.10 (see reference 6). For the gap lengths typically used with diaphragms required to operate at bass frequencies, a figure for E_{max} of 40 kV/cm is applicable.

To increase the available acoustic output level, or to permit a smaller loudspeaker to produce a given level, three possible techniques are:

(a) Mount a unit of the Fig. 3.1 type in a very thin and light plastic enclosure and fill this with a gas having a higher E_{max} value than air.
(b) Employ a multiple-diaphragm construction, such as the double-diaphragm one shown in Fig. 3.11.
(c) Use folded constructions with multiple small gaps as described in reference 3, in which the sound waves are emitted in a direction in line with the diaphragm surfaces.

All these schemes involve fairly formidable constructional and/or insulation problems, though (a) has been employed commercially by Dayton-Wright. A problem with it is that reflections are liable to occur, both because of the mass of the enclosure, if thick enough to be thoroughly gas-tight, and because of the different values of ρc for the gas and for air.

Thus the simple air-spaced scheme of Fig. 3.1 has been almost exclusively adopted in practice, and its use is assumed in the treatment given here.

The large-signal electric-field intensities for a loudspeaker as in Fig. 3.1 will first be considered at high frequencies, where the diaphragm amplitude is negligible compared with the plate spacing.

At $V_{sig} = 0$, and assuming the diaphragm to be accurately centred, the electric fields in both gaps are of intensity V_{pol}/d. The application of a signal voltage V_{sig} introduces an additional field component of magnitude $V_{sig}/2d$ and, since the diaphragm displacement may be neglected at high frequencies, the field components due to the diaphragm polarization remain virtually at the value V_{pol}/d. The total electric-field intensity E_{tot} in the gap where the fields are additive is therefore given by

$$E_{tot} = \frac{V_{pol} + V_{sig}}{d} \tag{3.40}$$

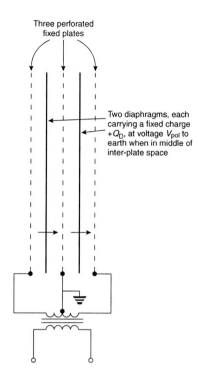

Three perforated
fixed plates

Two diaphragms, each
carrying a fixed charge
$+Q_D$, at voltage V_{pol} to
earth when in middle of
inter-plate space

Figure 3.11. Double-diaphragm scheme,
for increased total-force capability.

The design problem is to choose the value of V_{pol} so that the maximum diaphragm force can be developed without exceeding the maximum field intensity E_{max} that the air can withstand. Equation (3.40) thus becomes

$$E_{max} = \frac{V_{pol} + V_{sig_{max}}}{d} \tag{3.41}$$

where $V_{sig_{max}}$ is the maximum peak instantaneous signal voltage that can be withstood.

Rearranging equation (3.41) gives:

$$V_{sig_{max}} = 2(E_{max}d - V_{pol}) \tag{3.42}$$

Substituting this in equation (3.15) gives the maximum peak instantaneous signal force that can be produced, for a given value of V_{pol}, without exceeding the maximum permissible field intensity E_{max}, and the result is

$$F_{sig_{max}} = \frac{2\varepsilon_0 A V_{pol}}{d^2}(E_{max}d - V_{pol}) \tag{3.43}$$

The value of V_{pol} must now be chosen to maximize equation (3.43), for which purpose the equation must be differentiated with respect to V_{pol} and the differential coefficient equated to zero. This gives:

$$V_{pol} = \tfrac{1}{2}E_{max}d \tag{3.44}$$

where V_{pol} is in V, E_{max} (dielectric strength of air) is in V/m, and d (see Fig. 3.1) is in m.

Equation (3.44) thus gives the correct polarizing voltage to use for obtaining the maximum possible acoustic output when the diaphragm amplitude is small compared with the plate spacing.

Substituting from equation (3.44) in equation (3.43) gives the maximum peak instantaneous force that can be produced under these optimum design conditions:

$$F_{\text{sigmax}} = \tfrac{1}{2}\varepsilon_0 A (E_{\max})^2 \tag{3.45}$$

Taking $E_{\max}$ as $40\,\text{kV/cm}$ or $4 \times 106\,\text{V/m}$, equation (3.45) gives $F_{\text{sigmax}} = 70\,\text{N/m}^2$ approximately. This represents an absolute upper limit for units having the spacing d in the region of $2\,\text{mm}$ but, in practice, because of slight dimensional errors and field non-uniformities due to the perforations etc., a value of $50\,\text{N/m}^2$ is probably a more realistic one to assume.

Though the maximum permissible signal voltage at high frequencies is given by equation (3.42), it is convenient also to have formulae giving this voltage as a function of V_{pol} or $E_{\max}d$ only, for the condition where V_{pol} has been set in accordance with equation (3.44) for the maximum possible force and acoustic output. Thus, from equations (3.42) and (3.44):

$$V_{\text{sigmax}} = E_{\max}d \tag{3.46}$$

$$V_{\text{sigmax}} = 2V_{\text{pol}} \tag{3.47}$$

3.2.10 Maximum attainable forces at low frequencies

An accurate low-frequency analysis, fully taking into account all effects, seems hardly feasible, and that given here assumes that the diaphragm displacement is uniform over its area. In fact the displacement is far from uniform, both because the diaphragm is a resonant membrane fixed at its periphery, and also because the effective air-mass loading on it is non-uniformly distributed. The profile taken up by the diaphragm varies with frequency, making matters very complex. However, the mean velocity and displacement must nevertheless be related in a simple manner to the motional current in the Fig. 3.2 circuit.

The justification for the following analysis is felt to be that it gives considerable insight into some fundamental principles, but it must be borne in mind all along that the quantitative predictions concerning maximum signal input levels and peak diaphragm displacements show large discrepancies when compared with what occurs in practice, as discussed later.

At low frequencies, the diaphragm amplitudes are not negligible compared with the plate spacing, so that the gaps are now a function of both the applied signal voltage and the magnitude and phase angle of the total mechanical load on the diaphragm. As mentioned at the end of Section 3.2.2, the negative compliance as given by equation (3.14) should be included as part of this total mechanical load.

At the frequencies here involved, i.e. within an octave or two of the low-frequency resonance, the acoustic contribution to the total mechanical load is normally that of a certain almost-constant effective air mass, or 'accession to inertia'[5], represented by X_{ma} in Fig. 3.3.

The resistive loading due to sound radiation, represented by R_{ma}, is usually relatively small, but additional damping is normally provided, e.g. by fine-mesh cloth on one or both of the plates, to prevent the Q-value being far too high. Initially, however, all damping will be ignored in the following treatment, its effects being considered in Section 3.2.11.

To avoid possible ambiguity, the unqualified terms 'total mechanical load' and 'mechanical resonance frequency' will henceforth be avoided, for in some contexts there might be uncertainty as to whether the effect of the electrically produced negative compliance is included or not. Adjectives 'polarized' or 'unpolarized' will therefore always be added to indicate the respective inclusion or not of the effect of the negative compliance.

Thus, at low frequencies and ignoring all damping, the total unpolarized diaphragm impedance varies from a compliance below resonance to a mass reactance above resonance. With polarization on, this still applies, but the negative compliance now

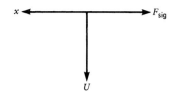

Figure 3.12. Diagram showing phasing of velocity and displacement relative to force, for electrostatic loudspeaker operating above the polarized resonance frequency, damping being neglected.

gives rise to a lower resonance frequency. Below the polarized resonance frequency, ω_p, the diaphragm displacement is in phase with the signal force; above resonance it is in antiphase with this force.

In order to be able to determine the maximum permissible signal voltages, and the maximum obtainable amplitudes, at these low frequencies, it is necessary to obtain a quantitative relationship between V_D, V_{sig} and x/d.

Now the magnitude of the diaphragm velocity is given by

$$|U| = F_{sig}/X_{tot} \tag{3.48}$$

where X_{tot} = the total polarized reactance. The phasing of this velocity for frequencies above ω_p is shown in Fig. 3.12.

The magnitude of the displacement is given by

$$|x| = F_{sig}/\omega X_{tot} \tag{3.49}$$

and since the displacement is in antiphase with respect to F_{sig} when X_{tot} is positive (mass reactance) it is given correctly, including its sign, by

$$|x| = -F_{sig}/\omega X_{tot} \tag{3.50}$$

Let m be the total effective mass loading per unit area on the diaphragm, including both air mass and the actual diaphragm mass. Then, if ω_0, is the unpolarized resonance frequency, it may be shown that

$$\text{total unpolarized reactance} = +\omega_0^2 mA \; [(\omega/\omega_0)^2 - 1] \tag{3.51}$$

From equation (3.14)

$$\text{negative compliance reactance} = +\frac{2V_{pol}^2\varepsilon_0 A}{\omega d^3} \tag{3.52}$$

Hence the total polarized reactance, X_{tot} is given by

$$X_{tot} = \omega_0^2 \frac{mA}{\omega} \; [(\omega/\omega_0)^2 - 1] + \frac{2V_{pol}^2\omega_0 A}{\omega d^3} \tag{3.53}$$

By substituting this for X_{tot} in equation (3.50), and by substituting for F_{sig} in equation (3.50) from equation (3.15), some minor algebraic manipulation yields an equation which, for present purposes, is most conveniently expressed in the form

$$\frac{1}{2}\frac{V_{sig}}{V_{pol}} = \frac{x}{d}\left\{ \frac{d^3 - \omega_0^2 m}{2\omega_0 V_{pol}^2} [(\omega/\omega_0)^2 - 1] + 1 \right\} \tag{3.54}$$

where ω_0 = unpolarized angular resonance frequency.

For a given loudspeaker, all the quantities within the curly brackets of equation (3.54) are constants, except for ω, so that the equation shows how the factor of proportionality between diaphragm position and signal voltage varies, in magnitude and sign, with frequency. The maximum signal input that can be tolerated may be limited either by the diaphragm hitting a fixed plate, causing the onset of distortion, or by reaching the maximum permissible voltage gradient E_{max}, a complication being that E_{max} is a function of the gap length (see Fig. 3.10). In practice, the limit is more likely to be set by hitting a plate than the idealized theory predicts, for the diaphragm amplitude at the centre considerably exceeds the mean amplitude.

However, pursuing the idealized analysis, the next requirement is to determine the voltage gradients in the gaps, and to do this it is necessary to derive, in addition to equation (3.54), an equation for the diaphragm voltage as a function of its position. This may be done by substituting from equation (3.54) in equation (3.11), the latter being appropriately rewritten as

$$\frac{V_D}{V_{pol}} = 1 - \left(\frac{x}{d}\right)^2 - \frac{1}{2}\frac{V_{sig}(x/d)}{V_{pol}} \tag{3.55}$$

The substitution then gives

$$\frac{V_D}{V_{pol}} = 1 + \left(\frac{x}{d}\right)^2 \left\{ \frac{d^3\omega_0^2 m}{2\varepsilon_0 V_{pol}} [(\omega/\omega_0)^2 - 1] \right\} \tag{3.56}$$

It is convenient at this point to introduce a parameter β, given by

$$\beta = \frac{d^3\omega_0^2 m}{2\varepsilon_0 V_{pol}^2}[(\omega/\omega_0)^2 - 1] \tag{3.57}$$

Equations (3.54) and (3.56) may then be shortened to

$$\tfrac{1}{2}V_{sig}/V_{pol} = -(x/d)(\beta + 1) \tag{3.58}$$

$$V_D/V_{pol} = 1 + (\beta x/d)^2 \tag{3.59}$$

Two particular values of the frequency-dependent parameter β are of especial significance. When $\beta = 0$, equation (3.57) shows that this corresponds to operation at the unpolarized resonance frequency ω_0, At this frequency the force F_{sig} due to V_{sig} may be said to be driving only the negative compliance, or, alternatively, it could be said that the total electrically produced force is zero. Of interest is the fact, evident from putting $\beta = 0$ in equation (3.59), that under these conditions the diaphragm voltage V_D remains constant at V_{pol}, even for large amplitudes. In practice, however, because of finite damping, here neglected, such constant-diaphragm-voltage operation is never fully realized, even at ω_0.

When $\beta = -1$, equation (3.58) shows that finite diaphragm amplitudes are obtained for infinitesimally small signal inputs, so $\beta = -1$ evidently corresponds to operation at the polarized resonance frequency ω_p. Here again, damping prevents this condition from being fully realized in practice, though the increased sensitivity may be usefully exploited, as described later.

Equation (3.57) may be expressed in a simpler form by invoking equation (3.14) and also the relationship

$$\omega_0^2 mAC_{md} = 1 \tag{3.60}$$

Substitution from these gives

$$\beta = -\frac{C_{me}}{C_{md}}[(\omega/\omega_0)^2 - 1] \tag{3.61}$$

As mentioned near the beginning of Section 3.2.7, a ratio $-C_{me}/C_{md} = 3.5$ is adopted by Quad.

From equations (3.58) and (3.59) the diagrams shown in Fig. 3.13 may be drawn. The lines, plotted from equation (3.58), represent the voltage of the left plate to earth, which is equal to $\tfrac{1}{2}V_{sig}$, the diaphragm-voltage parabolas being plotted from equation (3.59).

It is important to know how far up and down the Fig. 3.13 graphs the operating point can be allowed to swing without causing the field intensity in either gap to exceed the relevant value of E_{max}. To provide this information, an expression must

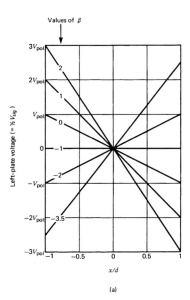
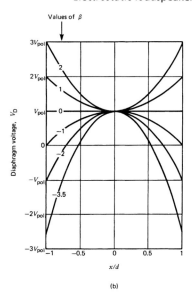

Values of β

Values of β

(a)

(b)

Figure 3.13. Graphs plotted from equations (3.58) and (3.59), for idealized electrostatic loudspeaker without damping. $\beta = 0$ corresponds to ω_0, $\beta = -1$ corresponds to ω_p. With $-C_{me}/C_{md} = 3.5$, $\beta = -3.5$ represents zero frequency.

be derived for the voltage gradient in a gap as a function of diaphragm position and ω.

Let the voltage gradient be called positive when the field direction is towards the right. Then the gradient E_R in the right-hand gap of Fig. 3.1 is

$$E_R = \frac{V_D + \tfrac{1}{2}V_{sig}}{d - x} \tag{3.62}$$

and substituting for V_{sig} and V_D from equations (3.58) and (3.59) leads to

$$E_R = \frac{V_{pol}}{d}[1 - \beta(x/d)] \tag{3.63}$$

Similarly

$$E_L = \frac{V_{pol}}{d}[1 - \beta(x/d)] \tag{3.64}$$

The graphs of Fig. 3.14 have been plotted from equations (3.63) and (3.64).

Normally, as explained in Section 3.2.9, the value of V_{pol} is chosen to maximize the attainable diaphragm force at high frequencies, and equation (3.44) shows that V_{pol}/d is then equal to $\tfrac{1}{2}E_{max}$, E_{max} here being the breakdown field intensity for a gap of length d.

Operation up to the beginning of the broken-line regions in Fig. 3.14 is theoretically possible without exceeding this value of E_{max}.

For $\beta = -1$ (i.e. $\omega = \omega_p$), the strongest field occurs in the smaller of the two gaps, so that a somewhat higher field intensity than $E_{max} = 2V_{pol}/d$ can be withstood, allowing diaphragm displacement up to $x/d = \pm 1$ even at frequencies a little lower than ω_p. However, for the Quad ratio of C_{me} to C_{md}, equation (3.61) shows that, at very low frequencies, β tends towards the value -3.5, and from Fig. 3.14 it is then evident that the diaphragm cannot move more than $1/3.5$ of the way from the middle

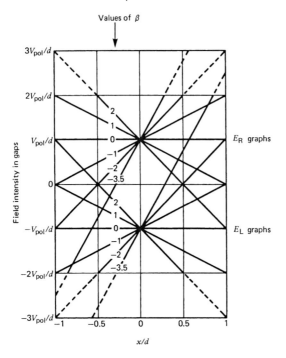

Figure 3.14. Gap voltage–gradient graphs plotted from equations (3.63) and (3.64), for idealized electrostatic loudspeaker without damping. $\beta = 0$ corresponds to ω_0, $\beta = -1$ corresponds to ω_p. With $-C_{me}/C_{md} = 3.5$, $\beta = -3.5$ represents zero frequency.

to a plate without the gap field intensity exceeding the value $2V_{pol}/d$. With such a relatively small displacement, the relevant value of E_{max} is fairly closely equal to $2V_{pol}/d$. Finally, reference to Fig. 3.13(a), or substitution in equation (3.58), shows that the value of V_{sig} required to obtain this displacement is $1.43V_{pol}$, or a plate-to-earth voltage of $0.71V_{pol}$. Note that this is considerably less than the permissible peak value of V_{sig} at high frequencies, which is $2V_{pol}$.

For $\beta = +1$, Fig. 3.13 shows that the left-plate voltage, for $x/d = -1$, is $2V_{pol}$, or $V_{sig} = 4V_{pol}$. This is twice the signal voltage permissible at high frequencies, but the result is of no practical significance, for, with a diaphragm not moving as a true piston, E_{max} would be reached near the edge of the diaphragm at the normal high-frequency signal-voltage limit.

As already mentioned, diaphragm damping is normally employed, reducing the Q-value to typically about 2 and, when this is done, Figs 3.13 and 3.14 are very misleading for frequencies close to ωp, corresponding to $\beta = -1$. A loudspeaker may be made, however, with little, if any, intentional damping, the resultant low-frequency peak being taken care of by electrical equalization – though some damping is actually always desirable for suppressing higher-frequency resonance modes in the diaphragm.

3.2.11 Effects of damping

In the previous section, where damping was neglected, the diaphragm displacement was always either in phase with V_{sig}, or in antiphase, depending on the frequency. With damping, the displacement has some intermediate phase angle, making analysis more complex. To avoid undue complexity, however, the following treatment will be restricted to the conditions applying at the polarized resonance frequency ω_p, for which there is pure damping control, resulting in the displacement being in quadrature with the signal voltage. This provides the most complete contrast to the conditions of Section 3.2.10.

Thus the present aim is to plot graphs, at the one frequency $\omega = \omega p$ and with damping, of signal voltage, diaphragm voltage and voltage gradient, against x/d, for comparison with Figs 3.13 and 3.14. These same quantities will also be plotted against time.

Let the instantaneous signal voltage be

$$V_{sig} = V_{sig}\sin\omega_p t \tag{3.65}$$

Then from equation (3.16)

$$F_{sig} = \frac{\varepsilon_0 A V_{pol}}{d^2} \times V_{sig} \sin \omega_p t \tag{3.66}$$

The instantaneous diaphragm velocity, which is in phase with F_{sig} at ω_p, is given by

$$U = \frac{F_{sig}}{R_M} = \frac{\varepsilon_0 A V_{pol}}{d^2 R_M} \times V_{sig} \sin \omega_p t \tag{3.67}$$

where R_m is the total mechanical damping resistance.

Now the instantaneous displacement, x, is given by

$$x = \int U \, dt \tag{3.68}$$

so that, integrating equation (3.67)

$$x = - \frac{\varepsilon_0 A V_{pol}}{\varepsilon_p d^2 R_m} \times V_{sig} \cos \omega_p t \tag{3.69}$$

Substituting from equation (3.57), and bearing in mind that $\beta = -1$ at ω_p, leads, after some algebraic juggling, to

$$x/d = \frac{\omega_p m A}{R_m} [1 - (\omega/\omega_p)^2] \times \frac{\frac{1}{2}\hat{V}_{sig}}{V_{pol}} \cos \omega_p t \tag{3.70}$$

i.e.

$$x/d = Q[1 - (\omega/\omega_p)^2] \times \frac{\frac{1}{2}\hat{V}_{sig}}{V_{pol}} \cos \omega_p t \tag{3.71}$$

Hence, for a given design, where V_{pol}, Q and ω_0/ω_p are known, a particular peak value of input signal V_{sig} may be chosen, and a table of values of x/d may be calculated from equation (3.71) for a series of values of the angle $\omega_p t$. From equation (3.65) the corresponding values of the instantaneous signal voltage V_{sig} may then be determined. Equation (3.11) may then be used to calculate the related values of diaphragm voltage V_D. From the data in the resultant table, the voltage gradients in the two gaps may also be determined.

For the ratio of diaphragm stiffness to negative stiffness of 3.5:1 adopted by Quad, the corresponding ratio $(\omega_0/\omega_p)^2$ in equation (3.71) is equal to 1.4, i.e. equal to the factor by which the loudspeaker input capacitance increases, at very low frequencies, on switching on the polarizing supply. The results shown in Figs 3.15, 3.16 and 3.17 have been calculated for this ratio, the Q-value being taken as 2.0.

It was established that at high frequencies (see equation (3.47)) the maximum signal voltage that can be applied without danger of air breakdown is given by

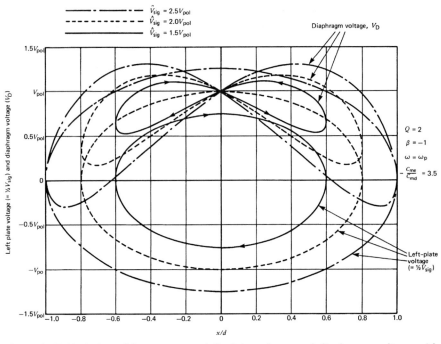

Figure 3.15. Variation of instantaneous left-plate voltage and diaphragm voltage with diaphragm position, at polarized resonance frequency, for idealized electrostatic loudspeaker with damping.

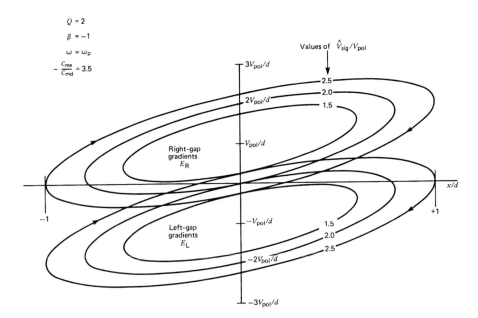

Figure 3.16. Variation of instantaneous voltage gradients in gaps with diaphragm position, at polarized resonance frequency, for idealized electrostatic loudspeaker with damping.

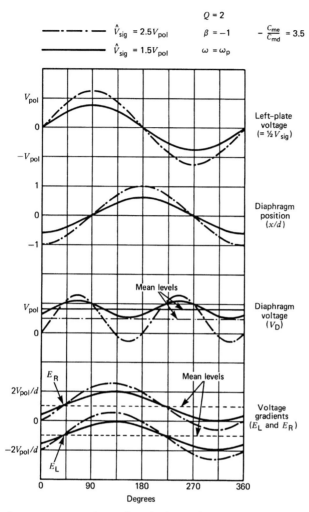

Figure 3.17. Waveforms for idealized electrostatic loudspeaker operating at the polarized resonance frequency, with damping. Conditions as for Figs 3.15 and 3.16 but, for clarity, only two signal levels are represented. In practice, the maximum acceptable value of V_{sig} is less than $1.5V_{pol}$.

$V_{sig_{max}} = 2V_{pol}$. But Fig. 3.16 shows that this input level now gives rise to a peak voltage gradient exceeding $2V_{pol}/d$. This occurs, however, in a gap spacing of only about $d/2$, for which Fig. 3.10 shows that the breakdown gradient is about 14% greater than for a gap of d. Consequently a signal input level very nearly as large as for high frequencies can be withstood, as far as electrical breakdown is concerned. Nevertheless, because the diaphragm movement at its centre is about twice the mean movement, the diaphragm is likely to hit the plates well before this electrical limit is reached. But since ω_p usually corresponds to a frequency not much above 40 Hz, at which normal music programme signals do not contain full-amplitude components, this restriction on the acceptable low-frequency signal level is seldom an embarrassment.

An interesting feature of Fig. 3.16, as of Fig. 3.14, is that the difference between the voltage gradients in the two gaps remains constant all the time at $2V_{pol}/d$.

Figure 3.17 presents the information of Figs 3.15 and 3.16 in a different form, now plotted against time. Two points that become evident are:

(a) The diaphragm voltage is a pure second-harmonic sine wave, with an amplitude proportional to the square of the signal-input voltage. This is shown in Fig. 3.17 for operation at the polarized resonance frequency with damping, and consideration of Fig. 3.13 shows that the same thing must occur at other signal frequencies, even when damping is not significant. The production of the second harmonic in Fig. 3.13 can be envisaged in terms of 'projecting' a sine-wave of x/d on to the relevant parabola, in the manner familiar in the context of transistor characteristics, etc.

(b) The mean diaphragm voltage shifts downwards by an amount proportional to the square of the signal-input voltage.

Point (a) is given further attention in Section 3.2.12, but point (b) is also worthy of more detailed consideration[34].

With an ideal constant-charge diaphragm, charged once and for all and then isolated, the downward shift in mean diaphragm voltage could be maintained indefinitely in the presence of a large and continuous low-frequency input. However, the diaphragm is in practice slightly electrically conductive, and is connected to the polarizing supply voltage. Consequently, with a steady and long-sustained low-frequency input, the mean diaphragm voltage must creep back to its original no-signal value.

In the absence of such creeping-back, the loudspeaker is theoretically absolutely linear and of perfectly constant sensitivity but, when the mean diaphragm voltage has crept back up to the voltage of the polarizing supply, there must, of course, be an increase in sensitivity, the value of V_{pol} to be inserted in the equations then being higher than that of the supply itself.

Figure 3.13 shows that for positive values of β, i.e. for frequencies above the unpolarized resonance frequency ω_0., the diaphragm-voltage parabolas are upward-facing, so that, with constant charge, the mean diaphragm voltage would be expected to rise rather than fall. Again, however, with a sustained signal input, the mean voltage must revert to that of the polarizing supply, in this case causing a reduction in output level. At ω_0 itself, no change would be expected to occur.

An experiment was done on a Quad ESL63 in which a constant low-level input signal was applied at 1000 Hz, a microphone being placed near the loudspeaker. The amplified microphone output was fed via a highly selective 1000 Hz band-pass filter to an oscilloscope and digital voltmeter. A low-frequency input was also fed to the loudspeaker at a level somewhat below the maximum acceptable level. The measuring

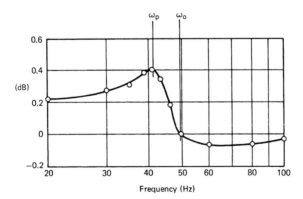

Figure 3.18. Variation in magnitude of 1000 Hz output from Quad ESL63 with frequency of sustained 7 V r.m.s. low-frequency input.

system was quite non-responsive to the low-frequency output as such, but registered the changes in sensitivity of the loudspeaker caused by the effects described above.

The result obtained is shown in Fig. 3.18, and is in satisfactory general agreement with the reasoning given. The ratio of the two marked frequencies agrees well with the predicted ω_0/ω_p) ratio of $\sqrt{1.4}$, 1.4 being the factor by which the low-frequency inter-plate capacitance increases on applying the polarizing voltage. The flattening-off at very low frequencies would be expected from Fig. 3.13, for there is almost pure stiffness control, giving constant x/d, and β tends towards a fixed value of -3.5. Lastly, it was found that the magnitude of the voltage increase at ω_p was accurately proportional to the square of the low-frequency input voltage, again as predicted.

In all normal circumstances, the Fig. 3.18 effect is of such very small magnitude that it is quite undetectable subjectively. It does not occur at all, of course, for short-duration bass transients, but only for sustained inputs.

3.2.12 Harmonic and intermodulation distortion[34]

In considering non-linearity distortion, the following facts, already established, are relevant:

(a) Under absolutely constant-charge conditions, an ideal electrostatic loudspeaker is virtually free from harmonic distortion, provided only that the diaphragm does not hit the fixed plates and that the voltage gradients in the gaps never exceed the breakdown value for air.

(b) The diaphragm-voltage variation with sinusoidal input and perfectly constant charge is of exactly sine waveform, at twice the signal frequency. This is shown in the Fig. 3.17 waveforms, and it also occurs under the zero-damping conditions of Fig. 3.13(b), since projecting a fundamental-frequency sine wave of x/d up on to the appropriate parabola generates a double-frequency sinusoidal variation of V_D.

(c) The force/voltage sensitivity of an electrostatic loudspeaker varies in direct proportion to the diaphragm charge.

(d) Purely from considerations of structural symmetry, it follows that the distortion with finite diaphragm-charging resistance must involve only odd harmonics, for the output waveform must also be symmetrical.

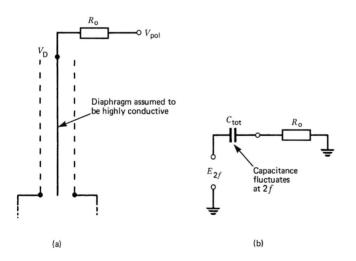

(a) (b)

Figure 3.19. Diagrams relating to explanation of distortion mechanism.

Because the diaphragm-charging resistance is distributed over the diaphragm surface, the true situation is much more complex than that shown in Fig. 3.19(a), but the latter is nevertheless useful for clarifying ideas.

As far as a.c. conditions in the polarizing circuit are concerned, the loudspeaker may be regarded as a generator of second-harmonic e.m.f., E_{2f}, in series with a capacitance C_{tot}, as shown in Fig. 3.19(b). This capacitance, however, is not constant, and fluctuates in value with a periodicity of $2f$ when the diaphragm displacement, x, occurs at a frequency f, the relevant relationship being that of equation (3.6).

When R_0 is large, such that most of E_{2f} appears across it, a current is caused to flow in R_0 of almost pure $2f$ sine waveform, and this current, flowing to the diaphragm, constitutes a fluctuation at $2f$ in the diaphragm charge Q_D. Since the sensitivity of the loudspeaker is directly proportional to Q_D, the output becomes amplitude-modulated at $2f$. Now with a fundamental at f modulated at $2f$, sidebands are produced at $f + 2f$ and $f - 2f$. The former represents third-harmonic distortion and the latter a change in fundamental amplitude, normally quite negligible.

As soon as R_0 becomes low enough for an appreciable proportion of E_{2f} to appear across C_{tot}, the $2f$ fluctuations in C_{tot} cause the current flowing in R to be a distorted $2f$ waveform, containing harmonics at $4f$, $6f$ etc. Then the signal at f becomes modulated not only at $2f$, but also at $4f$, $6f$ etc., producing additional output harmonics at $5f$. $7f$ etc.

However, R_0 is indeed made very large in practice, so that only weak modulation occurs, and almost purely at $2f$, producing a very low level of output harmonic distortion, which is almost entirely third-harmonic.

Though this topic could obviously be pursued in greater analytical detail, a full analysis would have to take into account the effect of the non-uniform diaphragm displacement and the distributed nature of the diaphragm resistance and capacitance, rendering any simple concept of a time-constant rather meaningless. It is doubtful whether such a complex analysis would be of any real value, for the influence of diaphragm resistivity on distortion can be determined experimentally. A value of 10^9 Ω per square is found to be adequately high in practice.

Nevertheless, it is useful to have some idea of the expected theoretical law of variation of distortion with diaphragm resistivity, and this may be deduced approximately by reasoning related to Fig. 3.19.

The percentage distortion is proportional to the percentage variation in Q_D. Now the mean voltage across C_{tot} is nearly enough constant at V_{pol}, except at very high low-frequency signal levels, so that the percentage variation in Q_D can be taken approximately as being related in a fixed ratio to the absolute $2f$ voltage appearing across C_{tot}. Hence it is evident, as would be expected, that the maximum distortion will occur when $R_0 = 0$ and the whole of E_{2f} appears internally across C_{tot}, none appearing at the diaphragm terminal. If the distortion under this condition, for some specific input voltage and frequency, is D_0, then the distortion with finite R_0 will be proportional to the fraction of E_{2f} that appears across C_{tot}, and it is therefore given by

$$D = D_0 \times \frac{X_{C_{tot}}}{\sqrt{X_{C_{tot}}^2 + R_0^2}}$$

(3.72)

$$D = D_0 \times \frac{1}{\sqrt{1 + (2R_0 \omega C_{tot})^2}}$$

(3.73)

where $\omega = 2\pi \times$ fundamental frequency.

The reactances in equation (3.72) are those applying at the second-harmonic frequency, $2f$. Equation (3.73) is represented graphically in Fig. 3.20.

The important feature of Fig. 3.20 is that, provided the distortion is already low, doubling R_0 gives a 6 dB reduction in distortion, and this would be expected to apply also to the surface resistivity of a resistive diaphragm.

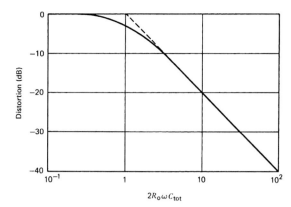

Figure 3.20. Variation of distortion in accordance with equation (3.73).

It is also of interest to consider how the percentage distortion varies with the signal input level. The significant fact in this context is that the second-harmonic component of diaphragm voltage has a magnitude proportional to the square of the signal input voltage. That this is so can be seen from Figs 3.13(b) and 3.17, but may be established formally from equation (3.11) by inserting $V_{sig} = \hat{V}_{sig} \sin \omega t$ and $x = \hat{x} \sin (\omega t + \phi)$ in this, ϕ being the phase angle between displacement and signal voltage, as discussed in Sections 3.2.10 and 3.2.11. Both terms in equation (3.11) then give rise to a second-harmonic component, and since $\hat{x}$ is proportional to $\hat{V}_{sig}$ both of these components are proportional to the square of $\hat{V}_{sig}$. (The two components are of equal magnitude and opposite sign under the conditions represented by $\beta = 0$ in Fig. 3.13(b).)

Thus, if V_{sig} is doubled, E_{2f} in Fig. 3.19(b) is increased by a factor of four, giving four times as much fluctuation in diaphragm charge and consequently a fourfold increase in the percentage distortion. In other words, the percentage distortion is proportional to the square of the input voltage, which is the normal state of affairs for most third-order, or cubic, distortion mechanisms.

Only harmonic distortion, with a single audio input, has been considered so far, but, of course, intermodulation distortion also occurs. From the modulation nature of the distortion mechanism, it is evident that large bass inputs will modulate the amplitude of high audio frequencies just as effectively as they modulate the amplitude of lower audio frequencies, when all these are handled by the same diaphragm. In the presence of two inputs at f_1 and f_2 the intermodulation products are of the $f_1 \pm 2f_2$ type.

In practice, with slight residual asymmetry of construction, some even-order harmonic and intermodulation distortion may sometimes be observed, but in general the distortion levels are much lower than with moving-coil systems.

A further cause of distortion is non-linearity of the diaphragm-stretching compliance, but such distortion should be very small with the correct choice of diaphragm dimensions and materials, as mentioned in Section 3.2.7. An important point, however, is that such non-linearity gives rise to low-frequency distortion only, and does not result in significant intermodulation between low and high frequencies.

Distortion of a sort can also result, in the presence of large low-frequency inputs, from 'rustling' of the plastic dust-covers within which electrostatic loudspeakers are often enclosed. This effect, and also the effect on output at very high frequencies, can be minimized by using the thinnest available plastic material.

Lastly, Doppler distortion, giving intermodulation products of an FM variety, occurs in electrostatic as in other loudspeakers but, even when the whole audio spectrum is handled by one diaphragm, the magnitude of such distortion has been found to be too small to be subjectively noticeable[26].

3.3 Radiating the sound

The maximum forces per unit area that can be developed by a single-diaphragm air-dielectric electrostatic drive system are very much smaller than those that can be produced using moving-coil principles.

As stated in Section 3.2.9, a practical limit, for a diaphragm-to-plate spacing in the region of 2 mm, is about 50 N/m² peak, equivalent to 0.5 millibar peak. To put this in perspective, a 100-turn 25 mm diameter voice coil in a flux density of 1 tesla produces a force of 39 N for a peak current of 5 A. If this drives a diaphragm of effective diameter 165 mm (a so-called 200 mm diaphragm), the peak force per m² of diaphragm area is approximately 1800 N/m². In a tweeter, where a probably somewhat smaller maximum force is applied to a very much smaller diaphragm area, the peak force per unit diaphragm area may be some tens of thousands of N/m².

Though the forces per unit area are so very much smaller than with moving-coil systems, there is the great compensating advantage that they are developed on an almost massless diaphragm, i.e. a diaphragm whose own mechanical impedance is very small. Such a low-impedance mechanical generator is inherently well suited to driving the low impedance of the air for sound radiation, but problems arise, both at low frequencies and at high frequencies, in exploiting this feature effectively.

3.3.1 Low frequencies

The obvious starting point for thoughts about electrostatic loudspeakers at low frequencies is probably the totally enclosed cabinet arrangement shown in Fig. 3.21.

Because the mass of the diaphragm, plus the mass of the air that effectively moves with it, is so very much less than the total mass in a moving-coil system, the resonance frequency will be much higher, so that at very low frequencies, such as 50 Hz, the diaphragm motion will be controlled almost entirely by the stiffness of the air within the cabinet in combination with the diaphragm-stretching stiffness and electrical negative stiffness. The latter may be calculated from equation (3.14), and the stretching stiffness, as explained in Section 3.2.7, is typically made 3.5 times the magnitude of the negative stiffness. The stiffness due to the enclosed air volume, assuming adiabatic operation, is also readily calculated and turns out to be the dominant contribution unless the cabinet is very deep. Hence, ignoring the small mass reactance, the diaphragm velocity for the maximum available peak signal force of 50 N/m² can be determined at, for example, 50 Hz.

Now at very low frequencies, where the cabinet dimensions are quite small compared with the wavelength, a system as in Fig. 3.21 constitutes a simple acoustic source of strength S, given by

$$S = UA \qquad (3.74)$$

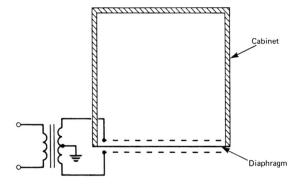

Figure 3.21. Basic cabinet-type electrostatic loudspeaker.

where S = strength of source (m³/s)
         U = diaphragm velocity (m/s)
         A = diaphragm area (m²)

The total power radiated by such a source in free space is given by[7]

$$W = \pi S^2 f^2 \rho / c \qquad (3.75)$$

where W = acoustic strength (W)
         S = source strength (m³/s)
         f = frequency (Hz)
         ρ = air density (kg/m³)
         c = sound velocity (m/s)

(If S is an r.m.s. quantity, W will be the mean power. If S is a peak quantity, W will be the peak instantaneous power.)

A calculation on the above basis for a cabinet of internal dimensions 0.3 m × 0.3 m × 0.3 m, using a diaphragm-to-plate spacing, d, of 2.5 mm and V_{pol} = 5 kV, shows that at 50 Hz and with a peak signal force of 50 N/m² the maximum obtainable mean output power would be only about 0.1 mW. However, the analysis shows that, for a constant cabinet depth, the maximum available power is proportional to the fourth power of frequency and to the square of the frontal area. Thus at 100 Hz, and with frontal dimensions 0.6 m × 0.6 in, a maximum mean output power of about 25 mW would be available – rather more if the beneficial effect of some mass reactance is allowed for. Increased loading resulting from the presence of nearby wall and/or floor surfaces would often give a further increase in practice, just about satisfying normal requirements. At 50 Hz, however, the maximum output in free space would be only about 1.5 mW.

The performance of a system basically of the Fig. 3.21 type can be improved by adding acoustic mass and damping behind the diaphragm to lower the resonance frequency to, say, 50 Hz, and schemes of this kind are considered in some detail in Part 2 of reference 2. The resulting designs unfortunately tend to be rather bulky and more suitable for building into walls etc. than for construction as independent units.

Another quite feasible approach is to use a multiple stack of diaphragms and plates to produce a greatly increased driving force. However, a loudspeaker made in this way is liable to have rather similar shortcomings to moving-coil designs, including cabinet resonances and diffraction effects at higher frequencies.

A fundamentally different approach to the problem of obtaining adequate bass output power is to employ a sufficiently large area of unbaffled diaphragm operating as a doublet radiator. This exploits a unique advantage of the electrostatic principle – the relative ease and economy with which very large diaphragm areas can be provided. Cabinet-work is then virtually eliminated, together with its complex effects at high frequencies. A doublet excites fewer room eigentones, but this is not necessarily an advantage, as plenty of evenly spaced eigentones are usually considered preferable to a few with large gaps[8,9].

Consider initially the case where the unbaffled diaphragm is of the same size as the diaphragm in Fig. 3.21. Because of the absence of the enclosed-air stiffness, a given electrically produced force, at some specific low frequency, will now produce a much larger diaphragm velocity. On the other hand, the resistive component of the air loading on the diaphragm will be much less than it is when a cabinet is used. Hence it is not immediately obvious whether the acoustic power will be greater or smaller.

To obtain a quantitative answer, information is required on the resistive and reactive air loading on an unbaffled diaphragm. The particular case of a circular diaphragm vibrating with uniform amplitude all over has been successfully analysed, and the information is given in graphical form in reference 7, on which Fig. 3.22 is

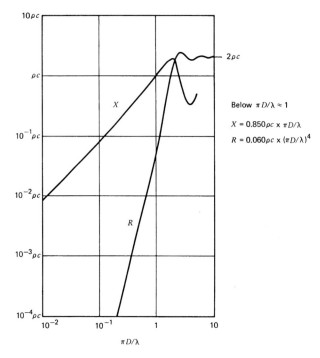

Figure 3.22. Mechanical resistance and reactance per unit area due to air loading on both sides of unbaffled piston of diameter D. The curves give the averaged effects for the whole area, but the loading is not uniformly distributed[5].

based. These graphs may be taken as applicable to a square diaphragm of equivalent area without serious error. At low frequencies, corresponding to values of $\pi D/\lambda$ of less than unity, calculations may be conveniently based on the low-frequency asymptote formulae given in Fig. 3.22.

From these formulae the following equation may be derived:

$$W = \frac{0.644 \, (F_{sig}/A)^2 D^4 f^2}{c^3 \rho} \quad (\pi D/\lambda < 1) \tag{3.76}$$

where $W =$ total acoustic output power from both sides of the diaphragm (W)
  $F_{sig}/A =$ force per unit area applied by the diaphragm to the total air load (N/m²)
  $D =$ diaphragm diameter (m)
  $f =$ frequency (Hz)
  $c =$ velocity of sound (m/s)
  $\rho =$ density of air (kg/m³)

(If F_{sig} is the r.m.s. force, W will be the mean power. If F_{sig} is the peak instantaneous force, W will be the peak instantaneous power, equal to twice the mean power.)

Taking F_{sig}/A as 50 N/m² peak, a 0.34 m diameter diaphragm, whose area is the same as that of a 0.3 m × 0.3 m square one, gives a mean output power at 50 Hz of 0.55 mW. Doubling the diameter to 0.68 m, equivalent to 0.6 m × 0.6 m, yields a 50 Hz mean output power of 8.8 mW.

The above figures are based on the notion that the force acting on the total air load is simply the electrically produced signal force of 50 N/m² peak. However, in the absence of the large stiffness due to the air in a closed cabinet, the total stiffness is very much reduced, being only that resulting from the diaphragm tension and electrical negative stiffness. By suitable choice of diaphragm-to-plate spacing d, these stiffnesses may be made such that the resonance frequency is somewhere in the 50 Hz region, and damping resistance must be added, as mentioned in Section 3.2.11, to control the Q-value. The maximum available force at the resonance frequency is then increased Q times, and the available power output by Q^2 times. For a Q-value of 2, the 0.68 m diameter diaphragm considered above would then be capable of a mean output power at 50 Hz of approximately 35 mW. In practice, the free-space power output at such a low frequency might be limited to rather less than this figure by the diaphragm hitting the plates.

It is thus evident that a bass-power capability comparing favourably with a cabinet design of similar frontal area is in fact achievable, and equation (3.76) shows, moreover, that the performance in this respect increases very rapidly with an increase in linear dimensions.*

An arrangement, seldom practicable, which is capable of a greatly enhanced performance at very low audio frequencies, is that in which an electrostatic loudspeaker diaphragm, say 0.6 m square, is mounted in a hole in the wall between two rooms. Mean output powers to each room in the region of 200 mW at 50 Hz would then be achievable.

3.3.2 Walker's equation

When designs are based on the use of an unbaffled diaphragm, it becomes possible to adopt a refreshingly simple approach to calculating the performance over a wide frequency range and with good accuracy. This very potent method, which exploits the reciprocity principle[7,10], appears to have been first conceived and put to practical use by Peter Walker, who gave details in a lecture to the Acoustics Group of the Institute of Physics and the Physical Society at Imperial College in December 1965. He derived a very simple and fundamental equation, which will here be stated first, the theoretical justification for it being given afterwards. The equation is:

$$P = I_{sig} \times \frac{V_{pol}}{2\pi crd} \tag{3.77}$$

where P = sound pressure at a large distance r on the axis of a plane unbaffled diaphragm of any size or shape (N/m²). The diaphragm stiffness, mass and intentionally introduced damping are neglected.

I_{sig} = signal current fed to plates (A)
V_{pol} = polarizing voltage (V)
c = velocity of sound (m/s)
r = measuring distance (m)
d = diaphragm-to-plate spacing (m)

By a 'large' measuring distance r is meant a distance such that if sound waves were emitted from the measuring point they would become sufficiently nearly plane waves, by the time they reached the diaphragm, for no significant phase differences to occur over the diaphragm area.

More than one approach can be adopted for proving the above equation, but the author, being always conscious that traps for the unwary exist in connection with the reciprocity principle[10], has a preference for an approach that involves applying this

---

* If the output is limited by the diaphragm hitting the plates, rather than by the maximum available force, doubling the linear dimensions gives an 18 dB increase in the maximum obtainable output, instead of the 12 dB given by equation (3.76).

principle only between terminal pairs of like kind, electrical if possible. The following proof is based on the exposition given at the above-mentioned lecture, and differs from that given more recently by Peter Walker[11].

Assume that at a large axial distance r from the loudspeaker a very small omni-directional capacitor microphone capsule of ideal performance is placed, d.c. polarized in the usual manner. As it is stiffness-controlled, a constant a.c. signal voltage applied to the capsule will give frequency-independent diaphragm displacement, and hence constant current fed to it will give frequency-independent diaphragm velocity. The microphone then constitutes an acoustic source of strength S (see equation (3.74)) in which S is related to the microphone current by a simple constant:

$$S = kI_{mic} \qquad (3.78)$$

When used as a microphone, an output voltage related in a frequency-independent manner to the sound pressure is, of course, obtained, and it is not difficult to show, by considering the internal parameters of the capsule, that the constant of proportionality is the same as in equation (3.78), so that

$$V_{mic} = kP \qquad (3.79)$$

where P = sound pressure. The gap spacing, polarizing voltage and diaphragm stiffness affect both operations equally, and so do not affect the equality of the constants in equations (3.78) and (3.79).

Now a source of strength S produces an air-particle displacement x at a distance r given by[7]

$$x = \frac{S}{4\pi rc} \qquad (3.80)$$

where x = displacement (m)
S = source strength (m^3/s)
r = distance from source (m)
c = velocity of sound (m/s)

With the unbaffled electrostatic loudspeaker diaphragm in place at this distance r, it is so light, and has so high a compliance, that over most of the audio-frequency spectrum it simply takes up the full particle displacement x, and hence produces an open-circuit output voltage which is related to x by equation (3.38) of Section 3.2.8. Omitting the irrelevant minus sign, and using V_{esl} instead of V_{sig} to avoid possible confusion, equation (3.38) may be written

$$V_{esl} = 2V_{pol}x/d \qquad (3.81)$$

From equations (3.78), (3.80) and (3.81), it follows that a current I_{mic} fed to the capacitor microphone capsule produces an open-circuit output voltage from the loudspeaker as given by

$$V_{esl} = \frac{kI_{mic}V_{pol}}{2\pi cd} \qquad (3.82)$$

The mutual impedance for signal transmission from the microphone to the loudspeaker is therefore given by

$$Z = \frac{V_{esl}}{I_{mic}} = \frac{kV_{pol}}{2\pi cd} \qquad (3.83)$$

The reciprocity principle simply states that, with passive reversible devices, such as normal transducers, this mutual electrical impedance will be the same if measured the other way round, a current I_{esl} being fed to the loudspeaker and the open-circuit voltage V_{mic} from the microphone being observed. Hence, under these conditions

$$\frac{V_{\text{esl}}}{I_{\text{mic}}} = \frac{kV_{\text{pol}}}{2\pi rcd} \qquad (3.84)$$

Substituting for k in this equation from equation (3.79), and rearranging, then gives equation (3.77), which the author feels could appropriately become known as Walker's equation! (I_{esl} is now called I_{sig}.)

At very low frequencies, equation (3.77) ceases to be directly applicable, owing to the effect of diaphragm stiffness. The response for constant input current rises above that for simple mass control but, as ω_0 and Q are known, a correction is easily calculated.

A significant point is that if the plates are fed from a current signal source of infinite impedance, then the linear negative compliance of equation (3.14) is no longer effective. The explanation for this is given in Section 3.2.8. The consequence is that, with current drive, the relevant resonance frequency is the unpolarized one, ω_0.

At very high frequencies, the mass of the diaphragm, together with some impedance due to the air in the plate holes, becomes significant, causing the sound pressure to fall slightly below that predicted by equation (3.77). However, over a range of at least five octaves, Walker's equation gives the long-distance axial sound pressure directly and without requiring any corrections.

It is instructive to consider how the flat long-distance axial response, with current drive, predicted by equation (3.77), fits in with other notions. First, if the diaphragm is large compared with the wavelengths in use, the effective air load on each side of it will be resistive and of value ρc mechanical ohms per unit area. In this case, constant-voltage drive, producing constant force, would give constant, i.e. frequency-independent, total acoustic output power. Hence constant-current drive would give a total output power inversely proportional to the square of the frequency. How does it come about that a total output power varying in this manner can produce a constant sound pressure and intensity at a large distance on-axis? It can only be because the directivity factor[7] of the large-diaphragm source, measured at a large distance, is proportional to frequency squared, or, in other words, the directivity index[7] rises at 20 dB/decade with frequency, and this is, indeed, always the case for any plane radiating surface at sufficiently high frequencies.

Second, if the diaphragm is small compared with the wavelength., and is of circular shape, the constant axial pressure for constant drive current predicted by equation (3.77) can be shown to be consistent with the data in Fig. 3.22, as follows.

Constant applied voltage would produce constant force, virtually all used, at low frequencies, in driving the reactive part of the Fig. 3.22 air-load impedance, and therefore giving a diaphragm velocity U proportional to $1/f$. Since R is proportional to f^4, the total acoustic output power U^2R is proportional to f^2. The diaphragm being small compared with the wavelength, the directionality is constant, so that the sound intensity on axis is also proportional to f^2 and the axial sound pressure is therefore proportional to f. Hence, for constant axial pressure the drive voltage must be attenuated at 20 dB/decade with rising frequency, and this attenuation is automatically provided when current drive is adopted.

Walker's equation may also be utilized for determining axial sound pressures at shorter distances than the 'large' ones referred to above, and at angular positions off-axis. This is because the total sound pressure at any point is the vector sum of the contributions, considered one at a time, from each small part of the diaphragm and, since the remainder of the diaphragm has negligible mass and stiffness, its presence does not affect the radiation from the particular elementary area being considered.

This elementary area, if small enough, therefore has a cosine, or figure-of-eight, polar characteristic at all frequencies, and the absolute measuring distance need no longer be large in order to satisfy, nearly enough over the small elementary area, the plane-wave requirement mentioned previously.

Thus by suitable integration, done numerically using a calculator or a computer, it is possible to predict, with good accuracy, the frequency responses at different distances and angles off-axis, enabling polar diagrams also to be drawn.

So far it has been assumed that the whole diaphragm area is to be driven by just one simple pair of fixed plates, giving polar characteristics, at different frequencies, that are determined purely by the size and shape of the diaphragm. However, it is quite practicable to divide the plate area into a number of electrically separate parts, and feed these parts with currents of unequal magnitudes and/or phases. In this manner the polar characteristic obtained for a given overall diaphragm size may be varied over a wide range and given desirable forms, as is done in the Quad ESL-63[11]. Equation (3.77) can still be used to determine the long-distance axial response, the current I in it now being the vector sum of the currents to the various parts of the electrode system. This statement assumes that the diaphragm-to-plate gap is the same everywhere.

3.3.3 Optimum directional characteristics

The fact that the maximum available forces per unit area are relatively small with simple air-spaced electrostatic systems imposes a certain restriction on the designer with regard to achieving desirable polar characteristics at high frequencies, because a simple electrostatic tweeter of size comparable with typical moving-coil ones is incapable of producing a sufficient acoustic output level.

Electrostatic principles, however, make possible various other solutions to the problem of high-frequency radiation – solutions that are either impracticable with electromagnetic designs, or cannot be achieved so elegantly.

Because the choice of possible arrangements is so wide, it seems particularly desirable to try to establish at the outset just what directional characteristics loudspeakers should have in order to give the most natural and pleasing reproduction of well-recorded music. No single recipe can be confidently stated to be universally right, partly because personal preferences are inherently involved, but also because the optimum directional characteristics are very considerably influenced by the acoustics of the room in which the loudspeakers will be used[35]. However, by using scientific reasoning whenever possible, combined with careful experimenting and critical subjective comparisons between reproduced and live music, a number of guiding principles have emerged.

The simplest circumstance to consider initially is that in which a single loudspeaker is listened to in anechoic surroundings. Clearly, if the listener is on-axis, then nothing but the loudspeaker's axial response is of any consequence. Ordinarily, however, some latitude of listener positioning is obviously desirable, and then, even under anechoic conditions, loudspeaker responses somewhat off-axis, particularly horizontally, become of importance.

Now it is an inescapable fact that any loudspeaker with a flat axial frequency response, but whose directional characteristics are frequency-dependent, must have a non-level frequency response for angles off-axis. If the directivity of such a loudspeaker becomes very great at high frequencies only, this is equivalent to saying that it will have a very non-level frequency response for angles only a little off-axis: and if the high-frequency polar characteristic has pronounced side lobes, then the frequency responses for angles well off-axis will be of peculiar and complex shapes.

Thus, even under these very simple conditions of use, the first guiding principle emerges, which is that if the directivity of a loudspeaker is to increase with frequency, it should do so only in fairly mild degree, so that excessive high-frequency directivity is avoided. Under these simple anechoic, monophonic, conditions, a perfectly omni-directional loudspeaker, equivalent to a point source of sound, or a perfect doublet loudspeaker, would be equally and ideally suitable.

For listening to stereo under anechoic conditions, omnidirectional loudspeakers would again appear, at first sight, to be an ideal choice, and are certainly much preferable to loudspeakers having excessive directivity at high frequencies only. In practice, however, even in anechoic surroundings some degree of directivity, if possible almost independent of frequency, can be usefully exploited to give satisfactory stereo imaging over a larger listening area[12,30].

When loudspeakers are used in normal, relatively reverberant surroundings, the off-axis performance assumes even greater significance, because even if the listener is situated exactly on the axis of a loudspeaker, a large part of the total sound heard is the result of energy emitted from the loudspeaker at angles well off-axis and reflected in complex ways from room surfaces and furniture[13,14].

The overall impression of musical quality and balance of a loudspeaker under such conditions is, indeed, very much influenced by the character of the sound emitted by it in non-axial directions. At one time some designers were inclined to think that the optimum loudspeaker performance, for a given listening room, would be that giving, as nearly as possible, a flat frequency response for the total sound as received by a listener. For ordinary loudspeakers with typical directional characteristics, this would necessitate having an axial response that rises with frequency. Experience has shown, however, that such a solution seldom sounds right, and it is now widely recognized that a level, or nearly level, axial frequency response should be provided.

Though a listener to music reproduced in live surroundings is certainly influenced by the overall sound balance as affected by the listening-room acoustics, he is also able, in some subtle way, to sense the direct sound as such, and a rising frequency characteristic in this is instinctively interpreted as unnatural when high-quality programme sources are used.

Thus the second guiding principle to emerge is that the axial frequency response of any loudspeaker at normal listening distances should be approximately level.

Turning now to stereo reproduction in reverberant surroundings, considerable directivity is desirable, not only for the reason mentioned above in relation to anechoic surroundings but also, probably more importantly, to minimize the influence of wall reflections in upsetting the process whereby positional information and a true sense of ambience are conveyed.

Clearly, to minimize the disturbing influence of wall reflections, loudspeaker output in directions well off-axis should, in general, be as small as possible, though the need for this feature is very dependent on the conditions of use. In large rooms, where the loudspeakers can be placed a long way from walls, but where listening distances are nevertheless reasonably short, the loudspeaker images in the walls are sufficiently weakened by distance, and transients from them are sufficiently delayed in time, for satisfactory stereo to be possible without the need for very much loudspeaker directivity.

Taking all aspects into account, it is evident that some loudspeaker directivity is always desirable, but that the nature of this directivity needs to be rather carefully controlled, both for good stereo imaging, and to avoid an unpleasant or unnatural quality in the overall sound received by listeners.

The directivity of a loudspeaker at very low frequencies cannot in practice be made very great, for any totally enclosed cabinet design has substantially zero low-frequency directivity, and a doublet design has a theoretical directivity index[7] of 4.8 dB. At higher frequencies, more directivity is desirable, so that the directivity must in practice be allowed to increase with rising frequency, but it is most important that this increase should be achieved in a smooth and continuous manner. It is also important that the off-axis frequency responses which accompany this increase in directivity should not have exaggerated and unnatural features, such as pronounced peaks or dips, nor be of a shelving or step-like form[13].

Gently falling high-frequency responses are frequently met in natural circumstances, for example when listening to sounds that have travelled round corners[13], so it is hardly surprising to find that the off-axis sound from loudspeakers should be arranged to have this kind of characteristic if the overall result is to be interpreted as natural-sounding.

Not only should the directivity increase smoothly and continuously with frequency above a certain frequency, but also the magnitude of the increase should be appropriately optimized – sufficient to give good stereo conditions, but not so great that listening positions become unduly critical. If any large increase in directivity is allowed to occur by 1000 Hz, the reverberant sound will appear to be bass-heavy and dull, and yet, for good stereo, some directivity at 1000 Hz is desirable. In this respect a doublet system has a real advantage, for directivity at 1000 Hz is then obtainable without having to suffer a variation in directivity below this frequency.

Non-uniformly increasing directivity, accompanied by complex off-axis frequency responses, tends to occur with many loudspeakers employing separate physically spaced units for different parts of the spectrum, the polar characteristic narrowing towards the high-frequency end of the woofer range, for example, and widening again when operation is transferred to the tweeter. The change occurs within quite a small frequency band, of course, when high-order crossover networks are employed.

Quite apart from such changes in polar characteristic width caused by different sizes of radiating element, any loudspeaker employing separate spaced units is bound to exhibit a frequency response markedly dependent on listening position at frequencies close to the crossover frequency, where approximately equal outputs are being emitted from different physical locations. This effect, mainly a vertical one with most designs, can be minimized by reducing the spacing between units, and by adopting symmetrical arrangements where high frequencies are radiated from an area flanked symmetrically by areas radiating lower frequencies.

Thus the third guiding principle to emerge is really an extension of the first, and is that the increase in directivity with rising frequency should not only be reasonably mild in degree, but should also occur in a smooth and continuous manner, and should be accompanied by off-axis frequency responses that are devoid of 'unnatural' features[14].

With electrostatic loudspeakers, crossover schemes may either be eliminated completely or may be of a simple low-order type, giving, in combination with suitably arranged radiating areas, an unusual degree of freedom from the undesirable effects mentioned above. Consequently, well-designed electrostatic loudspeakers give the feeling that the sound is coming from a naturally homogeneous source, and this, in combination with the inherently excellent transient response and very low distortion, is sufficient to account for their reputation for giving significantly more realistic reproduction of the best programme material than do other types of loudspeaker.

3.3.4 Near-field and far-field directivity

The polar characteristics and directivity indices normally given in books are accurate only if the measuring distance is sufficiently large in relation to the radiating element size and frequency. The distance must be large enough for the fall-off in intensity to follow the inverse-square law, i.e. pressure is inversely proportional to distance. If this condition is not satisfied, the effective directivity may differ greatly from published data.

Reference 15 gives helpful information on the near-field and far-field radiation from a circular piston in an infinite baffle, and shows that the axial pressure at very low frequencies begins to fall off approximately inversely with distance, beyond about one radius away from the diaphragm centre. At higher frequencies this distance increases progressively.

For a doublet radiator, an extra complication arises because the velocity and displacement in a spherical sound wave fall off with distance, at short distances and low frequencies, more rapidly than does the pressure[7,16]. This makes the displacement, at small distances from a point source, greater than is given by equation (3.80), leading to a correspondingly greater pressure from a doublet loudspeaker than is given by equation (3.77). This is the same effect that gives bass lift for close speech with ribbon microphones, and is really quite a separate issue from the main one being considered here. The lift is about 3 dB at 50 Hz for a distance of 1 m.

At short or medium distances at high frequencies, owing to the beaming of the radiation, the fall-off in axial intensity with distance is, in general, less rapid than the inverse square law would predict, but complex fluctuations with distance can occur because of interference effects. Thus, with a different law of attenuation of intensity with distance applying at low and high frequencies, it is clearly impossible to achieve an axial frequency response that is independent of distance.

As discussed in Section 3.3.3, there are good reasons for wishing to avoid too great a directivity at high audio frequencies, and if it were practicable to have a uniformly driven plane radiating surface small enough in all dimensions to avoid such excessive directivity, then far-field conditions would apply fully at all normal listening distances.

For example, page 105 of reference 7 shows that a small unbaffled diaphragm, 7 cm in diameter, has a directivity index at 10 kHz of approximately 13 dB, the response at 20° off-axis being about −6 dB. This is probably as much directivity as is really desirable. Reference 15 shows that far-field conditions, i.e. pressure falling off inversely with distance, accompanied by a constant polar characteristic, would apply even at this high frequency for any measuring distance exceeding a small fraction of a metre.

However, the problem is that so small an electrostatic diaphragm area would be incapable of radiating sufficient power, and this has led to the widespread adoption of radiating elements in the form of long, vertical strips. Even if such a strip is made narrow enough to give a satisfactory horizontal polar characteristic, its length is liable to result in the above-mentioned near-field complications being introduced, causing the vertical polar characteristics to differ greatly at high frequencies from those given in books[16] for a vertical untapered line source, and also making the axial frequency response somewhat distance-dependent.

Figure 3.23 shows the results of some measurements by the author on a straight, unbaffled, uniformly energized electrostatic loudspeaker strip, suspended in mid-air

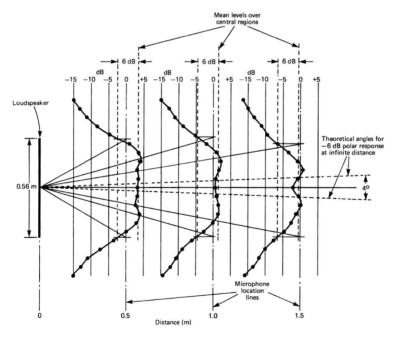

Figure 3.23. 10 kHz measurements on an unbaffled electrostatic doublet strip radiator. The sloping lines show the effective −6 dB polar response angles at microphone distances of 0.5, 1 .0 and 1.5 m. The effective directivity at 3 kHz is greater than that shown.

vertically. The measurements were made in an ordinary room using 10 kHz warble-tone, with 8 Hz triangular FM waveform covering a total bandwidth of one-third of an octave. Polyurethane foam sound-absorbing material was placed on the wall about 1 m behind the strip unit.

For each of the three curves, determined at three different horizontal distances from the loudspeaker, the measuring microphone was moved along a vertical line and the variation in its output was noted. The measuring system sensitivity was held constant; thus the changes in sound pressure with horizontal distance are correctly shown and follow approximately a law in which the pressure falls off by 3 dB per doubling of axial distance. This is what would be expected at high frequencies and moderate distances with a long and narrow strip radiator, because the sound waves expand normally in horizontal planes but do not expand very much vertically. At much lower frequencies there is almost equal expansion in both directions, and the pressure then falls off by 6 dB per doubling of distance.

Also shown in Fig. 3.23, in broken line, is the vertical polar diagram angle for 6 dB pressure loss, as predicted from the ordinary textbook data, which assumes an ideally infinite measuring distance. A most important point is thus clearly emphasized, which is that the effective polar response of a straight-strip radiator of moderate size, at normal domestic listening distances, is of much wider angle at high frequencies than the usual classical theory might lead one to suppose.

As a check on the validity of the Fig. 3.23 measurements made in non-anechoic surroundings, the response at 1.5 m was calculated for vertical positions within the limits of the strip height, and gave a curve agreeing with the measured one to within about ±1 dB and showing the same central dip flanked by maxima. For the calculation, the strip was replaced by 14 point sources, but the effects of the figure-of-eight polar characteristics of these, and the slightly different distances from them to the 'measuring' point, were neglected – a simplification which is justified within the confines of the strip height.

3.3.5 Curved-strip radiators

Vertical long-strip high-frequency radiating elements in electrostatic loudspeakers are sometimes formed into an arc of a circle (the same technique has also sometimes been used in the horizontal plane, for example in the Bowers and Wilkins DM70), with the aim of preventing the polar characteristic from becoming excessively narrow at very high frequencies, and the same technique has also been employed in moving-coil line-source loudspeakers for sound reinforcement[17].

Formulae, and sets of long-distance polar diagrams, for such circular-arc sources are given in reference 16, but some caution is necessary in interpreting this information if the right practical conclusions are to be drawn. Some general points are:

(a) If the frequency is high enough, the vertical polar diagram approximates closely to being sector-shaped, or wedge-shaped, with an included angle equal to that subtended by the curved strip about its centre of curvature.

(b) In determining whether the frequency in (a) is in fact 'high enough', the most significant parameter is Δ/λ, where Δ is as shown in Fig. 3.24 and λ is the wavelength. This is a better parameter to adopt than the often-quoted R/λ or L/λ values, for it applies in a much more nearly invariant manner over a very wide range of practical magnitudes of the angle ϕ. When $\Delta/\lambda = 0.5$, the polar diagram shows some semblance of its full-width shape; when the frequency is high enough to give $\Delta/\lambda = 1$, the approximation to the ideal shape is good enough for most practical purposes. Figure 3.25, based on a diagram in reference 16, shows one example – a 60° arc with $\Delta/\lambda \approx 1$, corresponding in this case to $R/\lambda = 8$.

(c) At frequencies low enough for Δ to be only a small fraction of a wavelength, forming the strip into a circular arc has very little effect on the polar characteristic, which nevertheless starts narrowing with increasing frequency approximately in accordance with the normal theory for a straight-line source. It is found that the long-distance polar diagram for a curved source always narrows down, with rising frequency, to a certain minimum width before beginning to open out again in conformity with (a) and (b) above. The minimum width occurs with a value of Δ/λ somewhere in the region of about 0.3.

(d) In most domestic circumstances, however, an important consideration is that, at the fairly short listening distances normally involved, the effective high-frequency polar-diagram width, even for a straight strip, is so much greater than the usual long-distance directivity theory would predict, that the amount of widening of this characteristic produced by typical amounts of strip curvature is liable to be much less significant than might at first be expected. This is considered in greater detail in Section 3.3.4.

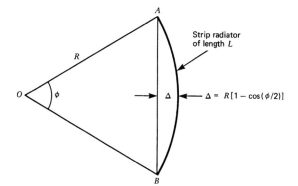

Figure 3.24. Geometry of a curved-strip radiator. The polar characteristic shape at very high frequencies tends towards that of the sector-shaped figure *OAB*.

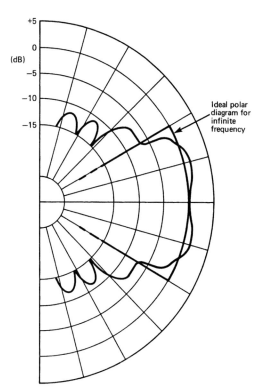

Figure 3.25. Polar diagram for uniformly energized curved-strip radiator, having parameters $\phi = 60°$, $R/\lambda = 8$, $\Delta/\lambda = 1.07$.

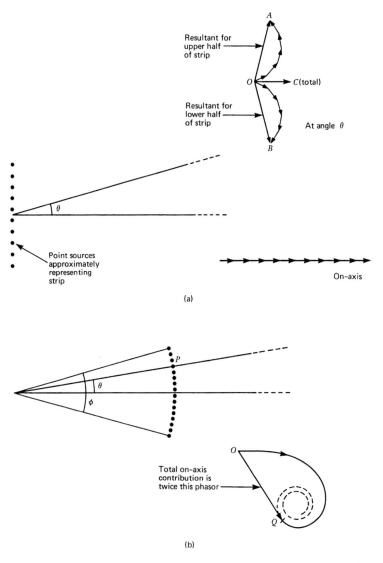

Figure 3.26. Approximate representation of strip radiators by arrays of point sources. In (a) the individual source phasors are shown, whereas in (b), for clarity and ease of drawing, this is not done. For a distant measuring point on-axis in (b), identical phasor diagrams, as shown in full-line, apply to both the upper and the lower halves of the strip. For θ as shown, the diagram for below-P contributions continues into the broken-line region, that for above-P contributions stopping before Q is reached.

(e) For a given length of strip, the larger the radius of curvature is made, the higher the frequency has to become before a semblance of the sector-shaped polar characteristic begins to be established.

To obtain some physical insight into how the above effects come about, it is helpful to regard a uniformly energized radiating strip, whether straight or curved, as an assemblage of closely spaced point sources. Thus, in Fig. 3.26(a), it is evident that a measuring point a very long distance away on the axis receives in-phase signals from all the point sources, whereas if the distant measuring point is off-axis, the contributions from the point sources will be at various different phase angles, giving a smaller resultant pressure.

Taking the contribution from the middle of the strip as being of reference phase, then, for a small angle off-axis as shown, the contributions from the upper half of the strip are of leading phase, giving rise to a resultant pressure from this half of the strip as represented by OA. For the lower half of the strip, the resultant is of equal magnitude but lagging in phase, as represented by OB. The overall resultant is therefore represented by OC.

For angles well off-axis, at high frequencies, the phasors for the upper and lower halves curl up into figures of small diameter, which become of smooth rather than angular form if the number of point sources and phasors is made very large. The overall resultant then fluctuates in magnitude as the polar angle is varied, giving rise to the familiar side lobes of the polar diagram, with interspersed zeros.

For a curved source, as in Fig. 3.26(b), it is evident that, even on-axis and at a large distance, the phasors can no longer line up in-phase, so that the large on-axis high-frequency polar response previously obtained no longer occurs. But for any polar direction within the included angle θ of the source arc, there are, on either side of the polar direction line, several of the closely spaced point sources that contribute very nearly in-phase, whereas on departing further from this line the point sources acquire different phase angles at an increasing rate, causing the phasor diagram to curl up and hence to contribute little further to the resultant pressure. Thus, at very high frequencies, an almost constant pressure level is obtained for all polar directions within the sector angle ϕ. Outside the sector angle, there is no tendency for any fairly closely spaced point sources to contribute in-phase, and the high-frequency response falls off greatly.

Clearly, if Δ in Fig. 3.24 is very much less than the wavelength, the total phase shifts built up for polar angles within the sector angle are quite small, and curving the strip then has little effect.

The various on-axis phasors for Fig. 3.26(a) cannot come into line at short measuring distances, for the point sources are then not equidistant from the measuring point. This broadens the polar diagram in a manner somewhat similar to that for long distances with a curved radiator. No true zeros, but only minima, now appear between the lobe maxima.

3.3.6 Speech-reinforcement applications

Reference 18 gives much interesting theoretical and practical information on large-scale speech-reinforcement applications of line-source loudspeakers, though the circular-arc variety is not included. Reference 19 is concerned with, among other topics, an effective technique for improving the bass performance of doublet-type line-source loudspeakers.

Some of the ideas in these papers are undoubtedly applicable to electrostatic line-source loudspeakers. Though the latter do not appear to have been used to any substantial extent for speech-reinforcement purposes, they certainly do have superior attributes to offer for this application. The present author has never accepted the notion that quality standards of the highest grade are irrelevant in such work.

3.3.7 Various schemes for complete loudspeakers

In Section 3.3.3 certain conclusions were reached with regard to the directional characteristics that a loudspeaker should possess in order to give the best stereo reproduction under normal domestic conditions. Briefly, the directivity should be fairly constant up to about 1 kHz, the effective directivity index increasing smoothly, by probably somewhere in the region of 5 dB, from 1 to 10 kHz.

In Section 3.3.1 it was shown that, quite apart from directional aspects, an unbaffled doublet system has technical advantages at low frequencies when compared with cabinet designs. The directional advantage is also, moreover, a very real one, for such a system starts off, at bass frequencies, with a directivity index of theoretically 4.8 dB. Thus higher directivity indices are permissible at medium and high frequencies than with moving-coil cabinet systems without exceeding the above-mentioned mild and smooth directivity increases. The resultant designs are sometimes criticized for having poor directivity characteristics, i.e. poor off-axis frequency responses compared with the best moving-coil designs, but it should be borne in mind that it is largely *because* of these directional characteristics that the stereo imaging is so good. Excessive high-frequency directivity, much beyond that mentioned above, is certainly to be avoided whenever possible, however.

Sections 3.3.4 and 3.3.5 emphasize that, at the fairly small listening distances involved domestically, a plane radiating element having at least one dimension that is quite large, say 1 m, is effectively much less directional at high frequencies than the usual far-field theory might lead one to expect. It does, however, inevitably have an axial frequency response that varies somewhat with distance and is liable to exhibit fluctuations due to interference effects.

One way to obtain a more nearly ideal performance is to make the radiating element very long in one dimension, in the form of a strip extending, as nearly as is practicable, from floor to ceiling. Then there is, ideally, no expansion of the sound waves vertically, the only cause of attenuation of the direct sound with distance being the horizontal expansion, giving an attenuation rate of 3 dB per doubling of distance.

Such a design gives a sound quality pleasingly independent of the vertical positioning of the listener's ears, but care must be devoted to providing a satisfactory horizontal polar characteristic. The great height, however, in combination with the improved acoustic loading resulting from there being only horizontal expansion of the sound waves, enables fairly adequate volume capability to be provided, over the whole audio spectrum, from just one such strip diaphragm, thus avoiding the need for crossover circuitry. Because the gap is large for high as well as low frequencies, the effective efficiency is rather low, but the economic availability of amplifiers rated at 100 W or more renders this not unacceptable.

The Acoustat 'Two + Two' loudspeaker[20] is based on the above concept, except that the diaphragm area is divided effectively into two vertical strips, angled at 9° with respect to each other, to improve the horizontal polar characteristic. It is a doublet design, and gives fully satisfactory results only when positioned at least 1 m away from the wall behind it.

Another way to accommodate a tall strip radiator in a room is to put it on a wall surface, or right in a corner. Doublet operation is, of course, then unsuitable, and it is necessary to provide a cavity of some kind behind the diaphragm, designed to provide a well-controlled acoustic impedance of suitable magnitude . When circumstances permit, a cavity-wall construction could be used. By developing forces of increased magnitude, through the use of a multiple-diaphragm scheme on the lines of Fig. 3.11, and/or by employing a dielectric with a much higher electrical breakdown strength than air, the size of the cavity may be minimized, permitting it to be made as a surface-mounted structure.

Though it is often said that loudspeakers should be spaced well away from walls[21], the arguments for this are largely nullified when a loudspeaker of very shallow depth is located at a wall surface or corner, because such arguments are concerned with interference effects produced by the combination of the direct sound with that coming from the acoustic images of the loudspeaker in the nearby walls. With a loudspeaker positioned really closely in a corner, for example, the nearest separately positioned images are those involving reflections from the far walls of the room – ignoring possible furniture reflections – and these images are so far away, and hence weakened and time-delayed, that the listener does not interpret their effect as a coloration of the direct sound. The absence of close wall-images should give improved stereo positioning without the need for much loudspeaker directivity, and the listening area for satisfactory stereo should be less restricted than usual.

However, the practical problems with such designs, not least that of providing a sufficiently well controlled impedance at the back of the diaphragm, are difficult to solve in a fully satisfactory manner, and this, together with the fact that most living-rooms do not have two symmetrically positioned corners conveniently available for mounting the loudspeakers, makes such schemes much less attractive, at least commercially, than they might at first sight appear to be. Somewhat similar remarks are applicable to schemes involving large diaphragms covering extensive wall areas[2].

Even free-standing loudspeakers of floor-to-ceiling height are not always accept-able domestically, and the main demand is clearly for electrostatic loudspeakers of reasonably small size that can be easily moved about. Some possible ways to obtain a satisfactory polar characteristic, combined with adequate volume capability, in such medium-sized loudspeakers, are:

(a) Employ a vertical, or nearly vertical, strip radiator, only several cm wide, for the highest frequencies, and arrange for crossover circuitry to introduce addi-tional area, on either side of this strip as the frequency falls, so that at the lowest frequencies at least the major part of a m[2] of doublet diaphragm area is in action. This is the solution adopted in the Quad Mark 1 design, discussed in greater detail in Section 3.4.1.

(b) Use a single diaphragm of fairly large area to handle the whole audio spectrum, but with the addition of an acoustic lens system to give a satisfactory polar char-acteristic. This rather expensive technique has been successfully exploited by Beveridge in a cabinet-type electrostatic loudspeaker which, not being a doublet, does not necessarily have to be spaced well away from the wall behind it[22,23].

(c) Divide the total diaphragm area into a large number of small areas, all fed with full-frequency-range signals in the same phase, but arranged to form a surface of appropriate contour for giving the desired polar characteristic as nearly as possible.

(d) Have a single large plane diaphragm, or the equivalent thereof, but divide the fixed-plate system into a number of sections driven by signals of such phases and amplitudes as to cause the diaphragm to radiate sound with the desired polar characteristic. This technique is more flexible than the others in that it enables the manner in which the polar characteristic varies with frequency to he more satisfactorily controlled, A particular version of this general scheme is used in the Quad ESL63.

3.3.8 Amplifier considerations

In evolving a practical electrostatic loudspeaker, problems of drive circuit design, sound volume capability, and efficiency, are closely tied in with the directional-char-acteristic considerations discussed in the previous sections.

Up to the present point, the input terminals of the loudspeaker have been regarded as being the fixed plates, i.e. the terminals at the left-hand side of Fig. 3.2. Even if attention is confined to electrostatic loudspeakers not containing built-in amplifiers, some passive circuitry (other than a transformer) is usually interposed between the plates and the actual connection terminals on the back of the loudspeaker.

Before outlining some specific design approaches, a couple of basic notions will first be considered.

The first is that, referring to Fig. 3.2 and ignoring the 'dead' capacitance C_s, the electrical input impedance of the basic loudspeaker, as far as practical drive require-ments are concerned, is nearly enough equal to the reactance of the plate-to-plate capacitance C_0, except at frequencies in the region of the fundamental diaphragm resonance and below. In other words, the motional component of the input imped-ance is fairly negligible except at low frequencies. This is so because the impedance

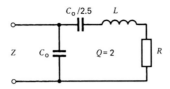

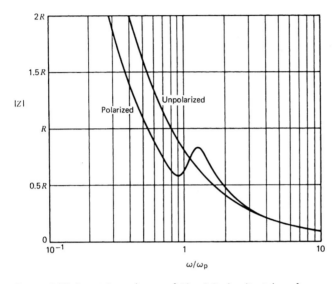

Figure 3.27. Input impedance of Fig. 3.2 circuit at low frequencies, ignoring C_s, for a diaphragm-stretching stiffness 3.5 times the negative stiffness, with $Q = 2$. R is the total series resistance, almost entirely due to intentional damping. The unpolarized curve is simply the reactance of C_0.

of the part of the Fig. 3.2 circuit to the right of $-C_0$ is inevitably much larger than that of $-C_0$ at high frequencies. Alternatively, with reference to Fig. 3.3, it may be said that the mechanical impedance of the diaphragm plus air loading is much higher than the reactance of the negative compliance, except at low frequencies.

Figure 3.27 shows the calculated impedance characteristic for the Fig. 3.2 circuit, ignoring C_s, when the damping is such as to give a typical Q-value of 2 for the series-tuned-circuit part of the network, the diaphragm-stretching stiffness being taken as 3.5 times the electrically produced negative stiffness.

The second notion is that the true power efficiency of a basic electrostatic loud-speaker is very high, because the only elements in Fig. 3.2 or Fig. 3.3 which can dissipate the signal power fed in are the damping resistance R_{md} and the air-load resistance R_{ma}, the latter representing the useful sound output.

The damping resistance per unit area is normally only a small fraction of ρc, whereas at medium and high frequencies a doublet diaphragm of moderate dimensions, for example, sees an air-load resistance per unit area of approximately $2\rho c$. Under these conditions the major part of the electrical input power appears as acoustic output power.

However, though the true power efficiency of the basic Fig. 3.2 loudspeaker is then very high, the main part of the signal input current, at medium and high frequencies, goes to C_0. This is a quadrature, or wattless, current, greatly increasing the volt-amps (VA) supplied and hence the size of amplifier needed. The size of amplifier required for supplying, say, 100 VA of reactive output is at least as great as is required for supplying 100 W of output to a resistive load. In fact a 100 VA reactive load results in a much larger dissipation of power in the amplifier output transistors than does a 100 W resistive load. Therefore, other things being equal, tougher transistors and/or larger heat sinks are necessary in the reactive case[24].

What really counts, of course, with a given loudspeaker, is the size of amplifier required to enable it to produce peak SPLs of a specified magnitude under normal conditions of use. This, however, is too vague to form the basis of a proper quantitative specification, for it requires 'normal conditions of use' also to be specified.

One widely adopted scheme is to specify the SPL produced at a stated distance on-axis, in anechoic surroundings, for a specific input voltage, and to accompany this information by an impedance/frequency curve. The specific input voltage is often taken as 2.83 V r.m.s. (1 W in 8 Ω) and the distance as 1 m, the SPL being given in decibels relative to the reference level of 2×10^{-5} N/m^2. Alternatively, and perhaps more elegantly, the sensitivity may be given in μbar/volt.

A point of interest is that if a closed-cabinet loudspeaker and a doublet one have the same sensitivities, measured as above, the latter is liable to sound less loud than the former when listened to at a moderately large distance in a living-room of average reverberation time. This is because the reverberant sound intensity is higher for the more nearly omnidirectional loudspeaker, and contributes more to the overall volume as judged by a listener. The difference can be quite appreciable subjectively.

With most electrostatic loudspeaker designs, the electrical-input impedance reaches its lowest value at very high audio frequencies but, since a transformer is normally used the designer can easily set the absolute level of the impedance characteristic at what he considers to be an appropriate value. The optimum choice in this respect is by no means always easy to decide upon, however.

Increasing the step-up ratio of the transformer, by reducing the number of primary turns, will increase the loudspeaker sensitivity, defined as above. But, if carried too far, the impedance at high frequencies may become much lower than that for which the intended type of amplifier can produce its full output voltage swing. Up to a certain point this may be argued not to matter, for music waveforms do not normally contain very high frequency components of as large a magnitude as those at lower frequencies. Indeed, with ordinary radio and analogue disc sources, it is found that the peak rates-of-change of programme voltage do not normally exceed that of a 2.5 kHz sine waveform having a peak-to-peak voltage equal to that which can just be handled at medium frequencies without overload[25].

However, when ordinary peak programme meters are used at the originating end, and when the radio or recording system employs pre-emphasis and de-emphasis, a result somewhat like that just mentioned is virtually inevitable, even if the original natural sounds themselves have much higher rates of change – the peaks will get squashed. By turning down the programme level at the control desk, sounds having these very high rates of change could, of course, be faithfully transmitted, but the noise performance of ordinary disc and radio systems makes this solution unattractive. However, with the advent of digital systems, which have superb noise performance, it has become quite feasible to preserve full rates of change even in music containing very loud cymbal clashes, etc.

Tests carried out by the author on his own and other digital recordings, using a special very fast responding peak programme meter circuit arranged to indicate both peak voltage and peak rate of change of voltage, have shown that rates of change corresponding to full-amplitude sine waves at much higher frequencies than the above can sometimes occur. The most extreme case was that of a close-up recording of drumkit rim-shots, which gave a frequency figure, on the above basis, of about 15 kHz. A cymbal clash in another recording gave 6.5 kHz. A recording of a contemporary work by Gordon Crosse, 'Wintersong', involving some quite intense bell sounds, gave 8 kHz. Applause can sometimes involve quite high peak rates of change, leading to a frequency figure of 5 kHz or more.

From evidence such as the above, it would seem that, in the digital era, an electrostatic loudspeaker should be designed so that its impedance at very high audio frequencies does not fall to quite such a low value as would previously have been thought to be innocuous; or, conversely, that an amplifier for use with an existing design of loudspeaker should be capable of giving full output voltage swing into a lower value of load impedance than would once have been considered necessary.

Actually, it is generally believed that some degree of momentary clipping can be allowed to occur that is not subjectively detectable, but related specifications really require to be reassessed in the light of digital developments. If, for economic or other reasons, occasional clipping has to be permitted, then it is important to ensure that recovery from clipping is very rapid and not accompanied by longer-duration, 'recovery transients'.

3.3.9 Some further design problems

The large majority of electrostatic loudspeakers are of the doublet type, so that Walker's equation (3.77), or the comparable equation (3.111) of Section 3.3.11, provide the best starting point for design purposes.

Initial thoughts, in the absence of practical experience, might involve the notion of a single uniformly driven diaphragm, say circular, not too large in order to avoid excessive high-frequency directivity, and with the listener at a sufficient distance to experience virtually the full high-frequency axial response directly predicted by equation (3.77). The plates would be driven with frequency-independent current via a large value of series resistance, from a voltage amplifier plus step-up transformer. However, such a design is rather impractical for several reasons, and it is instructive to consider these.

It was established in Section 3.3.8 that for frequencies well above the main resonance frequency, the electrical-input impedance is nearly enough that of the capacitance C_0; see Fig. 3.27. Under these conditions the peak plate current is related to the peak plate voltage by

$$\hat{I}_{\text{sig}} = \hat{V}_{\text{sig}} \times \frac{2\pi f}{C_0} \tag{3.85}$$

Substituting this in Walker's equation (3.77) gives

$$\hat{P} = \hat{V}_{\text{sig}} V_{\text{pol}} \times \frac{fC_0}{crd} \tag{3.86}$$

Substituting for $\hat{V}_{\text{sig}} V_{\text{pol}}$ and C_0 in this, from equations (3.15) and (3.31) respectively, then leads to

$$\hat{P} = \frac{\hat{F}_{\text{sig}} f}{2cr} \tag{3.87}$$

The physical significance of this last relationship is perhaps more easily appreciated if it is expressed in the form

$$\hat{P} = Af \times \frac{\hat{F}_{\text{sig}}/A}{2cr} \tag{3.88}$$

where $\hat{P}$ = peak instantaneous sound pressure at a large distance r on-axis (N/m²)
A = diaphragm area (m²)
f = frequency (Hz)
$\hat{F}_{\text{sig}}/A$ = peak instantaneous signal force per unit area (N/m²)
c = velocity of sound (m/s)
r = measuring distance (m)

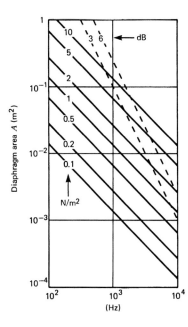

Figure 3.28. The full-line graphs show minimum areas of unbaffled doublet diaphragm required for producing, at 1 m on-axis, the peak instantaneous sound pressures indicated, assuming that the maximum instantaneous signal force that can be developed is 50 N/m². The diaphragm must be circular, or of squat proportions, so that the effect to which Fig. 3.32 relates is not invoked. The broken-line graphs apply accurately to a circular diaphragm only, and show the relationship between area and frequency for the directivity index to exceed its low frequency value of 4.8 dB (figure-of-eight) by 3 and 6 dB respectively.

As stated in Section 3.2.9, the practical maximum value of $\hat{F}_{sig}/A$ attainable in a simple air-dielectric loudspeaker of the Fig. 3.1(b) type is about 50 N/m² when the spacing d is made sufficient to permit operation at bass frequencies. Using this value of $\hat{F}_{sig}/A$, graphs may then be plotted from equation (3.88), as shown in full line in Fig. 3.28, giving the minimum necessary diaphragm area as a function of frequency for a series of peak instantaneous axial sound-pressure levels at some standard distance r, taken as 1 m.

If the diaphragm in a simple circular design, as is now being contemplated, is made large enough to produce a satisfactory volume, it will become excessively directional at high frequencies.

Reference 7, page 112, gives information relating to the directivity index of an unbaffled circular piston. At low frequencies this index is 4.8 dB, and it has increased by 3 dB by the time $\pi D/\lambda$ has risen to about 3.3, D being the diaphragm diameter. The corresponding frequency in Hz is easily shown to be

$$f_{3dB} = \frac{320}{\sqrt{A}} \tag{3.89}$$

where A = diaphragm area (m²).

The 3 dB broken-line graph in Fig. 3.28 is a plot of equation (3.89), the 6 dB graph being derived in a similar manner.

Suppose it is decided that a peak instantaneous axial sound pressure of 5 N/m² at 1 m should be obtainable at 100 Hz. This is approximately 3.5 N/m² r.m.s. (sine-wave), corresponding to an SPL of 105 dB. Then Fig. 3.28 shows that a diaphragm area of 0.69 m² is required. Figure 3.28 also shows that if this diaphragm area is kept the same at higher frequencies then the directivity index will have increased by 6 dB by the time the frequency has reached about 610 Hz. If, however, the effective diaphragm area is arranged to decrease as the frequency rises, so that it is always just sufficient to give the desired maximum sound pressure, then the frequency for a 6 dB increase in directivity index becomes 3.7 kHz. This frequency is nevertheless still considerably lower than is really desirable.

It should be noted that if, by suitable circuit arrangements, the effective diaphragm area is made inversely proportional to frequency at the higher frequencies, the total electrical impedance then ideally becomes independent of frequency, i.e. resistive.

It is evident that the following techniques may be employed to give a more satisfactory polar characteristic from a circular design, or one of comparable proportions:

(a) Tolerate a smaller maximum volume capability, permitting the use of a smaller diaphragm – not normally an acceptable solution, though very satisfactory when applicable.
(b) Use a much smaller gap for the parts of the diaphragm handling the higher frequencies. As shown in Fig. 3.10, higher field intensities are then permissible, giving increased maximum signal forces per unit area.
(c) Divide the diaphragm into small areas located on a cylindrical or spherical surface.
(d) Use an acoustic lens to widen the polar characteristic.
(e) Use a delay-line technique, as in the Quad ESL63, to cause different annular areas of the diaphragm to vibrate with appropriately different phasings, thus radiating a spherical wavefront rather than a narrow beam at high frequencies. This enables plenty of volume to be obtained without requiring small gaps, with their attendant constructional and tolerancing problems.

Except for (a), all these techniques are fairly tedious and expensive to carry out in a fully satisfactory manner, and the alternative approach of using long-strip radiators for the high frequencies tends to lead to easier and more economical designs.

So far the effect of diaphragm-stretching stiffness has been ignored, but in practice it is essential to take such stiffness into account at an early stage in the design procedure. If the diaphragm-stretching stiffness were zero, and there were no added damping, it would theoretically be necessary, for flat response, to maintain constant-current feed to the plates right down to the bottom of the audio spectrum. This would require a very high value of series resistance, and make the overall sensitivity of the loudspeaker undesirably low. Such a condition is a purely imaginary one, of course, because stability of the diaphragm could not be obtained.

By arranging for the total stiffness to be such that the resonance, involving the diaphragm-plus-air-load mass, occurs at a low bass frequency, with a Q-value of about 2, a larger effective value of F_{sig}/A may be produced at frequencies in the region of resonance, giving improved bass-power capability. Suitable damping must be introduced, by fine-mesh cloth on one or both of the plates or in some other way.

Walker's equation (3.77) indicates that the frequency response at low frequencies will be flat, for constant current fed to the plates, if the diaphragm-stretching stiffness is zero, and if there is no damping other than that due to the air loading.

Referring to Fig. 3.2 and ignoring C_s, the circuit, with current feed and no diaphragm stiffness or damping, becomes that shown in Fig. 3.29(a). The reactance of the diaphragm mass M_{md} is normally very much less than the air-load mass reactance X_{ma} at low frequencies, typically by a factor of the order of 100.

The combination of I_{sig} and C_0 is equivalent to a voltage $I_{sig}/\omega C_0$ acting in series with C_0. C_0 and $-C_0$ in series then give a zero impedance, which is equivalent to saying that with constant-current feed no negative stiffness is produced. Thus Fig. 3.29(a) is equivalent to (b), in which X_{ma}/α^2, the reactance of a virtually constant air-load mass, is very much the dominant impedance at low frequencies. With a driving voltage inversely proportional to ω, and an impedance proportional to ω, I_{mot} is inversely proportional to ω^2, i.e. the diaphragm velocity is inversely proportional to ω^2. This is the condition giving a flat frequency response at low frequencies, as may readily be verified by reference to Fig. 3.22.

When there is finite stretching stiffness and considerable extra damping, the plates still being fed with constant current, the relevant circuit becomes that of Fig. 3.29(c), which may be more conveniently shown as in (d).

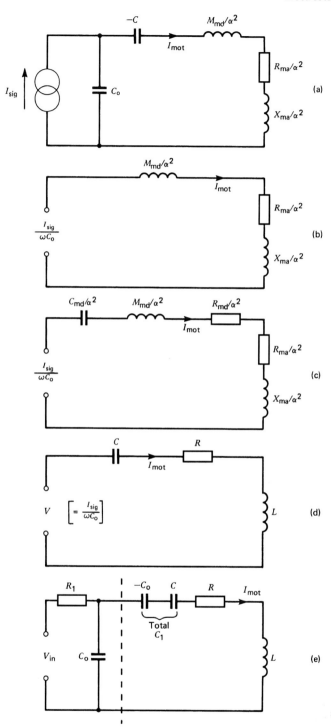

Figure 3.29. Electrical impedance circuits relating to the determination of the frequency response of doublet electrostatic loudspeakers at low frequencies.

With C and R absent, it has already been shown that the loudspeaker gives a flat response. The change in its response on introducing C and R is therefore the same as the change in response of the (d) circuit when C and R are introduced. For Fig. 3.29(d):

$$I_{mot} \text{ without } C \text{ and } R = V/pL \tag{3.90}$$

$$I_{mot} \text{ with } C \text{ and } R = \frac{V}{pL + 1/pC + R} \tag{3.91}$$

where $p = d/dt = j\omega$ (s may be used instead of the Heaviside operator p, if preferred.) Hence

$$\frac{I_{mot} \text{ with } C \text{ and } R}{I_{mot} \text{ without } C \text{ and } R} = \frac{p2LC}{1 + pCR + p^2LC} \tag{3.92}$$

$$= \frac{p^2T^2}{1 + pT/Q + p^2T^2} \tag{3.93}$$

where $T = 1/\omega_0 = \sqrt{LC}$ and $Q = \sqrt{LC}/R$.

This is a second-order high-pass filter response, with a cut-off frequency equal to the unpolarized resonance frequency ω_0, of the loudspeaker, and is thus the response for ideal constant-current feed with finite stretching stiffness and added damping.

It would be theoretically possible, by retaining constant-current feed and introducing sufficient damping, to obtain, say, a second-order Butterworth response. However, as already implied, increased power-output capability and greater practical sensitivity can be achieved by adopting a higher value of Q, typically about 2, and feeding the plates from a voltage source via a series resistance of only moderate value. When this is done, the relevant circuit becomes that of Fig. 3.29(e).

It is desired to determine in what manner I_{mot} departs from being inversely proportional to ω^2, which is the condition giving a flat acoustic response. Analysis of the Fig. 3.29(e) circuit gives

$$I_{mot} = \frac{V_{in}}{R_1} \times \frac{pC_1R_1}{1 + p(C_1R + C_0R_1 + C_1R_1) + p^2(LC_1 + C_0C_1R_1R) + p^3C_0C_1R_1L} \tag{3.94}$$

For a level acoustic response, I_{mot} must be inversely proportional to p^2T^2, so that with I_{mot} actually varying as in equation (3.94), it is evident that the acoustic response obtained must be of the same form as equation (3.94) but with a p^3 term in the numerator, giving a bass response falling off asymptotically at 60 dB/decade, with level response at higher frequencies.

In practice, the design approach may be much simplified by appreciating that the impedance of the part of the Fig. 3.29(e) network to the left of the broken line is normally much less than the impedance of the part to the right, so that the two parts may be treated separately with no very serious error; R is likely to be at least 10 times as large as R_1.

Practical designs are thus best carried out on this simplified basis, the response being checked by measurement later on, perhaps leading to a slight change in the value of R_1 used.

An effect likely to be quite significant is that, whereas Walker's equation (3.77) assumes an unbaffled diaphragm, in a practical doublet loudspeaker there is always a certain amount of baffling caused by the mouldings to which the fixed plates are attached, the means for supporting these mouldings, the dust covers, and the outer protective and decorative covering of the loudspeaker. Such 'vestigial baffling' can increase the bass response by several decibels, and is another reason why some experimental adjustment to the originally selected value of R_1 is likely to be necessary in practice.

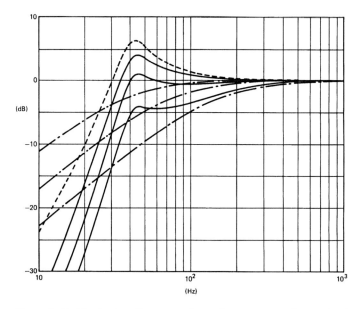

Figure 3.30. The broken-line curve shows the response of a constant- current-fed doublet electrostatic loudspeaker with a resonance frequency of 40 Hz and $Q = 2$, ignoring 'vestigial baffling'. The chain-dotted curves relate to the R_1C_0 part of Fig. 3.29(e), the full-line curves showing the overall responses for the corresponding values of R_1C_0.

The broken-line curve in Fig. 3.30 shows the response that would be given by a constant-current-fed unbaffled diaphragm with 40 Hz resonance frequency and a Q-value of 2. The three chain-dotted curves represent responses of the R_1C_0 part of the Fig. 3.29(e) circuit for three different corner frequencies. The full-line curves show the overall theoretical responses of the loudspeaker, ignoring 'vestigial baffling', with these three different values of R_1C_0 and illustrate how changing the value of R_1 modifies the bass-response characteristic. (Varying R_1 also affects the sensitivity, but this effect is not shown in Fig. 3.30: all curves are based on a common 0 dB level.)

An important point to appreciate is that a decision on the diaphragm size and resonance frequency predetermines the gap d that must be adopted, assuming the use of the full polarizing voltage as given by equation (3.44). The argument is as follows.

The total diaphragm compliance necessary can be determined once an estimate has been made of the air-mass loading per unit area, since acceptable bass performance requires that the resonance frequency should occur at an appropriately low frequency such as 40 Hz. The negative compliance must be numerically larger than this total compliance, by a factor that has been taken as typically 2.5 in earlier sections of this chapter, though some designers may choose to depart considerably from this particular value.

From equations (3.14) and (3.44) it follows that

$$C_{\text{me}} = -\frac{d}{E^2_{\text{max}}\varepsilon_0 A} \tag{3.95}$$

The required negative value of the compliance C_{me} having been determined as above, equation (3.95) gives the necessary value of d.

As already emphasized, a fairly large diaphragm area is necessary in a doublet design in order to permit the radiation of sufficient power at low frequencies. Even apart from power output capability, however, reducing the diaphragm size, while retaining the same frequency response, results in a large reduction in sensitivity. The explanation is as follows.

If the diaphragm diameter is halved, the air-mass loading per unit area is halved (see Fig. 3.22) and the total air-mass loading is therefore divided by eight. For the same resonance frequency as before, the total compliance must therefore be increased by eight times. With A in equation (3.95) reduced by a factor of four, d must be doubled to obtain the required eight times increase in C_{mc}. These changes greatly increase the power that must be supplied to the loudspeaker for a given sound-pressure level on-axis. That this is so may be appreciated by considering the behaviour of the Fig. 3.29(e) circuit, in which, at medium and high frequencies, most of V_{in} appears across R_1 giving a current of approximately V_{in}/R_1 and a power consumption of approximately V_{in}^2/R_1.

Now from Walker's equation (3.77), a certain definite current is required for a specific sound-pressure level at a large distance on-axis, independently of diaphragm size, and independently of the gap d, if it is assumed that V_{pol} is made proportional to d. Thus, if the diameter is halved, accompanied by a doubling of d as described above, the capacitance C_0 is divided by eight. The time-constant R_1C_0 must remain the same as before to give the same frequency response, so R_1 must be increased by a factor of eight. With the same current flowing, the required power input is therefore increased eight times.

It thus becomes clear that a uniformly driven full-frequency-range small-diaphragm design will inevitably have very low sensitivity.

Some practical solutions were outlined in Section 3.3.7, and the most widely adopted ones are based on the use of strip radiators, in some cases of floor-to-ceiling height.

Whereas sufficiently small radiating elements ideally need constant-current feed to give a flat axial response at reasonable listening distances, practically sized elements often require the signal current to increase at high frequencies. The designer needs to know by how much this current should increase, and a universal curve has been derived to provide this information.

3.3.10 A universal curve and some related ideas[32]

Consider a straight uniformly driven narrow strip radiator of height h, as shown in Fig. 3.31(a). Sound arriving at the listening point M from the point P on the strip has travelled an extra distance A compared with sound from the centre of the strip, and A is given by

$$\Delta = (r^2 + y^2)^{1/2} - 2$$
$$= r[(1 + y^2/r^2)^{1/2} - 1] \tag{3.96}$$

and if $y \ll r$, then

$$\Delta \approx y^2/2r \tag{3.97}$$

The approximate phase lag of the sound from P with respect to that from the centre is therefore given by

$$\phi = 360° \times y^2/2r\lambda \tag{3.98}$$

If the strip is considered as equivalent to a large number of equal point sources, each of these gives rise to a pressure component at M phased approximately in accordance with equation (3.98), the various component phasors then forming a diagram as in Fig. 3.31(b). At very low frequencies, all the phasors are very nearly in line.

The magnitude of the fall-off in frequency response at higher frequencies, caused by listening at a finite distance r, can be determined once the value $(h/2)^2/2r\lambda$ is known, i.e. once $h^2/r\lambda$ is known, since, from equation (3.98), this determines the angle

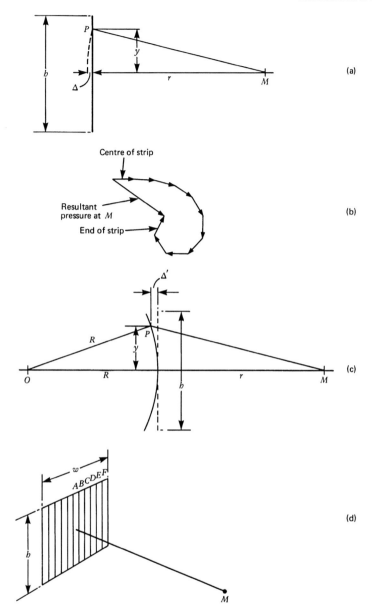

Figure 3.31. Diagrams relating to the derivation and use of the universal curve shown in Fig, 3.32.

of the last phasor in the diagram and thus dictates the shape of the rest of the diagram. From these considerations it follows that:

(a) If h is doubled, a given high-frequency loss will occur at four times the wavelength, i.e. at a quarter of the frequency.

(b) If r is doubled, a given high-frequency loss will occur at half the wavelength, i.c. at twice the frequency.

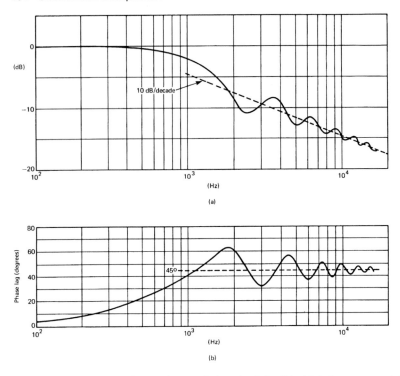

Figure 3.32. (a) Universal curve for predicting the fall-off in high-frequency axial response at short listening distances from a uniformly energized narrow-strip radiator, relative to the theoretical response at an infinite distance from a straight strip. For convenience, the frequency scaling relates to a straight strip of length 1 m and a listening distance of 1 m, but the accuracy is poor under these actual conditions. For a straight strip of length h, at a listening distance r, both in m, the frequencies shown above should be multiplied by r/h^2 and the accuracy will be satisfactory if r is at least twice h. If the narrow strip is curved, with radius of curvature R m, then the frequencies shown should be multiplied by $rR/h^2(R + r)$ (all in m). If the strip is not narrow, an additional high-frequency loss is introduced, and this additional loss may be predicted by using the above curve with the strip width w replacing h. The asymptotic high-frequency fall-off rate is then 20 dB/decade. (b) The corresponding phase curve, approximate only for frequencies above 1 kHz. The angle of lag given is that by which the axial sound pressure lags in phase relative to the phase it would have if only a very small, exactly central, part of the radiating system was operative.

Hence if a single curve of high-frequency loss against frequency is derived for a specific pair of values of r and h, it may be used, in accordance with (a) and (b) above, to deduce the loss for any other values of r and h. In Fig. 3.32(a), a frequency scaling corresponding to $r = h = 1$ m is conveniently adopted. The frequencies shown should therefore be multiplied by r/h^2 (r and h must be in m).

The phase angles for drawing the many phasor diagrams from which the Fig. 3.32(a) curve was derived were obtained from the approximate formula (3.98), each half of the radiating strip being represented by 20 point sources. The curve is therefore rather inaccurate if actually used for the condition $r = h$, but gives satisfactory accuracy under more typical listening conditions for a moderately sized electrostatic loudspeaker, where r is considerably greater than h. There are three causes of inaccuracy when r is not much greater than h:

(a) Use of equation (3.97) instead of (3.96);
(b) For doublet radiating points near the ends of the strip, the listening point is at angles well off the maxima of their figure-of-eight polar characteristics;
(c) The outer point sources are further from the listening point than are the more nearly central ones, introducing an inverse square law error.

All the above effects are reasonably negligible provided that $r \geqslant 2h$.

The use of the universal curve may be extended to cover curved-strip radiators. Referring to Fig. 3.31(c), the sound reaching the listening point M from the point P has now travelled an extra distance Δ', compared with the condition for a straight strip, Δ' being given approximately by

$$\Delta' = y^2/2R \tag{3.99}$$

Thus the total extra distance travelled by the sound from P, compared with sound from the centre of the strip, is given approximately by

$$\Delta_{\text{tot}} = \frac{y^2}{2}\left[\frac{1}{r} + \frac{1}{R}\right] = \frac{y^2}{2} \times \frac{R+r}{rR} \tag{3.100}$$

and the approximate total phase lag is

$$\phi_{\text{tot}} = 360° \times \frac{y^2}{2} \times \frac{R+r}{rR\lambda} \tag{3.101}$$

The shape of the phasor diagram, and therefore the magnitude of the high-frequency response loss, is now dictated by the value of

$$h^2 \left[\frac{R+r}{rR}\right] \quad \text{(all in metres)}$$

so that the frequencies shown in Fig. 3.32 should now be multiplied by

$$\frac{rR}{h^2(R+r)} \quad \text{(all in metres)}$$

Finally, use of the Fig. 3.32 universal curves may be extended to the situation where the radiating strip is not narrow. With reference to Fig. 3.31(d), a wide straight-strip radiator may be regarded as equivalent to a large number of narrow-strip radiating elements placed side by side.

At any specified high frequency, strip A gives a total sound-pressure phasor at M which is attenuated in magnitude in accordance with the notions given above. Ignoring figure-of-eight directivity and inverse square law errors as before, strips B, C, D etc. all give equal-magnitude phasor contributions to the pressure at M, but these contributions lag more and more in phase as strips progressively further off-centre are considered. Thus a second high-frequency-attenuating mechanism is introduced, of exactly the same nature as that applying to each strip, and the additional high-frequency loss thus introduced may be determined by using the universal curve, but with w in Fig, 3.31(d) replacing h.

A particular case is that of a square diaphragm, which gives twice the high-frequency attenuation, in dB, than is given by a straight narrow strip of height equal to the side of the square. The high-frequency asymptotic rate of attenuation is then 20 dB/decade, instead of the 10 dB/decade for a narrow strip that is shown in Fig. 3.32(a).*

---

* Because most of the triangles invoked in deducing the effect of horizontal displacement are slightly larger than those for vertical displacement along the central strip, the higher-frequency maxima and minima, of small amplitude, do not come at quite the same frequencies as for a single strip. This is normally of no practical consequence, however.

Obviously, perfect equalization of a response such as that of Fig. 3.32(a) is not very practicable, but the situation is more favourable than it might at first appear to be, for two reasons:

(a) The majority of reasonably portable electrostatic loudspeakers have strip heights not much greater than about 0.5 m, so that most of the wiggles in the frequency response corresponding to Fig. 3.32(a) occur above the audio band.
(b) When a floor-to-ceiling strip is used, the conditions that give rise to the wiggles in Fig. 3.32 do not occur. Such a strip, plus its images in the floor and ceiling, ideally simulates an infinitely long strip radiator, the listening distance therefore inevitably being very small in comparison. The value of r/h^2 for use in Fig. 3.32 is thus vanishingly small, so that all audio frequencies come beyond the right-hand limit of the curve, where the wiggles have died out. A perfectly smooth response is thus theoretically obtained, and it has the attractive feature of being independent of the listening position.

The argument in (b) above predicts that a very narrow floor-to-ceiling doublet loud-speaker, whose plates are fed with frequency-independent current, will have a frequency response falling with rising frequency at 10 dB/decade throughout the audio spectrum. To counteract this, the current must be made to rise at 10 dB/decade with frequency, corresponding to the plate voltage falling at 10 dB/decade.

It is interesting to observe that this latter conclusion is consistent with the fact that an infinitely long and very narrow unbaffled strip sees an acoustic load in which X is proportional to frequency and R (much smaller than X) is proportional to frequency cubed. Since the shape of the radiated field is ideally independent of frequency, a level frequency response at the listening position requires a frequency-independent power output. From the above X and R considerations, this requires a force inversely proportional to the square root of the frequency, i.e. a plate voltage falling at 10 dB/decade with rising frequency, as previously deduced.

A more direct and simple approach to the design of floor-to-ceiling electrostatic doublet loudspeakers, enabling the sensitivity to be very easily calculated, is given in the next section.

3.3.11 A basic equation for floor-to-ceiling doublets

Walker's equation (3.77) of Section 3.3.2 gives a simple and direct relationship between the sound pressure at a large distance r on-axis, and the signal current fed to the plates, for a plane unbaffled diaphragm of finite size. A comparable relationship may be established between the horizontally axial sound pressure at a distance r from a narrow vertical doublet strip radiator of infinite length, and the signal current per unit length fed to the plates.

The derivation of this new relationship follows generally similar lines to those adopted in deriving Walker's equation, though it involves the concept of a very narrow strip-shaped pressure capacitor microphone of height h, placed at a distance r from a very narrow strip-shaped unbaffled electrostatic doublet loudspeaker also of height h. The microphone is assumed to be at the maximum of the figure-of-eight horizontal polar characteristic of the loudspeaker, and h is assumed to be so large in relation to r and the wavelength that vertical expansion of the waves passing between these units may be ignored – in other words, ideal cylindrical waves are assumed.

Equations (3.78) and (3.79) are applicable to this modified form of microphone, but it is also necessary to establish the magnitude of the air-particle displacement at a distance r from the microphone when a current I_{mic} is fed to it. The required information for this is contained in reference 27, which gives the following equation for a radially pulsating cylinder of infinite length:

$$W_1 = \pi^3 \rho a^2 U_0^2 f \tag{3.102}$$

where W_1 = mean power radiated per unit length (W/m)
U_0 = peak instantaneous radial velocity (m/s)
a = radius of cylinder (m)
ρ = density of air (kg/m^3)
f = frequency (Hz)

Hence the peak instantaneous volume velocity per unit length is $2\pi a U_0$.

Now for the microphone used as a loudspeaker, equation (3.78) shows that its volume velocity per unit length is given by

$$S/h = k I_{mic}/h \tag{3.103}$$

If the microphone is of small enough cross-section, the fact that it is not cylindrical is of no significance.

By equating equation (3.103) to $2\pi a U_0$, and then substituting for U, from equation (3.102), the following equation is obtained:

$$W_1 = \frac{\pi \rho f k_2 \hat{I}_{mic}^2}{4h^2} \tag{3.104}$$

where W_1 = mean power output from microphone per unit length (W/m)
$\hat{I}_{mic}/h$ = peak instantaneous microphone current per unit length (A/m)

The mean power passing through unit area at a distance r is given by:

$$\text{power per unit area} = \frac{\pi \rho f k^2 \hat{I}_{mic}^2}{4h^2 \times 2\pi r} \tag{3.105}$$

If U_r is the peak instantaneous air-particle velocity at distance r, then

$$\tfrac{1}{2} U_r^2 \rho c = \frac{\rho f k^2 \hat{I}_{mic}^2}{8h^2 r} \tag{3.106}$$

The peak air-particle displacement is given by

$$\hat{x} = U_r/\omega \tag{3.107}$$

From equations (3.106) and (3.107) may be derived:

$$\hat{x} = \frac{k \hat{I}_{mic}}{4\pi h \sqrt{crf}} \tag{3.108}$$

and therefore, from equation (3.81):

$$V_{esl} = \frac{V_{pol} k I_{mic}}{2\pi d h \sqrt{crf}} \tag{3.109}$$

in which V_{esl} and I_{mic} may be peak or r.m.s. values as preferred.

Hence, by the reciprocity principle, a current I_{esl} fed to the loudspeaker plates produces an output voltage from the microphone given by

$$V_{mic} = \frac{I_{esl} k V_{pol}}{2\pi d h \sqrt{crf}} \tag{3.110}$$

Hence, from equation (3.79), and using the symbol I_{sig} in place of I_{esl}:

$$P = I_{sig} \times \frac{V_{pol}}{2\pi d h \sqrt{crf}} \tag{3.111}$$

where P = sound pressure at a distance r from the floor-to-ceiling strip, on the maximum of its horizontal figure-of-eight directivity characteristic (N/m^2)

I_{sig} = signal current fed to plates (A)
V_{pol} = polarizing voltage (V)
d = diaphragm-to-plate spacing (m)
h = height of strip (m)
c = velocity of sound (m/s)
r = measuring distance (m)
f = frequency (Hz)

Inspection of equation (3.111) shows that for a level frequency response the current must be made proportional to the square root of frequency, i.e. it must be increased at 10 dB/decade with rising frequency, as established via other arguments in Section 3.3. 10. Equation (3.111) also shows that the pressure falls off in inverse proportion to the square root of distance, which is what would be expected when the sound expands horizontally but not vertically. Finally, the equation shows that the pressure is proportional to the current per unit length, again as would be expected.

When a floor-to-ceiling doublet is insufficiently narrow, the axial sound pressure at normal listening distances is less, at high frequencies, than is given by equation (3. 111), owing to the effect discussed in detail in Section 3.3.10. The magnitude of the high-frequency response loss may be determined with the aid of the Fig. 3.32(a) universal curve, the strip width w replacing h.

3.3.12 Radiation impedance of baffled strip radiators

It has been shown in earlier sections that performance calculations for unbaffled doublet electrostatic loudspeakers may be very greatly simplified by making use of Walker's equation (3.77), or, for floor-to-ceiling strip designs, the equation given in Section 3.3.11.

Sometimes, however, as already mentioned, electrostatic loudspeakers are constructed in which the diaphragm is mounted in a plane baffle, or a hole in a wall, or in a cabinet of some kind, and in such cases the above simple design approach cannot be adopted. It then becomes desirable to have access to curves giving the reactive and resistive components of the air-load impedance, per unit area, for the shape of diaphragm to be used.

Such information is readily available when the diaphragm can be regarded as a circular piston, either mounted in an infinite plane baffle, or at the end of a tube, this latter case being relevant to that of a totally enclosed cabinet loudspeaker used well clear from room surfaces. But prior to the appearance of the first edition of this book in 1988, no satisfactory information had been published for the important case of a long-strip radiator in a plane baffle or in a cabinet, there being quite large discrepancies between the data in the references then available[2,28,29].

However, R. D. Ford had succeeded previously in deriving the resistive component of the acoustical loading on such a strip by rigorous mathematical analysis, the result being presented by him in Chapter 1 of the first and second editions of this book. The resistive curve in Fig. 3.33 was derived from this analysis.

No corresponding analysis yielding a reactive-loading curve had appeared at that time, and the reactive curve then given in Fig. 3.33 was largely based on an experimental investigation by the present author using a baffled steel strip vibrating at a very low frequency in water. More recently, however, Stanley Lipshitz has fully solved the problem analytically[36] producing the accurate results now given in full-line in Fig. 3.33. The broken-line curve is that based on the water experiment. (Both Lipshitz and Ford chose to plot their curves with $2\pi a/\lambda$ as abscissa, where the strip width is $2a$, so that their abscissa values need to be divided by π for presentation, as in Fig. 3.33, with w/λ as abscissa.)

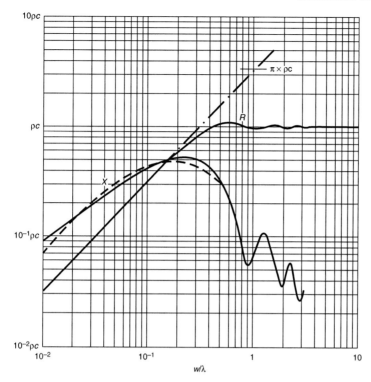

Figure 3.33. Mechanical resistance and reactance per unit area for air loading on one side of an infinitely long strip of width w in an infinite plane baffle. The full-line curves are based on an analysis by S. P. Lipshitz *et al.*, whose resistive curve agrees precisely with the independent analysis of R. D. Ford. The broken-line curve is an experimental one.

3.4 Practical designs

3.4.1 Quad Mark 1

The Quad Mark 1 electrostatic loudspeaker of 1957, which was in production for 29 years, underwent only very minor modifications during this period. The design provides an excellent illustration of the practical utility of many of the theoretical ideas already presented, and will therefore be discussed in some detail.

The circuit diagram of one of the later versions is given in Fig. 3.34 and some of the constructional details are shown in Fig. 3.35.

The bass-radiating units have Saran diaphragms (see Section 3.2.7), the low Young's modulus of this plastic permitting the diaphragm in each unit to be supported at two intermediate positions across its width, and yet still to have the required low resonance frequency. This construction greatly eases tolerances with regard to achieving a sufficiently uniform gap everywhere.

The fixed plates for these bass units are made of insulating material 2 mm thick, with the conducting coating on the outside surface, remote from the diaphragm. The gap between the inside surface of the plates and the diaphragm is also nominally 2 mm. With plates as thick as this, the bass units are largely self-protecting if excessively large signal inputs are applied, for if a spark occurs between the inner faces of the plates, the plate insulating material, over a localized area, temporarily develops

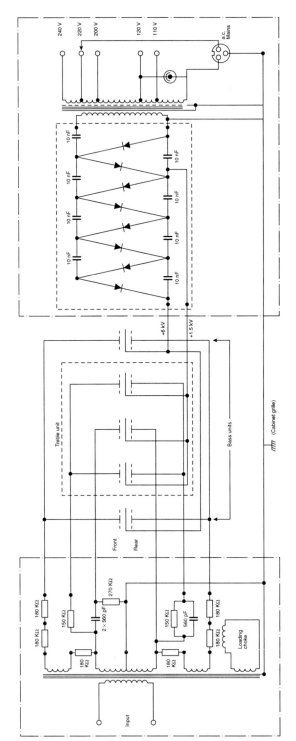

Figure 3.34. Complete circuit of Quad Mark 1 electrostatic loudspeaker.

a large voltage gradient across it, reducing the voltage across the air gap. Sparking through the holes, from one conductive coating to the other, is unlikely, but even if it did occur on a rare occasion, drastic damage would not be caused, because with the bass units fed via a very high value of series resistance, the energy available for the spark is only that stored in the inter-plate capacitance.

The treble unit has the conductive coatings on the inner surfaces of the plates, with only 0.5 mm gaps, and this reduced gap spacing, permissible at higher frequencies, minimizes the volt-amps required for a given output level, since the current demanded by equation (3.77) now produces a smaller voltage drop. The plates are thinner than in the bass units, to reduce the loss of high-frequency response, and the response irregularities, caused by the impedance of the air masses in the holes.

Whereas the bass units operate with a polarizing voltage of 6 kV, the treble unit requires only 1.5 kV, with correspondingly reduced peak signal voltages. These reduced voltages ease the problem of avoiding damage due to sparking, for the diaphragm plastic itself, Mylar in the treble units, is normally able to withstand the full peak signal voltage directly across it.

In versions of the loudspeaker prior to serial no. 16800, the capacitor and resistor associated with the signal feed to the central high-frequency strip were omitted. This was entirely satisfactory for use with the 15 W amplifier originally specified, but when more powerful amplifiers became readily available occasional breakdowns of the high-frequency strip diaphragm insulation occurred. The results tended to be catastrophic, for the full amplifier power could be poured into the breakdown spark, taking place from one conductive coating to the other. The capacitor and resistor allow virtually the full transformer output voltage to be applied to the high-frequency strip above about 1000 Hz, but reduce the voltage that can be applied at lower frequencies, where the highest signal levels normally occur. The presence of the series capacitor also limits the amount of energy that can be fed into a spark.

The above-mentioned capacitor and resistor have negligible effect on the overall frequency response of the loudspeaker and their presence will be ignored from now on.

The 'loading choke' shown in Fig. 3.34 will also be ignored. It consists merely of a few shorted turns round the transformer core and its purpose is to reduce the

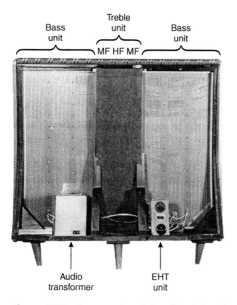

Figure 3.35. Constructional details of Quad Mark 1 electrostatic loudspeaker.

magnitude of the input-impedance peak that occurs somewhere in the 100 Hz region – the actual frequency is very dependent on the signal amplitude.

Transformers for use in electrostatic loudspeakers are critical components requiring considerable thought and care to be devoted to their design. The transformer in Fig. 3.34 has a step-up ratio to the whole secondary of approximately 1:290, the secondary shunt inductance at low frequencies and moderate signal levels being well over 5000 H. The peak instantaneous secondary voltage may exceed 8 W.

It is customary to represent the performance of an audio transformer by means of a fairly simple equivalent circuit, involving an ideal transformer in association with leakage-inductance, winding-capacitance and loss-resistance circuit elements. Information and formulae are given in many textbooks, but it is often overlooked that the performance of all transformers becomes much more complex than is represented by such equivalent circuits if the frequency is made high enough. To account fully for the observed behaviour, the equivalent circuit would have to be impracticably elaborate, incorporating a myriad of capacitances and leakage inductances between winding layers and even between individual turns.

In most audio transformers of low to moderate impedance, the performance remains simplex, i.e. in accordance with the normal type of equivalent circuit, up to well above the maximum working frequency. Indeed, by suitably choosing the values of terminating resistors and capacitors, the high-frequency response of such transformers may be made to follow, for example, a second- or third-order Butterworth characteristic. Nevertheless, if the measuring frequency is increased far enough above the filter cut-off frequency, the response curve is always found to develop a series of increasingly complex peaks and troughs.

The important point to be made here is that in an electrostatic loudspeaker transformer, where the total number of secondary turns is sometimes well over 10 000, complex behaviour, not fully predictable, is liable to occur within the audio spectrum.

One way to avoid such trouble is that adopted in the Quad ESL63, where two separate transformers, each of ratio 1:122.5, are used to drive the plates, rather than a single transformer with a centre-tapped secondary. Each transformer can then be of very simple design, involving a single-section primary winding next to the core, followed by a single-section secondary winding with plenty of air space all round it to minimize the shunt capacitance. This secondary winding, of course, requires only half the number of turns that would be needed if a single transformer were employed, and simplex behaviour is obtained up to about 50 kHz.

The transformer in Fig. 3.34, however, is considerably more elaborate than just mentioned, the winding geometry being shown diagrammatically in Fig. 3.36. Complex behaviour involving the high-impedance secondary sections does indeed occur at high audio frequencies, but is of no importance, because the full secondary voltage feeds only the bass units, and the relevant leakage inductance is intentionally made so large that the output voltage from this secondary is attenuated at high frequencies as detailed below. The audible effect of any such complex resonances is further reduced by the fact that the high-frequency response of the large bass units, at normal listening distances, is much reduced by the mechanism to which Fig. 3.32 relates.

The frequency response of the Fig. 3.34 transformer measured to the low-impedance secondary that feeds the treble unit is quite free from complex resonances over the full audio spectrum. This secondary is relatively tightly coupled to the primary, the leakage inductance referred to it being approximately 145 mH only.

In order to appreciate how the bass, medium-frequency and high-frequency diaphragms function in combination to produce a good approximation to a flat overall frequency response, it is convenient to derive the simplified, single-ended, equivalent circuit shown in Fig. 3.37, which is based on Fig. 3.34 and some measured transformer parameter values. As already mentioned, the CR elements associated with the high-frequency strip, and the 'loading choke', are ignored. Elements representing the shunt inductance and core losses of the transformer are also omitted, for they play no significant part in determining the frequency response.

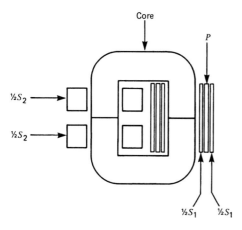

Figure 3.36. Transformer geometry for Quad Mark 1 electrostatic loudspeaker.

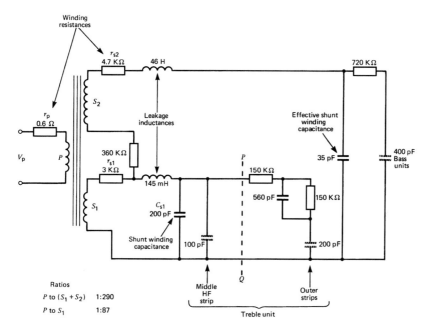

Figure 3.37. Simplified equivalent circuit for Quad Mark 1 electrostatic loudspeaker.

Though the Fig. 3.37 circuit is only an approximation to the full truth at high frequencies, it is an adequate representation for practical purposes. The total effective series resistance feeding the bass units is 1.14 MΩ, causing the plate current to fall off at low frequencies with a corner frequency of 350 Hz. The polarized resonance frequency is about 70 Hz, the Q-value being reduced to around 2.5 by suitable diaphragm damping.

The low-frequency response that would be obtained with constant current fed to the bass-unit plates, ignoring 'vestigial baffling' for the time being, is shown by curve 1 in Fig. 3.38(a) – see Section 3.3.9 for theory. The first-order fall-off in plate current at low frequencies, mentioned above, is represented by curve 2, and when this is

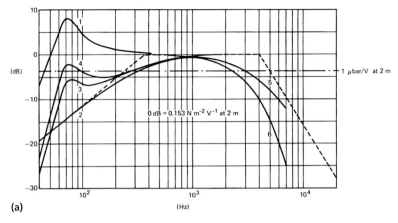

(a)

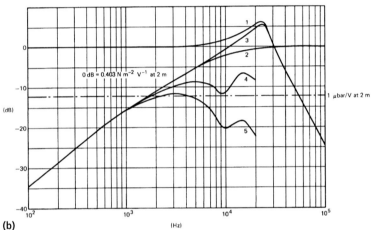

(b)

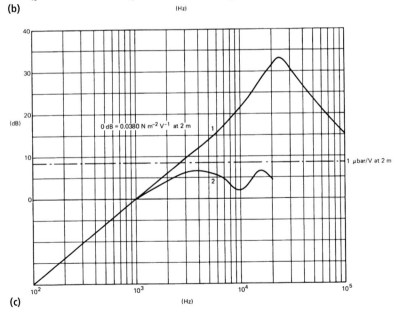

(c)

added to curve 1, curve 3 results. When allowance is made for 'vestigial baffling' (see Section 3.3.9), a low-frequency response more nearly that shown by curve 4 is obtained.

If the 46 H leakage inductance and 35 pF shunt capacitance were absent, the current would become frequency-independent at high frequencies. The sound pressure that would be produced at a large distance on axis, assuming flat diaphragms, is then readily calculable using Walker's equation (3.77). With $V_{pol}=6$ kV, $d = 2.3 \times 10^{-3}$ m (slightly greater than the physical gap to allow for the finite permittivity of the plate material), and $r = 2m$, direct substitution in equation (3.77) gives a sound pressure of 0.153 N/m² for a plate current of 254 µA, corresponding to 1 V across the primary winding.

At high frequencies, however, the pressure actually produced by the bass units is less than the above figures for two reasons:

(a) The leakage inductance and shunt winding capacitance attenuate the plate current.

(b) At a distance of 2 m, and with curved plates, the axial sound pressure is reduced below the figure given directly by Walker's equation, owing to the effects to which Fig. 3.32 relates.

Considering (a) first, the initial point to appreciate is that when the frequency is high enough for the leakage inductance and shunt capacitance to be significant, the reactance of the 400 pF plate capacitance is so much less than the series resistance that it has negligible influence on the frequency response of the current. The relevant circuit is therefore nearly enough that of Fig. 3.39. This has a second-order low-pass response with a resonance frequency of 4.0 kHz and a Q-value of 0.51. Note that the designer has the freedom to choose the Q-value over a limited range, by altering the ratio of the series and shunt resistor values, keeping their sum constant.

Allowing for effect (a) thus gives the high-frequency response shown by curve 5 of Fig. 3.38(a).

With regard to effect (b), the two bass units will be taken as approximately equivalent to a 0.6 × 0.6 m square diaphragm. All the units are curved, with a radius of curvature of approximately 3 in, though, in the case of the bass units, this is done for structural rather than acoustic reasons. From Fig. 3.32(a), with $h = w = 0.6$ m, $R = 3$ m and $r = 2$ m, the approximate resultant frequency response may be deduced, and when this is added to that for effect (a) above, the response shown by curve 6 of Fig. 3.38(a) is obtained. The curve has not been continued above 7 kHz, because, as already mentioned, the transformer exhibits complex behaviour at high frequencies, rendering such a continuation rather meaningless.

Attention will now be given to determining the acoustic output contribution from the medium-frequency strips in the treble unit, the initial step being to determine the magnitude and frequency response of the current fed to the approximately 200 pF total capacitance of these.

Referring to Fig. 3.37, there are effectively two circuits in cascade, one to the left of the broken line PQ and one to the right. The left-hand circuit has a resonance frequency of approximately 24 kHz, and in this region the impedance of the right-hand circuit is, nearly enough, simply that of the 150 kΩ resistor. Thus the left-hand circuit gives a second-order low-pass response corresponding to the presence of a series damping resistance of 7.5 kΩ and a shunt damping resistance of 150 kΩ, leading to a Q-value of approximately 2. This response is shown in curve 1 of Fig. 3.38(b). The current response of the network to the right of PQ is shown in curve 2. Curve 3, which is the sum of 1 and 2, thus shows the manner in which the current fed to

Figure 3.38 (opposite). Theoretical response curves, at 2 m on-axis, for Quad Mark 1 electrostatic loudspeaker. (a) Contribution from bass units. (b) Contribution from medium-frequency (MF) areas. (c) Contribution from central high-frequency (HF) strip.

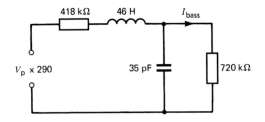

Figure 3.39. Equivalent circuit relating to high-frequency attenuation of current fed to bass units in Quad Mark 1 electrostatic loudspeaker.

the medium-frequency plates varies with frequency for a constant input voltage to the transformer primary.

Once again the response at 2 m on-axis is attenuated by the mechanism to which Fig. 3.32 relates though, because of the relatively narrow width of the treble unit, only the effects of the height and curvature need to be allowed for. Taking $h = 0.55$ m, $R = 3$ m and $r = 2$ m, the Fig. 3.32(a) curve is shifted to the right by two octaves, and when this effect is combined with curve 3 in Fig. 3.38(b), curve 4 is the result.

It is now necessary to take into account the additional high-frequency loss caused by the mass of the diaphragm plus the effective mass due to the inertia of the air in the plate holes. Equation (5.57) of reference 7 relates to this problem, but it is evident that a factor of 2.5 was inadvertently omitted in deriving it from equation (4) of reference 31. For present purposes the corrected equation is conveniently expressed in the form:

$$X_h \approx \frac{\omega \rho b^2}{\pi a^2}\left[t + 1.7a\left(1 - \frac{2.5a}{b}\right)\right] \tag{3.112}$$

where $X_h =$ mechanical reactance per unit plate area due to masses of air associated with the holes (MKS mechanical (Ω/m^2))

$\rho \ =$ density of air (kg/m³)
$a \ =$ radius of hole (m)
$b \ =$ distance between hole centres (m)
$t \ =$ thickness of plate (m)

The formula assumes that the holes are arranged as in Fig. 3.40(a). If, as is often the case, they are arranged as in (b), the effective value of b for substitution in equation (3.122) may be obtained from

$$b^2 = 100\pi a^2/x \tag{3.113}$$

where $x =$ percentage open area

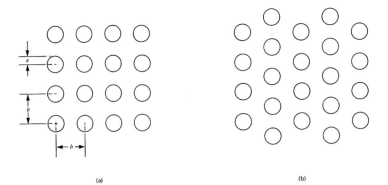

(a)

(b)

Figure 3.40. Alternative dispositions for holes in electrostatic loudspeaker plates.

Because there are two plates in a loudspeaker, twice the value of X_h given by equation (3.112) must be taken, and when the sum of this and the reactance of the diaphragm mass per unit area is equal to $2\rho c$ (840 MKS mechanical Ω/m^2), then there will be a theoretical response loss of 3 dB.

Equation (3.112), however, treats the air in each hole as a lumped cylindrical mass, the second term in the square brackets being an end correction. The compressibility of the air is ignored, and this leads to quite a significant error at the highest audio frequencies with the typical hole dimensions used in electrostatic loudspeakers. The behaviour is then more like that of a short length of transmission line, and the loss of response at, for example, 15 kHz is liable to be appreciably less than that predicted by the simple theory.

The above theory gives a loss of 3 dB at about 4 kHz for the Quad Mark 1 loudspeaker, and when this loss is included the predicted axial response for the medium-frequency radiating areas is as shown by curve 5 in Fig. 3.38(b).

The absolute sensitivity for the medium-frequency part of the system will now be deduced. 0 dB in Fig. 3.38(b) corresponds to a plate current, for 1 V r.m.s. input, of $87/150 = 0.58$ mA r.m.s. From Walker's equation (3.77), this would give an axial sound pressure at 2 m of 0.403 N/m^2, neglecting the effects represented by curves 4 and 5. In calculating this result, V_{pol} and d were taken as 1.5 kV and 0.5×10^{-3} m respectively.

Lastly, the acoustic output from the middle high-frequency strip will be determined. Because the low-impedance secondary voltage is applied directly to the high-frequency strip, the shape of the frequency response of the plate current will be as for curve 1 in Fig. 3.38(b) but tilted upwards with rising frequency by 20 dB/decade. This is shown as curve 1 in Fig. 3.38(c), with 0 dB arbitrarily chosen to be the level of the curve at 1000 Hz. The axial acoustic output response falls below this current response owing to the same mechanisms that give the reduced responses represented by curves 4 and 5 in Fig. 3.38(b), and taking these two effects into account then yields curve 2 in Fig. 3.38(c).

The reactance at 1000 Hz of the 100 pF high-frequency plate capacitance is 1.59 $M\Omega$, and the current to it for 1 V r.m.s. input is 54.7 μA r.m.s. Walker's equation (3.77) then gives an axial sound pressure at 2 m of 0.0380 N/m^2.

Having now derived the individual responses for the bass, middle and high-frequency radiators, the overall axial response of the loudspeaker must next be deduced.

The obvious technique is to draw the three individual response curves on a single sheet of paper with a common 0 dB reference level, which can conveniently be taken as 0.1 N/m^2 per volt (1 μbar/V) as indicated by the chain-dotted lines in Fig. 3.38(a), (b) and (c). However, in order to be able to derive the sum of these pressure responses with good accuracy, it is necessary to know their relative phasings.

The phasing problem may be considerably eased by appreciating the fact that because the medium and high-frequency strips have the same gaps and polarizing voltage, the same curvature, and negligible high-frequency loss related to width, their combined acoustic output on-axis is related in a simple and absolute manner to the sum of the currents fed to the plates. This total current may be calculated from the element values in the Fig. 3.37 circuit, or determined experimentally. Allowance must then be made for the Fig. 3.32 effect, and for the high-frequency loss caused by the mass of the diaphragm and the air in the holes, as already explained. This procedure yields the M17 + HF curve of Fig. 3.41.

Also shown in Fig. 3.41, with the same reference level, is the response of the bass units, taken from curves 4/6 of Fig. 3.38(a).

The MF + HF curve, ignoring the fluctuations in the 10–20 kHz region, approximates fairly closely to that of a tuned circuit with a Q-value of 0.5, centred on 3.3 kHz. This, at the crossover frequency of 1.4 kHz, gives a phase lead of 43°.

With regard to the phase lag of the bass-unit contribution at 1.4 kHz, this frequency is sufficiently far above the 70 Hz cut-off frequency to render phase effects related

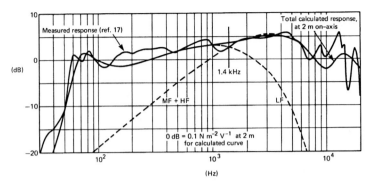

Figure 3.41. Calculated and measured responses for Quad Mark 1 electrostatic loud-speaker.

to the latter negligible. Thus what is required is the phase lag at 1.4 kHz associated with curve 6 of Fig. 3.38(a). This may be considered to have two components:

(a) A component corresponding to that of a second-order low-pass filter with a cut-off frequency of 4 kHz and a Q-value of, nearly enough, 0.5. This gives a phase lag of 38.5° at 1.4 kHz.
(b) A component due to the effect to which Fig. 3.32 relates.

To determine (b), the phase curve in Fig. 3.32 is utilized, the operation being carried out in two stages. First, for $h = 0.6$ m, $R = 3$ m and $r = 2$ m, the value of

$$\frac{rR}{h^2(R + r)}$$

is 3.33, so that the phase curve relating to vertical phasing effects is that of Fig. 3.32 shifted upwards in frequency by this factor. This gives a phase lag at the 1.4 kHz crossover frequency of 18.5°. Second, for horizontal phasing effects, the value of r/h^2 is 5.55, and the Fig. 3.32 phase curve, when shifted in frequency by this factor, gives a phase lag at 1.4 kHz of 11°. Thus the total phase lag at 1.4 kHz due to the Fig. 3.32 effect is 29.5°.

Taking into account, now, both phase-lag components given under (a) and (b) above, the relevant total 1.4 kHz phase lag is 38.5 + 29.5 = 68°.

The phase lead for the treble unit at 1.4 kHz, as already determined, is 43° so that the bass and treble outputs, at 2 m on-axis, have a relative phase angle of 68 + 43 = 111° at the crossover frequency. It is then easily deduced that the total sound pressure at this frequency exceeds that of either contribution alone by 1.1 dB.

Similar calculations may be carried out for frequencies on either side of the crossover frequency, and the resultant calculated response curve may then be drawn, as in Fig. 3.41. Also shown in Fig. 3.41 is a curve taken from reference 17 and based on measurements made by PTT, Bern. No absolute sensitivity information is given there.

In normal use, the effective response of this loudspeaker at frequencies well below 1 kHz is increased by the effect of floor reflections, counteracting the falling low-frequency response trend of the Fig. 3.41 curves to a useful extent.

A feature of the design not yet mentioned is that a sheet of fibrous material over 1 cm thick – removed when the Fig. 3.35 photograph was taken – is fixed behind the treble unit, spaced away from it by a similar distance. This serves two purposes:

(a) It provides damping, of about 600 MKS Ω/m^2, for the main resonance of the treble unit, which occurs at about 260 Hz. Without such damping, the Q-value would be very high, and undesirable buzzing sounds could then be produced by the diaphragm hitting the plates if the music contained a strong component at the resonance frequency.

(b) Because the velocity of sound in this damping material is several times less than in air, it introduces a time delay for the sound radiated rearwards from the back of the diaphragm, bringing it into an approximately antiphase relationship with the rearward-travelling sound coming from the front of the diaphragm. This gives the loudspeaker a cardioid type of polar characteristic at medium and high frequencies.

Looked at in full theoretical detail, the effects of introducing this fibrous material are rather complex, and design of this aspect has to be carried out to some extent on a cut-and-try basis. The effect of the material on the frontal response on axis is quite small. Whether suppression of the rearward radiation is considered to be subjectively advantageous depends a good deal on living-room acoustics and loudspeaker positioning as well as on individual preferences.

3.4.2 The Quad ESL63

The Quad ESL63 is a fully symmetrical doublet design, employing a delay-line technique to provide good medium- and high-frequency directional characteristics, both vertically and horizontally, combined with adequate volume capability,

As ideally conceived, the loudspeaker would have a single stretched diaphragm occupying the whole area but, for practical reasons, four equal-sized sub-units, each measuring about $60 \, cm \times 19 \, cm$, are employed. The gap between the diaphragm and the perforated fixed plates is the same everywhere and is nominally 2.5 mm. The plates are made of very thin material to minimize the effective mass of air in the holes, in the interests of good very-high-frequency response, and each plate is kept flat by being bonded with adhesive to the inner face of a multicellular plastic louvre moulding, of a rather similar nature to that often used in fluorescent lighting fittings.

The copper electrode coatings are on the outer surfaces of the insulating plates, and are the light-coloured areas in Fig. 3.42. Over the central part of the loudspeaker, these electrodes are in the form of a number of electrically separate annular areas or rings, the black approximately circular lines being the gaps between the rings.

The 12 small black circular 'blobs' in the photograph are where screws pass through pillars in the louvre mouldings to locate the louvre on one side of the diaphragm in an accurate and stable manner relative to the louvre on the other side. These pillars pass through clearance holes in the diaphragm, but because of the low-impedance acoustic loading on a doublet diaphragm at low frequencies, the volume-velocity of the air flow through the holes is quite small and has very little effect on the sound output. At high frequencies the holes merely reduce, by a negligible amount, the active diaphragm area.

The complete practical circuit is shown in Fig. 3.43, but the very considerably idealized and simplified circuit of Fig. 3.44 will be used for the initial explanation. The delay-line is here shown as an unbalanced line for convenience.

The relevant basic delay-line theory is represented by the following three equations:

$$Z_0 = \sqrt{L/C} \qquad (3.114)$$

$$f_c = \frac{1}{\pi\sqrt{LC}} \qquad (3.115)$$

$$T_s = \sqrt{LC} \qquad (3.116)$$

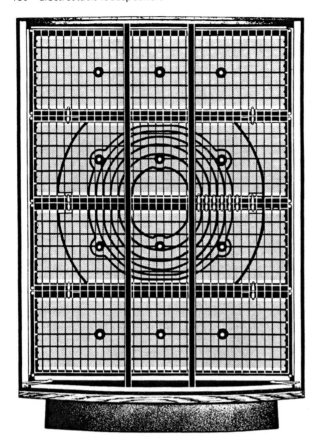

Figure 3.42. The inside of the Quad ESL63.

where Z_0 = characteristic impedence (Ω)
  L = inductance per section (H)
  C = capacitance per section (F)
  f_c = cut-off frequency (Hz)
  T_s = time delay per section (s)

A convenient additional formula, readily derived from the above, is

$$T_{tot} = Z_0 C_{tot}$$
(3.117)

where T_{tot} = total time delay (s)
  C_{tot} = total capacitance (F)

The L and C values shown in Fig. 3.44 lead approximately to $Z_0 = 900\,k\Omega$, $f_c = 16\,kHz$, $T_s = 20\,\mu s$, and $T_{tot} = 100\,\mu s$ between points A and B.

At very low frequencies, say 100 Hz and below, the time delays within the delay-line are insignificant, and all six ring capacitances are effectively simply in parallel. Moreover, because the reactance of the large outside area is much higher than R at these low frequencies, all the capacitances, including this outside one, have virtually the full secondary voltage across them so that the whole diaphragm is then almost uniformly voltage-driven.

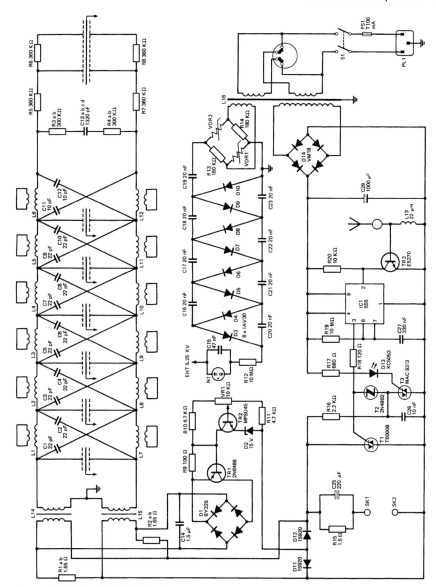

Figure 3.43. Quad ESL63 circuit.

The total electrode current at these low frequencies falls off, therefore, with reducing frequency, towards an asymptotic rate of 20 dB/decade, but because of diaphragm resonance and vestigial baffling the bass response is held tolerably level down to below 50 Hz. Hence the design principles at bass frequencies are as described in Section 3.3.9, being the same in essence as for the earlier Quad design. Ignoring vestigial baffling, the actual operating conditions for the ESL63 are fairly closely those represented by the lowest curves of Fig. 3.30.

In the idealized scheme of Fig. 3.44, it is evident that all the input current I_1 supplied by the transformer to the delay-line must find its way to electrode capacitances, and therefore, from Walker's equation (3.77), the long-distance axial response

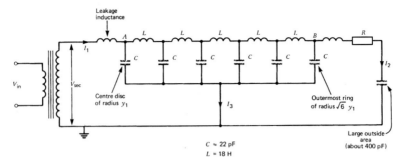

Figure 3.44. Simplified explanatory circuit for ESL63.

will be flat, over most of the audio spectrum, if this input current is frequency-independent, no matter how the current is divided between the various electrode areas.

To keep the delay-line input current constant for a constant input voltage to the transformer, and thus to give the loudspeaker a flat axial frequency response, the initial idealized notion is simply that the line must be terminated in a resistance equal to its characteristic impedance, this being achieved nearly enough, at medium and high frequencies, by making the resistor R in Fig. 3.44 equal to Z, The following conditions are then satisfied:

(a) $I_1 = I_2 = V_{sec}/R$ $\hspace{6cm}$ (3.118)

(b)   All the ring capacitances have a frequency-independent voltage V_{sec} across them.

(c) $I_3 + I_2 = I_1$ $\hspace{6.5cm}$ (3.119)

On progressing outwards from the centre disc to the outermost ring, the electrode voltages are increasingly delayed by successive time increments of T. The effect, so far as the sound radiated from one side of the diaphragm is concerned, is as if all the annular areas were connected in parallel and were thus fed with equal in-phase voltages, but were physically displaced in an axial direction as shown in Fig. 3.45.

At high frequencies such a physical arrangement would tend to radiate an expanding, approximately spherical, wavefront into the space to the right of it, as if coming from a point source at 0, but would radiate a contracting wavefront towards the left, tending to concentrate the radiation through a small region near 0.

With a planar system of rings plus delay-line, however, expanding waves are radiated from both sides symmetrically, each expanding approximately as if from a point source on the opposite side. The flat arrangement lends itself much more straightforwardly to practical construction, and has the further great advantage that the part of the total diaphragm area that is used in this manner for medium and high frequencies can have a low resonance frequency, allowing it also to contribute fully to the radiation of bass frequencies.

The minimum number of sections required in the delay-line follows logically once a decision has been taken on the outer ring diameter and the distance of the point 0 away from the diaphragm. These dimensions, which are related to the desired directional characteristics as discussed in more detail later, determine the distance cT_{tot} in Fig. 3.45, T_{tot} being the total time delay given by the delay-line and c being the velocity of sound. Thus the value of T_{tot} is established and, if N is the number of sections in the delay-line, then

$NT_s = T_{tot}$ $\hspace{6.5cm}$ (3.120)

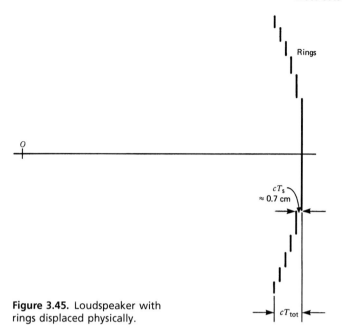

Figure 3.45. Loudspeaker with rings displaced physically.

and it immediately follows, from equations (3.115) and (3.116) that

$$N = \pi f_c T_{tot} \tag{3.121}$$

Hence, for $T_{tot} \approx 100\ \mu s$ and $f_c \approx 16\ kHz$ as in the ESL63, equation (3.121) dictates that five sections are necessary between points A and B.

Throughout that part of the working frequency range where the delay-line can be taken as being terminated nearly enough by the resistor R, equal to its characteristic impedance Z_0, the current supplied by the transformer is V_{sec}/Z_0, and the volt-amps supplied are V_{sec}^2/Z_0. From equation (3.117) it then follows that the volt-amps may alternatively be expressed in the form:

$$(VA)_{\text{with delay-line}} = V_{sec}^2 C_{tot}/T_{tot} \tag{3.122}$$

If, however, the rings are arranged physically as in Fig. 3.45 and are fed electrically in parallel, then the volt-amps required at high frequencies are given by

$$(VA)_{\text{no delay-line}} = V_{sec}^2 \times 2\pi f C_{tot} \tag{3.123}$$

Hence:

$$\frac{(VA)_{\text{no delay-line}}}{(VA)_{\text{with delay-line}}} = 2\pi f T_{tot} \tag{3.124}$$

The frequency $f = 1/T_{tot}$ corresponds to a 360° phase shift in the delay-line, the distance cT_{tot} in Fig. 3.45 then being one wavelength. Equation (3.124) shows that at this frequency the use of a delay-line reduces the volt-amps required by a factor of 2π. For $T_{tot} = 100\ \mu s$, as given below equation (3.117), the frequency for this condition is 10 kHz. Strictly speaking, equation (3.124) is correct for the part of the delay-line lying between A and B, only if the first capacitance C, at A, is not included in C_{tot}.

It is thus evident that a further theoretical advantage which can be claimed for employing delay-lines in electrostatic loudspeakers is that it enables the required electrode currents to be obtained with an amplifier of smaller volt-amps rating.

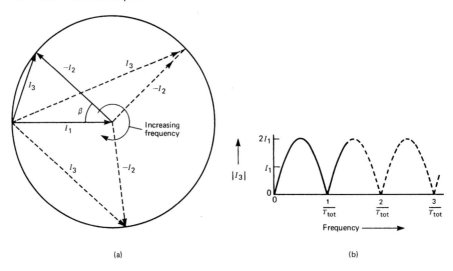

(a)   (b)

Figure 3.46. Diagrams relating to long-distance axial response of ring system with undamped delay-line terminated in $R = Z_0$.

In some designs[2] the saving can, indeed, be very substantial, but it is unfortunate that in the ESL63, owing to the wide plate spacing used over the whole diaphragm area, the impedance level is so high that the theoretical advantage is largely lost because of the adverse effect of stray capacitances. This aspect is considered in detail later.

On the assumption that the Fig. 3.44 delay-line is lossless, and correctly terminated in $R = Z_0$, all the capacitances will ideally have voltages of equal magnitudes across them throughout the working frequency range. However, this state of affairs would not lead to a very satisfactory overall performance, for reasons that will become apparent.

When $R = Z_0$, equation (3.118) applies, so that I_1 and I_2 are ideally equal in magnitude independently of frequency. Equation (3.119) applies too, and because the phase angle between I_1 and I_2 varies greatly with frequency, the magnitude of I_3 also varies greatly, as shown by the phasor diagram of Fig. 3.46(a). The angle β is the angle by which the output voltage of the delay-line lags on the input voltage, and is given by

$$\beta = 360° \times T_{tot} f \tag{3.125}$$

where T_{tot} = the total time delay through the delay-line.

From the geometry of Fig. 3.46(a), and equation (3.125), is obtained:

$$|I_3| = 2I_1 \sin (180° \times T_{tot} f) \tag{3.126}$$

and Fig. 3.46(b) is based on this equation. The graph is only meaningful, of course, up to the maximum frequency for which the delay-line has an approximately level frequency response, and for the values given below equation (3.117) the cut-off frequency is at $1.6/T_{tot}$, so that the broken-line part should be ignored.

Now the current I_3, by Walker's equation (3.77), is a measure of the total axial acoustic output pressure, due to the ring system alone, at an adequately large measuring distance r.

It is obviously very undesirable that the output from the ring system should be allowed to fluctuate in the violent manner of Fig. 3.46(b), though it must be emphasized that, even then, the total axial response, represented by I_1 (see Fig. 3.44), and including the contribution from the diaphragm area outside the ring system, does not

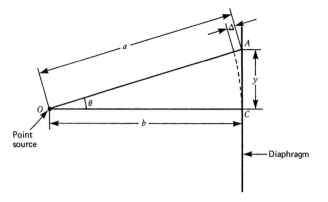

Figure 3.47.
Diagram relating to determination of signal voltage distribution for making an infinite massless diaphragm simulate a point source.

fluctuate. But at angles off-axis, for which the high-frequency contribution from these outer parts is greatly attenuated, the large fluctuations with frequency in the output from the ring system become dominant.

If the resistor R in Fig. 3.44 was taken to earth instead of to the outer electrode area, the full result of the Fig. 3.46(b) nulling would be observed on-axis.

To prevent the occurrence of undesirably large fluctuations in the off-axis response and directivity index with frequency, it is necessary to introduce losses, or damping, into the delay-line, causing the signal voltage on the ring electrodes to fall off with increasing radius, especially at the higher frequencies. Such damping is also necessary for achieving the wanted type of polar characteristic, as explained later.

To gain a greater understanding of the principles involved, the design problem will now be considered from a different starting point[11]. So many interrelated electrical and acoustic principles are involved in the functioning of a loudspeaker of this type that it is only by thinking around them at length from various points of view that a fully coherent and rational overall picture can be obtained, revealing the various design options and limitations in their most vivid and straightforward form. It is hoped that the following pages will be found to complement the treatment given in reference 11 in a helpful manner.

The idealized concept now introduced involves a massless insulating diaphragm of infinite extent, uniformly charged, and the question to be answered is just how this diaphragm should be driven, by means of signals fed to electrodes suitably disposed over its surface, to make it radiate sound with a specific polar characteristic as desired.

One very simple such polar characteristic to consider first is that of an ideal omnidirectional point source. Suppose such a point source is placed a distance b away from the infinite diaphragm, as represented in Fig. 3.47. Because the diaphragm is assumed to be massless and sufficiently slack, it is obvious that it cannot have any influence on the propagation of the spherical waves from the point source. Nevertheless, it is true to say that the waves in the space on the remote side of the diaphragm are caused purely by the vibration of the diaphragm. Therefore, if the diaphragm can be made to vibrate in this same manner by electrical means, with the point source removed, a listener situated on this remote side will receive exactly the same sound as with the point source present.

The obvious initial move would therefore appear to be to establish just what velocities need to be given to the diaphragm, as a function of position; though a more enlightened approach, as will shortly be shown, is to think in terms of pressures rather than velocities. First, however, the velocity approach will be pursued a little way, because it gives a helpful insight into some of the acoustic principles involved.

With reference to Fig. 3.47, let the strength of the omnidirectional point source be S (m³/s). Then, at the point A, the radial air-particle velocity U (m/s), acting along the line OA, is given by[7]

$$U = Sf/2ac \qquad\qquad (3.127)$$

where c = velocity of sound (m/s). This radial velocity at A may be regarded as having two components, one acting in the plane of the diaphragm, and the other, called U_{90}, acting normally to the surface of the diaphragm. The former component has no effect on the diaphragm motion; the diaphragm at A therefore vibrates with the velocity component U_{90}, which is $U \cos \theta$, and hence is given by

$$U_{90} = Sfb/2a^2c \qquad\qquad (3.128)$$

or, from the Pythagoras relationship in Fig. 3.47:

$$U_{90} = \frac{Sfb}{2(y^2 + b^2)c} \qquad\qquad (3.129)$$

(It is evident that this same value of U_{90} must apply at any point on the diaphragm circle with centre C and radius y.)

Thus, if every small area δA of the diaphragm can be made to vibrate with a velocity as given by equation (3.129), the diaphragm as a whole will radiate sound waves as if they were coming from the omnidirectional point source. The wavefront shape will be spherical, with uniform directivity. Similar waves, of opposite phase, will of course be radiated from the other side of the diaphragm, coming as if from a point source located at the position of the image of the original point source in the diaphragm plane.

If the effect of vibration, in accordance with equation (3.129), of just one small area δA by itself is considered, the rest of the diaphragm being taken as having zero velocity, then this small area will emit a hemispherical wave, of uniform directivity, known as a Huygens wavelet. Since the other parts of the diaphragm are being considered as stationary, they must be regarded, in this context only, as constituting an infinite plane baffle if the magnitude of the Huygens wavelet is to be calculated. The total output from the diaphragm, on this basis, may be regarded as being due to the sum of a large number of Huygens wavelets.

Some puzzlement may at first arise on contemplating the fact that the Huygens wavelets from remote parts of the diaphragm must arrive at the listening position time-delayed relative to those arriving from the nearer parts, perhaps suggesting that a short transient signal will be elongated and smudged over by the arrival of the later wavelets. The reason this does not occur is that the numerous wavelets, whose waveforms are always bipolar, neutralize each other except during the arrival of the initial transient. Some further clarification of this aspect is given later in connection with Fig. 3.49.

A design approach directly based on considering the diaphragm one small area at a time, and endeavouring to calculate the electrical drive conditions required for making this alone move in accordance with equation (3.129), would be analytically very tedious. This is because, if the electrical drive is applied only to the area in question, the acoustic output from it will cause other parts of the diaphragm to move, so that to achieve the stipulated condition of having these other parts stationary it would be necessary to apply drive voltages to all of them to prevent them moving.

If just one small area of the massless diaphragm is electrically driven, then it emits a wavelet of doublet type, not a uniform-directivity hemispherical wavelet. It is quite legitimate, however, to regard the total output from the diaphragm as consisting of a large number of such doublet wavelets – the resultant wave may be synthesized in terms of component wavelets of either type, provided their magnitudes and phasings are correct.

Though the above complexities are undoubtedly instructive and fascinating, they may fortunately be totally circumvented by approaching the design problem on a basis of pressures rather than velocities. When the omnidirectional point source is in action. it gives rise to a pressure P, at the point A in Fig. 3.47, given by

$$P = Sf\rho/2a \tag{3.130}$$

and from the geometry of Fig. 3.47 this may be expressed as

$$P = \frac{Sf\rho}{2[y^2 + b^2]^{1/2}} \tag{3.131}$$

This is therefore the sound pressure very close to the diaphragm surface on the side of it remote from the source. Hence, with the point source removed, the requirement is to drive the massless diaphragm in such a manner that it produces surface acoustic pressures everywhere in accordance with equation (3.131). If this is done, the diaphragm will radiate waves as if from the original point source.

When a voltage is applied to electrodes covering just a small area of massless and slack diaphragm, the whole of the force developed acts on the air load on that area, half the force being used in producing a pressure increase on one side of the diaphragm and half in producing a pressure decrease on the other side. Thus, from equation (3.15), the acoustic pressure P is given by

$$P = \frac{1}{2} \times \frac{\varepsilon_0 V_{pol} V_{sig}}{d^2} \tag{3.132}$$

so that

$$V_{sig} = \frac{2Pd^2}{\varepsilon_0 V_{pol}} \tag{1.133}$$

Substituting for P in this last equation from (3.131) gives

$$V_{sig} = \frac{Sf\rho d^2}{\varepsilon_0 V_{pol}[y^2 + b^2]^{1/2}} \quad \text{(omnidirectional point source)} \tag{3.134}$$

Equation (3.134) thus shows how the signal voltage must be made to vary with the radial distance y from the centre of an infinitely large massless and sufficiently slack diaphragm, to make it radiate waves as from an omnidirectional point source located at a distance b away from the centre on the other side of the diaphragm.

If, instead of an omnidirectional point source, a doublet, or figure-of-eight, point source is assumed, then equation (3.130) becomes

$$P = \frac{Sf\rho}{2a} \cos \theta \tag{3.135}$$

where $S =$ the strength (m³/s) of an omnidirectional source giving the same pressure at a given distance as the doublet source does on-axis. Both a and $\cos \theta$ may be expressed in terms of other quantities in Fig. 3.47, leading to

$$P = \frac{Sf\rho b}{2(y^2 + b^2)} \tag{3.136}$$

Substituting for P in equation (3.133) from (3.136) gives

$$V_{sig} = \frac{Sf\rho b d^2}{\varepsilon_0 V_{pol}(y^2 + b^2)} \quad (\cos \theta \text{ point source}) \tag{3.137}$$

If the point source has a $\cos^2 \theta$ directional characteristic, it follows similarly that

$$V_{sig} = \frac{Sf\rho b^2 d^2}{\varepsilon_0 V_{pol}[y^2 + b^2]^{3/2}} \quad (\cos^2 \theta \text{ point source}) \tag{3.138}$$

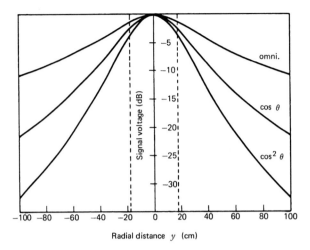

Radial distance y (cm)

Figure 3.48. Variation of signal voltage with radial distance for making an infinite massless diaphragm simulate point sources having omnidirectional, $\cos \theta$ and $\cos^2 \theta$ polar characteristics, each point source being 30 cm behind the diaphragm. The broken lines show the limits of the ring/delay-line system in the ESL63.

Figure 3.48 is based on equations (3.134), (3.137) and (3.138), and emphasizes the point that increased directivity requires a more rapid attenuation with radial distance of the signal voltages applied to the ring electrodes.

Equations (3.134), (3.137) and (3.138) give no information about the phasing of the voltages fed to the various elementary areas or rings. From Fig. 3.47, however, it is evident that to simulate a spherical wavefront emanating from 0, the voltage fed to a point such as A must be delayed relative to that fed to the centre C of the diaphragm by a time corresponding to the distance (shown i.e. by a time Δ/c). Δ is given by

$$\Delta = [y^2 + b^2]^{1/2} - b \tag{3.139}$$

and if $y^2 \ll b^2$, then this becomes

$$\Delta \approx y^2/2b \tag{3.140}$$

Now if the electrode system, as in the ESL63, has a central disc surrounded by a series of rings, all of the same area, and therefore capacitance, as the disc, then it is easily shown, from purely geometrical considerations, that the outside radii of the rings must be made proportional to the square root of the ring number, numbering the rings outwards from the centre and counting the centre disc as ring number 1. Thus

$$y \propto \sqrt{n} \tag{3.141}$$

where n = the ring number.

Now with these equal-capacitance rings forming the capacitances of an L-C delay-line, the time delay is proportional to n, and therefore, from (3.141), to y^2. This, fortunately, is approximately what is required, in accordance with equation (3.140), for giving the correct phasings for a spherical wavefront, provided, as already mentioned, that $y^2 \ll b^2$.

If the equal-capacitance ring system is continued outwards to radii such that this last condition is not satisfied, then the time delays in the feeds to the outer rings must be made smaller than would be obtained by using a straightforward extension of the delay-line scheme. In the ESL63 design, the maximum value of y^2/b^2 is just over 0.3, giving a departure from the ideal time-delay value at this radius of about 8%.

Before considering the consequences of continuing the concentric rings outwards to only a certain limited radius, some further attention will be given to equation (3.137), to see how it ties in with other notions.

First, to obtain a constant value of S, equation (3.137) shows that V_{sig} must be made proportional to f. But from equation (3.127) a constant value of S causes an air-particle velocity proportional to f. Therefore, to obtain an air-particle velocity independent of f, and hence a flat frequency response, V_{sig} must be made independent of f, requiring a frequency-independent voltage to be applied to all the rings. This is for frequency-independent $\cos\theta$ directivity. The same applies, via equations (3.134) and (3.138), for obtaining a flat response with frequency-independent omni-directional or $\cos^2\theta$ directivity respectively.

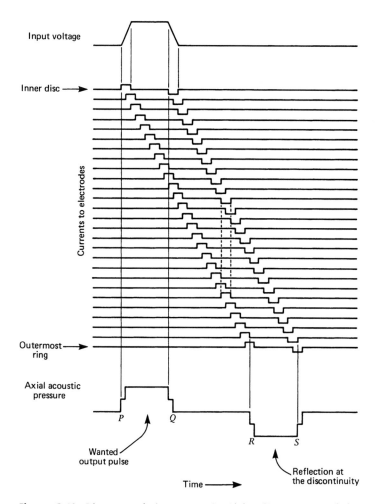

Figure 3.49. Diagram relating to a ring/delay line system of finite radius without damping, showing how a reflection at the discontinuity is generated. The time delay between the diaphragm and the measuring point is ignored.

Second, it is interesting to note that if b is made zero, equations (3.134), (3.137) and (3.138) predict $V_{sig} = \infty$ at $y = 0$, and $V_{sig} = 0$ for any finite value of y. Though perhaps initially puzzling, this is actually perfectly reasonable, being merely the limiting case of a situation where b is made small but finite. V_{sig} is then very large when $y = 0$, i.e. at the centre of the ring system, and falls off very rapidly as y increases. To limit the magnitude of V_{sig}, at the maximum desired sound-output level, to a value that can be achieved in practice without gap breakdown, it is necessary to adopt an adequately large value of b. This leads to the utilization of a large area of diaphragm for the radiation of sound of all frequencies.

It is not practicable, or economic, to continue the rings-plus-delay-line scheme out to, or beyond, the largest radii shown in Fig. 3.48, and it is important to understand, as clearly as possible, just what effects to expect if this system is terminated at a relatively small radius as shown by the broken lines.

Reference 11 states that 'there will be interference waves generated by the discontinuity'. How best can one envisage the mechanism whereby these waves are produced, and how can their adverse effects be minimized?

The author has found that regarding matters on a pulse basis brings considerable enlightenment. Imagine a ring system, of finite radius but with a large number of rings, fed from a lossless delay-line directly terminated by a simple shunt resistor of value equal to the characteristic impedance. Suppose that the input voltage waveform applied to the delay-line is a single pulse, with finite rise and fall times as shown in Fig. 3.49, and that the delay-line cut-off frequency is high enough for this pulse-voltage shape to be reproduced with little distortion, but with time delay, at each ring electrode. Then the currents fed to the various electrodes will be as shown. By Walker's equation (3.77), each such current will cause a proportional pressure contribution at a distant measuring point on-axis, and the total pressure waveform at this point will be the sum of all these contributions, reproducing approximately the signal-input voltage pulse, as shown.

Throughout the QR time interval, a positive-going contribution is always accompanied by a negative-going one, for example where indicated by the broken lines, so that the total pressure is zero. After time R, however, because of the absence of further rings, there are no longer any positive-going pulses to accompany the negative-going ones, so that cancellation does not occur and a long resultant negative-going pulse appears at the measuring point. This is the 'reflection at the discontinuity'. It should be noted that no corresponding reflection occurs in the delay line, because of the terminating resistor, here connected directly across its output.

Clearly, if the rings were to continue outwards indefinitely, the condition for the production of a reflection would not arise at all. More significantly, however, if losses are introduced into the delay-line, so that the pulse amplitudes shown in Fig. 3.49 become progressively less as the outermost ring is approached, the discontinuity-reflection pulse will also be of reduced amplitude.

As mentioned earlier, when the current in the delay-line-terminating resistor is fed to additional electrode areas outside the ring system, as in Fig. 3.44, the on-axis response fluctuations given by a ring system of limited radius are eliminated. Off-axis, however, they are not eliminated, the interference waves from the discontinuity causing large fluctuations in frequency response, and unsatisfactory polar characteristics, unless appropriate losses are introduced into the delay line.

In the ESL63 these losses are obtained by having a few short-circuited turns in each of the inductors, as shown in Fig. 3.43. The attenuation per section introduced in this way increases with rising frequency, leading, as would be expected from Fig. 3.48, to increased directivity at high frequencies. Above about 1 kHz, the directivity index increases, in a smooth and gentle manner, from its low-frequency doublet value of 4.8 dB to about 10 dB at 10 kHz. This was argued in Section 3.3.3 to be a desirable characteristic for stereo listening in normal surroundings.

In order to be able to make the directivity index start increasing at a frequency as low as 1 kHz, the ring system must have an adequately large radius in relation to

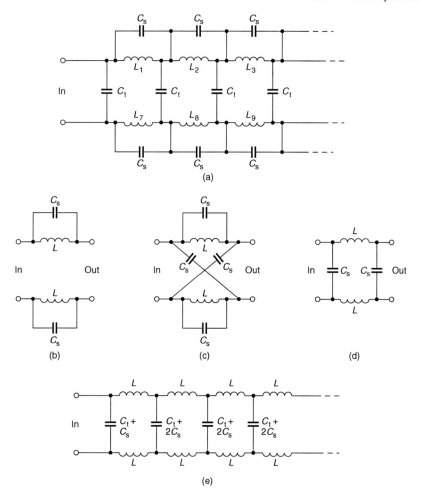

Figure 3.50. Circuits relating to the practical delay-line.

the wavelength at this frequency, and this is the main consideration determining the outer ring radius adopted.

The distance b in Fig. 3.47 has already been determined, as previously explained, by the need to be able to obtain a specified maximum sound-pressure level, at say 1 m on-axis, without gap breakdown. With both b and the outer ring radius thus dictated, the number of rings, and therefore delay-line sections, follows logically – see text preceding equation (3.121).

The centre disc of the ESL63 ring system is about 15 cm in diameter. This is much larger than could be adopted for the diaphragm of a normal tweeter, and it may at first seem surprising that the ring system can be made as 'coarse' as this and yet still give a satisfactory high-frequency polar characteristic. A clue, however, resides in the fact that the side lobes in the polar characteristics of a ring source[16] are at a much higher level, in relation to the axial response, than is the case for a disc source. See also reference 11, page 797.

Attention will now be directed to explaining why the 22 pF lattice capacitors are necessary in the practical delay-line of Fig. 3.43. The need for these capacitors, and the components R_3, C_{13} and R_4, arises from the fact that the shunt winding capacitances of the inductors, stray capacitances across the outside between rings of different sizes, and wiring capacitances, are unfortunately of the same order as the inter-electrode capacitances directly associated with the wanted transduction mechanism.

Hence, ignoring stray capacitances from one ring to rings other than its immediate neighbours, the effective electrical circuit, in the absence of the lattice capacitors, and ignoring the shorted turns, is as shown in Fig. 3.50(a). The capacitances labelled C are the wanted transduction ones.

The unwanted capacitances C_s across the inductors prevent the correct delay-line behaviour from being obtained. The principle that is exploited to overcome this is that if, to a simple balanced circuit as shown in (b), is added a pair of lattice capacitors C_s as in (c), then this latter circuit may easily be shown to have exactly the same behaviour, with regard to input and output impedances and signal transmission, under the relevant condition of balanced operation, as does the circuit shown at (d). Hence, by adding lattice capacitors C_s all along the line, the equivalent delay-line circuit becomes that shown at (e).

In the ESL63, $C_s \approx C_t$, so that the effective shunt capacitance value per section of balanced line, within the main body of the line, is about 66 pF. Now such a balanced line can be regarded as having a top half and a bottom half, each constituting an unbalanced line with shunt capacitance per section of value $2C_t + 4C_s$, i.e. about 132 pF. To obtain the same time delay per section as for the unbalanced line shown in Fig. 3.44, the inductor value has to be reduced from 18 H to $18 \times (22/132) = 3$ H, and this is the nominal value for the self-inductance of each inductor used in the ESL63. It gives the balanced line a characteristic impedance of approximately 300 kΩ.

Only about a third of the current supplied by the transformer to the delay-line is usefully employed in driving the transduction capacitances in the ring system. Therefore, if the whole of the delay-line output current were to be fed to the large surrounding electrode areas, these would give a disproportionately large acoustic output level. For this reason the components R_3, R_4, and C_{13} are added and take the major portion of the delay-line output current.

Whereas in Fig. 3.44 the electrode area outside the ring system is shown, for simplicity, as a single capacitance, it is divided into larger and smaller parts in the practical design of Fig. 3.43. This arrangement aids the smoothness of the transition from figure-of-eight directivity at low frequencies to increased directivity at high frequencies.

Arriving at a truly optimum design for a system as described above is, indeed, a long and difficult undertaking, for not only are there many evident parameter values to be chosen, but also the true circuit is a good deal more complex than has been described. The transformer leakage inductance and secondary shunt capacitance very significantly affect conditions at the input end of the delay-line, and there are various stray capacitances not taken into account in the above description but which have significant effects. Furthermore, the air-cored inductors are sufficiently closely spaced for mutual inductance between them to be very significant, increasing the effective inductance, time delay per section and characteristic impedance at the lower frequencies.

Fortunately, the introduction of losses makes the termination of the delay-line, for adequate freedom from reflections, less critical than it would otherwise be, but it is most important to make sure that at no point along the line, at any frequency, does the voltage exceed the input voltage, otherwise the likelihood of breakdown is increased.

If the value of the lattice capacitors is made slightly larger than the ideal value derived above, the effect is as if a negative capacitance were to be connected across each inductor, causing a fall in effective inductance with rising frequency. Herein lies a further possibility for adjustment of performance.

In the ESL63 the plate electrodes are deposited on quite thin insulating material, and the loudspeaker operates with very high signal and polarizing voltages over the whole area. This makes it imperative to incorporate auxiliary protective circuitry, to prevent the occurrence of serious damage if the loudspeaker is overdriven.

Two separate protective schemes are employed. The first, involving the diode bridge D_1, diodes D_{11} and D_{12}, and transistors TR_1 and TR_2, functions as a symmetrical shunt clipper to prevent the peak instantaneous voltage across each transformer primary from exceeding about 40 V, even though the input to the loudspeaker may do so. The novel circuit arrangement adopted results in much less power dissipation in the power transistor TR_1 than would be the case with a more conventional scheme.

The second protective circuit normally comes into operation only if a spark actually occurs somewhere, despite the existence of the above circuit. Such a spark generates RF interference, picked up by the aerial (a few cm of wire) on the base of TR_3. This triggers the type 555 timer circuit, which then generates a several-second positive pulse on its terminal 3, triggering the triac T_1 into conduction and thus short-circuiting the signal fed to the transformer primaries. The diac T_2 does not conduct under these conditions.

Obviously the circuit can operate as described only if there is a d.c. supply voltage for the 555, and this will not be present if the loudspeaker mains supply has inadvertently not been switched on. The user may then instinctively turn the volume control up high in the attempt to get some sound. The possibility of serious damage under these conditions is guarded against by the part of the circuit involving T_2 and T_3. In the absence of a d.c. supply, triac T_3 is not made conductive, and the application of a signal input then develops a signal voltage across C_{26} which, via diac T_2, puts T_1 into the conducting state and short-circuits the signal.

Capacitor C_{25} provides a small amount of frequency-response correction.

3.5 A general principle

As a result of much thinking about the fundamental theory underlying the ESL63, a valuable general principle has become evident, and may be stated in the form of a theorem as follows.

The long-distance polar characteristic of any plane, unbaffled, massless and slack diaphragm, driven by means of multiple electrodes with signal voltages varying with position in a specified manner, is cos θ times the long-distance polar characteristic, at the same frequency, of a plane diaphragm of the same shape and size in a rigid and infinite plane baffle, the vibrational velocity of this diaphragm varying over its surface in the same specific manner as the above voltages. Uniform diaphragm-to-electrode spacing is assumed, and θ is the angle of off-axis.

Thus, for example, published polar characteristics for a piston in an infinite baffle have only to be multiplied by cos θ to give the directivity information for an unbaffled massless flexible diaphragm of the same diameter and uniformly driven electrostatically over its whole surface.

3.6 Safety

Polarizing circuits are normally designed to have insufficient output-current capability to be lethal, but the secondary circuits of the signal transformers, delay-lines and plates are very hazardous indeed, and extreme care must therefore be taken when servicing or experimenting with electrostatic loudspeakers.

Acknowledgements

The author would like to thank Peter Walker for his willingness in supplying all the information that was requested on Quad loudspeakers, and for his patience in discussing various related topics.

References

1. HUNT, F V, *Electroacoustics*, John Wiley, Chichester (1954).
2. WALKER, P J, 'Wide range electrostatic loudspeakers', *Wireless World*, **61**, Nos 5, 6 and 8, May, June and August (1955). Errata: p.265, line 18 should be 'In practice the compliance will be considerably less than . . .'. p.266, line 23 should be 'velocity of motion will vary inversely with frequency'. p.265, lowest curve in Fig. 2 is incorrectly copied from p. 127 of reference 7 below.
3. WALKER, P J and WILLIAMSON, D T N, British Patent No. 815978. Appn date 20 July (1954).
4. WILLIAMSON, D T N, 'The electrostatic loudspeaker', *JIEE*, **3**, No. 32, 460–463, August (1957).
5. MCLACHLAN, N W, *Loudspeakers*, Clarendon Press, Oxford (1934).
6. WHEATCROFT, E L E, *Gaseous Electrical Conductors*, Clarendon Press, Oxford (1938).
7. BERANEK, L L, *Acoustics*, McGraw-Hill, New York (1954). Errata: The graphs on pp. 119, 120, 122 and 123 should appear on pp. 122, 123, 119 and 120 respectively; the captions stay where they are.
8. GILFORD, C L S, 'The acoustic design of talks studios and listening rooms', *Proc. IEE*, **106**, Part B, No. 27, 245–258, May (1959).
9. GILFORD, C L S, 'Acoustics for radio and television studios', *IEE Monograph*, Peter Peregrinus Ltd, Bristol (1972).
10. BERANEK, L L, *Acoustic Measurements*, John Wiley, Chichester (1949).
11. WALKER P J, 'New developments in electrostatic loudspeakers', *JAES*, **28**, No. 11, 795–799, November (1980). Erratum: Fig. 1 diagram should say 'force, proportional to square of voltage . . .'.
12. BRITTAIN, F H and LEAKEY, D M, 'Two-channel stereophonic sound systems', *Wireless World*, **62**, No. 5, 206–210, May (1956).
13. * SHORTER, D E L, 'A survey of performance criteria and design considerations for high-quality monitoring loudspeakers', *Proc. IEE*, **105**, Part B, No. 24, 607–623, November (1958).
14. SHORTER, D E L, 'High quality loudspeakers', *JIEE*, **9**, 253–256, June (1963).
15. * KEELE, D B, 'Low-frequency loudspeaker assessment by nearfield sound-pressure measurement', *JAES*, **22**, No. 3, 154–162, April (1974).
16. OLSON, H F, *Elements of Acoustical Engineering*, Van Nostrand, New York (1947).
17. GAYFORD, M L, *Acoustical Techniques and Transducers*, Macdonald and Evans, London (1961).
18. TAYLOR, P H, 'The line-source loudspeaker and its applications', *J. Brit. Kinematograph Soc.*, **44**, No. 3, 64–83, March (1964).
19. BAXANDALL, P J, 'A bi-directional line-source loudspeaker with Von Braunmuhl and Weber Baffle', preprint at AES 50th Convention, London (1975).
20. COLLOMS, M, 'An American electrostatic', *Hi-Fi News*, **29**, No. 2, 46–49, February (1984).
21. HARWOOD, H D and GILFORD, C L S, 'Monitoring-loudspeaker quality in television sound control rooms', *BBC Eng. Mono.* No. 78, Part 11, 10–16, September (1969).
22. BEVERAGE, H N, US Patent No. 3 668 335. Appn date 17 June (1969).
23. MCKENZIE, A A, 'American letter', *Hi-Fi News*, **22**, No. 8, 29–30, August (1977).
24. BAXANDALL, P J, 'High-fidelity amplifiers', chapter 14 in Radio, *TV and Audio Technical Reference Book*, Ed. by SW Amos, Butterworth, London (1977).
25. BAXANDALL, P J, 'Audio power amplifier design', *Wireless World*, **84**, No. 1505, 53–57, January (1978).
26. * ALLISON, R and VILLCHUR, E, 'On the magnitude and audibility of FM distortion in loudspeakers', *JAES*, **30**, No. 10, 694–700, October (1982).
27. MORSE, P M, *Vibration and Sound*, McGraw-Hill, New York (1948).
28. OLSON, H F, *Acoustical Engineering*, Van Nostrand, New York (1957).
29. HUETER, T F and BOLT, R H, *Sonics*, John Wiley, New York (1955).
30. * KATES, J M, 'Optimum loudspeaker directional patterns', *JAES*, **28**, No. 11, 787–794, November (1980).
31. INGARD, U, 'On the theory and design of acoustic resonators', *JASA*, **25**, No. 6, 1037–1061, November (1953).
32. LIPSHITZ, S P and VANDERKOOY, J, 'The acoustic radiation of line sources of finite length', paper presented to the AES 81st Convention, Los Angeles (1986).
33. STRENG, J H, 'Sound radiation from circular stretched membranes in free space', *JAES*, **37**, No. 3, 107–118, March (1989).

---

* References 13 and 15 are reprinted in *Loudspeakers; An Anthology*, Vol. 1, AES (1953–1977). References 26 and 30 are in Vol. 2 (1978–1983).

34. STRENG, J H, 'Charge movements on the stretched membrane in a circular electrostatic push-pull loudspeaker', *JAES*, **38**, No. 5, 331–339, May (1990).
35. WALKER, P J, 'The sound of the room', *Hi-Fi News*, **36**, No. 8, 33–37, August (1991).
36. LIPSHITZ, S P, SCOTT, T C and SALVY, B, 'On the acoustic impedance of strip radiators', AES Preprint No. 2977, 89th Convention, Los Angeles, September (1990).
37. SCHROEDER, M R, 'Progress in architectural acoustics and artificial reverberation: concert hall acoustics and number theory', *JAES*, **32**, No. 4, 194–203, April (1984).

4 The distributed mode loudspeaker (DML)

Graham Bank

4.1 Introduction

It is only rarely that a paradigm shift happens in any branch of science. It is more than a little surprising, therefore, that a new class of acoustic radiator should come into being. What is not too surprising is that it should come from a field that is outside the traditional loudspeaker and headphone arena.

When Ken Heron at DERA (Defence Evaluation and Research Agency) applied for his new invention[1], which describes a very high stiffness aluminium honeycomb panel designed to be used as a limited bandwidth public address loudspeaker, he showed that a 'resonant multi-modal radiator' with limited bandwidth could be used as a loudspeaker. At the time Dr Heron was working on composites for use in helicopters, and DERA were hoping to replace heavy internal structures within the helicopter with lightweight panels. In doing the trials it was found that the panels re-radiated significant output in the audio band – not desirable in an aircraft – but it sparked the idea that this could be put to use as a loudspeaker in its own right.

4.2 Historical background

The historical introduction to this new technology will be brief, since it only came into the public arena in the early 1990s.

4.2.1 DERA

DERA had specified that, from its early beginnings as a public address possibility, the panel would only be an efficient radiator above its coincidence frequency. This frequency was set in their patent by choosing the panel stiffness and areal density within specific limits. DERA, as far as they were concerned, were sure that such a panel would only be used in public address applications, because of its limited bandwidth.

Having started their work, it became clear to DERA that they needed an established loudspeaker company to progress the technology. In 1994 Verity Group plc negotiated a licence from DERA to develop the technology, and bring it to a commercial realization.

4.2.2 Verity

With the licence in place, Verity Laboratories, headed by Henry Azima, started work with a small team of engineers in January 1995, leading to the first patents being filed in September 1995. By early September 1996 the work undertaken during 1995 and 1996 was consolidated into a total of 21 European and International Patents.

The technology was made public at a number of venues, where technology demonstrations were given. First of these was for the World Press Launch on 27 September 1996, at UBS in London, England. It was closely followed by similar events in Tokyo, Japan (October 1996), Frankfurt, Germany (November 1996) and at the US Consumer Electronic Show, Las Vegas (January 1997).

Verity Group plc subsequently changed its name to NXT plc and currently has close to 1000 patents world-wide in various stages of application or grant. The technology is licensed, and NXT plc can be contacted directly[2], or via their website[3].

4.3 Traditional loudspeakers

In earlier chapters, the traditional loudspeaker has been fully described. Since its inception in the 1920s it has been the subject of a substantial amount of research work in all areas of its operation. Because it can be easily modelled, if some assumptions are made, it has found favour among all the disciplines involved with loudspeaker design.

Simple modelling, and the assumptions made, can hide the fundamental limitations of the traditional moving-coil (strictly speaking, *moving-mass*) loudspeakers. These limitations make the use of such loudspeakers quite deterministic, but they have difficulties which all arise because the radiated sound is correlated from a uniformly vibrating (i.e. pistonic) diaphragm. Early work by Frederick[4] and Thuras[5] in the 1930s shows the need to provide an enclosure for such loudspeakers, demonstrating a complete understanding of the equivalent circuit methods for modelling behaviour.

In all modelling approaches to traditional loudspeakers the characteristics of the piston lead to an on-axis output which is a maximum for all frequencies. However, the off-axis output depends upon the frequency being reproduced, as well as the piston radius (a), giving rise to an ever-narrowing major pressure lobe as the frequency increases. This forces the designer to use ever-smaller diameter drive units at increasingly higher frequencies. In most cases, two or three drive units are needed in order to approach a nearly constant radiated acoustical power. This inevitably leads to extra cost and complexity in crossover design.

A plot of the directivity of a rigid piston in an infinite baffle, as shown in Fig. 4.1, illustrates the problem for a 100 mm diameter drive unit. This characteristic for a rigid piston is made worse by the interaction of such sources when placed close to

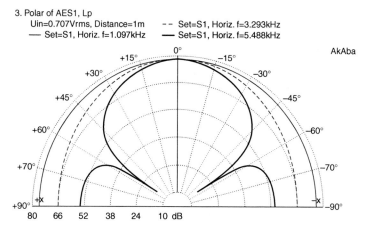

Figure 4.1. Polar response for a rigid piston in an infinite baffle, for 1.097 kHz (———), 3.293 kHz (– – –) and 5.488 kHz (▬▬).

reflecting boundaries. Traditional loudspeakers often suffer marked variation in response when placed in normal listening rooms, and this subject is fully discussed in Chapter 8.

4.4 Bending waves in beams and plates

Rayleigh[6] recounts the story of Sophie Germain, who first presented the solution to the problem of the modes of a plate in 1815, in response to a prize of 3000 francs offered by Napoleon. However, her choice of boundary conditions proved incorrect, and Kirchhoff gave a more accurate theory in 1850. In his treatise on sound, Rayleigh described the problem as 'one of great difficulty', since there is no analytical solution.

It is instructive to start the analysis of plates by considering a beam, and this can be extended to plates. Both Cremer, Heckl and Ungar[7], and Fahy[8], derive the equation for the motion of an elemental section of a beam, and Fahy's expression is shown in

$$B \frac{\partial^4 \eta}{\partial x^4} = -m \frac{\partial^2 \eta}{\partial t^2} \tag{4.1}$$

where B = bending stiffness
m = mass per unit length
η = transverse displacement

This expression can be extended to describe a thin plate[8], infinite in extent, lying in the xz plane, where we can now write, in rectangular Cartesian coordinates;

$$B \left(\frac{\partial^4 \eta}{\partial x^4} + 2 \frac{\partial^4 \eta}{\partial x^2 \partial z^2} + \frac{\partial^4 \eta}{\partial z^4} \right) = - \mu \frac{\partial^2 \eta}{\partial t^2} \tag{4.2}$$

where μ = mass per unit area. Since the derivative of the spatial coordinate in equation (4.2) is a different order from that with respect of the time, the phase velocity of the wave will be frequency dependent (i.e. dispersive).

A solution that satisfies equation (4.2) is given by

$$v(\omega) = \sqrt[4]{B/\mu} \, \sqrt{\omega} \tag{4.3}$$

where ω = angular frequency ($\omega = 2\pi f$)
f = bending wave frequency
$v(\omega)$ = bending wave velocity

This important result confirms that the bending wave is dispersive, with a phase velocity given in equation (4.3).

Although there is no analytical solution to equation (4.2), it is possible to use Rayleigh–Ritz variational methods, as well as Finite Element Analysis (FEA), and some semi-analytical solutions. By using one of the various methods the modal frequencies of a plate can be calculated. In order to design a DML loudspeaker it is necessary to optimize these modal frequencies to produce a modal density that is sufficient to give the illusion of a continuous spectrum.

For a typical panel whose lowest mode is set at f_0, then useful output is available from $2.5f_0$, as shown in the typical power output measurement of Fig. 4.2. The panel response is a direct result of the interleaving of the modal frequencies, solved from the partial differential of equation (4.2). The spectrum of a typical panel is shown in Fig. 4.3, where each vertical line represents a panel mode. Modal density can be seen to increase rapidly above 150 Hz, and this effect can be clearly seen in corresponding features in the power response.

At first sight the modal nature of the radiator usually raises eyebrows, since traditional loudspeaker designers have tended to avoid modal behaviour, especially

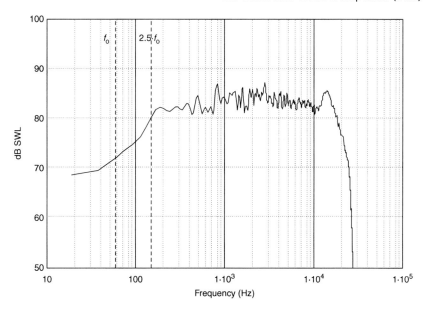

Figure 4.2. Power response of a typical DML loudspeaker, showing f_0 and $2.5f_0$.

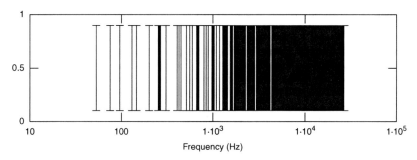

Figure 4.3. Calculated modes for the DML panel above.

low-order modes, since they can produce large fluctuations in measured output. What makes the DML a unique radiator is the density and spacing of these modes, producing the same effect as a continuous spectrum.

4.5 Optimizing modal density

Clearly, the selection of the right modal density is an important part of the design process, and correct selection of panel size and parameters will produce a known modal spectrum, suitable for a DML.

4.6 Early work

When Verity took the exclusive licence from DERA, little was known about how to make a broadband radiator using prior art. It was not until Henry Azima, Neil Harris and Martin Colloms[9] (the co-inventors of DML) started working on both the theoretical and practical aspects of the DML as a loudspeaker that their invention began to take shape.

By using beam functions as a starting point, Harris set about analysing plate modes with a number of analytical and numerical tools, to be able to correlate the smoothness of the acoustic output, both objectively and subjectively, with the evenness of the modal density. Although this produced a heuristic approach at first, it quickly became clear that the smoothness of the response was related to the smoothness of the driving point impedance.

A further solution from equation (4.2) is the panel's mechanical impedance at any arbitrary point. The natural modes of a DML panel need to be driven in such a way as to encompass as many of the modes as possible.

4.7 Current methodologies

By having the wave speed in the panel frequency-dependent, it is possible to design the panel such that the force at a particular drive point is de-correlated from the returning wave, which has been reflected from one or more of the boundaries. This de-correlation makes the panel *appear* infinite in size in respect of the drive point.

For a thin, isotropic panel of infinite size, Cremer et al.[7] give the analytical solution for the drive point impedance, as shown in equation (4.4):

$$Z_m = 8 \sqrt{B\mu} \tag{4.4}$$

To all intents, albeit driving a modal panel, the drive point looks like a mechanical resistor with a mean value given by equation (4.4).

This is illustrated in Fig. 4.4, where the actual measured mechanical impedance of a DML panel is compared to the theoretical value given in equation (4.4). Although the measured mechanical impedance shows the panel modality, the value of the impedance soon settles to that of the infinite plate. This confirms that the value for an infinite plate can be conveniently used as the load for the driving force for a DML panel.

The impedance-measuring rig is only accurate up to about 5 kHz, due to the finite size and effective mass at the measuring point. The measured impedance compares

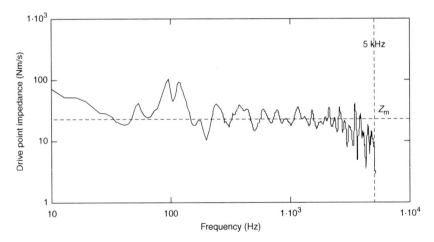

Figure 4.4. Comparison between measured mechanical impedance and infinite plate impedance, at the drive point, up to 5 kHz.

well with the analytical solution, since it is taken over quite a small area. When a larger area is used to drive the panel, then the diameter of the drive patch will affect this impedance at higher frequencies.

4.8 Panel mechanical measurements

Both static and dynamic properties of the panel will dominate the panel characteristics, and so some method is needed in order to measure these mechanical aspects of the panel.

4.8.1 Static measurements

Static measurements can be used to evaluate the values of bending stiffness for a panel. In order to calculate the bending stiffness of a panel, the flexural or tensile modulus of the panel material needs to be measured. The flexural modulus of a material is more suitable for the flat panels used for DMLs, because these operate in bending. However, the tensile modulus data may also be used and is more widely available from materials suppliers. If available, the data provided by a panel material supplier can be used directly in equation (4.6) to calculate the panel bending stiffness (if in monolithic form).

For bending stiffness measurements, it is recommended that a test standard be used as a reference in order to ensure accurate, repeatable measurements. There are many test standards available covering the measurement of flexural and tensile properties of materials. To calculate the bending stiffness, the sample dimensions, span, deflection and loading need to be measured. This technique can be used for both monolithic and sandwich construction panels. A sample is simply subjected to a force and the resulting displacement measured at the mid-point of the sample. Figure 4.5 shows the typical set-up.

A simple graph of the force versus deflection (where force = load (kg) × 9.807) will allow the user to identify the linear region of the curve. The gradient of this elastic region can then be used to calculate the flexural modulus of the panel material. A typical curve is shown in Fig. 4.6. It should be noted that the initial portion of the curve might be non-linear, caused by the test piece bedding-in and since it does not represent the performance of the material, it should be ignored.

The flexural modulus of the panel material is determined from the gradient (m) of the elastic region of the graph, using equation (4.5):

$$E_p = \frac{L^3 m}{4wt^3} \tag{4.5}$$

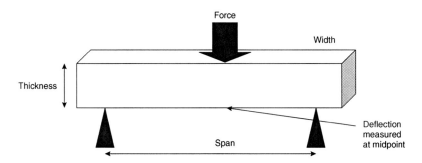

Figure 4.5. Three-point bending test set-up.

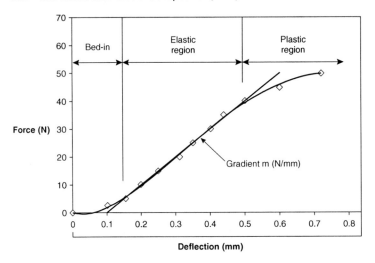

Figure 4.6. Typical force versus deflection for a beam sample.

where E_p = flexural modulus of panel (Pa or N/m²)
L = sample span (m)
m = gradient of linear region of force versus displacement graph
w = sample width (m)
t = sample thickness (m)

This value of flexural modulus can be used in equation (4.6) to calculate the bending stiffness of the panel using

$$B = \frac{E_p t^3}{12(1 - \nu^2)} \qquad\qquad (4.6)$$

where B = bending stiffness (Nm)
ν = Poisson's ratio

The sample dimensions may be restricted by the availability of the material but the relevant test standards make certain restrictions on the sample dimensions. For many of the materials used for flat panel speakers, the most relevant standard is ASTM D790[10] which covers the flexural properties of plastics. The following list covers the restrictions detailed in this standard for composite laminated materials:

- a ratio of span to sample thickness of between 16 and 40
- the sample also needs to overhang the supports by 10% of the span, e.g. if the span is 100 mm, the sample must overhang by 10 mm at either end.
- sample width should not be greater than one quarter of the span for specimens that are greater than 3.2 mm thick. For thinner samples the specimen should be a standard size of 50 mm long, 12.7 mm width and tested on a 25 mm span.
- for each material type, five samples should be measured in order to ensure a statistically accurate result.

Clearly, the bending stiffness of a sandwich panel consists of the three contributions from the skin, core and glue line. To determine the bending stiffness of a sandwich panel, the individual parts of the panel can be measured separately and these parts summed to obtain a value for the complete panel.

4.8.2 Dynamic measurements

It is possible to use a direct measurement of the panel mechanical impedance with an impedance head. This is a standard technique, and commercial impedance heads are available, for example from Bruel & Kjaer[11], who describe such an instrument. Details of an instrument designed and built at the BBC are given by Mathers et al.[12], for those who want to build their own.

Two piezoelectric transducers are used together. A force transducer measures the force applied to the panel, and an accelerometer measures the movement of the panel. The transfer function of force/velocity is directly related to the mechanical impedance. The outputs of the two piezoelectric transducers are charge. These are therefore conditioned by a high-impedance charge amplifier to convert to a voltage, which is fed into an FFT analyser. This can be stand alone, or a PC integrated system. This analyser is two-channel and measures the transfer function, $T(\omega)$:

$$T(\omega) = \frac{\text{Force } (\omega)}{\text{Acceleration } (\omega)}$$

However, the force measured by the force transducer is not the true force, but has added to it the inertial forces of the sensing tip. This force therefore needs to be subtracted when calculating the mechanical impedance:

True force (ω) = force (ω) − $m \times$ acceleration(ω)

where m is the mass of the sensing tip.

The velocity is found from the acceleration by scaling by angular frequency:

$$\text{Velocity } (\omega) = \frac{\text{acceleration } (\omega)}{i\omega}$$

The mechanical impedance Z_m is therefore given by:

$$Z_m(\omega) = i\omega(T(\omega) - m)$$

$$Z_m(\omega) = \frac{i\omega[\text{force } (\omega) - m \times \text{acceleration } (\omega)]}{\text{acceleration } (\omega)} \qquad (4.7)$$

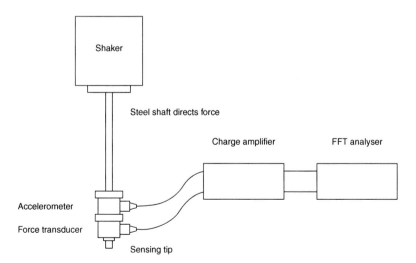

Figure 4.7. Schematic for impedance head set-up.

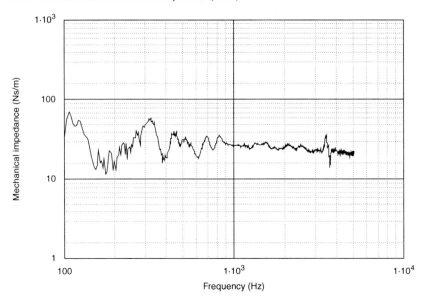

Figure 4.8. Modulus of mechanical impedance for a typical panel.

The mechanical impedance can therefore be obtained from a measurement of the transfer function and a knowledge of the tip mass, and evaluating equation (4.7) (see Figure 4.7).

The tip mass determines the upper frequency limit of the measurement. When the inertial forces of the tip are comparable to the forces of interest, then it becomes progressively more difficult to extract the true force from the combined force. An estimate of the upper frequency limit is therefore given by setting the inertial forces equal to twice the panel forces:

$$\text{Force}_{\text{inertial}} = 2 \times \text{Force}_{\text{panel}}$$

$$\Rightarrow m \times \text{acceleration} = 2 \times Z_{\text{m}} \times \text{velocity}$$

$$\Rightarrow m\omega = 2 \times Z_{\text{m}}$$

$$\Rightarrow \text{frequency}_{\text{upper}} = 2\,\frac{Z_{\text{m}}}{m}$$

Note that the mechanical impedance can become very small at the frequency of a particular panel mode, for a low-loss panel. This can lead to the measurement being sensitive to the peaks in the Z_{m} between panel modes, but cutting off the minima in Z_{m} in between.

Figure 4.8 is a typical measurement of the modulus of mechanical impedance of a panel driven at one of its optimum positions, showing the limit of measurement at about 5 kHz, given by the particular tip mass.

4.9 Drive points

For an infinite panel, the location of the drive point is meaningless, but not so for a finite panel. To give optimal results, the drive should be at a point that couples to all the panel modes, and while ensuring de-correlation between panel velocity and drive point.

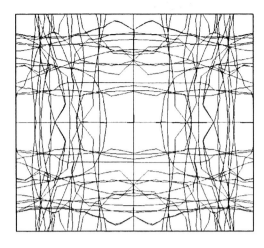

Figure 4.9. Nodal map for the first 20 modes of a DML panel.

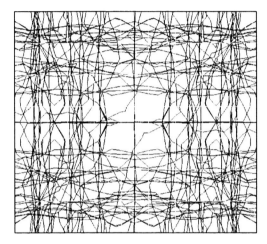

Figure 4.10. Nodal map for the first 30 modes of a DML panel.

There are a number of locations on the panel that achieve this objective, and these can be determined by a number of methods. FEA is one such approach, and the eigenvalues for a free panel can be determined, along with the related mode shapes. These mode shapes are converted in contours of nodal lines, where the displacements are a minimum. Regions of no nodal lines represent potential positions to drive the panel. It is usually only necessary to look at the first 20 or so sets of modes in the nodal map in order to find suitable candidates as drive points. These drive points are already known for complete sets of panel sizes and aspect ratios, but for optimal performance the ratios of panel length to width, as well as the most appropriate target values for B and μ are already known, as an integral part of the invention.

By way of example, an FEA model was constructed for a typical DML panel, and the model solved. By extracting the panel displacements associated with each mode, it is possible to generate composite maps for 20 modes (Fig. 4.9) and 30 modes (Fig. 4.10). Care should be taken with higher mode orders because of the accuracy of the FEA solvers at higher eigenfrequencies.

4.10 Mechanical model

Since the radiation impedance is small compared to the mechanical impedance of a typical DML, its effect on the acoustical performance is also small. The mechanical losses in a DML are typically very low, the primary damping action is due to acoustic radiation. Indeed, the power ratio, given in equation (4.8), can be as high as 90%, depending on the material chosen.

$$\gamma = \frac{2\eta_{\text{rad}}}{\eta_{\text{mech}} + 2\eta_{\text{rad}}} \times 100\% \tag{4.8}$$

where η_{mech} =   mechanical loss factor
        η_{rad} =   radiation loss factor each side of panel
        γ =   power ratio (radiation loss/total loss)

A good approximation, therefore, could be to model the mechanical behaviour of the DML, and assume that all the power is radiated. The approximation holds for frequencies at which the loudspeaker is truly modal, and is large enough to be self-baffling. At lower frequencies, a simple diffraction model will provide a fairly accurate extension to the high-frequency case.

The radiation from a DML can be considered to be that radiated from a randomly vibrating surface, and the existing motion of the panel will be uncorrelated to any new input being applied. Therefore, it looks like an infinite plate at the drive point. The radiation intensity from such an area is shown by Morse and Ingard[13] to depend on the square of the mean velocity, and hence the pressure at the panel surface is proportional to the panel velocity. In order to achieve a constant velocity with a constant force, the mechanical impedance must be resistive, and an infinite panel operating in bending waves meets this criterion[14].

These assumptions have been shown to give useful results, and measurements confirm that, to calculate the radiated acoustic power, we need only calculate the mechanical power delivered to the panel[15]. Given that the DML is a resistance-controlled device, and that we do not need to consider the acoustic radiation in detail, we can develop an equivalent circuit from Fig. 4.11. This represents a simplified version of the 'inertial magnet driver' application, as described in the NXT White Paper, published in 1996[16].

The coupled equations of motion are given in the following equations:

$$M_m \frac{d^2 x_m}{dt^2} + r\left(\frac{dx_m}{dt} - \frac{dx_p}{dt}\right) + k(x_m - x_p) - F = 0 \tag{4.9}$$

$$Z_p \frac{dx_m}{dt} + r\left(\frac{dx_p}{dt} - \frac{dx_m}{dt}\right) + k(x_p - x_m) + F = 0 \tag{4.10}$$

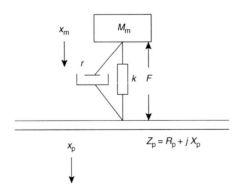

Figure 4.11. Mechanical elements and forces for a panel driven by a damped mass-spring oscillator.

If the force is assumed to be sinusoidal with angular frequency, and using the same symbols to refer to the peak values of variables, then

$$F(t) \equiv F\mathrm{e}^{j\omega t} \text{ (and similarly for } x_{\mathrm{m}} \text{ and } x_{\mathrm{p}})$$

$$-\omega^2 M_{\mathrm{m}} x_{\mathrm{m}} + j\omega r(x_{\mathrm{m}} - x_{\mathrm{p}}) + k(x_{\mathrm{m}} - x_{\mathrm{p}}) + F = 0 \tag{4.11}$$

$$j\omega Z_{\mathrm{m}} x_{\mathrm{p}} - j\omega r(x_{\mathrm{m}} - x_{\mathrm{p}}) - k(x_{\mathrm{m}} - x_{\mathrm{p}}) - F = 0 \tag{4.12}$$

or in matrix form, separating the stiffness, mass and resistance matrices

$$(\mathbf{K} - \omega^2 \mathbf{M} + j\omega \mathbf{R})\mathbf{x} - \mathbf{F} = 0 \quad \text{or} \quad \mathbf{x} = (\mathbf{K} - \omega^2 \mathbf{M} + j\omega \mathbf{R})^{-1}\mathbf{F} \tag{4.13}$$

where:

$$\mathbf{M} = \begin{pmatrix} M_{\mathrm{m}} & 0 \\ 0 & 0 \end{pmatrix} \quad \mathbf{K} = \begin{pmatrix} k & -k \\ k & k \end{pmatrix} \quad \mathbf{R} = \begin{pmatrix} r & -r \\ -r & r \end{pmatrix} + \begin{pmatrix} 0 & 0 \\ 0 & Z_{\mathrm{p}} \end{pmatrix} \tag{4.14}$$

$$\mathbf{x} = \begin{pmatrix} x_{\mathrm{m}} \\ x_{\mathrm{p}} \end{pmatrix} \quad \mathbf{F} = F\begin{pmatrix} -1 \\ 1 \end{pmatrix} \tag{4.15}$$

So the specific velocity, or mobility, Y_p in the panel is given by

$$Y_{\mathrm{p}} = \frac{u_{\mathrm{p}}}{F} = \frac{j\omega x_{\mathrm{p}}}{F} \tag{4.16}$$

$$Y_{\mathrm{p}} = j\omega \begin{pmatrix} 0 & 1 \end{pmatrix} \begin{pmatrix} k-\omega^2 M_{\mathrm{m}}+j\omega r & -k-j\omega r \\ -k-j\omega r & k+j\omega r+j\omega Z_{\mathrm{p}} \end{pmatrix}^{-1} \begin{pmatrix} -1 \\ 1 \end{pmatrix} \tag{4.17}$$

$$Y_{\mathrm{p}} = \frac{\omega^2 M_{\mathrm{m}}}{(\omega^2 M_{\mathrm{m}}(Z_{\mathrm{p}} + r) - Z_{\mathrm{p}}k) - j\omega(kM_{\mathrm{m}} + rZ_{\mathrm{p}})} \tag{4.18}$$

By inspection, noting that the velocity in the spring and damper is the difference between the velocities in the mass and panel, the equivalent circuit using the impedance analogue can be drawn as in Fig. 4.12.

It is a relatively straightforward task to verify that the ratio of panel velocity, u_{p}, to force, F, matches that given by the reciprocal of equation (4.18), i.e.

$$Z_{\mathrm{m_{eff}}} = Z_{\mathrm{p}}\left(1 - \frac{k}{\omega^2 M_{\mathrm{m}}}\right) + r - \frac{j}{\omega}\left(k + \frac{rZ_{\mathrm{p}}}{M_{\mathrm{m}}}\right) \tag{4.19}$$

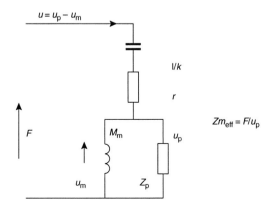

Figure 4.12. Impedance analogue model of DML panel.

4.11 Implementation for a practical moving-coil exciter

Figures 4.11 and 4.12 represent a panel driven by an idealized point source and, if we consider the motor system to be a moving-coil device, then M_m represents the mass of the magnet, cup and pole piece. The spring/damper represent a means of attachment of the motor to the panel. To account for the effect of the coil, we must add in series with Z_p a mechanical mass, M_c, equal to the mass of the voice coil. Additionally, the impedance Z_p is only real for a point source, and will generally be a complex quantity for a finite diameter voice coil. Remember that the pseudo-random fluctuations in X_p due to modal behaviour are not considered in this simple model.

The high-frequency component of X_p is systematic, however, and therefore of importance. Figure 4.13 shows the variation of both real and imaginary parts of typical panel impedance, due to the drive point from a finite coil, of radius r. This is termed the aperture effect, and Fig. 4.14 shows the mechanical model of such a system. The corresponding equivalent circuit is shown in Fig. 4.15.

The effective mechanical impedance relating u_p to F for Fig. 4.15 is

$$Z_{m_{\text{eff}}} = Z'_p \left(1 - \frac{k}{\omega^2 M_m}\right) + r - \frac{j}{\omega}\left(k + \frac{rZ'_p}{M_m}\right) \tag{4.20}$$

where: $Z'_p = R_p + jX_p + j\omega M_c$

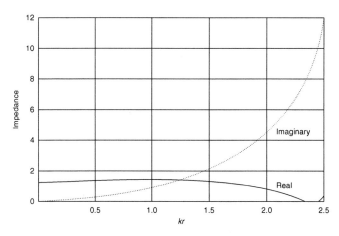

Figure 4.13. Typical panel impedance presented to the driving point; k is the wavenumber, r the coil radius.

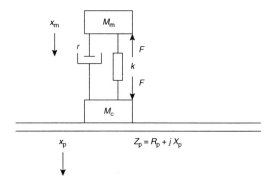

Figure 4.14. Mechanical elements and forces for a panel driven by a practical moving-coil exciter.

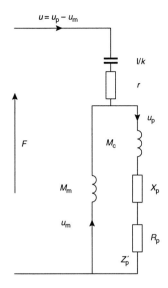

Figure 4.15. Impedance analogue model of DML panel with moving-coil motor system.

At high frequencies, equation (4.20) simplifies considerably. If all the terms involving negative powers of ω are ignored, we have $Z_{m_{eff}} = Z'_p + r$, i.e.

$$Z_{m_{eff}} \approx (R_p + r) + j(X_p + \omega M_c) \approx R_p + j\omega M_c \tag{4.21}$$

which gives the high-frequency limit for the DML as

$$f_{max} \approx \frac{R_p}{2\pi M_c} \tag{4.22}$$

A similar simplification gives the low-frequency limit which, if we ignore k, is given by:

$$Y_p \approx \frac{1}{R_p} + \frac{1}{j\omega M_m} \quad \text{so} \quad f_{min} \approx \frac{R_p}{2\pi M_m} \tag{4.23}$$

Alternatively, if we assume the influence of M_m is small and that stiffness dominates, it is evident that

$$f_{min} \approx \frac{k}{2\pi R_p} \tag{4.24}$$

4.11.1 Complete electromechanical model

Ideally, an electromechanical model is needed which will enable acoustic engineers to use familiar software modelling tools. This will allow the application of DML technology to any specific acoustic requirement. Given that a stiff, light panel can be designed to have optimal modal distribution and low loss, it has been shown that in order to model the acoustic pressure or acoustic power, it is only necessary to calculate the mean velocity in the panel.

The exciter and panel mechanical model can be transformed into an equivalent electrical impedance circuit by a method proposed by Bauer[17]. This has the added advantage that multiple exciters and panel/exciter compliances can be also added to model any level of DML implementation.

If we write the laws of motion for any mechanical system comprising masses, stiffness and damping:

- Force = $(j\omega$ Mass$) \times$ Velocity
- Force = Resistance $\times$ Velocity
- Force = Velocity/$(j\omega$ Compliance$)$

Then, using the voltage–force–pressure analogy, called EFP for brevity, we can write;

- Voltage = $(j\omega L) \times$ Current
- Force = $R \times$ Current
- Force = Current $/ (j\omega C)$

where L, R and C are the electrical analogues of the Mass, Resistance and Compliance in the mechanical domain.

Mechanical structures and motions have magnitudes and directions, i.e. they are vector quantities, while the electrical variables in a circuit are scalar. Therefore, to be able to model using electrical circuits, we need to restrict the analysis to motion in a single axis. The mechanical network components that are used are shown in Fig. 4.16, for a mass. The circuit implies that the mass is considered as a single lump, with all parts having the same velocity.

When it comes to representing stiffness, this has a capacitor as its equivalent, as shown in Fig. 4.17, where the velocities are represented by the two currents in the two branches. Although shown with potentials, one should always consider the terminals connected to a zero impedance voltage source. The equivalent for a dashpot (damping) is shown in Fig. 4.18, and the currents in the two legs again represent the velocities at the two ends of the dashpot.

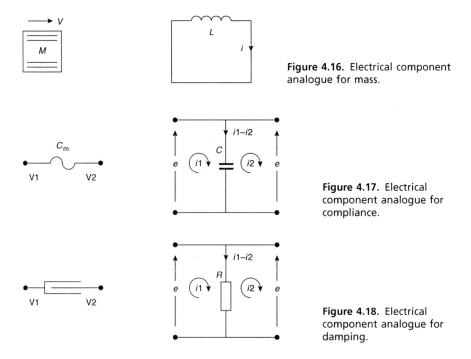

Figure 4.16. Electrical component analogue for mass.

Figure 4.17. Electrical component analogue for compliance.

Figure 4.18. Electrical component analogue for damping.

4.11.2 Basic exciter model

The mechanical model can be transformed into an equivalent electrical impedance circuit by the method proposed by Bauer mentioned in the previous section. In considering the basic inertial drive for a DML panel, we can use these analogues to construct a complete equivalent circuit, and thereby determine the velocity characteristics of the loudspeaker.

Starting with a simple free exciter, the mechanical arrangement is shown in Fig. 4.19. Transforming this into its impedance equivalent gives the diagram of Fig. 4.20.

The transformers are there, at the moment, as a method of ensuring the circuit is correctly drawn, in accordance with Bauer's method. Removing the transformers gives the schematic shown in Fig. 4.21, and adding the moving-coil gyrator element gives the complete analogue for a moving-coil exciter, as shown in Fig.4.22.

The final circuit has the mechanical elements accessible via nodes w, x and y, while the electrical side connects to the nodes s and t. This basic exciter equivalent can therefore be coded into any electrical circuit simulator, for example AkAbak[18], where the currents and voltages can be evaluated. Exciter design, and the significance of the various parameters described are illustrated by Roberts[19].

In order to use this approach for a complete DML application, it is necessary to extend the method to include panel, frame and any compliance used. This method has been employed by Tashiro *et al.*[20] for calculation of the terminal impedance, mechanical impedance at the driving point and mean panel velocities for a multimedia application. This reference compares the modelled panel velocity, with the measured SPL actually achieved (with 1/3 octave smoothing), as shown in Fig. 4.23.

Notice that the acoustical output for this multimedia panel falls below 250 Hz, since the panel cannot support bending waves at this low frequency. In fact, the first mode of the panel is set at 116 Hz (f_0), and useful output is available at $2.5f_0$.

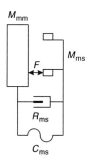

Figure 4.19. Mechanical arrangement for inertial drive exciter.

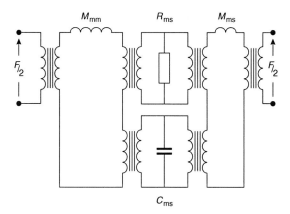

Figure 4.20. Impedance equivalent, basic exciter.

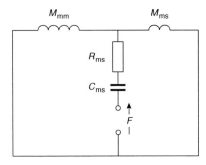

Figure 4.21. Impedance analogue for the mechanical components of an exciter.

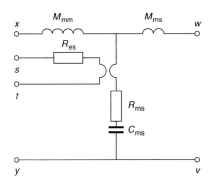

Figure 4.22. Complete electromechanical impedance analogue for an exciter.

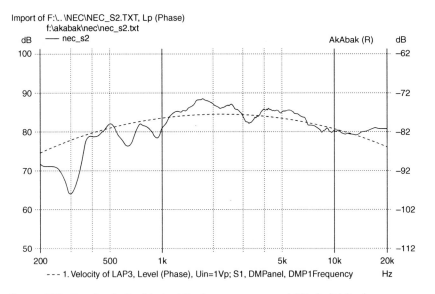

Figure 4.23. Panel velocity (dashed line) versus measured SPL (solid line).

4.12 Radiation simulation modelling

In order to appreciate the radiation from a DML, it is necessary to construct some form of mathematical model. Commercially available software for DMLs is available only to DML licensees, and such a modelling tool is described in the next section. However, any suitable commercial FEA solver can be used, and this is mentioned later in the section. Panzer and Harris[21] describe a modelling method which gives good agreement with measured DML panels. They start by considering an infinite panel and then move onto panels of finite size.

4.12.1 Infinite panel

In one dimension and normal to the panel, a single plane bending wave can have the form given by Cremer et al.[22];

$$v(x) = A \, e^{-jk_b x} \tag{4.25}$$

where $k_b = \sqrt[4]{\omega^2(\mu/B)}$ bending wave number (m^{-1})

   B bending stiffness (Nm)
   μ mass surface density (kg/m^2)
   A arbitrary constant

Such a plane transversal surface wave can radiate only if it matches the projected longitudinal acoustic wave. This is illustrated in Fig. 4.24 for the case where the wavelength in the panel (λ_b) is less than the trace wavelength in the air (λ_0), and there is no effective radiation. The trace of the acoustical wave is the projection onto the panel surface.

Compare this with Fig. 4.25, where λ_b is greater than λ_0, and radiation takes place at angle ϑ_c (the coincidence angle) where

$$\sin(\vartheta_c) = \frac{\lambda_0}{\lambda_b} = \frac{c_0}{\sqrt{\omega}} \sqrt[4]{\frac{\mu}{B}} \tag{4.26}$$

The far-field response is shown to be proportional to the velocity wave-number spectrum, by Morse and Ingard[23], who also show that this is the two-dimensional Fourier transform of the radiating velocity field due to the bending wave. From this

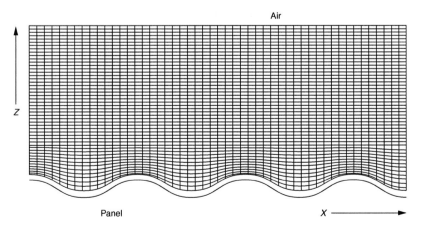

Figure 4.24. No effective radiation, λ_b less than λ_0.

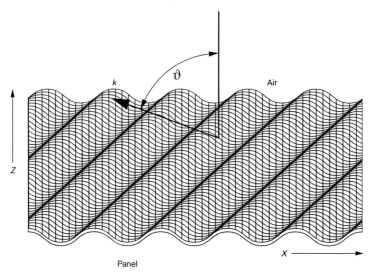

Figure 4.25. Radiation, λ_b greater than λ_0.

spatial spectrum only those wave number components whose k-factor is smaller than the wave-number, k_0, radiate into the far-field.

For radiation from a finite-sized panel, they first consider an infinite panel behind an aperture cut into an infinite baffle. Using this approach they calculate the radiation patterns for an infinitely large panel with various aperture sizes, and conclude that for this case the aperture size controls the directivity. A similar approach is adopted by Fahy[24], who illustrates the sound radiation from flexural waves in plates by investigating an infinitely long strip set into an infinite baffle.

4.12.2 Finite panel

When the panel is finite in size it has to satisfy the boundary conditions, and such a panel comes with a built-in aperture function (its overall size). As such, a finite vibrating panel has no external forces and moments applied at its edges.

By selecting the correct mechanical parameters it is possible for a finite sized panel to radiate well below coincidence, and thus provide a much more useful bandwidth. Results of modelling using this technique give reasonable agreement with measured panels, as shown in Figs 4.26 and 4.27, and confirm that a DML can be expected to have useful bandwidth, as well as wide directivity.

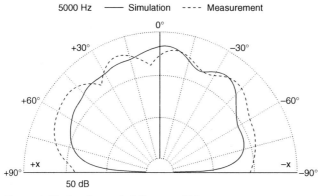

Figure 4.26. Small panel, 260 mm × 230 mm.

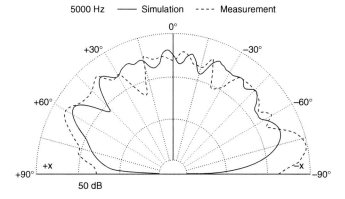

Figure 4.27. Large panel, 600 mm × 600 mm.

4.12.3 FEA solvers

The same results can be produced using commercial FEA programs, provided they have suitable extensions in order to model the air in contact with the panel (this should include the infinite fluid region). Bank[25], describes such a technique, using a commercial solver (PAFEC[26]) to show that the polar response of a DML is substantially independent of size, but will depend, of course, on where coincidence occurs. At this frequency typical panels will exhibit a degree of beaming, as described in the work of Panzer and Harris[21].

4.13 Performance

There are many examples of DML loudspeakers available, with a host of different specifications and levels of performance to suit each licensee's requirements. However, a number of aspects of the performance of DMLs are universal, and different from the traditional moving-coil (mass) counterpart.

A traditional loudspeaker, acting for the most part as a single, finite-sized radiator, has correlated output. These sources are typified by having measurements of amplitude and phase that vary smoothly and in a predictable way as small changes are made to the measurement position. Measurements made a short distance apart from each other will give smoothly varying results, until the wavelengths are of the order of the distances involved.

However, because of the nature of DML loudspeakers, the output is for the most part uncorrelated, and predicting the level and phase at points near to each other becomes very difficult. The DML loudspeaker gives the listener the impression of a continuous spectrum, although measuring instruments indicate a fully modal, uncorrelated output. This said, we can use traditional techniques to make measurements, but must bear in mind that these will need some interpretation.

4.14 Acoustical measurements

In order to put the previous sections into proper perspective, a typical panel has been designed and measured to give a complete signature. The panel was not chosen because it represented the ultimate in DML technology, but it does show all the characteristics that are to be expected when measuring DMLs using typical loudspeaker testing techniques. Other techniques are available, of course, and these results are not meant to be exhaustive, but it is hoped that they give an insight into the diffuse nature of the radiation from a DML. They are presented in an order that is typical of how a DML loudspeaker engineer might analyse a panel.

For the record, the panel is a composite, made from glass fibre/resin skins bonded onto a 3 mm NOMEX[27], honeycomb core, with a 90 grams per square metre (gsm) adhesive layer. Bending stiffness, $B = 6.99$ N.m; areal density, = 0.704 kg/m^2, giving mechanical impedance, $Z_m = 17.75$ Ns/m. A standard NEC Authentic[28] exciter, type NX33 (19 mm diameter voice-coil) was used for all measurements, except for the distortion measurements, where a type NX43 (25 mm diameter voice-coil) was used.

Acoustical measurements were made in a large test room (900 m^3), using standard measuring techniques. The measuring distance was set to 1 metre, and the input voltage for SPL measurements set to 1 volt. The input voltage for the distortion measurements was set to give the test SPL.

4.14.1 On-axis impulse response

Although not usually of primary interest to the loudspeaker engineer, the on-axis impulse response is shown first to illustrate the unusual behaviour of a DML in the time domain. After an initial fast rise time, the initial impulse *sees* an infinite panel (the impulse has not had time to reach the panel edges). Shortly after that is a region of diffuse radiation, which emanates from the pseudo-randomly vibrating panel, such that subsequent impulses are decorrelated from the initial impulse.

After about 13 milliseconds the first reflection, from the floor, is picked up by the microphone, while the panel is still radiating. As with normal room measurements, the impulse response would be truncated beyond the first reflection (Fig. 4.28).

4.14.2 Terminal impedance

This is a standard measurement, made by inserting a known series resistor between the loudspeaker terminals and the power amplifier. Voltage and current are measured simultaneously, and modulus of impedance extracted. Notice, in Fig. 4.29, the low-frequency region, where the panel is sparsely modal, and the pseudo-random behaviour of the panel can be seen. As the frequency rises into the densely modal region, then the deviations in the impedance become much smaller.

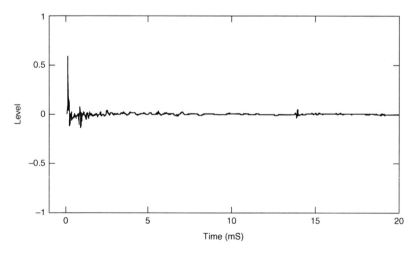

Figure 4.28. On-axis impulse response.

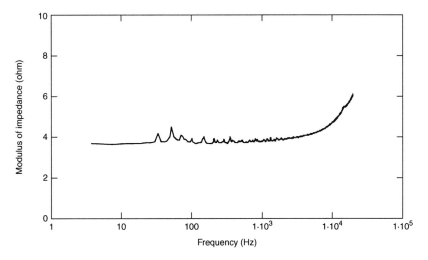

Figure 4.29. Terminal impedance.

4.14.3 SPL on-axis

SPL is measured, first on-axis at 1 metre distance, for a 1 volt r.m.s. input level. Figure 4.30 shows an unsmoothed response for the test panel. The modal nature of the panel radiation means that some smoothing is normally applied to measurements, but they are presented here without smoothing.

4.14.4 SPL 30° off-axis

Of course, a slight change of microphone position will give substantially the same overall output, but the detail will change markedly with even small changes to the microphone position. In Fig. 4.31 a measurement at 30° off-axis shows the change in response detail when the data is unsmoothed, although the overall response is very similar to Fig. 4.30.

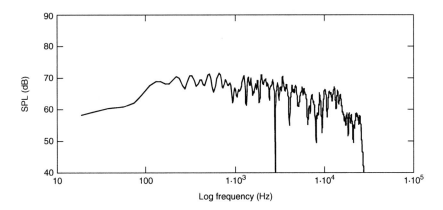

Figure 4.30. On-axis, 1 m 1 volt input, unsmoothed response.

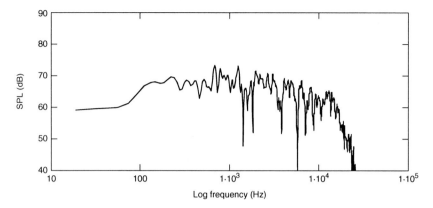

Figure 4.31. 30° off-axis, 1 m 1 volt input, unsmoothed response.

4.14.5 2D Polar response

Polar responses indicate the directivity for any loudspeaker, and results from such measurements are useful for calculating the directivity Q of the DML. Measurements are taken using a revolving turntable, synchronized to the measurement equipment.

Measurements are usually taken at single frequencies, whilst the panel is rotating or, more normally, a wide-band response is taken at set angles, and the polar responses subsequently calculated from this data. Either method can be adopted, depending on the preferences of the user. Figure 4.32 shows some typical DML polar responses at various frequencies.

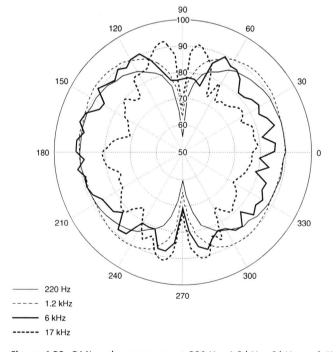

Figure 4.32. DML polar response at 220 Hz, 1.2 kHz, 6 kHz and 17 kHz.

4.14.6 3D Polar response

Because the DML's behaviour and subsequent radiation pattern has no axis symmetry, it is necessary to map the polar response of the panel in both axes to characterize its output fully. This requires a large amount of data storage, and is not often measured routinely, for this reason. The panel is first set in landscape orientation for a complete 2D polar response, and then rotated by increments into the portrait orientation, with polar responses made at the increments. To reduce the data set to manageable proportions, the number of angular steps can be reduced, at the expense of resolution at higher frequencies.

Manipulating the data into a 3D presentation is done with a mathematical analysis tool, like MathCad[29], where results for 200 Hz and 1.2 kHz are shown in Figs 4.33 and 4.34. The front of the panel is to the right, rear to the left, with the plane of the panel visible as a dip in the otherwise spherical response.

At 6 kHz the radiation pattern can still be resolved reasonably well, as illustrated in Fig. 4.35. Results at higher frequencies are less useful, unless measurements are made at smaller intervals, in order to preserve the detail.

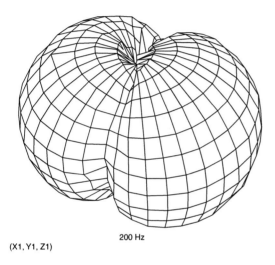

(X1, Y1, Z1)   200 Hz

Figure 4.33. 3D polar, 200 Hz.

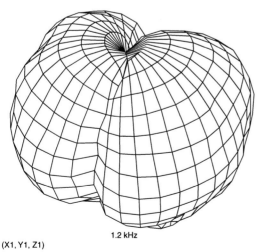

(X1, Y1, Z1)   1.2 kHz

Figure 4.34. 3D polar, 1.2 kHz.

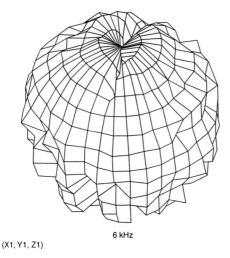

6 kHz

(X1, Y1, Z1)

Figure 4.35. 3D polar, 6 kHz.

4.14.7 Sound power response

In Fig. 4.36 the sound power can be seen to give a satisfactory low-frequency extension compared to the on-axis measurement shown earlier, with only a slight reduction in level. Depending on the panel size, the self-baffling effect for a DML can give a good extension in the power output to low frequencies.

An alternative method for measuring the sound power has been developed by Gontcharov, Hill and Taylor[30], since they confirm that a single point measurement cannot characterize fully the output from a DML. By employing an array of 16 microphones, however, they show it is feasible to characterize the polar characteristics of a DML, or any loudspeaker for that matter, fully by making a limited number of SPL measurements.

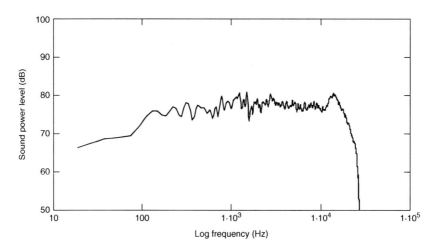

Figure 4.36. Sound power response, calculated from 2D polar data.

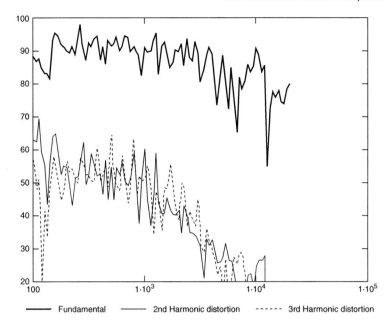

Figure 4.37. Second and third harmonic distortion products, fundamental set at 90 dB SPL.

4.14.8 Harmonic distortion

Harmonic distortion was measured using CLIO[31], a PC based FFT type analyser. Input level was adjusted to give an average level of 90 dB SPL. Results are shown in Fig. 4.37.

4.15 Psychoacoustics

By the very nature of its operation, the DML loudspeaker produces acoustic output that is diffuse and uncorrelated. It is no surprise then, that the subjective performance of such a loudspeaker differs from traditional loudspeakers in a number of areas. Early research work has already indicated the areas of perceived differences, as well as the likely causes.

4.16 Loudness

When auditioning DMLs, listeners have remarked on the position-independence of sound quality. This section compares the subjective loudness properties of DML and conventional loudspeakers. Flanagan and Harris, in their AES paper[32], use auditory modelling of objective measurements and the results of psychometric experiments to determine the perceived loudness of loudspeakers. They also discuss psychoacoustic mechanisms, in order to explain the observed phenomena.

4.16.1 Method

Measurements were made to compare the SPL levels produced by a normal cone loudspeaker and a DML, similar to that used in earlier work[33]. The data consisted

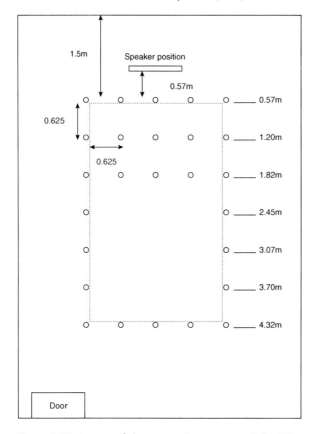

Figure 4.38. Layout of demonstration room and the 35 measurement points.

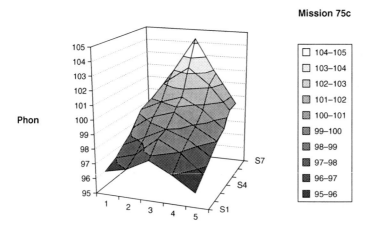

Figure 4.39. Traditional centre channel, loudness level, in phons.

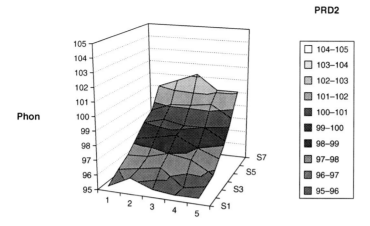

Figure 4.40. DML centre channel loudness level, in phons.

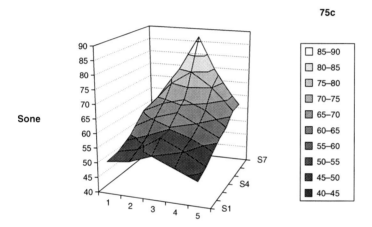

Figure 4.41. Traditional centre channel, loudness in sones.

of an array of 35 points (5 × 7) at which full spectrum, unweighted, 1/3 octave measurements were taken. The measurements were made in the listening room shown in Fig. 4.38.

4.16.2 Modelling and results

The Zwicker procedure for calculating loudness is quite complex and the ISO document[34] is filled with a series of charts. A BASIC program was produced by Zwicker, Fastl, and Dallmayr[35]. An MS-DOS version of this named zw.exe is available as shareware[36]. By editing the raw data files, these can be made to conform to the required ASCII input format, using a simple text editor. The output of the program copies the input data to the screen along with loudness level in phons and loudness in sones.

Calculations were done for two loudspeakers; a commercially available centre channel loudspeaker for Home Theatre use (Mission 75C), and a prototype 50 cm × 60 cm DML loudspeaker for surround and centre channel duty (PRD2).

prd2_centre

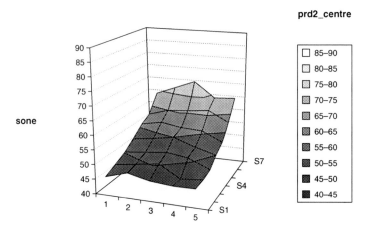

Figure 4.42. DML centre channel, loudness in sones.

Phon values for each set of results were entered into a 5×7 table in a similar orientation to the physical layout of the measurement grid. This allows a three-dimensional map to represent the loudness level in the modelled demonstration room, as shown in Figs 4.39 and 4.40. Repeating this for the sone values gives corresponding maps of loudness, as shown in Figs 4.41 and 4.42.

4.16.3 Conclusions

From these graphs a distinct difference between panel and cone can be seen. Loudness level drop in phons from the centre front to the centre back of the room is approximately equivalent for traditional cone and panel. Loudness drop in sones from front to back of room is 31 sones for the cone and 22 sones for the panel, an improvement of 9 sones. Note also the improved distribution of loudness across the room for a DML, compared to a traditional loudspeaker.

4.17 Stereophonic localization

Experiments on sound localization using specular radiators are often performed under near-anechoic conditions that do not correspond closely to normal listening environments. Harris, Flanagan and Hawksford[37] propose a hypothesis that 'localization precision as a function of room acoustics is minimized by the use of diffuse acoustic radiators such as DML panels'. They test this proposition and present the results from a series of psychometric tests to establish the conditions under which this hypothesis is valid.

4.17.1 Method

The subjective effects of the listening environment on listening tests are well documented[38,39]. In order to avoid these variations, image localization tests are usually carried out under near-anechoic conditions, with a single, well-defined listener location. In normal use, however, loudspeakers are listened to in rooms with reflective boundaries, and by more than one person at a time. Harris *et al.* argue that conventional loudspeakers, as specular acoustic radiators, are particularly susceptible to acoustic room modes and boundary interference, and have a limited 'sweet-spot' for optimal listening. In contrast, diffuse acoustic radiators such as the DML are cited as having 'sympathetic boundary interactions' and 'providing a more even power distribution across a large listening area and over a wide frequency band'[40].

Harris *et al.* carried out a series of psychometric listening tests, and went on to compare the ability of these two classes of radiator to localize a stereo image in an untreated room.

4.17.2 Results and conclusions

Table 4.1 Extract from summary of overall results.

Experiment	Description	Ranking by significance		
		Better	*Worse*	*Rank*
2	Tiled DML in wall	0	2	3
3a	Mission 2 way	0	3	4
4	Tiled DML free space	2	1	1.5
5	Tiled DML + subwoofer	2	1	1.5

Results of a series subjective listening tests seem to confirm the hypothesis that localization precision as a function of room acoustics is minimized by the use of diffuse acoustic radiators such as DML panels. A summary of results is given in Table 4.1. In these tests, the diffuse radiators performed at least as well as a quality two-way cone-in-box loudspeaker system, even in the 'sweet-spot' which should favour the latter.

Later work by Harris *et al.*[41] shows the results of a large number of subjective listening tests, using a more refined experimental procedure. The results from this series of tests confirm the earlier hypothesis that DMLs lessen the degradation caused by room acoustics on stereophonic localization.

4.18 Boundary reaction

Early work by Azima and Harris[40] on the reaction of diffuse sources to boundaries used FEA models as well as measurements, to compare DMLs with traditional loud-

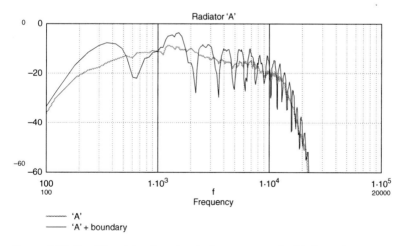

Figure 4.43. Traditional loudspeaker in free space and with boundary.

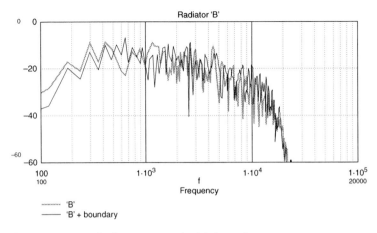

Figure 4.44. DML in free space and with boundary.

speakers. Later work, by Azima and Mapp[42], uses a traditional two-way loudspeaker with smooth amplitude and directivity behaviour, and a DML some 58 cm × 66 cm in size. Using SPL measurements, they show that the response from a traditional loudspeaker is subject to the comb-filtering associated with a nearby boundary, as indicated in Fig. 4.43. For the case of the DML the measured response shows much less correlation between the direct and reflected outputs, as illustrated in Fig. 4.44. Azima and Mapp extended the normal interpretation of the measured curves by using inter-aural cross correlation (IACC) techniques. Other researchers[43] have reported the preference of listeners when the value of IACC for left and right ears is low, allowing a more objective analysis of the measured data to be made. Measured results for the most critical angle (40°) indicate that the DML shows almost no correlation between direct and reflected sounds.

4.19 Acoustic feedback margin

Mapp and Ellis[44] investigated the potential use of DML in improving the feedback margin in sound reinforcement systems. The wide directivity and diffuse radiation results in a much more uniform sound field distribution within a room, compared to traditional cone loudspeakers. Using two test cases, they show a 4.7 dB gain before feedback for the first case, of a simple sound reinforcement system. The second case simulated a teleconferencing set-up, where duplex operation allows an open microphone to be in proximity to the system loudspeakers. Using a pair of loudspeakers, the gain advantage from using DMLs was 7.4 dB, over the conventional cone loudspeakers. They conclude that the more sound sources present, the greater the advantage of using DMLs.

4.20 Sound reinforcement applications

The radiation characteristics of DMLs would seem to lend themselves to use in sound reinforcement applications. Mapp and Colloms[45] investigated the improvements to intelligibility from using DMLs, and concluded that the radiation patterns are of a form which showed (in a single source experiment) considerable gain over and above the inverse square law for sound power fall-off with distance of a conventional loudspeaker. In later work, Mapp and Gontcharov[46] describe three installation case studies, where DMLs have been effectively used:

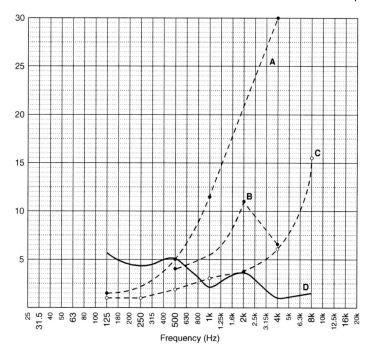

Figure 4.45. Comparison of typical Q-values: A – short column, B – two-way, C – 4-inch cone, D – DML (0.36 m²).

- A large architectural design office
- A large reverberant hall
- A transportation concourse

They concluded, from these test cases, that the directivity index for a typical DML has advantages in all three cases.

To illustrate the directivity characteristics, Fig. 4.45 shows typical Q-values for a number of loudspeaker types, calculated from the expression given in

$$Q = \frac{180°}{\text{arc sin } (\sin \alpha/2 \cdot \sin \beta/2)} \tag{4.27}$$

where α and β are the $-6\,\text{dB}$ horizontal and vertical coverage angles.

In this comparison the DML clearly stands out, since its Q-value decreases with frequency, in contrast to all other types. Although unexpected for such a large loudspeaker, the Q-value again confirms the wide directivity and diffuse nature of the radiation from the DML. In good acoustic conditions, they report that the DML appears to be able to offer better performance in terms of coverage and intelligibility than conventional cone devices. Caution should be exercised, however, when using the reflected field under highly reverberant conditions, since this represents a departure from traditional design techniques.

4.21 Distortion mechanisms

In any new transduction device the level of non-linearities is an important issue. Colloms et al.[47] report on distortion measurements made on typical DMLs. Because

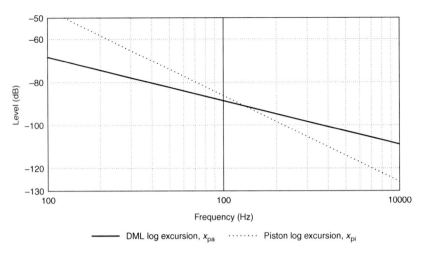

Figure 4.46. Comparison of displacement with frequency, for conventional cone ($\cdots$), versus DML(——).

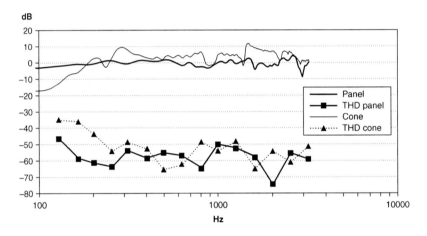

Figure 4.47. Comparison of THD with frequency, cone versus DML, both at 90 dB SPL.

the DML is a velocity device, the sound pressure is proportional to the mean panel velocity. This contrasts markedly with a conventional moving coil/mass loudspeaker, where sound pressure is proportional to the rate of change of velocity (or acceleration).

Therefore, comparing the displacements of the two, as shown in Fig. 4.46, illustrates that at low frequencies, where high coil displacement gives rise to distortion, the DML displacement is always less than in a conventional loudspeaker. This should give an improvement in DMLs, especially at LF, as illustrated in Fig. 4.47. The slight difference in SPL levels is to take account of the greater radiated power from the DML, so that in reverberant conditions the levels are correctly matched. Colloms *et al.* conclude that for optimized exciter designs, the distortion products from DMLs would be typically 10 dB lower than an equivalent cone piston loudspeaker.

4.22 The future

With such an exciting new technology, it is expected that the future will reveal many more applications which will benefit from the diffuse characteristics of DMLs. This will inevitably mean that any attempt to cover the use of DMLs will soon be dated. In order to avoid this to some extent, the introduction given in this chapter is meant as a primer in DML technology, enabling the reader to understand the underlying principles and to follow its progress with that knowledge.

Acknowledgements

The writer is indebted to New Transducers Ltd, and its parent NXT plc for access to its Intellectual Property, and its kind agreement to release as much material and data as possible in order to illustrate and explain the technology. Thanks also go to all those past and present employees of New Transducers Ltd, and its many consultants and friends, who have made valuable contributions to the knowledge base already in existence.

References

1. HERON, K, Panel Form Loudspeaker, Defence Research Agency, DRA Farnborough (1991).
2. NXT plc, Registered Office, 2nd Floor, Ixworth House, 37 Ixworth Place, South Kensington, London, SW3 3QH, UK.
3. http://www.nxtsound.com
4. FREDERICK, H A, Acoustic Device, US Patent No. 1,955,800 (1934).
5. THURAS, A L, Sound Translating Device, US Patent No. 1,869,178 (1932).
6. RAYLEIGH, J W S, *The Theory of Sound*, Vol. 1, Dover Publications, New York (1945).
7. CREMER, L, HECKL, M and UNGAR E, *Structure-Borne Sound*, 2nd edition, Springer-Verlag, Berlin (1988).
8. FAHY, F, *Sound and Structural Vibration*, Academic Press, London (1987).
9. AZIMA, H, HARRIS, N and COLLOMS, M, International Patent Application, WO 97/09842, International Filing Date: 2 September (1996).
10. ASTM Standards, Designation D 790–66.
11. NEILSEN, L F, WISMER, N J and GADE, S, Improved Method for Complex Modulus Estimation, The Technical University of Denmark and Bruel & Kjaer, Denmark.
12. MATHERS, D and ACKROYD, D M, 'Measuring the properties of loudspeaker materials for low-frequency drive units', 82nd AES Convention, March 1987, pre-print 2422.
13. MORSE, P and INGARD, U, *Theoretical Acoustics*, Section 7.4. McGraw-Hill, New York (1968).
14. MORSE, P and INGARD, U, *Theoretical Acoustics*, Section 5.3.19. McGraw-Hill, New York (1968).
15. GONTCHAROV, V, HILL, N and TAYLOR, V, 'Measurement aspects of distributed mode loudspeakers', 106th AES Convention, September (1999).
16. 'NXT White Paper', (C) New Transducers Ltd (1996).
17. BAUER, B B, 'Equivalent circuit analysis of mechano-acoustic structures', *JAES*, **24**, No.8, 643–655.
18. AkAbak (TM), formerly Panzer & Partner, now supported by New Transducers Limited, Huntingdon, UK.
19. ROBERTS, M, 'Exciter design for distributed mode loudspeakers', 104th AES Convention, May 1998, pre-print 4743.
20. TASHIRO, M, BANK, G and ROBERTS, M, 'A new flat panel loudspeaker for portable multimedia', 103rd Audio Engineering Convention, September 1997, preprint 4527.
21. PANZER, J and HARRIS, N, 'Distributed-mode loudspeaker radiation simulation', 105th AES Convention, San Francisco, September 1998, preprint 4783.
22. CREMER, L, HECKL, M and UNGAR, E, *Structure-Borne Sound*, second edition, Springer-Verlag, Berlin (1988).
23. MORSE, P and INGARD, U, *Theoretical Acoustics*, McGraw-Hill, New York (1968).
24. FAHY, F, *Sound and Structural Vibration*, Academic Press, London (1987).

25. BANK, G, 'The intrinsic scalability of the distributed mode loudspeaker (DML)', 104th AES Convention, May 1998, preprint 4742.
26. PAFEC, Program for Automatic Finite Element Calculations, SER Systems Ltd, Strelley Hall, Nottingham, England, NG8 6PE.
27. NOMEX, is a registered trademark of DuPont.
28. NEC Authentic Ltd, 3–14–1 Hisamoto, Takatsuku, Kawasaki-City, Kanagawaken 213 Japan.
29. MATHCAD, MathSoft Inc., 101 Main Street, Cambridge, MA 02142, USA.
30. GONTCHAROV, V, HILL, N P R and TAYLOR, V J, 'Measurement aspects of distributed mode loudspeakers, 104th AES Convention, May 1998, preprint 4970.
31. CLIO, Audiomatica SRL, Via Faentina 244/G, 50133 Florence, Italy.
32. FLANAGAN, S and HARRIS, N, 'Loudness: a study of the subjective differences between DML and conventional loudspeakers', 106th AES Convention, Munich, May 1999, preprint 4872.
33. COLLOMS, M and ELLIS, C, 'Diffuse field planar loudspeakers in multimedia and home theatre', 103rd AES Convention, September 1997, preprint 4545.
34. ISO Standard 532, Acoustics – Method for calculating loudness level. International Organization for Standardization (1975).
35. ZWICKER, E, FASTL, H and DALLMAYR, C, 'BASIC program for calculating the loudness of sounds from their 1/3 octave band spectra according to ISO 532 B', Acustica, 55, 1984, 63.
36. www.measure.demon.co.uk/Acoustics_Software/loudness.html
37. HARRIS, N, FLANAGAN, S and HAWKSFORD, M J, 'Stereophonic localisation in the presence of boundary reflections, comparing specular and diffuse acoustic radiators', 104th AES Convention, Amsterdam, May 1998, preprint 4684.
38. BARRON, M, 'The subjective effects of first reflections in concert halls – the need for lateral reflections', Journal of Sound and Vibration, 15(4), 475–494 (1971).
39. OLIVE, S E, SCHUCK, P L, SALLY, S L and BONNEVILLE, M E, 'The effects of loudspeaker placement on listener preference ratings', JAES, 42, No. 9, 651–669 (1994).
40. AZIMA, H and HARRIS, N, 'Boundary interaction of diffuse field distributed-mode radiators', 103rd AES Convention, September 1997, preprint 4635.
41. HARRIS, N, FLANAGAN, S and HAWKSFORD, M J, 'Stereophonic localisation in rooms, comparing conventional and distributed-mode loudspeakers', 105th AES Convention, San Francisco, September 1998, preprint 4794.
42. AZIMA, H and MAPP, P, 'Diffuse field distributed-mode radiators and their associated early reflections', 104th AES Convention, May 1998, preprint 4759.
43. ANDO, Y, 'Subjective preference in relation to objective parameters of music sound field with a single echo', JASA, 62, No.6, December (1977).
44. MAPP, P and ELLIS, C, 'Improvements in acoustic feedback margin in sound reinforcement systems', 105th AES Convention, San Francisco, September 1998, preprint 4850.
45. MAPP, P and COLLOMS, M, 'Improvements in intelligibility through the use of diffuse acoustic radiators in sound distribution', 103rd AES Convention, New York, September 1997, preprint 4634.
46. MAPP, P and GONTCHAROV, V, 'Evaluation of diffuse mode loudspeakers in sound reinforcement and PA systems', 104th AES Convention, Amsterdam, May 1998, preprint 4758.
47. COLLOMS, M, PANZER, J, GONTCHAROV, V and TAYLOR, V, 'Distortion mechanisms of distributed mode (DM) panel loudspeakers', 104th AES Convention, Amsterdam, May 1998, preprint 4757.

5 Multiple-driver loudspeaker systems

Laurie Fincham

5.1 Introduction

Ideally, a loudspeaker system should possess an extended and smooth frequency response which is maintained over a sufficiently large radiation angle to cover the intended listening area. In addition it should be capable of reproducing all types of programme material without any audible non-linear distortion. In practice, a single drive unit cannot meet these requirements, and so the majority of loudspeaker systems employ two or more drive units, each of which reproduces only part of the audio spectrum. The electrical network used to split the audio signal into the required frequency bands and assign each band to the appropriate drive unit is called a *crossover* or *dividing* network.

In the early days of multiway loudspeakers, little attention was paid to the design of the crossover networks. As a result, the advantage of increased frequency range obtained by the use of multiple driver units was often offset by gross irregularities in the response around the crossover frequency – which is usually located in that part of the spectrum where the ear is most sensitive to response changes.

The need to tailor crossover networks to suit the individual drive units more accurately was beginning to be appreciated by the 1960s. However, manufacturing variations between one unit and another existed to such an extent that elaborate networks were needed having 'adjust-on-test' components.

The situation changed in the mid-1970s, thanks to more consistent manufacturing standards, the introduction of digital techniques in acoustic measurement and filter network optimization, and a better understanding of the effect of crossover networks and drive unit positioning on the directional characteristics of a loudspeaker. These developments permit a more systematic approach to the design of such filters as an integral part of the system. Each drive unit can meet the numerous quality requirements over only a restricted bandwidth. It is therefore essential that the output of each drive unit outside its working range is reduced to a sufficiently low level by adequate attenuation in the crossover filter.

Practical filters cannot have an infinitely sharp cut-off and so there is inevitably an overlap region. The design must therefore ensure a flat response over this transition region, as well as tailoring the response for each individual drive unit.

In multiway loudspeaker systems there are two main categories of crossover networks – *high-level* and *low-level*. In systems employing high-level crossover networks, each drive unit is fed via its own passive filter section – high-pass for HF (high-frequency) units, bandpass for MF (mid-frequency) units and low-pass for LF (low-frequency) units. These various filter sections are normally connected in parallel and fed from a single, low source impedance (constant-voltage) power amplifier (see Fig. 5.1). With low-level crossover networks, each drive unit is fed via its own power amplifier, each of which has a low-level active filter network connected at its input (see Fig. 5.2).

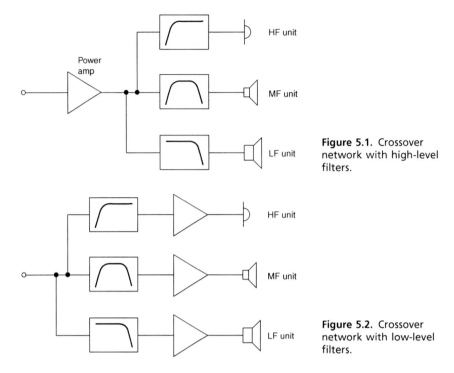

Figure 5.1. Crossover network with high-level filters.

Figure 5.2. Crossover network with low-level filters.

The performance requirements for both high- and low-level crossover networks are the same, but the design procedures are different. High-level networks have the additional requirement that the individual filter sections must have the correct response when loaded by the complex input impedance of each drive unit, while at the same time presenting an acceptable overall load impedance to the power amplifier.

5.2 Crossover networks – theoretical design criteria

5.2.1 Target function approach

The discovery in the early 1970s that the majority of loudspeaker drive units are so-called minimum-phase-shift systems[12] allowed a more systematic approach to crossover or dividing network design. This was called the Target Function Approach in which the frequency response of a drive unit, mounted in its enclosure, was made to conform to some desired target function by means of a suitable filter network. If the filter network was also minimum-phase-shift, then only the magnitude of the filter response need be considered during the design phase.

The target function approach to crossover network design requires that the overall electro-acoustic response of each drive unit and its corresponding filter section must be considered together, and made to conform as closely as possible to some desired filter shape referred to as the target function, $T(f)$[10]. The frequency response of the drive unit measured under working conditions is represented by $S(f)$ and the filter by $H(f)$ where $T(f) = H(f)S(f)$ (see Fig. 5.3(a)).

Normally, $T(f)$ is defined and $S(f)$ can be found by measurement. Therefore, the design problem reduces to devising a filter whose frequency response or transfer function most closely approximates $H(f)$, where

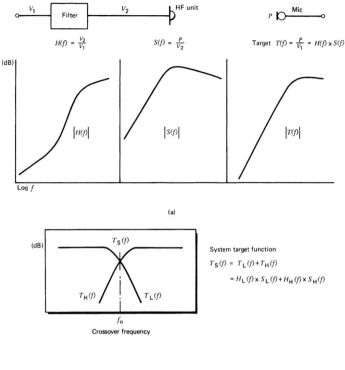

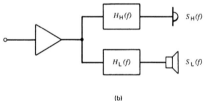

Figure 5.3. Target functions.
(a) Drive unit. (b) System.

$$H(f) = T(f)/S(f)$$

The overall system response is given by the vector sum of the individual unit target responses. For example, for a two-way system (see Fig. 5.3(b)),

$$T_S(f) = T_L(f) + T_H(f)$$

where $T_S(f)$ is the overall system target response and

$$T_L(f) = S_L(f) H_L(f)$$

$$T_H(f) = S_H(f) H_H(f)$$

$S_L(f)$ and $S_H(f)$ represent the measured frequency response of the low- and high-frequency drive units and $H_L(f)$ and $H_H(f)$ are the corresponding low- and high-pass filter sections which together make up the complete crossover network.

5.2.2 Choice of unit target functions

The following unit target functions may be used when a flat overall system response is required. The results given are for a two-way system and apply for a measuring point which is equidistant from the effective acoustic centre of each drive unit[4,5,9].

(a) Butterworth or maximally flat

First-order Butterworth

$$T_L(f) = \frac{1}{1 + s_n}$$

$$T_H(f) = \frac{s_n}{1 + s_n}$$

where $s_n = s/2\pi f_0$ is the normalized complex frequency variable,
   $s = j2\pi f$
   f_0 is the crossover frequency.

With both units connected in-phase (same polarity) the system response is

$$T_S(f) = T_L(f) + T_H(f) = \frac{1 + s_n}{1 + s_n} = 1$$

i.e. flat amplitude and phase (see Fig. 5.4(a)).

For out-of-phase connection (opposite polarity):

$$T_S(f) = T_L(f) - T_H(f) = \frac{1}{1 + s_n} - \frac{s_n}{1 + s_n} = \frac{1 - s_n}{1 + s_n}$$

The overall response $T_S(f)$ has an all-pass characteristic, i.e. flat amplitude, but with a frequency-dependent phase or delay characteristic[8] (see Fig. 5.4(b)).

Note that the low- and high-pass outputs are in phase quadrature at all frequencies, and are each attenuated by 3 dB at the crossover frequency. The total power output for either polarity is:

$$|T_H|^2 + |T_L|^2 = \frac{1}{1 + s_n^2} + \frac{s_n^2}{1 + s_n^2} = 1$$

Hence, for an omni-directional system there is constant acoustic-power output with frequency. Despite its apparently ideal characteristic when connected in-phase, the first-order target function is not used in practice due to its very slow attenuation rate of only 6 dB/octave[9,10].

Third-order Butterworth

$$T_L(f) = \frac{1}{1 + 2s_n + 2s_n^2 + s_n^3} = \frac{1}{(1 + s_n)(1 + s_n + s_n^2)}$$

$$T_H(f) = \frac{s_n^3}{1 + 2s_n + 2s_n^2 + s_n^3} = \frac{s_n^3}{(1 + s_n)(1 + s_n + s_n^2)}$$

For in-phase connection:

$$T_S(f) = T_L(f) + T_H(f)$$

$$= \frac{1 + s_n^3}{(1 + s_n)(1 + s_n + s_n^2)}$$

$$= \frac{(1 + s_n)(1 - s_n + s_n^2)}{(1 + s_n)(1 + s_n + s_n^2)} = \frac{(1 - s_n + s_n^2)}{(1 + s_n + s_n^2)}$$

$T_S(f)$ has a second-order all-pass characteristic with a flat amplitude[8] (see Fig. 5.5(a)).

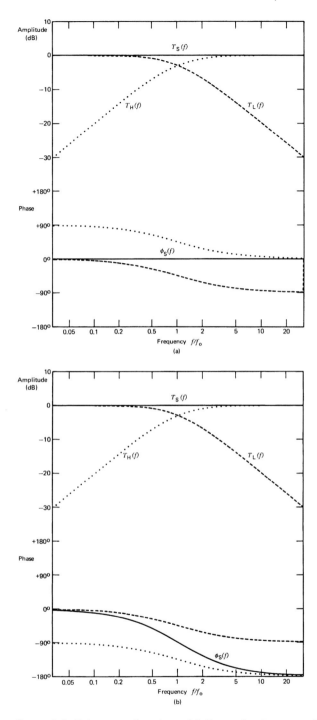

Figure 5.4. Unit target functions. (a) First-order Butterworth (in-phase). (b) First-order Butterworth (out-of-phase).

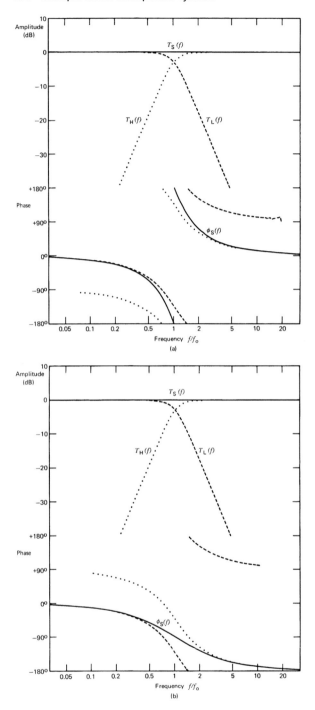

Figure 5.5. Unit target functions. (a) Third-order Butterworth (in-phase). (b) Third-order Butterworth (out-of-phase).

For out-of-phase connection:

$$T_S(f) = T_L(f) - T_H(f)$$

$$= \frac{1 - s_n^3}{(1 + s_n)(1 + s_n + s_n^2)}$$

$$= \frac{(1 - s_n)(1 + s_n + s_n^2)}{(1 + s_n)(1 + s_n + s_n^2)} = \frac{(1 - s_n)}{(1 + s_n)}$$

Again the total system output has an all-pass characteristic, but this time only of first order when the two outputs are connected out-of-phase (reversed polarity) (see Fig. 5.5(b)). Note that its delay characteristic is identical to the out-of-phase connected first-order Butterworth target function (see Fig. 5.4(b)). As for the first-order case, the outputs are -3 dB at the crossover frequency and are in phase quadrature at all frequencies. Reversing the polarities, however, while maintaining the all-pass characteristic, worsens the delay. For either polarity the total power output remains constant[8].

The third-order Butterworth is one of the most useful target functions as it combines a reasonably high attenuation rate, 18 dB/octave, with only a modest frequency-dependent delay if the unit outputs are connected in opposite polarity[5,8].

Fifth-order and higher odd-order Butterworth filters can also be shown to have all-pass characteristics but are seldom used in passive crossover designs because of their complexity, cost and insertion losses[6,7,9,11].

(b) Linkwitz–Riley

Another class of filter, which can also be used as a target $T(f)$ to provide a flat system response, is the so-called Linkwitz–Riley, which consists of two Butterworth sections in cascade[6,7,9,11].

.

Second-order Linkwitz–Riley

$$T_L(f) = \frac{1}{(1 + s_n)^2}; \quad T_n(f) = \frac{s_n^2}{(1 + s_n)^2}$$

When connected out-of-phase, the second-order Linkwitz–Riley gives an all-pass characteristic (see Fig. 5.6(a)).

$$T_S(f) = T_L(f) - T_H(f)$$

$$= \frac{1 - s_n^2}{(1 + s_n)^2} = \frac{(1 - s_n)(1 + s_n)}{(1 + s_n)(1 + s_n)}$$

$$= \frac{1 - s_n}{1 + s_n} \quad \text{(first-order all-pass)}$$

Fourth-order Linkwitz-Riley

$$T_L(f) = \frac{1}{(1 + \sqrt{2}s_n + s_n^2)^2}; \quad T_H(f) = \frac{s_n^4}{(1 + \sqrt{2}s_n + s_n^2)^2}$$

For the fourth-order Linkwitz–Riley target shapes, an in-phase connection gives a flat frequency response (see Fig. 5.6(b)):

$$T_S(f) = T_L(f) + T_H(f)$$

$$= \frac{1 + s_n^4}{(1 + \sqrt{2}s_n + s_n^2)^2}$$

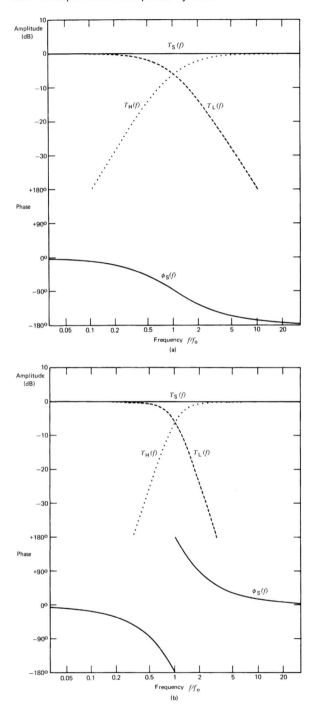

Figure 5.6. Unit target functions. (a) second-order Linkwitz–Riley. (b) Fourth-order Linkwitz–Riley.

$$= \frac{(1 - \sqrt{2}s_n + s_n^2)(1 + \sqrt{2}s_n + s_n^2)}{(1 + \sqrt{2}s_n + s_n^2)^2}$$

$$= \frac{1 - \sqrt{2}s_n + s_n^2}{1 + \sqrt{2}s_n + s_n^2} \quad \text{(second-order all-pass)}$$

Both second- and fourth-order Linkwitz–Riley filters have the property that the low and high outputs are in-phase at all frequencies and are $-6\,\text{dB}$ at the crossover frequency[1,6,11] (see Figs 5.6(a) and 5.6(b)). They do not have the constant-power characteristic of the odd-order Butterworth filters; instead there is a $-3\,\text{dB}$ dip in the power response at the crossover frequency[8].

(c) Linear-phase or constant-voltage target functions

Of the target functions described in the previous section, only the in-phase connected first-order Butterworth has a linear-phase (or uniform-delay) characteristic – all the others have an all-pass response. Several other linear-phase target functions have been identified, and some of these are described below[2,3,4,10].

The general requirement for a linear-phase or constant-voltage[10] target function is that the system target function

$$T_S(f) = T_L(f) + T_H(f) = 1$$

for any given shape of $T_L(f)$ and $T_H(f)$ may be calculated from the relationship

$$T_H(f) = 1 - T_L(f)$$

or where $T_H(f)$ is defined:

$$T_L(f) = 1 - T_H(f)$$

Asymmetrical constant voltage
If $T_L(f)$ has a third-order Butterworth characteristic[10], then the system target function $T_S(f)$ will be linear phase if

$$T_H(f) = 1 - T_L(f)$$

$$= 1 - \frac{1}{1 + 2s_n + 2s_n^2 + s_n^3} = \frac{2s_n + 2s_n^2 + s_n^3}{1 + 2s_n + 2s_n^2 + s_n^3}$$

The resulting target shapes $T_L(f)$ and $T_H(f)$, which are asymmetrical, are shown in Fig. 5.7(a), from which it can be seen that $T_H(f)$ has an attenuation rate of only 6 dB/octave and peaks around the crossover frequency. It can be shown that, for all such asymmetrical designs, the attenuation rate of one of the filters will be only 6 dB/octave, regardless of the order of the other[10].

Symmetrical constant voltage
Symmetrical designs are also possible[10], a second-order design (attenuation rate 12 dB/octave) being given by

$$T_L(f) = \frac{1 + 3.7s_n}{1 + 3.7s_n + 3.7s_n^2 + s_n^3}$$

and

$$T_H(f) = \frac{3.7s_n^2 + s_n^3}{1 + 3.7s_n + 3.7s_n^2 + s_n^3}$$

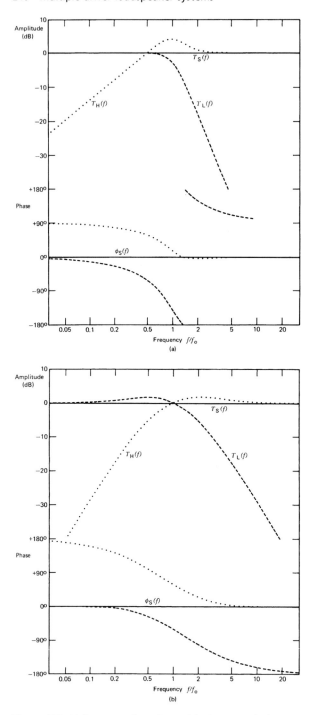

Figure 5.7. Unit target functions. (a) Asymmetrical constant-voltage. (b) Symmetrical constant-voltage.

The response of both sections is peaked by about 2 dB in the pass-band and there is a 120° phase shift between the two sections (see Fig. 5.7(b)).

5.2.3 Filler-driver designs

In filler-driver designs, the missing term required to make the system response linear-phase is provided by a so-called filler-driver unit[4]. For a second-order system where $T_L(f)$ and $T_H(f)$ target functions are second-order Butterworth type, then

$$T_L(f) = \frac{1}{1 + \sqrt{2}s_n + s_n^2}$$

and

$$T_H(f) = \frac{s_n^2}{1 + \sqrt{2}s_n + s_n^2}$$

The required response of the filler-driver is given by

$$T_{FD}(f) = 1 - (T_L(f) + T_H(f)) = \frac{\sqrt{2}s_n}{1 + \sqrt{2}s_n + s_n^2}$$

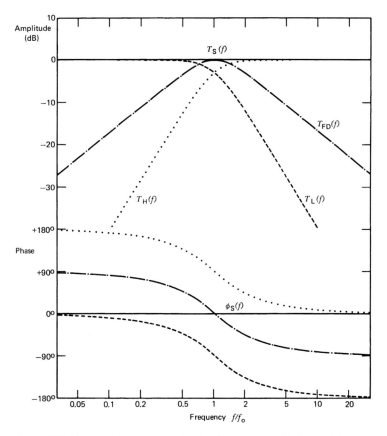

Figure 5.8. Linear phase – second-order Butterworth high- and low-pass, plus a filler-driver.

From Fig. 5.8 it can be seen that $T_{FD}(f)$ takes the form of a first-order band-pass filter, with centre frequency f_0 having upper and lower attenuation rates of 6 dB/octave.

5.2.4 Delay-derived high-slope filters

Delay-derived high-slope filters have the form[2]:

$$T_H(f) = \exp(-\tau(f)) - T_L(f)$$

where τ represents pure time delay. In effect, instead of achieving the desired high-pass response $T_H(f)$ by subtracting $T_L(f)$ from 1 (see Fig. 5.9(a)), $T_L(f)$ is subtracted from a time-delayed version of the input signal (see Fig. 5.9(b)).

5.2.5 Delay-derived target functions

$$T_L(f) + T_H(f) = \exp(-\tau(f))$$

By making the time delay τ equal to the zero-frequency phase and group delays of the low-pass target function $T_L(f)$, the attenuation rate of the derived high-frequency target function $T_H(f)$ may be greater than that achievable from the simpler constant voltage designs.

5.2.6 Avoiding interference

A further consideration in the choice of the target function concerns the directional characteristics of the loudspeaker system and their effect on the quality of sound heard under practical conditions. If, in a multiway loudspeaker, the acoustic centres of all the drive units are not equidistant from the listener, then the additional phase shift between the drive units resulting from the differential time delay may be sufficient to cause audible response irregularities in the crossover region due to interference. This can only be avoided completely with so-called coincident driver designs in which the high-frequency unit is mounted co-axially with respect to the

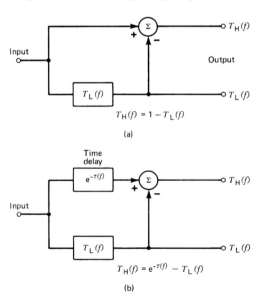

Figure 5.9. Linear phase. (a) Constant voltage. (b) Time-delay derived.

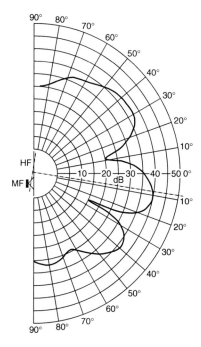

90° 80° 70° 60° 50° 40° 30° 20° 10°

HF
MF

−10 −20 −30 −40 −50 0°
dB

Axis:
zero inter-unit
time delay − − − −

10° 20° 30° 40° 50° 60° 70° 80° 90°

Figure 5.10. Vertical polar characteristic at crossover frequency. Two-way system with unit target functions. Third-order Butterworth.

low-frequency unit such that their acoustic centres are effectively coincident. In non-coincident designs, the effects of unwanted interference can be minimized through suitable drive unit layout and choice of target function.

For maximum horizontal distribution of sound without interference, the drive units should be mounted one above the other. Because of the unavoidable separation between the units, some interference effects must then occur when the listener is located above or below the design axis, and thus no longer equidistant from the different sound sources. The amount of this interference sets a limit to the angle above and below the axis within which the response can be maintained substantially constant.

This situation is further complicated by the phase shift necessarily associated with the high- and low-pass characteristics of the individual filter/unit combinations. The high-frequency drive unit, which at crossover normally has a phase lead over the low-frequency unit, is commonly mounted above the latter. What happens then is illustrated by the polar diagram in Fig. 5.10, which shows how the loudspeaker response at crossover varies with angle in the vertical plane. In this example, the high- and low-pass filter functions are of third-order, for which the inter-unit phase difference is 90°, and the crossover frequency is 3 kHz. The vertical distance between the axes of the units is 170 mm, which at 3 kHz is 1.5 times the wavelength of the sound. It can be seen that the main lobe of the polar characteristic, instead of coinciding with the axis of zero inter-unit time delay, is tilted downwards and has a maximum amplitude 3 dB above the on-axis response. A great deal of sound energy is thus directed away from the listening area and towards the floor, producing unwanted frequency-dependent reflections which modify the relationship between the direct and reflected sound in the room. Worse still, there is a region just above the axis, where the outputs from the two units are beginning to get out of phase, and at one angle almost cancel each other. As a result, a small vertical displacement produces a large change in the response of the system around crossover, and hence in the spectrum of the reproduced sound. This effect is illustrated in Fig. 5.11, which

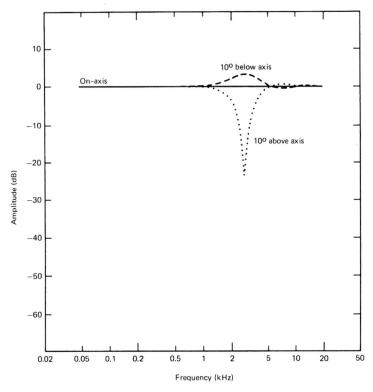

Figure 5.11. Two-way system with third-order $T(f)$. Variation in frequency response with vertical angle.

shows how the frequency characteristic of the loudspeaker referred to in Fig. 5.10 varies with angle above and below the design axis. It may be noted in passing that a similar phenomenon occurs with a first-order crossover. In this case the slower rate of cut-off gives a wider range of overlap between the two sound sources, and the tilt in the polar characteristic extends over at least two octaves.

One way of dealing with this situation is to mount the low-frequency drive unit (or midrange unit in the case of a three-way system) above the high-frequency unit, thus turning the polar diagram upside down. The main lobe is then directed away from the floor and the cancellation region placed where it can do little harm.

The in-phase relationship between the high- and low-frequency sections of a Linkwitz–Riley system design ensures that the main lobe of the polar curve remains symmetrical about the axis of zero inter-unit time delay. In addition, the variation in frequency response with vertical angle is also reduced (see Fig. 5.12). For this reason, Linkwitz–Riley target shapes are widely used for high-quality loudspeaker systems.

5.2.7 Target functions for design axes having non-zero inter-unit time delay ('Wiggle Factor method')

The ideal layout for a practical two-way system, having a fourth-order Linkwitz–Riley target function and a crossover frequency of 3 kHz, is shown in Fig. 5.13(a). Note that the LF and HF sections sum to a flat overall system response and that the phase characteristics overlay at all frequencies (see Fig. 5.13(b)). With some speaker

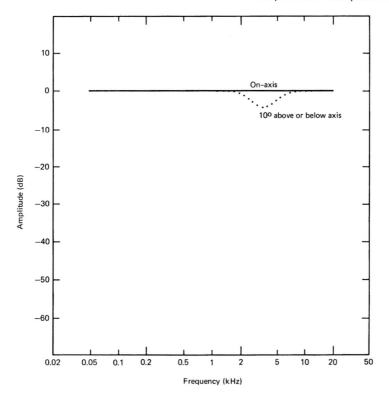

Figure 5.12. Two-way system with fourth-order Linkwitz–Riley $T(f)$. Variation in frequency response with vertical angle.

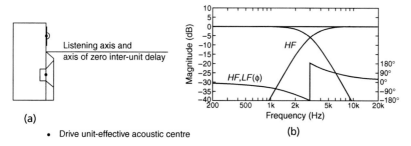

Figure 5.13. Two-way system with zero inter-unit time delay on listening axis, (a) system layout, (b) system response.

systems it is not always practical to arrange for the axis of zero inter-unit time delay to coincide with the desired listening axis (see Fig. 5.14(a)). The resulting inter-unit time delay on the listening axis (in this example estimated at 88 μs due to a physical offset of 30 mm) results in additional phase difference between the drive unit outputs of 90° at the crossover frequency. The resultant system response, instead of being flat, has a dip of nearly 4 dB at 3.6 kHz (see Fig. 5.14(b)). The most direct way to correct this would be to introduce an additional delay in the feed to the HF unit of 88 μs, which could be implemented by means of cascaded all-pass sections or a

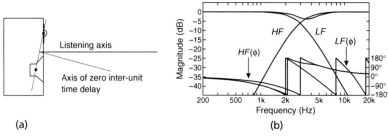

(a)   (b)

Figure 5.14. Two-way system with 88 μs inter-unit time delay on listening axis, (a) system layout, (b) system response.

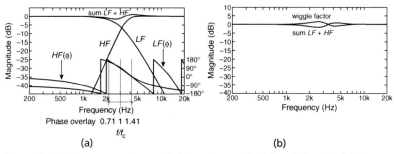

(a)   (b)

Figure 5.15. Two-way system with 88 μs inter-unit delay, (a) revised HF target to give best phase overlay, (b) system response and calculated 'Wiggle Factor'.

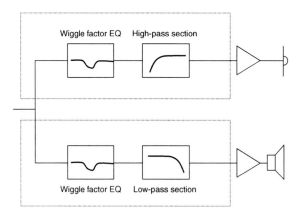

Figure 5.16. Block diagram of two-way crossover with 'Wiggle Factor'.

digital delay line. Where the complexity and cost of such an approach is not justi-fied, the following simpler and more cost effective method can be used. The additional differential phase-shift around the crossover frequency, due to the additional inter-unit time delay, can be partially compensated by adjusting both the slope and cut-off frequency of either the LF or HF target shapes[17,19]. It can be seen from Fig. 5.14(b) that the inter-unit phase difference, at the crossover frequency (3 kHz), is about 90°. By changing the HF target from a 3 kHz, fourth-order Linkwitz–Riley to a 3.6 kHz, fifth-order Butterworth, the phases of the LF and HF sections overlay for about one half-octave above and below the crossover frequency, f_c (see Fig. 5.15(a)).

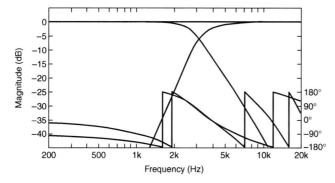

Figure 5.17. Two-way system response on listening axis with 'Wiggle Factor' correction.

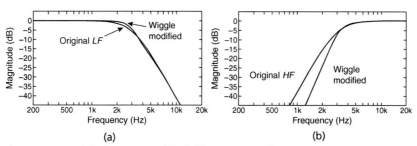

Figure 5.18. Original-versus-modified filter sections for two-way system with 'Wiggle Factor' correction. (a) LF, (b) HF.

The revised target shapes, although substantially in-phase around the crossover point, still do not sum to a flat response (see Fig. 5.15(b)). However, if the minimum-phase-shift inverse of this response irregularity, called the 'Wiggle Factor' , is applied to both LF and HF sections of the dividing network (see Fig. 5.16), then the resulting system response will be flat (see Fig. 5.17). The revised target shapes for the low- and high-frequency sections are compared in Figs 5.18(a) and 5.18(b), from which it can be seen that the effect of the 'wiggle' is slightly to increase the peaking of the LF section at the crossover frequency and increase the low-frequency roll-off of the HF section but without modifying its response significantly above 3 kHz. Wiggle Factor is a very versatile design technique and is applicable to any non-coincident driver system design, employing high- or low-level crossovers.

5.2.8 Summary

None of the target functions discussed so far is ideal in every respect. In practice, some compromises must be made, and these are summarized below:

First-order Butterworth:

● not practical, attenuation rate too low

Third-order Butterworth:

● simple and practical
● asymmetric off-axis response for non-coincident drive units
● widely used

Second-order Linkwitz–Riley:

- good polar response
- simple
- dip in power response around crossover frequency

Fourth-order Linkwitz–Riley:

- good off-axis response
- excellent attenuation in stop-band
- dip in power response around crossover frequency
- widely used in high-quality systems

Linear-phase/constant-voltage:

- the low attenuation rates in either high- or low-pass sections are impractical
- poor off-axis performance
- seldom used

Filler-driver:

- complicated
- second-order design requires first-order bandpass response from filler-driver, which is usually impractical
- poor off-axis performance
- used in the early 1970s, seldom used since

Delay-derived:

- to date there is no conclusive scientific evidence to show that linear phase is a necessary requirement for a high-quality reproduction system. Improvements in both source material and transducers, however, may yet show that linear-phase designs are superior. That being so, the delay-derived designs appear to offer a practical solution. These are readily implemented using DSP along with other suitable techniques employing FIR filters[16].

5.3 Practical system design procedures

5.3.1 Two-way system

(a) The chosen drive units should be mounted in the proposed enclosure. For optimum uniformity of response in the horizontal plane, the units should be mounted one above the other. In addition, the axis of zero inter-unit time delay should coincide with the reference or design axis. In a typical listening environment, the loudspeaker system is placed so that the design axis points directly to the listener's head. Note that with the usual arrangement, in which the high-frequency unit is mounted above the woofer, the axis of zero inter-unit time delay will be directed downward (see Fig. 5.19). Where this axis does not pass through a convenient listening point, it may be tilted upwards by sloping or stepping the front baffle, or by reversing the positions of the two units so that the HF unit is below the woofer as shown. If for aesthetic or cost reasons a complicated baffle arrangement is not suitable, then nearly equivalent results may be obtained for a planar vertical front baffle with the Wiggle Factor target function design approach..

(b) Measure the frequency response of both drive units both on-and-off axis, both horizontally and vertically with respect to the reference axis, keeping the microphone

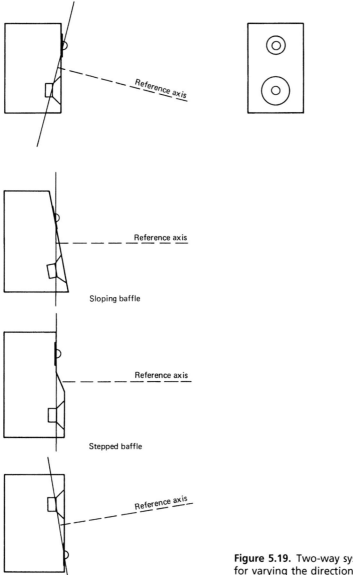

Figure 5.19. Two-way system – methods for varying the direction of the reference axis.

at the same distance from the enclosure. Normally a measuring distance of 2–3 m is used, as this is close to that used for normal domestic listening. This measurement will usually need to be carried out in an echo-free or anechoic room, although alternative measuring methods using gating techniques can enable satisfactory measurements to be made in semi-reverberant rooms (as discussed in Chapter 12).

(c) Select a suitable crossover frequency. Ideally the measured response of the low- and high-frequency units should be smooth for at least an octave above and below the proposed crossover frequency. It is also important to ensure that the crossover frequency is not so high that the LF unit has become highly directional, or so low that the HF unit distorts due to overloading under normal conditions of use. Typically for a 200 mm LF unit and a 20 mm or 25 mm dome HF unit, the crossover frequency will be between 2.5 kHz and 4 kHz.

(d) Next select a suitable target function from those given in the previous section. Where either drive unit has an irregular response close to the crossover frequency, select a target function having a relatively high slope, e.g. 18 dB or 24 dB per octave.

(e) Calculate the required low- and high-pass filter shapes, $H_L(f)$ and $H_H(f)$, that are needed to give the chosen target functions $T_L(f)$ and $T_H(f)$:

low-pass $$H_L(f) = \frac{T_L(f)}{S_L(f)}$$

high-pass $$H_H(f) = \frac{T_H(f)}{S_H(f)}$$

where $S_L(f)$ and $S_H(f)$ are the measured responses of the low- and high-frequency units respectively.

(f) The final part of the design consists of devising suitable filter sections which approximate as closely as desired to the calculated shapes $H_L(f)$ and $H_H(f)$.

A common misconception about crossover network design is that practical filters will have response shapes which correspond to that of an ideal target function. This would be true only if each drive unit had a perfectly flat response but in practice, of course, they do not. In addition, their response shapes will also change according to the size and shape of the enclosure in which the units are mounted. For any chosen target function, therefore, a successful crossover filter design must combine the dual functions of frequency division and response equalization.

5.3.2 Crossover design procedure

A sound knowledge of electrical circuit theory, plus perseverance, experience and some luck, will usually be needed to achieve a successful design. The more complex crossover designs are only practical using computer-aided optimization. Fortunately today, affordable PC-based design packages are readily available, many with built-in optimizers (see Filter Shop from Linear-X, for example). This greatly reduces the time and effort needed to achieve the intended result.

(a) Equalization and frequency division

Even though in practice these two intrinsic functions of a correctly designed crossover network will usually be combined, it is often beneficial during the initial design phase to consider them separately. In this way the choice of target function, for example, can be changed and its performance compared without the need to recalculate the equalization (see Fig. 5.20). Later on, when the design is complete, the two can be

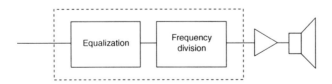

Figure 5.20. Crossover network – idealized block diagram.

combined. This is the only practical course of action in a passive network design where each section in a ladder network will interact with all the others unless costly and complicated impedance compensating sections are employed. For low-level crossover networks it is optional. An important distinction between active and passive crossover networks is that equalization in passive networks can only be achieved through *attenuation* whereas active equalizers can *boost* or *attenuate*. In any practical passive design, therefore, a balance must be found between the need to flatten the response while maintaining acceptable overall system efficiency. The next section on equalizer design will discuss the more general case of active equalizers. The additional constraints imposed by practical passive crossovers will be covered in a later section.

(b) Equalization

If the measured and smoothed response data for each drive unit are equalized so that they are flat over a sufficiently wide range, say 1–2 octaves above or below the intended crossover frequency, then the frequency dividing filter shapes will correspond quite closely to those of a classical target function.

The measured frequency response of a given drive unit, on any particular axis, often contains many irregularities due to both interference effects, resulting from secondary radiation from the edges of the enclosure, and resonant phenomena in the diaphragm and enclosure. The equalization element of any crossover network should not attempt to compensate for all of these, partly for practical reasons, but more importantly because the peaks and dips caused by interference depend on the measuring axis, whereas those due to resonance do not. Compensating for an interference dip could lead to an improvement on one axis only to make matters worse on another. It is important therefore to average the on- and off-axis response data spatially before starting the crossover design. This procedure will tend to smooth out the effects of interference without obscuring the peaks or dips due to resonance. For drive units that have a very uneven response, some frequency response smoothing – 3–6 points/octave is usually sufficient – can often simplify the design procedure.

The first step is to devise a series of cascaded equalization sections which will flatten the smoothed response of each drive unit over a frequency range extending at least 1–2 octaves above and below the intended crossover frequencies. Secondly the frequency division sections, which in cascade approximate the desired drive unit target functions, are added to complete the crossover section for the low- and high-frequency bands.

A selection of second-order equalization sections (also known as biquads)[14], which will accommodate most practical crossover design requirements, are shown in Figs 5.21(a–h) and 5.22(a–c) together with their associated transfer functions. Any of these biquad sections may be realized, in an active implementation, using the state-variable configuration. A single op-amp implementation can be achieved by using the Friend biquad[14]. Note that these convenient building blocks can also be utilized for the implementation of digital dividing networks. By performing an s-to-z transform of the transfer functions of each of the biquad sections, using the bilinear transform for example, then a satisfactory design can be achieved using IIR (infinite impulse response) filter sections. Many other digital filter designs based on other topologies and design approaches, including genuine linear phase implementations, can be implemented[15,16,17,18].

(c) LF and HF shelves – boost or cut

The second-order LF shelves may be used to raise or lower the existing cut-off frequency (f_1) and modify the Q-factor (Q_1) of any second-order high-pass section to some desired new values f_2 and Q_2 (see Fig. 5.21(b) and (d)). The resulting second-order high-pass section, with cut-off frequency f_2 and Q-factor Q_2, can then form

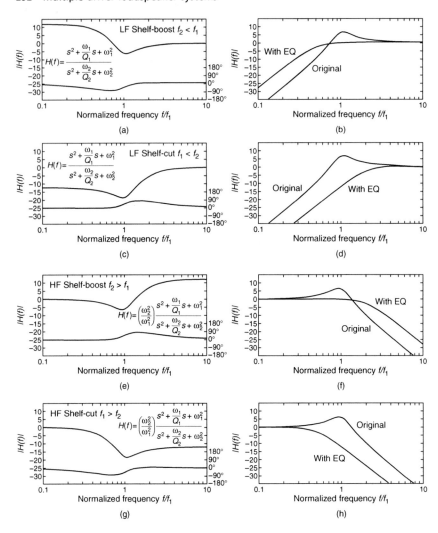

Figure 5.21. (a)–(h) Second-order (biquad) equalizer sections – shelving.

part of the desired system target function. For example, if the required MF target function were a fourth-order Linkwitz–Riley high-pass with a cut-off frequency of 400 Hz, then suitable values for f_2 and Q_2 would be 400 Hz and 0.707. The corresponding electrical high-pass section of the crossover would also have an f and Q of 400 Hz and 0.707 respectively, since a fourth-order Linkwitz–Riley is equivalent to two cascaded second-order sections having the same cut-off frequency and a Q of 0.707.

In a similar way, the HF shelves can be used to extend or lower the existing cut-off frequency of a middle- or high-frequency unit.

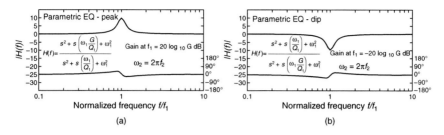

Figure 5.22. (a)–(b) Second-order (biquad) equalizer sections – parametric.

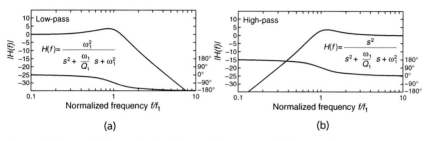

Figure 5.23. (a)–(b) Second-order (biquad) frequency division sections.

(d) Narrow-band boost and cut (parametric equalizers)

These sections can be used to correct for unwanted peaks or dips in the smoothed drive unit responses and are normally specified in terms of the peak or dip frequency, Q-factor and gain (see Fig. 5.22(a–b)).

(e) Frequency division

Shown in Fig. 5.23 are the frequency division elements, and their corresponding transfer functions, which are used to band limit the response of each drive unit. Any desired drive unit target function can be achieved by cascading a suitable combination of first- and second-order sections with the equalized drive unit response.

5.3.3 Design example 1: active MF crossover

The spatially and frequency smoothed response data for a horn-loaded MF unit is shown in Fig. 5.24(a). The desired MF target function had a bandpass characteristic consisting of a sixth-order, 268 Hz, Linkwitz–Riley high-pass section in cascade with a fifth-order, 2.65 kHz, low-pass Butterworth filter section (see Fig. 5.24(b)). The block diagram for the final crossover (see Fig. 5.25) consisted of four equalization sections (two peak, one dip and one HF boosting shelf) and three frequency division sections (two second-order high-pass and one second-order low-pass). The individual equalization and frequency-division sections are shown in Figs. 5.26(a) and (b) and cascaded in Fig. 5.27. The final optimized-versus-desired target shapes are shown in Fig. 5.28.

5.3.4 Design example 2: passive MF crossover

A typical example showing both the passive circuit topology and the corresponding response shape for the MF section of a three-way system is given in Figs 5.29–5.33.

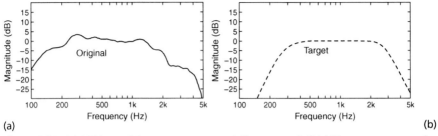

(a)   (b)

Figure 5.24. (a) MF horn driver response – spatially averaged, (b) MF target response for horn driver.

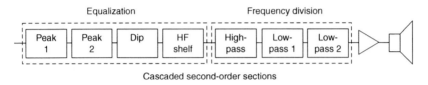

Figure 5.25. Block diagram: MF crossover network (Active).

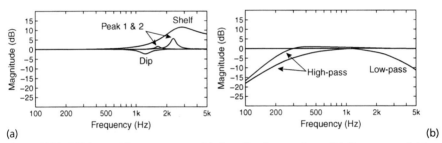

(a)   (b)

Figure 5.26. MF horn driver crossover. (a) Equalization sections. (b) Frequency division sections.

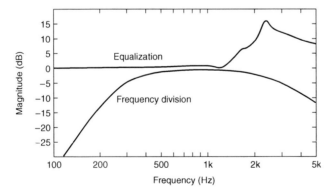

Figure 5.27. MF horn driver crossover. Overall equalization and frequency division responses from cascaded second-order sections.

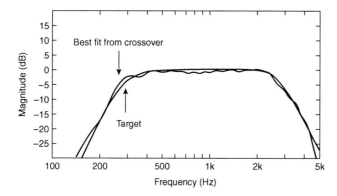

Figure 5.28. MF horn driver crossover. Target-versus-actual acoustic response via optimized crossover.

The target function in this example is a fourth-order Linkwitz–Riley bandpass with crossover frequencies at 400 Hz and 3 kHz.

The design of passive crossover networks is complicated by the need to ensure that the desired target crossover filter shapes are achieved when the network is loaded by the complex input impedance presented by each drive unit when mounted in the intended enclosure. For this reason it is first necessary to measure the input impedance of each driver, when mounted in the system enclosure, and then model that

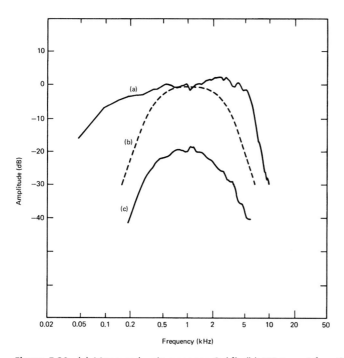

Figure 5.29. (a) Measured unit response $S_M(f)$. (b) MF target function $T_M(f)$. (c) Required filter response $H_M(f)$.

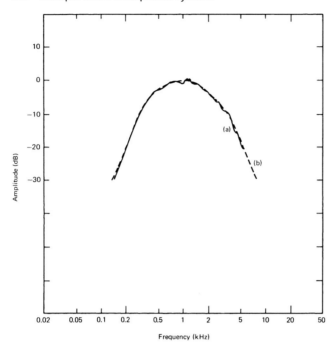

Figure 5.30. (a) Required filter response $H_M(f)$. (b) Measured response of optimized MF filter section.

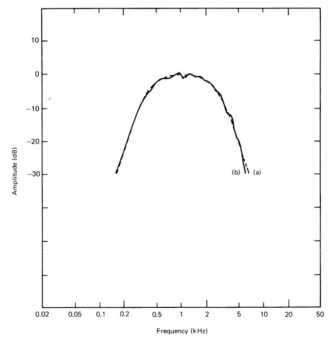

Figure 5.31. (a) MF target function $T_M(f)$. (b) Measured MF response via optimized filter network.

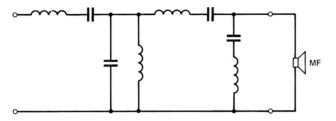

Figure 5.32. Passive MF section – circuit topology.

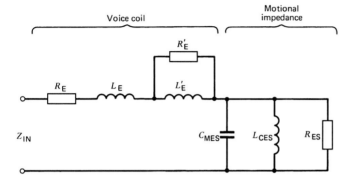

Figure 5.33. Electrical circuit topology used to model moving-coil drive unit input impedance.

impedance characteristic by means of an equivalent electrical circuit. The circuit given in Fig. 5.33 was used for the example given above.

The circuit values are obtained by using a computer-aided optimizer to determine those component values which give the best fit between the measured and calculated input impedance. This same topology may be used, with suitable changes of component values, to represent accurately the input impedance of most moving-coil drive units.

5.3.5 Load equalization

With complex passive designs, it is often useful to connect equalizing sections in parallel with the drive unit so that a more uniform and nearly resistive load is presented to the filter. Using the circuit topology shown in Fig. 5.34, it is possible to design a circuit such that, when it is connected in parallel with a drive unit, its input impedance is absolutely flat and equal to the d.c. resistance of the voice coil. Variations of this circuit are often used for the crossover sections between the LF and MF drivers in three-way designs, in order to minimize the effect of interaction between the motional impedance of the drivers and the passive filter network. The circuit in Fig. 5.34, although capable in theory of providing an exact solution, often requires impractical component values, or ones which are expensive and not always necessary. One simplification is to use a series CR network connected in parallel. By a suitable choice of C and R it is possible to flatten the input impedance curve effectively above the peak at the resonance frequency.

Load compensation

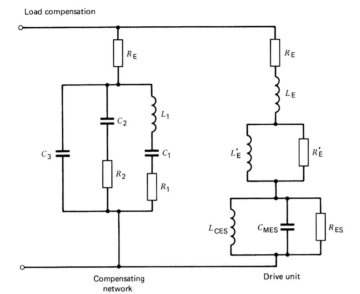

Compensating
network

Drive unit

$$L_1 = C_{MES} R_E^2 \qquad C_2 = L'_E / R_E^2$$
$$C_1 = L_{CES} / R_E^2 \qquad R_2 = R_E^2 / R'_E$$
$$R_1 = R_E^2 / R_{ES} \qquad C_3 = L_E / R_E^2$$

Figure 5.34. Network for compensating loudspeaker impedance. Input impedance, $Z_{i-n} = R_E$ (after Thiele).

5.4 Summary

Successful crossover design for multiway loudspeaker systems need not depend on cut and try methods. The target function approach, together with computer-aided circuit optimizers, can always be used to provide a solution which is both accurate and cost effective. The same procedure, with slight variations, may be used for both high- and low-level crossover designs.

In the future it may be expected that DSP-based digital filters, with their ability to provide linear-phase systems, time-delay correction between non-coincident drivers and complex but practical driver equalization, will come to dominate crossover design and narrow the gap between loudspeakers and the rest of the audio chain.

References

1. KEFTOPICS, 'A target function approach to the design of filters', KEF Electronics Limited. Vol. 2, No. 1 (1978).
2. LIPSHITZ, S P and VANDERKOOY, J, 'A family of linearphase crossover networks of high slope derived by time delay', *JAES,* **31,** No. 1/2, 2 (1983).
3. LIPSHITZ, S P and VANDERKOOY, J, 'Is phase linearization of loudspeaker crossover networks possible by time offset and equalization?' *JAES,* **32,** No. 12, 946 (1984).
4. BAEKGAARD, E, 'A novel approach to linear phase loudspeakers using passive crossover networks', *JAES ,* **25,** No. 5, 284 (1977).
5. ASHLEY, J R and HENNE, L M, 'Operational amplifier implementation of ideal electronic crossover networks', *JAES ,* **19,** No. 1, 7 (1971).
6. GARDE, P, 'All-pass crossover systems', *JAES,* **28,** *No.* 9, 575 (1980).

7. LINKWITZ, S H, 'Active crossover networks for noncoincident drivers', *JAES*, **24**, No. 1 (1976).
8. ASHLEY, J R, 'On the transient response of ideal crossover networks', *JAES*, **10**, No. 3, 241 (1962).
9. THIELE, A N, 'Optimum passive loudspeaker dividing networks', *Proceedings of the IREE*, July (1975).
10. SMALL, R H, 'Constant voltage crossover network design', *JAES*, **19**, No. 1, 12 (1971).
11. KEFTOPICS, 'Crossover filters an integral part of overall system engineering', KEF Electronics Limited, Vol. 4, No. 2.
12. KEFTOPICS, 'Reference Series Model 105', KEF Electronics Limited, Vol. 3, No. 1.
13. BERMAN, J M and FINCHAM, L R, 'The application of digital techniques to the measurement of loudspeakers', *JAES*, **25**, No. 6, 370 (1977).
14. SEDRA, A S and BRACKETT, P O, *Filter Theory and Design: Active and Passive*, Pitman Publishing, London, pp. 392, 591–628 (1979).
15. WILSON, R, 'Filter topologies', *JAES* , **41**, No. 9, (1993).
16. WILSON, R, ADAMS, G and SCOTT, J, 'Application of digital filters to loudspeaker crossover networks' *JAES*, **37**, No. 6 (1989).
17. HAWKSFORD, M O J, 'Asymmetric all-pass crossover alignments', *JAES*, **41**, No. 3, (1993).
18. HAWKSFORD, M O J, 'Digital signal processing tools for loudspeaker evaluation and discrete-time crossover design', *JAES*, **45**, No. 1/2 (1997).
19. HILLERICH, B, 'Acoustic alignment of loudspeaker drivers by nonsymmetrical crossovers of different orders', *JAES*, **37**, No. 9 (1989).
20. SCHUCK, P L, 'Design of optimized loudspeaker crossover networks using a personal computer', *JAES*, **34–35** (1986–7).
21. WALDMAN, W, 'Simulation and optimization of multi-way loudspeaker systems using a personal computer', *JAES*, **36**, No. 9, (1988).

Bibliography

AARTS, R M, 'A new method for the design of crossover filters', *JAES*, **37**, No. 6 (1989).
AARTS, R M and KAIZER, A J M, 'The simulation of loudspeaker crossover filters with a digital signal processor', preprint 2471, 82nd AES Convention, London (1987).
ADAMS, G J and ROE, S P, 'Computer-aided design of loudspeaker crossover networks', *JAES*, **30**, No. 7/8, 496 (1982).
AUGSPURGER, G L, 'Electrical versus acoustical parameters in the design of loudspeaker crossover networks', *JAES* , **19**, No. 6, 509 (1971).
BLINCHIKOFF, H J and ZVEREV, A I, *Filtering in the Time & Frequency Domains*, John Wiley, Chichester (1976).
BULLOCK, R M, 'Loudspeaker crossover systems: an optimal crossover choice', *JAES*, **30**, No. 7/8, 486 (1982).
BULLOCK, R M, 'Satisfying loudspeaker crossover constraints with conventional networks old and new designs', *JAES*, **31**, No. 7/8, 489 (1983).
BULLOCK, R M, 'Passive three-way all-pass crossover networks', *JAES*, **32**, No. 9, 626 (1984).
CHEN, W, 'Performance of cascade and parallel IIR filters', *JAES*, **44**, No. 3 (1996).
CROWHURST, N H, 'Crossover design', *Audio Magazine*, **26**, September (1963).
FINK, D G, 'Time offset and crossover design', *JAES*, **28**, No. 9, 601 (1980).
DE WIT, R P, KAIZER, A J M, and OP DE BEEK, F J, 'Numerical optimization of the crossover filters in a multiway loudspeaker system', *JAES*, **34**, (1986).
JACKSON, L B, *Digital Filters and Signal Processing*, Kluwer Academic, London (1989).
KEFTOPICS, 'The Calinda and Cantata loudspeakers', KEF Electronics Limited, Vol. 3, No. 2.
KEFTOPICS, 'A target function approach to the design of filters', KEF Electronics Limited, *Vol.* 2, No. 1 (1978).
LINKWITZ, S H, 'Passive crossover networks for non-coincident drivers', *JAES*, **26**, No. 3, 149 (1978).
MITRA, S K, DAMONTE, A J, FUJIL N and NEUVO, Y, 'Tunable active crossover networks', *JAES*, **33**, No. 10, 762 (1985).
OPPENHEIM, A V and SCHAFER, R W, *Digital Signal Processing*, Prentice Hall, Englewood Cliffs, NJ (1975).
POTCHINKOV, A, 'Frequency-domain equalization of audio systems using digital filters. Part 1: Basics of filter design', *JAES*, **46**, No. 11, (1998).
RABINER, L R and GOLD, B, *Theory and Application of Digital Signal Processing*, Prentice Hall, Englewood Cliffs, NJ (1975).

SEDRA, A S and BRACKETT, P O, *Filter Theory and Design: Active and Passive*, Pitman Publishing Limited, London (1979).

SHPAK, D J, 'Analytical design of biquadratic filter sections for parametric filters', *JAES*, **40**, No. 11 (1992).

VANDERKOOY, J and LIPSHITZ, S P, 'Power response of loudspeakers with noncoincident drivers – The influence of crossover design', *JAES*, **34**, No. 4, (1986).

6 The amplifier/loudspeaker interface

Martin Colloms

6.1 Introduction

The loudspeaker does not stand alone. For it to operate, it must be fed electrical power corresponding to audio signals. Given that loudspeaker efficiency is low, when compared with other systems such as an electric motor, fairly large amplifiers are necessary to produce realistic sound intensities. It is fortunate that our great aural sensitivity does not require high acoustic power for good subjective loudness. High-quality domestic and monitoring loudspeaker systems show a typical conversion efficiency of 0.5–1.5% for the conversion of electrical input to radiated acoustic power, while stage, public address and sound-reinforcement systems using horn-loaded drivers may offer an increased efficiency of the order of 5–15%. For example, a loud-speaker rated at 92 dB/W at 1 m is only 1% efficient under free-field conditions. Consequently substantial electrical inputs are required, corresponding to a range of 20–400 W of amplifier rating, according to the system size and application. Most of this power is dissipated as heat in the drive units.

Traditional power amplifier design is based on considerations of a continuous power rating into a defined resistive load, usually 8 Ω. Loudspeakers and loudspeaker systems are also generally specified in terms of a given loading, most commonly an 8 Ω 'impedance'. The use of the word impedance is revealing, since it includes the possibility of reactive as well as resistive components and is a clue to the often complex nature of a loudspeaker load. It is this complexity which can make demands on a power amplifier far beyond the nominal 8 Ω matching requirement. Such demands make the amplifier-to-speaker interface a subject worthy of closer investigation in addition to general issues of interconnection practice.

The amplifier is commonly located at some distance from the loudspeaker and the two need to be linked by an appropriate choice of electrical cable. Specifications need to be set here if the designed performance is not to be compromised. Considerations of loudspeaker connector and switching practice should also be included as links in the transmission chain. For lower-quality applications where very long speaker cables are required, such as for sound distribution, the system can be made efficient by operating at higher voltage, lower current like an electrical power transmission network, using nominal '100 V' lines and matching transformers at source and load.

While the use of a power amplifier remote from the speaker is usual, an alternative method may also be employed, namely the 'active' loudspeaker. Here the passive, power level, crossover network which may divide the frequency range between the various drivers in the system is replaced by an 'active' or electronic line level filter. From this filter, the divided ranges are fed to separate power amplifiers for each drive unit. While this 'active' electronic assembly can be remotely positioned, it is usual to combine it, adjacent to, or built into the loudspeaker enclosure to form a complete active speaker system. The on-board electronics may then offer extra facilities such as special user-selectable low-frequency alignments, or incorporate control techniques such as motional feedback, a servo method for linearizing and/ or extending the low-frequency output of a speaker system.

An active system is driven with analogue signals at line or low level via a screened cable, preferably balanced for lowest system noise. A variant is also possible whereby a digital-to-analogue converter is incorporated into the loudspeaker system. Feed of encoded digital audio is thus possible, direct from PCM systems, CD players and digital studio mixing desks. Via a computer or infrared remote control data link, such a loudspeaker may be programmed digitally with respect to many audio parameters, such as output level and frequency response. The latter can be controlled and compensated to near perfection in real time and thus be programmed to take account of, and compensate for, the effects of local boundary reflections. Design refinement may extend to DSP executed crossover networks operating in the digital domain with separate DACs for each operating driver channel.

Blurring the traditional distinction between amplifier and loudspeaker, the computer industry has come up with a new data interface for peripheral components which has implications for the loudspeaker. Developed from the concept of an amplifier-equipped speaker system with a digital input capability, the more recent generations of million-selling digital audio sound cards designed for computers have created new design opportunities for the speaker engineer.

While drawing up the specifications for an advanced intelligent computer port, one with still more compact connector practice, sufficient thought was given to include digital audio data with both send and receive capability. Other data handling duties include memory/backup, keyboard, mouse and printing. Called 'USB' for Universal Serial Bus, the new standard forms an important part of Microsoft's 'plug and play' strategy whereby peripherals may be simply plugged in as required, online, system active, and they are automatically identified and configured by the operating system, immediately ready for use.

For an active 'computer' loudspeaker system to be recognized it must have some measure of Bus intelligence and USB loudspeakers effectively include a new generation digital audio sound card. Here the physical interface between the loud-speaker and the computer for data handling/storage/replay comprises a compact cable fitted with a miniature multiple contact telephone style connector. The Bus data rate is up to ten times that of the older serial computer port and two data rates are specified, 1.5 Mbps and 12 Mbps (about 1.5 Mbyte/s) ; and the latter higher speed is necessary for digital audio of near CD quality and the like. High-quality multi-channel audio and video is beyond this Bus capability. The lower rate is, however, suitable for slower peripherals and perhaps MP3 and related highly data-reduced formats.

A USB compatible speaker will include a USB-equipped 'sound card' which gener-ally offers analogue and digital inputs as well as analogue and digital outputs. Thus the mic and/or line inputs to a computer are now usually available at the master speaker location and not, as is usual, inconveniently tucked away at the back of the computer.

USB-equipped speakers need not be costly. A good-quality example system with a central subwoofer powered by a 140 mm low-frequency driver of respectable power output, partnered by two robustly constructed satellites, is currently priced near the upper range of computer/desktop multi-media speaker systems.

USB-interfaced speakers may have some basic controls accessible by the user, while the linked intelligence means that the usual on-screen software allows for a virtual control panel for speaker system operation, via the DSP in the speaker enclosure. In addition to the usual facilities – volume, balance and tone – more sophisticated signal processing is readily available for ambience, image manipulation, including 3D sound field simulation and graphic frequency equalization. Music audio editing pack-ages are increasingly popular and much broadcast program editing, for example radio production, is now done on a PC platform. The appeal of good USB speaker systems is strong for such applications. Speakers are also assuming greater importance as audio related aspects of the PC expand. With digital audio available at several levels of replay quality over the Internet – e.g. MP3 (MPEG layer 3) data-reduced audio coding, games software now at the 16-bit level for the audio data component, and the PC capable of CD, DVD, television and radio reception, including digital radio (DAB) – versatile local speaker systems have become a vital component of a comprehensive personal computing installation.

6.2 The electrical load presented by the loudspeaker

For power amplifiers, test and design practice has tended to assume that a laboratory 8 Ω or 4 Ω fixed resistor is a valid load. Amplifiers are often specified in terms of their continuous capability into such a load, the output being expressed in watts, W, with a zero power factor, i.e. the voltage and current remain in phase for such a 'resistive' load. Loudspeakers, whether single drivers or multi-way systems, are also specified in similar terms, i.e. as a nominal impedance of 8 Ω, or 4 Ω or rarely 16 Ω. The IEC standards[1,2] suggest that the allowed modulus of impedance over the frequency range should not vary by more than 20% of the nominal rating, i.e. not fall below 6.4 Ω for an 8 Ω system and 3.2 Ω for a 4 Ω system. In practice, excursions to higher than nominal excess values are regarded as less harmful since they reflect easier loading for both amplifier and cable.

With very few exceptions, the load impedance of a practical loudspeaker is non-uniform, and these variations are generated by reactive electrical components resulting from the operation of a coupled dynamic transducer, even a single drive unit. Take the commonest example, a moving-coil driver mounted in a sealed box (Fig. 6.1). The impedance curve shows the 8 Ω nominal value at only three points. The peak is at 56 Hz and corresponds to the system resonance, produced by a combination of the moving mass of the diaphragm and the compliance contribution of the driver suspension plus the trapped air within the box. Below resonance, the phase angle of the current drawn relates to a corresponding complementary component of reactive loading, of typically 200 μF. This corresponds to the 'stiffness'-controlled region of the loudspeaker system. At resonance, the impedance is high, and resistive. Above resonance, the speaker is now in the mass-controlled region, dominated

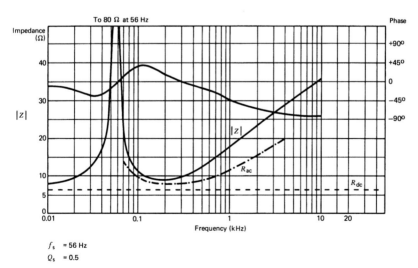

f_s = 56 Hz
Q_s = 0.5

Figure 6.1. Impedance characteristic of a single moving-coil driver, $|Z|$, R_{dc}, phase R_{ac}.

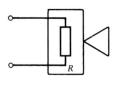

(a)

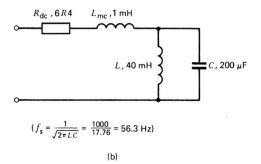

$(f_s = \dfrac{1}{\sqrt{2\pi LC}} = \dfrac{1000}{17.76} = 56.3 \text{ Hz})$

(b)

Figure 6.2. (a) Notional 8 Ω loudspeaker load. (b) Example equivalent electrical circuit of a single 8 Ω moving-coil driver.

by the effective moving mass of the driver and showing a parallel or shunt, inductive input impedance of around 40 mH. With increasing frequency, the load value settles near to the d.c. resistance of the voice coil and becomes almost purely resistive again at around 6.5 Ω. With increasing frequency, the input impedance now begins to rise, approaching a 6 dB/octave rate, this due predominantly to the electrical inductance of the voice-coil winding which in this example is about 1 mH. These values are represented in the equivalent circuit (Fig. 6.2(b)). Additionally, small resistive losses in the suspension and the magnetic circuit are also present in practical drivers.

These reactive components of electrical impedance make greater demands on the power amplifier than a purely resistive load; the significant power factor increases the power dissipation in the output stage and imposes additional demands on the stability criteria for the amplifier. When several drivers are combined in multiway systems, classic book design practice may suggest that the required filter networks should be designed for constant resistance loading to benefit the amplifier. However, in practice the networks or crossovers are generally designed to provide the desired acoustic frequency response for the system, hopefully consistent with a tolerable variation in speaker input impedance., The crossovers are often deliberately mismatched in respect of impedance in order to provide driver equalization, frequently at the cost of a severely non-uniform frequency response with respect to impedance.

For example, a two-element, 12 dB/octave roll-off Butterworth design of filter network for an 8 Ω nominal load set at a crossover frequency of 1.6 kHz, ($\omega = 10^4$) will in theory use a combination of a 10 μF capacitor and a 1 mH inductor (of typically 0.3 Ω series resistance) (Fig. 6.3). With the low-pass filter form complemented by a reciprocal high-pass network, the input impedance for this textbook example is essentially resistive at close to 8 Ω, provided that the loudspeaker driver terminations are themselves 8 Ω.

When such a network is misterminated, e.g. by a real drive unit, the resulting network can then reflect a substantial variation in input loading constituting an overall mistermination[3] (Fig. 6.4).

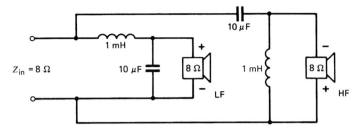

Figure 6.3. Two-pole passive two-way crossover (theoretical), 12 dB/octave Butterworth at 1.6 kHz.

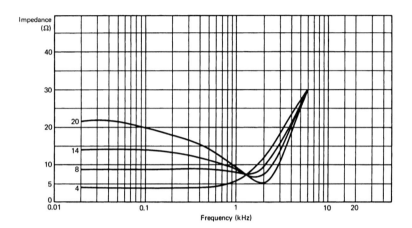

Figure 6.4. Misterminated two-pole filter, effect on input impedance (after Kelly).

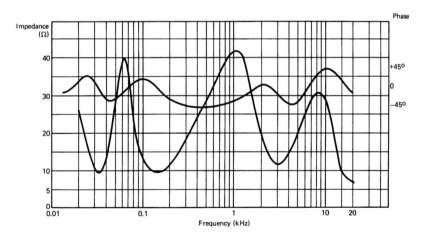

Figure 6.5. Impedance/phase characteristic of three-way crossover with equalizer. Crossovers at 3 kHz, reflexed at 34 Hz.

Simplifying, if a 20 Ω driver load were to be present, the input impedance will actually fall to 5 Ω near the crossover point due to excessive 'undamped' current flowing in the filter loop. With multi-way systems, which also frequently use additional equalizing networks to take account of the drive unit responses, the overall load impedance presented by a speaker system can become very complex (Fig. 6.5).

The impedance curve is conventionally and conveniently measured by feeding the loudspeaker from a frequency sweep generator with a sufficiently high output imped-ance (e.g. > 1 kΩ) (see Chapter 12), to give a good approximation to a constant current signal. The voltage variation at the speaker terminals corresponds to the modulus of the impedance, and is preferably plotted on a linear amplitude scale for clarity. Additionally, the phase angle may be plotted, recording the phase angle differ-ence between the generator source and that measured at the load. The resistive and reactive components may then be extracted and plotted separately.

One objection to the current sweep method for assessing impedance is that only one frequency is stimulated at a time, whereas music programme drive over a given spectrum can simultaneously excite many frequencies. There may also be a complex, short-term time history of input power and current in the speaker system. Thus judgement of the overall impedance rating can be made on only an approximate basis, that of correspondence with the ±20% tolerance defined by the published standard.

An alternative method with greater relevance consists of a broad-band pink noise stimulus, recording the r.m.s. voltage and current, and applying Ohm's law to calcu-late the nominal overall impedance value. Questions of spectral weighting may also arise for the noise stimulus to try and describe the loading effect on an amplifier more precisely. Neither of these methods tells the full story and, additionally, under dynamic or music conditions the instantaneous impedance can be substantially lower than the steady stimulus results suggest.

Research[4,11] has shown that on a simulated programme commercial 4–8 Ω rated multiway speakers can draw peak currents corresponding to a dynamic impedance as low as 2 Ω. Some worst case examples even fall below this. The problem arises due to the parallel nature of the sections of multiway systems. For example, a low-frequency pulse may excite the bass section, and its highly reactive nature may then require negative current from the amplifier following the pulse. If a mid- or treble-frequency transient, negative-going, immediately follows the low-frequency signal, the two currents may sum to a negative peak much higher than the swept steady-state impedance curve suggests. This peak current, which may be assessed using a variety of synthesized pulses related to observed music transients, may be used to define more precisely the effective impedance of the system.

A conflict of opinion arises over the test signal used to explore these peak demands and its likelihood of occurrence on a real programme. A statistical analysis of typical programme may show the incidence of such peak demands as a percentage proba-bility. It is up to the amplifier designer whether or not to design his product to cope with the worst case probability of stimulus and commercial load, or to arrive at some compromise.

Practical tests by the author have provided considerable backing for the above view of the concept of 'dynamic impedance'. In a large review test project for some 55 amplifiers covering a wide price range, extensive double-blind subjective evalua-tion was matched by laboratory testing. While several areas of correlation were apparent between the two domains of test, objective and subjective, analysis of the whole report revealed a strong link between a good subjective result at specifically higher power levels and one particular laboratory test parameter. This turned out to be the peak current capability, assessed on short-term pulsed power delivery into 2 Ω and with simulated reactive loads. The results suggested that for amplifiers rated from 30 to 50 W into 8 Ω, a peak current capability of 10–15 A or more is desirable, rather higher than book values indicate. Up to 100 W, 25 A is required, while for

200 W, 50 A is a desirable design target. This takes into account the real life loading imposed by the more complex high-fidelity loudspeaker systems.

It is also possible to compensate fully the input impedance of a loudspeaker, though this can be expensive and, after strong commercial promotion, has not found wide adoption (see Section 6.3). Other classes of transducer present different loads to the moving-coil type considered so far. For example, the 'isodynamic' or stretched-film speakers, using a spaced surface conductor pattern and an acoustically open array of magnets, are moving-coil in principle but their moving mass is so low compared with their total resistive losses, acoustic and mechanical, that the reactive component of impedance at low frequencies is small. Given good crossover design in the case of multiway types, such systems can have a predominantly resistive input impedance and are comparatively easy to drive.

Electrostatics provide another class, these capacitor-principle transducers operate on the high voltages (up to 5 kV) produced by a matching step-up transformer. While from a theoretical viewpoint the electrostatic is highly efficient and hence in the mid-range presents a high impedance, nevertheless the overall load impedance can still be difficult to drive at the frequency extremes (Fig. 6.6). At low frequencies, the finite primary inductance of the matching transformer allows the impedance to approach that of the d.c. winding resistance, usually a low value. At high frequencies, depending on the crossover arrangements, the intrinsic capacitance loading of this speaker (up to 2 μF) makes itself felt as a falling modulus with a phase lead, this also tending to destabilize the usual design of negative feedback amplifier. For this type of load, the matching amplifier must have a good stability margin and be capable of decent current output at the frequency extremes. With some amplifiers, a series buffering resistor of 1–2 Ω is a wise precaution for an electrostatic load and the subsequent loss in output level at 20 kHz is a necessary condition. Note that many amplifiers will also show some overshoot and response peaking with the shunt capacitance of an electrostatic speaker load. The series resistance may not be a bad thing after, all since it may well restore the correct response over the required bandwidth.

Ribbon speakers, if of the single-conductor type, will also require transformer matching for their exceedingly low diaphragm resistance (as low as 0.05 Ω) and the complex behaviour of the transformer deserves consideration. Piezo-electric drivers, sometimes used as small high-frequency units, generally offer such a small load capacitance as to impose only a very mild loading. They are typically as little as 0.1 μF and are used with a series resistor of 50 to 200 Ω, this forming a simple low-pass equalizing filter.

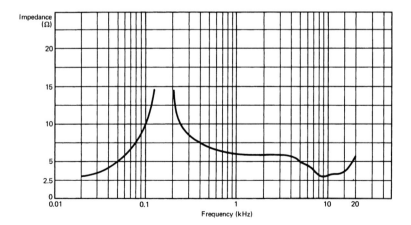

Figure 6.6. Impedance of electrostatic loudspeaker.

The commonest load problem for the amplifier appears to rest with the dominant moving-coil type and its relatively high mass diaphragms, where inertia results in back e.m.f. The tighter the electromagnetic coupling, i.e. the lower the system Q-factor, the greater is the potential negative peak current demand. With the recent loudspeaker trends towards higher rated sensitivity, lower system Q and generally lower impedance, the demands on amplifiers have correspondingly increased.

6.3 Impedance compensation

It is possible to compensate for the variations in input impedance of a speaker by applying appropriate networks in parallel with the crossover whose electrical components are conjugate with those of the system. The simplest of these is the Zobel network, used to control the rising impedance of a moving-coil driver above the resonance range, this due to coil inductance. The network (Fig. 6.7) consists of a resistor R_{dc} and capacitor C. The capacitance is given by:

$$C = \frac{L}{R^2}$$

where R is the effective a.c. resistance (includes losses, eddy currents, etc.) in the region of compensation and L is the motor coil inductance. The network resistor typically corresponds to the d.c. resistance of the voice coil.

For a common driver type, $L = 0.5\,\text{mH}$, $R_{dc} = 6.4\,\Omega$, and the a.c. resistance at 1–3 kHz is around $10\,\Omega$. The calculated capacitance is 6 μF and the beneficial effect on the driver impedance is clearly apparent (Fig. 6.8).

Such compensation can improve the termination for a particular crossover network and for the amplifier. In another example, the crossover point chosen for a given

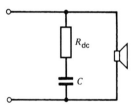

Figure 6.7. Zobel compensation of driver inductance.

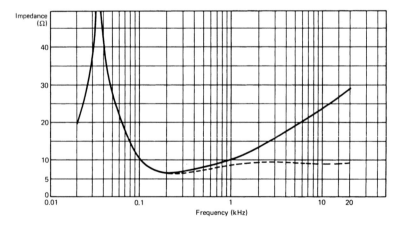

Figure 6.8. Compensation of coil inductance, effect on impedance curve (see Fig. 6.7).

high-frequency unit may be rather close to its fundamental resonance, in the capacitative region of the impedance curve. In this case the previously chosen capacitance for the Zobel network may be replaced by an inductor, though here compensation will hold only down to the resonance frequency. Completion of this compensation may be achieved by a full conjugate impedance matching, whereby the rise at resonance is almost wholly countered by an appropriately damped parallel resonant circuit, working with the above Zobel network covering the upper frequencies.

6.4 Complete conjugate impedance compensation

In Fig. 6.9 such a circuit for a 19 mm dome, plastic foil high-frequency driver is shown, while the compensated and uncompensated impedance curves are illustrated in Fig. 6.10. Here the final circuit provides a virtually uniform 6.4 Ω resistive loading. Full conjugate impedance compensation may also be applied to moving-coil speaker systems over the whole frequency range.

In theory the compensation could be applied at the input terminals of the completed speaker equipped with crossovers but, in practice, it is worth while to try to compensate each section of the loudspeaker separately as in the case of the commercial example illustrated in Fig. 6.11).

Taking the low-frequency 'way' first, here a two-pole filter network supplies the two paralleled bass drivers. Incidentally, these are loaded by a combination of a sealed box at the diaphragm rear and a ported, tuned cavity at the front, a bandpass enclosure form. The resulting low-frequency motional impedance is nulled by a circuit representing the inverse characteristic of the driver /enclosure combination. This

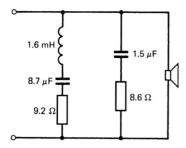

Figure 6.9. Conjugate impedance compensation of 19 mm plastic foil HF unit.

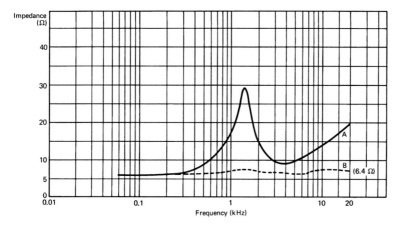

Figure 6.10. Compensated impedance curve of 19 mm plastic foil HF unit (KEF T27).

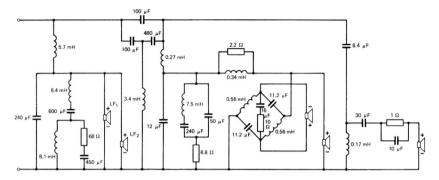

Figure 6.11. Commercial three-way network with conjugate impedance compensation (KEF R104:2).

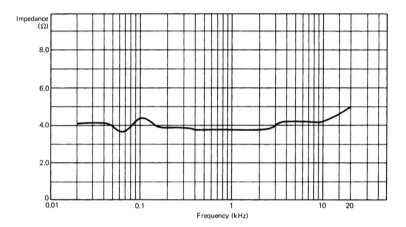

Figure 6.12. Fully compensated impedance curve of three-way system. Note neglible variation at low frequencies (see Fig. 6.11).

compensation comprises the five-element network shown which of necessity must employ components of substantial power headroom and value. They must match the dynamic, peak current capability of the system if the compensation is to be maintained over the specified power range. Premature overload of the network will simply result in unnecessary harmonic distortion and still more severe amplifier loading, negating the entire principle. The next section, bandpass for the mid-frequencies, begins with a third-order high-pass network which also includes some frequency response equalization. It leads to a third-order low-pass at the upper crossover frequency. A conjugate matched network is present, namely the four-element network which compensates for the simple, sealed box resonance of the back loaded pair of matched, mid-range drivers. Incidentally note that one mid-range unit is fed via an all-pass delay network to help widen the vertical polar response of the system.

The final treble arm, high-pass, is also third-order but does not require a compensating network. This is mainly because this high-frequency driver has ferrofluid in the magnet gap, of sufficient viscosity to damp the motional impedance and make compensation unnecessary at the chosen crossover point. Also in this case the slightly rising inductive impedance with increasing frequency has been left in circuit to compensate for a possible mild increase in source resistance, which is often present with some combinations of amplifier and speaker cable.

In this particular example, the overall impedance is specified at 4 Ω and, as can be seen in Fig. 6.12, the compensation is very successful. In effect, the power amplifier working this speaker is dealing with an almost laboratory-grade 4 Ω resistance. Setting the speaker to a lower than usual 4 Ω value (not 8 Ω) was considered to be acceptable by the designers on the grounds that most amplifiers are designed with tolerance of loading below 8 Ω, and that most were also capable of their continuous power maximum into 4 Ω, perhaps twice their normal rating. Taking advantage of this, the compensated form of speaker can then claim a higher voltage sensitivity; namely 92 dB at 1 m for a 2.83 V input (in fact 2 W into 4 Ω) compared with the 89 dB which would hold if it had been designed to the 8 Ω standard.

CAD (computer aided design) was employed for both the crossover and the compensation networks. The latter needed to be closely toleranced to be effective, and likewise the loudspeaker system tuning and Q factors, both bass and mid-range, needed to be held constant in manufacture and be maintained over the designed power range if the compensation were to work as intended.

Several side effects of compensation are possible. Some designers have ascribed, at the highest quality level, some loss of clarity and dynamic expression to the technique. Given that power level passive filter components are generally large, and are to some degree imperfect, and that compensation can greatly add to the number of components loading the amplifier terminals, some level of unwanted interaction may occur between crossover components, e.g. stray magnetic fields and circulating currents in the tracking or wiring, and untoward subjective effects might well be anticipated.

There is also a view that, while an uncompensated speaker might suffer from a couple of possibly prejudicial dips in impedance, the mean value remains fairly high and the mean power demanded from the amplifier is thus quite moderate. However, if such a speaker is fully compensated to a constant low value, it will draw full current at all parts of the frequency spectrum. It is possible to show statistically that the amplifier will run hotter, its power supplies will be more heavily loaded and, since distortion is generally proportional to current draw, the average distortion level will also be higher.

When seeking a particular advantage in speaker design, it is worth considering all the implications for the amplifier/speaker interface and not just the benefit to the speaker specification. A case in point is the present regrettable trend to claim and sometimes provide higher sensitivity in order to produce favourable A/B comparisons in a point-of-purchase showroom. A number of commercial loudspeakers are not only significantly optimistic concerning rated sensitivity, +2 dB and more, but claim 8 Ω loading where in fact the mean value may be as low as 5 Ω and some showing minima at 2.8 Ω. Such a speaker, which might by calibrated measurement offer a 91 dB /W 'sensitivity' with the usual industry '8 Ω watt', actually offers a true watt sensitivity of only 87 dB and will only attain the higher figure if the amplifier employed has sufficient thermal and current capacity. In one test example, a prestigious three-way monitor design which is rated at 1 kW (8 Ω) peak program power (unclipped speech and music) was tested on a suitably powerful amplifier. It met the specified sensitivity and drew upwards of 58 A peak at full level, due to its lower than nominal impedance loading. This author feels that this is an unreasonable demand to make on cable, connectors and, not least, amplifiers.

6.5 Sound level and amplifier power

The maximum sound level required is the main criterion in defining amplifier ratings. The acoustic environment has to be taken into account – its size, absorption and reverberation characteristic, as well as size of audience and the spacing from the

Table 6.1 Room SPL versus amplifier power

Room level (dBA)	94	100	106	112
Power per channel (watts)	9	36	150	600

speakers to that audience. Stage sound systems for large audiences of between 10 000 and 50 000 persons can use up to 20 kW of amplification (see Chapter 10). Conversely, in the case of sensitive, high-quality loudspeakers used in a domestic living-room, satisfactory peak levels may be obtained with as little as 10 W per channel. In a typical domestic room of 0.5 second reverberation time, 80 m³ volume and dimensions 2.8 m high by 6.7 m long and 4.2 m wide (the IEC mean standard), a pair of speakers of 90 dB/W sensitivity, will produce around 85 dBA SPL at the listening position for a 1 W stereo input, a comfortable level for music reproduction.

An SPL of 100 dBA is a respectably loud listening level, while 106 dBA is as loud as most expensive audio systems can provide; 112 dBA is approaching rock concert and studio monitoring levels. Given a modern loudspeaker system with a 90 dB/W axial sensitivity at a 1 m measuring distance, the amplifier powers required per channel for these sound levels described are as shown in Table 6.1. This simple table readily shows how quickly one can run out of amplifier power in the quest for still higher sound levels.

For non-demonstration level domestic reproduction of classical music, 30 W per channel is more than adequate for a 90 dB/W loudspeaker system. However, if the system sensitivity is rather lower, say 84 dB/W, then 120 W will be necessary for equivalent results. A very few loudspeakers, such as the large, pure foil ribbons and the 'pulsing sphere' electrodynamic, offer 80 dB/W sensitivities or even less.

A further factor is the dynamic range of the programme. Much commercial programme material has a fairly compressed nature, with the high peak transient levels of natural sounds strongly compressed or squashed by tape saturation and/or by the deliberate use of electronic compressors or limiters. The peak-to-mean level of such programme may be only 5–10 dB. Thus, for a given average-level subjective loudness, an amplifier would have to deliver a maximum of only 10 dB more than the average requirement in order to handle the peaks. Suppose a mean level of 90 dBA spl was considered sufficient, supplied by a mean or equivalent continuous power of 5 W; a peak headroom of 50 W would be sufficient to ensure freedom from amplifier clipping or voltage overload.

However high quality programme signals from direct-cut analogue disc, and many digital sources including compact disc, may well have a greater dynamic range, which may be expressed as a wider peak-to-mean ratio of up to 20 dB. With such a programme, an amplifier would need to be increased to a 500 W rating if the same 90 dBA average subjective loudness rating were to be maintained without peak clipping. Practical considerations generally deter such a choice and 100 W is just about the maximum realistic power for domestic amplifiers on grounds of performance and price. Only two solutions are therefore available. Either the speaker sensitivity must be raised by the necessary 7–8 dB, or alternatively the overall sound level setting must be reduced. The first solution is unrealistically difficult since much larger loudspeaker systems would be required if the low-frequency response were to be maintained at the previous standard. A 98 dB/W domestic loudspeaker sensitivity is in any case a rather difficult target, at present achieved by only a few specialist hybrid horn designs. So we are left with the second solution, namely to turn down the volume to a lower mean level.

Thus, with the best-quality programme, a 100 W amplifier will be working at a mean level of just 1 W. In this context, digital and similar quality sources can be said to impose less demand on the speaker in thermal terms than poorer quality material reproduced at higher mean levels. Given this peak-to-mean relationship, and assuming that in normal use amplifiers are not overdriven beyond clipping, then the continuous power rating need not be very great for speaker or amplifier – of the order of 10 W at a conservative estimate. What is needed is an ability to provide large short-term peaks free from voltage or current clipping or, in the case of the speaker, freedom from network saturation and/or driver/system mechanical overload. Of course some operators will abuse the available headroom, and sensible allowance needs be made for a degree of irresponsible overdrive. In professional systems, music PA and announcement installations designed for public safety, power margins are precisely calculated, peak headroom is predicted and limiters are invoked as necessary.

The traditional view of maximum current demand for a 100 W 8 Ω rated amplifier would follow the requirements of a 6.4 Ω minimum impedance value:

$$I_{rms} = \frac{V}{R} = \frac{28.3}{6.4} = 4.42 \text{ A} \quad (W = 100 \text{ W})$$

and

$$I_{pk} = \sqrt{2 \times I_{rms}} = 6.25 \text{ A}$$

However, high-quality multiway loudspeakers can impose an impedance loading equivalent to 1/3 or less of their rated value, which gives typically:

$$I_{pk} = 16 \text{ A}$$

A 25 A peak capability for this amplifier would provide some reserve for the most awkward commercial speaker loads, and also provide for the 20–30% margin most amplifiers can handle above their rated power.

Amplifier specifications based on a continuous rating are thus inappropriate for most music and speech programme sources driving loudspeaker loads. In engineering terms, they should instead be designed for a low duty cycle with a large peak current capability. Nevertheless, the peak currents demanded by a speaker system may persist for up to 200 ms on modern programme, and not the rather short 20 ms indicated in the Institute of High Fidelity recommendation concerning amplifier testing (IHF 202).

6.6 Remote crossovers, remote or built-in amplifiers?

The power amplifier may be remote from a high-quality loudspeaker and linked to it via a heavy-duty speaker cable or, alternatively, it may be built into the speaker or fitted adjacent to it, with the main audio signal line now running at a lower 'line level'. It may use balanced or unbalanced input connection.

It has also been proposed that the passive crossover of a multiway loudspeaker should be remotely sited with the power amplifier, to remove it from possible interactions with the drive units in the enclosure. In this configuration, the demands made upon the remaining cables from the crossover to the loudspeaker are reduced since they carry band-filtered signals. Certainly there is some degree of microphony present in the large film dielectric capacitors used in better-quality crossover filters while the drivers also produce both static and fluctuating magnetic fields in operation. All in all, the inside of a speaker enclosure is a fairly hostile environment for a passive electrical filter network.

Several variations on the local amplifier theme are possible, including active speaker systems, which will be covered in Section 6.10. With the power amplifier inbuilt, provision needs to be made for mains power supply to the speaker, plus some means for placing them in a standby mode or even switching them off when not in use. This is an additional complication. Ideally, balanced lines would be used to link the audio signals to a speaker to conform with studio practice. With a good separate ground connection, this studio practice three-wire technique provides the best immunity to interference, and also fully exploits the cable performance.

Certainly, a built-in power amplifier removes the need to take account of losses in the speaker cable, which may be an important consideration over long cable runs. Balanced line working can operate at low loss over long cable runs of 100 m or more, driven by a suitably low source impedance. In such circumstances, this solution will cost rather less in higher-quality installations than long runs of copper-intensive, low-loss speaker cable used with a distant power amplifier.

6.7 Damping factor and source resistance

Historically, much attention has been paid to one particular amplifier parameter, the damping factor. This is defined as the ratio of the load impedance to the amplifier output or source resistance:

$$DF = \frac{R_L}{R_S}$$

For an 8 Ω load used with an amplifier having a low source resistance of 0.04 Ω, a typical value with the customary negative feedback circuits, and with no allowance for the speaker cable, the damping factor is 200. Depending on the circuit, figures up to 1000 are possible, though they become increasingly difficult to measure. Indeed, it is possible to make the amplifier output impedance negative, taking the damping factor beyond infinity. However, such impressive-sounding figures have little meaning in terms of practical usage.

By implication, the term damping factor suggests that the small, finite output resistance of an amplifier affects loudspeaker damping, particularly at resonance. In fact, the circuit combining the amplifier and the loudspeaker contains considerable resistance of which the amplifier contribution is usually quite small (see Fig. 6.13). In this circuit, a total loop resistance of 8.44 Ω is present at low frequencies, of which value the amplifier's 0.04 Ω represents only 0.5%. Clearly, small changes in amplifier damping factor, by even ten to one at the 200 level, will have little effect on the overall electromechanical damping of the speaker.

Output resistance is a more sensible term than damping factor, and a useful criterion would read: 'source resistance less than 5% of the overall loop resistance including the speaker', i.e. sensibly less than 0.4 Ω for 8 Ω systems. This criterion is useful for both the amplifier and the speaker cable. Only a few low-feedback, generally vacuum tube, amplifiers have output impedances significantly greater than 0.5 Ω. The odd example has been found as high as 2 Ω, and this does result in significant interaction with the loudspeaker load impedance, specifically the variation of the latter with frequency, resulting in a mild alteration in the audible and measured frequency response with such a source (see also Fig 6.24).

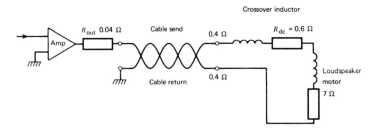

Figure 6.13. Amplifier/loudspeaker connections.

6.8 Level rather than watts

Given that a modern power amplifier is expected to act as a device of negligible output impedance and high peak current rating, it can be said to approximate to a voltage source.

The concept of continuous power, $V \times I$ (volts times amps), is valueless where loudspeakers are concerned. This is especially true in view of a loudspeaker's non-uniform impedance. The important criterion with regard to loudness calculations is voltage level expressed in dB. If adopted more widely, this would conveniently combine with loudspeaker sensitivity ratings, the latter in any case being referenced to a nominal 2.83 V input which is 1 W into 8 Ω. If amplifiers were thus rated in dBW (1 W, 8 Ω = 0 dBW) for *level* rather than power it would be easy to calculate the resulting system SPL. For example, a 20 dBW amplifier (100 W) used with a 90 dB/W sensitivity loudspeaker would generate a maximum of 90 + 20, or 110 dB at 1 m. The necessary condition is simply that the amplifier should be able to sustain its rated *level* into the chosen load, under peak or transient conditions.

With 20 dBW (29 dBV) as a benchmark, it becomes much easier to scale the relative size of power amplifiers. A 50 W amplifier is 17 dBW, and a 200 W amplifier is 23 dBW. A better feel of the sound-level difference is also portrayed by this method, and it also reveals the nonsense of many domestic amplifier ranges, where several models may be marketed in ascending order of price and power in a narrow range from 40 to an 80 W. The price may well be proportional to the quoted 'horsepower', but the sound level difference between adjacent models is so small as to be almost inaudible in practice.

6.9 Axial SPL and room loudness

For loudspeakers with a relatively uniform frequency response, the A-weighted sound level for a stereo pair in a typical room is generally 5 dB below the 1 m axial free field result taking into account reverberation and the typical direct-to-reflected contribution. For a normal listening position, the maximum level in the above example (20 dBW + 90 dBW sensitivity) would reach 105 dBA (110 dB − 5 dB). On programme drive, the actual sound level reading will depend on the spectral content, meter response averaging, the peak-to-mean ratio and the reverberant characteristics of the room, i.e. whether it is either more lively or more damped than average.

6.10 Active loudspeaker systems

An 'active' loudspeaker employs an electronic crossover, with separate (usually local) power amplifiers feeding each drive unit (see Fig. 6.14). It has a number of attractions for the engineer, allowing many of the more difficult aspects of system design to be brought into the province of active filter electronics which may be readily synthesized. Such circuitry is highly effective and can offer special advantages, while the frequent complicating interactions of passive filters with speaker drivers are avoided. With single drive-unit speakers or simple inexpensive designs, the 'active' question does not arise. However, with increasing loudspeaker system complexity, the size and cost of a passive crossover network increases. If executed in a low-loss, high-linearity form, it can constitute a considerable part of the final cost of any given design. Except in those rare cases where impedance compensation is employed, we have to accept that the system input impedance will usually be complex. Substantial amplifiers may be required in order to cope with the potentially high peak current demands.

At this point an active speaker design becomes more attractive. At the highest subjective quality level, there is still some doubt concerning the sound quality of

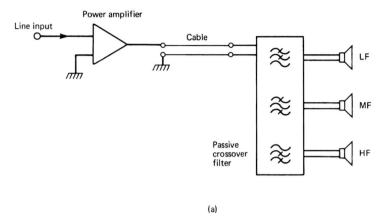

(a)

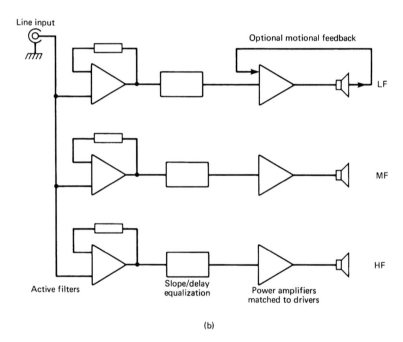

(b)

Figure 6.14. Comparison of passive and active multiway systems.

high-power passive crossovers. Reversible electrolytic capacitors and relatively cheap cored inductors are used more often than expected, components for which many critical amplifier designers would express considerable doubt. Still more costly constructions with transformer-cored inductors and plastic film capacitors also have notable losses, and may also saturate on peak programme levels.

On the other hand, the implementation of the active electronics for a speaker cannot be taken for granted if high-fidelity performance is the objective, and this must be carefully designed to meet the quality level anticipated for the product. Book designs for active filters using inexpensive integrated-circuit operational amplifiers may not be of sufficiently high sound quality. In fact from one viewpoint an active filter can be said to represent a greater challenge than a top quality preamplifier.

Component quality, layout, grounding concepts and power supply design are crucial factors. The designer of an active filter and amplifier should be fully aware of high-fidelity practice if the final results are not to be compromised.

One can nonetheless put forward a list of strong advantages for active design.

(a) The use of individual power amplifiers for each drive unit brings considerable rewards. The first is the much-reduced complexity of loading on each amplifier, which may allow the use of simpler, speaker-optimized amplifier design, e.g. in terms of power bandwidth and spectral/duty cycle power demand. It should be remembered, however, that even a single moving-coil driver remains a more difficult load than a laboratory test resistor.

(b) With separate amplifiers working over separate frequency bands, intermodulation distortion for the system is greatly reduced, including that which may be generated at the interface of the amplifier to the passive crossover.

(c) An active system may be designed to provide a higher sound-pressure level with a given drive unit resource. Several factors are at work here. The power rating of a loudspeaker is, as mentioned earlier, a vague specification generally based on the assumption of a typical programme energy distribution, a typical peak-to-mean ratio of 10 or 15 dB, and the estimated power ratings of the drivers and crossover parts. The result is a peak programme rating corresponding to the maximum size of amplifier which may be used to provide unclipped programme. The continuous sine wave drive rating is generally much lower, particularly in the high-frequency range. Thus a 200 W rating may be specified even if the sine wave rating for the high-frequency unit may be only 10 W. With an active system, the power amplifier size may be scaled to suit the individual driver sensitivity and power ratings, allowing their working limits to be exploited fully. Taking into account the fact that overload protection may also be precisely tailored to each driver, including modelling of thermal time constants, the potential for higher system sound output for a given rated power may be realized.

(d) It is also true that active systems may be driven to still higher sound levels with comparatively slight subjective impairment, particularly if mild clipping can be tolerated. When an amplifier driving a passive system clips, the clipping distortion is wideband. Assuming the usual condition of overload occurring in the bass to mid-range, the higher-order amplifier harmonics are readily reproduced by the high-frequency unit and are consequently easily heard. In the case of an active system, clipping harmonics of the bass amplifier are restricted to the lower frequency units, and are largely filtered by their natural acoustic roll-off. The high-frequency system can continue unimpaired, driven by its own presumably unclipped amplifiers. In a given arrangement, such short-term clipping of the system by a few dB is generally considered acceptable subjectively and, in practice, allows an active system to play up to 4 dB louder than the passive equivalent, which is equivalent to more than doubling the amplifier power. From another viewpoint, the separation of the amplifier/speaker system into filtered 'ways' will consequently reduce intermodulation distortion, whether in clip/overload conditions or not.

(e) Direct connection of amplifiers to loudspeakers offers further benefits. Drivers may have resonances outside of their normal working ranges, which may intrude on the pass-band of the driver in an adjacent frequency range, and be acoustically excited by that driver's output. With a passive crossover, the termination or source impedance offered to a driver outside its working range is often high and variable. Of course, the fundamental resonance of a driver is susceptible to electromagnetic damping, controlled by the source impedance. Thus a directly connected power amplifier offers the benefit of a low source impedance, for example helping to control the out-of-band resonance of a high-frequency unit. Where still greater damping is required, the amplifier output impedance can easily be made negative (see Fig. 6.15) up to the value of the total resistance in the driver. (More than this results in positive feedback and destructive oscillation will result.) Such freedom for driver control can notably reduce

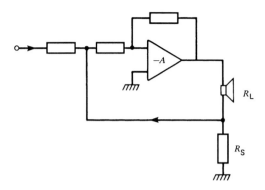

Figure 6.15. Control of amplifier negative output resistance for variable damping.

system coloration. In one commercial example, a fairly high Q resonance for a 34 mm dome unit, which was located at 900 Hz, gave a noticeable 'nasal' coloration in a passive form of speaker despite compensation for its motional impedance. The active version, when completed, avoided this coloration because the in-band output of the mid-drivers was no longer capable of acoustically exciting the dome at its resonance, the latter being well damped by the direct amplifier connection.

Loudspeakers are essentially voltage-controlled devices, i.e. the sound output is proportional to the applied voltage, while the current drawn by a moving-coil driver may be significantly non-linear. Where the source impedance is low, this does not pose much of a problem but, in the case of the higher source impedance imposed by the interposition of a passive crossover network, the current component of distortion now appears in series with the applied voltage according to the additional source resistance. This may be seen as a rise in distortion for the driver's acoustic output. Two-way systems often show this as a rise in distortion to 1% or more in the mid-range, generally due to the uncontrolled fundamental resonance of the HF unit. Direct amplifier connection often results in reduced distortion from this cause, the benefit being in addition to the advantage of essentially negligible distortion contribution from active filters, compared with the passive, high-power equivalents.

(f) Active connection readily permits the application of motional feedback to the drivers, then allowing both magnetic and mechanical non-linearities to be taken within the control of the power amplifier via negative feedback (see Fig. 6.16).

(g) With active crossovers, the system designer is also given considerable freedom to exploit the higher classes of low-frequency system alignment. Passive system design generally stops at the fifth order, for example a reflex tuned system with a series capacitor of 300–500 µF. Higher orders of electronically assisted alignments may be easily devised. For enhanced low-frequency system design, a popular class of sixth-

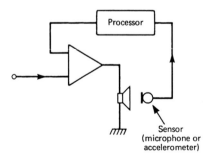

Figure 6.16. Simple motional feedback principle.

order alignment uses a fourth-order, somewhat overdamped bass reflex, augmented by a second-order 'electronic' boost to maintain and extend the response, the result providing a valuable combination of reduced enclosure size, increased efficiency and boosted low-frequency extension.

(h) Active or multiple-amplifier speaker designs may also benefit from the use of electronic delay networks which can be adjusted to compensate for differential time delays between drivers and thus facilitate more accurate crossover design. In addition, with active design, differences in driver sensitivity and production tolerances are easily trimmed out via pre-set gain controls in the amplifiers.

Comparative tests which have been carried out on basic repeatable systems show important subjective differences between active and passive versions of the same speaker; when the full potential of active connection has been exploited in a specific design, the superiority of the active method becomes obvious. Reported gains are of the order of 20–30% for overall sound quality, particularly for clarity and dynamic range. Close control of the amplifier/loudspeaker interface is the main foundation for the improved speaker performance to be achieved via active system design.

6.11 A typical active speaker system

Sophisticated active speaker systems are undoubtedly complex creations in their own right. In Fig. 6.17 a fully equipped three-way active system is outlined. The line-level input signal, nominally 1 V, is initially buffered and generally band-limited to the working frequency range of the system. Thereafter come the three active filter stages, namely high-pass for the high-frequency range, band-pass for the mid-range, and low-pass for the low frequencies. Stages for driver equalization follow, these compensating for any anomalies in their acoustic outputs taking into account delay and enclosure diffraction. If required, delay networks may be included to bring the acoustic centres or radiating planes of the mid- and high-frequency units into synchronism with the low-frequency driver. The use of electronic delay compensation allows the drivers to be conveniently mounted on a common baffle plane, with considerable benefits of diffraction control, reduced costs and, finally, convenience (see also Section 6.15).

Pre-set gain controls allow adjustment of absolute sensitivity for the low frequencies, followed by optimum driver level adjustment for the middle and high frequencies. Protection systems are included for the three power amplifiers, and ideally these are programmed to follow the safe operating area of the given driver. The dimensions of the safe operating area are bounded by absolute voltage and current, by the time-integrated thermal history, and by the maximum diaphragm excursion. For example, a treble unit may have a continuous rating of just 10 W and yet cope happily with its share of a 100 W peak programme, assuming undistorted speech and music drive. It can readily absorb short, high-power impulses, integrating the energy via the voice-coil thermal time constant. The protection circuit can be designed to follow these factors and thus protect the unit from damaging overload, while simultaneously allowing exploitation of its maximum output level. An optional motional feedback section has been included in the low-frequency part of this active system example.

6.12 Driver equalization and motional feedback (MFB)

In theory, any driver may be equalized to a desired frequency response in any box by means of an appropriate electrical shaping network. For example, a typical 300 mm driver mounted in an undesirably small 30-litre enclosure was found to produce an overdamped bass response which measured 9 dB down at the system resonance frequency of 45 Hz.

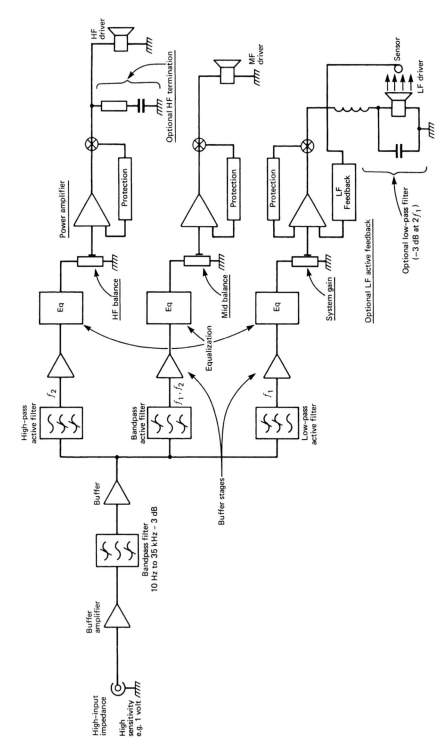

Figure 6.17. Fully equipped active system.

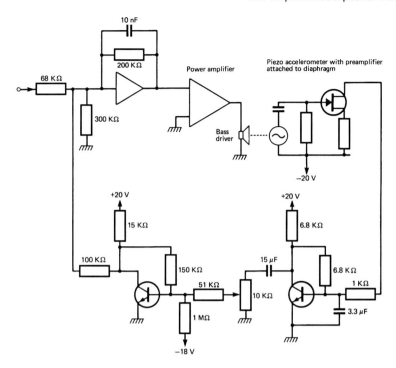

Figure 6.18. Motional feedback circuit (after Philips 4882).

An additional active filter, providing a second-order boost of 12 dB at 38 Hz, restores the system to a flat Butterworth frequency response, and the output now extends satisfactorily to −3 dB at 33 Hz.

The main requirement for such equalization is an amplifier large enough to produce the additional power. This requirement is assessed in conjunction with the likely power spectrum of the programme, the speaker's thermal power-handling and mechanical excursion limits, and the maximum sound level required. An alternative to such calculated corrective pre-equalization is the application of feedback (MFB) techniques. A sensor is attached, either to the diaphragm itself or at a position in front of the driver. It provides a signal which can be processed into a form suitable for applying electrical feedback to the driving amplifier. By negative feedback, errors in the speaker output may be actively corrected via the production of compensating changes in the amplifier's output. All aspects of the transfer characteristic of the driver are linearized, including the piston mode frequency response and the usual non-linear distortion (generally due to a combination of mechanical and magnetic factors).

In a successful commercial MFB design series (Fig. 6.18), a simple piezo-electric accelerometer was attached to the bass cone apex and equipped with a small inte-grated-circuit preamplifier to improve the local signal-to-noise ratio. Diaphragm acceleration is directly proportional to sound power over the reference low-frequency range and the sensor output, suitably scaled and band-limited, forms the basis of the electrical feedback signal to the driving amplifier. Distortion reduction will occur only over the frequency range where the harmonic products fall within the necessarily filtered pass-band of the feedback signal. Stability difficulties may arise if the frequency range is too wide. Furthermore, the available amplifier headroom and the excursion limits of the driver need to be matched. With reducing frequency, below the driver cabinet system's normal pass-band, a point will be reached where the rising distortion and extra power required are so great as to exceed the design limits. At this point, the system may overload dramatically; effectively, of course, the feedback loop is then broken and gross distortion results. The MFB designer must therefore ensure that the maximum input level and the combination of required low-frequency bandwidth do not allow the system to enter this region of either feedback-related or power-limit failure.

This class of design is now increasingly popular for subwoofers which have become commonplace for home theatre applications, with or without MFB control. Avoidance of gross overload is achieved by safe area monitoring, working in conjunction with a form of subjectively matched electronic limiting, in fact an active compressor, which backs off the gain as the system approaches overload and prevents the distortion exceeding pre-set limits, e.g. 5% at 30 Hz. In one exotic example, a solid state inertial/position sensor allows the control system to be informed of the absolute position of the motor coil in the available gap height and dead centre correction is possible under active drive, further increasing headroom.

6.13 Full-range feedback

Traditional methods for feedback control of drivers have concentrated on the low-frequency range, where lumped parameter theory holds, with the aim of extending the bass response and reducing the usual non-linearity. However, full-range response and linearity correction is also possible. In one interesting, if rather complex, design, a multi-way loudspeaker was fitted with full-range sensors for each drive unit, these sensors being capable of integrating the bulk of the radiated acoustic output by sensing the movement of the entire diaphragm assembly regardless of potential breakup modes (see Figs 6.19 and 6.20).

The sensing method employs the variable-capacitor principle, where one electrode is a fixed, acoustically porous conductive mesh placed close to the diaphragm. The

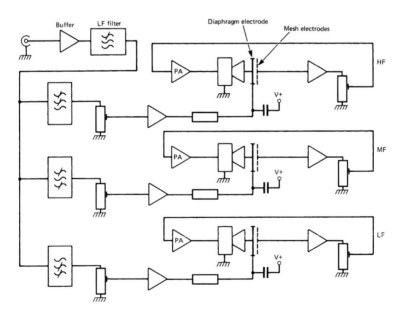

Figure 6.19. Schematic diagram for full-range MFB via electrostatic sensors (Backes and Muller).

Figure 6.20.
Demonstration
model using
full-range MFB as
in Fig. 6.19.

second electrode is the diaphragm itself, rendered conductive by a suitable coating. A low current, safe polarization of 400 V is applied between the electrodes, and the output of the mesh electrode forms the basis of an averaged feedback signal used to linearize the driver's acoustic output.

With constant-charge operation, the diaphragm displacement is largely proportional to the sensor output voltage. The respectively filtered input signals are mixed with the feedback at the electrodes, and the sensor output is used to drive the power amplifiers. While successful working models have been demonstrated, this design has proved rather costly to produce.

An alternative route involves the accurate measurement and modelling of a speaker system and then providing the appropriate complex pre-equalization of linearity, dynamic characteristics and response/acoustic power, ideally via an inverse algorithm operating via a DSP unit.

6.14 Speaker adaptability

Often a speaker may be designed to produce a uniform axial output in a laboratory test environment such as a 2π or 4π anechoic chamber but it will actually be used in real rooms or halls. This final location will modify both the measured and the perceived frequency response due to the proximity of immediate boundaries such as the floor, the side and rear walls, and not least the ceiling (see Chapter 8). Sound quality will also be influenced by the reverberant character of the environment.

A 'bright' room with little mid- and high-frequency absorption will tend to make the speaker sound bright in tonal balance and an over-absorbent room will have the opposite effect. At low frequencies, strong standing wave modes are often present and the listener will hear major variations in low-frequency loudness and frequency balance as he or she moves about in the room. In addition, the boundaries local to the speaker directly augment the output in an irregular manner, producing up to 9 dB of response lift at the lowest frequencies. This may begin from 90 Hz, according to path lengths, and reach a maximum in the 25–40 Hz range for typical rooms of 80 m³. The low-frequency gain is dependent on building construction, these larger values pertaining for brick and concrete forms, while the US style of open-plan timber frame house may show much less lift and a consequently shorter LF reverberation.

Assuming sensible placement of the loudspeaker in the room, i.e. by providing asymmetric spacing from floor, back and side walls, a fairly uniform low-frequency distribution is possible over a realistic stereo listening space, this being confirmed by multiple spatial averaging and also corroborated in listening tests. A loudspeaker may advantageously be fitted with controls for low-frequency output which allows for level and shape adjustment, and which may be acoustical and/or electrical.

A more sophisticated concept of adjustment is also possible. Conventional response equalization falls into the steady-state category – a third-octave band equalizer or alternatively a parametric equalizer is placed in the chain to try and account for the response variations imposed on a speaker's output by the listening environment (see Chapter 9). This is called steady-state equalization, since all signals suffer continuous modification in the attempt to compensate the blend of direct and reverberant sound in the listening room. Unfortunately, modification of the response will include the axial direct sound and will alter the tonal balance of the loudspeaker. The ear can readily distinguish between direct and reverberant sound by assessing the leading edges of music transients.

What is ideally required is a method which leaves the first arrival direct sound from the loudspeaker untouched, but which then equalizes, compensates, even cancels the reverberant/reflective aberrations in the room following the transient. The speaker's output must remain untouched for the first millisecond or few, after which the real-time equalizer may generate its programmed corrections.

The steady evolution of digital audio processors and microcomputers has made such a real-time equalizer practicable though so far it has not found much favour for hi-fi applications. Automobile sound is another matter and better quality radio-electronics systems for cars, which include CD multi-players and surround-sound processing, are now frequently equipped with quite powerful DSP. Speaker systems installed in the vehicle are exhaustively calibrated, and the required corrections for good stereo and a natural sound balance may be pre-programmed into the DSP.

In the domestic listening room application, a calibration microphone is placed at the optimum listening point, and one speaker of a stereo pair is excited by a computer-generated test signal. The result is stored and analysed, the computer recording the reflected room modes and their amplitudes. The processors are dual-channel, and the calibration is then repeated for the second loudspeaker channel. In the 'operate' mode, a fast processor continuously computes corrections in real time, these corresponding to the inverse of the room modes, preserving their delay timing and providing a near-perfect 'anechoic' compensation at the listening position. At present it is possible to compute in a frequency range up to several kHz, with sufficient resolution to cancel very complex room modes.

The tonal quality of the programme is essentially unaltered, save for the secondary effect of the altered reverberation, but the subjective result is the obvious elimination of local boundary modes at lower frequencies. Interestingly, where a good room is concerned, together with a sensible placement of boundary matched/compatible loudspeakers and additionally a favourable position for the listener, the subjective improvement is actually quite small. However, for non-ideal conditions where aberrations of ±9 dB are possible in the range up to 500 Hz, the ADSP (adaptive digital signal processor) can provide a major correction which is then easily audible and certainly preferred.

One aspect of digital equalization is the potential for accurate linearization of the amplitude and phase response of any particular loudspeaker in addition to, or instead of, the room. In this circumstance the loudspeaker may be specifically designed for the best engineering and physical characteristics, diffraction control, driver mounting and appearance, efficiency, etc., largely ignoring the usual premise of a neutral tonal balance/flat response. Provided that tolerances and consistency are good, the resulting axial response, or some preferred summation of the acoustic output in the frontal region, may then be equalized to very high precision using an inverse digital filter calculated from a measurement of the real object. This digital filter may be realized in a small electronic unit inserted into the audio system, for example digitally linked to a DAC via the popular SPDIF interface.

6.15 Digital loudspeakers

This section covers two types of digital loudspeaker in which the signal in digitally coded form is brought closer to the transducing process, where the electrical signals are converted into air-pressure variations, or sound energy.

Theoretical work is proceeding on a loudspeaker whose sound generating section is a matrix of small 'digital' or on/off elements which are energized by an array of electromagnetic solenoids or alternatively by piezo actuators. Modelling indicates that significantly greater excursion may be generated by spiral forms of piezo exciter. By suitably driving the array from a digitally processed signal, preferably to the full 16-bit 44/48 kHz PCM resolution or better, the matrix of elements is intended to reconstruct the original acoustic wavefronts in time, frequency and amplitude. Problems of ultrasonic and beat-frequency noise generation will be considerable and in this basic form the system may never be commercially realized.

Direct conversion from digital signals to sound pressure is the ultimate objective and may be possible with micro-machine pumps. In an alternative earlier proposal, the voice-coil of a direct radiator driver is subdivided and each element is driven digitally. Data integration occurs over the complete coil. Half the bit range – the upper, easier half – is handled by the split coil windings while the lower half, in this case the remaining 8 bits, operates in the analogue domain, with a simple DAC and the requirement for only a low-power, linear amplifier for this part.

An intermediate form of 'digital speaker' (Fig. 6.21) which is successfully in production begins with an analogue or conventional speaker of active design. Here the designer can choose to place a D/A (digital-to-analogue) converter in the loudspeaker electronics. A digital audio processor may also be present. Digitally coded signals can be fed directly to the system, avoiding losses in intermediate analogue control electronics as well as in the line-level linking cables. Control of the signal replay level, for example, is achieved via another digital line, in this case leading to a managing microcomputer built into the speaker. In a domestic realization, the latter would ideally also be interfaced via a hand-held infrared remote control. The concept of remote D/A and A/D converters is likely to find increasing application. The more that audio signals are handled in digitally coded form, the less danger there is in theory that degradation will occur in the transmission chain compared with analogue electronic interfaces and cables. This trend is apparent in digital studio practice, for example with the introduction of remote A/D converters at the microphone position.

Digital speaker systems can be as sophisticated and versatile as the designer wishes. In one example, the equalization and crossover functions are located in the digital domain and conversion to band-limited, filtered audio is accomplished in separate DACs placed close to the power amplifiers for the respective drive units. Research

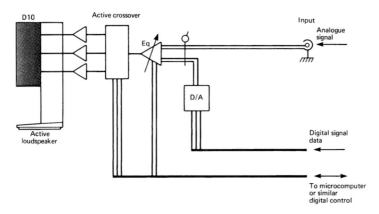

Figure 6.21. Intermediate analogue/digital loudspeaker arrangement (courtesy Meridian).

is also continuing on a new class of Delta Sigma power DAC operating at the tens of watts level, and essentially capable of direct conversion of the digital input to electrical power at loudspeaker level with a wide dynamic range. Obvious complications relate to ultrasonic and EMC filtering and the need to comply with today's strict radiation requirements. Another important question concerns the dynamic range, which may be a problem when viewed over a range of speaker/driver sensitivities; the consequences of audible background noise from the more sensitive drive units of a digital loudspeaker system may be serious.

For general applications, it is considered good practice for a power amplifier to provide a signal-to-noise ratio of 110 dB or better, to ensure inaudible backgrounds with a range of loudspeakers. Because noise in digital systems has a tendency to be correlated and/or have side tones in the audible band, this is a serious issue. A typical 18-bit DAC directly connected via a power amplifier to a speaker will not necessarily deliver an inaudible background with drivers of higher than average sensitivity without the introduction of some analogue gain ranging between the DAC and the power amplifier. Good practice suggests up to 20 or 30 dB of fine stepped control of gain digitally, this augmented by analogue ranging in 20 dB steps inserted after the DAC. These gain stages are synchronized under microprocessor control. DSP loudspeakers are likely to benefit from textbook design crossovers, thanks to the ability to offset physical delays between the driver outputs; the polar responses are correspondingly uniform and well behaved which can improve the overall sound quality.

A problem which besets the loudspeaker and amplifier designer in this field is that of designing for the optimum combination of maximum output level, power handling and distortion and damage limits. This question is complicated by the variation of distortion and power handling with frequency, a feature common to most loudspeakers. A neater solution may be found in the digital loudspeaker where the incoming musical programme is assessed by the digital processor for its spectral and power content. This can then be compared with a predictive model for the dynamic envelope for the loudspeaker components, and the DSP may then calculate the spectral content in real time so that it can comfortably fit the specification for distortion, low frequency extension and power. For example, a given loudspeaker may be capable of 25 Hz at 5% distortion at 90 dB, but with a damage limit of 96 dB. In the mid-range it may be happy to cruise at 106 dB. Where the programme is likely to drive the speaker beyond the low-frequency limit, a progressive filtering or cut back of low-frequency extension could occur to maintain a subjective performance which remains free from significant distortion. The speaker may then be said to have automatic, optimized adaptation to the bandwidth and level demands made upon it.

6.16 Cables and connectors

When the signal source is remote from the loudspeaker, and whether the power amplifier is separated from the loudspeaker or built into it, a length of audio cable must be used to transmit the audio power. Traditionally the view has been taken that well-specified cable runs impose negligible loss in terms of audio quality. However, recent developments in metallurgy have provided us with advanced grades of copper conductor which indicate that a basic cable design and performance can no longer be taken for granted for critical applications.

For normal practice, two classes of cable can be considered; a high-current type for the loudspeaker-to-amplifier connection, and a low-current type 'interconnect' for line-level signals between the signal source and the amplifier input. Speaker cable ideally needs to be of low resistance, to minimize power loss in the low-impedance speaker circuit, and is thus of suitable large cross-section area (CSA).

Interconnect cable can generally be of much higher resistance, though the ground return may be significant where stray hum currents may run between source and load devices. Significant ground or 'signal minus' resistance will add noise and hum to the circuit as the stray current resolves to an interfering voltage seen by the load or termination. For a given connection, cable resistance over 10 ohms may give trouble, while still lower values may add secondary effects not directly audible as noise. Two-wire interconnection is called single ended or SE and often uses coaxial cable to help screen the inner live or hot signal conductor from induced electrical noise. Here the source impedance is generally of low impedance, less than 1 k Ω and often less than 100 Ω, with the termination at the power amplifier input at 10 k Ω or more, and typically in the 20–100 kΩ range.

The usual alternative to the lower cost SE connection, eminently useful for shorter lengths, is the balanced form, originally specified to a 600 Ω termination to define the optimum loading for the balanced mode signal coupling transformers used. Popular in broadcast and recording studios for low-noise line-level connection, a three-wire balanced system allows for the effective subtraction of common mode noise, i.e. hum induction and the like, entering the cable. The design consists of a twisted pair, already conferring significant immunity for the balanced/differential signal, while the chassis/ground currents are relegated harmlessly to a separate 'power ground' wire, generally an external braided shield. Professional grade active speaker systems normally have balanced inputs and many top-quality audio amplifiers also conform to this connection practice.

Cable has the fundamental properties of capacitance, inductance and resistance which all exert some measurable influence on sound quality. The designer chooses a cable that is appropriate for the application, to minimize these basic influences, but may also exercise some freedom to voice the installation by additional selection of cable by sound.

6.16.1 Speaker cable practice, normal and bi-wiring

Here currents are high while the source impedance is low. Radiation and induced interference problems are minor in most applications, and such cable is generally of an unscreened twin-conductor form. The conductor cross-sectional area and resistivity, coupled with run length, together determine the resistance. The inductance is partially controlled by the geometry, increasing as the conductors are spaced further apart, and conversely reducing when the conductors are closely positioned, for example as a twisted pair. The rarely used coaxial construction gives minimal inductance and deserves wider notice. The more closely the conductors are spaced, the higher the inter-conductor capacitance, with the coaxial form generating the highest values. Note that loudspeaker cable can act as an aerial to radio frequency fields and often delivers EMC to the output terminals of an amplifier. Some amplifier designs are susceptible and may demodulate RF presented at these terminals, resulting in audible interference. Even when the effect is not directly audible, subtle changes in sound quality may still be present. Double blind testing of a given type of cable with and without a designed RF termination (for example, with maximum loss predominately at the peak broadcast spectrum level at around 1 MHz) showed audible gains with the termination in place, the subjective result being not dissimilar to a reduction in jitter for a digital audio chain. This also suggests that a screened speaker cable may be of value under difficult EMC conditions, for example near transmitters.

6.16.2 Transmission lines and cables

With longer runs, some engineers have considered whether audio cables have significant transmission line properties. Transmission lines are designed to transmit complex waveforms accurately over a wide frequency range, generally above 0.5 MHz, and they require appropriate source and termination impedances. If these are not matched, reflections will occur, distorting the waveform, impairing the frequency response and reducing power transfer.

Taking 20 kHz as the highest audio frequency, the electrical wavelength is 16 km, which is very long compared with normal audio cable runs. Viewed alternatively, the maximum likely run of 100 m has its first transmission resonance or reflection at 3 MHz, which is well above the audio band. Speaker cable runs are normally about 3–30 m, and therefore line properties may be considered negligible. On this basis, the transmission line approach may be dismissed, and instead a cable may be represented by an equivalent circuit employing a lumped-parameter approach (see Fig. 6.22(a)).

In this circuit the parallel capacitance is shown, together with the mutual or loop inductance, and finally the total or loop resistance. The loop includes the send and return conductors. For a remote cable run of twin conductors of 10 m, typically spaced by 0.9 mm, the inductance L is 8.0 μH, the capacitance C is 600 pF, and the resistance R is 100 mΩ.

Modern amplifiers have a low output resistance of typically 50 mΩ, plus a small series inductance of around 2 μH, to stabilize behaviour on possible capacitive

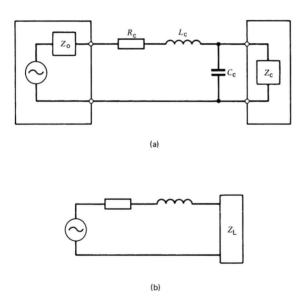

(a)

(b)

Figure 6.22. (a) Speaker cable equivalent circuit. (b) Simplified equivalent circuit.

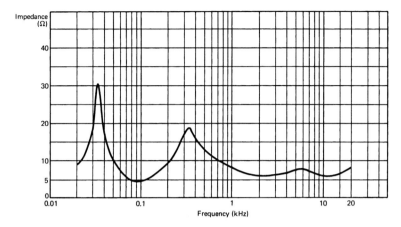

Figure 6.23. Modulus of impedance of a typical multiway loudspeaker (Yamaha NS1000).

loads. This amplifier inductance amounts to 0.25 Ω impedance at 20 kHz, where it has maximum effect, and may be summed with the cable inductance contribution of 1 Ω, giving a total of 1.25 Ω at 20 kHz. The impedance of typical loudspeakers lies in the 4–8 Ω range and, taking 4 Ω as a pessimistic reference, the example cable will impose a minor loss of 0.32 dB at low to middle frequencies rising to a significant −2.4 dB at 20 kHz. With a 30 m run, the amplifier power at the load will be reduced severely by 20 kHz. It is worth noting that several hi-fi speaker cables of the spaced twin variety have a larger 2 cm conductor spacing with a substantially higher inductance; audible changes in high-frequency response will become significant at just 10 metres even with 8 Ω speaker loading. Note that cable capacitance is a small factor; even at 30 m, it represents only a 352 Ω reactance at 20 kHz, which represents a negligible shunt loss. The inductance is clearly the critical factor here, and indicates that heavy duty coaxial cable could well be preferable for long runs. Industry standard RG9 coaxial is a possible choice for speaker lines, with an inductance of typically 10% of the commercial speaker twin cable, combined with a relatively low 13 mΩ/m loop resistance. Capacitance is also low at 30 pF/m for this design.

Fortunately, loudspeaker systems with an impedance of 4 Ω or less at 20 kHz are uncommon. The impedance curve for a typical multiway system (Fig. 6.23) shows the high-frequency impedance still rising at 20 kHz, due to the component represented by the voice-coil inductance of the high-frequency drive unit. For runs up to 10 m, the inductive loss with twin cable is often negligible. However, if the cable conductor is too thin, its series resistance will result in both power loss and, to a lesser extent, changes in the frequency response of the system. Reference to Fig. 6.23 will show that load impedance is a complex parameter which varies with frequency. Significant source resistance, working in conjunction with these variations, imposes errors in the frequency response.

In a given installation it is possible to terminate the speaker connection end of a long cable by adding a shunt capacitor, usually damped by a series resistor. A 500 nF/2 Ω combination can restore the level at 20 kHz without harm to the overall arrangement or threat to the power amplifier.

6.16.3 Resistance effect

Taking an example loudspeaker with an impedance range from 3.5 Ω to 80 Ω (and noting, if relevant, the dynamic minima, i.e. peak current equivalent impedance), let us choose an unsuitable loudspeaker cable of the type commonly used to wire bell

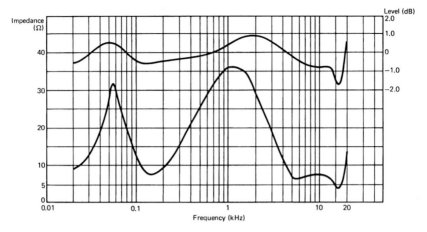

Figure 6.24. Effect on frequency response (upper curve) due to interaction with loud-speaker impedance (lower curve).

circuits. With a loop resistance of 0.1 Ω/m, the total resistance of a 10 m length is 1 Ω. Inductance is 6.5 µH, amounting to an additional 0.8 Ω at 20 kHz. Its effect on frequency response may be seen in Fig. 6.24. The result is certainly audible and not especially memorable. For the long speaker cable runs met in sound-reinforcement applications, a higher voltage distribution system is therefore recommended (see Chapter 10). Here the so-called 70 V or 100 V lines are driven and terminated by matching step-up/step-down transformers, keeping the losses of even inexpensive cable to a tolerable minimum. Both the cost and insertion loss of these transformers is a factor; with modest powers and moderate audio bandwidth, they are economic, saving in the copper content for long cable runs but, where high power, wide frequency range transmission is required, suitable transformers prove to be expensive and will still impair the resulting fidelity to some degree.

6.16.4 General rules: speaker cables

Ideally the loudspeaker cable should be short, with the power amplifier placed in close proximity to the loudspeakers. For runs of up to 10 m, the cable should be a closely spaced or twisted pair type, of loop inductance preferably < 0.5 µH/m, and loop resistance < 20 mΩ/m.

Above 10 m, a heavy-duty coaxial should be considered, of capacitance < 60 pF/m, resistance < 10 mΩ/m and inductance < 0.1 µH/m. For very long cables of over 50 m, the transformer-matched, high-impedance transmission system should be used, noting that some quality loss will occur.

In critical audio applications, it has been found that several secondary aspects of cable design and construction influence sound quality, such aspects not necessarily materially altering the basic properties of capacitance, resistance and inductance. Specific design details may include the choice of dielectric or insulation, e.g. PVC, polyethylene, polypropylene, PTFE, silicon rubber; whether the conductors are made solid or stranded and, if stranded, in what geometrical form they are distributed; and also whether they are separately insulated (Litz). Other factors include mechanical rigidity of the overall cable assembly and the grade of conductor, i.e. metal type and/or alloy.

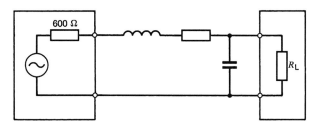

Figure 6.25.
Interconnect equivalent
circuit.

6.16.5 Signal interconnect cable

The value of '600 ohms' is normally associated with balanced lines and derives from traditional impedance matched, transformer coupled telephony usage. Modern audio sources tend to have a low source impedance of below 100 ohms and generally work best with a high termination impedance of 10 K ohms or more; true 600 ohm working is generally reserved for very long cable runs where the transmission line properties of the cable need to be taken into account. A recent application for matched balanced lines is for the transmission of digital audio to the AES-EBU standard where the impedance is strictly defined at 110 ohms. An XLR style connector is used, though there are now some concerns about the quality of the matching impedance of this connector from an RF quality viewpoint. This is because the bandwidth required for the best transmission of digital audio exceeds 25 MHz, beyond the anticipated design of this universal audio balanced connector.

With long runs of around 200 m, capacitance in audio cables will become important in the equivalent circuit (Fig. 6.25). A 20 m length of common coaxial cable imposes a shunt capacitance of 2 nF with a −3 dB break point at 133 kHz. This is comfortably high if the source can tolerate this load (at 20 kHz it represents 4 kΩ). However, note that the increased loading may still impair the peak output level of an amplifier and increase distortion, even if clipping or gross overload does not occur.

At 500 m the loading for this example is 50 nF or 160 Ω, which is very significant. A special low-impedance output buffer or line driver is one answer, which can offer a 10 Ω or lower source impedance with the ability to drive up to a 0.2 μF without distress. Twisted-pair cable is a popular lower capacitance alternative for SE connections, while the addition of an external coaxial braid, terminated at one end only, provides shielding from electrical interference. Ordinary two-conductor coaxial is also widely used. With longer runs, cable quality is a definite factor, and the choice of insulator, dielectric, copper and construction all affect the sound quality to a small degree. Copper-clad steel conductors are not recommended due to ferro-magnetic distortion effects and, on very long runs, polythene insulation has been found to be superior to PVC, showing better dielectic properties.

The key feature of balanced working is the use of differential mode signals for the two equal twisted pair signal lines, surrounded by a third conductor, the coaxial shield. Differential working endows the system with the ability to reject signals in the common mode, i.e. interference and hum induction relative to ground. With good design, such rejection can exceed 80 dB at low- and mid-frequencies and is a valuable feature where very low level signals are involved such as microphone circuits.

6.16.6 Copper grade and its effect on sound quality

Until recently, the properties of copper as a conductor of electrical signals have not been given much attention. Conductivity is traditionally regarded as the main factor and, in commercial purities of 99% or better, the absolute conductivity varies little. In progressive terms of increasing purity, the copper grades are known as 'bar refined' 99.5% (also known as TPC or tough pitch copper), 'electrolytically refined' 99.9%

(the commonest electrical grade) and 'high purity' better than 99.99%, possessing 100 p.p.m. of residual oxygen. If 'oxygen free', it may contain only 3–5 p.p.m. of oxygen.

Recent tests, using master-quality programmes and associated equipment of high quality, suggest that in addition to the R, L and C properties of a cable, the above-listed grades of copper conductor are also influential. Sound quality appears to show improvement with increasing copper purity, in particular with a continued reduction in oxygen content. Such tests on the sound of conductors may be verified by comparing realistic lengths of electrically matched cables, for example between 10 and 50 m, with negligibly short lengths, using the substitution method. Given high standards of programme quality and critical judgement, the general conclusion reached is that the shorter the cable, the better the sound, not surprisingly. This finding is essentially independent of any consideration of the normal electrical parameters. The subjective losses then associated with long cable runs are described as a loss of clarity, with an impaired reproduction of subtle, low-level musical information. Definition on music transients appears to be mildly impaired throughout the frequency range.

Microscopy at a moderate ×200 magnification reveals the structure of the conductor. Bar-refined copper shows a highly crystalline makeup, of some 150 000 crystals per metre. An analysis of the structure indicates that the crystals have a pure interior, while the impurities congregate at the crystal boundaries. The oxygen content is present in the reduced form of Cu_2O, a semiconductor. Considering the conductive path between crystals, the boundary has the properties of a junction diode, a junction capacitor and a low shunt resistance, the latter being the dominant feature. Increasing copper purity results in a lowered Cu_2O content with fewer crystals, those remaining being of larger size. Oxygen-free copper has around 50 000 crystals per m.

A test[6] was devised to examine the effect of conductor crystals on sound quality. A non-crystalline conductor was made and compared with standard copper. The test conductor was judged to be superior, closely resembling the very short bypass. This reference 'wire' was made with 10 m of hollow 0.6×10^{-3} m diameter polythene tubing, filled with mercury, a non-crystalline metal, liquid at room temperature. The results indicated that conductor crystal properties were a significant factor.

To improve conductive performance for real wire, an annealing process was applied to oxygen-free copper which encouraged the growth of larger crystals; these were of 500 μm diameter as compared with the usual 30 or 10 μm of bar-refined copper. When drawn into wire, these large crystals are greatly elongated to 50 mm in length, providing a much-reduced crystal content – as low as 20 per m for this drawn 'LC' linear crystal or 'mono crystal' grade. The draw used is 10:1, i.e. from 1.6 mm diameter to 0.16 mm. Independent listening tests[7,8] have been reported as confirming the superiority of this new grade of copper conductor, specifically where high sound quality is important. High-quality loudspeakers were said to be improved by a noticeable step in quality when LC is substituted for normal cable over a 10 m run.

Such cable also improves performance on shorter runs, such as in an interconnect role of just 1 m length. The lower the signal level, the greater is the improvement, and some recording studios have experimented with LC cable for use on long microphone lines. Studio tests have shown that lengths of more than 100 m of conventional balanced audio twin cable do impose a significant impairment in sound quality. A leading Japanese studio has noted only relatively mild degradation when using up to 250 m of LC cable. This result is also relevant to long runs of audio cable used to feed active loudspeakers. At present it is difficult to explain the subjective results, but it is suspected that the equivalent circuit of the bulk conductor is relevant. The lowered crystal content of LC wire does result in a small reduction in the measured bulk capacitance of the manufactured wire. This is rather small at 1–2%, but verifiable by precision measurement at 1 kHz. Note that the capacitance change is not itself responsible for the alteration in subjective performance; it is merely an indicator of the altered internal structure of the conductor.

Recently a critical analysis has been published[9] which looks more closely at the theoretical aspects of conductivity. Working from Maxwell's equations for an electromagnetic wave, it was shown that, for normal speaker cable diameters of 1–2 mm, the very low resistivity of the usual conductive metals – copper or aluminium – will result in a significant propagation effect in the audio frequency range. Surprisingly, it turns out that the skin effect is involved, where the higher frequencies travel at one velocity, mainly travelling in the skin, and thus fail to penetrate the conductor. Lower-frequency signals do penetrate fully, and the very low bulk resistivity of good metal conductors results in a small, lagging phase angle between the low and high frequencies, this constituting a differential delay. If the cable is resistively driven and terminated, the effect is very minor but, when audio cable is used conventionally, i.e. misterminated, the delay can result in reflected waves which can potentially interact with the advancing wave via minor wire non-linearities. It was suggested that the subjective improvement due to 'linear crystal' and similar 'massed crystal' grades of copper conductor was due to a reduced non-linearity, working in association with the differential frequency delay phenomenon. The analysis indicated practical constructions and considerations which would result in a cable offering superior sound quality when required in audio installations of the highest calibre, e.g. studio mastering and audiophile grade domestic.

For long interconnect or line cables, balanced, matched and terminated working is to be preferred on the above basis. In general, cable runs should be no longer than is necessary. Smaller conductor diameters fall below the differential delay threshold, and 0.6–0.1 mm are considered ideal, so the heavier duty, thicker cables should be kept short. The metallurgical grade of the conductor is significant and reduced crystal forms in copper and aluminium are capable of better results. Hand in hand with such attention to detail, superior dielectrics are also advisable, with polythene and PTFE being prefered to the usual vinyl. The overall geometry and physical construction also play a part.

In recent years, where the very high cost can be borne, pure silver has shown substantially increased use as a conductor for high quality audio cables. Interestingly, research at Netherlands Radio resulted in a patent involving the use of carbon in audio cables, usually as a jacket surrounding a metallic conductor. This pioneering work has also resulted in the successful production of audio interconnect cable made entirely without metallic conductors. The configuration is in the form of insulated micro-fibres of conductive graphite in a Litz form and superior audio qualities are ascribed to its use. It has a substantially higher resistance than metallic conductors and this may need to be taken into account where chassis or ground current flow may be a factor.

6.16.7 Fuses and protection

Fuses are often employed as protection devices in the interface between amplifier and loudspeaker. Ideally they should act as a low-value linear resistor but, in practice, they are rather non-linear and their resistance is significantly modulated by the heating effect of the current passing through them. Generally, a fast-blow, simple tube fuse is employed. At 60% of its full or blow current, the resistance is double the rest value and, as the rated level is approached, it may increase by between three and four times.

Intermodulation and harmonic distortion measurements taken close to the blow current can reach as much as 4%. Figure 6.26 shows the changing V/I curve measured for a 20 Hz signal large enough to cause burn-out in about 0.8 s. The trace uses a 5 kHz pilot tone with the 20 Hz power fundamental filtered out. Cyclic changes in resistance are evident. As Greiner[10] points out, provision for separate fuses for individual drivers in a multiway system will help to reduce such intermodulation. In practice, this is desirable since a fuse large enough for adequate LF driver protection is unlikely to protect a much lower powered HF unit. Fuses may also be placed

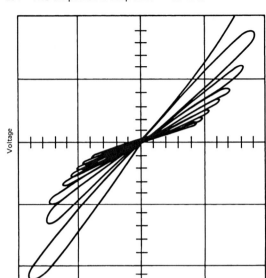

Figure 6.26. *V/I* characteristic of a fuse up to burnout (after Greiner).

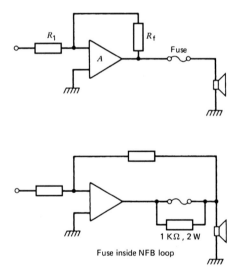

Figure 6.27. Use of feedback to reduce fuse non-linearity.

within the feedback loop of a power amplifier, which then minimizes the non-linearity of the protection component (Fig. 6.27).

A novel device has become available which exhibits quite good linearity as well as a convenient self-resetting behaviour. Called a 'posistor', it is finding favour among some loudspeaker designers and may be left permanently wired into circuit. Figure 6.28 shows the *V/I* characteristic for the device, which may be obtained in various current ratings suitable for a range of drive units. When subject to overload, e.g. 10 dB above maximum rated level, the resistance increases by 100 times or more in

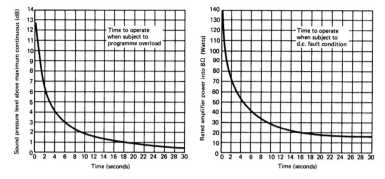

Figure 6.28. Characteristics of a polymer 'self reset' circuit breaker.

less than one second as the device heats up. The time constant is not dissimilar to the coil of a high-frequency unit. A small degree of hysteresis means that it does not return fully to its original value when the overload is removed. Some critics feel that these devices may influence the audible dynamics of the system. This criticism is only valid for high quality systems and they are most effective for general-pupose audio use. Resettable miniature circuit breakers (MCBs) have also been used for loud-speaker protection. Ideally the matching amplifier could be fitted with a well-designed electronic protection system, which may be programmed to suit the characteristics of the speaker in use.

6.16.8 Connectors

The connectors used to terminate loudspeaker cables are also important. Signal current levels cover a very wide range (over 80 dB in the case of digital programme), representing a maximum peak of 10 A to a minimum of 500 μA. In theory at least, contact specifications for switches and connectors are stretched to their limits by this working range. Switch and connector contacts need to cope with high inductive peak currents and yet they must provide contact resistances low enough to ensure freedom from semiconducting effects due to contact contamination, as the latter will degrade the clarity and reliability of low-level signals. Contacts rated at 1 A continuous, with less than 30 mΩ resistance, are suitable for most applications. Special care is required for high-power applications. This criterion relates to both the speaker plugs and sockets as well as to speaker line switching. The latter should be of the 'break before make' variety, to avoid unduly loading the amplifier during switching.

The XLR series of plugs and sockets is satisfactory for speaker practice, though it has generally proved too costly for domestic applications. Two-pin DIN connectors have frequently been used in the past, but are now considered to be inadequate for high-quality connections. They have therefore been supplanted by the use of binding posts accepting spade terminals, bare wire or alternatively 4 mm single-pole plugs. This is unfortunate, since the possibilities for misconnection as well as unreliable terminations are legion. In addition, EEC safety legislation has now outlawed equipment and cables fitted with the unshrouded 4 mm plugs on grounds of its potential compatibility with some European mains wall sockets. Binding post connections may not remain tight over long periods; soldered connections are perhaps the ideal solution, eliminating all mechanical contacts. Spade terminals have recently been introduced which have a laminated spring arrangement helping to maintain more constant contact force. The 'Speakon' multi-pole plug from Neutrik has found increasing acceptance for professional applications due to its contact performance and mechanical reliability.

XLR plugs and sockets are well suited to signal interconnection at 'line' level and are almost universally employed in recording and broadcasting studios. Domestically, both DIN and phono type plugs are used, but the latter are rather variable in both dimensions and quality and, as a result, are not very reliable except when made to high precision. Furthermore, the live signal pin of a phono plug usually makes contact before the shell or ground connection, which may give rise to a potentially damaging hum if the system is already energized when the plug is inserted or removed. The '5 pin' DIN plug is in fact better toleranced and has superior audio performance, but requires care and skill when making a reliable soldered connection to the small internal contact pins. It is also not physically compatible with some of the larger audiophile cable constructions. In the case of line-level interconnect links, connector contact resistance as well as spurious contact effects, for example due to diodic non-linearities, are even more important, and contact resistance values of 5 mΩ or less are to be preferred, these to be maintained over long periods. For critical applications, superior gold-plated phono plugs and sockets are available, and some versions do make the ground contact before the signal.

It should also be noted that, in the presence of a sound field, cabinet surfaces and structures absorb significant sound energy and consequently vibrate. The result, for the usual low contact force push-on connectors, is some degree of physical movement giving rise to variable contact effects. Both for this reason and for long-term reliability, locking-type connectors are preferred in more critical appplications.

References

1. IEC 581-1 (BS 5492, 1980) Audio Systems.
2. IEC 268-5 (BS 5428 1977) Loudspeakers.
3. KELLY, S, 'Network niceties', *Hi-Fi News*, August (1984).
4. MARTIKAINEN, I, VARLA, A and OTALA, M, 'Input current requirements of high quality loud-speaker systems', AES Preprint no. 1187(D).
5. OTALA, M and HUTTUNEN, P, 'Input current requirements of loudspeakers', AES 75th Convention, March (1984).
6. KAMEDA, O, 'Recent progress in oxygen free copper conductor for audio wire and cable', *Journal of Japan Audio Society*, **23**, No. 9, September (1983).
7. COLLOMS, M, 'Linear crystals', *Hi-Fi News*, October (1984),
8. COLLOMS, M, 'Crystals, linear and large', *Hi-Fi News*, November (1984).
9. HAWKSFORD, M J, 'The Essex echo', *Hi-Fi News*, August (1985) (with correction in October 1985).
10. GREINER, R A, 'Amplifier loudspeaker interfacing', *Loudspeakers: An Anthology*, Vol. 2, 135–140, AES (1980).
11. VANDERKOOY, J and LIPSHITZ, S, 'Computing peak currents into loudspeakers', AES Preprint No. 2411, 81st Convention (1986).
12. MCMANUS, J A and EVANS, C, 'The dynamics of recorded music', AES Preprint No. 3701, 95th Convention (1993).

Bibliography

ADAMS, G J and YORKE, R, 'Motional feedback in loudspeaker monitor', *Proc. IREE*, **85**, March (1976).
COLLOMS, M, *High Performance Loudspeakers*, John Wiley, Chichester, fifth edition (1996).
CORDELL, R, 'Open loop output impedance and interface intermodulation in amplifiers', AES 64th Convention (1979), Preprint 1537.
STAHL, K E, 'Synthesis of loudspeaker mechanical parameters by electrical means: new method for controlling low frequency behaviour', *Loudspeakers: An Anthology*, Vol. 2, 241–250, AES (1980).
WERNER, R E and CARRELL, R M, 'Application of negative impedance amplifiers to loudspeaker systems', 9th AES Convention 1957, in *Loudspeakers: An Anthology*, Vol. 1, 43–46, AES (1980).

7 Loudspeaker enclosures

Graham Bank and Julian Wright

7.1 Introduction

The earliest loudspeaker enclosures were cabinets in the true sense – crafted, decorative wooden furniture providing protection and cosmetic effect. Briggs[1] shows such cabinets as the HMV Victor gramophone horn of 1905 (Fig. 7.1 complete with audience!) and the BTH loudspeaker and amplifier of 1926 (Fig. 7.2). Hunt[2] provides a fascinating insight into the early history of loudspeaker enclosures, chronicling the confusion generated by the patent 'goldrush' of the 1920s.

Loudspeaker designers realized at a very early stage that frequency-dependent reinforcement and cancellation effects, due to the interference of radiation from the front and rear surfaces of the diaphragm, were rarely advantageous. A number of different methods were employed to avoid this problem, perhaps the most obvious being the airtight enclosure encasing the rear radiation – the closed box. Several designers used a vent, although some failed to appreciate the low-frequency enhancement that this could provide, using the vent merely as a form of pressure equalizer. However, a few notable individuals had the insight to appreciate the acoustical benefits of closed and vented boxes, including Frederick[3] and Thuras[4], who were among the earliest designers of such enclosures for enhancement of low-frequency radiation. The origins of the acoustic labyrinth can be traced to Olney[5].

Although the primary engineering function of a loudspeaker enclosure is to enhance low-frequency radiation, there are additional phenomena affecting performance at higher frequencies, such as mechanical and acoustical resonances and diffraction. The modern loudspeaker systems designer must consider a multiplicity of factors, some of which are counteractive.

We attempt here to introduce and discuss the major considerations in the design of a loudspeaker enclosure. Where possible, the reader is referred to texts covering specific topics in greater detail. Section 7.2 discusses the lumped-parameter method of modelling the behaviour of loudspeaker enclosures. Section 7.3 introduces the more common enclosure types. Sections 7.4 and 7.5 cover the mechanical and acoustical design considerations, and Section 7.6 discusses possibilities for the future in the form of finite element analysis.

7.2 Lumped-parameter modelling

Since the arrival of low-cost computer processing, theoretical analysis of the behaviour of loudspeakers has become a very powerful tool for the designer. Computer modelling provides a highly cost-effective way of optimizing a design prior to any construction, although the fine details are currently still left to testing and modification of physical prototypes.

Figure 7.1. HMV Victor gramophone, 1905. (Reproduced with the kind permission of Wharfedale International Ltd)

Figure 7.2. BTH 'Panatrope' loudspeaker and amplifier, 1926. (Reproduced with the kind permission of Wharfedale International Ltd)

Lumped-parameter loudspeaker models have been developed over many decades[6]: the technique itself even predates the moving-coil loudspeaker. Pioneered by, among others, Olson[7], this technique matured with the writings of Novak[8], Thiele[9] and Small[10], and continues to undergo refinement[11].

The lumped-parameter approach requires many simplifying and limiting assumptions, although it nevertheless provides a very useful method for obtaining a qualitative understanding of the behaviour of loudspeakers in their enclosures. The premise of this model is that each of the variables affecting the behaviour of a loudspeaker system can be treated as a lumped or bulk parameter of the system. By analogy with passive electrical components, an electrical network can be constructed which represents a simplified model of the behaviour of the electro-mechano-acoustical loudspeaker system.

Referring to Chapter 2, Fig. 2.54, we take as a starting point the general mobility analogue electrical circuit model of a moving-coil loudspeaker. This form of the network is preferred because the input terminals relate directly to the real loudspeaker terminals, i.e. the voltage source is equivalent to the output of an amplifier. In practice, the 'voice coil inductance', L_c, is replaced by a more accurate model of the motor impedance[12]. However, for the purposes of our analysis we shall neglect this motor impedance on the understanding that its effect at low frequencies is small but not insignificant. For clarity, we write:

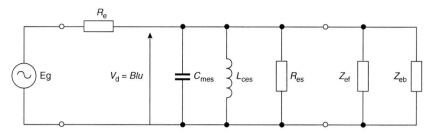

Figure 7.3. General mobility analogue electrical circuit model of a moving-coil loud-speaker.

$$L_{ces} = C_m(Bl)^2 \tag{7.1}$$

$$C_{mes} = M_m/(Bl)^2 \tag{7.2}$$

$$R_{es} = (Bl)^2/R_m \tag{7.3}$$

$$Z_{ef} = \frac{(Bl)^2/Z_{af}}{S_d} \text{ (front radiation impedance)} \tag{7.4}$$

$$Z_{eb} = \frac{(Bl)^2/Z_{ab}}{S_d} \text{ (back radiation impedance)} \tag{7.5}$$

giving the circuit shown in Fig. 7.3.

By defining the input voltage Eg (normally assumed to be of the form Eg = |Eg| exp ((t), i.e. sinusoidal) the network can be solved for the voltage V_d, and hence surface velocity u or volume velocity $U = u \cdot S_d$.

Note that R_e, C_{mes}, L_{ces} and R_{es} are defined only by the properties of the loud-speaker drive unit. The system design defines Z_{ef} and Z_{eb}, the mobility analogues of the acoustic impedance (radiation impedance). Therefore, in order to implement a lumped-parameter model, the parameters of the drive unit must first be obtained[10].

7.2.1 Piston range

It must be remembered that the lumped-parameter model is only valid when the drive unit vibrates in a piston-like manner, i.e. all radiating mechanical parts should vibrate with the same velocity magnitude and phase. Such behaviour is dependent upon the geometry and materials employed in the driver design. Often the cone of a high-fidelity bass or midrange unit will be designed to bend at very low frequencies in order to maximize control over resonant behaviour. In short, there is no substitute for an intimate knowledge of the mechanical performance of the actual driver[13]. However, Beranek[14] asserts that, for a conical diaphragm of height (or depth) b, piston behaviour can be expected when b is less than approximately $\lambda/10$, where λ is the wavelength of sound in air.

7.2.2 Point source approximation

Having solved the network for the required volume velocities, the far-field acoustic pressure p can be calculated by assuming that the system behaves as a point source. If the wavefront is truly spherical, the following equation can be used[15]:

$$p = j\frac{\rho c k}{2\pi r} u S_d \exp j(\omega t - kr) \tag{7.6}$$

where ρ is the density of air
         c is the velocity of sound in air

ω is the angular frequency (where $\omega = kc$)
k is the wavenumber (where $k = 2\pi/\lambda$)
r is the distance from the source to the observation point

Note that this expression assumes radiation into 2π steradians, i.e. the source lies in an infinite plane. To model radiation into 4π space, the denominator above should be $4\pi r$. Equation (7.6) can be simplified to:

$$p = j\frac{\rho f U}{r} \exp j(\omega t - kr) \tag{7.7}$$

If the system is not acting as a point source (e.g. at higher frequencies) it will be necessary to subdivide the various radiating surfaces into elements which are small enough to be treated as point sources, and calculate some surface integrals[15]. This is clearly more complicated, but can extend the validity of the model. It would be easier to treat the system as a point source, but to do this we must be confident that the radiated wavefront is spherical. As a guide, the upper frequency limit for such behaviour is taken to occur when the circumference of the driver is equal to one wavelength[14], i.e.

$$\pi d = \lambda \quad \text{or} \quad ka = 1, \quad \text{or} \quad f_{max} = c/\pi d \tag{7.8}$$

where d and a are the effective piston diameter and radius of the driver and f_{max} is the upper frequency limit of point source behaviour.

Taking $c = 345$ m/s. Table 7.1 shows some typical values for drivers of different diameters.

Table 7.1 Calculated loudspeaker point-source upper frequency limits

Nominal diameter (in.)	Effective piston diameter (m)	f_{max} (Hz)
6.5	0.135	813
8	0.165	666
10	0.21	523
12	0.26	422
15	0.33	333

It should be stressed that these frequencies are only a rough guide, as the maxim requires that the source acts as a piston (see Section 7.2.1).

7.2.3 Radiation impedance

The most difficult problem facing the user of this model is the assignment of values to Z_{af} and Z_{ab}, the radiation impedances acting on the front and back of the diaphragm respectively. For the purposes of a general discussion of lumped-parameter modelling, we can make some simple approximations to obtain values of Z_{af} and Z_{ab} which should be correct to within an order of magnitude. The best summary to date was provided by Beranek[14] in 1954.

The rear radiation impedance will be dominated by the air compliances and vent masses, although there will be some mass-loading effects due to the proximity of enclosure walls and some resistances, i.e. losses due to heat and cabinet motion. These effects are complex and work in this field is still in its infancy. Furthermore the lumped-parameter model for these phenomena will cease to be valid near and above the first standing wave frequency[16-18]. See also Section 7.5.

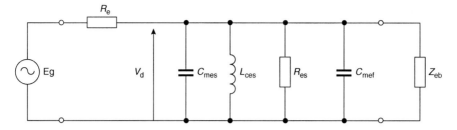

Figure 7.4. Simplified analogue electrical circuit.

The radiation impedance acting on the front of the diaphragm is slightly easier to define and, depending on the frequency of interest, is usually simplified to one of two basic models: the loudspeaker is assumed to behave either as a flat disc vibrating in an infinite baffle (radiation into 2π steradians), or as a flat disc vibrating in the end of a long tube (radiation into 4π steradians)[14]. These simplifications have been made in the absence of more accurate models. Radiation impedance data for cones and domes are now available[19,20].

The matter is further complicated in the case of multiple sources by mutual radiation impedance effects, that is, the impedance of the motion of a source due to the radiation from another source[17,21,22]. Note also that any source of radiation, including vents, etc., will also have an associated 'self' radiation impedance (acoustic impedance due to its own motion). In the case of a vent, this. is often approximated as an 'end-correction' where the effective length of the vent is increased in accordance with the associated additional volume of air required to give the appropriate additional mass.

Chapter 1, Fig. 1.7, shows the radiation resistance and reactance on one side of a flat circular disc vibrating in an infinite baffle. Below about $ka = 0.5$ the radiation resistance is very low, and the radiation reactance rises at a constant 6 dB/octave. This is equivalent to a constant mass of value $8\rho a^3/3$[14]. So, for the purpose of this model we take the front radiation impedance to be a simple constant mass, based upon the assumption that the loudspeaker is acting as a flat disc in an infinite baffle, with the low-frequency radiation resistance neglected. This gives the circuit shown in Fig. 7.4, where the mobility analogue of the constant mass is the capacitor, C_{mef}. We therefore limit the frequency range over which the radiation impedance model is valid to $ka < 0.5$. This range could be extended by inserting, at any frequency, the appropriate values of radiation impedance[14]. For simplicity, the rear radiation impedance is taken to be predominantly air compliances and vent masses. It must be emphasized that, in any attempt to use the lumped-parameter model for loudspeaker enclosure design, *designers must clearly specify their assumptions about radiation impedances and solid angle of radiation*.

7.3 Enclosure types

Concise Oxford Dictionary: **Baffle-board**, device to prevent spread of sound, esp. round loudspeaker cone to improve tone(!).

7.3.1 Infinite baffle

The true infinite baffle is a theoretical concept. The loudspeaker is mounted in an inert flat plane which extends to infinity (Fig. 7.5). In this way, the forward radiation is completely isolated from the rear radiation, and there are no baffle edges to cause diffraction, nor are there any enclosure walls to generate mechanical or

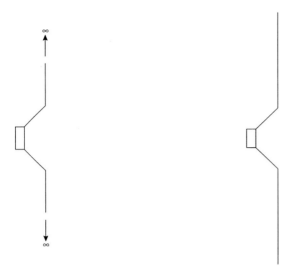

Figure 7.5. Infinite baffle.   **Figure 7.6.** Finite baffle.

acoustical resonances. Furthermore the solid angle of radiation from either face of the diaphragm is undisputably 2π steradians. Such a baffle allows a greatly simplified theoretical analysis, and has occupied many writers from the time of Rayleigh to the present[15,23,24].

7.3.2 Finite baffle

Arguably the simplest 'enclosure', the finite, or open, baffle (Fig. 7.6) serves as a practical mounting mechanism. Considering the loudspeaker diaphragm as a point source, the resulting acoustic radiator is a dipole where the path length between the two point sources is given by the physical distance around the baffle from the front to the rear of the diaphragm. In this case the on-axis pressure response is that of a comb filter with a roll-off at low frequencies, and the directivity has a figure-of-eight nature (Fig. 7.7).

Olson[24] gives the equation describing far-field pressure as:

$$p = \frac{\rho f u}{r} \cos\left(t \sin\left(kD/2\right) \cos\theta\right) \tag{7.9}$$

Note that low-frequency output can be extended by increasing the value of D, the source separation. The practical interpretation is to increase the size of the baffle. In practice the loudspeaker is not a point source, but has finite dimensions. The resulting combination of numerous front-to-rear acoustic paths of slightly varying lengths gives rise to a considerable corruption of both the idealized comb filter and the directivity. This phenomenon can be emphasized by offsetting the loudspeaker from the centre of the baffle. In this way the uniformity of the acoustic output can be greatly improved. Directivity can be further controlled by using arrays of dipole sources[25,26].

Although isolating the front radiation from the rear radiation is usually desirable, there are advantages to the dipole system. First, the low-frequency roll-off is only first order, i.e. 6 dB/octave, which is more conducive to electronic equalization than other enclosure configurations, because the electronic gain demands less increase in loudspeaker displacement than in higher-order systems. Second, the reduced radiation

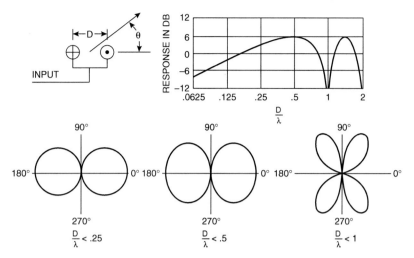

Figure 7.7. Schematic diagram, on-axis and polar frequency response characteristics of a dipole source. (Reproduced with permission from 'Gradient Loudspeakers' by H. F. Olson in *Loudspeakers: An Anthology*, Vol. 1, 305, AES (1980).)

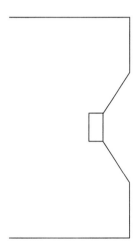

Figure 7.8. Open-backed cabinet.

from the sides of the source can be used to advantage for sound reproduction in enclosed spaces. There exists, in comparison to the other enclosure types, an extra variable in optimizing the location of the source for high-fidelity reproduction, i.e. one can rotate the source and alter the in-room response.

The most common practical version of the finite baffle is the open-backed cabinet (Fig. 7.8) where protection and stability are provided, while maintaining the dipole behaviour. Standard specifications exist for finite baffles[27]: these are normally used for testing purposes.

7.3.3 Closed box

The closed box is also known as sealed box or acoustic suspension[28]. It is also sometimes erroneously referred to as an infinite baffle. It is indeed the nearest practical embodiment of an infinite baffle, but the finite nature of the enclosure gives rise to resonance and diffraction effects and the solid angle of radiation is no longer clearly defined.

The rear radiation is completely isolated by means of a sealed enclosure (Fig. 7.9). The enclosed air provides a compliance (Fig. 7.10) which also serves to limit the low-frequency excursion of the drive unit, hence improving longevity. The resultant acoustic pressure output has the nature of a second-order high-pass filter, the asymptotic roll-off being 12 dB/octave (Fig. 7.11). The rear radiation must be dissipated as heat (usually by placing absorbent materials in the box) or by motion of the cabinet walls. The latter is highly undesirable (see Section 7.4).

Note that it may be ill advised to seal the enclosure hermetically: any changes in atmospheric pressure would cause changes in the equilibrium position of the driver. This problem is normally overcome by allowing air to pass through the cabinet, by leakage around or through the panels.

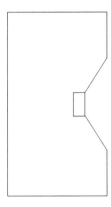

Figure 7.9. Closed box.

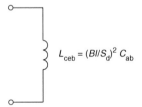

$$L_{ceb} = (Bl/S_d)^2 C_{ab}$$

Figure 7.10. Equivalent circuit for simplified radiation impedance of a closed box.

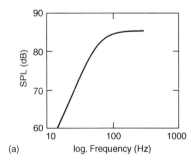

(a)

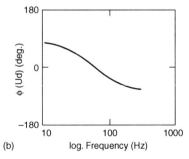

(b)

Figure 7.11. (a) Sound pressure level, and (b) volume velocity phase ϕ (Ud) of a typical closed box alignment.

In the normal closed-box design, the box is large and the compliance of the enclosed air is high: the compliance of the drive unit itself predominates, and the enclosure serves merely to isolate the rear radiation. However, in the acoustic suspension design, the enclosure is small enough to exert the controlling influence upon the effective compliance of the system. The only source of acoustic radiation is the front face of the driver: the required volume velocity is obtained easily from a calculation of V_d.

7.3.4 Vented box

The vented box is also referred to as ported box, bass-reflex, phase inverter[9,29]. The rear radiation is added to the direct radiation, over a limited bandwidth at low frequencies, by means of an aperture which is usually located on the front or rear of the cabinet (Fig. 7.12). This aperture may be a simple hole, but is often a tube or short tunnel. A Helmholtz resonator is formed by the mass of air in the vent resonating with the compliance of the air in the box[15] (Fig. 7.13). Superficially, this configuration provides 'more bass' than the closed box, but the fourth-order (24 dB/octave) low-frequency roll-off (Fig. 7.14) generally results in greater output in the upper bass (*circa* 80 Hz) but less 'deep bass' (*circa* 30 Hz). Placing the vent at the rear of the cabinet may prove beneficial in reducing the audibility of high-frequency spuriae radiated through the vent.

The total acoustic output will be governed by the (complex) sum of the volume velocity of the front face of the driver and the (out-of-phase) volume velocity of the vent. Hence the nett volume velocity is proportional to $(V_d - V_v)$. The Appendix shows a listing of a QBasic Personal Computer program to model vented box behaviour (QBasic is shipped with most Microsoft operating systems).

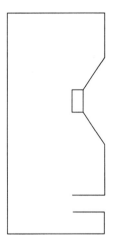

Figure 7.12. Vented box.

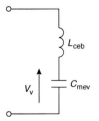

Figure 7.13. Equivalent circuit for simplified radiation impedance of a vented box.

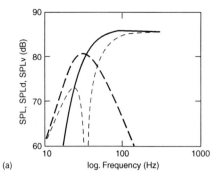

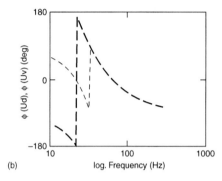

Figure 7.14. (a) Sound pressure levels, and (b) volume velocity phases of a typical vented box alignment. (– – –, driver; – – –, vent; ——, system).

7.3.5 Passive radiator

The passive radiator is also known as auxiliary bass radiator (ABR) or drone cone phase inverter[7,30]. Here the vibrating air mass of the vent is replaced by a suspended mass of solid material (Figs 7.15 and 7.16). The resultant capability for very high mass allows the use of very low tuning frequencies. Olson[7] claims several additional advantages over the normal air-mass vent, among them the feasibility of a large radiating area of the phase inverter (hence lower velocities for improved linearity). To achieve a large area with a normal vent would require a long tube, introducing viscous losses along the tube walls. Some additional design flexibility is provided by the possibility of controlling the mechanical compliance of the ABR. Typical realizations include cone loudspeakers without a motor system (i.e. no magnet and voice coil) and polystyrene 'plugs'.

As with the vented box, the total acoustic output will be governed by the (complex) sum of the volume velocity of the front face of the driver and the (out of phase) volume velocity of the passive radiator. Hence the nett volume velocity is proportional to $(V_d - V_p)$.

7.3.6 Transmission line

The transmission line enclosure is also called acoustic labyrinth[17,31,32]. Here the rear radiation is directed through a labyrinth which is usually open-ended, allowing enhancement at low frequencies (Fig. 7.17). The labyrinth can be thought of as an

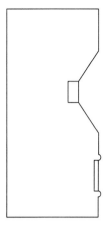

Figure 7.15. Passive radiator system.

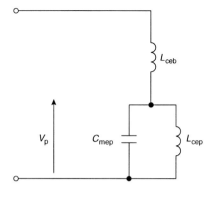

Figure 7.16. Equivalent circuit for simplified radiation impedance of a passive radiator enclosure.

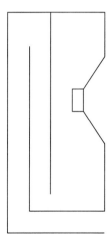

Figure 7.17. Transmission line.

organ-pipe, which supports resonant behaviour. At the frequency where the length of the transmission line is equal to $\lambda/2$, the velocity at the opening is in-phase with the forward radiation from the driver and maximum reinforcement occurs. By lining the labyrinth with a suitable absorbent material, higher frequencies can be attenuated and the significance of the resonances can be controlled. This material can also reduce the velocity of sound in the line[33] and hence reduce the required line length. Of course, there are other frequencies at which organ-pipe resonances occur, and the positive or negative contributions of these resonances must also be considered. The practical performance of the system is also affected by the distance between the driver and the opening of the transmission line: if they are separated by a distance greater than about $\lambda/4$, the combination of volume velocities will not be purely additive (this applies to any multi-source system). The resulting output has a subjective character significantly different from that of a vented system, although the theoretical asymptotic roll-off will still be 24 dB/octave.

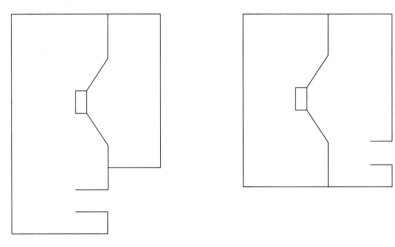

Figure 7.18. Conceptual bandpass system.   **Figure 7.19.** Practical bandpass system.

To determine the radiation impedance acting on the rear of the driver, we consider the labyrinth as a pipe in which plane waves are propagated[15]. If the pipe has length l, the resulting acoustic impedance at the drive unit end is:

$$Z_{ab} = \frac{\rho c}{S} \frac{(Z_{al} + j(\rho c/S) \tan kl)}{((\rho c/S) + jZ_{al} \tan kl)} \tag{7.10}$$

where S is the cross-sectional area of the pipe and Z_{al} is the acoustic impedance at the end of the pipe.

If the end of the line is open and therefore radiates, Z_{al} can be approximated at low frequencies by treating the radiator as a piston and hence $Z_{a} = j\omega \cdot 8\rho a^3/3S^2$ (see Section 7.2.3 above).

In the case where the line is rigidly terminated and is used simply as a superior mechanism for suppression of the rear radiation, $Z_{al} \to \infty$ and:

$$Z_{ab} = \frac{\rho c}{S} \frac{1}{j \tan kl} \tag{7.11}$$

Practical designs often incorporate a tapered labyrinth, where the cross-sectional area of the line decreases with distance along the line. Here the analysis requires a revised treatment[34].

7.3.7 Bandpass enclosures

Returning to Fig. 7.14, if the output of the vent alone is used the result is a band-pass filter. Conceptually this can be achieved by taking a vented system and enclosing the forward radiation from the driver (Fig. 7.18). A more practical realization is shown in Fig. 7.19. Because the bandwidth is restricted, the most common application for a bandpass system is a sub-woofer. The drive unit is mounted internally, on the baffle dividing the two chambers. The result is effectively a closed box on the rear of the driver and a vented box on the front, with no direct radiation from the drive unit. This system has been described in detail by Fincham[35]. This configuration produces a bandpass filter with second-order high-pass and second-order low-pass roll-offs, sometimes called a fourth-order bandpass. By venting the rear chamber as well, a 4th + 4th, or 8th-order, bandpass response results. Geddes[36] examines both

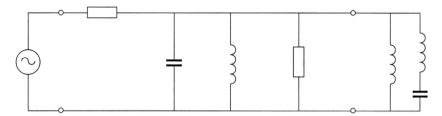

Figure 7.20. Simplified electrical analogue circuit for fourth-order bandpass system. Acoustic output is due entirely to vent volume velocity.

systems. Such bandpass enclosures can be tuned to provide greater pass-band output in comparison to the normal high-pass alignments.

Logically, on the basis of the assumptions concerning the rear radiation impedance in Sections 7.3.3 and 7.3.4 above, the circuit model for the fourth-order bandpass would be as shown in Fig. 7.20. Note that C_{mef} has been removed. However, we have now made two significant approximations concerning radiation impedance of enclosures, and the accuracy of this bandpass model requires careful appraisal.

7.3.8 Multi-chamber systems

Figure 7.21 shows an arbitrary multi-chamber arrangement. Such systems are rare, but have relevance for highly specialized designs[37]. Their increased complexity renders a lumped-parameter approach questionable.

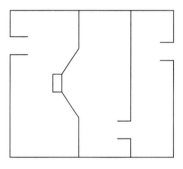

Figure 7.21. Multi-chamber enclosure.

7.4 Mechanical considerations

7.4.1 Choice of material

A loudspeaker enclosure should contribute nothing to the output that could possibly have a negative influence upon, or colour, the reproduced sound. Coloration could occur if the panels of a cabinet vibrate and then reradiate into the sound field, or if they are acoustically 'transparent' at any or all frequencies. The traditional method of cabinet design is to regard the enclosure stiffness at low frequencies as of paramount importance. Middle- to high-frequency panel vibrations are then treated by bracing and/or damping techniques. Alternative methods have used very stiff materials where the first panel resonance is pushed up higher in frequency, leaving the designer with the need for a modest, if any, amount of panel damping. The vast selection of different materials available for constructing an enclosure has been the source of numerous inventions and claims throughout the history of loudspeakers and the literature abounds with references, most of which can be found in an excellent treatise by Hunt[2].

Materials that could be used for enclosures can all be characterized by quantifiable properties: Young's modulus, density (and/or mass), internal self-damping. Harwood and Mathews[38] have made a number of studies of the mechanical properties of various cabinet materials, and these are reproduced in Fig. 7.22. Briggs in one of his early books[39] starts his section on materials by saying that 'Acoustically speaking, a good motto for cabinets is "the denser the better" ...', but this statement was made before some of the more recent man-made materials had come onto the market. Recent fabrication techniques have allowed aluminium honeycomb, foamed plastic and synthetic resins to be adapted for enclosures. Some of these materials may not seem to be ideally suited on first examination, but designers' ingenuity has pressed them into service with excellent results. No ideal cabinet material exists, just one ideal requirement that it should contribute nothing audible to the sound.

Whatever material is chosen, the mechanical properties, details of construction, panel shape and panel curvature will determine the final behaviour of the enclosure. Flat rectangular panels still dominate the designs of most enclosures, but advances in new materials mean that alternative techniques, like those mentioned above, can be used to manufacture cabinets. These materials can be shaped into forms which are inherently stiffer than flat panels. Local bracing can easily be incorporated into a design by using ribs, which do not add extra mass, and changes in panel thicknesses at structurally weak points are very easily accomplished. It is important when trying to make a rigid enclosure to choose a method of construction which will make rigid joints at the corners and edges. Barlow in 1970[40] described how the bass response he obtained from his 'stout experimental cabinet' was less than he had expected (at that stage it was common practice to fit such enclosures with removable fronts, and often removable backs as well). In this instance, a loss of some 3 dB from 60–100 Hz was apparent. In view of this, Barlow changed to an all-glued construction, with the bass unit screwed in from the outside and access to the cabinet being obtained via the bass unit mounting hole, and the output was restored. This is now the universally accepted method for the construction of wooden cabinets.

7.4.2 Coincidence effect

The situation is further complicated by the coincidence effect described by Beranek[41] where, at a particular frequency (critical frequency), sound impinging on a panel at a particular angle will have the same wavelength projected onto the panel surface as the flexural wave in the panel. Although there may be no mechanical resonance, the panel will radiate strongly at this frequency. Care must be taken with especially lightweight panels, where the coincidence frequency may occur in the sensitive mid-band area. Fortunately this particular phenomenon is amenable to treatment by a small amount of damping material.

7.4.3 Panel modes

A rectangular box may be thought of as having its edges pinned by the action of the adjacent panels, and some simple calculations can be made to give the panel resonance frequencies and mode shapes. Leissa[42] gives expressions for these panel modes, and shows some of the low order mode shapes. Figure 7.23 shows some typical panel vibrational modes that can be measured either by accelerometers or, in this case, directly measuring velocity using a Laser Vibrometer. The calculations for simply supported and clamped panels are given by Iverson[43] and Moir[44], and the expression for a clamped panel is given for reference:

Material	Density	Max Q	Elastic modulus $E \times 10^{-9}$ N/m^2							
			62.5 Hz	125 Hz	250 Hz	500 Hz	1 kHz	2 kHz	4 kHz	6 kHz
16 mm Three layer chipboard 0.600 density	0.60	50	*2.08	2.02	1.95	1.8	1.52	1.15	0.83	0.67
18 mm Three layer chipboard 0.600 density	0.61	45	*2.42	2.37	2.2	1.95	1.6	1.17	0.74	0.53
9 mm Resin bonded Birch 7 ply along outer grain	0.72	67	*12.2	12.2	12.0	11.1	9.8	7.8	5.4	4.1
9 mm Resin bonded Birch 7 ply across outer grain	0.72	63	*5.7	5.5	5.3	4.8	4.2	3.55	2.8	2.4
9 mm Casein bonded Birch 7 ply along outer grain	0.67	77	*11.1	11.1	11.0	10.2	8.8	7.0	4.9	3.9
17 mm Resin bonded Birch 13 ply along outer grain	0.74	54	*7.5	*7.5	7.3	6.6	5.6	4.1	2.55	1.8
17 mm Resin bonded Birch 13 ply across outer grain	0.73	54	*7.0	*7.0	6.9	6.3	5.3	3.95	2.55	1.8
10 mm Mahogany 5 ply along outer grain	0.61	97	*8.5	8.3	7.9	7.2	6.1	4.7	3.2	2.2
10 mm Mahogany 5 ply across outer grain	0.61	82	*3.8	3.8	3.6	3.2	2.65	2.1	1.63	1.38
12 mm Birch faced Red Pine core blockboard along surface grain	0.535	68	*9.0	*8.4	7.4	6.0	4.3	2.55	1.5	1.25
12 mm Birch faced Red Pine core blockboard across surface grain	0.535	75	*4.9	4.9	4.9	4.75	4.3	3.65	2.95	2.4
G.R.P. 20% random fibres	1.315	31	5.8	5.8	5.7	5.2	4.8	4.9	5.5	—
G.R.P. 50% random fibres	1.55	47	10.7	10.7	10.3	9.2	8.5	8.7	10.0	—
G.R.P. 70% random fibres	1.51	54	9.4	9.3	8.9	8.1	7.7	7.8	8.3	—
G.R.P. 50% woven rovings	1.59	46	12.8	12.8	12.5	11.8	11.1	10.8	10.5	—
Carbon Fibre/Honeycomb Sandwich	0.255	170	—	—	—	7.2	5.2	3.0	—	—

* Extrapolated

Figure 7.22. Properties of various manufactured materials. (Reproduced from *Factors in the Design of Loudspeaker Cabinets* by H. D. Harwood and R. Mathews[38].)

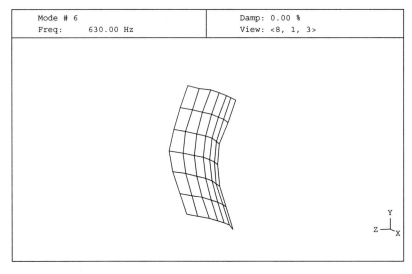

(a)

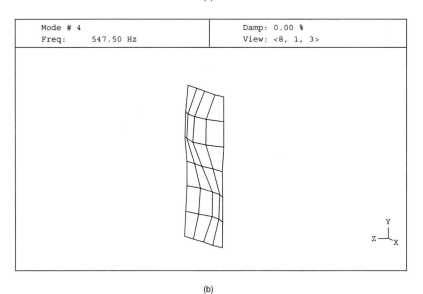

(b)

Figure 7.23. (a) Unclamped panel, 630 Hz, (b) clamped panel, 547.5 Hz.

$$f_c = \frac{12\sqrt{D}}{2\pi\rho} \sqrt{\frac{7}{2}\left(\frac{1}{a^4} + \frac{4}{7}\frac{1}{a^2b^2} + \frac{1}{b^4}\right)} \qquad (7.12)$$

where $D = Eh^3/12 \ (1 - \nu^2)$
$f_c =$ fundamental frequency
$a,b,h =$ length, width and thickness of panel
$\rho =$ mass per unit area of the panel
$E =$ modulus of elasticity
$\nu =$ Poisson's ratio

7.4.4 Panel bracing

It is necessary to balance the various parameters of a chosen cabinet material, since taking any one single feature to its limit (density, for example) may not yield the best overall result. However, it is possible to change the behaviour of any panel by mechanical methods such as bracing, where little extra mass is added, and Tappan[45] has shown typical improvements from bracing cabinet panels. Results from a method described by Wright[46] are shown in Fig. 7.24. Here a sequence of velocity measurements made on four panels of the same size are shown for comparison.

They are, in order:

(a) normal medium density fibreboard (MDF) cabinet
(b) braced MDF
(c) double-braced MDF
(d) aluminium honeycomb sandwich (Aerolam™), taken from a Celestion SL600 cabinet

The addition of bracing and finally a change to a different panel material clearly show the improvements in behaviour for panels of the same nominal dimensions. Bracing is a very worthwhile technique for cabinet designers to consider, since it can make dramatic improvements to performance, without a severe penalty on cost. The effectiveness of such methods is usually confined to lower frequencies, and Iverson[43] concludes that, if panel resonances occur at high frequencies, then little vibration results. Stiff lightweight materials could be chosen for this case, providing due care is taken over the coincidence effect.

7.4.5 Driver decoupling

As well as the detailed construction of the cabinet affecting the cabinet behaviour, the coupling of drive units to the baffles can also have a dramatic effect upon the complete system. Recent techniques for decoupling the drive unit from the front baffle, or the magnet from the chassis, have resulted in much lower panel vibrations. For example, the KEF 104/2 uses a system of carefully controlled synthetic isolation bushes between magnet and chassis, effectively reducing the vibrations transmitted to the enclosure. This technique does, however, require an isolation system that has good long-term stability.

7.4.6 Damping

Other techniques for minimizing the effect of cabinet resonances have relied on using additional damping. There are two main methods of application. The first often takes the form of a single unconstrained layer of material added to the inside walls of the cabinet, where this layer experiences higher strains and therefore dissipates more energy. Damping material is either painted on or glued and stapled, depending on its formulation. In either case the solvents that are trapped may evaporate from the surface, and cause long-term problems, if they interact with the drive unit materials. The second method is to use two thin panels bonded together with a damping layer between them. This constrained-layer technique does however have the disadvantage of needing to cut through the damping layer during manufacture, usually coating the cabinet-maker's saw with a sticky deposit, but it is intrinsically more effective than the unconstrained layer method.

7.4.7 Audibility

What is the optimum panel behaviour? A number of researchers have tried to answer this question, and diagrams and charts of audibility abound. For example, Harwood and Mathews[38] published such a chart in 1977, reproduced here in Fig. 7.25, which

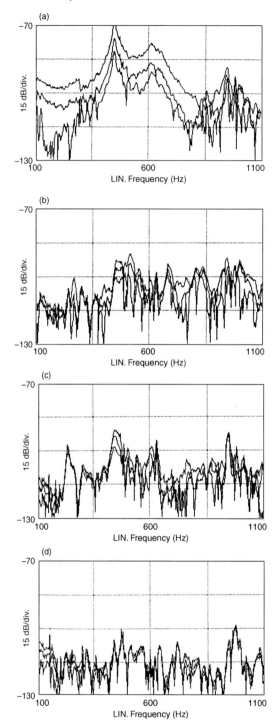

Figure 7.24. Velocity magnitudes measured near the centres of four panels of similar size and different materials.

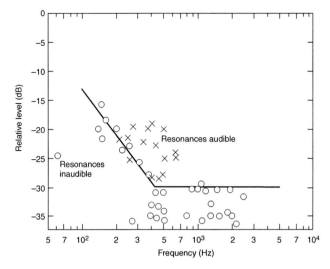

Figure 7.25. Criterion for perceptibility of panel resonance. (Reproduced from *Factors in the Design of Loudspeaker Cabinets* by H. D. Harwood and R. Mathews[38].)

indicates the target level for audible resonances. The chart is only a guide, but it does illustrate that high-frequency resonances, i.e. above 400 Hz, are less audible than those below this frequency. Toole[47] has presented some results on listening tests and perception, whilst later work by Lipshitz *et al.*[48] illustrates a computational technique for predicting the radiation from enclosure panels.

7.5 Acoustical considerations

In designing a loudspeaker enclosure, due regard should be given to both the forward and rearward sound power. Sound emitted from the front of the diaphragm will be modified (diffracted) by the shape of the front of the enclosure, while that from the back of the diaphragm will be modified by the detailed construction inside the cabinet. Apart from the cases where this rear radiation is harnessed to augment the very low frequencies by some phase-changing mechanism, we can assume that the rear radiation from the diaphragm should be completely absorbed.

7.5.1 Standing waves

The degree to which the rear radiation affects performance will be determined by the geometry of the enclosure and any standing waves generated within it. These are described in detail in Chapter 8, Sections 8.2.1 and 8.2.2, and can be calculated for a small loudspeaker enclosure in exactly the same way as for a room. However, their effect in cabinets can be reduced in ways which are impractical with a room: although there is little flexibility in the geometry of a room, it is possible to change the internal shape of a cabinet. This can be achieved with non-parallel walls, angled internal baffles, or panel spacings which are not harmonically related.

Alternatively, some absorbing material can be placed inside the cabinet. This should be positioned in a region where maximum particle velocity occurs rather than at a pressure maximum, since energy is transferred to the absorbent by a kinetic process. This implies that the absorbent should be fitted in the central void of the enclosure rather than on the walls. The absorbent should be firmly retained, since any movement may cause the damping action to be non-linear.

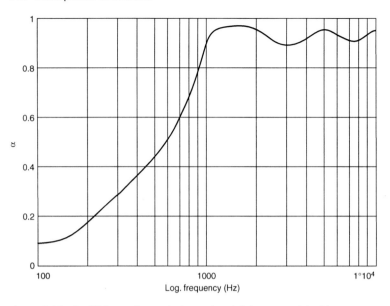

Figure 7.26. Coefficient of sound absorption (α) for normal incidence upon a sample of polyester foamed plastic.

7.5.2 Absorbers

Failure to absorb the rear-radiated power will give rise to excitation of the cabinet panels as well as standing waves within the cabinet space and, since the diaphragm mass is likely to be relatively low, re-radiation back through the cone itself to combine with the original radiated sound. A number of researchers have studied this particular phenomenon, including Barlow[40], who concludes that most cabinet designs will need some absorbent filling to reduce this effect.

The different effects of various filling materials have been studied by Bradbury[33], Small[28], Leach[11] and others, and these findings are still the subject of debate, especially when trying to predict accurately low-frequency behaviour. For mid-frequencies their use is always recommended for damping standing waves within the enclosure. The wide variety of filling materials do have diverse absorption properties, and Fig. 7.26 shows the performance of a 25 mm thick sheet of a typical polyester foamed plastic. While attenuation is good at high frequencies, performance is dramatically reduced at lower frequencies.

7.5.3 Partitioning

Even large amounts of absorbent may not completely eradicate the standing waves within the cabinet. In these cases elaborate forms of baffling or partitioning, for example B&W Matrix™, have been used successfully. The inside of the cabinet is effectively broken up into small interconnected chambers and so low-frequency standing waves are suppressed. Since the individual chambers have to be filled with some absorbent, this can make the whole cabinet quite difficult to assemble. Other less ambitious designs have one or two connected enclosures, where the first is small to prevent standing waves, and the second is joined by way of a hole or tube where only low frequencies are coupled.

7.5.4 Diffraction

The radiated wavefront will be affected by the shape of the cabinet due to the phenomenon of diffraction[51]. Olson[49] shows (Fig. 7.27) a number of different cabinet shapes and their relative merits in terms of on-axis Sound Pressure Level (SPL). It is worth bearing in mind that the sound power output of the system should be considered along with the on-axis SPL, to get the correct balance between direct and indirect sound when listening in a normal room. Although some of the shapes are more practical and acceptable than others, they could all suffer from diffraction at the first boundary between the front baffle and cabinet side panel. Kates[50] shows how a mathematical model can be used to predict simple reflections from mechanical features on the cabinet front panel, and how to suppress some of these effects by adding a sound absorbing blanket to the cabinet baffle. Recent publications on diffraction have been confusing and contradictory. For clarification see Vanderkooy[52] and Wright[53].

7.5.5 Vents

Enclosures which are not sealed box types, but which use the rear radiated sound as part of the low-frequency output of the loudspeaker, will need some form of opening or vent to couple to the outside of the cabinet. Such openings must be considered carefully in terms of dimension, shape, position and air flow. Small[29] gives an expression for the minimum area and diameter as follows:

$$S_v \geqslant 0.8 f_B V_D \tag{7.13}$$

$$d_v > \sqrt{f_B V_D} \tag{7.14}$$

where S_v is the area of the vent
   d_v is the diameter of a circular vent
   V_D is the peak displacement volume of the driver diaphragm
   f_B is the resonance frequency of the vented enclosure

The shape of the vent is also important, since rapid changes in section will interrupt the smooth flow of air and this will give rise to distortion. Salvatti et al.[54] have conducted extensive research into vent geometry, and Pederson and Vanderkooy[55] discuss measurement techniques. Although the lumped parameter analogue circuit used for vented enclosures does not usually contain any detailed modelling of the vent behaviour, Fincham[35] shows a transmission line-type equivalent circuit which gives good correspondence with the measured performance. Recent work has shown that the position of the vent in a cabinet can dramatically affect the coupling to the low-frequency standing wave within the enclosure, giving rise to very marked changes in performance with different vent positions. This effect can cause perturbations in the pass-band of the system even if the first standing wave is an octave higher than the system's nominal low-pass cut-off. Figure 7.28 shows two SPL curves from a band-pass enclosure, where the position of the port in the first case causes a severe resonance to be reproduced. Moving the port to a second position not only reduces the effect dramatically, but restores the output in the pass-band region.

7.6 Finite element modelling

The advantage of the lumped-parameter model for a combination of loudspeaker and enclosure is one of simplicity and speed. A computer model can easily be developed from simple principles and a good approximation can be obtained for most cases. However, it is limited by a number of simplifications: the model does not account for standing waves in the enclosure, resonances within the loudspeaker diaphragm structure, diffraction or cabinet panel modes. The model is ideal for low

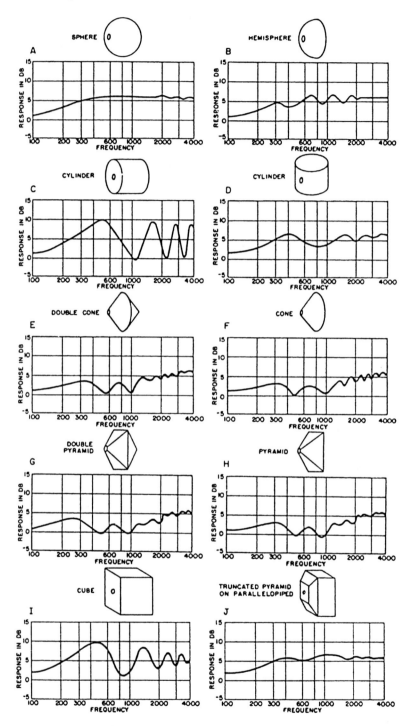

Figure 7.27. Ten different enclosures tested for their effects on the sound pressure response of a loudspeaker. (Reproduced from *Acoustical Engineering* by H. F. Olson[7].)

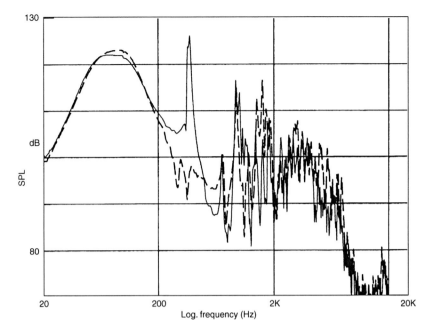

Figure 7.28. Effect of vent location upon sound pressure output from a bandpass enclosure (– – –, original vent position; - - -, optimized location).

frequencies, but of limited use at higher frequencies. It is possible to improve the lumped-parameter model by adding extra parameters, but these can make any such model very cumbersome, as well as limiting its flexibility.

We require, therefore, a completely flexible method for modelling the mechanical and acoustical parts of the whole loudspeaker system. Such a method is that of finite element (FE) analysis. Advances in the power of desktop computers have led to a revolution in the availability of such techniques to loudspeaker design. The FE technique originated as a method of predicting the stress in large engineering structures. It was developed to determine the natural resonance frequencies of mechanical structures, and from there to model the behaviour of coupled structures and fluids.

7.6.1 Mechanical finite element modelling

Frankort[56] illustrated the technique in 1978 by dividing a loudspeaker cone into small annular rings, or sections, which are joined together to form a complete cone. In his model the cone was represented by six elements only, and the simultaneous equations were solved for a complete range of frequencies. In more modern techniques using general purpose FE computer models, the loudspeaker moving parts are modelled by subdividing them into small elements which are joined together at connecting points (nodes) to form a complete mesh. The mechanical properties of each small area determine how the whole matrix of elements will behave as a complete structure, and both the free (natural) vibration and forced vibrational responses can be determined. This can give a very accurate model of the vibrational behaviour, predicting the velocity of the radiating surface. The next difficulty lies in predicting what the sound pressure distribution might be from all these separate elements vibrating with different magnitudes. This stage is much more difficult.

7.6.2 Acoustical finite element modelling

The use of an FE model which includes the fluid (in our case, air) is a fairly recent development, and Macey[57] and Kaizer *et al.*[58] describe the methods available. The difficulty with modelling the air in front of the loudspeaker is caused by the need to represent an infinite region of air. Since an infinite region of air would need an infinite number of elements to model it, we must represent this region in some other way, by a finite number of elements. We must also have a model which ensures that the solution represents only outward travelling waves[58].

The solution relies on replacing the infinite fluid by a boundary element which satisfies the Helmholtz equation at the boundary between the finite elements and the infinite fluid. We now need to model the loudspeaker voice coil, diaphragm, surround and cabinet details to be able to predict the complete soundfield produced. The normal methods for modelling the mechanical parts are employed, using the properties of the various materials. Since the majority of loudspeaker drive units are round and cabinets are often rectangular, symmetry in the model is used wherever possible. This technique can reduce the size and complexity of an FE model quite considerably, and is always recommended for consideration. When the structure has been suitably represented by a mesh of connected elements, the region directly in front of any radiating part is meshed with pressure-based acoustic finite elements, where the pressure at the nodes is the only degree of freedom. At a suitable distance from the radiating structure, a boundary element is coupled to the surface of the acoustic finite elements, and the model is then complete. The relevant boundary conditions are applied, along with the appropriate forcing function. This can be derived from a knowledge of the force *Bli* exerted on the coil in a magnetic field of known flux density *B* when a current *i* flows in a length of conductor *l*.

The solution to the FE problem will yield the velocity, acceleration and displacements of the loudspeaker diaphragm and/or cabinet walls, as well as the pressure at the nodes within the acoustical finite elements and pressures anywhere within the infinite fluid region modelled by the boundary element. Directivity patterns can also be generated, and these are defined as the directivity at $r = $ infinity.

To illustrate the results of the method, Fig. 7.29 shows the output from a half-model of a loudspeaker, including drive unit, air within the cabinet, the cabinet panels

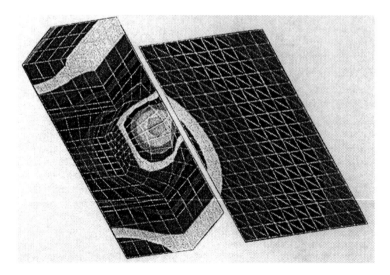

Figure 7.29. Pressure magnitude contours of a complete drive unit, enclosure and surrounding infinite region of air (courtesy PAFEC Ltd and Celestion International Ltd).

and the infinite region of air outside the cabinet. The region outside the cabinet has a series of display patches which give a graphical representation of the pressure magnitude on a selected plane in front of the loudspeaker.

The technique is now gaining acceptance in the repertoire of loudspeaker design tools. It is possible to predict how a complete drive unit, vibrating cabinet panels and diffracting front baffle will interact. Such a powerful tool enables loudspeaker designers to raise their ingenuity to higher levels, accurately predicting how a loudspeaker will behave even before it exists.

Appendix

```
'Computer model of a vented enclosure using QBasic Version 1.0
'Last revision: October 1993, J.R. Wright
'N.B. Code is optimized for clarity, not for speed!

DECLARE SUB ComplexParallelImpedance (Zreal1!, Zimag1!, ZReal2!, ZImag2!,
ZoutReal!, Zoutlmag!) DECLARE SUB PotentialDivider (VinReal!, VinImag!, Zreal1!,
Zimag1!, ZReal2!, ZImag2!,VoutReal!, VoutImag!)
DECLARE FUNCTION log10! (value!)

CONST PI = 3.14159
CONST RHO =118      'density of air

'Define driver parameters
Sd = .03

Bl = 7.7
Re = 2.8

'Generally, calculate component values: here they are preset
Lces = .01956: Cmes = .00061: Res = 31.7
Cmef= .00006: Lceb =.0175: Cmev = .00078
'N.B. The model can be changed to Closedbox by setting Cmev very high, e.g. 10000

'Set input voltage: 2 V gives 1 W into nominal 4 ohms
Eg=2

'Initialise graphics
SCREEN 9      'Colour EGA
VIEW (200, 100)(400, 225), , 14
WINDOW (0, 70)(50, 100)      '30 dB vertical range for plot

'Loop for required frequency range, say 20 500 Hz, logarithmic spacing
FOR i = 0 TO 50
  index = log10(20) + (i * log10(500 / 20) / 50)
  frequency =10 ^ index
  omega = 2 * PI * frequency 'angular frequency

  CALL ComplexParallelImpedance(0, (omega * Lceb) (1 / (omega * Cmev)), 0, 1 / (omega *
Cmef),
ZpReal, ZpImag)
  CALL ComplexParallelImpedance(ZpReal, ZpImag, Res. 0, ZpReal, ZpImag)
  CALL ComplexParallelImpedance(ZpReal, ZpImag, 0, 1 / (omega * Cmes), ZpReal,
ZpImag)
  CALL ComplexParallelImpedance(ZpReal ZpImag, 0, omega * Lces, ZpReal, ZpImag)
  ' now we have the effective impedance of the parallel network, ZpReal + j* ZpImag

'Calculate Vd (complex)
CALL PotentialDivider(Eg, 0, Re, 0, ZpReal, ZpImag, VdReal, VdImag)
```

```
'Calculate Vv (complex)
CALL PotentialDivider(VdReal, VdImag, 0, omega * Lceb, 0, 1 / (omega * Cmev), VvReal,
VvImag)

'Calculate nett volume velocity magnitude
UReal = (Sd / BI) * (VdReal VvReal)
UImag = (Sd / B1) * (VdImag VvImag)
UMag = SQR(UReal * UReal + UImag * Uimag)

'Calculate Sound Pressure Level at 1 m, assuming point source behaviour
Pressure=RHO * frequency * UMag
Spl = 20 * log10(Pressure / .00002)

'Plot SPL
IF.i > 0 THEN LINE (i 1, previousSpl)(i, Spl), l 1

    previousSpl = Spl        'Hold SPL for next LINE command
NEXT i

END

SUB ComplexParallelImpedance (ZReall, Zimag1, ZReal2, ZImag2, ZoutReal, ZoutImag)
A = ZReall * ZReal2 ZImagl *ZImag2
B = ZReall + ZRea12
C = Zreall * ZImag2 + ZReal2 * ZImagl
D = ZImagl +ZImag2
Denominator = B * B + D * D
ZoutReal = (A * B + C * D) / Denominator
ZoutImag = (B * C A * D) / Denominator
END SUB

SUB PotentialDivider (VinReal, VinImag, Zreall, ZImagl, ZReal2, ZImag2, VoutReal,
VoutImag)
A = ZReall + ZReal2
B = ZImagl + ZImag2
C = Zreal2 * VinReal ZImag2 * VinImag
D = ZImag2 * VinReal + ZReal2 "' VinImag
Denominator = A * A + B * B
VoutReal = (A * C + B * D) / Denominator
VoutImag = (A * D B * C) / Denominator
END SUB

FUNCTION log10 (value)
log10 = LOG(value) / LOG(10)      'Convert natural logarithm to log. base 10
END FUNCTION
```

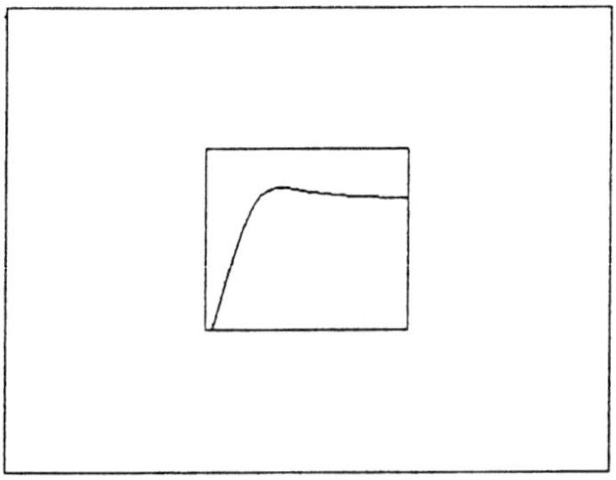

Figure 7.A1. Screen output of above program.

References

1. BRIGGS, G A, *Audio and Acoustics*, Wharfedale Wireless Works Ltd, Bradford (1963).
2. HUNT, F V, *Electroacoustics*, Am. Inst. Phys. (1982).
3. FREDERICK, H A, *Acoustic Device*, US Patent No. 1,955,800 (1934).
4. THURAS, A L, *Sound Translating Device*, US Patent No. 1,869,178 (1932).
5. OLNEY, B, *Sound Reproducing System*, US Patent No. 2,031,500 (1936).
6. BERANEK, L L, 'Some remarks on electro-mechano-acoustical circuits', *J. Acoust. Soc. Am.*, **77** (4) (1985).
7. OLSON, H F, *Acoustical Engineering*, Professional Audio Journals Inc., Philadelphia (1991).
8. * NOVAK, J F, 'Performance of enclosures for low-resonance high-compliance loudspeakers', *JAES*, **7** (1) (1959).
9. * THIELE, A N, 'Loudspeakers in vented boxes', *JAES*, **19** (5) and (6) (1971).
10. * SMALL, R H, 'Direct radiator loudspeaker system analysis', *JAES*, **20** (5) (1972).
11. LEACH, W M, Jr, 'Electroacoustic-analogous circuit models for filled enclosures', *JAES*, **37** (7/8) (1989).
12. WRIGHT, J R, 'An empirical model for loudspeaker motor impedance', *JAES*, **38** (10) (1990).
13. BANK, G and HATHAWAY, G T, 'A three-dimensional interferometric vibrational mode display', *JAES*, **29** (5) (1981).
14. BERANEK, L L, *Acoustics*, Am. Inst. Phys., New York (1986).
15. KINSLER, L E and FREY, A R, *Fundamentals of Acoustics*, John Wiley, New York (1962).
16. MEEKER, W F, SLAYMAKER, R H and MERRILL, L L, 'The acoustical impedance of closed rectangular loudspeaker housings', *J. Acoust. Soc. Am.*, **22** (2) (1950).
17. IH, J G, 'Acoustic wave action inside rectangular loudspeaker cabinets', *JAES*, **39** (12) (1991).
18. SAKAI, S, KAGAWA, Y and YAMABUCHI, T, 'Acoustic field in an enclosure and its effect on sound-pressure responses of a loudspeaker', *JAES*, **32** (4) (1984).
19. SUZUKI, H and TICHY, J, 'Radiation and diffraction effects by convex and concave domes', *JAES*, **29** (12) (1981).
20. WRIGHT, J R, 'Radiation impedance calculation by Finite Element Analysis', *Inst. Acoust. Bulletin*, Nov./Dec. (1994).
21. SUZUKI, H, 'Mutual radiation impedance of a double-disk source and its effect on the radiated power', *JAES*, **34** (5) (1986).
22. * JACOBSEN, O, 'Some aspects of the self and mutual radiation impedance concept with respect to loudspeakers', *JAES*, **24** (2) (1976).
23. RAYLEIGH, J W S, *The Theory of Sound*, Dover Publications, New York (1945).
24. CRANDALL, I B, *Theory of Vibrating Systems and Sound*, Macmillan, London (1926).
25. * OLSON, H F, 'Gradient loudspeakers', *JAES*, **21** (3) (1973).
26. HAWKSFORD, M O J, 'The Essex echo: within these walls', *Hifi News & Record Review*, June (1988).
27. BRITISH STANDARDS INSTITUTION, BS 6840 Part 5. Sound System Equipment (1990).
28. * SMALL, R H, 'Closedbox loudspeaker systems', *JAES*, **20** (10) and **21** (1) (1973).
29. * SMALL, R H, 'Ventedbox loudspeaker systems', *JAES*, **21** (5), (6), (7) and (8) (1973).
30. SMALL, R H, 'Passive radiator loudspeaker systems,' *JAES*, **22** (8) and (9) (1974).
31. BAILEY, A R, 'The Transmissionline Loudspeaker Enclosure', *Wireless World*, May (1972).
32. LETTS, G S, *A Study of Transmission Line Loudspeaker Systems*, honours thesis, School of Electrical Engineering, University of Sydney (1975).
33. * BRADBURY, L J S, 'The use of fibrous materials in loudspeaker enclosures', *JAES*, **24** (3) (1976).
34. ROBERTS, M, 'An acoustical model for transmission line woofer systems', *Proc. Inst. Acoust.*, **12** (8) (1990).
35. FINCHAM, L R, 'A band-pass loudspeaker enclosure', Audio Eng. Soc. convention preprint No. 1512 (1979).
36. GEDDES, E R, 'An introduction to band-pass loudspeaker systems', *JAES*, **37** (5) (1989).
37. BOSE, A G, *Multiple Porting Loudspeaker Systems*, US Patent No. 4,549,631 (1985).
38. HARWOOD, H D and MATHEWS, R, *Factors in the Design of Loudspeaker Cabinets*, Research Department Engineering Division, British Broadcasting Corporation, BBC RD 1977/3 (RD 621.395.623.7) (1977).
39. BRIGGS, G A, *Cabinet Handbook*, Rank Wharfedale Ltd, Idle, Bradford (1962).
40. * BARLOW, D A, 'The development of a sandwich-construction loudspeaker system', *JAES*, **18** (3) (1970).
41. BERANEK, L L, 'The transmission of radiation of acoustic waves by structures', *Proc. Inst. Mech. Eng*, **173**, 12 (1959).

42. LEISSA, A W, *Vibrations of Shells*, NASA SP288 (1973).
43. * IVERSON, J K, 'The theory of loudspeaker cabinet resonances', *JAES*, **21** (3) (1973).
44. MOIR, J M, 'Structural resonances in loudspeaker cabinets', *J. Brit. Sound Recording Assoc.*, **6** p. 183 (1961).
45. * TAPPAN, P W, 'Loudspeaker enclosure walls', *JAES*, **10** (3) (1962).
46. WRIGHT, J R, 'Automatic vibration analysis by laser interferometry', 88th Audio Eng. Soc. Convention Preprint No. 2889 (1990).
47. TOOLE, F E and OLIVE, S E, 'The modification of timbre by resonances: perception and measurements', *JAES*, **36** (3) (1988).
48. LIPSHITZ, S, HEAL, M and VANDERKOOY, J, 'An investigation of sound radiation by loudspeaker cabinets', Audio Eng. Soc. Convention Preprint No. 3074 (1991).
49. OLSON, H F, 'Direct radiator loudspeaker enclosures', *JAES*, **17** (1) (1969).
50. KATES, J M, 'Loudspeaker cabinet reflection effects', *JAES*, **27** (5) (1979).
51. EVEREST, F A, *The Master Handbook of Acoustics*, 3rd edition, TAB (1994).
52. VANDERKOOY, J, 'A simple theory of cabinet edge diffraction', *JAES*, **39** (12) (1991).
53. WRIGHT, J R, 'Fundamentals of diffraction', *JAES*, **45** (5), (1997).
54. SALVATTI, A, BUTTON, D, and DEVANTIER, A, 'Maximizing performance from loudspeaker ports', Audio Eng. Soc. Preprint 4855 (1998).
55. PEDERSON, J A and VANDERKOOY, J, 'Near-field acoustic measurements at high amplitudes', Audio Eng. Soc. Preprint 4683 (1998).
56. FRANKORT, F J M, 'Vibration pattern and radiation behaviour of loudspeaker cones', *JAES*, **26** (9) (1978).
57. MACEY, P C, *Acoustic and Structure Interaction Problems using Finite and Boundary Elements*, PhD thesis, University of Nottingham, November (1987).
58. KAIZER, A J M and LEEUWESTEIN, A 'Calculation of the sound radiation of a non-rigid loudspeaker diaphragm using the finite element method', *JAES*, **36** (7/8) (1988).

Note: Items marked * can also be found in *Loudspeakers: An Anthology*, Vol. 1, 2nd edition, AES (1980).

8 The room environment: basic theory

Glyn Adams
revised by John Borwick

8.1 Introduction

Any comprehensive work on the subject of loudspeakers would be incomplete without a section devoted to the environments into which loudspeaker systems are placed for their operation. In general terms, loudspeakers are most commonly used in large auditoria or open-air theatres for sound reinforcement (see Chapter 10), or in smaller rooms of typically domestic living-room size for reproducing recorded music. This chapter discusses the second category of environment, namely that of rooms of modest dimensions.

The room environment can be considered as the final link in the sound-reproduction chain, in that it relays the sound output of the loudspeaker system to the listener's ears. The transfer characteristic of this link is certainly not a simple function, but if we accept that the sound waves which are radiated by the loudspeaker system in three dimensions are received by the listener either directly or via reflections off the walls, floor and ceiling, then the dimensions of the room, the nature of its boundaries, and the positions of both the source loudspeaker and the receiving listener are all influential factors in this link. The effects of these factors on the reproduced sound come under the general heading of room acoustics, the basics of which are presented in Section 8.2.

Most loudspeaker systems radiate sound to a greater or lesser extent in all directions, and thus the sound waves received by the listener which have undergone one or more reflections from the room boundaries are a function of the directional characteristics of the loudspeaker system. Any discussion of the effects of the room environment on the sound reproduction of a loudspeaker system must therefore include the influence of the loudspeaker system's own directional characteristics, even though this is not strictly part of the 'room link'. This influence, and that of the position of the loudspeaker system relative to the room boundaries, are the subjects discussed in Section 8.3.

Although in some cases it is not always possible to explain satisfactorily why a room colours the sound in a particular way, it is generally found possible to determine the nature of the colorations, and often their sources, by carrying out a set of acoustic measurements of the room together with an examination of its construction and surrounding environment. The most important of these measurements are discussed in Section 8.4, while the effects of particular room constructions and features are dealt with in Section 8.5. The latter section should also prove useful to those considering the design and construction of dedicated or special-purpose listening rooms.

8.2 Basic room acoustics

Room acoustics is a science which deserves the in-depth treatment that only a complete book on the subject can offer; however, the short presentation of the fundamentals of room acoustics given here should enable the reader to understand and explain most of the common phenomena observed in listening rooms. The four main topics are standing waves, room modes, reverberation and flutter echoes.

8.2.1 Standing waves

The phenomenon of standing waves is observed when two similar wave motions, having the same wavelength but travelling in opposite directions, combine together in the same medium. Considering each travelling wave by itself as it passes some fixed point P in its path, it can be seen from Fig. 8.1 that the wave motion observed at this point will have the same peak amplitude A as that of the travelling wave. The wave motion observed at some other arbitrarily fixed point Q will also have this amplitude but in general will have a different phase of motion from that at point P.

When the two travelling waves combine together, the resultant wave no longer has the same amplitude of motion at all points, but has an amplitude which depends on the point of observation. At some points, described as nodes and spaced at equal intervals of half the wavelength, the amplitude falls to a minimum, while at intermediate points, described as antinodes, the amplitude reaches a maximum equal to the sum of the amplitudes of the individual travelling waves.

Because this variation in amplitude of the resultant wave is fixed with respect to the space coordinate along the travelling axis, the resultant wave is called a stationary or standing wave. Another fundamental difference between travelling waves and standing waves is the behaviour of the phase. For example, if the amplitudes of the two travelling waves are equal, then the wave motions observed at all points between

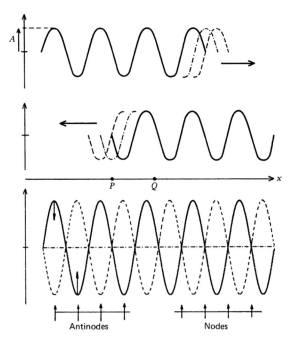

Figure 8.1. Standing waves occur when two similar wave motions having the same wavelength, but travelling in opposite directions, combine in the same medium.

one pair of adjacent nodes have the same phase. Furthermore, the phase changes by 180° in passing from between one pair of adjacent nodes to the next, such that only two possible phases of motion can be observed anywhere along the standing wave. In the following sections of this chapter, the maximum amplitude of motion which is observed at an antinode is simply referred to as the amplitude of the standing wave.

In a listening room, standing wave motion occurs because the sound waves travelling outwards from the sound source are, in general, wholly or partly reflected at the room boundaries back into the room. The incident and reflected waves thus fulfil the conditions required for the formation of a standing wave as explained above. However, most loudspeaker systems radiate travelling waves in more than one direction; furthermore, these waves, unlike those shown in the simple example of Fig. 8.1, spread out and thereby decrease progressively in amplitude as the distance they have travelled increases. To explain the formation of standing waves under these conditions requires a little more knowledge of the nature of the sound radiation from the source and how the sound waves are reflected at the room boundaries.

Consider a typical closed-box type loudspeaker system employing a single low-frequency drive unit. The sound radiation from such a system is rather difficult to calculate for the general case (see Chapter 1); however, for low frequencies, where the wavelength of sound in air is much greater than the physical dimensions of the loudspeaker system, it is found that the sound propagated in free space takes the form of spherical waves radiating outwards uniformly in all directions as depicted in Fig. 8.2. If the loudspeaker system is driven by a sine-wave input signal such that the diaphragm of the loudspeaker drive unit vibrates sinusoidally with an acceleration given by $\sqrt{2}A_D \sin(\omega t)$, then the sound pressure at some point O a distance r from the sound source is given by[1]:

$$p = \frac{\rho}{4\pi r} S_D \sqrt{2}A_D \sin(\omega t - kr) \qquad (8.1)$$

where S_D is the effective projected surface area of the loudspeaker diaphragm, and ρ is the density of air. If the loudspeaker diaphragm is considered to be equivalent to a flat circular piston at low frequencies, then equation (8.1) is not valid for values of r less than about $2S_D/\lambda$. as shown by Pierce[2].

The term 'free space' means that the loudspeaker system must be suspended in free air away from any boundaries or objects. In this rather artificial situation, equation (8.1) shows that the sound-pressure amplitude falls off smoothly as the distance r from the source is increased. Furthermore, for given values of r and S_D, the sound-

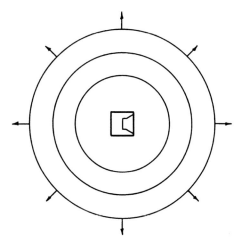

Figure 8.2. For a typical closed-box type loudspeaker system, the sound radiation into free space at low frequencies takes the form of spherical waves which spread out uniformly in all directions.

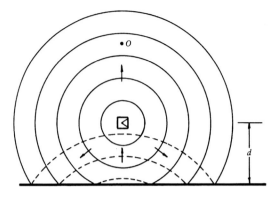

Figure 8.3. When a loud-speaker system is placed near a solid boundary, some of the waves which would otherwise radiate away from the listener at point O are reflected back onto the source and towards the listener.

pressure amplitude/frequency response indicated by equation (8.1) is dependent only upon the variation of A_D as a function of frequency. Within the upper frequency limit defined above, the r.m.s. amplitude A_D in many loudspeaker systems shows an inherent independence of frequency for a given amplitude of input signal down to a lower limit described as the low-frequency cut-off. A loudspeaker system which shows this behaviour will therefore provide a flat sound-pressure amplitude/frequency response in free space between these frequency limits.

In practice, loudspeaker systems are not used in free space but are placed near to boundaries such as the walls of a room. Consider then the effect of placing the loud-speaker system above a single solid boundary as shown in Fig. 8.3. The spherical waves radiating down towards the boundary are reflected back towards the source and the listener at point O. If an input signal is suddenly applied to the loudspeaker

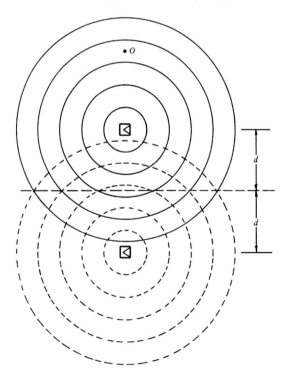

Figure 8.4. The wave reflected at the boundary in Fig. 8.3 arrives at the listener some time later and with a smaller amplitude than the wave which he receives directly from the source. This delayed signal can be represented by a second loudspeaker system, identical to the first, which is placed behind the boundary.

system, the listener at point O will first receive a sound signal when the direct sound wave has travelled the distance r. At some time later, the observer then receives a second sound signal when the waves which were at first travelling away from him have been reflected at the boundary and travelled to point O. The second signal arrives with a time delay of $2d/c$ seconds after the first, where d is the distance of the loudspeaker system from the boundary. Because the second signal has travelled a distance of $2d$ more than the first signal, it is also of smaller amplitude.

If the listener were able to distinguish between the two separate sound signal arrivals, he might interpret the second arrival to be an echo of the first resulting from a reflection. Alternatively, he might imagine that there were two separate sound sources, each fed with the same input signal, but positioned at different distances away from him as illustrated in Fig. 8.4. Here the second source has been placed a distance of $2d$ further away from the listener than the first, so that the second signal arrives with the same time delay and amplitude as the real reflection shown in Fig. 8.3.

Comparison of Figs 8.3 and 8.4 illustrates the concept of using an image of the sound source to replace the effect of the reflection at a solid boundary. The image source is positioned exactly as if it were the mirror image of the sound source that the listener would see if the boundary were replaced by a mirror. Once the image source has been introduced, the reflecting boundary can be removed altogether, leaving the sources radiating into free space. We are then at liberty to make use of the relationships, such as equation (8.1), which apply to sound radiation in free space. Thus, under steady-state conditions, the sound pressure at point O in Figs 8.3 and 8.4 is given by:

$$\frac{\rho S_D \sqrt{2} A_D}{4\pi} \left\{ \frac{1}{r} \sin(\omega t - kr) + \frac{1}{r + 2d} \sin(\omega t - kr - 2kd) \right\} \tag{8.2}$$

The amplitude of the sound pressure given by equation (8.2) as a function of frequency is shown in Fig. 8.5 for values of r and d equal to 1.5 m and 0.5 m respectively. The amplitude of the diaphragm acceleration was taken to be independent of frequency. The sound pressure due to the wave coming directly from the source, which is represented by the first term in equation (8.2), is shown as a broken curve. The second term, which represents the image source and hence the reflected wave, adds constructively to the first term at some frequencies, but destructively at others because of the phase shift introduced by the different distances of the two sources from the listener. The peaks and dips in the frequency response introduced by the reflection are equally spaced along the linear frequency axis, giving a characteristic similar to that of a comb filter. This simple example demonstrates how reflections can so easily modify the frequency response of the sound source.

When a loudspeaker system is placed within a rectangular room, it is enclosed by six boundaries consisting of three pairs of parallel surfaces. To remove each pair of parallel boundaries requires the substitution of an infinite number of image sources, as can be verified by considering the analogous optical situation where an object is placed between two parallel mirrors facing each other. The primary image resulting from the first reflection in the front mirror is reflected in the back mirror to give a secondary image, which is in turn reflected in the front mirror, and so on.

Replacement of all six boundaries thus requires a triple infinity of image sources. Fortunately, real walls like real mirrors are not perfect reflectors; some of the incident wave energy is absorbed at the boundary and thus the high-order images become steadily weaker as they undergo successive reflections. The images close to the source are consequently of most importance, although the number which must be included to represent the effects of the boundary adequately will depend on the degree of absorption.

Now consider the geometry of the set of images which is formed when the loudspeaker system is placed in a corner of a rectangular room as shown in Fig. 8.6.

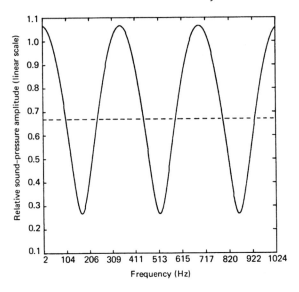

Figure 8.5. The amplitude of the sound pressure at point O in Figs 8.3 and 8.4 as a function of frequency for $r = 1.5$ m, $d = 0.5$ m. The sound pressure due to the direct sound wave only is also shown as a broken curve. The reflected wave, which arrives some time after the direct wave, gives rise to dips in the frequency response similar to the characteristic of a comb filter.

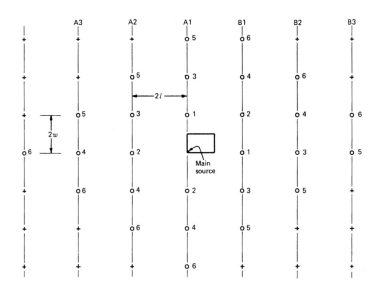

Figure 8.6. The image-source model of a sound source positioned in the corner of a rectangular room. This plan view shows images up to sixth order (numbered circles), but the regular pattern of images extends *ad infinitum* on all sides. The complete pattern is further repeated above and below the room at intervals equal to twice the room height.

Images up to and including the sixth order only are shown, but a regular pattern is clearly established within this number, making the addition of further orders simply a matter of continuation of the pattern *ad infinitum*. The pattern of images shown in the plan view of Fig. 8.6 likewise repeats itself above and below the room at intervals equal to twice the room height.

The regular pattern of images which is formed when a sound source is placed within a rectangular room gives rise to an important phenomenon – the generation of plane waves. Consider the row of image sources labelled A1 in Fig. 8.6. Because the pattern of images is repeated at intervals above and below the room, this row is in fact a section of an entire vertical plane array of image sources. If an input signal is suddenly applied to the loudspeaker system, all the images begin to radiate spherical waves at the same instant. At some time later, the spherical wavefronts radiating from each source interfere with each other as shown in Fig. 8.7.

The wavefronts combine together to give a resultant wavefront which shows much less curvature than the individual spherical wavefronts. Because the image sources are arranged in a plane, the resultant wave behaves rather like a plane wave in that the wavefront remains approximately in the shape of a plane as it travels away from the array of image sources. Unlike the spherical wave, which diverges and therefore weakens in amplitude as it travels away from the source, the plane wave does not spread out and thus remains of the same amplitude everywhere along its path.

Returning now to the phenomenon of standing waves, it is evident from the simplified image-source model shown in Fig. 8.6 that the spherical waves radiating out from the loudspeaker system are reflected at the boundaries of a rectangular room in such a way that plane waves are effectively generated travelling in a number of different directions within the room. Consider, for example, the plane waves generated by the arrays labelled A1 to A3. If the loudspeaker system is driven with a sine-wave signal

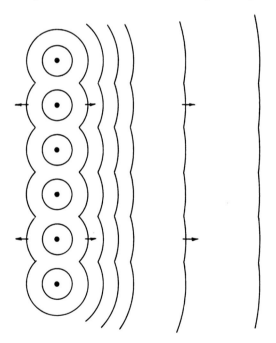

Figure 8.7. A row of images from the image-source model of Fig. 8.6 in fact represents a section of an entire vertical array of image sources. Because all of these images radiate in unison with the main source in the room, their wavefronts combine to give an almost plane wavefront as shown.

then, under steady-state conditions, the three travelling waves travelling into the room from the left-hand side will combine together to form a single resultant travelling wave moving across the room from left to right. Likewise the arrays of image sources labelled B1 to B3 give rise to a single resultant travelling wave moving across the room from right to left. The conditions for the formation of a standing wave are now exactly as shown in Fig. 8.1. Because the travelling waves are approximately in the form of plane waves, the nodes and antinodes of the standing wave exist as flat planes within the room of minimum and maximum sound pressure respectively, which lie parallel to the arrays of image sources.

8.2.2 Room modes

If the sound-pressure distribution throughout a typical domestic listening room is investigated using a hand-held sound-level meter, while the loudspeaker system is excited with a sine-wave signal, it will be found that the amplitudes of the standing waves pass through several strong maxima as the frequency of the sine-wave signal is slowly increased from, say, 20 Hz. These increases in sound-pressure level at certain frequencies are examples of the phenomenon of room resonance, which occurs when the room is excited at the frequencies of its natural modes of vibration. The standing waves which are set up at these natural frequencies of vibration are referred to as room modes (eigentones).

An understanding of the mechanism by which these room resonances occur can be gained by further consideration of the image-source model shown in Fig. 8.6. To simplify the analysis, consider again the resultant travelling waves entering the room from the left- and right-hand sides due to the arrays A1-A3 and B1-B3 respectively. If the room boundaries are assumed to be perfect reflectors, then all the image sources are of the same strength and thus the amplitudes of the two travelling waves will be equal. Because the amplitude of the standing wave is equal to the sum of the amplitudes of the two travelling waves, we need only examine the frequency dependence of one of the travelling waves in order to see the resonance behaviour.

If we consider the arrays of image sources A1, A2 and A3 to be the sources of three plane waves travelling into the room then, under steady-state conditions, the resultant sinusoidal pressure variation at some point in the room is given by:

$$p = \sin(\omega t) + \sin(\omega t - 2kl) + \sin(\omega t - 4kl) \tag{8.3}$$

where $\sin(\omega t)$ has been arbitrarily chosen to represent the pressure variation at the given point due to array A1. The contributions from arrays A2 and A3 are of the same amplitude but have phase shifts appropriate to their distance away from A1.

The amplitude of the sinusoidal pressure variation given by equation (8.3) represents the amplitude of the resultant travelling wave entering the room from the left-hand side. This amplitude is plotted as a function of frequency in Fig. 8.8 for a room having a length l of 5 m. The sound-pressure amplitude of the travelling wave, and hence that of the standing wave, passes through several maximum values as a function of frequency, demonstrating the phenomenon of room resonance. The maximum values of sound-pressure amplitude are equal to three times the value of the contribution due to array A1 only, and occur at frequencies given by:

$$f = \frac{cn_1}{2l} \quad n_1 = 1, 2, 3, \ldots \tag{8.4}$$

The resonance frequencies given by equation (8.4) are also easily deduced by examination of equation (8.3); clearly, the amplitude of the resultant travelling wave will reach a maximum if the individual travelling waves arrive in phase in the room. For this to be the case, the frequency must be such that the distance of $2l$ between adjacent arrays is equal to an integral number of wavelengths. At frequencies which lie between the resonance frequencies, the individual travelling waves arrive more or

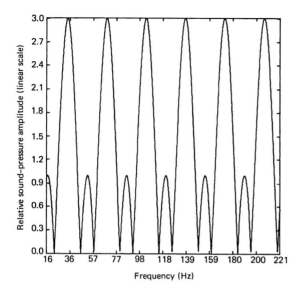

Figure 8.8. The amplitude of the resultant travelling wave entering the room from one side, as a function of frequency, due to the contributions of the arrays Al, A2 and A3 of Fig. 8.6. The amplitude of this travelling wave, and hence that of the standing wave, shows a number of maximum values demonstrating the phenomenon of room resonance.

less out of phase with each other and thus partly cancel to give a resultant travelling wave of smaller amplitude than at resonance.

In the simplified treatment given above, only three image-source arrays were considered, but of course in practice there will be many more arrays of higher orders which will contribute to the resultant travelling wave. In general we can consider the contributions from N arrays by rewriting equation (8.3) in the form of a summation:

$$p = \sum_{n=1}^{N} \sin \left[\omega t - (n-1)2kl \right] \tag{8.5}$$

The amplitude of the travelling wave given by equation (8.5) is shown in Fig. 8.9 for $N = 30$, $l = 5$ m. By comparison with Fig. 8.8, we can see that when the contributions from a larger number of arrays of image sources are taken into account the resonances become more sharply defined and of greater amplitude.

In real rooms, where the boundaries are not perfect reflectors, the contributions to the summation of equation (8.5) will become steadily weaker with increasing n because of the partial absorption of energy which occurs at each reflection. If the effect of boundary absorption is included in the summation, then it can be truncated after a finite number of terms which will depend on the degree of absorption. An analysis of this kind is beyond the scope of this text but, from the simple examples illustrated by Figs 8.8 and 8.9, we can expect the following observations. In rooms with little boundary absorption, such that a large number of terms must he included in the summation, the room resonances will be strong and have sharply defined frequencies. As the degree of boundary absorption is increased, such that the number of terms to be included in the summation is decreased, the room resonances become weaker and the resonance peaks become broader.

The amplitude of the sinusoidal pressure variation at different points along the standing wave can be calculated by combining the resultant travelling waves entering the room from opposite directions. Continuing with our previous example, the sound

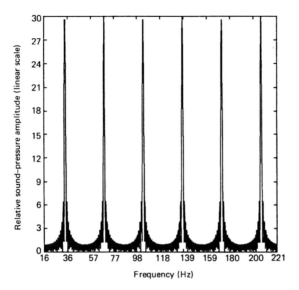

Figure 8.9. The amplitude/frequency of the resultant travelling wave entering the room from one side due to the contributions of 30 adjacent plane arrays of image sources. Compared to Fig. 8.8, the resonances are both stronger and more sharply defined as a result of considering a larger number of arrays.

pressure at a point a distance d from the left-hand wall due to the contributions from arrays A1–A3 and B1–B3 at a resonance frequency can be written:

$$p = A \sin(\omega t - kd) + B \sin(\omega t - 2kl + kd) \qquad (8.6)$$

where A represents the strength of the images in arrays A1–A3, and B represents the strength of the images in arrays B1–B3. If the boundaries are perfect reflectors, the images will all be of the same strength as the source and hence A and B will then be equal. The amplitude of the sound pressure along the standing wave for this condition is shown in Fig. 8.10 for the first three resonance frequencies. Because the amplitudes of the resultant travelling waves entering from the left- and right-hand sides are equal, the pressure falls to zero at the nodes of the standing wave.

Now consider the effect of introducing some absorption into the room boundaries. The images are then no longer of the same strength but become weaker with increasing order. For example, if the boundary absorbs 50% of the incident wave energy at each reflection, then a primary image will be half the strength of the source, a secondary image a quarter of the strength, and so on. In the image-source model of Fig. 8.6, the orders of the images are shown, and close examination reveals an important trend in the strengths of the images making up the plane arrays. Comparing rows A1 to B1, A2 to B2 and A3 to B3, it is evident that the images are of one order higher to the right-hand side of the room than they are in the corresponding rows on the left-hand side. This means that the amplitudes of the resultant travelling waves entering the room from these opposite sides will not be equal, resulting in only partial cancellation of the sound pressure at the nodes of the standing wave.

Figure 8.10 (*opposite*). The variation of the amplitude of the sound pressure along a standing wave set up in the length direction of a room having perfectly reflective boundaries. The pressure variation is shown for the first three axial modes ($l = 5$ m): (a) $n_l = 1$, $f = 34.5$ Hz; (b) $n_l = 2$, $f = 69$ Hz; (c) $n_l = 3$, $f = 103.5$ Hz.

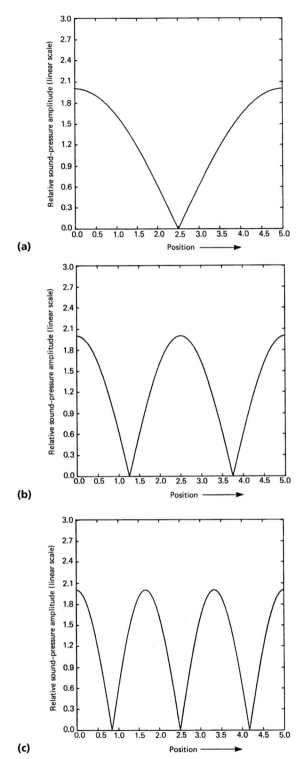

(a)

(b)

(c)

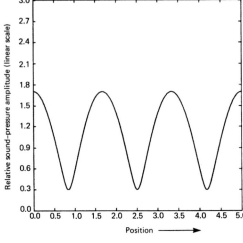

(a)

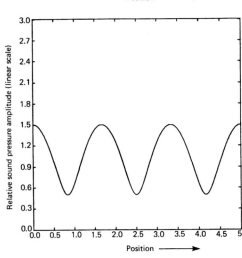

(b)

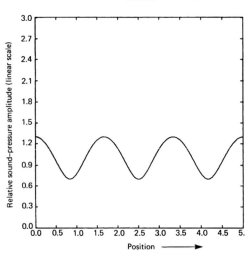

(c)

We can make a rough calculation of the amplitude of the sound pressure along the standing wave when the boundaries are partially absorbent by modifying the value of B relative to A in equation (8.6). Figure 8.11 shows the pressure variation in the room calculated at the third resonance frequency for values of B of $0.7A$, $0.5A$ and $0.3A$, corresponding to boundaries which absorb respectively 30%, 50% and 70% of the incident wave energy at each reflection. As the boundaries are made more absorbent, the difference in the amplitude of the sound pressure between the nodes and antinodes of the standing wave becomes less pronounced. The ratio of the pressure amplitude at an antinode to that at a node is described as the 'standing wave ratio' and is easily measured using a sound-level meter to provide an indication of the degree of boundary absorption.

So far in this section we have considered only the resonant standing waves that can be set up along the axis of the length of a rectangular room. Clearly, the analysis applies equally well to the arrays of image sources lying beyond the other pair of opposite side walls and the floor and ceiling. Two further sets of room modes can thus be added to the set defined by equation (8.4), whose frequencies are given by:

$$f = \frac{cn_w}{2w}, \quad n_w = 1, 2, 3, \ldots$$

$$f = \frac{cn_h}{2h}, \quad n_h = 1, 2, 3, \ldots \qquad (8.7)$$

where w is the width of the room and h is the height. Because these modes are standing waves which are set up along the axes of the length, width and height of the room they are classified as 'axial modes'.

We have seen that the phenomenon of room resonance can be explained by considering planes of image sources which lie parallel to each other and are spaced at equal intervals such that their contributions to the sound pressure in the room all arrive in phase at the resonance frequencies. This geometrical feature does not apply only to planes of image sources which lie parallel to each of the room boundaries, as is shown in Fig. 8.12.

Here there are four sets of plane arrays, each lying at an oblique angle to two of the room axes but parallel to the third, which give rise to four resultant travelling waves entering the room. The standing wave so formed has nodal planes which are parallel to two of the three pairs of room boundaries rather than to only one pair, as in the case of an axial mode. This type of standing wave pattern is described as a 'tangential mode'. The sound-pressure distribution throughout the room for one of the many possible tangential modes is illustrated in Fig. 8.13.

If you can imagine the image-source model of Fig. 8.6 in three dimensions, you will appreciate that it is also possible to define sets of equally spaced plane arrays of image sources which lie at an oblique angle to all three axes of the room. Eight sets of equally spaced plane arrays can be identified in this case, thus giving rise to eight resultant travelling waves entering the room. The resonance modes defined by these arrays are described as 'oblique modes', and are characterized by the presence of nodal planes which are parallel to all three pairs of room boundaries.

Because the spacing of the parallel planes of image sources determines the resonance frequencies, the tangential and oblique modes have resonance frequencies which are related to combinations of the room dimensions. An expression which gives the frequencies of the axial, tangential and oblique modes of a rectangular room was derived by Rayleigh[3] using a more rigorous mathematical treatment than has

Figure 8.11 (*opposite*). The approximate variation of the amplitude of the sound pressure along a standing wave set up in the length direction of a room having partially absorbent boundaries. The pressure variation is shown at the third axial mode for different degrees of absorption of the incident wave at each reflection: (a) 30% absorbed; (b) 50% absorbed; (c) 70% absorbed.

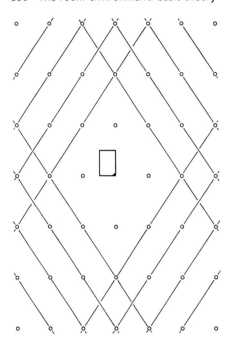

Figure 8.12. The image-source model of a sound source positioned in the corner of a rectangular room showing sets of plane arrays of image sources lying at an oblique angle to two of the room axes. These arrays give rise to four travelling waves in the room which produce a standing wave pattern described as a tangential mode.

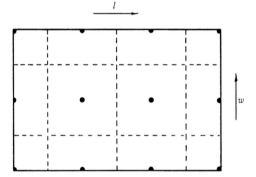

• Antinodes, maximum amplitude
--- Nodal planes, minimum amplitude

Figure 8.13. The distribution of the amplitude of the sound pressure throughout a room at the tangential mode described by $n_l = 3$, $n_w = 2$, and $n_h = 0$.

been presented here. By solving the wave equation for a plane wave in three dimensions, and imposing the necessary conditions at the room boundaries, he showed that the frequencies of the room modes are given by:

$$f = \frac{c}{2} \sqrt{\left[\frac{n_l}{l}\right]^2 + \left[\frac{n_w}{w}\right]^2 + \left[\frac{n_h}{h}\right]^2} \tag{8.8}$$

where n_l, n_w, and n_h are integers. By comparison with equations (8.4) and (8.7) it is clear that the frequencies of the axial modes are given by setting any two of n_l, n_w, and n_h to zero. The frequencies of the tangential modes are related to two of the room dimensions and are thus given by setting any one of n_l, n_w, and n_h to zero. Lastly, the resonance frequencies of the oblique modes, which are related to all three room dimensions, are given by equation (8.8) when n_l, n_w, and n_h are all non-zero. It is also useful to note that the integers n_l, n_w, and n_h indicate respectively the number of nodal planes that are formed perpendicular to the length, width and height axes of the room. The example shown in Fig. 8.13 illustrates this feature.

8.2.3 Reverberation

The use of image sources to replace the reflective boundaries of a rectangular room was shown in the previous sections in order to investigate the steady-state behaviour of the sound-pressure distribution throughout the room. Fortunately, the image-source model is also valid for the analysis of the room response to time-varying input signals, such as the sound output of the loudspeaker system fed with a music signal or the impulsive sound produced by a hand-clap.

If we take the latter example of a hand-clap as being the sound source in the image-source model of the rectangular room shown in Fig. 8.6, then one can argue that all the image sources are in effect further 'pairs of hands' which clap in unison with the main source. The sound perceived in the room when a hand-clap takes place thus consists of an initial component, due to the sound wave which travels directly from the source to the listener, followed by a succession of further hand-claps which arrive at later times related to the distances of the images from the listener. Because the room boundaries are not perfect reflectors, the images get weaker with increasing order and thus the sound contribution from the distant images eventually dies away to nothing. The effect of the image sources, and hence the reflections which they represent, is thus a continuation of the sound in the room for some time after the source sound signal has stopped. This phenomenon is called reverberation.

The extent to which the sound in the room is maintained after the source is stopped or switched off will depend on the degree of absorption of sound energy which occurs at each reflection. Following the concepts developed in Section 8.2.2, if the boundary absorption is small – such that a large number of image sources will make significant contributions to the sound pressure in the room – then the sound level will take a long time to die away after the source is switched off. As the degree of boundary absorption is increased, so that only those image sources nearest to the room become significant, the time taken for the sound to die away shortens. Because the distances of the image sources from the room are dependent on the room dimensions, it is also clear that the time taken for the sound to die away will be related to the room size.

This relationship between the degree of boundary absorption, the room size, and the length of time for which a sound reverberates in a room was investigated by Sabine[4]. In quantifying the relationship, Sabine defined the 'reverberation time' of a room to be the time taken (after the sound source is switched off) for the sound level to reduce by 60 dB below some steady level established prior to the switching. The sound level generally decays in an approximately exponential manner after the source is switched off, and thus a plot of sound level in dB versus time falls more or less along a straight line with a slope which indicates the reverberation time. This facilitates estimation of the reverberation time in cases where the steady level established before switching is less than 60 dB above the background noise level, as shown in Fig. 8.14.

The reverberation time of a listening room has an important bearing on the quality of the sound reproduction which can be obtained. Thus, for the purposes of design of such rooms, we need to relate the reverberation time to the absorption of the room boundaries in a predictable way. The degree of absorption which takes place when a sound wave is reflected at a boundary can be defined by an absorption coefficient α, which represents the fraction of randomly incident sound energy which is absorbed. Thus, if the boundaries of the room modelled in Fig. 8.6 have an absorption coefficient of α, and the strength of the source is W, then the primary images will have a strength of $(1 - \alpha)W$, the secondary images a strength of $(1 - \alpha)^2W$, and so on.

Using the image-source model of a rectangular room with image strengths modified in this way to represent the effect of boundary absorption, Eyring[5] derived an

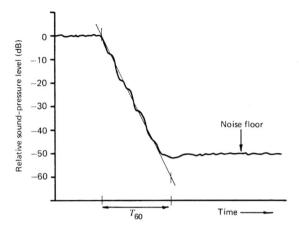

Figure 8.14. The decay in the sound-pressure level after the sound source has been switched off is generally exponential, giving a constant decay slope when plotted on a logarithmic scale. This facilitates estimation of the reverberation time T_{60} in cases where the background noise level is high, or where the decay is ragged.

expression for the decay of the sound energy in a room after the sound source had been switched off. The value of the reverberation time predicted by this expression is given by:

$$T_{60} = \frac{0.161V}{a} \tag{8.9}$$

where V is the volume of the room in m³ and a is the total room absorption given by:

$$a = -S \ln (1 - \alpha) \tag{8.10}$$

S is the total area of the interior surfaces of the room in m³.

In practice, a typical room will have several different types of building construction and decorative finish for the walls, floor and ceiling, which each have their own value of absorption coefficient. Equation (8.10) is not strictly applicable to this case because, in its derivation, all the room boundaries were considered to have the same absorption coefficient. However, an estimate of the reverberation time can be obtained from equations (8.9) and (8.10) in such cases by replacing the absorption coefficient with an average absorption coefficient defined as follows:

$$\bar{\alpha} = \frac{\sum \alpha_i S_i}{S} \tag{8.11}$$

where α_i and S_i are the absorption coefficient and area of each surface making up the interior room boundaries. The absorption coefficients of some typical surface finishes are shown in Table 8.1.

An alternative approach to calculating the reverberation time for rooms which have a mixture of absorbing materials present was investigated by Millington[6] and Sette[7]. Their analysis shows that the total room absorption can in some cases be given more correctly by:

$$a = \sum -S_i \ln (1 - \alpha_i) \tag{8.12}$$

Table 8.1 Absorption coefficients of some common surface treatments

Frequency	125 Hz	500 Hz	2 kHz	4 kHz
Brick wall, untreated	0.02	0.03	0.05	
Plaster on brick	0.01	0.02	0.04	0.05
Plywood panelling	0.28	0.17	0.10	0.11
Glass	0.04	0.05	0.05	
Glass window	0.35	0.18	0.07	0.04
Concrete floor	0.01	0.02	0.02	0.02
Carpet on concrete (not glued)	0.12	0.18	0.28	0.45
Heavy drapes	0.10	0.50	0.82	
Felt	0.13	0.56	0.65	
Rock wool	0.35	0.63	0.83	
Velcro material over 40 mm thick rock wool	0.29	0.32	0.50	0.56
Varnished wooden floor	0.08	0.12	0.15	0.16
False ceiling made from 16 mm rough painted chipboard tiles suspended 0.5 m below concrete ceiling	0.19	0.23	0.25	0.30

Sources: Kinsler and Frey[1], Pierce[2], Vogelaar[33].

Table 8.2 Absorbing-power increments due to persons, metric sabins

Frequency	125 Hz	500 Hz	2 kHz	4 kHz
Person standing	0.12	0.59	1.13	1.12
Musician sitting	0.60	1.06	1.08	1.08

Source: Kutruff[8].

The reverberation time follows from equation (8.9).

The introduction of persons or pieces of furniture into a room will in general give rise to a reduction in the reverberation time, because the wave motion in the room is partially absorbed by these items. With the aid of equation (8.9), the measured values of reverberation time before and after a single item is introduced into a reverberant room can be used to calculate the value of the increment by which the total room absorption is increased[2]. The effect on the reverberation time of introducing a number of such items into the room can then be found by adding a corresponding number of increments to the total room absorption given by equation (8.10) or equation (8.12). The increments for persons are given in Table 8.2.

8.2.4 Flutter echoes

A flutter echo is a distinct type of reverberation characterized by the perception of a periodic train of sound impulses or echoes in the decaying sound level. If the impulses arrive after equal time intervals of about 25 ms or more, the ear is able to resolve the impulses as separate events occurring in the time domain[8]. This 'fluttering' in the sound level imparts a rough quality to music signals reproduced in the room, which can be likened to the introduction of a tone having a sharp, metallic timbre[9].

The impulsive sound produced by a hand-clap is a good test signal for investigating the presence of flutter echoes. In the previous section we saw that when an impulsive sound of this sort is produced in a rectangular room there are many components which arrive after the direct sound with time delays that depend on the distances of the image sources from the listener. If all the boundaries of the room have similar absorption coefficients, the decaying sound of the hand-clap will contain a mixture of several periodic trains of decaying sound impulses having similar strengths but with different periods relating to the room dimensions. The intervals between reception of successive sound impulses will not be equal in this case and will often be too small for the ear to resolve the separate impulses, particularly if the room dimensions are small. The perception of a flutter echo under these conditions is unlikely.

Now consider a rectangular room which has one pair of parallel boundaries that are much less absorbent than the other room boundaries. The decaying sound of a hand-clap will in this case contain a dominant periodic train of sound impulses which is much stronger than the other components. This arises from the contributions made by the set of strong image sources formed by reflection of the source in the pair of parallel boundaries having low absorption. Because the sound decay is now dominated by one periodic train of impulses, and provided that their time interval is sufficient, the ear will perceive a flutter echo.

We can conclude that flutter echoes are most likely to occur between two parallel surfaces placed opposite each other when their absorption coefficients are smaller than those of the other surfaces making up the room boundaries. The flutter echo can thus be reduced by increasing the absorption coefficient of one or both of the offending pair of surfaces, or alternatively by making the other surfaces of the room boundaries more reflective. If these remedies are unacceptable, then the flutter echo can often be removed by angling one of the offending surfaces so that it is no longer parallel with the other.

8.3 Loudspeaker placement

The positions in which we place loudspeaker systems within a room are, more often than not, dictated by the convenient spaces which happen to be available rather than by any scientific premise. However, if the best possible quality of sound reproduction is desired from some given loudspeaker systems, they need to be positioned with a little more care.

This section discusses the effects on the low-frequency sound reproduction of the proximity of the loudspeaker system to the room boundaries, and its position relative to the nodes and antinodes of the room modes. Consideration is also given to the influence of the directional characteristics of the loudspeaker system on these effects. The section concludes with a brief study of the additional factors which apply when positioning a stereo pair of loudspeakers.

8.3.1 Proximity to room boundaries

In the study of basic room acoustics presented in Section 8.2 we were chiefly interested in understanding the intrinsic acoustic properties of the room, and only brief mention was therefore made of the properties of the sound source. In fact, a loudspeaker sound source cannot be regarded as being entirely independent of the room acoustics, because the efficiency with which it converts electrical input power into acoustic output power is influenced by the sound waves which are reflected back on to the loudspeaker diaphragm from the room boundaries.

In examining the influence of boundary reflections on the efficiency of a loudspeaker system, the analysis can be simplified by first considering the effect of the early reflections from the boundaries nearest to the source. Let the loudspeaker system be positioned near one corner of the room at some distance away from the three nearest mutually perpendicular boundaries. In Section 8.2.1 and Figs 8.3 and 8.4 the concept of using an image of the sound source to replace the effect of the reflection at a solid boundary was introduced. In applying this concept to the present case, we see that the influence of the three room boundaries can be represented by seven images of the loudspeaker system as shown in Fig. 8.15. Because we are ignoring the influence of the other room boundaries for the moment, there are no further images introduced by successive reflections between parallel boundaries. For the

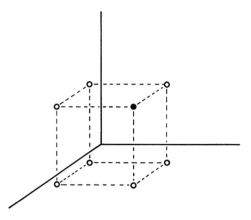

- Main source
- ⊙ Image sources

Figure 8.15. The images of a sound source caused by the three nearest room boundaries. These seven image sources represent the early reflections which combine with the direct wave to form the 'early sound wave'.

purposes of analysis, we can thus consider the three boundaries to have been removed, leaving the loudspeaker system and its seven images to radiate into free space.

If we consider the loudspeaker system to be non-directional, so that it radiates spherical waves, then the sound pressure at some point O in the room can be found by summing together the individual sound pressure contributions from each of the eight sources. From the similar example given in Section 8.2.1, which considered the summation of the contributions from two sources, we can expect the resultant sound pressure at point O to have a strong frequency dependence. The amplitude of the resultant sound pressure at any frequency clearly depends on the amplitudes and phases of the individual sound-pressure contributors arising at point O from each source. Because the amplitude and phase of the contribution from each source depend on its distance from point O, the frequency dependence of the resultant sound pressure will change as point O is moved throughout the room.

This variation in the sound-pressure amplitude with position is an important factor in itself, but it means that we cannot see the overall effect of the presence of the boundaries by considering the resultant sound-pressure amplitude at just one point. To determine an overall effect of the early reflections, we need to consider the sound-pressure amplitude at all points throughout the room and combine them in some form of summation or average. This laborious averaging procedure can fortunately be avoided by making use of the relationship which exists between the mean square pressure averaged throughout the room and the acoustic power output of the sound source. The influence of the boundaries near the source on the overall sound level throughout the room can thus be determined by calculating that part of the total acoustic output power of the loudspeaker system (and its seven images) which enters the room.

An analysis of this type was carried out by Waterhouse[10] for a non-directional source placed near solid perfectly reflective boundaries. He shows that the relative power output of the source positioned at perpendicular distances of x, y and z from the three boundaries is given by:

$$\frac{W}{W_f} = 1 + j_0(2kx) + j_0(2ky) + j_0(2k(x^2 + y^2)^{1/2} +$$

$$j_0\{2k(x^2 + z^2)^{1/2}\} + j_0\{2k(y^2 + z^2)^{1/2}\} + j_0\{2k(x^2 + y^2 + z^2)^{1/2}\} \tag{8.13}$$

where W_f is the power output of the source when it is in free space, and $j_0(a) = (\sin a)/a$. The amplitude of vibration of the source is here assumed to be independent of the environment.

In equation (8.13) the unity term represents the power contribution of the main source, while the remaining seven terms relate to the contributions of the image sources. If the main source is placed far away from all three boundaries such that x, y and z are very large, then the j_o terms tend to zero, leaving $W = W_f$. This is simply the limiting case of radiation into free space. Moving to the opposite extreme, if the main source is placed right into the corner of the three boundaries so that x, y and z are zero, then the j_o terms become equal to unity, giving $W = 8 W_f$. The power output of the source is thus increased by eight times, which is equivalent to a power gain of approximately 9 dB.

This increase in the power output of the source is a little difficult to explain without first introducing the concept of radiation resistance. If we consider the low-frequency drive unit of a typical moving-coil direct-radiator type loudspeaker system, then the efficiency with which it converts the electrical input power into acoustic output power will be of the order of 1% in free space. This poor efficiency is due to the relatively small mechanical load which the air presents to the diaphragm of the drive unit, compared with the mechanical load of the diaphragm itself and its associated enclosure. Thus most of the input power is dissipated in the energy losses involved in just moving the diaphragm, while the remaining small percentage is dissipated in working against the resistive part of the air load to generate acoustic radiation. The resistive part of the air load, or radiation impedance, is known as the radiation resistance (see also Chapter 1).

The acoustic output power of the drive unit operating in a closed box at low frequencies is given by[1]:

$$W = V_D^2 S_D^2 R \tag{8.14}$$

where R is the radiation resistance presented to the front side of the diaphragm, and V_D is the r.m.s. value of the axial velocity of the diaphragm.

When the loudspeaker system is placed near one or more room boundaries, a part of the sound wave radiated by the loudspeaker diaphragm is reflected off the boundaries back onto the diaphragm. If the wavelength of the sound radiation is much larger than the distance between the loudspeaker diaphragm and the boundary, then the combination of the outgoing and reflected waves constitutes an increased air loading on the diaphragm, compared to free-space loading, which takes the form of an increase in the radiation resistance.

Because the radiation resistance is a very small part of the total mechanical impedance of the system, this increase in radiation resistance has very little bearing on the magnitude of the total mechanical impedance. This means that the amplitude of the diaphragm velocity, which is dependent on this magnitude and that of the electrical input signal, will also be virtually unchanged by the increase in radiation resistance.

The assumption that the amplitude of the diaphragm velocity is independent of the environment was made in the derivation of equation (8.13), and it would appear to be valid in the case of the typical loudspeaker system cited above. However, it is clear that the increase in acoustic output power is gained only because the value of the diaphragm velocity in equation (8.14) remains the same when the radiation resistance is increased. If the loudspeaker system had been highly efficient, such that the radiation resistance represented a considerable part of the total mechanical impedance of the system, then an increase in radiation resistance would give rise to a reduction in the amplitude of the diaphragm velocity. The power output given by equation (8.14) would then increase by a smaller factor than that predicted by equation (8.13).

Most of the loudspeaker systems which we use today fall into the highly inefficient category, and thus we can make use of equation (8.13) with little likelihood of error. Under this condition, the power output is directly proportional to the radiation resistance and so we can write:

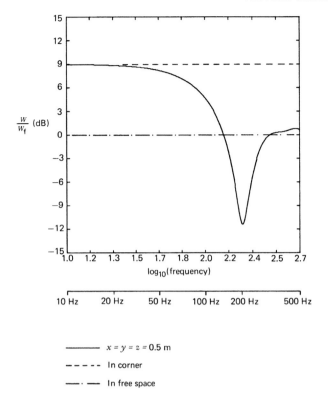

Figure 8.16. The sound power output of a non-directional sound source placed near three mutually perpendicular room boundaries relative to the power output in free space. The source is positioned 0.5 m away from each boundary ($x = y = z = 0.5$ m).

$$\frac{W}{W_f} = \frac{R}{R_f} \qquad (8.15)$$

where R_f is the radiation resistance for free-space loading.

Now consider the power output versus frequency given by equation (8.13) for the loudspeaker system placed some distance away from the three boundaries. Figure 8.16 shows the relative power output obtained when the loudspeaker diaphragm is placed 0.5 m away from each boundary. The power output for free space loading and for the corner position are also shown on the same plot for reference. Unlike these last two conditions, the power output for $x = y = z = 0.5$ m shows a strong variation as a function of frequency, with the power output even becoming less than the free-space value at some frequencies.

This variation in the power output can be explained as follows. At low frequencies, where the wavelength is very large compared with the values of x, y and z, the waves reflected back onto the loudspeaker diaphragm are virtually in phase with the outgoing wave, and the radiation resistance is increased by 9 dB as for corner mounting. As the frequency is increased, the wavelength becomes comparable with the distances of the loudspeaker diaphragm from the boundaries so that the phase difference between the reflected and outgoing waves increases. The reinforcement of the air load is therefore diminished, and the radiation resistance falls towards the free space value. In particular, when the wavelength is equal to roughly four times the values of x, y and z, the reflected waves, represented by the seven image sources shown in Fig. 8.15, all arrive at the loudspeaker diaphragm approximately in antiphase to the outgoing wave. The air loading is then severely diminished, resulting in a dip in the power output.

At higher frequencies, where the values of $2kx$, $2ky$ and $2kz$ are greater than π, the component of the pressure on the diaphragm due to the outgoing wave which is in phase with the diaphragm velocity becomes more significant than the corresponding components of the sound-pressure contributions arriving at the diaphragm due to the image sources[11]. Because these components determine the radiation resistance experienced by the diaphragm, the contributions of the image sources become less significant at high frequencies and are then only evidenced by small increases or decreases of the power output about the free-space value.

We can conclude from these examples that, when the loudspeaker system is positioned some distance away from the boundaries, the power output will be increased by 9 dB at low frequencies but will fall to the free-space value at high frequencies. The frequency range over which this transition takes place is centred on $f = c/4x$ when x, y and z are approximately equal. At frequencies below the transition range, the distance of the source away from the boundaries is so small compared with the wavelength that the source behaves exactly as if it were right in the corner of the three boundaries. Above the transition range, the source is far enough away from the boundaries compared with the wavelength to behave as if it were in free space.

The variation of the relative power output in the transition range amounts to about 20 dB in the example shown in Fig. 8.16. The 9 dB boost at low frequencies is basically unavoidable and must be taken into account in the design of the loudspeaker system. However, the dip of about 11 dB is more serious because it occurs over a narrow frequency range, which makes compensation of the loudspeaker power response impractical. Fortunately, the magnitude of this dip can be reduced considerably by positioning the loudspeaker system so that the values of x, y and z are different from each other. The distances between the main source and the image

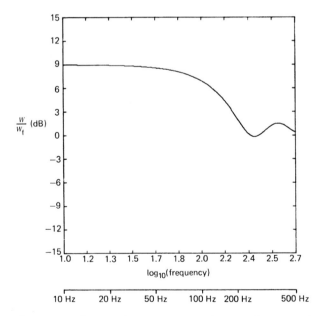

Figure 8.17. The sound power output relative to free-space loading, as for Fig. 8.16, but for asymmetrical placement ($x = 0.5$, $y = 0.35$, $z = 0.15$ m).

sources are then spread over a wider range, making the cancellation effect less pronounced.

Figure 8.17 shows the relative power output calculated from equation (8.13) for a source positioned at $x = 0.5$ m, $y = 0.35$ m, $z = 0.15$ m. The dip in the power output can clearly be reduced to an acceptable amount by selecting this type of asymmetrical position relative to the near room boundaries. Having removed the dip in this way, the remaining boost in power output at low frequencies can be compensated for by making suitable choices of the cut-off frequency and the characteristic of the low-frequency roll-off of the loudspeaker system. Alternatively, if these parameters are fixed, a particular asymmetrical position can often be chosen which best complements the power-output/frequency response of the loudspeaker system[12].

Determination of the optimum position for the loudspeaker system is in practice best achieved by experiment, using the analysis of this section only as a guide. The changes in the quality of sound reproduced by the loudspeaker system when it is moved from one position to another are often clearly audible, which enables the optimization to be carried out by ear. This technique also allows for the selection of a position which may not be optimum in the true sense, but which provides pleasing results by virtue of the introduction of some bass boost.

When carrying out experiments in this way, it is useful to know what relative change in power output of the source we can expect when it is moved from one position to another. Our judgement of the success or failure of each experiment will be based on this relative change rather than on the absolute power-output/ frequency response. If we consider an asymmetrical position relative to the three boundaries as a starting point, then the movement to a second asymmetrical position nearer or further from the boundaries would be a good experiment to try first. By calculating the power output for both positions of the source using equation (8.13), the relative power gain obtained by moving the source is easily found by subtracting the power responses expressed in dB.

Figure 8.18 shows the relative power gain calculated in this way when the source is moved from $x = 1$ m, $y = 0.7$ m, $z = 0.3$ m to a position nearer the boundaries given by $x = 0.33$ m, $y = 0.23$ m, $z = 0.1$ m. The power gain peaks to about 8 dB in the frequency range 70–250 Hz. This example illustrates the general finding that the relative power gain obtained by moving the source from one position to another is confined to a limited frequency range defined by the transition ranges at the two positions. The power output at very low frequencies (and at high frequencies) thus remains unchanged. In practice, the power output at low frequencies is also influenced by the excitation of the room modes in a way which is dependent on the position of the loudspeaker system, as discussed in the next section.

As a further example of the relative power gain obtained by moving from one position to another, consider the effect of raising the loudspeaker system from a position near the floor. Figure 8.19 shows the power gain obtained by moving from $x = 1$ m, $y = 0.7$ m, $z = 0.3$ m to a point 0.5 m higher at $x = 1$ m, $y = 0.7$ m, $z = 0.8$ m. The power gain in this case is negative, showing a broad dip of about 6 dB around 110 Hz. This reduction might well compensate for an excess in power output at the lower position, thereby improving the sound reproduction. Conversely, if the power output of the loudspeaker system was lacking in this frequency range, the sound reproduction would probably benefit from the additional reinforcement gained by placing the system near the floor.

Before leaving this section, it is important to say a few words about the special position defined by $x = y = z = 0$. Because the values of x, y and z are the distances of the loudspeaker diaphragm to the three nearest room boundaries, it is not possible to realize this position in practice due to the obstruction of the loudspeaker enclosure. However, as has previously been shown, this position offers the attractive benefit of providing 9 dB of power boost at all frequencies. In realizing this benefit, some loudspeaker engineers have developed loudspeaker enclosures which allow the loudspeaker diaphragm to be placed as close as possible to the junction of the room

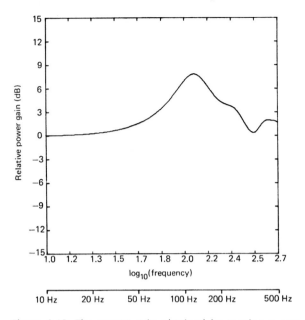

Figure 8.18. The power gain obtained by moving a non-directional source from one position ($x = 1$, $y = 0.7$, $z = 0.3$ m) to a second position nearer the corner ($x = 0.33$, $y = 0.23$, $z = 0.1$ m).

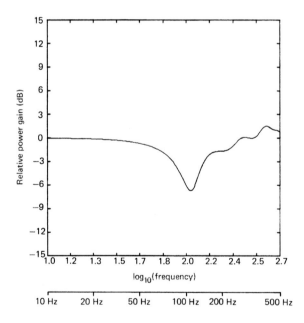

Figure 8.19. The power gain obtained by raising a non-directional source from one position ($x = 1$, $y = 0.7$, $z = 0.3$ m) to a second position 0.5 m higher ($x = 1$, $y = 0.7$, $z = 0.8$ m).

boundaries. The work carried out by Allison[13,14] has been a major contribution to this interesting area of loudspeaker design.

8.3.2 Coupling to room modes

In the previous section we considered the influence on the power output of the source of only those three mutually perpendicular room boundaries which lie nearest to the source. When the sound source is placed in a rectangular room having three pairs of parallel boundaries, there are many more image sources formed in addition to those shown in Fig. 8.15 for the simple three-boundary case. These additional images represent the successive reflections which occur between opposite pairs of parallel boundaries, and give rise to the phenomenon of room resonance discussed in Section 8.2.2. The analysis of Section 8.3.1 still applies to this case, but the introduction of further images into the model means that the expression for the power output of the source given in equation (8.13) will gain some additional terms.

In the three-boundary case, the change in power output relative to the free-space value was shown to be due to the waves reflected back on to the loudspeaker diaphragm by the room boundaries. These reflected waves can change the radiation resistance experienced by the loudspeaker diaphragm and hence the power output as given by equation (8.14). When we consider the complete image-source model of a sound source in the corner of a rectangular room, given in Fig. 8.6, we can see that there will be many more reflected waves falling onto the loudspeaker diaphragm.

In general, these reflected waves will arrive at the sound source with different phases and will thus tend to cancel out; however, at the frequencies of the room modes, many of the reflected waves arrive at the source in phase with each other and in phase with the source. Unless the room boundaries are highly absorbing, these in-phase components can combine with the outgoing wave in such a way that the radiation resistance experienced by the diaphragm is increased. For this corner position of the source, we can therefore expect an increase in power output of the source at the frequencies of the room modes.

The calculation of the power output of a sound source placed within a rectangular room requires the use of some complicated mathematics which are beyond the scope of this text. However, the results of such an analysis have been published by Salava[15], and these clearly demonstrate that the radiation resistance presented to the loudspeaker diaphragm is considerably influenced by the room modes.

Salava calculated the radiation resistance of a vibrating rigid circular piston placed in one wall near the corner of a rectangular room. The circular piston was chosen to represent the behaviour of the loudspeaker diaphragm at low frequencies. He chose a room of dimensions $5.5 \times 3.5 \times 6$ m, having a reverberation time of 0.5 s, as an example of a typical listening room. The radiation resistance as a function of frequency calculated by Salava for this environment shows a number of narrow peaks centred on the frequencies of the room modes. The magnitudes of these peaks are not all the same but fall within a range of between 6 to 24 times the free-space value of the radiation resistance. From equation (8.14), the power output of the source would therefore increase by between 7 and 14 dB above the free-space value at the frequencies of room resonance if the amplitude of the piston velocity was maintained constant.

At frequencies which lie between the room resonances, Salava's calculations show that the radiation resistance never falls below the free-space value, but that it does sometimes fall below the value given by equations (8.13) and (8.15) for the simple three-boundary case. Without embarking on a more detailed analysis, we can only postulate from this that the additional terms to be included in equation (8.13), which account for the influence of the other three room boundaries, can make both positive and negative contributions, depending on the frequency.

Because these additional terms represent the reflections which occur after the first reflections from the three nearest room boundaries, we can consider the analysis and

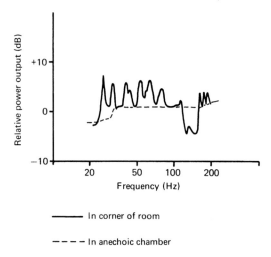

In corner of room

- - - - In anechoic chamber

Figure 8.20. The measured power output of a closed-box loudspeaker system placed in the corner of a room ($x = 0.7$, $y = 0.7$, $z = 0.5$ m). The power output measured in an approximately free space (an anechoic chamber) is shown by the broken curve. The augmentation of the power output due to the room modes is clearly shown, as is a broad dip around 150 Hz due to the early reflections.

results of the previous section to be a valid representation for the time period of the 'early sound wave' which comprises the direct sound wave and the first reflections. For the time period following the propagation of this early sound wave up to the establishment of a steady-state condition, the power output of the source given by equation (8.13) is progressively modified such that the amplitudes of the modal frequencies are selectively augmented relative to other frequencies. From the results of some measurements made by the author[16] of the radiation resistance of a loud-speaker diaphragm at low frequencies, this steady state augmentation is typically of the order of 10 dB, as shown in Fig. 8.20.

So far in this section we have considered only a corner location of the source. This position was also chosen for the image-source model of Fig. 8.6 and the analysis of room modes given in Section 8.2.2 because it turns out to be the position at which the source can most effectively excite all the room modes. If the source is moved to some other position in the room, then certain modes are still excited to the same extent but others are excited less or not at all. Because the power output of the source is related to the mean square pressure averaged throughout the room, this change in the strength of the modes is also evidenced by a corresponding change in the radiation resistance experienced by the source.

The change in excitation of a particular room mode as a function of the position of the source is most easily illustrated by redrawing the image-source model of Fig. 8.6 for the new source position. Consider, then, the resulting image-source model shown in Fig. 8.21 for a source positioned in the centre of the room. Compared with Fig. 8.6 we can see that the positions of the image source have changed such that, for example, the plane arrays of image sources which generate travelling waves along the length axis are now spaced at intervals equal to one room length instead of two.

At the frequency of the first axial mode in the length direction ($n_1 = 1$) given by equation (8.8), the wavelength is equal to twice the room length, and thus the plane arrays in question are spaced at intervals of half a wavelength. The resultant travel-ling waves entering the room from the left and right are therefore both formed from the combination of a series of travelling waves which arrive alternately in-phase and

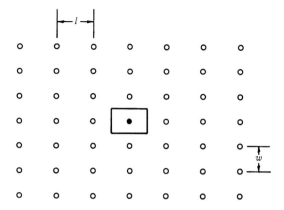

Figure 8.21. The image-source model of a sound source placed in the centre of a rectangular room showing the first few orders of images. Compared to the model for a corner location of the source shown in Fig. 8.6, the image spacing has been halved.

out-of-phase. If the room boundaries are perfect reflectors, then the in-phase and out-of-phase components are of the same strength and thus they cancel completely. The amplitude of the standing wave is then zero, and the mode is said to be not excited. When the room boundaries are partly absorbing, the in-phase and out-of-phase components do not quite balance, and thus a standing wave of small amplitude is still established. The mode is then said to be weakly excited.

At the frequency of the second axial mode in the length direction ($n_1 = 2$) the components making up the resultant travelling waves combine together in a different way than for the first mode. Here the wavelength is equal to the room length and thus all the components from the plane arrays of image sources arrive in-phase, so that they combine constructively. A standing wave of similar amplitude to that obtained for corner placement of the source is therefore set up at this frequency. The mode is then said to be strongly excited.

From these two examples it is easily seen that placement of the source in the centre of the room will cause the axial modes having even values of the integer n_1 to be strongly excited, while the axial modes having odd values of n_1 will be only weakly excited or not excited at all.

Continuing with our examples of the first and second axial modes, it is interesting to note from the variation of the amplitude of the sound pressure along the length axis of the room, shown in Fig. 8.10, that weak or strong excitation of a mode corresponds respectively with the existence of a node or antinode at the position of the source. The study of further examples of particular room modes, and how they are excited for different positions of the source, confirm that this correspondence holds in general. Thus, for any particular mode, we can expect that the mode will be most strongly excited when the source is placed at an antinode, but only weakly excited when placed at a node.

This dependence of the strength with which a mode is excited on the position of the source leads to the concept of the 'coupling' of the source to a room mode. When the source is placed at an antinode, the mode is strongly excited, which is interpreted as good coupling to the mode. Conversely, poor coupling is said to be obtained when the source is placed at a node because the mode is then weakly excited. As the source is moved from a node to the nearest antinode, the coupling progressively changes from poor to good in much the same way as the amplitude of the sound pressure measured with a microphone increases as the microphone is moved along a given standing wave from a node to an antinode.

In general, for some arbitrary position of the source, each mode will be excited to a greater or lesser extent depending on the position of the source relative to the nodes and antinodes of the corresponding standing wave. Because the positions of the nodes and antinodes are different for each mode, we can expect some modes to be excited more strongly than others, and some modes to be only weakly excited. The only positions of the source which ensure that all modes are strongly excited are at the corner points of the room, because all modes have antinodes at these places.

An understanding of how the coupling to various room modes can be influenced by changing the position of the source is useful in deciding on the optimum position in which to place a loudspeaker system for best results. Unfortunately, the degree of coupling which yields the best results is not clearly defined, because it depends on a number of factors which are often different in each case. We have seen, for example, that the power output of the source is increased at the frequency of a room mode if the source is positioned so that it is strongly coupled to the mode. Thus strong coupling to room modes, particularly in the low-frequency range, is a means of obtaining acoustic amplification of the sound output of the loudspeaker system. This may well prove welcome if the loudspeaker system has a relatively poor low-frequency power output.

However, acoustic amplification of this sort is confined to narrow frequency ranges at low frequencies which can sometimes lead to audible coloration; musical sounds are then impaired by the impression that all low-frequency notes have the same frequency content. This form of coloration is particularly noticeable in small rooms which have widely spaced modes in the low-frequency range and/or when the loudspeaker system has an extended or excessive power output/frequency response at low frequencies. In these cases it will probably be best to select a loudspeaker position which has weaker coupling to the low-frequency room modes so that the coloration is reduced.

In practice, there is regrettably no substitute for some experimentation with loudspeaker positioning if the best results are to be obtained. The degree of coupling to the low-frequency room modes which proves to be optimum depends on the characteristics of the loudspeaker system, the reverberation time of the room as a function of frequency, the room size, and finally, but not least, the subjective preferences of the listener and his position in the room.

For rooms of typically domestic living-room size which have fairly lively acoustics ($T_{60} = 0.5$ s at 500 Hz) it is usually found necessary to position the loudspeaker systems so that they are fairly well coupled to the room modes at low frequencies in order to achieve a satisfactory balance between the low and middle frequency ranges. Some low-frequency coloration is then inevitable, but may well be accepted in preference to the alternative of weak coupling with insufficient bass.

The importance of the position of the listener must not be overlooked. If the listener is positioned at an antinode of a mode, he will experience a greater sound-pressure level at the frequency of that mode than if he were positioned at a node. The 'coupling' of the listener to a particular room mode determined by his position is therefore as important as the coupling of the loudspeaker system. For the typical listening room mentioned above, the optimum position of the listener is generally found to be where there are fewest nodes at low frequencies, because some advantage must be taken of the amplification provided by room resonance. This optimum position is thus very often near to the room boundaries, rather than in the central part of the room where nodes are more likely to occur.

A few minutes spent mapping out the positions of nodes and antinodes for the modes in the low-frequency range, particularly the first three or four axial modes in each direction, can provide a considerable insight into how the coupling of both source and listener can be changed to advantage. The correct use of this information can in many cases prove to be a most effective form of sound system equalization[17].

8.3.3 Influence of source directivity

The study of the effects of the near room boundaries and the room modes on the power output of a loudspeaker system presented in Sections 8.3.1 and 8.3.2 assumed that the source was non-directional. This condition is satisfied fairly well in practice, particularly at low frequencies, by many of the closed-box and vented-box type loudspeaker systems which are in most common use for sound reproduction in the home. These systems normally only begin to become directional for frequencies above a few hundred Hz, where the wavelength becomes comparable with the box dimensions. However, there are some types of loudspeaker system design which also show directional behaviour at low frequencies. The best known example is the open-backed box or baffle which, at low frequencies, exhibits a radiation pattern similar to that of a dipole or doublet source (see, for example, the electrostatic loudspeaker, Chapter 3).

If a loudspeaker drive unit is mounted in an open-backed box or on a flat open baffle, the front and back surfaces of the vibrating loudspeaker diaphragm can be thought of as being two individual sound sources vibrating in antiphase with each other. At low frequencies, where the dimensions of the open box or baffle are much smaller than a wavelength, both sources can be treated individually as being non-directional. Because the radiation from each source must travel some distance d to reach the edge of the baffle before it spreads out in all directions, the two sources are effectively spaced apart by this distance. The combination of two such sources having equal amplitude but opposite phase is described as a dipole.

The radiation pattern of a dipole source takes the form shown in Fig. 8.22, often described as a 'figure-of-eight' characteristic (see also Chapter 1). The positive and negative signs indicate the relative phase or polarity of the sources, as well as the relative phase of the sound radiation from front and back. At any point in the plane of the baffle the sound-pressure contributions from the two sources are of equal amplitude but are out-of-phase and thus the resultant sound pressure is zero. At any point in space which does not lie on this plane, the sound-pressure contributions still tend to cancel but are prevented from doing so completely by the amplitude and phase differences introduced by the separation d.

The image-source concept used previously in this chapter is also applicable to the investigation of the excitation of the room by a directional source, but the analysis becomes a little more difficult. The images in this case are also characterized by the directional behaviour of the main source as defined by a true mirror image. Thus, in computing the combined effect of a number of image sources, the directional properties of each image source must be taken into account. Despite this difficulty, the image-source model can still provide some understanding of the influence of source directivity without recourse to too much mathematics.

As an example, consider a dipole source positioned some distance in front of a single boundary as shown in Fig. 8.23. The main source and the single image source

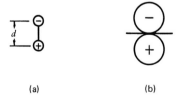

(a)   (b)

Figure 8.22. A loudspeaker drive unit mounted in an open-backed box or on an open baffle behaves as a dipole source represented at (a) by two out-of-phase non-directional sources separated by a distance d. The directional characteristic of this combination shown by (b) has lobes of radiation directed mainly to the front and back which are out of phase with each other.

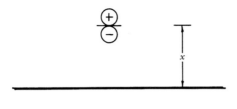

Figure 8.23. The image of a dipole sound source in a single solid boundary.

which is formed in this case are each represented by their figure-of-eight radiation pattern, but they could equally have been shown as spaced pairs of out-of-phase non-directional sources as in Fig. 8.22. Remembering that the boundaries are effectively removed by introducing the image sources, we can see that the sound radiation which enters the room arises from the positive forward radiation of the main source together with the negative backward radiation from the image source. At low frequencies, where the wavelength is much greater than the distance x of the dipole source in front of the boundary, these two components are very nearly out-of-phase and therefore largely cancel each other. This effect is clearly quite different from that obtained with a non-directional source, where the image would provide reinforcement of the output of the main source at these frequencies.

The relative power output of a dipole source positioned in front of a solid, perfectly reflecting boundary was analysed by Waterhouse[10] for various orientations of the source. When the plane of the baffle is placed parallel to the boundary, he showed that the power output W of the source is given by:

$$\frac{W}{W_\mathrm{f}} = 1 - 3j_\mathrm{o}(2kx) + \frac{6}{(2kx)^2}\{j_\mathrm{o}(2kx) - \cos(2kx)\} \tag{8.16}$$

where W_f is the power output of the source in free space. The amplitude of vibration of the source is again assumed to be independent of the environment, as it was for equation (8.13).

The variation of the value of equation (8.16) as a function of frequency is shown in Fig. 8.24 for distances between source and boundary of 0.5 m and 1.5 m. These curves confirm that the power output at very low frequencies is indeed reduced relative to the free-space value when the baffle is placed parallel to a boundary. However, the power output does become reinforced around the frequency range where the distance x is approximately equal to a quarter wavelength. The contribution from the image source, and hence the reflection which it represents, then arrives in phase with that of the main source. By careful choice of the distance x it may in some cases be possible to centre this boost in a frequency range where reinforcement is desirable. In general, it is clear that an open-baffle type loudspeaker system should not be placed too close to the parallel boundary which lies behind it, otherwise low-frequency output will be lost.

Because a dipole source is directional, it is not surprising to find that its power output is affected in a different way than that shown above when the baffle is placed at right angles to a solid boundary. In this case the relative power output is given by[10]:

$$\frac{W}{W_\mathrm{f}} = 1 + \frac{3}{(2kx)^2}\{j_\mathrm{o}(2kx) - \cos(2kx)\} \tag{8.17}$$

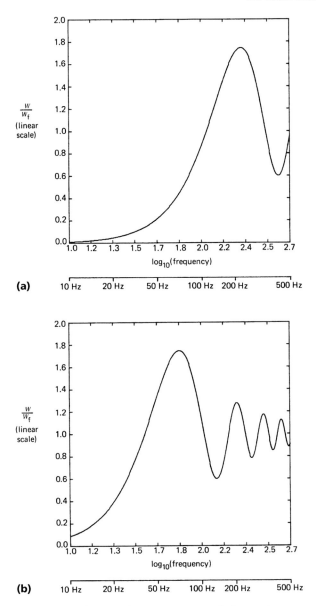

(a)

(b)

Figure 8.24. The sound power output relative to free-space loading of a dipole sound source placed with the baffle parallel to a single solid boundary at distances of 0.5 m (a), and 1.5 m (b).

The value of this equation as a function of frequency is plotted in Fig. 8.25 for $x = 0.5$ m and $x = 1.5$ m. In contrast to the previous orientation, this arrangement provides positive reinforcement of the power output of the source at very low frequencies. The power boost amounts to 3 dB which is similar to that obtained when a non-directional source is placed near a single boundary.

From the results shown in Figs 8.24 and 8.25 it is evident that the power

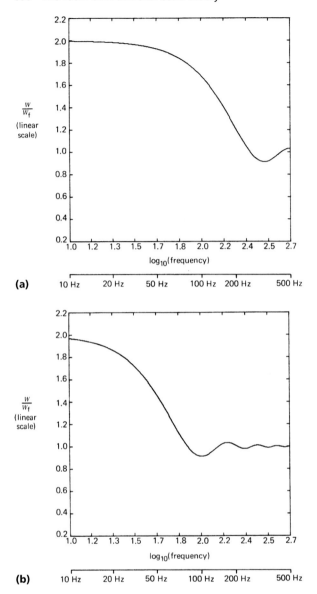

Figure 8.25. The sound power output, as Fig. 8.24, for a dipole source with the baffle placed at right angles to the boundary at distances of 0.5 m (a), and 1.5 m (b).

output/frequency response of a dipole source in the period of the early sound wave is capable of being optimized for best results by careful selection of the position of the source relative to the near boundaries. However, unlike the case of the non-directional source, the directivity of the dipole source requires that attention must also be paid to the orientation of the source relative to the room boundaries. Thus, in a typical listening room arrangement where the baffle is placed parallel to the wall in front of the listener, the maximum power boost at low frequencies will probably be obtained by bringing the baffle forward away from the parallel wall, while at the

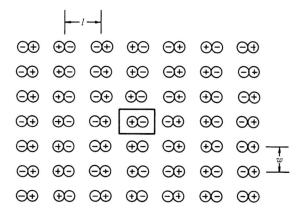

Figure 8.26. The image-source model of a dipole sound source placed in the centre of a rectangular room for the first few orders of images.

same time placing it as near as possible to the floor and side wall boundaries which are at right-angles to it.

In carrying out experiments with the position of the loudspeaker system in this way, we must remember that the position of the source also influences the degree of excitation of the room modes at low frequencies. When the source is directional, the room modes will be excited differently than they would be by a non-directional source. A rough indication of these differences can be obtained from a cursory study of the image-source model of the directional source placed in a rectangular room. Figure 8.26 shows the plan view of such an image-source model for the first few orders of images of a dipole source positioned in the centre of the room. This position is hardly typical, but it facilitates a useful comparison with the observations made about Fig. 8.21 for a non-directional source in the same position.

For the orientation of the source chosen in Fig. 8.26, it is interesting to note that the plane arrays of dipole image sources (which generate travelling waves that enter the room from the left and right sides) alternate in polarity from one plane to the next. Thus, at the frequency of the first axial mode in the length direction ($n_1 = 1$), where the wavelength is equal to $2l$, the contributions from the plane arrays lying for instance on the left-hand side, all arrive at the room in phase. The resultant travelling wave entering the room from the right-hand side is similarly formed constructively. These travelling waves combine to produce a standing wave having a nodal plane lying in the centre of the room, as illustrated in Fig. 8.10.

At the frequency of the second axial mode in the length direction ($n_1 = 2$), the wavelength is equal to l and hence equal to the spacing of the planes of images to the left and right sides. Because of the polarity reversal between alternate planes, the resultant travelling waves sum to zero amplitude if the boundaries are perfect reflectors, and the mode is therefore not excited. If the boundaries are partly absorbing, the mode is then weakly excited.

These results are exactly the opposite to those found for the non-directional source considered in Section 8.3.2. The coupling of the dipole source to a mode appears to be most effective when the source is placed at a pressure node, as it was for the above example at the first mode. Conversely, the coupling has been shown to be poor when the source is placed at an antinode, as applies when the source is placed at the centre of the room at the second mode (refer to Fig. 8.10).

In a standing wave, the velocity of air motion reaches a maximum where the pressure reaches a minimum, and vice versa. Thus a velocity node is formed at a pressure antinode, and a velocity antinode is formed at a pressure node. We can therefore conclude that, for the axial modes considered, the coupling of the dipole source to a mode is greatest when the source is placed in the region of maximum velocity amplitude, i.e. at a pressure node.

Referring again to Fig. 8.26, we can see that the plane arrays of image sources which will give rise to standing waves along the width and height axes of the room do not show the alternating polarity from one plane to the next that applies in the length direction. These features are, of course, related directly to the orientation of the source relative to the room boundaries. For the orientation shown, the coupling of the source to axial modes in the width and height directions will therefore follow the findings of Section 8.3.2 for a non-directional source, i.e. the coupling is greatest when the source is positioned at an antinode. However, because the dipole image sources are poor sound radiators in the directions which are parallel to the baffle, the travelling waves entering the room along the width and height axis are considerably weaker than those entering along the length axis.

From this brief examination of the simplified image-source model, we can expect that a dipole source will excite those axial room modes which propagate in a direction which is parallel to the baffle by a lesser degree than would a non-directional source. Also, because the coupling to the axial modes which propagate in a direction at right angles to the baffle is weak in the vicinity of a pressure antinode, we can expect the first few of these modes to be only weakly excited if the source is positioned fairly near a corner, away from the pressure nodes which form in the central area of the room. Thus, in applications where the acoustic amplification offered by the excitation of room modes at low frequencies is not required, the dipole source provides some advantage over the non-directional source.

Before leaving this topic, it is interesting to consider the possibility of selecting a directional characteristic for the source which provides coupling to the modes in a way which is independent of position. We have seen that the non-directional and dipole sources are in some ways complementary, in that one is able to excite a mode in a position where the other cannot. It is therefore reasonable to suppose that a combination of these sources would provide similar excitation of the mode at any point along the standing wave. In microphone design, just such a combination of non-directional and dipole characteristics is used to provide a unidirectional or cardioid characteristic as described by Kinsler and Frey[1] and illustrated here in Fig. 8.27.

A sound source with a cardioid directional characteristic radiates principally in the forward direction, with very little sound radiation emerging from the back. Although the radiation normal to the forward axis is weaker than from the front, it is still very much stronger than that of the dipole source and should not be overlooked. If a cardioid source is substituted into the image-source models of Figs 8.6 and 8.21 with

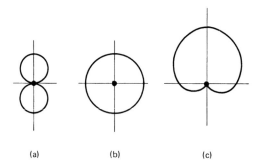

(a)   (b)   (c)

Figure 8.27. As with microphones, the directional characteristics of a dipole source (a) and a non-directional source (b) can be combined to give a source having a cardioid directional characteristic (c).

the forward axis directed to the right, then a brief study will confirm that this source, unlike the non-directional and dipole sources, is able to excite strongly all the axial modes in the length direction from both corner and centre locations. Unfortunately, the same does not apply to the axial modes in the other two directions, which still behave as if the source were non-directional.

8.3.4 Stereo imaging

When a pair of loudspeaker systems is used for the reproduction of stereo recordings, there are some considerations in addition to those already covered regarding their positioning within the listening room. These are primarily concerned with the mechanism by which the listener perceives stereo images. This mechanism of stereo image perception has been studied by many authors and the interested reader can refer to the bibliography (and also Chapter 14). The treatment given here is highly simplified, but it serves to illustrate and explain most of the commonly observed phenomena.

Let us first consider a hypothetical situation in which the listener and two identical loudspeaker systems are arranged in free space as shown in Fig. 8.28. If only one loudspeaker system is driven, then the sound wave reaching the listener must travel a greater distance to reach one ear than the other. This additional path length gives rise to a time difference between the signal arrivals at the two ears. Also, at high frequencies where the wavelength is smaller than the head dimensions, the shadowing of one ear by the head can introduce an appreciable level difference between the signals received by the two ears. Both of these time and level differences enable the brain to locate the position of the source by comparing the signals received by the left and right ears. The time difference appears to be the most important, particularly at low frequencies, while the level difference plays a supporting role at frequencies above about 1 kHz. It is also likely that the shape of the ear pinna and the influence of unconscious head movements are important contributors to our ability to localize sound sources.

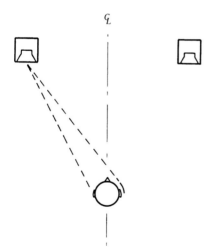

Figure 8.28. Our facility to localize sound sources relies on detection of the differences in the sound signals received by our two ears. The path-length difference from the source to each ear, and the shadowing of one ear by the head, are two of several important differences we can detect. When listening to a stereo sound system, the signals received by each ear from the left and right sound sources are partially combined before comparison, which enables phantom sound sources to be perceived at points lying between the two sound sources.

When the second loudspeaker system is also driven, each ear receives two signals, one from each loudspeaker system, making four signals in total. Because the ear demonstrates a degree of integration of incoming signals, each pair of signals is partially combined prior to the comparison of the electrical output signals from both ear mechanisms by the brain. The partially combined signals are compared by the brain as they would be for the incoming signals from a single source, with the result that we perceive a single phantom image lying somewhere between the two loud-speaker systems. Thus, for example, if the listener is positioned on the central axis between the loudspeaker systems, as shown in Fig. 8.28, then a centrally placed phantom image is perceived when both loudspeaker systems are driven with the same signal.

The position of the phantom image is quite sensitive to an imbalance in the sound level output of the two loudspeaker systems, a difference of about 10–15 dB being sufficient to localize the image completely over to the louder side. A time difference between the left and right sound signals can also give rise to a similar shift in the position of the phantom image. Both of these time and amplitude differences are used in most stereophonic recordings to create phantom images at any desired locations between the loudspeaker systems.

We have seen in the preceding sections of this chapter that when a loudspeaker system is placed in a room instead of in free space, the resulting reflections at the room boundaries can be represented by mirror images of the loudspeaker system lying outside the room. In common with the mechanism of perception of flutter echoes discussed in Section 8.2.4, only the contributions from distant image sources will be resolved by the ear as being separate from the output of the main source, the loudspeaker system. These contributions are perceived as echoes or reverbera-tion. However, the contributions from the main source and the image sources which arrive at the listener within an interval of about 15 ms are not resolved by the ear as separate signals, but are combined together and treated as a single signal arrival. As a result, the perception of stereophonic images in a real room environment involves the interpretation of the combined contributions of the main left and right sources and their near image sources which represent early reflections.

As an example of how these early reflections can influence stereo imaging, consider the nearest primary images formed in the boundaries lying behind and to the sides of the loudspeaker systems as shown in Fig. 8.29. If only one loudspeaker system is driven, the listener now perceives a sound image positioned somewhere between the

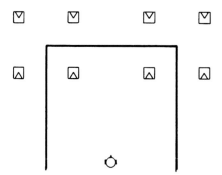

Figure 8.29. The nearest image sources of a stereo pair of loudspeaker systems in the walls behind and to the sides representing the early reflections which fuse with the direct sound wave. If these boundaries are good reflectors the image sources are strong, resulting in an effective displacement of the localized position of each source towards its nearest images. When stereo sound signals are reproduced these displacements cause the width and depth of the 'sound stage' to be increased.

corresponding cluster of four sources comprising the loudspeaker system and its three nearest images. The location of the sound image within the cluster depends on the strengths of the image sources relative to the main source. These strengths will clearly depend on the degree of absorption presented by the boundaries to the incident sound waves. Reflective boundaries will give rise to strong images and hence to a noticeable shift in location of the sound image towards the boundaries.

Because the orientations of the image sources towards the listener are different from that of the main source, the strengths of the contributions from these sources, and hence the amount of image shift, are also dependent upon the directional characteristics of the loudspeaker system. These are often dependent on frequency but, at any given frequency, we can consider each loudspeaker system and its nearest image sources as a single sound source displaced both sideways and away from the listener. When both loudspeaker systems are used to reproduce stereo recordings, this effective displacement of the left and right sound sources enables stereo images to be perceived at points which lie outside the area between the two loudspeaker systems.

From this simple analysis we can see that the positions of the left and right loudspeaker systems relative to the near room boundaries will have a bearing on the location of stereo images, particularly if the boundaries are reflective. Because the strength of the nearest image sources of each loudspeaker system determine its effective displacement, both the directional characteristic of the loudspeaker system and the absorption of the near boundaries as functions of frequency influence the location of stereo images. When the sound of a single musical instrument composed of several frequency components is reproduced, it is thus easy to understand how these components might be localized at different points. This undesirable effect gives images which are difficult to localize and have poor clarity[18,19].

When positioning a given pair of loudspeaker systems for stereo sound reproduction, we are seldom at liberty to make changes in their directional characteristics or to make modifications to the treatment of the room boundaries. However, to keep in line with the expectations of the recording engineer, we can at least ensure that the left and right sound channels are as nearly identical as possible. Because of the important role played by the early reflections in determining the strength and effective position of each loudspeaker system, the mechanism by which these reflections occur should be made the same for both channels. Thus the position of both systems relative to their nearest room boundaries should be the same, and, further, the construction and wall treatments of these boundaries should be symmetrical about the central listening axis. Items of furniture or wall treatments which imbalance this symmetry will have a similar effect on the stereo image location. For instance, if one loudspeaker system is placed in front of a window while the other is placed in front of a bookcase, there will be a noticeable shift of stereo images towards the window.

8.4 Measurement of room acoustics

As a science, room acoustics theories offer the engineer some understanding of the acoustic phenomena experienced in listening rooms, as well as the opportunity to make predictions based on calculation. The success of such predictions, and hence the confirmation of the theories, can be assessed by carrying out acoustic measurements of the sound field within the listening room. Most of the acoustic phenomena discussed in this chapter have been extensively investigated by measurement, and the reliable ties so formed between theory and measurement have established a basis for the diagnosis of acoustic problems by measurement. Together with a knowledge of room acoustics theories, these measurements can offer considerable guidance towards the realization of a cure for such problems.

In the preceding sections, the dependence of the sound field in terms of both time and frequency have been discussed and demonstrated. To simplify analysis, it is common practice to consider variations in the sound field as a function of either time or frequency, leaving the other one of these variables fixed. This practice is often carried through to measurements, and thus two of the most commonly used room measurements consist of the measurement of frequency response under steady-state conditions, and the time decay of one or more fixed-frequency components. A brief study of these measurements is given below (see also Chapter 12).

8.4.1 Measurement of reverberation time

In Section 8.2.3 reverberation time was defined as the time taken after the sound source is switched off for the sound level to reduce by 60 dB below some steady level established prior to the switching. Sabine[4] in his experiments made use of an organ pipe as a steady sound source, and his original choice of a pipe sounding the octave of middle C, i.e. 512 Hz, seems to have set the standard of reverberation time measurement at 500 Hz. If a single value of reverberation time is quoted for a room, it is generally understood to be for this frequency. When the reverberation time is required as a function of frequency, it is usually measured in frequency bands which cover an octave or third of an octave. Reverberation time is rarely measured or quoted for frequencies below 50 Hz or above 10 kHz.

For measurement of the reverberation time of listening rooms of modest dimensions, the resident loudspeaker systems can often be used to provide the sound source. As a rule, the loudspeaker system is driven with a pink noise signal which is filtered by an octave or third-octave wide filter centred on the frequency range of interest. Alternatively, a sine-wave signal may be used (if one particular frequency is of interest) or the sine-wave frequency can be modulated (or wobbled) to cover a certain range. Whichever signal is chosen, it must first be applied until the sound level in the room reaches a steady value before switching the signal off to observe the decay.

The decay of the sound level after the source has been switched off can be heard by ear, but it is normal practice to trace the decay of the amplitude of the sound level in dB versus time on some type of recording instrument. The sound-pressure level at some point of interest in the room is measured with a microphone whose electrical output signal is rectified to produce an r.m.s. value (strictly speaking the 'true r.m.s.' value should be used) which is in turn converted to a logarithmic level for recording against the time base. High-speed level recorders can be used to trace the decay on chart paper, or computer graphics, in addition to providing the means of rectification and log conversion. Alternatively, the decay curve can be measured and recorded using a storage oscilloscope.

Although the definition of reverberation time refers to a drop in sound level by 60 dB, the reverberation time can still be estimated roughly in cases where the background noise level is less than 60 dB below the steady level established by the sound source. Since the general nature of the decay in sound level is exponential with time, the time expected for the sound level to drop by 60 dB in the absence of background noise can be estimated by extrapolating the decay curve as shown in Fig. 8.14. However, if the signal-to-noise ratio is less than about 30 dB, such estimates are unlikely to be reliable and steps should be taken to reduce the background noise level and/or increase the power output of the sound source. The power output of a loudspeaker sound source can often be beneficially increased by positioning the loudspeaker system right into a corner of the room. Alternatively, the noise level relative to the signal can usually be reduced by inserting a second filter between the microphone and the recording apparatus.

In rooms which contain relatively little absorption, the decay curve will generally be ragged, particularly at low frequencies where the frequencies of the room modes are widely spaced. The estimation of reverberation time from such curves is rather difficult, and is often made doubly so by their poor repeatability. In such cases it is normal to record several decay curves (say 10) and make estimates of the reverberation time by drawing a straight line through each ragged decay curve which best

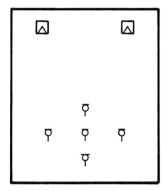

Figure 8.30. When making reverberation measurements of a listening room, it is important to average measurements taken for different positions of both sound source and microphone. The normally used positions of the loudspeaker systems, and microphone positions within the listening area, should be the first choices.

represents the trend of the decay. The values of reverberation time corresponding to the slopes of these straight line decays then provide the estimates which can be averaged together.

This time-consuming and laborious process has prompted the development of special instrumentation which is able to collect and average the decay curves automatically. In addition, the instruments are usually capable of fitting the 'best' straight line to the averaged decay curve and so compute the reverberation time.

The values of reverberation time, particularly at low frequencies, are seldom found to be independent of the position of the microphone within the listening room. Thus, to obtain a meaningful impression of a room's reverberation, it is wise to carry out reverberation time measurements for several microphone locations and combine these by averaging. The position of the sound source can also influence results. For these reasons, and because of the need for the matching of left and right channels of a stereo system, it makes sense to use the loudspeaker systems in their normal positions as alternative sound sources driven one at a time, and to use several microphone positions within the normal listening area. A typical arrangement of sound source and microphone locations is illustrated in Fig. 8.30. The reverberation times versus frequency collected in this way can be combined into several averages to indicate differences in the left and right channels, and variations with microphone position. The latter may help to suggest an optimum listening position, or provide some indication of the cause of some acoustic problem.

8.4.2 Measurement of frequency response

The variations of sound pressure as a function of frequency at given points within the listening room are of fundamental interest in making an assessment of the room's acoustics. The measurement of 'frequency response' implies measurement of the variations of both amplitude and phase (see Chapter 12) but in room measurements it is normal to measure the amplitude/frequency response only.

In electrical engineering it is common practice to use a sine wave signal whose frequency is slowly varied to measure the frequency response. However this method is not convenient in rooms because the relatively long delay time of the room resonances requires an unusually slow sweep.

Because of the random nature of music signals, the room modes, which take a finite time to build up and decay, are not excited as strongly as they are under steady-state conditions. To simulate this effect, the sine wave signal can be modulated or

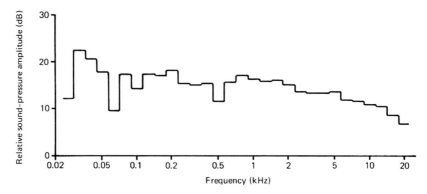

Figure 8.31. The third-octave spectrum of the sound-pressure amplitude measured at the listening position of a typical domestic listening room. The loudspeaker system was excited by a pink noise signal.

wobbled rapidly about the measurement frequency so that the room modes never become completely established. Alternatively, a random noise signal can be used to excite the sound source, and the microphone signal analysed with some form of frequency-response analyser.

The frequency analysis is most commonly achieved by filtering the microphone signal through a band-pass filter of octave or third-octave width with an adjustable centre frequency. Some instruments employ a multiplicity of such filters, with fixed centre frequencies to cover the entire audio range and so provide a 'real-time' display of the frequency response. These instruments are particularly valuable when investigating the variation of frequency response as a function of microphone position within the room. A pink noise excitation signal is normally employed in this type of measurement because its octave or third-octave frequency spectrum is flat.

Figure 8.31 shows the third-octave spectrum of the sound-pressure amplitude measured at a point located in the normal seating area of a typical listening room. The spectrum analyser contained 30 third-octave filters in this case, covering the range from 25 Hz to 20 kHz. Most analysers provide some means of adjusting the averaging time of the output signal from the filters; a short averaging time, and hence fast response, is useful in observing spectrum changes as the microphone is moved around the room, while a long averaging time is more suited to the generation of hard-copy output for fixed microphone positions.

In assessing the acoustics of a room and the performance of the loudspeaker systems, it is helpful to measure the frequency responses at the same positions as those suggested for the reverberation time measurements, shown in Fig. 8.30. The variation of frequency response with microphone position is of major importance in suggesting the optimum listening positions and in the identification of acoustic problems. In the latter case, it is helpful to measure the frequency response at increments of say 0.5 m along a line joining the source loudspeaker and the listening position. By plotting this series of frequency responses versus position as an isometric projection, many interesting features of the sound field can be identified. By measuring the frequency responses in the listening area for excitation by the left and right loudspeakers driven one at a time, an indication of the channel matching and symmetry of a stereo system can also be obtained.

8.4.3 Other measurements

The measurements of reverberation time and frequency response for different positions of both sound source and microphone can often provide adequate information for the assessment of a room's acoustics or the identification of acoustic problems. However, the correct interpretation of such measurements requires a degree of experience, and even then soundly based conclusions are sometimes proved to be in error. It is commonly felt that such measurements reveal only part of the story, and that a different approach which more nearly approximates our mechanism of hearing and perception should be sought.

One approach which recognizes the importance of examining the frequency content of individual reflections has led to the use of a technique developed by Heyser[20] known as time delay spectrometry (see also Chapter 12). This measurement system allows the frequency content of the microphone signal to be analysed at different times after the excitation of the sound source. The technique thus provides a measurement of variations of the sound field as a function of both time and frequency simultaneously. An interesting application of the technique is to make use of its 'time windowing' of the frequency response to examine the frequency spectrum of the early sound wave, which includes the direct sound wave and the first reflections from the nearest room boundaries. Reflections which arrive at the microphone after this wave can similarly be examined by delaying the time window as required.

The emergence of digital signal processing has resulted in the digitizing of a large number of acoustic measurements. Typically measurements can be made with a card of electronics plugged into a PC, thereby enabling measurements to be easily stored and manipulated. One such widely used device is MLSSA (Maximum-Length Sequence System Analyser)[31]. Although this system can be used for loudspeaker measurements in general, it is also applicable to room acoustics. MLSSA measures the impulse response of a system and manipulates that data to display other information. Examples of relevant displays are A-weighted, octave or third-octave spectra thereby emulating real time spectrum analysers. The decay of a room can also be displayed as a three-dimensional energy versus time versus frequency plot. The system measures reverberation via the Schroeder decay curve. The reverberation time is given by the best fit to the decay curve. The early decay time can also be calculated, equivalent to the slope over the first 10 dB of decay, it being suggested that this figure is more representative of the subjective experience of reverberation.

In recent years, interest has also grown in the use of measurement microphones placed inside a dummy human head[21], and in the processing of these microphone signals[22] to imitate the discrimination of the time arrivals of various reflections which is made by our hearing mechanism.

8.5 Listening room design

In our own homes, the scope for creating a good environment for listening to recorded music is generally limited to making a choice from the rooms available, and perhaps to the selection of wall treatments and room furnishings. However, even with these constraints, there is a very wide range of possible sound qualities. Experimentation with loudspeaker and listener positions can also yield considerable improvements, as discussed in detail in Section 8.3.

In the professional audio industry it is not uncommon to construct rooms specifically for the purpose of monitoring live performances and/or recorded material. In these circumstances the acoustic consultant has much more scope than is normally available in the domestic case. As a result, many diverse approaches to dedicated listening room design have been tried (see Chapter 9), but opinions differ as to the optimum acoustic conditions which should be aimed at. In this section only the basic rules of room design are discussed briefly and details of the construction of some simple types of absorber are given.

8.5.1 Room size and shape

The choices of room size (i.e. volume) and room shape depend to some extent on the intended application. Apart from the obvious requirements of being able to accommodate the sound equipment comfortably and the seating of listeners at a sensible distance from the loudspeaker systems, the minimum room volume which will prove acceptable depends on the nature of the sound signal to be reproduced. For the uncoloured reproduction of speech, for example, where the fundamental frequencies rarely extend below 100 Hz, a room volume of 40 m³ or greater will generally be satisfactory. For music signals, on the other hand, which can contain frequencies extending down to 20 Hz and below, the natural reproduction of low frequencies is difficult to realize in rooms of less than about 100 m³ in volume.

These minimum recommended volumes are related to the lowest frequencies which are required to be reproduced without significant coloration, because of the way in which room modes are distributed with respect to frequency. By way of an example of this distribution, Fig. 8.32 shows the frequencies of the first 51 room modes calculated using equation (8.8) for a rectangular room of dimensions $l = 4.7$ m, $w = 3.2$ m, $h = 2.5$ m (volume = 37.6 m³). Each room mode is here shown as a single vertical line positioned along the frequency axis at its resonance frequency. It is clear from this example that adjacent room modes are more widely spaced at the low-frequency end of the resonance range, and that the number of modes in a given bandwidth increases in proportion to the square of the frequency[1].

The relevance of this variation in the frequency separation between adjacent modes was investigated by Gilford[23,24]. He defined coloration as a selective reinforcement of sound at a particular frequency in relation to the spectrum as a whole. In this context he found that for an individual room mode to be audible as coloration, a separation of the order of 20 Hz from the adjacent modes was necessary. Applying this rule to the example illustrated by Fig. 8.32 we can see that in this case the first few low-frequency modes which lie below 70 Hz are indeed separated by this order and are thus likely candidates for the production of colorations, if the spectrum of the sound signal extends below 70 Hz. For frequencies above about 70 Hz, adjacent room modes in this example become spaced by less than 20 Hz, and these modes are therefore less likely to be perceived individually as colorations.

To reduce the severity of colorations at low frequencies due to room resonance, there are a number of courses open, the most obvious one being to increase the room volume. However, to achieve the necessary close spacing of adjacent room modes down to the lowest frequencies to be reproduced would often require a room of impracticably large size. In practice we have to make a compromise between the room volume which is available and the degree of coloration which is acceptable for

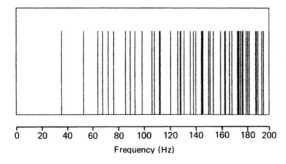

Figure 8.32. The first 51 resonance frequencies of a rectangular room of dimensions $l = 4.7$ m, $w = 3.2$ m, $h = 2.5$ m, plotted on a linear frequency scale. The modes at low frequencies are more widely spaced than those at higher frequencies, which makes their perception as a form of coloration more likely.

the intended purposes of the room. After the choice of room volume has been made, the remaining courses involve the selection of the room shape and the degree of low-frequency absorption that is introduced into the room.

Rooms of rectangular shape are most commonly used for listening room construction, as there appears to be little acoustic virtue in making use of angled or curved walls. Impracticably large angles are required to influence the resonance behaviour at low frequencies significantly, and curved or undulating boundaries are usually detrimental, being the cause of undesirable focusing effects. An often helpful exception is the introduction of a small angle, say 5°, between an otherwise parallel pair of opposite walls in order to reduce flutter echoes (Section 8.2.4). The choice of room shape thus rests mainly in the selection of the length, width and height of a basically rectangular shell. These are normally expressed in terms of a ratio of the three dimensions.

For a given room volume, the resonance frequencies and hence the spacing of adjacent modes is determined by this ratio, and several studies have been carried out to determine the 'best' ratio which provides the most even spacing between adjacent room modes at low frequencies. For example, Louden[25] calculated the spacings of the first 216 room modes for 125 different dimension ratios. By computing a figure which represented the evenness of these spacings, Louden was able to sort the dimension ratios into a preferred order. The most even distribution was given by the ratio 1:1.4:1.9. In a more recent study, Bonello[26] adopts a slightly different approach by considering the number of room modes which fall in third-octave wide bands of the frequency spectrum.

Although these studies are helpful and important, the simpler rule proposed by Gilford[23] is probably adequate and may well be more relevant. He found that the axial modes are generally of most importance in determining colorations, and he therefore recommends making a list of all axial modes up to about 350 Hz. This list should be examined for groups of modes which have nearly the same frequency, or for adjacent modes which are separated by 20 Hz or more. If these features are found, the chosen dimensions should be adjusted in an effort to make the frequency spacings between adjacent axial modes more even.

The optimum choice of dimension ratio for a given room size will generally still leave some widely spaced room modes at low frequencies which will be sources of coloration if excited by the sound source. However, the excitation of these modes, and hence the degree of coloration, can be reduced by increasing the absorption coefficient of the room boundaries in the low-frequency range. As indicated in Section 8.2.2, the introduction of damping in this way both reduces the excitation of the modes and broadens their resonance peaks.

8.5.2 Reverberation time characteristic

The reverberation time of a room as a function of frequency has come to be regarded as the most important single feature in determining the subjective quality of the acoustics of a room. In the author's experience, the reverberation time is not sufficient in itself to define all the subjective qualities of a room; however, it does still provide a very useful yardstick which allows the acoustics to be classified in a general way. Thus, based on the results of reverberation time measurements, an engineer might class a room as being 'very dead', or at the other extreme, as being 'very lively'. For listening rooms of modest dimensions, the reverberation time at 500 Hz would cover the range from about 0.1 s to 1 s respectively between these two extremes. A typical European domestic living-room would have a reverberation time of around 0.5 s.

While the value of reverberation time in the mid-band provides a general indication of the room's acoustics, its value as a function of frequency has an additional bearing on the tonal balance between the low, middle and high frequencies being reproduced. An increase in reverberation time over part of the frequency range will usually give the subjective impression of an increased output from the source in this range, and thus the reverberation time versus frequency characteristic should ideally be flat. In practice it is difficult to provide the same absorption for low frequencies as for middle and high frequencies, and some increase in reverberation time at low frequencies is usual. A reverberation time at 50 Hz equal to twice that at 1 kHz is typical, and represents the type of reverberation characteristic under which most loudspeaker systems are designed to operate. If additional low-frequency sound absorption is introduced into a room to obtain a flat reverberation time characteristic, most loudspeaker systems will then appear to be lacking in low-frequency output.

To obtain an indication of the tonal balance which will be imparted by the reverberation time characteristic of an existing room, it is probably adequate to make reverberation time measurements for octave-wide frequency bands as described in Section 8.4.1. The use of narrower frequency bands, such as third-octave, provides more detail but is usually essential only to investigations of acoustic problems. Similarly, if the acoustic treatment of a room is being calculated in advance of its construction using the relationships given in Section 8.23, then it would be usual to start off from the desired reverberation time characteristic specified at octave intervals.

The requirement for an environment which is acoustically comfortable has been part of the natural evolution of domestic living quarters. These show worldwide regional variations as well as local variations, such as between tower-block flats and single houses. Because most domestic audio equipment is designed to operate in these environments, there is often a need for duplicating a 'typical' domestic room acoustic for the evaluation and development of new equipment (see Chapter 13). For European homes, the recommended reverberation time characteristics that should be duplicated for such tests are given in an IEC Standard[27] for listening tests on loudspeakers. This Standard recommends a reverberation time of 0.4 ± 0.05 s between 250 Hz and 4 kHz, and a maximum value of 0.8 s at low frequencies. A room volume of 80 m(3) is also recommended. Similar recommendations are also to be found in the DIN Standard[28].

8.5.3 Room layout

The reverberation time of a given room depends largely on the total room absorption, as shown in Section 8.2.3. However, similar values of total room absorption can be achieved with very different arrangements of the absorbing and reflecting areas which make up the room boundaries. For example, the differing surface treatment of two opposite boundaries, such as the floor and ceiling, could be exchanged without changing the total absorption. Thus, within the specification of some desired reverberation time, there are a number of possible approaches to the layout of the acoustic treatment which is applied to the room boundaries. Because the layout of this treatment, relative to the positions of the loudspeaker systems and listener, influences the strength and frequency content of reflections received by the listener, each particular approach can yield a different subjective impression.

In European domestic listening rooms, the treatment of the ceiling and all four side walls is quite often the same, being a fairly reflective finish such as paint or paper on plaster, while the floor is generally carpeted. This typical layout is also recommended by the IEC Standard[27], but the addition of an absorbent treatment, such as heavy drapes, to the wall behind the listener, is specified to prevent the occurrence of strong reflections at middle and high frequencies from this wall back towards the listener. The side walls in this arrangement may prove to be too reflective compared with the mean absorbing power of the other two pairs of parallel boundaries, resulting in the generation of flutter echoes. These can usually be cured by partly draping the side walls and/or by making the surface of the side walls less regular. Irregularities can be introduced most easily by placing items of furniture in front of the walls, or by hanging several pictures having thick frames on each side

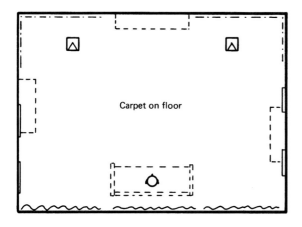

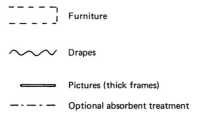 Furniture

Drapes

Pictures (thick frames)

Optional absorbent treatment

Figure 8.33. Recommended layout of a typical domestic listening room. By placing the loudspeaker systems along one of the longer walls, they can be positioned away from the room corners at different distances from the nearest boundaries. The listener is then also able to position himself near the boundary opposite the loudspeaker systems away from the central part of the room where nodes occur at low frequencies.

wall. The introduction of an absorbent treatment to the wall areas immediately behind and to the side of the loudspeaker systems is a matter for personal taste, depending on the type of stereo imaging desired. However, if the loudspeaker systems have a dipole-type radiation pattern, e.g. when the systems are of the electrostatic type, some moderate absorbent treatment having a mid-band absorption coefficient of about 0.25 is recommended for these areas.

Before deciding upon the surface treatment of the various room boundaries, it is clear from the above example that the positions of the listener and the loudspeaker systems must first be selected. The effects of changing these positions on the reproduction of low frequencies are discussed in detail in Section 8.3. As a general rule, one should aim to keep the loudspeaker systems away from room corners, and to place the listener near the wall opposite the loudspeaker systems rather than in the central part of the room. For a room which is appreciably longer than it is wide, this rule is best satisfied for stereo listening by placing the loudspeaker systems against one of the longer walls. The loudspeaker systems can then be separated by a distance similar to the room width, making a listening position near the rear wall practicable.

This approach to loudspeaker and listener placement, combined with the recommendations for wall treatments given above, is illustrated in Fig. 8.33. In addition, for correctly balanced stereo images, the room layout should be kept as symmetrical as possible about the central listening axis. The treatment of the boundaries nearest to the loudspeaker systems is of most importance in this respect, as shown in Section

8.3.4. Where possible, the central listening axis should be defined centrally in the room so that the left and right loudspeaker systems are placed equidistant from their nearest side wall.

8.5.4 Sound absorbers

In a typical domestic listening room there are many elements which act as sound absorbers, such as the furniture, the surface treatments of the room boundaries, and the actual construction of the boundaries in cases where these are not completely rigid. These absorbing elements are generally most effective over specific frequency ranges, as indicated by the absorption coefficients given in Table 8.1 (page 341). A typical combination of these elements is often able to provide a reasonably smooth reverberation time versus frequency characteristic, although the absorption of low frequencies is usually less than at middle and high frequencies.

In professional applications, where the listening room is dedicated to the purpose of sound reproduction, the desired reverberation characteristic is usually obtained by introducing purpose-made sound absorbers into the room. These can be located within the boundary constructions of the room and thus leave most of the floor area free to provide a working space. Sound absorbers can take many forms and the interested reader is referred to Kutruff[8] and to Ford[29] for details of the basic types. Some further references of interest are also given in the bibliography. In this section only the simplest type of absorber, namely the panel or membrane absorber, is described.

The panel absorber provides a very simple means of introducing absorption into the room over specific frequency ranges. The basic construction of such an absorber is shown in Fig. 8.34. Here a flexible panel of mass M kg/m² is fixed along its edges to a framework attached to one of the internal surfaces of the room, such that a sealed air space of depth D is defined behind the panel. The mass of the panel and the stiffness of the enclosed volume of air together form a simple resonant system having a resonance frequency given by:

$$f_0 = \left(\frac{c}{2\pi}\right)\left(\frac{\rho}{m_D}\right)^{1/2} \tag{8.18}$$

where c is the speed of sound in air and ρ is the air density (see Porges[30]). When sounds are produced in the room around this frequency, the panel is excited into resonance and thereby absorbs some of the sound energy in the room by the dissipation of internal losses within the panel, and by the radiation of sound from the enclosed side of the panel into the air space. The damping of the resonance of the absorber can be adjusted by placing a blanket of glass fibre material within the air space as shown in Fig. 8.34.

If the absorber is underdamped, the resonance of the panel may become audible, giving a characteristic 'hang-over' at the resonance frequency. Conversely, overdamping can render the absorber inefficient. In the example shown in Fig. 8.34, the damping blanket is purposely fixed to the room surface so that it does not come into contact with the vibrating panel. This scheme usually provides adequate damping, whereas the attachment of the blanket to the vibrating panel or the filling of the complete air space with absorbent reduces the effectiveness of the absorber.

When correctly damped, the panel absorber provides absorption over a fairly broad frequency range centred on the resonance frequency. The absorption coefficient for low-frequency panel absorbers is then typically of the order of 0.5 around resonance. By using panel absorbers of several different resonance frequencies, it is thus possible to adjust the total room absorption as a function of frequency. Panel absorbers are ideally suited to the absorption of frequencies in the range 50–350 Hz, where the construction proves to be simple using readily available materials. For example, a panel of 6 mm thick plywood ($M = 3$ kg/m²) fixed against an air space of 50 mm in depth provides absorption centred on 160 Hz.

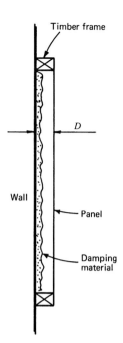

Timber frame

D

Wall

Panel

Damping
material

Figure 8.34. A simple panel absorber construction. To be most effective these absorbers must be correctly damped and the air space properly sealed.

Panel absorbers may also be used for absorbing frequencies below 50 Hz, but heavier panels and larger air spaces are then required. One way of providing these is to construct a second room within the first using a timber framework and heavy panels. By making this inner shell smaller than the outer room, the required air spaces behind the panels can be defined. A typical form of construction of such panels is shown in Fig. 8.35. The resonance frequency of this construction is of the order of 20 Hz. Low frequencies effectively pass through the inner shell and reflect off the more rigid boundaries of the outer shell. The dimensions of the outer shell thus still define the frequencies of the room modes at low frequencies. At frequencies above about 50 Hz, the inner shell starts to become an effective boundary, and thus the comments made earlier in this section regarding the choice of room size and shape also apply to the inner construction. The internal surfaces of this type of construction provide a convenient means of fixing the lighter panel absorbers required for the upper bass frequencies.

Absorption at middle and high frequencies is most easily provided by using surface treatments such as fabric covering, drapes and carpet. If greater absorption at middle frequencies is required, some areas of the walls can be covered with a blanket of fibreglass, which is in turn covered with thin cloth to improve the appearance. The thickness of the blanket determines the frequency above which the absorption is most effective. For a blanket of 30 mm thickness, the absorption coefficient is approximately 0.8 for frequencies above about 400 Hz. The absorption of the blanket at high frequencies can be reduced if necessary by covering it with a perforated panel[8], or thin polythene sheeting.

As well as absorption, it is necessary to consider the question of diffusion. This is the process of sound being reflected from irregularly shaped surfaces, so that it is diffused uniformly throughout the room. This helps to break up the early reflections and prevent echoes, thereby reducing coloration. It can also add a greater sense of spaciousness and liveness to the music. A good diffuser has a roughness of the same order of magnitude as the wavelength being reflected, lower frequencies being scattered only by large surfaces or deep recesses. A broadband diffuser therefore has

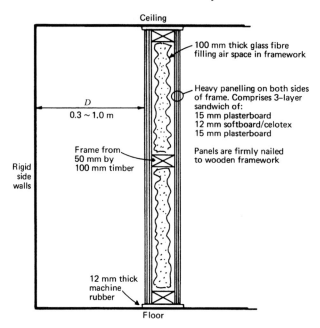

Ceiling

100 mm thick glass fibre
filling air space in framework

Heavy panelling on both sides
of frame. Comprises 3-layer
sandwich of:
15 mm plasterboard
12 mm softboard/celotex
15 mm plasterboard

Panels are firmly nailed
to wooden framework

D
0.3 ~ 1.0 m

Frame from
50 mm by
100 mm timber

Rigid
side
walls

12 mm thick
machine
rubber

Floor

Figure 8.35. Panel absorber construction for providing absorption at low frequencies. This form of partitioning, often referred to as 'Camden Partitioning', can be used to construct a complete self-standing room shell which defines the airspace between the panels and the outer rigid room shell.

surfaces of different dimensions. Good diffusion also requires a uniform distribution of diffusers around the room. However, studies suggest that a uniform diffusion may not be wholly necessary, and it is more important that diffusion is present to the side and behind the listener, to create a sense of space. Absorbers, diffusers and reflectors are all used in control room design[32].

8.6 Frequency response equalization

In both domestic and professional sound-reproduction systems, equalizers are often employed to alter the tonal balance of the reproduced sound. In domestic systems, the equalizer usually takes the form of bass and treble controls on the amplifier, while in professional applications it is common practice to make use of graphic equalizers which provide adjustment of the frequency response in octave or third-octave wide intervals. Because these variations in the frequency response are caused by the interference of sound waves which arrive at different times and with different frequency contents at the listening position, they cannot be untangled and thereby corrected by means of a simple time-invariant equalizer.

Because acoustic problems in listening rooms are unfortunately a common feature, the desire to effect a cure to such problems without recourse to alteration of the room construction has led to the development of a number of digital signal processing (DSP) systems for use as room equalizers. Systems of this type are able to equalize correctly the sound-pressure response at the listening position by taking into account the arrival times and spectra of the direct and reflected sound waves. The growing interest in digital audio technology may make the use of such adaptive digital filters commonplace in the future.

Acknowledgements

The author is indebted to Gert Jan Vogelaar of PolyGram Recording Services, Sound Labs, Baarn, The Netherlands, for his helpful comments on the text and for his patience in introducing the author to many of the practical aspects of listening room design.

Thanks are also due to B&W Loudspeakers Ltd for their provision of computing and measurement facilities, and for their encouragement to experiment with the design of listening rooms.

References

1. KINSLER, L E and FREY, A R, *Fundamentals of Acoustics*, 2nd edition, John Wiley, New York (1962).
2. PIERCE, A D, *Acoustics*, McGraw-Hill, New York (1981).
3. STRUTT, J W (Baron Rayleigh), *The Theory of Sound*, Vol. 2, 2nd edition, Macmillan & Co., London (1896).
4. SABINE, W C, *Collected Papers on Acoustics*, Harvard University Press, Cambridge, MA (1922).
5. EYRING, C F, 'Reverberation time in dead rooms', *JASA*, **1**, No. 2, part 1, 217–221, Jan. (1930).
6. MILLINGTON, G, 'A modified formula for reverberation', *JASA*, **4**, No. 1, part 1, 69–82, July (1932).
7. SETTE, W J, 'A new reverberation time formula', *JASA*, **4**, 193–210, Jan. (1933).
8. KUTRUFF, H, *Room Acoustics*, 2nd edition, Applied Science Publishers, London (1979).
9. BILSEN, F A, 'Thresholds of perception of repetition pitch', *Acustica*, **19**, 27–32 (1967/68).
10. WATERHOUSE, R V, 'Output of a sound source in a reverberation chamber and other reflecting environments', *JASA*, **30**, No. 1, 4–13, Jan. (1958).
11. JACOBSEN, O, 'Some aspects of the self and mutual radiation impedance concept with respect to loudspeakers', *JAES*, March (1976).
12. BALLAGH, K O, 'Optimum loudspeaker placement near reflecting planes', *JAES*, **31**, No. 12, Dec. (1983).
13. ALLISON, R F, 'The influence of room boundaries on loudspeaker power output', *JAES*, June (1974).
14. ALLISON, R F, Influence of listening rooms on loudspeaker systems', *Audio*, **63**, No. 8, 36–40, Aug. (1979).
15. SALAVA, T, 'Performance criteria for loudspeakers in rooms', AES preprint No. A3, presented at the 47th AES Convention, Copenhagen (1974).
16. ADAMS, G J, 'Adaptive control of loudspeaker frequency response at low frequencies', AES preprint No. 1983, presented at the 73rd AES Convention, Eindhoven (1983).
17. GROH, A R, 'High-fidelity sound system equalization by analysis of standing waves', *JAES*, **22**, No. 10, 795–799, Dec. (1974).
18. QUEEN, D, 'The effect of loudspeaker radiation patterns on stereo imaging and clarity', *JAES*, **27**, No. 5, 368–379, May (1979).
19. QUEEN, D, 'Speaker directivity', *Audio*, **63**, No. 9, 37–41, Sept. (1979).
20. HEYSER, R C, 'Acoustical measurements by time delay spectrometry', *JAES*, **15**, 370 (1967).
21. STAFFELDT, H, 'Measurement and prediction of the timbre of sound reproduction', *JAES*, **32**, No. 6, 410–414, June (1984).
22. SALMI, J, 'A new psychoacoustically more correct way of measuring loudspeaker frequency response', AES preprint No. 1963, presented at the 73rd AES Convention, Eindhoven (1983).
23. GILFORD, C L S, The acoustic design of talks studios and listening rooms', *JAES*, **27**, No. 1/2, 17–29, Jan./Feb. (1979).
24. GILFORD, C L S, *The Acoustics of Small Rooms at Low Frequencies*, PhD thesis, University of London (1964).
25. LOUDEN, M M, 'Dimension-ratios of rectangular rooms with good distribution of eigentones', *Acustica*, **24**, 101–104 (1971).
26. BONELLO, O J, 'A new criterion for the distribution of normal room modes', *JAES*, **29**, No. 9, 597–605, Sept. (1981),

27. IEC Publications 268–13, *Sound System Equipment, Part 13: Listening Tests on Loudspeakers* (1983).
28. DIN Standard 45 573, Tiel 4, *Loudspeaker Listening Tests* (1979).
29. FORD, R D, *Introduction to Acoustics*, I, Elsevier Publishing, London (1970).
30. PORGES, G, *Applied Acoustics*, Edward Arnold, London (1977).
31. MLSSA *Reference Manual: Version 6.0*, DRA Laboratories, Douglas D Rife.
32. NEWELL, P, 'Monitor systems – control rooms', *Studio Sound*, March (1991).
33. VOGELAAR, G J, Polygram Recording Services, Sound Labs, Baarn, The Netherlands.

Bibliography

Stereo imaging

LEAKEY, D M, 'Stereophonic sound systems', *Wireless World*, 154–160, April (1960).
VANDERLYN, P, 'Auditory cues in stereophony', *Wireless World*, 55–60, Sept. (1979).
BERNFIELD, B, 'Psychoacoustics of sound localization – a new approach', AES preprint No. J5, presented at the 47th AES Convention, Copenhagen (1974).
BERNFIELD, B, 'Simple equations for multi-channel stereophonic sound-localization', presented at the 50th AES Convention, London (1975).
BERNFIELD, B, 'Computer-aided model of stereophonic systems', AES preprint No. 1321, presented at the 59th AES Convention, Hamburg (1978).
BLOOM, P J, 'Creating source elevation illusions by spectral manipulation', preprint No. E1, presented at the 53rd AES Convention, Zurich (1976).
COOPER, D H, 'On acoustical specification of natural stereo imaging', preprint No. 1616, presented at the 65th AES Convention, London (1980).
GARDNER, M B and GARDNER, R S, 'Problem of localization in the median plane: effect of pinnae cavity occlusion', *JASA*, **53**, No. 2, 400 – 408 (1973).
GARDNER, M B, 'Some monaural and binaural facets of median plane localization', *JASA*, **54**, No. 6, 1489–1495 (1973).
DE BOER, E, 'Auditory physics, physical principles in hearing theory, I', *Physics Reports* (Holland), **62**, No. 2, June (1980).
BLAUERT, J, *Spatial Hearing*, The MIT Press, Cambridge, MA (1983).

Absorbers

BURD, A N, 'Low-frequency sound absorbers', BBC Research Dept Report, No. 1971/15, May (1971).
BBC, 'Guide to Acoustic Practice', May (1975).

Equalization

ADAMS, R W and TYLER, L B, 'Equalizing system', UK Patent Application, No. 2,068,678, Aug. (1981).
TRAON, J A, 'Systeme de correction acoustique automatique', French Patent, No. 2,413,009, Dec. (1977).
MORTON, R A, 'Method and equalization of home audio systems', US Patent, No. 4,306,113, Dec. (1981).
PAUL, J E, 'Automatic digital audio processor – ADAP', AES preprint No. 1266, presented at the 58th AES Convention, New York (1977).
FRYER, P A and LEE, R, 'Use of tapped delay lines in speaker work', AES preprint No. 1588, presented at the 65th AES Convention, London (1980).
CHASE, P, 'The equalization of room acoustic response in sound reproduction systems', Undergraduate Project Report, Institute of Sound and Vibration Research, University of Southampton, May (1985).
FARNSWORTH, K D, NELSON, P A and ELLIOT, S J, 'Equalization of room acoustic responses over spatially distributed regions', *Proc. IOA*, **7**, Part 3, 245–252 (1985).

9 The room environment: problems and solutions

Philip Newell

9.1 Introduction

The previous chapter has indicated the many ways in which the acoustic properties of any given room can affect the performance of any loudspeaker placed in it, whether this performance is being measured or judged by ear. Certainly in the domestic situation, which is where the vast majority of loudspeakers end up, the complex acoustic problems are probably beyond solution. Each listener, ordinary consumers and dedicated audiophiles alike, will need to locate the loudspeaker(s) and the chosen listening position for the best compromise between conflicting standing wave effects, flutter echoes, bass boosts and the rest.

Chapter 13 describes how special attention to loudspeaker and listener positioning, in a room, having carefully controlled acoustics and dimensions in line with the recommendations of the relevant IEC and AES Standards, can raise the subjective evaluation of loudspeakers almost to the precision levels of objective measurements. Here we look at another special type of room, that used for monitoring purposes by professional broadcasting and recording engineers – using the specialized types of loudspeaker and working practices described in Chapter 11.

9.2 Room equalization

Early approaches to the problem of bringing under control the loudspeaker/room interface for programme quality monitoring began with relatively simple electronic devices. Somewhere around 1970, the Altec Lansing Company began selling their Acousta-Voice systems of room equalization and, within a short space of time, the practice of 'Voicing' a room came into common usage in recording studios. By the mid-1970s, the majority of top studios around the world were using monitor equalization to try to achieve a more standardized frequency response at the listening position. This was almost invariably set up by feeding a pink noise signal into each loudspeaker system in turn, and adjusting the equalizers to give the desired response at the listening position (usually flat, with a slight top-end roll-off) as checked on one of the newly available third-octave, real-time analysers.

The development of solid-state electronics had made possible the production of portable analysers, spanning 20 Hz to 20 kHz in third-octave bands, which displayed the output from a test microphone or an electronic input in real-time (i.e. as it happened). Previous methods of room measurement had often involved the use of bulky equipment which measured each band sequentially. The overall response usually had to be analysed some time after the event (hence not in real-time). However, the apparent simplicity of operation of this portable equipment led to its use by non-acousticians, who often misused it in totally inappropriate circumstances.

By the late 1970s, the warning bells began to sound when engineers and producers

started to realize on a widespread basis that rooms which were supposedly equalized to within very tight limits were still sounding very different from each other.
Over the next five years, the whole concept of monitor equalization came into serious
question, but by that time the idea of professional studios using room equalization
had become so deep in the folklore of audio that the misuse of such devices continues
to this day. Although very few top-line studios now use room equalizers, they can
still be found lower down the scale of things in recording studios staffed by people

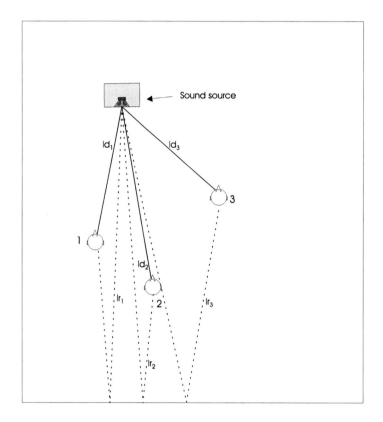

Examples for 344 Hz and 100 Hz

Position 1	ld=5.7 m	difference=10.2 m
	lr =15.9 m	relative phase@344 Hz=172°
		@100 Hz=267°

Position 2	ld=7.5 m	difference=6.7m
	lr =14.2m	relative phase@344 Hz=252°
		@100 Hz=158°

Position 3	ld=5.7 m	difference=12.5 m
	lr =18.2 m	relative phase@344 Hz=180°
		@100 Hz=228°

Figure 9.1. Position-dependent interference effects due to phase relationship of direct and
reflected waves in a reflective room. (ld = direct pathlength, lr = reflected pathlength.)

of lesser experience. Rather perversely, the studios which do still use these devices are often the ones with less acoustically controlled room designs, and it is in these rooms where the use of equalizers is least appropriate, as we shall see!

9.3 Correctable and non-correctable responses

The general rules for the use of loudspeaker/room equalization are: (a) when a room affects the response of a loudspeaker directly, such as by loading the diaphragm, then corrective equalization *can* be used, (b) when the effect of a room is indirect, such as by the superimposition of complex reflections and resonances on to the direct signal, then equalization will *not* tend to be corrective in any overall manner, and will be as likely to create as many response disturbances as it solves. The two situations may be described as minimum-phase, and non-minimum-phase, respectively.

A response modification such as the low-frequency boost experienced when flush-mounting a loudspeaker in a wall is a modification which takes place more or less simultaneously with the propagation of the sound waves from the source. In the case of flush-mounted loudspeakers, the wall provides a block to the radiation which, at low frequencies, would otherwise tend to be omnidirectional. If the pressure wave cannot travel behind the loudspeaker, then the pressure is concentrated in the forward direction, so the normal forward propagation is augmented by the would-be rear radiation. What is more, though, the constrained radiating space restricts the ability of the air to move out of the way of the diaphragm movements, and thus increases the load on the diaphragm, which in turn increases the work done, and thus increases the radiated power as compared to free-space radiation. In either case the effect is virtually instantaneous, it is equal throughout the space into which it radiates, and it can be considered to be a part of the actual loudspeaker response in those given conditions. A response modification of this nature is minimum-phase and *can* be equalized.

On the other hand, a non-minimum-phase response modification would be produced by the multiple reflections from room boundaries superimposing themselves upon the direct sound from a loudspeaker. In such cases, every different listening position will receive a different balance of direct and reflected sound from the boundaries, and no causal inverse equalization filter could correct the response at all locations. Each frequency will also exhibit a different reflected wave phase relationship to the direct signal at each position, as shown in Fig. 9.1.

Note that the term 'minimum-phase' relates to how the amplitude and phase responses track each other and has no relevance to the absolute quantity of a phase change. Essentially, a minimum-phase response is one where every change in the amplitude response has a corresponding change in the phase response, and vice versa. When the restoration to flatness of either response does *not* restore the other, the response is said to be 'non-minimum-phase', and cannot be corrected by a causal inverse filter. The degree of non-minimum-phase deviation is known as 'excess phase' and tends to build up with the summation of many types of time-shifted signals, or the re-combination of filters.

Figure 9.2 shows a typical response modification from a minimum-phase low-frequency boost caused by the flush-mounting of a loudspeaker. The relative responses of frequencies x and $2x$ will be the same throughout the room, and so can be corrected by equalization of the loudspeaker drive signal. Figure 9.3, on the other hand, shows the response of a non-minimum-phase effect, and furthermore, the relative responses of frequencies x and $2x$ are different at each position. Clearly, if they are different at each listening position, they cannot be universally corrected by equalization of the loudspeaker drive signal.

Two other observations can be made from Figs 9.2 and 9.3. Clearly the former can be modelled by relatively simple analogue filter circuits which can produce a mirror image of the amplitude *and* phase characteristics of the disturbed response, and return the transfer function to its original free-field response. However, the response of Fig.

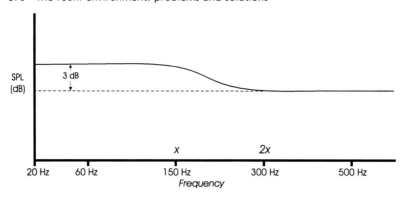

Figure 9.2. Example of a minimum-phase effect such as flush-mounting a loudspeaker, showing the boost to be expected at low frequencies when a loudspeaker designed for free-standing is flush-mounted. The dotted line indicates the low-frequency output when free-standing. The relative amount of boost will be the same at any position in the room, irrespective of any other irregularities the room may add.

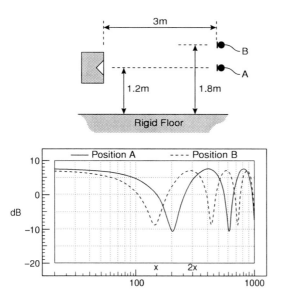

Figure 9.3. Example of a non-minimum-phase effect such as created by boundary reflections, showing the measured response at two different microphone positions in a reflective room. The signal from the loudspeaker had a flat response. At Position A the level difference between frequencies x and 2x is about 3 dB, whereas at Position B it is about 15 dB. No loudspeaker equalization can correct for both differences simultaneously.

9.3 is very difficult to mirror with analogue filters, and even perfect amplitude correction could not restore the original phase response. Different problems are posed by continuous (steady-state) musical sounds and sharp transients, and it is in such areas that simple third-octave equalization of monitors fails badly, though a closer approach to the desired result becomes possible with digital signal processing.

9.4 Digital correction techniques

Adaptive digital filtering can very accurately model the inverse responses necessary to correct either the minimum- or non-minimum-phase components of transient or steady-state signals, within certain limits. By means of measuring microphones, modelling delays (to allow the implementation of acausal (i.e. effect before cause) correction filters) and adaptive filtering processes, digital systems can be made to 'learn' what a given room will do to a loudspeaker response, and apply suitable corrections. Digital filters can be adjusted to give almost absolute correction in the amplitude, phase and time domains at one point in the room, or produce less accurate correction over a wider area. However, they cannot correct to a high degree of accuracy over a large area and, in any case, all corrections in one area are gained at the expense of response degradation elsewhere. The confusion set up in the reflective field of a room is ultra-complicated, and can only be dealt with properly at the boundaries themselves. This is why acoustic engineering is such a fundamental part of good listening room design. An example of correction by digitally adaptive filtering is shown in Fig. 9.4.

9.5 Related problems in loudspeakers

It should also be noted that there are often both minimum- and non-minimum-phase problems in loudspeakers, themselves, which are then added to the room problems. Loudspeakers inevitably exhibit internal reflection and resonance problems within both the cabinets and the drive units, and any non-minimum-phase disturbances can lead to narrow-band response irregularities which cannot be equalized out for the same reasons that room boundary effects cannot be equalized. The summation of the outputs of multiple drive units in multi-driver loudspeaker systems is a particularly common source of phase problems due, for example, to group delays in the crossover network filters (see Chapter 5). As with room response problems, however, the gradual boosts or roll-offs caused by the electromechanical characteristics of the drivers or their enclosures are usually of a minimum-phase nature, and *can* be equalized effectively.

Any time response (waveform) can be represented by a unique combination of amplitude and phase responses (as in the Fourier Transform) so any disturbance in either of these will inevitably affect the time response, and is particularly serious in the case of transients. Figure 9.5 shows an example of how a transient spike can be absolutely destroyed simply by manipulating the relative phases of the component frequencies. This shows very clearly why attempts to correct responses in the amplitude domain only, whilst allowing phase shifts to occur, can be disastrous to the transient response of any system to which it may be applied. Any gradual roll-off in the high- or low-frequency response of a loudspeaker does not involve reflections or response delays, but will still involve phase shifts, and will consequently affect the transient response. As these are minimum-phase problems, however, correction of the amplitude response by electronic means will apply a phase shift in the opposite sense, and thus tend to restore the phase and time responses to their correct values.

9.6 Equalization in auditoria

Stages and auditoria are also rooms and, though different priorities may dominate the control measures taken during live events, the same laws of physics are still at work. The stage monitor engineer may use graphic equalizers to help avoid feedback, but each correction will be highly dependent upon the relative position of the loudspeakers and microphones. For fixed microphones, such corrections can often be effective but, for a roving (hand-held) microphone, every movement will change

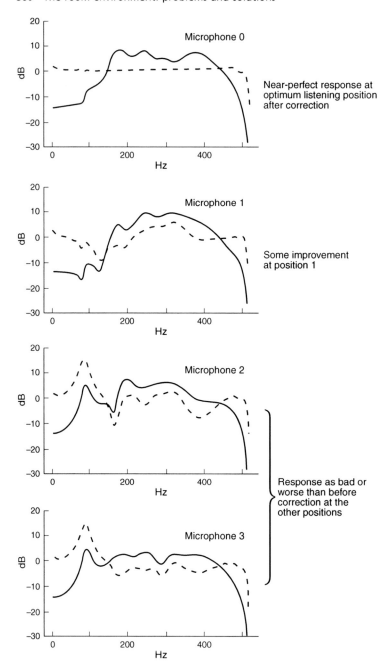

Figure 9.4. Room correction by means of digital active signal processing (after Elliott and Nelson). At position 0, the designated listening position, the correction (shown by the broken line) can be made almost perfect. However, for other positions in the room, the response tends to be worsened as position 0 is corrected.

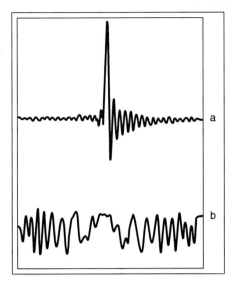

Figure 9.5. One period of waveform made up of 31 equal-amplitude harmonics. Top: all harmonics in approximately zero phase. Bottom: same amplitude spectrum, but phase angles 0 or π rad selected to minimize 'peak factor' (both waveforms are drawn to the same scale). These two waveforms sound very different at fundamental frequencies below 200 Hz, in contradiction to Ohm's Law of Acoustics which maintains that the ear is 'phase deaf'. The author has succeeded in playing simple melodies by changing phase angles of selected harmonics of the top waveform while keeping their amplitudes fixed ('phase organ'). (The figure and description are taken from Schroeder's *Models of Hearing*).

the pattern of nodal and anti-nodal regions at various frequencies. This can sometimes lead, during sound checks, to so many 'problem' frequencies being cut that what remains is very unnatural sounding and unhelpful to the musicians. This factor has led to an increase in the use of in-ear monitoring. The majority of musicians probably do not realize why their old stage monitor loudspeakers were so problematical, but they have nonetheless been glad to dispense with them.

In auditoria, equalizers should be used only sparingly because, as in any room, their effects are position dependent. All too frequently, using a measuring microphone, spectrum analyser and equalizer at the mixing position leads to a great sound for the front-of-house mixing engineer, but produces a poor sound balance for the 20 000 people who paid for seats in other parts of the hall. If a 'difficult' hall acoustic is encountered, then judicious and well-considered correction may, on many occasions, benefit all concerned (finding another hall may be out of the question) but such corrections are only approximate attempts at reducing the problem and will only partly succeed.

Unfortunately, the blame often lies with the architect of the hall, who may have had only a small knowledge of acoustics. Multi-use halls are particularly difficult to design and can be prohibitively expensive when it is remembered that a choral performance may demand 2–3 seconds reverberation time but a rock concert would, ideally, need only about one second. Such changes in the acoustic performance of large spaces are not easily achieved. Sound reinforcement engineers must often therefore do what they can with what they have been given, so one should not be too critical of their efforts (see Chapter 10).

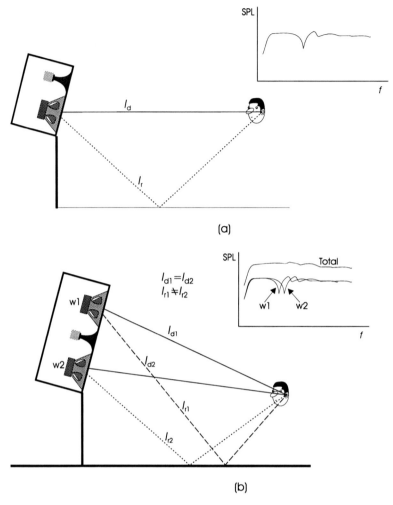

Figure 9.6. Floor reflection effects: (a) two-way loudspeaker response shows irregularities; (b) mounting a second low-frequency driver above the tweeter displaces the two irregularities due to the different reflected path lengths. The total response shows an almost complete absence of these disturbances.

9.7 An example of simple acoustic equalization

Before leaving this subject, we can look at a practical, acoustic engineering fix to a common problem. Figure 9.6(a) shows the effect of a floor reflection on the response of a two-way monitor system. The irregularities in the response are due to the combination of a time-shifted reflection with the direct signal. Figure 9.6(b) shows the smoothing effect of a second woofer, mounted vertically above the lower one. This is an example of an acoustic solution to an acoustic problem. The use of a third-octave equalizer on the response of the single woofer system would, at best, produce a response as shown in Fig. 9.7, which is less satisfactory and would create more disturbance of the time response waveform. Of course, another acoustic solution would be to make the floor out of a pile of mattresses, but perhaps the use of a second bass driver is a more practical solution[1,2].

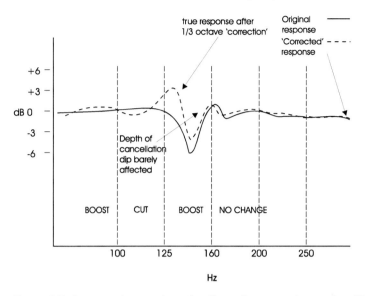

1/3 octave band centre frequency

Figure 9.7. Attempted correction of a floor dip to produce a flat 1/3 octave analyser reading is not a solution to the problem. Finer analysis will reveal a true response more like that shown by the broken line. The response would read flat on a 1/3 octave representation because there is equal total energy around each centre frequency.

9.8 Acoustic solutions

Now that we have looked at some of the problems, and the limitations of electrical solutions, we can begin to consider the ways in which acoustic engineering can be used to try to deliver as 'desirable' a loudspeaker signal as possible to the ears of the listener(s). The word 'desirable' is used deliberately here because total accuracy of the transfer function is not always what is required.

Here we are principally considering rooms of up to around 1000 cubic metres in volume. When rooms become *much* larger than this, however, say above 20 000 cubic metres, entirely new rules come into force. In very large rooms, statistical analysis can often be used in terms of the calculation of the quantities of diffusive, reflective, or absorbent surfaces which can be used to create the desired reverberation times. The exact positioning of the different surfaces is not always very important in such spaces, but, in smaller rooms, the localization of the surface effects become very noticeable to the ear. Traditionally, the control of reverberation time has been a prime objective in acoustic design, but true spatially independent reverberation can never exist in small listening rooms, either because of the localized surface effects or because the room mode density ceases to be evenly distributed well within the audible frequency range. For true reverberation to exist, the response in all parts of the room must be uniform: a totally diffuse sound field.

Figure 9.8 shows the modal distribution in two rooms of similar proportions but different dimensions. Note that in Fig. 9.8(a), depicting a room of $4.7 \times 3.2 \times 2.5$ m, there are only four modes below 70 Hz. If we increase each dimension of the room by a factor of three, as shown in Fig. 9.8(b), then the frequency scale for room modes is divided by three. Such a room, with dimensions of $14.1 \times 9.6 \times 7.5$ m would have over 50 modes below 70 Hz, which would result in the perceived response shown by the dashed line. It can clearly be seen by the comparison of the dashed lines in

(a)

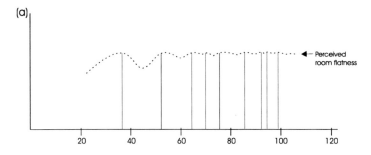

20	40	60	80	100	120

(b)

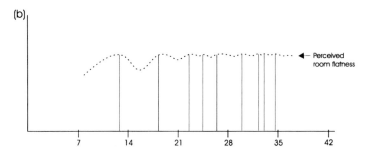

7	14	21	28	35	42

Figure 9.8. Perceived response in rooms with significant modal activity: in (a) a roller-coaster effect begins below 60 Hz; room (b) has similar proportions but dimensions multiplied by three, and the onset of the irregularity is delayed until below 20 Hz.

(a) and (b) (relative to the frequency scales) that the smoothness of the low-frequency response of a non-anechoic room is greatly improved as the size increases.

However, if merely changing the size of a room can have so great an effect on its response, changing the position of a loudspeaker in a room can be even more critical. In fact, there are cases where changing the position of a loudspeaker in a room can alter the relative balance of low and middle frequencies by 15 dB, or more. The situation is even further complicated by the radiation pattern of the particular loudspeaker, whether it is monopole or dipole, and the directivity of the middle and high frequency drive units. In fact, because of the magnitude and number of these variables, there will always be an artistic/subjective aspect to the ultimate choice of solutions, at least when listening to music. Some of these positional effects are described in more detail in Chapter 10.

Nevertheless, these effects have one saving grace; they are often of a minimum-phase nature, at least at the lower frequencies. This means that the acoustic boosts can be equalized, and this may yield a useful increase in the headroom of the loudspeakers and amplifiers. It tends to be the low frequencies that are power hungry, and a 9 dB electrical drive reduction at low frequencies, to compensate for the 'room gain', as Martin Colloms has called it, represents an 8-times power reduction in the demand from the amplifiers, and a huge cut in the working temperature of the voice coils, both of which tend to improve the dynamic response of a system.

Figure 9.9 shows the modal response of a typical room, and the dotted line indicates how the perceived response of the room tends to follow the average height of the modes. Because the modes are due to reflections and resonances, their energy sums with the direct radiation from the loudspeaker, and hence the overall response tends to be louder than would be perceived from the same power input to the loudspeaker in free-field (anechoic) conditions. A reasonably even modal spread can both make a loudspeaker sound louder, and extend its useful low-frequency response. The

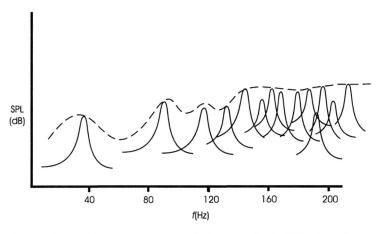

SPL
(dB)

40   80   120   160   200

f(Hz)

Figure 9.9. Modal response of a typical room: the individual modes are shown with typical frequency spreading and different amplitudes. This is more representative of reality than Fig. 9.8, which shows the modes as spot frequencies. A typical partially damped mode will be active for around 10 Hz either side of its nominal frequency.

price to be paid is a reduction in the flatness of the response, dips where they are not wanted, and a decay time of the response which can mask fine detail. However, the overall effect, whilst not technically accurate, may not necessarily be unpleasing. We shall return to this point later, when we make a clear differentiation between the room requirements for pleasant listening and quality control. The two may not always demand the same acoustic conditions.

9.9 Source pattern differences

So far in this chapter it has been assumed that the radiation pattern of the loud-speaker is that of a monopole, i.e. omnidirectional. However, other patterns exist, as shown in Fig. 9.10, where the figure-of-eight pattern of a dipole source, such as a typical electrostatic loudspeaker or an open-backed baffle (a) is compared with the circular (spherical in three dimensions) pattern of a monopole source (b) and the pattern to be expected from most cabinet loudspeakers, which become directional above about 300 Hz (c).

There are also certain types of loudspeaker, such as some of those produced by Bose and Canon, which maintain an omnidirectional 360° radiation pattern through the middle-and high-frequency ranges. All these radiation pattern differences produce radically different sound reproduction in conventional rooms, and all have their strengths and weaknesses. Even though the on-axis, anechoic responses may be virtu-ally identical, the presence of boundaries affects their responses in very different ways, and this becomes even more pronounced when the loudspeakers are used in stereo pairs. Stereo programme material generally has most of its low-frequency content distributed within the mix as central phantom images. Most cabinet loud-speakers radiate the low frequencies in all directions and, for relatively steady-state signals, unique standing wave patterns will be produced by the interference field (see Chapter 1) for each of the two separate sources. This will effectively excite many, if not most, of the room modes to varying degrees.

By contrast, the bi-directional, figure-of-eight radiation patterns of a pair of dipole loudspeakers principally tend to excite modes in one axis only. They also excite the modes most strongly when they are on the velocity antinodes (the pressure nodes)

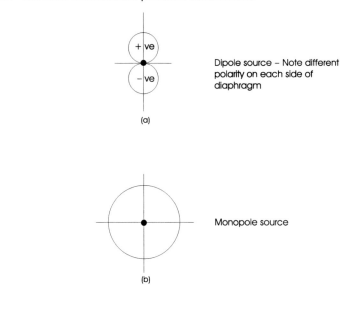

Dipole source – Note different
polarity on each side of
diaphragm

(a)

Monopole source

(b)

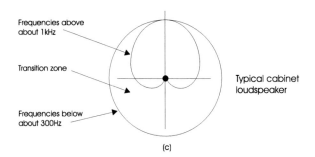

Frequencies above
about 1kHz

Transition zone

Typical cabinet
loudspeaker

Frequencies below
about 300Hz

(c)

Figure 9.10. Loudspeaker radiation patterns.

which is the exact opposite of monopole source behaviour. The modal excitation
pattern is thus very different for the two types of loudspeaker.

A further aspect of dipole radiation is that, because in almost all practical cases
the loudspeakers present their side-nulls towards the other loudspeaker in a pair,
virtually no radiation from one loudspeaker can impinge directly upon the diaphragm
of the other. Mutual coupling, as described in Chapter 1, can therefore take place
only via the less energetic reflective pathways. The far-field response of dipoles, as
compared to monopoles, therefore tends to be characterized by less modal density,
but stronger excitement of some of modes. Furthermore, except in central positions
some distance in front of the loudspeakers, the mutual coupling boost at the lower
frequencies, which is characteristic of monopole sources when radiating coherent
signals, will be absent from the combined output of the dipole pair. This can lead to
less low-frequency content in the ambient response of the room.

These factors together, plus the fact that the dipole sources cannot drive a pressure zone (see Section 9.17) no doubt contribute greatly to the widely held belief
that electrostatic loudspeakers have a tendency to be bass light, even when their
individual on-axis anechoic responses compare closely with a given monopole source.

It can therefore be concluded that care must be taken in the siting of dipole loud-speakers with regard to room boundaries and the modal distribution within a room if a relatively uniform response is to be produced at the listening location. In fact these considerations may well *dictate* the listening position.

In some circumstances, however, these complications can actually lead to advantages for dipole sources, for example in some difficult rooms their reduced acoustic coupling can be beneficial. Conversely, lack of understanding about dipole radiation and careless placement can lead to unsatisfactory in-room responses for which the loudspeakers themselves are not to blame. Injudicious placement of dipoles can result in only a few modes being strongly excited, which would produce a highly irregular response. On the other hand, the reduced modal excitation achieved by careful placement can lead to a subjective bass response that is both even and very well defined.

9.10 Listening rooms

When rooms are to be used for listening to music for serious enjoyment, almost all of the concepts discussed so far in this chapter must be taken into account. Simply placing loudspeakers in any convenient location will not suffice. It must be understood too that relatively small amounts of furniture, plus plastered walls and ceiling, will produce a highly reverberant room in which it is unlikely that pleasant listening conditions can be achieved, no matter what type of loudspeakers are used, or however carefully they are positioned. What is needed, when listening to music for pleasure, is a relatively flat amplitude response from the room/loudspeaker combination, and a quantity of reverberation which is appropriate for the music.

Except for the purposes of the quality control of recordings, it is probably true to say that all recorded music benefits from a little reverberation in the listening environment, although surround-sound systems, which can provide different ambiences for different types of recording, probably benefit from being sited in rooms with decay/reverberation times slightly lower than the shortest reverberation time in any recordings to be reproduced in them. However, for stereo reproduction in relatively small rooms, reverberation times (or more correctly in these cases, decay times) should be between about 0.3 and 0.5 second from 100 Hz to 1 kHz, with allowable reductions outside this range. In general, rock music and electronic music with high transient content tend to favour the shorter decay times, whilst acoustic music tends to favour the upper limits.

Control of the decay time can be accomplished by furnishings or by specific acoustic treatments, as described in Chapter 8, but the situation is greatly affected by the structure of the room. Bearing this in mind, some loudspeaker manufacturers, with their principal markets in their own countries, specifically tailor the low-frequency response of their products to take into account the absorption characteristics of the typical structures and furnishings of local homes. Flat anechoic chamber responses are most definitely not the objective in such cases. A subjectively flat response in the end-use environment is what is being sought.

One of the main requirements, whatever the nominal reverberation time, is that it should not vary significantly between adjacent third-octave bands, and that it should show a slight monotonic roll-off above about 100 Hz. In smaller rooms, the decay time below 100 Hz will usually fall, but in larger rooms it can be allowed to rise, though the 50 Hz reverberation time should not be more than twice that at 1 kHz. If the reverberation/decay time is not smooth, usually due to poorly damped resonances, it will colour the reproduced sound as certain bands of frequencies hang on when the rest of the music has decayed. In purpose designed listening rooms separate internal room structures can be built if the main structure is unsuitable.

9.11 Critical distance

Once the desired reverberation/decay time has been achieved, the distance between the loudspeaker(s) and the listener can be adjusted in order to achieve the desired balance of direct and reflected sounds. In fact, when different types of music require different quantities of room sound, the listening distance can be adjusted to change the balance. The critical distance is defined as that where the direct and reflected energies are equal. During critical listening, such as when searching for recording flaws, one would need to be much closer than the critical distance because reflected sounds can mask small details in the recording. However, choosing such a listening position may be possible only when the modal activity of the room is well distributed.

9.12 Control rooms

Although the function of almost all studio control rooms is to produce recordings (or live broadcasts) which will ultimately be heard in domestic rooms, there are other demands made of the monitoring systems and acoustics which lead to some very different design solutions. Control room monitoring usually needs to be capable of a degree of resolution which will show up any errors to such a degree that the recording personnel will always be one step ahead of consumers. Control rooms must also be practical working environments, where several people can work in different places in the room, yet all hear substantially the same musical and tonal balances. This implies that positionally dependent differences within the working area should be minimized.

Music mixing also tends to take place at higher sound pressure levels than that used for most domestic reproduction or standard listening tests, and this can further alter the balance of compromises. The reasons for this are numerous, and many are discussed in Chapter 11. However, in general, it can be said that most recording studio control rooms, for reasons which will be discussed in the following sections, are much more acoustically dead than most domestic living-rooms. Monitoring loudspeakers therefore need to be able to supply the extra power that this requires.

In fact, loudspeakers for recording use may need to supply up to 100 times the power of domestic loudspeakers. One hundred times the power corresponds to only four times as loud (a tenfold power increase only doubles the subjective loudness), and much of this can be soaked up in highly controlled rooms. The 100 times (20 dB) power increase poses many more electro-mechanical design problems than 'only' four times the loudness would suggest. Increased power usually means increased size, both to move more air and to lose the additional waste heat. As many loudspeakers in professional use are typically only 0.5–5% efficient, then 500 watts into such a loudspeaker would produce between 475 and 497.5 watts of heat in the voice coil. This must be dissipated in the metal chassis and by means of ventilation. Such problems can also call for different construction materials, which again moves studio loudspeakers one step further from their domestic counterparts.

9.13 The advent of specialized control rooms

Back in the days when most recording was done in mono, judicious movement of the loudspeakers and/or listening positions could often achieve a more or less acceptable sound, both in terms of critical distance factor and the avoidance of troublesome nodes or anti-nodes. When stereo arrived, however, a new set of restrictions arrived with it. The listening position became a function of the loudspeaker distances and the subtended angles between them. It was thus no longer a relatively simple matter of adjusting the location of a single loudspeaker and/or listening position, but moving a whole triangle, formed by the two loudspeakers and the listener. In other words, all three items needed to be moved not only in relation to their individual responses in the room, but also whilst maintaining relative angles and distances between them. There was also the necessity to maintain the stereo imaging, which can be very fragile. Moving the loudspeakers away from troublesome modes may disrupt the stereo balance or push them into impractical locations. Improving the siting of one loud-speaker may, through the need to maintain the triangulation, move the other loudspeaker into an acoustically worse position, or into an inconvenient place – perhaps where it needs to be moved every time the door is opened. These require-ments indicated the need for effective suppression of the temporally and spatially dependent characteristics of control rooms, but satisfactory solutions took a long time to develop.

Commercial recording studios put up with mainly poor rooms until the early 1970s, when serious efforts began to be made on an international scale to find control room designs which could be relied upon to produce recordings that would travel well; both to the outside world, and from one studio to another. This was an era when recordings really began to travel between studios, and even from country to country, during the production sequence.

One of the first big commercial efforts to produce acoustically standardized 'inter-changeable' rooms was by Tom Hidley at Westlake Audio in California, USA. These room designs spread world-wide, and were designed to have reverberation times of less than about 0.3 second. They incorporated large volumes of 'bass traps' in an effort to bring the low-frequency reverberation time into relative uniformity with the mid-band times, and also avoid the build-up of low-frequency standing waves, or resonant modes. An attempt to reduce the resonant mode problem further was effected by the use of entirely non-parallel construction to inhibit the formation of the more energetic axial modes. Monitor equalization was *de rigueur*, as the rooms were promoted on their flat frequency response at the listening position, and third-octave equalization was employed to achieve that goal. The rooms were generally quite well received at the time, and were a significant improvement on much of what was then current. Nevertheless, their responses were by no means as subjectively similar as their pink-noise, real-time spectrum analysis led people to expect. It soon became widely apparent that the control of reverberation time and the resulting combined direct/reflected (loudspeaker/room) response was not sufficient to describe the sonic character of a room. Some people already knew this, but they were a small minority, and few of them held any significant sway in the recording world. A typical design of an early Westlake-style room is shown in Fig. 9.11.

In the late 1970s, Don and Carolyn Davis were keenly investigating many acoustic and psycho-acoustic phenomena with the then newly developed TEF/TDS (Time Energy Frequency/Time Delay Spectrometry) measurement systems (see Chapter 12). TDS measurements made at Wally Heider Studios, Los Angeles, USA, and at RCA and Capitol Records, in Hollywood, had given them a lot to think about, leading them to the concept of a 'Reflection Free Zone' and the 'Live End, Dead End' control rooms.

At about the same time, Carolyn (Puddie) Rodgers was presenting new ideas about how certain room reflection characteristics could confuse the ear by giving rise to response filtering which closely mimicked the pinna (outer ear) transformations used by the brain to facilitate spatial localization[3]. This provided very explicit explana-tions of the psycho-acoustic relevance of the ETC (Energy/Time Curve) responses in the above-mentioned room measurements, and reinforced the Davis' LEDE proposals. Don Davis and Chips Davis (no relation) then wrote their seminal paper on the LEDE concept of control room design[4], and rooms based on this idea came into widespread use in the subsequent years. The concepts were further developed by Jack Wrightson, Russel Berger, and other notable designers. Sadly, however, Puddie Rodgers died before seeing the fruits of her labours fully mature.

The LEDE rooms rely on such psycho-acoustic criteria as the Haas effect (see Chapter 10) and the directional aspects of human hearing. Figure 9.12 shows the

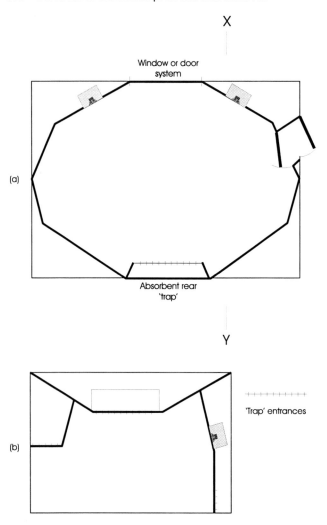

Figure 9.11. Typical Westlake/Hidley style room design of the late 1970s. (a) Plan view; absorbent bass traps strategically located to reduce the build-up of the most problematic modes. (b) Vertical cut through X-Y.

LEDE arrangement, where the front half of the room is largely absorbent, and its geometry is designed to produce a reflection-free zone around the principal listening position. The idea is to allow a clean first pass of the sound, directly from the loudspeakers, and then to allow a suitable time interval before the first room reflections return to the listener's ear. The rear half of the rooms are made diffusively reverberant, allowing the perception of a room 'life' which should not unduly colour the perception of the directly propagated information. The rooms require a diffuse reverberation effect, and proprietary diffusers such as those developed by Dr Peter D'Antonio, and marketed by his company RPG, are widely used in such rooms[5]. Strong, discrete, specular reflections are to be avoided, as they produce position dependent coloration and general response flatness irregularities.

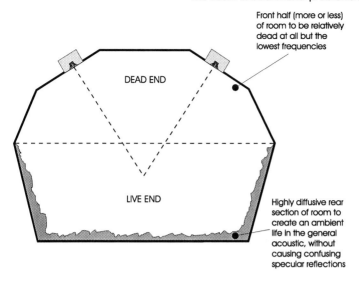

Front half (more or less) of room to be relatively dead at all but the lowest frequencies

DEAD END

LIVE END

Highly diffusive rear section of room to create an ambient life in the general acoustic, without causing confusing specular reflections

Figure 9.12. Concept of a Live-End, Dead-End control room in which a reflection-free zone is created around the listeners.

Concurrent with the introduction of the LEDE concept, Jensen in Europe was producing rooms as depicted in Fig. 9.13. These rooms used 'sawtooth' absorbers, which tended to absorb much of the incident wave from the loudspeakers, but had reflective surfaces at angles where they could reflect back sounds created within the room, such as by the speech and actions of the personnel. They would not, however, cause reflections directly from the monitors. Total absorption of the incident wave was not intended, because both the Jensen and LEDE rooms still sought to maintain a room decay time on the low side of the standard domestic range.

After growing criticism of his 1970s Westlake designs, Tom Hidley took a break from studio design between 1980 and 1983. During this lay-off, he thought over the problems and came up with a new concept that he called 'The Non-Environment' room principle. In these rooms, he made the front wall maximally reflective and, except for a hard floor, made all the other room surfaces as absorbent as possible[6]. The principle sought to drive the loudspeakers into something approximating an anechoic termination. Since the monitors were set into the front wall, the latter would act as a baffle extension, but it could not reflect sound from the loudspeakers because they were radiating away from it. However, along with the floor and equipment surfaces, the wall added life to the speech and actions of people within the room, thus reducing any tendency for the rooms to feel uncomfortably dead. The principles were not unlike those of the aforementioned Jensen rooms, but were taken to a greater extreme.

By introducing more absorption into the room and limiting reflections, the ratio of direct to reflected sound is increased, and hence the levels of coloration are reduced. This is achieved, however, at the expense of any consideration being given to domestic listening acoustic conditions. The consistency between such rooms is probably greater than that between most other types of room. One very famous recording engineer/producer, who loves these rooms for mixing, freely admits that for certain types of music recording, especially when the musicians are creating music live inside the control room (as is often the case these days), he must use rooms with a more inspiring ambience. Such things are very subjective, however, and he also produces artists who love the rooms for the whole recording process. A typical "Non-Environment" construction is shown in Fig. 9.14.

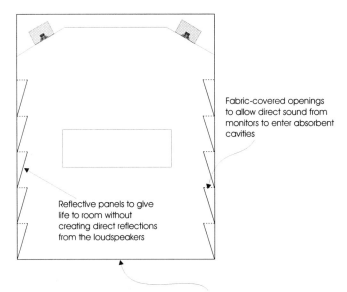

Fabric-covered openings
to allow direct sound from
monitors to enter absorbent
cavities

Reflective panels to give
life to room without
creating direct reflections
from the loudspeakers

Absorbent/reflective nature of
rear wall adjusted to suit desired
room conditions

(a)

(b)

Figure 9.13. Typical room in the style of Jensen. (a) The reflective side panels are relatively lightweight and act as low-frequency absorbers. Their angling also prevents 'chatter' between the hard-surfaced side-walls. (b) The hard front wall and flush-mounted monitors in the Jensen designed room at Sintonia Studios, Madrid, Spain.

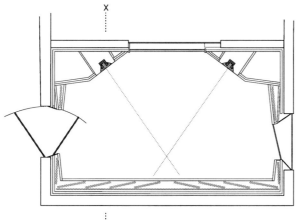

(a)

Shaded areas are wide-band absorber systems

Figure 9.14. The Non-Environment concept. (a) Plan of 'Non-Environment' control room; the shaded areas are wide-band absorber systems. (b) Vertical cut through X-Y. (c) Capri Digital, Italy: a Tom Hidley-designed control room using the Non-Environment principle. Note the front wall concept, which is essentially similar to the rooms shown in Figs 9.13(b) and 9.16, which are the inverse of LEDE.

(b)

(c)

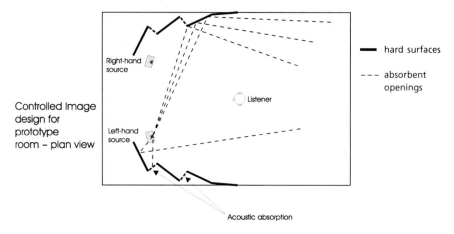

Figure 9.15. Controlled Reflection room, in the style of Bob Walker, for free-standing loudspeakers.

Figure 9.16. A Sam Toyashima control room at The Townhouse, London, having a hard front wall and highly absorbent rear wall.

Variations on the above themes are shown in Figs 9.15 and 9.16, which illustrate room design concepts by Bob Walker, at the BBC[7] and Sam Toyashima[8]. These, and others[9], are all highly successful concepts which perform well.

Note that all these room concepts must be fully understood if they are to be applied to their greatest effect, and that the principles involved are not necessarily interchangeable. Each one is a system in itself. It is the misapplication of the components of these philosophies which has led to some poor monitoring room designs. Nonetheless, there are some features which are more or less common to all of the rooms described above, and it is worthwhile looking at some of them in more detail.

9.14 Built-in monitors

In top-level control rooms, it is general practice to build the monitor loudspeakers into the front wall. Flush mounting of loudspeakers ensures that all the sound radiates in a forward direction. This avoids the problems discussed earlier, shown in reverse in Fig. 9.1, where the rear-radiations strike the wall behind the loudspeaker and return to the listener with a delay, giving rise to response irregularities due to the relative phase differences between the direct and reflected waves. None of the above control room concepts have front walls which are sufficiently absorbent at low frequencies to avoid such effects from free-standing loudspeakers, although the Walker room of Fig. 9.15 does go some considerable way towards limiting the damage.

Incidentally, flush-mounting is sometimes, wrongly, described as 'soffit' mounting. In many 1970s designs, the loudspeakers were flush mounted above soffits – a soffit being the underside of an overhang, or the ceiling of a recess. These overhangs were often built above windows, or to create recesses in which tape machines could be housed. This was never a good idea from an acoustic point of view, because the recesses could create resonant cavities. They also allowed a certain amount of rearwards sound propagation, leading to the problems described in the previous paragraph. Nonetheless, the concept was widespread and somehow, by erroneous extension, the term 'soffit mounting' has come into common use, especially in the USA, referring to flush mounting in general. Figure 9.17 shows a genuinely soffit-mounted loudspeaker, whereas the photographs in this chapter all show flush-mounted loudspeakers. Anyhow, loudspeaker positioning is a source of variability of room response, so it is a good idea for serious control rooms to standardize on a monitoring practice which has more in common from room to room, and flush-mounting fulfils this purpose.

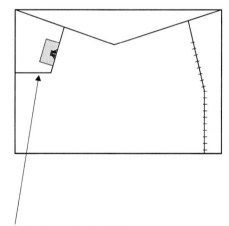

Soffit (underside of architectural overhang, such as balcony, etc.)

Figure 9.17. Soffit-mounted loudspeakers. In the example shown, the loudspeaker is mounted above the soffit. The alcoves below are often used to house tape recorders.

9.15 Directional acoustics

All the leading philosophies of control room design are directional in that the rear half of the room is acoustically different from the front. It has been found, almost universally, that such a practice is necessary for rooms to exhibit the most detailed stereo images, smoothest amplitude responses and best low-level perception (at least whilst creating a pleasant acoustic to work in), but this poses a dilemma for the design of surround-sound rooms. Building two 'front halves' of a good stereo room simply facing each other would leave the loudspeakers facing inappropriate far boundaries. However, this 'mirror imaged front-half' approach was tried for some quadraphonic rooms in the late 1970s and produced very poor results. Good surround rooms are beginning to emerge[10-13], but they are by no means universally accepted.

9.16 Scaling problems

Size differences greatly affect the response of rooms. Wavelengths in air are strictly related to frequency, and do not scale with room size. It therefore follows that a small room will exhibit a greater reflection density that a larger one of similar nature, because the sound will encounter more boundaries per second. What is more, in smaller rooms the first reflections will return with more energy because they have travelled less distance. Therefore, when control rooms are designed, the physical dimensions of the room dictate some of the treatments needed, and it is impossible simply to scale the various room design philosophies beyond certain limits. A Live End, Dead End room depends on the existence of reflection-free zones and a suitable time lapse between the arrival of the direct signal and the first returning energy from any room boundaries. If the room is too small, the required time intervals cannot be achieved, and the design will fail to function.

Non-Environment rooms, on the other hand, do not depend upon such carefully timed psycho-acoustic phenomena, but they do require considerable space for their absorber elements. The absorbers' sizes are wavelength related so, in very small rooms, especially if the structural walls are heavy and rigid and hence highly reflective, there may be insufficient space in the room to provide enough working area once the trapping is installed. An example at the other extreme is shown in Fig. 9.18.

Very small listening rooms of any design will produce problems due to the effects mentioned above, but they can also complicate issues by imposing additional loading on the fronts of the loudspeakers. The designs of most loudspeaker cabinets assume relatively free air loading. In small rooms, especially in cases where the low-frequency absorption is less than optimal, reflections can return with considerable energy. This can affect the performance of low-frequency drivers and, more especially, any tuning ports. Pressure zone loading may also have implications here, as discussed next.

9.17 The pressure zone

Below the frequency of the lowest mode which any room can support (i.e. the wavelength is twice the longest dimension of the room), there lies a region known as the pressure zone. Transition into the pressure zone is gradual, and depends to some extent on the Q of the lowest mode. Within the pressure zone travelling waves cannot exist, and the whole room is pumped up and down in pressure as the loudspeaker diaphragms move forwards and backwards respectively.

Many misconceptions exist with regard to loudspeaker response in the pressure zone. Some writers[14] say that below the pressure zone frequency, also known as the room cut-off frequency, the environment loads the sound source and the effect of this loading is to reduce the ability of the source to radiate sound into the rooms, resulting in reduced sound levels at these frequencies.

Elsewhere it is stated 'In rooms that are both physically and acoustically small, the pressure zone may be useful to nearly 100 Hz'[15]. It has also been suggested that control rooms, for example, should have their longest dimension at least equal to half a wavelength for the lowest frequency to be reproduced in the room; which for a '20 Hz' room would be around 8.6 metres. The implication often seems to be that

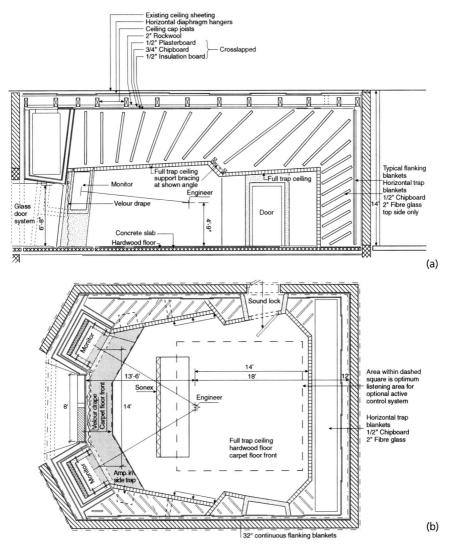

Figure 9.18. An extreme approach, aimed at controlling the room environment throughout the entire audible range, showing (a) the elevation and (b) plan view of a typical 'non-environment' absorbent control room (courtesy of Tom Hidley).
(Note: all dimensions are in feet and inches.)

in the pressure zone there will be no significant loudspeaker output. Indeed, this impression is reinforced by the use of the term 'room cut-off frequency'.

The origins of this misconception are clearly shown in Fig. 8.20 in Chapter 8, where the modal support of the combined loudspeaker/room response shows almost 10 dB of 'gain' over the anechoic response. It can also be seen that the response below the lowest mode falls back to the level of the rolling-off anechoic response, which would result in the pressure zone response being perhaps only about one-sixth as loud as the modally supported response at higher frequencies.

Obviously, however, in rooms where the modes are substantially damped, the differential between the modally supported and pressure zone regions will be greatly reduced. What is more, half-space (2π) loading, such as is approximated by flush-mounting the loudspeakers, will often serve to reinforce the output in the roll-off region and reduce the difference even more.

As for the concept of the pressure zone loading the loudspeakers, there are many things to consider. Perhaps if the room were very small, hermetically sealed, and had infinitely rigid walls, the statement would be true. Taken to an extreme, if the room were to be the same size as the cabinet loading the rear of the loudspeaker, then the loading effect could be readily understood. The reality is that the loading effect of the room is inversely proportional to the room/cabinet volume ratio. This means that a room with a cubic volume only ten times that of the loudspeaker cabinet would have only one tenth of the loading effect, which is hardly significant.

However, the mechanisms at work here are very interdependent. Below the resonance of a sealed box loudspeaker, the velocity of the cone for a given input power falls with decreasing frequency in a way that exactly offsets the tendency for the cone excursion to increase, and the effect is that the displacement stays constant with frequency. In the pressure zone, the pressure is displacement dependent, so a sealed box loudspeaker whose resonance was set at, or higher than, the frequency of the room cut-off (pressure zone frequency) would exhibit a flat response in the pressure zone. If the loudspeaker resonance is significantly below the pressure zone frequency, then the response will tend to rise with falling frequency until the resonance is reached, when the response will flatten out. Ported loud-speakers will tend to exhibit a 12 dB/octave (second order) roll-off below resonance, as opposed to their normal 24 dB/octave (fourth order) roll-off, when the port resonance is below the pressure zone frequency, although loading effects on the ports can complicate matters.

All of this assumes that the room is sealed and highly rigid. Non-rigid (diaphragmatic) walled rooms, such as are found in many timber-framed houses and most purpose-designed sound control rooms, present a totally different set of circumstances. Most timber-framed structures are, to varying degrees, transparent to low frequencies, and the pressure zone onset is defined by whatever high rigidity containment shell exists outside the inner structure. The general situation in rooms, however, is highly variable and a thorough understanding of the given structure is needed before any accurate pressure zone response prediction can be made.

Nevertheless, one or two definite statements *can* be made. Dipole radiators, including most electrostatic loudspeakers, cannot radiate sound below the room cut-off frequency, and only at very small distances from the front or rear of the diaphragm will any sound be perceived. A dipole source (see Chapter 1) will merely paddle back and forth, without creating any net change in the overall pressure within the room. The short circuit round the sides of the diaphragm will cancel out any local pressure changes due to the opposite polarity on the front and rear radiating surfaces – see Fig. 9.10(a).

The very fact that dipole loudspeakers are an exception in itself implies that the room 'cut-off' does not mean that conventional sealed box or ported loudspeakers cannot radiate useful output below that frequency. In fact, one characteristic of the response in the pressure zone is that the perceived response not only exists, but is extremely uniform both in level and with respect to position, because no modes exist to create spectral or spatial variations. Nonetheless, a relatively flat extension of the general low frequency response into the pressure zone can only be achieved in very damped (absorbent) rooms.

For these reasons, no loudspeaker system can be absolutely optimized for low-frequency performance without knowledge of the room in which it will be used. Good monitor loudspeakers must perform in real control rooms, and do the job for which they were designed. Good domestic loudspeakers must give pleasing results in less controlled circumstances. There are so many things that a room can do to modify a loudspeaker response that no simple set of idealized test specifications can truly represent a loudspeaker in normal use.

9.18 The general behaviour of loudspeakers in rooms

After discussing so many different situations and consequences relating to the use of loudspeakers in rooms, perhaps a little order can be restored by examining the behaviour of single and multiple sources in the two extremes of anechoic and highly reverberant rooms. If we also look at the behaviour of transient and steady-state signals in the same circumstances, then it should be possible to use these principles to interpret the performance of loudspeakers in most 'normal' rooms, whose properties are between the two extremes.

9.18.1 Mono sources

Figure 9.19 shows a typical set of response plots for a loudspeaker in an anechoic chamber. The individual plots illustrate the response on-axis, and at 15°, 30°, 45° and 60° off-axis. It should be evident from the plots that the high frequencies will progressively reduce in level as one moves further off-axis. If one were to move behind the loudspeaker, there would not be much high-frequency sound at all, and a glance at Fig. 9.10(c) will clearly show why.

If the high and low frequencies are equal in level on-axis, then it should be clear that in the room as a whole there will be a predominance of low-frequency energy. In fact, the areas of the high- and low-frequency patterns in Fig. 9.10(c) give a reasonable idea of the relative proportions of the radiated power in each frequency band. Nevertheless, a reasonably uniform response can be achieved within a restricted arc of ± 15°, directly in front of the loudspeakers.

If we now transfer the same loudspeaker into a reverberation chamber, the diffuse field that it creates will serve to integrate all the responses in all directions and, since the low frequencies are radiating in all directions, they should also be radiating more total power into the room. Figure 9.20 bears this out, and shows the significantly higher level of low frequencies that would be perceived in the far-field at any point in a highly reverberant room. In fact, on-axis at quite short distances, the response would still largely be as shown in Fig. 9.19. Figures 9.19 and 9.20 represent actual measurements of the same loudspeaker taken by Dr Keith Holland in the anechoic and reverberation chambers, respectively, at the Institute of Sound and Vibration Research in Southampton, UK. All music listening rooms fall somewhere between these extremes.

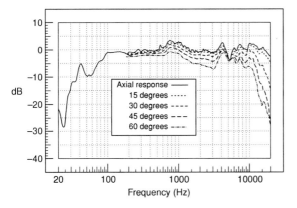

Figure 9.19. On-axis and off-axis responses of a loudspeaker in an anechoic chamber.

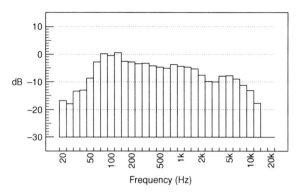

Figure 9.20. Total power response of the same loudspeaker as in Fig. 9.19, but made in a reverberation chamber.

9.18.2 Stereo sources

For the sake of avoiding confusion by introducing too many variables, let us consider two ideal point monopole sources, representing a stereo pair of perfectly flat, omni-directional loudspeakers. When reproducing a central mono image, the combined directivity would be as shown in Fig. 9.21(a). For a single, mono source, the radiation pattern is spherical, and equal at all frequencies (Fig. 9.21(b)) and the frequency response at any point in the room would be flat. Unfortunately, as shown in Fig. 9.22, the *off-axis* frequency response of a phantom central image from a stereo source is anything but flat[16] (it could not be expected to be so) but has in fact a comb-filter shape. In Fig. 9.23 the central in-room image created by the four speakers in a surround system[17] takes things a whole stage further. It can be seen that adding more sources can make the situation rapidly worsen. Only on-axis in an anechoic chamber can a phantom central stereo image mimic a discrete central image from a centrally positioned loudspeaker. In a reflective room this can never be, because *all* off-axis radiation will exhibit a comb filtered response. All reflections from points other than on the central plane between the loudspeakers will therefore also exhibit comb filtered response. This means that the reflective field will exhibit a different frequency balance from a central mono source than from a stereo phantom central image.

In rooms with significant modal activity, it must also be remembered that the modal pattern is very much dependent on the point(s) from which the room is driven. A central mono source has one drive point, whereas a phantom central image has two points of origin, separated by the distance between the loudspeakers, and neither source corresponds to the position of the phantom image. The two sources generating a central image thus drive the room very differently from a single mono source and, in semi-reverberant spaces, cannot be expected to sound the same as a mono source at any point in the room. Only on, or close to, the central plane in anechoic spaces will the discrete and stereo sources produce identical results. As we shall see in the next section, even these two cases may not be perceived to *sound* identical.

9.18.3 Steady-state versus transient performance

On steady-state signals, the situation can be very different between anechoic and reverberant rooms. In anechoic conditions, the interference pattern from the left and right loudspeakers will sum by 6 dB on the central plane and for a distance on either side which will depend on wavelength. The 3 dB of 'extra power' superimposed upon the 3 dB radiated power increase is gained at the expense of lower SPLs elsewhere in the room. Away from the central plane, the interference patterns will

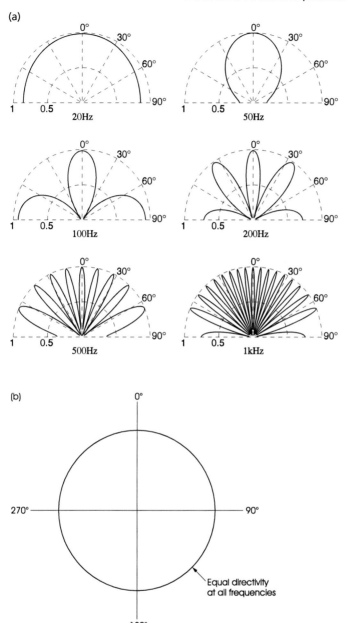

The above plot represents the polar response of a perfect, discrete, mono source.

Figure 9.21. Discrete versus phantom source patterns. (a) Combined directivity of a stereo pair of loudspeakers driven by a common signal – half-space shown for brevity. (b) Polar response of a perfect loudspeaker in an anechoic chamber.

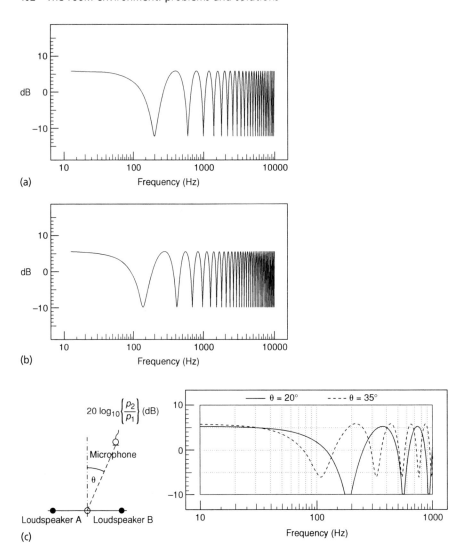

Figure 9.22. Comb filtering: off-axis frequency response of a pair of perfect loudspeakers when producing a central phantom image in an anechoic chamber. (a) 1 m off-axis; (b) 1.5 m off-axis; (c) at 20° and 35° off-axis.

produce comb filtering, as shown in Fig. 9.22, the nature of which is position dependent. Off-axis also there will be additional low-frequency power due to mutual coupling. That this is not perceived on-axis as solely an LF boost, but as an overall boost, can be considered to be a result of an overall directivity change when the two spaced drivers are operating together.

In highly reverberant conditions, a totally diffuse sound field will build up. As we are still considering perfectly flat omnidirectional loudspeakers here, a similar frequency distribution will be radiated in all directions, and so the response at any point in the room will be the same, and corresponds to a simple power summation of the two sources, producing a 3 dB rise in level, plus whatever low-frequency boost

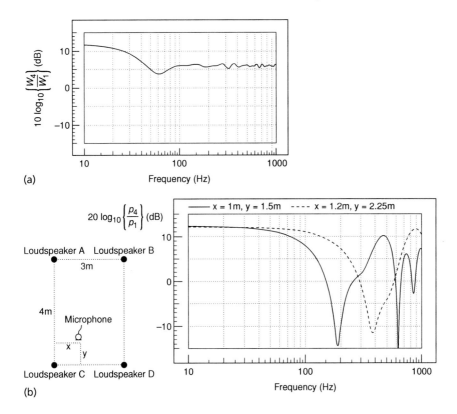

Figure 9.23. Frequency response of four perfect loudspeakers radiating the same signal. (a) Total power response in a highly reverberant room. (b) Combined frequency response of four omnidirectional loudspeakers at two different positions in an anechoic chamber.

results from mutual coupling. Audibly, though, things will sound very confused, with one note blurring into the next. In practice, many listening rooms of a reflective nature tend towards the 3 dB central summation, and some repercussions of this fact will be discussed in Section 9.18.4, but we should first look closely at the area of 6 dB summation in the more anechoic conditions, because the width of this region is very frequency dependent. At 20 kHz, the region of perfect summation will be only about 1 cm (half a wavelength), but at low frequencies it could rise to many metres. At around 2 kHz, for a listener located on the centre line, there would be an effective cancellation due to the spacing of the ears. Figure 9.24 shows how this is caused, as the path length differences are not the same from each loudspeaker to each ear. This is yet another mechanism by which a centrally panned image, from a pair of loudspeakers, differs in the way that it arrives at the ears, compared to the arrival of the sound from a discrete, central loudspeaker.

9.18.4 Transient considerations

For transient signals, on the central plane of a stereo pair of loudspeakers, the pressures will sum to produce a single pulse of sound, 6 dB higher than that emitted by each loudspeaker individually. At all other places in the room, different distances from the two loudspeakers create arrival time differences, and double pulses will result. This effect is clearly shown in Fig. 9.25. Although it may seem to an observer

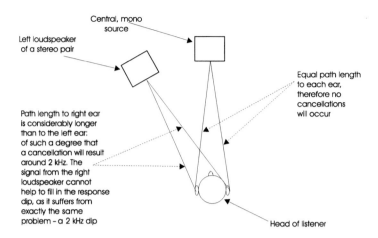

Central, mono source

Left loudspeaker of a stereo pair

Equal path length to each ear, therefore no cancellations will occur

Path length to right ear is considerably longer than to the left ear: of such a degree that a cancellation will result around 2 kHz. The signal from the right loudspeaker cannot help to fill in the response dip, as it suffers from exactly the same problem – a 2 kHz dip

Head of listener

Figure 9.24. Path length anomalies for phantom central image, comparing a central, mono source and the left speaker of a stereo pair.

on the central plane that four times the power (+6 dB) of a single loudspeaker is being radiated at all frequencies, to an observer *off* the central plane this would apply only at low frequencies, where mutual coupling occurs and the combined radiation from the pair of loudspeakers is essentially from a single, combined, omni-directional source. At higher frequencies, the effect of the constructive and destructive super-imposition of the signals results in an average increase of 3 dB. The zones of coupling are shown diagramatically in Fig. 9.26. However, this still apparently leaves us with a magic extra 3 dB of power on the centre line which, for transients, cannot be described in terms of radiation impedance, because radiation impedance is a frequency domain concept, and transients exist in the time domain. So, we need to look at this further.

In the case of a perfect delta function (a unidirectional impulse of infinitesimally short duration) the points of superimposition would only lie on a two-dimensional, central plane, of infinitesimal thickness. As this would occupy no perceivable space, then no spatial averaging of the power response would be relevant. With a transient musical signal, which has finite length, there would be positive and negative portions of the waveform. At the places along the central plane where the transients cross, they would not only meet at a point, but would 'smear' as they interfered with each other over a central area, either side of the central plane. The pressures would thus superimpose and produce a pressure increase of 6 dB over a finite region on each side of the centre line. As they crossed further, they would produce regions of cancellation and show overall power losses equal to the power gain in the central region of summation. The total power would thus remain constant when area-averaged. This effect is shown in Fig. 9.27, where the average height of all the transients occurring at any one time in the room would be the same as that of a single transient emitted by one loudspeaker, but increased in number as there are two sources.

On transient signals, therefore, because of their existence as separate bursts of energy, the performance of anechoic and reflective conditions differs only in that the reflective rooms will add an ever-increasing number of reflected energy bursts of ever-decreasing energy, which will add to the overall perceived loudness. However bearing in mind that a reflective/reverberant room must, by definition, be anechoic until the first reflection arrives, the subjective perception of the transient could change over time from the anechoic to the reverberant state, depending on the room size

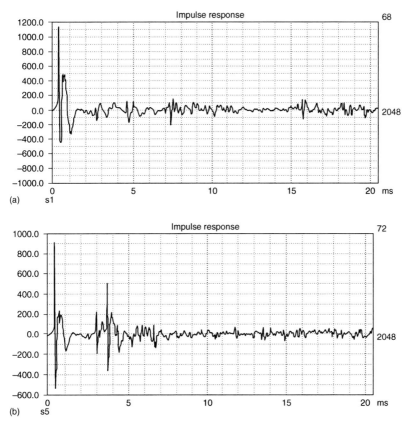

Figure 9.25. Multiple impulse effect. (a) Impulse response received from a centrally panned image from a pair of loudspeakers. Measurement taken on the common axis of both loudspeakers (centre line). The response from a central, mono loudspeaker would be essentially similar; (b) impulse response received from a position 1 metre behind (a) and 1 metre to the left of the common axis (centre line). There are *two* clear impulses, with the one arriving from the right-hand loudspeaker later in time and lower in level.

and the length of the transient. Continuous sounds (e.g. bass guitar notes) would, by contrast, be perceived more consistently in accordance with the type of room in which they were being reproduced.

9.18.5 The pan-pot dilemma

The perception of a 3 dB overall central summation (as tends to exist in most reverberant spaces) was at the root of the old pan-pot dilemma: should the electrical central position of a pan-pot produce signals which are 3 dB or 6 dB down relative to the fully left or fully right positions? Mono electrical compatibility of the stereo balance requires constant voltage, therefore each side should be 6 dB down (half voltage) in the centre, in order to sum back to the original voltage when added electrically. (Pan-pots, are potential (voltage) dividers, so it is the voltages which must sum.) On the other hand, in the case of an acoustic stereo central image in a reasonably reverberant room, it is the power from the two loudspeakers which must sum to unity in the centre, which therefore requires a condition whereby the output

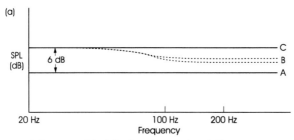

A: Response of single loudspeaker, anywhere in the room.

B: Response of a stereo pair of loudspeakers, each receiving the same input as 'A', anywhere in the room, except on the central plane: precise response may be position dependent.

C: As in 'B', but measured on the central plane.

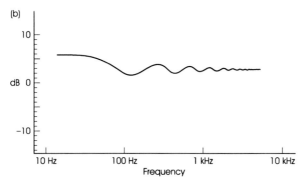

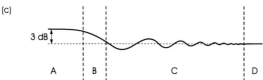

Zone A: Region where the separation distance between the loudspeakers is less than half a wavelength and where wholly constructive mutual coupling is effective.

Zone B: Region where the separation distance between the loudspeakers is less than half a wavelength, but where the mutual coupling is becoming less effective as the frequency rises.

Zone C: Region where the separation distance is greater than half a wavelength, and where the mutual coupling alternates, as the frequency rises, between being constructive or destructive.

Zone D: Region where the mutual coupling has ceased.

Figure 9.26. Mutual coupling effects – omnidirectional sources. (a) Pressure amplitude responses in an anechoic room. (b) Frequency response of a pair of loudspeakers at any position in reverberant chamber (combined power output). (c) Zones of loading: general response as in (b).

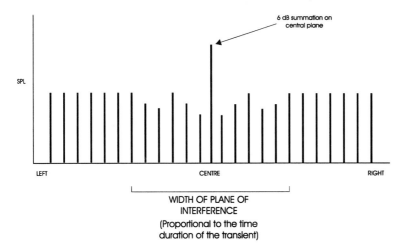

Figure 9.27. Pulse superimposition. The 6 dB (double voltage) summation of the power superimposition (3 dB) is gained at the expense of adjacent, destructive interference.

from each loudspeaker would be only 3 dB down (half power) when producing a central image. Nevertheless, below the mutual coupling frequency, Fig. 9.26(b) shows how the low-frequency response would still approximate to the −6 dB requirement, as the power summation tends to be augmented by a further 3 dB due to the mutual coupling between the two loudspeakers.

In stereo radio drama, where voices are often panned across the sound stage, −3 dB pan-pots produce a uniform level as the voice is panned from left, through centre, to right. However, if the programme is then broadcast in mono, the voice is perceived to rise by 3 dB as it passed through the centre position in the stereo mix. If the same panning movement were to be repeated for a bass guitar, then in the mono broadcast the guitar would still be subject to the same, uniform, 3 dB rise as it passed through the centre. Somewhat inconveniently, though, when heard in stereo, it would be perceived to increase only in its low-frequency content, due to the coupling between the two woofers. The degree of boost, and its frequency of onset, would depend on the distance between the left and right loudspeakers, introducing yet another variable.

As there is usually little dynamic panning of low-frequency instruments in stereo music recording, a −3 dB centre position used to be considered optimum. Yet for radio drama, where mono compatibility is an important priority for many listeners, a −6 dB centre position is preferred. Many mixing console manufacturers produce consoles with different pan-pot laws for different applications, but they usually opt for a −4.5 dB compromise, which produces only a 1.5 dB worst-case error in either instance. This compromise fits nicely with most real-life rooms in which music will be heard, because their acoustics are somewhere between reverberant and anechoic. In truly anechoic conditions, the acoustical sum on the central plane is identical to the electrical sum, at least in the region where the axial response holds true; but this is a central plane-only condition. Elsewhere in the room, the summation approximates to the panning effect at all points in a more reverberant space, though with less confusion in the sound. The repercussions of these anechoic/reverberant and steady-state/transient phenomena play total havoc with attempts to find subjectively reliable means of producing electrical fold-down systems for 5-channel/4-channel/3-channel/stereo/mono compatibility from surround mixes.

9.18.6 Limitations, exceptions and multi-channel considerations

The theoretical concept of perfect omni-directional loudspeakers breaks down badly when the question arises of where to put them. Within any room which is neither perfectly anechoic nor perfectly reverberant, boundary reflections produce an irregular frequency response. Perfectly reverberant conditions would be useless for listening to music, as much of the detail would be swamped by the reverberation. Anechoic conditions, in which fine detail is most readily perceived, provide no reflected energy, so the only sound heard by a listener is that which passes directly from the loudspeaker. Under such circumstances, however, there is no perceivable difference between a perfectly omni-directional loudspeaker and one which radiates a uniform frequency balance on an axis pointing directly towards the listener. A uniform response for ± 20° or 30° off-axis allows for some movement about the central listening position, and this is quite easy to achieve in practice. So, in anechoic conditions omni-directional loudspeakers would do nothing except waste power by radiating sound in unnecessary directions.

Just as the perceived response of a central image generated by two dynamic loudspeakers will be different from that generated by a single centre loudspeaker, an image which is generated by four loudspeakers (see Fig. 9.23) is likely to differ from that generated by two, because there could be a further mutual coupling rise of up to 3 dB. With electrostatic loudspeakers, however, this build-up is likely to be less because, as mentioned in Section 9.7, they behave less like volume velocity sources, and more like pressure sources, i.e. coupling to pressure *nodes*, and not anti-nodes. This situation can pose even more problems for anyone trying to design electronic systems to 'fold-down' multi-channel mixes into stereo or mono, while still attempting to maintain the original musical balance. As well as all the fold-down compromises caused by the effects of the room and combined loudspeaker directivity, it can now be seen that the desired fold-down can also depend on the type of loudspeakers on which the music was mixed, as well as those on which it will be reproduced. To make matters worse, the compatibility of mixes is also affected by the fact that the optimum fold-down must take into account the frequency dependence of the directionality of human hearing. The subjectively desirable level of a rear channel sound may be considered to be excessively loud, or bright, when reproduced from a frontal direction after fold-down, because the ears are much less sensitive to higher frequencies arriving from behind the head. Thus, when assessing claims about the degree of compatibility of mixes from surround to stereo or mono, one must ask, 'Compatibility with what?' The electrical fold-down equation may differ greatly from the purely acoustic, electro-acoustic or the perceived psycho-acoustic fold-down requirements. The acoustic nature of the listening environment will also add its own variables – the room coupling. Once again, however, relatively anechoic monitoring conditions would seem to offer the fewest complications for the electrical fold-down requirements, and, as the feeling of spaciousness will be in the surround channels, there is little need for the room to add any more.

So, even if all the other chapters in this book could enable a perfect loudspeaker to be designed and built, the question would still remain as to exactly in what sort of a room they would behave perfectly as a stereo pair. Only on the central plane in anechoic conditions, it would appear. In all other cases, the room will impart its influence upon the perceived response and, even in anechoic chambers, there are aspects of the weaknesses inherent in stereophonic reproduction which cannot precisely reconstruct the sound from a single source by means of a phantom image generated by two sources. The situation is that we have imperfect loudspeakers *and* imperfect rooms, trying to reproduce an imperfect notion.

This produces a strong argument in favour of three or five frontal loudspeakers, not only in terms of image stability when moving off-axis, but also from the point of view that fewer phantom sources and more discrete sources means that reflection, absorption and diffusion will all be more uniform in terms of frequency. True, in real situations, the reflection, absorption and diffusion will not all be uniform. Nonetheless, less confusing interaction between multiple sources for single-image positions means, at the very least, a more predictable set of starting conditions. Unfortunately, this still assumes that we have omni-directional, point source, loud-speakers with uniform frequency responses.

The job of acousticians and electro-acousticians is to get the best out of any given set of circumstances, and clearly this is no easy task. Only by a very careful balancing of all the parameters can optimum end results be realistically hoped for, but parameters can only be balanced if they are understood, and many of the points made in this chapter are not widely appreciated. The reason why no single principle of listening room design is universally accepted is because of the truly vast number of variables involved, of which the problems that have been discussed here form only a very small part. All of this does little, however, to change the conception in the minds of the public at large that room acoustics is a black art. In fact, in many instances, the object of an acoustician's work is to keep the variables within defined, acceptable limits, rather than to seek an as yet unachievable perfection.

9.19 Summing up

Rooms form the final link in the sound chain from the loudspeaker to the ear. They are also the most variable links in the chain, and are the most difficult to standardize. In rooms with poor acoustics, no loudspeakers can be expected to perform well. Given rooms with reasonably controlled acoustics, the general tendencies are for rooms with some acoustic ambience to sound more musical, and to support stereo images over a somewhat wider listening area. Very dead rooms allow more accurate perception of timbre and detail, but are sometimes unsatisfactory for the enjoyment of acoustic music. They do, however, allow very precise stereo imaging from the designated listening positions.

As the dead rooms provide no support by means of reflections, they tend to require more power output capability from the amplifiers and loudspeakers when compared to more lively rooms of comparable size. On the other hand, because they contribute very little to the total perceived sound field, they are generally more consistent in their performance.

Considering the wide range of sound control room concepts, and the great weight of experience which has been applied to their designs, the continuing existence of such variability of implementation suggests that there is no simple solution to the problem of room standardization which is consistent with the provision of all the desired acoustics. Of course, the desired acoustic can be very personal, and most of the generally accepted control room philosophies have their partisan followers. There is not one design of tennis racquet which is used to win all championships. Different weights, different string tensions and different designs suit different styles of play. Room effects are somewhat similar; there is no single room environment that is correct for all purposes.

Room effects become significantly more problematical as the number of sound sources increases. In fact, the problems can multiply so rapidly that the four, five or six sources commonly advocated for surround systems will make things so complex that there may be no room design which will deliver the same degree of fidelity as can be achieved in the better stereo rooms. If surround mixing is not carried out with due regard to the acoustic limitations, then trading quality for quantity may be a fact of life. The directional reflection philosophies, used in many stereo control room concepts, will not function when an array of sound sources are used to cover a 360° field. Surround-sound problems can be solved by highly damped rooms in which all the ambience is in the surround channels, but this can lead to very unpleasant room acoustics for normal living when the music is switched off. There would seem to be no easy solution to some of the questions raised by surround.

Acknowledgements

Thanks to Sergio Castro and Keith Holland for preparing the figures, and to Janet Payne for her excellent work on the text.

References

1. KINOSHITA, S, 'What innovation has the Kinoshita Monitor by Rey Audio brought to the studios?' Rey Audio Ltd, Sailama, Japan (1984).
2. NEWELL, P R, *Studio Monitoring Design*, Chapter 19, Focal Press, Oxford (1995).
3. PUDDIE RODGERS, C A, 'Pinna transformations and sound reproduction', *JAES*, **29**, 226–234, April (1981).
4. DAVIS, D and DAVIS, C, 'The LEDE concept for the control of acoustic and pychoacoustic parameters in recording control rooms', *JAES*, **28**, No. 9, 585–595, Sept. (1980).
5. D'ANTONIO and P, KONNERT, J H, 'The RFZ/RPG approach to control room monitoring', Preprint No 2157, presented at 76th AES Convention, New York (1984).
6. NEWELL, P R, HOLLAND, K R and HIDLEY, T, 'Control room reverberation is unwanted noise', *Proceedings of the Institute of Acoustics*, **16**, Part 4, 365–373 (1994).
7. WALKER, R, 'The control of early reflections in studio control rooms', *Proceedings of the Institute of Acoustics*, **16**, Part 4, 299–311 (1994).
8. TOYOSHIMA, S M and SUJUKI, H, 'Control room acoustic design', Presented at the 80th Convention of the Audio Engineering Society, preprint No 2325 May (1986).
9. VOELKER, E J, 'Room modes and deep tone reproduction in control rooms for music monitoring', Presented at the 80th Convention of the Audio Engineering Society, preprint No 2356, May (1986).
10. WALKER, R, 'A controlled reflection listening room for multi-channel sound', *Acoustics Bulletin*, **24**, No. 2, 13–19 (1999).
11. NEWELL, P R, 'From mono and stereo, through quadrophony to surround – a review of control room requirements and practices', *Proceedings of the Institute of Acoustics*, **19**, Part 6, 135–154 (1997). Reprinted in *Audio Media*, European Edition (December 1997 and January 1998).
12. NEWELL, P R, 'Surround monitoring, Part One', *Audio Media*, European Edition (May 1998). US Edition, (July/August 1998).
13. CHASE, J, 'Surround monitoring, Part Two', *Audio Media*, European Edition, 122–126, (July 1998).
14. HOWARD, D M and ANGUS, J, *Acoustics and Psychoacoustics*, 2nd edition, Focal Press, Oxford (2001).
15. DAVIS, D and DAVIS, C, *Sound System Engineering*, 2nd edition, Focal Press, Oxford (1997).
16. HOLLAND, K R and NEWELL, P R, 'Loudspeakers, mutual coupling, and phantom images in rooms', Presented at the 103rd AES Convention, New York, preprint No 4581 (1997).
17. HOLLAND, K R and NEWELL, P R, 'Mutual coupling in multi-channel loudspeaker system', *Proceedings of the Institute of Acoustics*, **19**, Part 6, 155–162 (1997).

Bibliography

HOWARD, D M and ANGUS, J, *Acoustics and Psychoacoustics*, 2nd edition, Focal Press, Oxford (2001).
DAVIS, D and DAVIS, C, *Sound System Engineering*, 2nd edition, Focal Press, Oxford (1997).
NEWELL, P R, *Studio Monitoring Design*, Focal Press, Oxford (1995).
KUTRUFF, H, *Room Acoustics*, 2nd edition, Applied Science Publishers, London (1979).
WALKER, R, 'A simple acoustic room model for virtual production', presented at the 106th AES Convention, Munich, Preprint No 4937 (1999).
ISHII, S and MIZUTANI, T, 'A new type of listening room and its characteristics', presented at the 72nd AES Convention, Anaheim, preprint No 1887 (1982).

10 Sound reinforcement and public address

Peter Mapp

10.1 Introduction

Sound reinforcement and public address form two different applications for loud-speaker technology, but they have enough in common to enable them to be discussed together. Sound reinforcement, as its name suggests, has to do with the boosting and distribution of live or natural sound and is usually connected with observable events in theatres, conferences, churches, auditoria, lecture rooms, etc. The term public address (PA) can be taken to include all sound systems in which there is no live sound, or observable natural event to reinforce, such as paging, general and emergency announcements in airports, railway stations, shopping and leisure centres, stadia, industrial complexes, public buildings, etc.

The two types of system have quite different functions to perform. Sound reinforcement, for example, can range from the very subtle reinforcement of a musician or speaker (so that only when the sound system is switched off can the audience tell that they were listening to 'amplified sound') to the reinforcement of live events in very large auditoria (where the reinforcement signal has to be considerably louder and more widely dispersed than the natural sound for the audience in distant areas to hear an intelligible signal). Nonetheless, the sound in such circumstances should be natural sounding, of high quality and with correct directionality and apparent time of arrival (i.e. the sound should appear to originate from the person speaking etc.). Cinema sound systems also fit into this category since, although there is no original sound to reinforce, the sound system is there to reinforce the visual event. We can see, therefore, that the prime design target for a sound-reinforcement system is naturalness of sound, i.e. a high quality of sound reproduction together with other factors such as direction and level, and of course intelligibility.

The prime object of a public address system, on the other hand, is that of intelligibility. The listener has no visual cues to rely on and must depend purely on the sound signal he hears. Interestingly, naturalness of sound and high intelligibility do not always go hand in hand as, in many situations, the lower bass frequencies may mask the higher-frequency components of speech, namely the consonants, which determine the intelligibility of the received speech. This is particularly true in both highly reverberant buildings and high-noise environments such as industrial complexes or noisy spectator sports, where either crowd noise or the event noise causes speech masking. Under such circumstances, the PA system frequency response is often deliberately tailored to maximize intelligibility by filtering both the lower and upper frequency ranges of the audible spectrum. The resultant quality of sound reproduction may be no longer 'hi-fi' but it should be intelligible. All too many PA systems, however, suffer from poor response tailoring or the use of inappropriate loudspeakers. The latter may provide insufficient control of the radiated sound or lack a sufficiently extended and smooth response. This can lead to beaming, poor coverage, excessive excitation of the reverberant field or transmission into unwanted areas.

Successful sound-system design is very much a question of using the right type of loudspeaker for the job. It requires a knowledge and understanding of the acoustic environment in which the loudspeaker is to be used, and detailed information regarding the loudspeaker's sound-radiation characteristics and Q (directivity) factor. Unfortunately, such information is often unobtainable – particularly in Europe, where there is a proliferation of small loudspeaker companies manufacturing products of which they have no real knowledge or detailed technical information.

Apart from the all-encompassing subject titles of sound reinforcement and public address, a number of other associated loudspeaker applications will also be discussed in this chapter which do not quite fit within these categories. These include sound conditioning, reverberation enhancement and cinema systems. We shall look at the types of loudspeakers involved – ranging from the simple single-cone driver to the new form of loudspeaker – the DML, complex arrays of drivers, and the use of compression drivers and hornflares. Also the fundamentals of speech intelligibility and the whole signal chain from microphone, mixer and signal processing to power amplifier, transmission line and the loudspeaker itself will be considered.

10.2 Loudspeakers and signal distribution

Public address and sound reinforcement systems may be divided into two groups associated with their sound distribution and coverage patterns. First, there are *high-level* distribution systems where either a single loudspeaker cluster sound source, or a few such sources, are used to cover an area, with each source radiating a relatively high sound-pressure level to do so (e.g. a theatre auditorium sound system with a main loudspeaker cluster located above the proscenium and covering most of the auditorium). Second, there are the *low-level* distribution systems, where a large number of loudspeaker sources are used, each operating at a relatively low level of sound output. These systems are often used in areas with relatively low ceiling heights and flat floors, such as conference rooms, exhibition suites or shopping centres and malls. Low-level distribution systems are also widely used in large churches and similar buildings. Here the coverage is achieved either from loudspeakers distributed along the structural columns on either side of the congregation or from a localized pew-back loudspeaker arrangement. In such cases, the signals feeding the loudspeakers may be electronically delayed, so that sound arrivals are 'synchronized' to improve intelligibility and also to improve perceived directionality. (See Section 10.5.)

Many sound systems in fact make use of both types of distribution. There is often no clear-cut reason for using one type, and so other considerations such as architectural constraints, accessibility and installation costs are often the deciding factors.

Distribution of the signal to the loudspeakers can be carried out in one of two ways:

(a) by standard low-impedance connection, i.e. 2–8 Ω;
(b) by high-impedance (nominally 100 or 70 V line) constant voltage distribution using step-up and step-down transformers.

Each method has its advantages and disadvantages, and tends to suit either the high- or low-level sound distribution system. Low-impedance signal distribution generally offers a wide frequency and dynamic range capability, but cable lengths must be kept short to minimize power losses due to cable resistance. Multiple connection of loudspeakers on to one common output can also become unwieldy, often requiring quite complex combinations of series-parallel connections to provide a reasonable load impedance for the amplifier (see Fig. 10.1). Installations with widely distributed loudspeakers can present a considerable wiring problem, particularly if one of the units should fail.

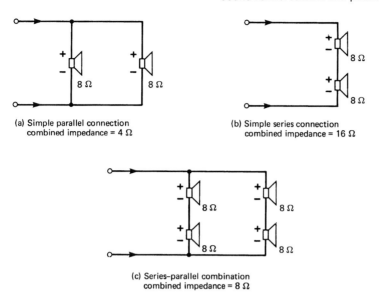

(a) Simple parallel connection
combined impedance = 4 Ω

(b) Simple series connection
combined impedance = 16 Ω

(c) Series–parallel combination
combined impedance = 8 Ω

Figure 10.1. Loudspeaker connections: (a) parallel; (b) series; (c) series-parallel.

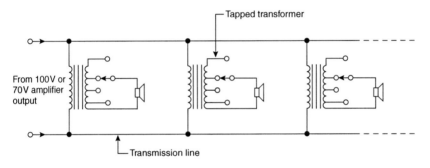

Figure 10.2. High impedance 'constant voltage' distribution system.

Constant-voltage, high-impedance (100 or 70 V line) distribution is ideal for large PA installations associated with either long cable runs or large numbers of loudspeakers. It is in essence very similar to electrical mains power distribution. In practice, a step-up transformer is fitted to the power amplifier, rated to take the maximum output capability of the amplifier, e.g. 100 W. Distribution to each loudspeaker is then a simple matter of making numerous parallel connections onto matching transformers (see Fig. 10.2). Typically, the secondary winding of the matching transformers will be fitted with a number of power tappings, e.g. 1, 2, 5, 10, 20 W, so that the signal level fed to each loudspeaker can be individually adjusted. This allows the loudspeaker coverage sound-pressure level to be accurately set, taking into account any local acoustic or background noise level conditions, etc.

Theoretically, one can continue to add loudspeakers until the maximum output capability of the amplifier is reached. However, in practice some contingency/spare power reserve must be left (e.g. 20%). Also, the resistance losses of the distribution cables themselves must be taken into account for, although transforming the audio signal to a higher voltage for transmission decreases the voltage drop seen at the end

Table 10.1 Typical two-wire cable lengths for 0.5 dB loss in SPL (metres)

Conductor size (areas in mm²)	AWG	Resistance Ω/305 m (1000 ft)	Low-impedance			High-impedance systems				
			4 Ω	8 Ω	16 Ω	200 W 100 V or 100 W 70 V (50 Ω)	100 W 100 V 50 W 70 V (100 Ω)	50 W 100 V 25 W 70 V (200 Ω)	10 W 100 V 5 W 70 V (1000 Ω)	2 W 100 V 1 W 70 V (5000 Ω)
5.26	10	1.00	120	240	480	1500	3000	6000	30 000	150 000
3.3	12	1.59	75	150	300	940	1800	3800	18 000	94 000
2.08	14	2.50	48	96	190	600	1200	2400	12 000	60 000
1.3	16	4.02	30	60	90	370	740	1500	7 400	37 000
0.82	18	6.39	19	38	76	230	460	920	4 600	23 000
0.52	20	10.1	12	24	48	150	300	600	3 000	15 000

of the line, this will still occur to some extent. The most common nominal line distribution voltages are 70 V (USA and Japan) or 100 V (UK and Europe) but other distribution voltages are also employed, such as 50 V, 30 V and 25 V, to meet certain building and safety codes.

Table 10.1 provides some typical cable length/transmission loss data. In practice, cable losses should be kept to within 0.5 dB (or 1 dB as an absolute maximum): remember that 0.5 dB represents a 12% loss of power. Electrically speaking, the actual cable length involved will be twice the apparent physical distance, i.e. the return path must not be neglected. It is important to note that not all power amplifiers will operate satisfactorily with a transformer load. Therefore, unless a 70 V or 100 V output is specifically provided, the amplifier manufacturer should be consulted as to the suitability of his product for this application. Certain cable types (e.g. mineral insulated cable (m.i.c.c.) can also present highly reactive loads – particularly at high frequencies where the coupling capacitance can become significant and the load can effectively increase.

In the past, 70 V and 100 V line-distributed PA systems have often been associated with poor sound quality. This has been due to the use of poor-quality line transformers or the low quality of the loudspeaker itself – or often a combination of the two. However, this need not be the case. Modern transformers can readily provide a frequency response extending from around 40 Hz to 15 kHz ± 1 dB. However, care must be taken when selecting a transformer to ensure that it has a low insertion loss and does not saturate prematurely under load. Otherwise more power will be lost in the transformer itself than in the line. An insertion loss of 0.5–1.0 dB maximum should be readily achieved with a good-quality transformer. Several power amplifier designs are now available which do not use a transformer output stage, and are therefore capable of matching the specification of any good-quality low-impedance device.

High-impedance line-distribution cables should be twisted in order to reduce both electromagnetic radiation and radiation pick-up. In installations where the input signal lines have to run in close proximity to the loudspeaker lines for extended distances, signals in the loudspeaker line may be picked up by the input lines, leading to hum and crosstalk or even oscillation in some cases. Balanced output lines should be used in such cases. It may also be necessary to ground the centre-tap terminal, rather than leaving the output floating. If hum or other interference is encountered with a balanced line output as described above, then it may be necessary to run a screen of (shielded) two-conductor cable to the loudspeakers and to ground the screen at the amplifier end. Fully screened and balanced low-impedance lines should be used for any long input circuit cable runs, and should be kept well clear of the loudspeaker lines, and other electrical cables or services. Minimum separation should be 300 mm, or 1000 mm for long parallel runs. (Where cables have to be crossed, this should be arranged such as to be at right angles in order to minimize potential interference.)

10.3 Loudspeaker coverage

10.3.1 Low-level distributed systems

The most common form of low-level distributed loudspeaker system is that using overhead ceiling-mounted loudspeakers. Such overhead units are widely used in low-ceiling conference venues, meeting rooms, offices, hotel paging systems, airports, shopping centre coverage, etc. The loudspeakers may be mounted directly into the ceiling itself, attached to it or suspended from it.

A number of common problems can be encountered with such installations, resulting in poor-quality sound. The first is the quality of the loudspeaker drivers themselves. Figure 10.3 shows the frequency-response characteristic of a widely used ceiling loudspeaker drive unit. Second, instead of being mounted in back boxes, the

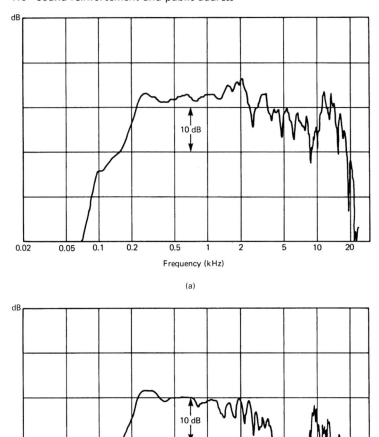

Figure 10.3. Typical ceiling loudspeaker frequency response curves: (a) on-axis; (b) off-axis.

units are often left open to operate into the ceiling void. Not only does this not provide the correct acoustic loading for the loudspeaker driver, but the ceiling void can add an unwanted reverberant sound component which filters through the ceiling (via the ceiling panels or gaps in or around light fittings, etc.). This delayed reverberant sound does nothing to enhance either the quality or the intelligibility of the reproduced sound. Conversely, for fire code regulation purposes, loudspeakers mounted in ceilings often have to incorporate a steel or high melting-point rear enclosure. This is frequently undamped and highly resonant. Furthermore, the enclosure is often of insufficient volume to load the cone drive unit correctly – resulting in reduced and uneven bass/lower-mid output. (The acoustic power response of many devices used in distributed sound systems is far from ideal and can adversely affect potential speech intelligibility unless appropriately equalized – see Section 10.6.) Third, an over-estimated coverage angle is very often assumed for the loudspeaker, which results in poor high-frequency coverage of the area with a resultant loss in both quality and intelligibility. A 200 mm (8-inch) cone loudspeaker is probably still the common device used for this application; and all too frequently a total coverage angle of 90° is assumed. However, 150 mm (6-inch) and even 100 mm (4-inch) cones are becoming increasingly popular – primarily for architectural and aesthetic reasons, though the smaller diameter brings with it improved high-frequency dispersion, albeit with some loss of efficiency and extended low-frequency response. In large PA and Voice Alarm (VA) systems, loudspeaker efficiency and sensitivity can be of critical importance – particularly if the system has to operate from a secondary battery-powered back-up supply. Here, the difference of 3 dB in sensitivity (e.g. from a 4-inch to a 6- or 8-inch device) is effectively equivalent to a doubling of the number

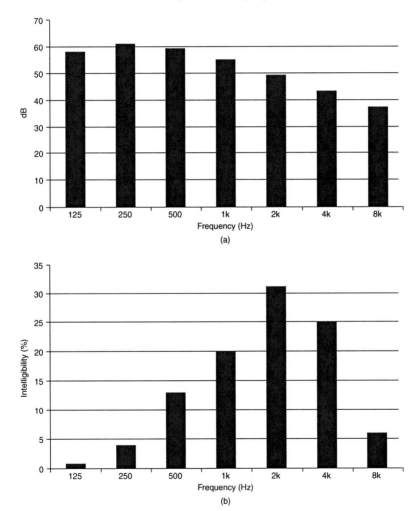

Figure 10.4. (a) Typical speech spectrum; (b) octave band frequency contributions to speech intelligibility.

Polar measurements in landscape plane

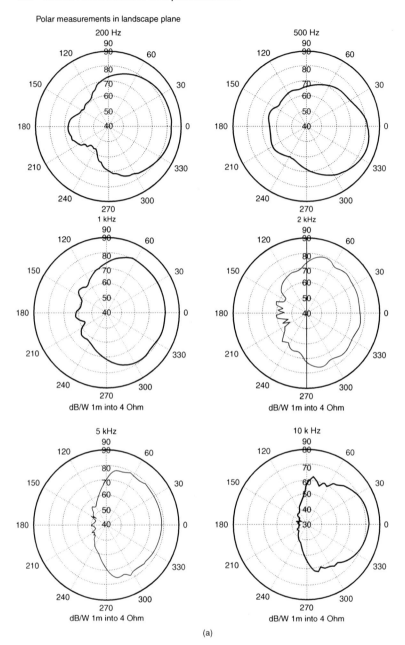

(a)

Figure 10.5 Distributed Mode (DML) ceiling loudspeaker: (a) baffle-mounted polar responses; (b) typical *in-situ* room frequency response.

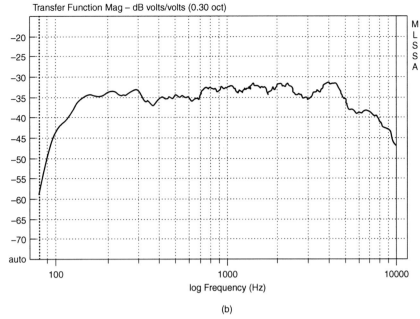

(b)

Figure 10.5 (continued)

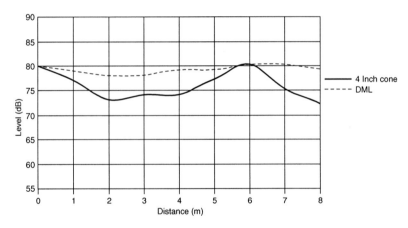

Figure 10.6 Office ceiling loudspeaker installation – variation of SPL between loud-speakers at 6 metres separation; comparison of 4-inch (100 mm) cone and DML.

of power amplifiers required. In terms of the overall system design and cost, the knock-on effect can be very significant when the additional costs for the required extra cabling/system monitoring/battery capacity, equipment racks and space are considered.

For speech intelligibility coverage calculations, the coverage at 4 kHz is the criterion that should be worked to, as the 4 kHz octave band contains much of the energy of the speech consonants which are essential to speech intelligibility. (The range 1–8 kHz, although containing only 5% of the power of the voice, is responsible for more than 75% of the intelligibility. (See Fig. 10.4.) Adequate high-frequency coverage is therefore essential both for good quality of reproduction and good intelligibility. Taking a 90° angle, or the coverage angle at 1 kHz, is not a suitable criterion, although it is often still used. As a rule of thumb, a 60° angle should be taken – but, as examination of simple piston radiation will show, even this is optimistic at frequencies much above 4 kHz (see Chapter 1). Note that a 200 mm diameter loudspeaker corresponds to a wavelength, λ, equivalent to a frequency of 1.7 kHz. When the diameter of the loudspeaker becomes 2λ or 4λ, severe restriction of the angle of radiation occurs (4 kHz has a wavelength of 85 mm). Loudspeakers with dual cones or coaxial drivers are therefore frequently used to improve the high-frequency coverage pattern. Alternatively, 100 mm drive units may be used, but these often lack the low-frequency extension and power-handling characteristics of 200 mm units. Active equalization can be used, however, to improve the former, though the resultant increase in the required audio power must be fully taken into account.

A new form of loudspeaker has recently been developed which effectively overcomes the coverage angle problem. This is the Distributed Mode Loudspeaker (DML) as overviewed in Chapter 4. This device inherently has an extended angle of coverage. Although naturally dipolar, when mounted in a ceiling or when fitted with a rear enclosure, wide dispersion still occurs – even at very high audio frequencies (see Fig. 10.5). The wide dispersion generally leads to an improved off-axis high-frequency performance. This feature together with the fast transient response and reduced boundary interaction effects has been shown to give rise to improved speech intelligibility and clarity in typical sound distribution systems [1]. Figure 10.6 shows the variation in coverage of a typical open-plan office distributed ceiling loudspeaker system at 4 kHz. The inter-speaker spacing is 6 m. The coverage variation which occurs between two high-quality 4-inch (100 mm) cone devices is compared with the corresponding variation for two DML commercial panels mounted at the same locations. A variation of 7 dB can be seen to occur with the 100 mm cones, whereas a variation of only < 2 dB was found to occur with the DMLs [2].

Armed with the loudspeaker's polar diagram or coverage angle information, the next step in the design of a sound system is to decide on the maximum variation in the overall coverage of the area which can be permitted. A coverage variation of within ±3 dB would be a reasonable standard to aim for, although ±2 dB is a more usual standard for high-quality conference and auditorium systems. As the quoted dispersion coverage angle of a loudspeaker is generally bound by its −6 dB points, it can readily be seen that, by allowing a slight overlap between adjacent units, extra coverage can be obtained due to the random addition of the contributions from each unit. A number of loudspeaker layouts for optimum coverage have been formulated over the years, some examples of which are presented in Fig. 10.7.

However, a useful rule of thumb is to use a separation between loudspeakers of 0.6h. Note that h is not the height of the loudspeaker above floor level, but the distance between the listeners' ears and the loudspeaker. (For seated listeners an ear height of 1.2 m is generally taken.) It should not be forgotten that, although a particular point is either within or on the −6 dB coverage angle boundary, the effects of the inverse square law must also be taken into account (Fig. 10.8). As the coverage angle employed increases, so too does the possible error. For example, let us assume that the loudspeaker-to-listener distance h (point a) is 2.5 m (Fig. 10.9). Then the distance to point b on the −6 dB coverage angle boundary will be:

$$\frac{h}{\cos \theta} = \frac{h}{\cos 30} = \frac{2.5}{0.866} = 2.89 \text{ m}$$

Path difference = 0.39 m
Difference in SPL = 20 log (2.5/2.89) = −1.26 dB

For an effective 90° coverage LS, the corresponding error would be 3 dB.

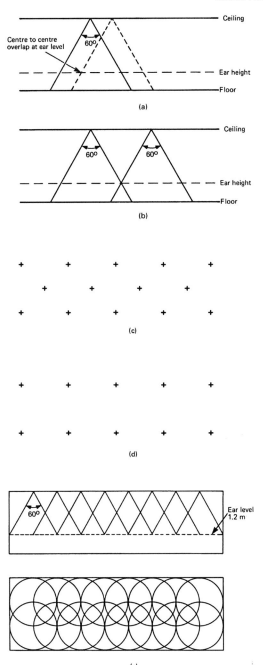

Figure 10.7. Examples of loudspeaker coverage patterns and layouts. (a) Loudspeaker dispersion and overlap for good coverage and minimum variation in SPL; (b) loudspeaker overlap for reasonable coverage, but slightly greater variation in SPL; (c) criss-cross layout; (d) simple rectangular layout; (e) Actual coverage from simple layout with centre-to-centre overlap.

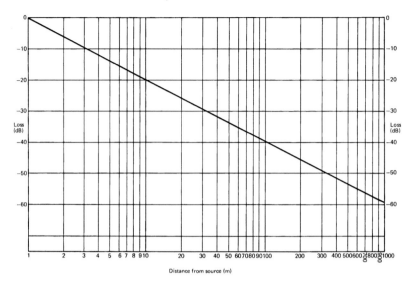

Figure 10.8. Inverse square law – attenuation of sound level with distance (direct sound).

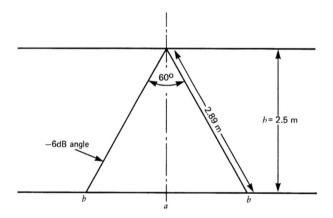

Figure 10.9. Coverage angle and SPL distribution from ceiling loudspeaker.

The above discussion relates to the Direct Sound component only. However, the overall SPL will of course be the resultant combination of the Direct and Reverberant sound fields. In highly reverberant spaces (e.g. >1.5 seconds) the ratio between the Direct and Reverberant components becomes critical with respect to speech intelligibility. (See Section 10.7.)

10.3.2 High-level distributed systems

The coverage from a high-level distribution system may be calculated in a corresponding way, though the coverage in the horizontal and vertical planes has to be individually calculated. Consider the loudspeaker source in Fig. 10.10(a) and assume its coverage angle to be 70° and its height to be 9 m. The discrepancy here between the two 6 dB coverage angle boundaries is immediately obvious. The variation caused

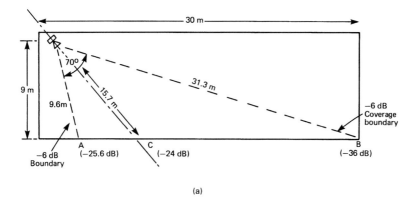

(a)

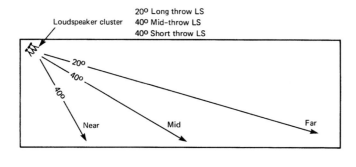

(b)

Figure 10.10. (a) SPL distribution and coverage from high-level PA system; (b) central cluster loudspeaker system.

by the combination of inverse square law and off-axis radiation is 12 dB, i.e. the system will sound more than twice as loud at the front of the room than at the back. Note that, although point A is on the −6 dB boundary, the increased path to point C means that the variation when these two effects are combined is only 1.6 dB.

To overcome this large variation in received SPL, it is usual to cover such a room by using a number of loudspeakers with well-controlled dispersion/radiation patterns, both to provide an even coverage and to minimize overspill onto unwanted areas or reflecting surfaces (Fig. 10.10(b)). Typically one would use a 40° × 20° (H × V) device to cover the far-field, a 60° × 40° device to cover the central area, and a 90° × 40° device to cover the area immediately in front of the loudspeaker. Depending on the exact geometry of the situation (i.e. height of LS, distance to be covered and width of room) it is often possible to use just two loudspeakers having appropriate coverage angles.

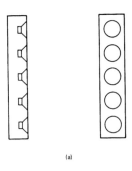

(a)

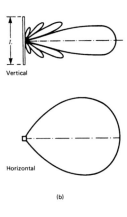

(b)

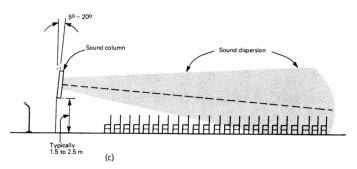

(c)

Figure 10.11. Column/line source loudspeaker principles: (a) construction; (b) generalized sound radiation pattern; (c) installation/coverage (note that it is essential to tilt the column downwards in order to achieve effective coverage).

10.3.3 Loudspeaker types

With high-level distributed systems, particularly in reverberant or acoustically difficult conditions, it is essential to use loudspeaker systems which will cause minimum excitation of the reverberant field. Essentially this is achieved by carefully directing the sound only to where it is wanted, i.e. onto the audience or congregation, and by minimizing overspill onto other sound-reflecting surfaces which could cause long path

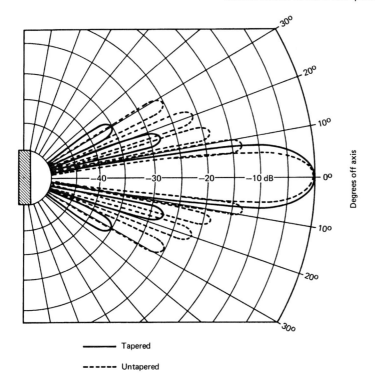

Degrees off axis

——— Tapered

------ Untapered

Figure 10.12. Polar response graph showing vertical sound radiation pattern for 3.3 m column loudspeaker at 1000 Hz. (After Parkin and Taylor.)

reflections or multiple echo sequences as well as general excitation of the reverberant field. Increasing the directivity Q of the devices employed also ensures that a higher ratio of direct-to-reverberant signal is achieved.

Traditionally, in Europe column or line-source loudspeakers have tended to be used, whereas in the United States compression driver and horn systems are the norm. However, over the past ten years or so, Constant Directivity devices have become increasingly widespread in Europe – particularly in the form of compact full-range combination devices (i.e. CD horn and bass cabinet combined into a single enclosure package). Usage of the column loudspeaker has correspondingly diminished – though it still has its uses where controlled radiation is required in a slim and in-expensive format. Recently, the electronically controlled line array has produced a resurgence of interest in the format – particularly in reverberant and architecturally sensitive buildings such as churches and cathedrals.

The purpose of the column loudspeaker is to produce a well-controlled narrow beam of sound in the vertical plane, while maintaining wide coverage horizontally (see Fig. 10.11). This is achieved by forming a vertical line of cone drivers inside a long narrow cabinet, so producing a sound source which is relatively large in the vertical plane in comparison to the wavelengths of sound involved, resulting in a narrowing down of the vertical dispersion angle. The small horizontal dimension ensures that coverage in this plane is not reduced.

The primary advantage of traditional column loudspeakers is that they are generally relatively simple and cheap to make – indeed, the vast majority on the market today consist of nothing more than a few inexpensive cone drivers mounted in a simple wooden, metal or plastic cabinet, frequently without any form

Table 10.2 Comparison of coverage angles for typical commercial column loudspeakers

Frequency (kHz)	Vertical coverage angle in degrees				Typical horizontal coverage
	Col. A	Col. B	Col. C	Col. D	
0.25	105	76	90	–	> 180
0.5	54	40	22	52	180
1	33	20	45	30	150
2	20	10	21	30	100
4	54	20	72	33	80
8	24	18	70	30	60
Column length	0.95 m	1.2 m	1.2 m	1.2 m	Width 200 mm

of internal acoustic damping material. It is therefore little wonder that such an approach gives rise to products with poor response curves. Even where an attempt is made to improve the 'on-axis' frequency response, the dispersion pattern and off-axis performance are usually neglected, giving rise to severe lobing and beaming at high frequencies.

If carefully designed, however, column loudspeakers can be made to provide both a reasonably well-controlled dispersion and an adequate frequency response. A number of techniques are used to achieve this, including power tapering and the use of secondary arrays or high-frequency elements. Power tapering is a method whereby the level and/or frequency content of the signal fed to each drive unit is progressively decreased from the centre of the column outwards, resulting in a reduction of the radiated side lobes and a more consistent frequency/dispersion angle characteristic (see Fig. 10.12 – the polar plot for a 3.3 m column LS with and without tapering).

Table 10.2 further illustrates the variations in coverage angles/patterns in typical column loudspeakers by comparing the vertical dispersion angles for four commonly used commercial products. A typical horizontal coverage angle is also included at each frequency for comparison. As Table 10.2 shows, the vertical coverage angle is hardly consistent, resulting in large variations in the resulting frequency response both within the audience area and in the excitation of the reverberant field.

Commercial column loudspeakers tend to be between 900 mm and 1200 mm in length – though a range of shorter models is available for use in low-level distributed systems. In the past, columns up to 3.3 m long have been built for particular applications, e.g. St Paul's Cathedral, London[3], but today this approach is rarely adopted. However, the recent introduction of DSP-controlled line source devices is bringing a new awareness of the technology and potential application[4-7]. The use of digital signal processing allows the radiated beam of sound to be 'steered'. This enables both the direction of the beam and the vertical radiation angle to be either locally or remotely adjusted. It also enables the physical loudspeaker cabinet to be installed vertically upright instead of being mechanically tilted down to cover the desired area – the tilt being produced electronically by delaying and shaping the signals fed to the individual drive units. DSP technology has also been applied recently to larger, full-range high-performance arrays intended for stadium and similar applications [8] (see Fig. 10.13). An advantage of arraying devices is that a narrower vertical (or horizontal) dispersion can be achieved with greater power handling and hence greater SPL.

In the USA, a range of horn loudspeaker formats has been developed over the years, including multicellular, radial and constant-directivity types. The advantage of being able to direct the sound output of a loudspeaker precisely and consistently is immediately obvious – not only in enabling an audience area to be uniformly covered at all frequencies of interest (i.e. so that almost every seat should be potentially capable of receiving an almost identical signal) but also in controlling overspill onto

Figure 10.13. DSP-controlled loudspeaker array (Courtesy EAW).

undesirable reflective surfaces and/or stage microphones, thereby improving both intelligibility and the potential feedback margin.

The latest types of constant-directivity horn enable precise control to be achieved from around 500 or 600 Hz up to 16 kHz and beyond, the lower frequencies being covered by separate conventional bass cabinets.

Table 10.3 gives the coverage data for a typical modern $60° \times 40°$ CD horn and shows that true control of the sound-dispersion pattern is achieved and maintained over the majority of the audio band. Needless to say, there is a rapidly growing trend to convert to this type of device. However, it should be remembered that a full-range MF/HF CD horn and driver costs at least three times as much as a good (but non-electronic) column loudspeaker, and probably six or seven times as much by the time the bass driver, cabinet and crossover are included. Figure 10.14 shows a typical compact CD horn and bass cabinet, while Fig. 10.15 shows a cluster of separate CD horns and bass cabinets employed to provide wide-range directional control in a

Table 10.3 Response of a $60° \times 40°$ CD horn

Frequency (kHz)	Vertical coverage (degrees)	Horizontal coverage (degrees)
0.6	45	55
1	40	50
2	40	60
4	45	60
8	45	60
16	40	50

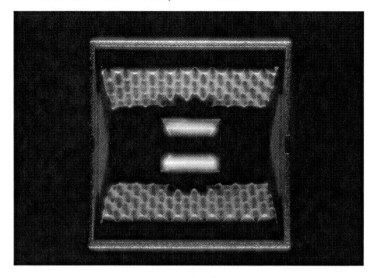

Figure 10.14. Compact full-range CD horn and bass cabinet enclosure (Courtesy d + b Audio).

Figure 10.15. Large CD horn cluster covering indoor ice rink (Courtesy Peter Mapp Associates).

reverberant ice rink environment. An interesting variant of the CD horn is the asymmetric horn which can be formed to produce different horizontal coverage angles at different vertical angles within the same device. This enables greater precision of coverage of a space by compensating for the increasing angular coverage that naturally occurs at greater angles of inclination. An example of such a device is shown in Fig. 10.16. This is a full-range fibreglass moulded outdoor device designed for

Figure 10.16. Full-range asymmetrical horn loudspeaker (Courtesy Community Loudspeakers).

Table 10.4. Comparison of sensitivities of typical devices

Device	Typical sensitivity (1 w/1 m dB spl)
100 mm cone	87–88
150 mm cone	89–90
200 mm cone	90–92
1000 mm column LS	92–95
1200 mm column LS	100
Full-range, cone-loaded PA horn LS.	100
Full range compact (CD horn based) cabinet	109
Re-entrant horn	105
120 × 40 CD horn	107
90 × 40 CD horn	109
60 × 40 CD horn	111
40 × 20 CD horn	114

stadia and theme park use. It has a nominal vertical dispersion of 40° whilst the horizontal coverage angle changes from 60° to 90° (or 40°–70°).

The greater efficiency and sensitivity of CD horns enables either higher SPLs to be generated or greater distances to be covered. Table 10.4 compares the sensitivities of a range of devices typically employed in sound reinforcement and PA systems.

To put the above values into perspective, consider the need to achieve a direct sound pressure level of 95 dB at the rear of a large hall or typical outdoor stadium stand at a distance of 30 m. This would require an input power of 11 watts for the 40 × 20 CD horn, whereas 282 watts would be required for the 1200 mm column. The equivalent power required for the 100 mm cone is 4.46 kW – assuming that such a power could be applied! The need to select the correct device for the job becomes immediately apparent.

The third type of loudspeaker, particular to public address systems, is the re-entrant horn. Figure 10.17 shows the basic operation of the device. As can be seen, the effective air path of the horn is increased by directing the sound-pressure output round a folded waveguide. The re-entrant horn can be used for a wide variety of applications in both high- and low-level distributed systems, but particularly in those where its increased sensitivity over a conventional cone unit (e.g. 100–105 dB compared with 90 dB for 1 W) and its inherently more rugged and weatherproof construction are required, e.g. paging and announcements/PA in industrial complexes or in outdoor use.

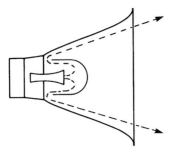

Figure 10.17. Re-entrant horn loudspeaker – principles of operation.

Table 10.5 Typical re-entrant horn coverage angles

Frequency (kHz)	250 mm Round horn	187 mm Round horn	280 × 153 mm Rectangular horn		508 × 254 mm Rectangular horn	
	Coverage angle					
			H	V	H	V
0.5	200°	175°	150°	126°	80°	160°
1	110°	100°	105°	72°	48°	120°
2	52°	72°	65°	48°	46°	50°
4	36°	40°	38°	48°	38°	30°

Figure 10.18. Typical stage monitor wedge loudspeaker (Courtesy Martin Audio).

The sound quality of most re-entrant horns is generally much poorer than that of a conventional cone unit. For example, most re-entrant horns cut off at around 300–400 Hz and do not extend much beyond 5 kHz. Indeed many do not even achieve this. Many also suffer from the presence of strong resonant peaks within this band. However, the restricted low-frequency response can be advantageous in noisy or difficult acoustic environments where speech intelligibility rather than quality is the primary objective. Although many manufacturers' descriptions would have you believe otherwise, the coverage angle of most re-entrant horns varies significantly with frequency, as Table 10.5 shows. In practice, few re-entrant horns seem to achieve much more than 40–50° coverage at 4 kHz, though some narrower rectangular models extend this to around 70° in the horizontal plane.

When planning to cover a specific area, it is obviously essential to have complete data of the loudspeaker's coverage and performance characteristics – yet all too often this vital information is either not available or incomplete.

A fourth type of loudspeaker, particular to sound reinforcement systems, is the stage monitor or foldback loudspeaker. These often take the form of low-profile 'wedges' and usually consist of either a 12- or 15-inch bass unit and high-frequency horn (see Fig. 10.18). As these loudspeakers are deliberately used in close proximity to live microphones, it is important that they exhibit a smooth frequency response – both on- and off-axis. A number of coaxial/dual concentric designs have been produced for this application.

10.4 Sound systems for auditoria

Many people, architects and interior designers included, seem to forget that one of the main functions of an auditorium, whatever its size, is to enable people to hear. This should therefore be reflected both in the acoustic design of the building and in the provision made for the installation of a sound system. The building should therefore have a reasonable value of reverberation time, be free from serious acoustic faults, such as long delay path echoes and sound focusing, and have room for the installation of a sizeable PA system in a useful position – which can, if necessary, be covered with a visually opaque but acoustically transparent material. Thus the internal designer/architect should be satisfied that the appearance of the building has not been compromised, while the sound-system designer should be able to produce a workable sound system of the desired quality, capable of performing the job required (assuming that an adequate budget is available). Sadly this often is not the case, with the result that many buildings and auditoria start out with an inadequate sound system. When starting a sound-system design from scratch, the first two questions which should be asked are 'what is the *system* going to be used for?' and 'what is the *building* going to be used for?' Obviously a theatre system will be very different from that for a sports stadium or an ice rink. The latter two systems will need to be designed to overcome high background noise levels, with the prime goal of producing an intelligible sound within what is likely to be an extremely reverberant and difficult acoustic environment. Furthermore, it must be established at the outset what types of programme material the system will be expected to handle. For example, does the ice rink require good music reproduction for ice dance, etc., or is a 'speech only' system required ? Likewise, will the sports stadium be used for other purposes, e.g. public meetings, or will an indoor space be used to stage musical or dramatic performances? The theatre application is perhaps the easiest to describe, though many auditoria today are built as multipurpose spaces needing to cater for a wide range of events and performances, including dramatic productions, concerts, conferences, conventions and indoor sports events such as boxing or snooker. A large number of concert halls are nowadays also used to stage large conferences – a requirement which is almost at the opposite end of the acoustic spectrum for which such buildings are primarily designed.

The size, shape and layout of the building will very much influence the design and location of the main sound-system components. For example, in Fig. 10.19 we see a fairly conventional theatre/auditorium shape. For events/performances occurring on the stage, it is important that the system should not only provide an adequate level of sound, but should also reinforce the impression that the sound is coming from the direction of the stage. Here the optimum location for the sound system is in the centre of the proscenium arch. This will give good coverage to nearly all areas and, by careful design and use of different loudspeakers to cover different areas, achieve an even distribution of sound. Furthermore, the system will be capable of reinforcing the impression that the sound is coming from the stage. The ear's ability to identify direction in the vertical plane is not particularly acute, so that the listener will localize

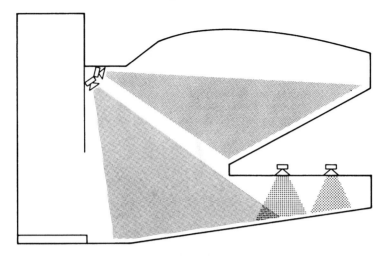

Figure 10.19. Typical auditorium with proscenium loudspeaker system.

on the stage rather than on the proscenium. (This effect may be further enhanced by the installation of a number of low-level 'image shift' loudspeakers along the stage front – a slight delay may be added to these such that listeners will hear the natural sound from the stage first.) This is further discussed in the next section.

A single central position for the sound system will also give rise to less unwanted reverberant excitation of the space, and the associated excitation of potentially long-path echoes. Although many European buildings have traditionally tended to be covered by loudspeakers mounted at the two sides of the stage, there is a growing trend towards a central cluster arrangement. Side positions are useful, however, for musical productions and special effects where distinct left and right (stereo) sound images are required. However, it is very disconcerting to see someone speaking on one side of the stage but hear them from the nearer loudspeaker on the opposite side. The side loudspeakers should therefore be used for distinct left and right effects or music reinforcement, etc. and the central cluster should be used for speech or vocals and for focusing attention towards the centre of the stage (see Fig. 10.20).

Furthermore, with the traditional 'side of stage' loudspeaker system, a listener first hears the sound from the nearer loudspeaker and will localize on this (Fig. 10.21). The listener then hears the direct sound of the person speaking followed by sound from the further loudspeaker. In large auditoria, or even in moderately sized ones where the stage is more than 10–15 m wide, sound received from the further loud-speakers can produce confusion rather than enhancement, the delay time being in excess of 30 ms.

The central cluster system may be made up from a number of possible elements, e.g. column loudspeakers, directional compact cabinet loudspeakers or even a horn and bass cabinet arrangement. As the object of the design is to produce as even a coverage as possible throughout the area, devices exhibiting only small changes in response within the nominal coverage pattern should be employed – a constant-directivity horn and bass cabinet would therefore be a good choice. The final config-uration of the loudspeakers will greatly depend on the interior design of the auditorium – e.g. the presence of boxes and the extent of the balcony. It may be possible to cover the main seating area with just two devices in the vertical plane, but at least two or three units will be required to provide adequate lateral coverage.

In Fig. 10.19 it can be seen that the seating under the balcony cannot be adequately covered from the central proscenium position because it lies in the shadow of the

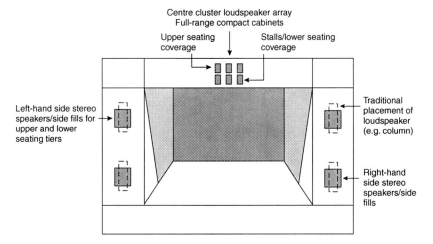

Figure 10.20. Typical stage and loudspeaker arrangements, showing central cluster and left–right stereo/side fills and traditional locations for superseded column speaker approach.

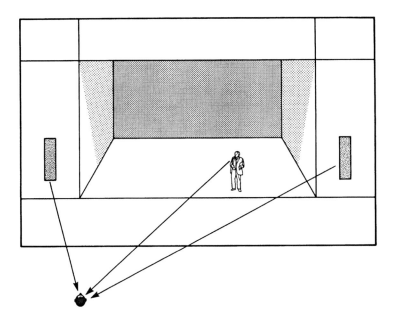

Figure 10.21. Traditional two-source loudspeaker arrangement for small hall or auditorium.

balcony itself. The solution here is to install local under-balcony loudspeakers to reinforce the coverage. These may either be ceiling mounted overhead types or units mounted just under the lip of the balcony and directed back underneath. This latter approach is generally preferred as it enables the direction of the sound to be better matched to the main proscenium loudspeakers. Ceiling units need to have wide dispersion due to the generally low ceiling height. A generous overlap will produce an even coverage – though the loudspeaker density needs to be relatively high. However, the loudspeaker units need not cover the full audio range, as it is only the middle and high frequency ranges which will be lacking. Power handling should not present a problem, because of the short loudspeaker listener distances involved (probably only 2–3 m or so for a seated listener). Although such a system will provide appropriate coverage and intelligibility, some further work is still required for it to become an acceptable reinforcement system.

10.5 Time delay and the Haas effect

First, as the system currently stands, a listener just under the balcony will hear both the main cluster and the local infill loudspeaker; i.e. he is in the overlap zone. However, he will hear the local under-balcony loudspeaker first, followed some time later by the main cluster due to the different propagation distances. If we assume the distance from the listener to the overhead loudspeaker to be 2 m, and the distance to the main cluster to be 30 m, a path difference of 28 m exists between the two sets of sound arrivals. This corresponds to a time delay of 81 ms, which will be perceived by the ear as a distinct echo, i.e. the listener will hear the same thing twice – but the second version will become confused with the first, resulting in loss of intelligibility. Figure 10.22 illustrates the effect.

As a general rule of thumb we can say that speech sounds which arrive within 30 ms of each other will be interpreted by the brain as constituting just one (louder) sound (i.e. they will fully fuse or integrate). However, as the delay between the sounds increases, they no longer fully integrate and become detectable as two separate events. With delays of 60–70 ms, quite distinct echoes are heard which serve only to confuse and degrade the intelligibility of speech. Figure 10.23 further illustrates this.

The solution to this coverage and arrival time delay problem is to delay the signal being sent to the overhead loudspeakers artificially (i.e. electronically) and

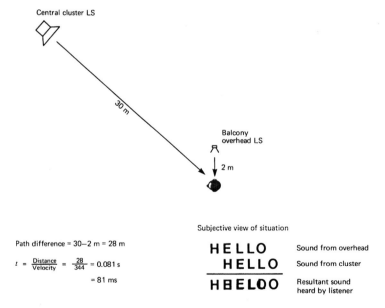

Figure 10.22. Effect of loudspeaker sound path differences on signal arrival times and intelligibility.

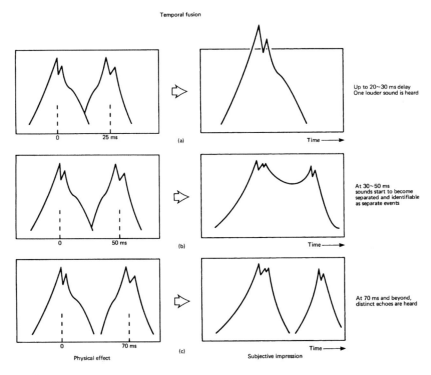

Figure 10.23. Subjective effect of time delay between sound arrivals.

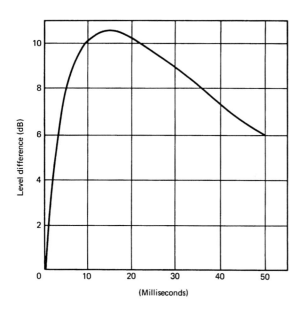

Figure 10.24. Haas effect curve.

synchronize it to the arrival of the main cluster signal. As we can see from Fig. 10.23, it is not necessary to align the arrival times exactly, as we have a 30 ms 'window' in which all the speech sound arrivals will fuse and integrate together, resulting in enhanced intelligibility. However, we should also make use of another psycho-acoustic trick, that of the precedence effect – commonly known as the Haas effect. Some confusion often surrounds the true meaning of this effect, but essentially there are two results which are of use to the sound-system designer.

Haas[9] found that a delayed secondary source could have a greater acoustic output than the primary source and yet not be judged subjectively as being as loud. He showed that, with time delays of between 10 and 25 ms, a secondary signal had to be about 10 dB higher in output level than the primary for it to be judged as being equally loud. This is quite an astonishing result and further explains the integration characteristics of the ear. The now famous 'Haas curve' is shown in Fig. 10.24. The curve is of great importance to the sound-system designer, but it should not be confused with the true precedence effect.

The precedence effect, studied by Wallach[10] and others[11,12], describes the phenomenon whereby human listeners lock onto the direction of the first arriving sound and ignore (either partially or completely) subsequent short-term delayed sounds or reflections; i.e. the delayed sounds integrate or fuse completely with the initial or direct sound as per the 30 ms rule. Thus we have two useful phenomena which we may use to advantage in sound-system design:

(a) Listeners tend to lock on to the first arriving sound and its direction, ignoring short-term delayed secondary sounds.
(b) Secondary sounds arriving within approximately 30 ms of the initial or primary sound fuse with it to produce one apparent sound of increased loudness. Furthermore, the secondary sounds may be up to 10 dB louder than the primary sounds before being judged equally loud (depending upon the time of arrival).

Unfortunately, a widely held and often quoted misconception has arisen as a result of the above findings. This implies that a secondary or delayed signal can be up to 10 dB louder than the primary or direct sound before it is perceived as a secondary source and localization on the primary source is lost. The inference is that one can increase the local sound level in a sound-system design by up to 10 dB from a secondary delayed loudspeaker before it is detected and heard in its own right, with a shift from the primary source. This is not the case. At +10 dB Haas actually stated that the secondary source would sound 'equally loud' which, by definition, means that the secondary source is clearly being detected as such. In fact, for the secondary source or echo to be just imperceptible for time delays of between around 10 to 25 ms, the secondary signal can be only about 4 to 6 dB higher in level than that of the primary signal at the reception point.

In Fig. 10.25, the dotted line (after Meyer and Shodder) presents a curve of 'echo' perception against delay and source-level difference. This very useful curve shows that echoes or delayed sounds become readily discernible at delays in excess of 35 ms. For example, at 50 ms delay, a secondary signal has to be more than 10 dB down before it becomes imperceptible, or greater than 20 dB down at 100 ms. The 'solid' curve in Fig. 10.25 provides us with a further very useful piece of information, namely that delayed sound is actually perceived as a sound source – i.e. the brain is no longer fooled into thinking that only one sound source exists. The curve (researched by Lochner and Burger) closely resembles the echo perception curve, except at very short time delays of less than 20 ms, where it was found that a rather higher output could be accommodated before a secondary, delayed loudspeaker was detected as a separate sound source; i.e. in this range one can tell that a secondary source is operating, but localization is not lost.

A third and final piece of useful data relating time delay and a secondary source of echo perception is shown in Fig. 10.26 and is again the work of Haas. The graph

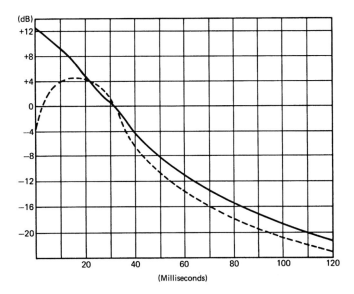

Figure 10.25. Echo perception curves (after references 11 and 12).

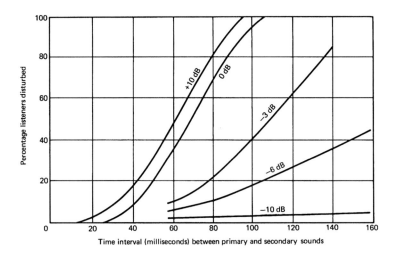

Figure 10.26. Haas echo disturbance curves.

provides an insight into the likelihood of an echo or secondary source causing disturbance to the listener. For example, when the primary and secondary sources were
of equal level, with 30 ms delay between them, less than 5% of listeners were
disturbed. At secondary-source levels more than 10 dB down, only a few per cent of
listeners were disturbed, although from the data we can see that an echo would
clearly be heard. When using the above data, it must be remembered that the results
were obtained under fairly abnormal laboratory conditions. In practice, listeners do
not generally set out to listen to the sound system in isolation, but will usually receive
other information and strong visual cues. Apart from outdoors, or in a very large
arena, it is unusual to be presented with just one or two discrete reflections. Usually,

a complex reflection sequence will be generated, adding further masking and increasing secondary energy build-up. This will modify the laboratory perception curves which, to give manageable results, were by their very nature simple and somewhat artificial.

Returning to the overhead loudspeaker under a balcony problem, we have seen that delaying the signal to the overhead loudspeaker by 81 ms will exactly synchronize the two sound arrivals. However, from our knowledge of the Haas/precedence effect, it can be seen that adding a further 15 ms or so delay to the overhead loudspeaker will ensure that the main cluster (stage) sound is heard first and its direction 'locked onto', while the secondary overhead sound will reinforce this by up to 4 or 6 dB without loss of localization.

10.6 Response shaping

A further adjustment which can be made to the under-balcony system to improve the effect, and hence the naturalness still further, is to tailor the shape of the overhead loudspeaker output frequency response so that it blends in with the spectrum from the main cluster which it is reinforcing. The introduction of a separate high-quality equalizer into the under-balcony circuit allows this to be readily carried out.

Figure 10.27 shows a simplified schematic diagram for the auditorium system we have developed. Note the inclusion of a digital delay line into both the main cluster and the side-fill loudspeaker circuits. This is to improve the overall naturalness and localization of the system. Delaying the main cluster and side-fills to synchronize with the sound of the person speaking/performing on stage (and adding 10–15 ms further delay) allows the audience to hear the natural direct sound component first, followed by the reinforcement component. Although this is often considerably louder than the direct sound, typically +10 dB, such delaying of the loudspeaker signal is well worth while as other cues are present, including a strong visual one which enhances the impression of apparent localization, i.e. 'I can see the person speaking and I can hear

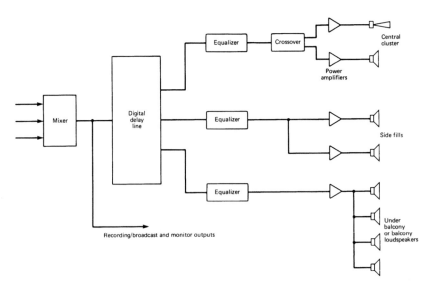

Figure 10.27. Simplified schematic diagram of theatre sound system using audio signal delays. (In practice the side fills would normally be operated as a L/R stereo system and the delayed balcony loudspeakers would take a combined Left + Right + Centre signal mix.)

the person speaking; therefore, the sound must be coming from the person speaking'. Therefore, provided that the overall quality of the sound is high, i.e. a wide smooth frequency response, no audible distortion or strong coloration, and sufficient level, then the use of a delay line stands a good chance of success, with gains in excess of 4 dB often being achieved.

Obtaining a smooth frequency response throughout the audience area is not always as straightforward as one might think. Furthermore a 'flat response' as measured with a real-time spectrum analyser will not sound natural, but will probably seem harsh, with too much emphasis on the high frequencies. It is customary to equalize, to some degree, all professional sound-reinforcement systems. There are two reasons for this – namely overall response shaping and control of feedback. The overall response may be made smoother, for a more natural effect, through the use of broadband equalization (e.g. 1/3 octave) and through the proper choice and installation of the loudspeaker drivers/components themselves. Where high system gain is required, narrow-band notch filters (e.g. 1/10 octave or narrower) may be necessary to tune out the first ring modes (feedback) of the system, once an overall smooth response has been achieved.

The response of the system is measured in the seating areas, taking an average of several measuring points. Before equalization can begin, however, it is essential that the system has first been correctly set up and installed. For example, the high-frequency horns must be correctly aimed, aligned and level matched; no obstructions should be in the way of the loudspeakers (a frequent cause of apparent HF loss) etc.

Figure 10.28 shows the effect of placing a high-quality loudspeaker with an extremely flat response adjacent to a large reflecting surface. The alteration to the unit's frequency response speaks for itself. There is no way with conventional (frequency domain) equalization that these deep interference notches can be overcome. Similarly, equalization is often used in an attempt to compensate for the poor response of the loudspeaker itself. Although some quite dramatic improvements can be achieved, this is not the correct way to design a sound system.

There are a number of ways of equalizing a sound system. Apart from the traditional 'time blind' Real Time Analyser (RTA) a number of acoustic analysers are becoming more generally available that enable the time domain aspects of the sound

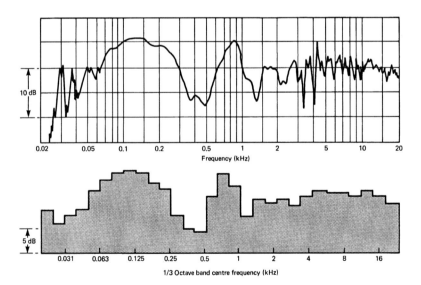

Figure 10.28. Effect of placing loudspeaker with a nominally flat response against a sound-reflecting surface (after Mapp).

field to be resolved. Via time windowing, such analysers allow the Direct, Early Reflected, Late Reflected and Reverberant sound field spectra to be viewed and analysed. With well-behaved devices (i.e. smooth axial and acoustic power responses) the author has found that correction of the direct and early reflected response can be particularly beneficial both in terms of naturalness of response, intelligibility and gain-before-feedback margin. The RTA however, still remains a powerful analysis and sound system equalization tuning tool.

Traditionally, the sound system is equalized by feeding pink noise into it and measuring the response on a 1/3 octave real-time analyser (RTA) or similar device. The system's acoustic response is then adjusted to fit a preferred house curve response, using the system equalizers (see Fig. 10.29): 1/3 octave graphic equalizers tend to be the most commonly used, though octave, 2/3 octave or combinations of 1/6, 1/3 and 1/1 octaves are also available. DSP-based devices including parametric equalizers are becoming increasingly common and offer the advantages of greater precision and control and generally incorporate a signal delay effectively for free as an inherent part of the digital processing.

The measuring microphone should initially be placed on the main axis of the loud-speakers, well away from any reflecting surfaces, and the response checked. If wide variations are found, then the loudspeaker installation itself should be investigated. (It is always worth initially checking the coverage of the system in the 4 kHz octave or 1/3 octave band – an equalizer cannot make up for poor coverage design.)

When using HF horn loudspeakers with compression drivers, a gradual fall-off is to be expected, following the Newman criteria (approximately 6 dB/octave, typically above 2–4 kHz). For this reason many compact full-range loudspeaker systems incorporate dedicated signal processors to compensate for this effect while also providing crossover and driver protection facilities.

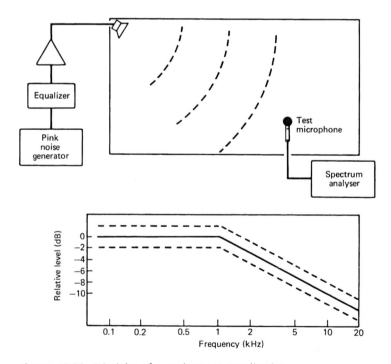

Figure 10.29. Principles of sound-system equalization.

It should be remembered when equalizing a system that in most cases the measuring microphone will be in the reverberant field of the loudspeaker (i.e. the reverberant component will dominate). It is therefore tending to measure the power response of the loudspeaker rather than the direct axial response, and cannot readily distinguish between the direct and reverberant components (see Chapter 12).

If the loudspeaker's own power response is smooth, then equalization should be fairly straightforward, with the direct sound and the measured responses tracking each other. However, if older forms of loudspeaker are being used which have an uneven power output characteristic, then equalization is more difficult and could even make some things worse, e.g. typical column loudspeakers or when an older type radial horn is being used with a bass enclosure. Figure 10.30 shows the on-axis direct field and reverberant power responses. Equalizing the system for a smooth power response (i.e. carrying out a standard equalization procedure) will cause the on-axis direct component to be uneven. Although the measuring microphone and RTA will not detect this, the ear will, causing a rather different subjective impression to be obtained from that indicated by the spectrum analyser. It is therefore essential that music and speech be regularly played through the system as a check during the equalization procedure. Loudspeakers exhibiting a flat power response are therefore preferable for PA and sound reinforcement systems – unfortunately this response information is rarely provided by manufacturers. Figure 10.31 shows the before and after equalization response curves for a set of good-quality 500 mm column loudspeakers installed in a very reverberant church (RT = 2 seconds). Before equalization, the sound quality was very 'muddy' with poor intelligibility and clarity. After equalization, the intelligibility was significantly improved and provided acceptable speech clarity. Although the high-frequency response above approximately 2.5–3 kHz drops off a little faster than might generally be desired for a distributed sound system, increased equalization to reduce the slope of the roll-off resulted in an unacceptable performance with a harsh, over-bright sound quality with emphasized sibilance. This was due to the differences between the direct and reverberant fields as illustrated in Fig. 10.30.

In practice, the type of house curve response presented in Fig. 10.32 has been shown to work well for distributed sound systems, though it should not be regarded as an absolute target which must be achieved at all costs. Figure 10.33 presents a suggested house curve for a rock music type of reinforcement system.

From the simplified sound-system schematic diagram shown in Fig. 10.27 it can be seen that each set of loudspeakers has its own equalizer. This enables the responses of the different loudspeaker types to be appropriately equalized, the required response curve for one set of loudspeakers being unlikely to match the requirements of another. Furthermore, where different acoustic conditions are met (e.g. under the balcony) a different compensation curve may well be required. Apart from producing a smoother response and more natural sound quality, correct equalization will also improve the gain-before-feedback margin by removing any sharp peaks in the response and ensuring that the full speech frequency bandwidth is available for use (see Fig. 10.34).

Once an appropriate house curve for the system has been achieved, a stage microphone should be set up and the gain-before-feedback checked. (Connecting this microphone to the RTA can be most instructive.) If sufficient gain is not achieved before feedback, the relative positions of microphone and loudspeaker should be checked to ensure that the microphone is not sitting under a loudspeaker side lobe, etc. Narrow-band notch filters can also be employed to improve the feedback margin by minimizing the excitation of the first few ring modes of the sound system. All too frequently, the prime cause of feedback is incorrect use of the microphone. Either a number of unnecessary microphone channels are left open (each doubling of the number of open microphones reduces the potential gain of the system by 3 dB) or, alternatively, the person speaking is standing too far away from, or off the axis of, the microphone. Where foldback (monitor) systems are provided for the musicians/

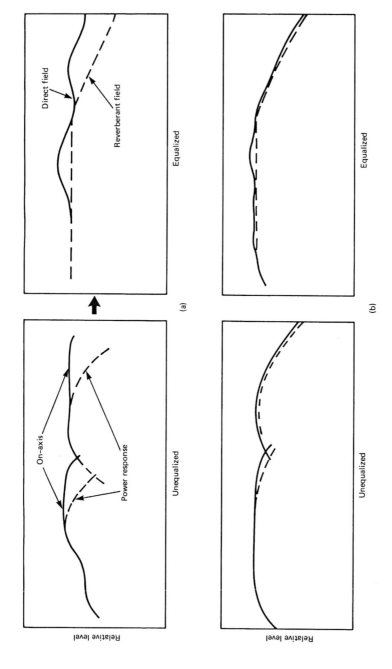

Figure 10.30. Sound-system equalization – effect of loudspeaker power response: (a) an uneven response; (b) a smoother response.

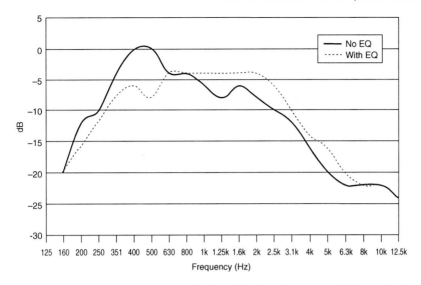

Figure 10.31. Frequency response of column loudspeaker system before and after equalization.

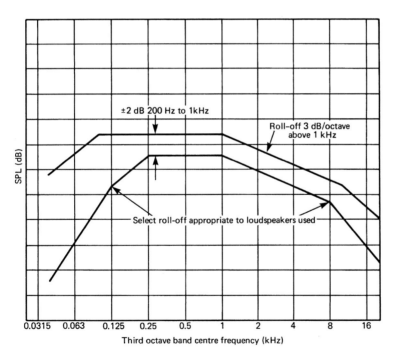

Figure 10.32. Typical response curve for speech reinforcement systems.

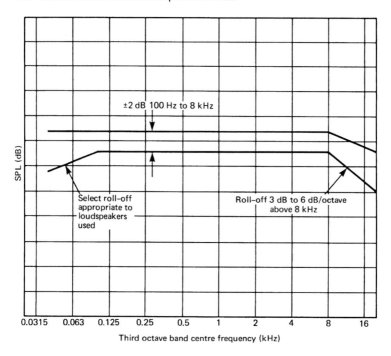

Figure 10.33. Typical response curve for high-level rock music system.

presenters etc. correct equalization of the response is essential if premature feedback is to be prevented. Also, in order to better hear particular instruments or other performers, an equalization curve may be set which emphasizes particular frequency bands. An equalizer with at least 1/3 octave resolution should be used.

10.7 Speech intelligibility

One of the most important aspects of sound reinforcement and PA (and VA) system design is that of achieving adequate speech intelligibility and clarity. In the past many systems appear to have been designed with either little or no attention being paid to this aspect or with scant regard to the basic parameters that affect intelligibility. Over the past 20–25 years our understanding of how loudspeakers and buildings interact acoustically, and how this affects perceived intelligibility, has grown considerably. Nowadays it is possible to predict the potential intelligibility of a sound system with a good degree of accuracy – provided that the appropriate acoustic/ electro-acoustic information is available. Whereas 10 years ago this was generally unlikely, this has changed significantly in the past 5 years – particularly with the introduction of computer-based acoustic and sound system design packages. However, in the final analysis, it does not matter how well a sound system is designed if it is not used or operated correctly. Intelligibility and clarity can be degraded surprisingly easily.

The main factors affecting speech intelligibility are:

● sound system bandwidth and frequency response
● loudness and more significantly S/N ratio
● reverberation time of the space

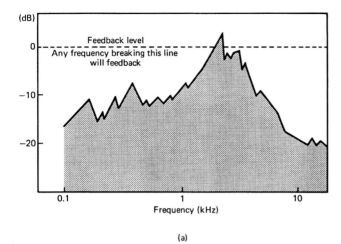

(a)

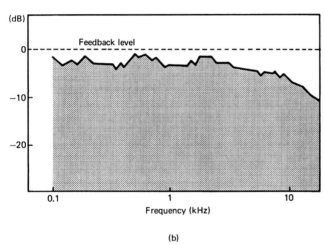

(b)

Figure 10.34. Equalizing a sound system to improve frequency response and gain before feedback: (a) before; (b) after equalization.

- the volume (and size and shape) of the space
- the distance between the listener and the loudspeaker(s)
- the directivity of the loudspeaker(s)
- the direct-to-reverberant ratio (which is dependent on the last four parameters)
- the number of loudspeakers operating

Other factors include

- listener acuity
- talker enunciation and rate of delivery
- talker voice type (male/female)
- system distortion
- system equalization
- uniformity of coverage
- sound focusing and presence of any discrete reflections

- direction of sound arrivals
- direction of interfering noise
- vocabulary and context of speech.

From the foregoing list of general factors, it can be seen that many aspects have to be considered and dealt with when designing and predicting the potential intelligibility performance of a sound system.

When designing sound systems for buildings or auditoria which possess well-controlled acoustics, or acoustics specifically designed for speech, e.g. with reverberation times of 1.5 s or less, producing adequate intelligibility should not present a major problem. Speech syllables typically occur three to four times a second and so, for reverberation times of 1.5 s or less, the effect of the reverberant overhang is small. Highly reverberant buildings such as ice rinks, churches, large sports halls and swimming pools, where reverberation times of 4–6 s can be expected, can present a considerable problem – particularly if a relatively high level of noise is present. However, it is possible to predict with reasonable accuracy the intelligibility of a sound system. Alternatively, by rearranging the equations given below, it is possible to determine the sound-system parameters and performance required to achieve a given degree of intelligibility within a given acoustic environment. The most commonly used prediction method is the 'articulation loss of consonants' (% Alcons) method of intelligibility prediction formulated by Peutz[13].

In reverberant environments the clarity of speech has been found to be a function both of reverberation time and the ratio of direct-to-reverberant sound. As a listener moves further away from a talker (decreasing the direct-to-reverberant sound ratio) articulation loss of consonants increases. That is, the intelligibility of speech reduces as the direct-to-reverberant ratio decreases. However, this relationship is maintained only to a certain distance, beyond which no further change takes place. The boundary corresponds to a direct-to-reverberant ratio of -10 dB (i.e. $3\times$ critical distance).

Figure 10.35 presents graphically a chart relating the articulation loss of consonants, the reverberation time of a space and the ratio of direct-to-reverberant sound-pressure levels. Peutz found that a 15% loss of consonants was the maximum loss which could be tolerated before the majority of listeners found the intelligibility of speech to be unacceptable. Therefore, this is generally taken as the minimum design target. (Although for many purposes, depending upon the nature of the system and use, a minimum value of 10% is required.) The following equations give the relationships between the room and sound-system parameters and the intelligibility in % Alcons:

$$\% \text{ Alcons of system} = \frac{200(D_2)^2(RT_{60})^2(n + 1)}{VQM} \tag{10.1}$$

$$\text{Maximum distance to furthest listener } D_2 = \sqrt{\frac{15VQM}{200(RT_{60})^2(n + 1)}} \tag{10.2}$$
for 15% Alcons

$$\text{Maximum allowable RT for 15\% Alcons} = \sqrt{\frac{15VQM}{200(D_2)^2(n + 1)}} \tag{10.3}$$

$$\text{Minimum loudspeaker } Q \text{ for 15\% Alcons} = \frac{200(D_2)^2(RT_{60})^2(n + 1)}{15VM} \tag{10.4}$$

where D_2 =   distance in m from the sound-system loudspeaker to the most distant listener
         Q =   directivity factor (axial)
         $(n + 1)$ =   total number of like loudspeaker groups contributing to the

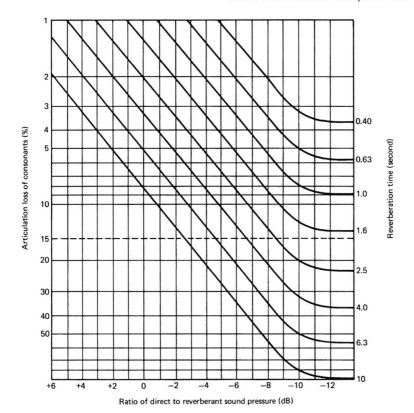

Figure 10.35. Articulation loss of consonants versus RT and direct-to-reverberant ratio. (Adapted from Peutz.)

reverberant field. ('1' represents the group contributing direct sound at the measurement point.)

M is a critical distance modifier, taking into account the higher than average absorption of the floor with an audience, for example,

$$M = (1 - \bar{a})/(1 - \bar{a}c) \tag{10.5}$$

where $\bar{a}$ = average absorption coefficient
$\bar{a}c$ = absorption in the area covered by the loudspeaker

If the Q of the loudspeaker is not known, it may be found from the following equation:

$$Q = \frac{180°}{\text{arc sin } (\sin \alpha/2 \sin \beta/2)} \tag{10.6}$$

where α and β are the -6 dB horizontal and vertical coverage angles of the loudspeaker.

Note the above articulation loss equations are valid only when the furthest listener distance D_2 is less than three times the critical distance, i.e. when $D_2 < 3.16D_c$.

For distances greater than three times the critical distance, the following relationship may be used to determine the % Alcons:

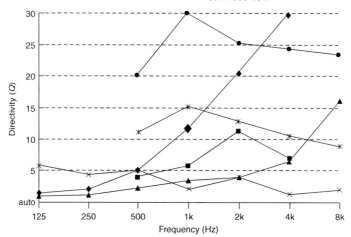

Figure 10.36. Directivity Q characteristics for a range of typical PA loudspeakers.

For $D_2 > 3D_c$,   % Alcons $= 9RT_{60}$ \hfill (10.7)

The critical distance D_c may be calculated from the following formula:

$$D_c = 0.03121 \sqrt{\frac{QVM}{RT_{60}(n+1)}}$$ \hfill (10.8)

or, if the room constant R is known:

$$D_c = 0.141 - \sqrt{\frac{QRM}{n+1}}$$ \hfill (10.9)

From the above equation it can be seen that D is proportional to the square root of Q and R:

$$D_c = 0.141 \sqrt{QR}$$ \hfill (10.10)

This shows that a 6 dB increase in directivity (four times increase in Q) corresponds to a doubling of the critical distance. Thus, by increasing the Q of the loudspeaker, we can increase the maximum distance at which sound will be intelligible in a reverberant space. Figure 10.36 compares the Q of a number of common loudspeaker types ranging from a single 100 mm cone to a short column speaker and CD horn. As can be seen, the Q factor for most loudspeakers increases with frequency. Exceptions to this are the two-way unit, where at crossover the radiation widens as a smaller (less directional) transducer takes over and the Distributed Mode Dipole loudspeaker which unconventionally becomes less directional at middle to high frequencies. (Also note the variations in directivity of practical Constant Directivity horns.) The above simple Alcons equation (10.1) has been developed to take account of background noise and redefined to operate with the direct and reverberant sound components more directly:

$$\% \text{Alcons} = 100(10^{-2((A+BC)-ABC)} + 0.015)$$ \hfill (10.11)

where $A = -0.32 \log\{L_R + L_N\}/\{10L_D + L_R + L_N\}$ for $A > 1$, $A = 1$
$B = -0.32 \log\{L_N\}/\{10L_R + L_N\}$ for $B > 1$, $B = 1$
$C = -0.50 \log\{RT_{60}\}/\{12\}$

where L_D is the direct sound level
L_R is the reverberant sound level
L_N is the ambient noise level.

This is known as the 'long form' % Alcons equation. A development of this is shown in equation (10.12) and is based on an algorithm developed for the TEF analyser.

$$\% \text{ Alcons} = 100(10^{-1.7/IParr}) \tag{10.12}$$

where $IParr = (ISN + IT) - (ISN*IT)$
ISN is the reduction in speech information due to S/N ratio
IT is the reduction in speech information due to reverberation
RT is the early decay time

From equation (10.1) it can be seen that the intelligibility of a sound system is related to the number of loudspeaker groups not directly contributing to the sound field at any given listening position – i.e. these loudspeakers contribute to the reverberant field and serve to reduce the ratio of direct to-reverberant sound. Furthermore, these distributed loudspeakers can also give rise to long-path echoes either from their own radiation or, more likely, from multiple reflection sequences which further serve to reduce intelligibility.

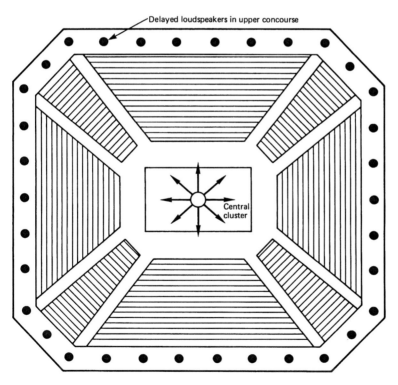

Figure 10.37. Central cluster system with delayed low-level distribution 'infill' system.

The advantages of a single 'point source' or central cluster system immediately become obvious, as the *n* factor in the intelligibility prediction equations is eliminated with a corresponding improvement in performance (although the combined *Q* factor for the cluster of individual components has to be calculated). However, a central cluster is not always the complete answer. For example, restricting the coverage of the cluster to just cover the audience area reduces undesirable reverberant excitation from reflective wall surfaces etc., but may fail to cover auxiliary galleries or walkways. The solution here is to cover these with a local low-level distributed system, fed with a suitable signal delay to synchronize sound arrivals in these areas with the main cluster. Figure 10.37 shows a typical layout of this kind.

From an examination of equation (10.1) it can be seen that the lower the loss of consonants, the better the intelligibility. Subjectively, values of 0–5 % are generally regarded as excellent, 5–10% as good and 10–15% as acceptable. A number of acoustic measuring systems incorporate facilities to measure the equivalent loss of consonants directly. An alternative and more sophisticated objective assessment approach is the Speech Transmission Index (STI and Rasti) method developed by Steeneken and Houtgast[14,15]. Here the scale ranges from 0 to 1 as shown in Fig. 10.38. The scale is rather more stringent than the % Alcons equivalent and is regarded by some as being too pessimistic or onerous for many sound system applications. (The Rasti/STI concept was originally intended for assessing natural speech intelligibility in auditoria, whereas it is arguable that a different expectation and consequent scaling might apply to that for a general paging system at a train station, for example.) Despite this, the method has been widely adopted within Europe to evaluate and set the design criteria for a wide range of sound systems ranging from emergency PA systems in public buildings, shopping centres, football stadia and transportation terminals to theatre and concert hall auditoria[16]. An advantage of the STI system is that it automatically takes into account any background noise and hence the prevalent S/N ratio as well as the effects of reverberation time and reverberant level (i.e. D/R ratio). Again, however, like the % Alcons method, the STI does not account for discrete, late reflections. System non-linearities and some forms of signal processing can also severely disrupt the method.[17,18] A disadvantage of the Rasti/STI method is that the potential result cannot be readily predicted without the use of a complex computer model.

It should be noted that the % Alcons method of intelligibility prediction is statistically based and assumes the existence of a true reverberant/diffuse field. Other factors which affect speech intelligibility, such as discrete echoes, cannot be catered for within the equations and must be separately checked, remembering that strong echoes occurring after approximately 50–60 ms will severely affect intelligibility.

The other major factor which affects the intelligibility of speech in both indoor and outdoor systems is that of masking produced by ambient background noise levels. This reduces the speech signal-to-noise ratio. Ideally, a signal-to-noise ratio of 25 dB is required to ensure optimum intelligibility. While it is possible and customary in theatres, multipurpose auditoria, conference rooms, etc. to achieve a low level of background noise around 25–30 dBA, many systems have to operate in far higher ambient levels. Since the optimum level for reproduced speech in the absence of significant background noise is approximately 70 dBA[19,20], this ideally means that the

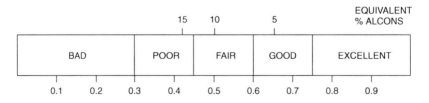

Figure 10.38. STI/RASTI Intelligibility Scale.

background noise level (with an audience) should not exceed 40–45 dBA. In many situations this is not possible – particularly in public address applications with loud crowd noise or considerable noise from the event itself.

In practice it has been found that, with higher PA levels, a signal-to-noise ratio of 10–15 dBA is usually acceptable [21], though, as the noise level increases and the level of the PA system is correspondingly increased, the intelligibility begins to decrease. Furthermore, the PA system could begin to become intolerably or even dangerously loud. For general PA announcements etc. under such conditions, a maximum level of 95 dBA is often recommended. However, in special circumstances, such as emergency warning and evacuation systems in noisy industrial environments, considerably higher levels are required, particularly if hearing protectors are being worn.

For good intelligibility it is vital that the high-frequency end of the speech range is adequately transmitted. The PA system must therefore be capable of responding up to at least 4 kHz and preferably 6 kHz without a significant fall-off in response. It is surprising, however, how many PA systems fall short of this goal, which is quite narrow-band by hi-fi standards. (Figure 10.4 illustrates the concept by presenting the individual octave band contributions to speech intelligibility.) From the figure it can clearly be seen that the 2 and 4 kHz octave bands alone contribute over 55% of the total.

Two of the main culprits of poor system response are the cheap type of re-entrant horn, which is often used, or column loudspeakers with a poorly controlled dispersion pattern, leading to excessive beaming (i.e. narrow coverage) at the higher frequencies and uneven sound power characteristic (see Sections 10.3.3 and 10.6). Whereas the falling response of the re-entrant horn loudspeaker can be improved by the introduction of appropriate equalization into the system (assuming that the horn and amplifiers can take the extra power required), little can be done to improve the coverage of the column loudspeaker without either physically or electrically modifying its design. However, the overall spectral response and hence the resultant potential frequency masking that often occurs with these devices when used in a reverberant environment can be noticeably improved by means of appropriate equalization [22].

It is worth noting that, in difficult acoustic environments, either with high noise levels or highly reverberant conditions (or both), shaping the overall response of the system to reduce the low frequencies and accentuate the higher ones can significantly improve the overall intelligibility of a system, though at the cost of naturalness.

10.8 Outdoor PA systems

In outdoor PA systems, although there is no reverberation to contend with, other problems arise which must be recognized and appropriately tackled if a worthwhile and intelligible PA system is to result. Both high-level long-throw and local distributed systems are used outdoors, depending on the nature of the event, etc. Local coverage should not present any major problems so long as adequate coverage is provided over the frequency range 500–4000 Hz. Care should be taken with the spacing and aiming of the loudspeakers to ensure that repeat echoes are not produced. Thus the main loudspeaker should be aimed into the crowd, and extra loudspeakers should be used to cover given areas. This is better than increasing the spacing and firing down the line one to another, as this can produce a regular series of distinct echoes, and even with a short spacing can result in a reverberant-like sound quality.

The effects of the inverse square law should not be forgotten, so that elevating the loudspeakers or standing them a little way off will lead to a more even distribution of sound. Constant voltage (100/70 V line) distribution is generally essential for such systems, because of the long cable runs. The ability to adjust the level of individual loudspeakers readily by altering their matching transformer tappings is a very useful and necessary feature.

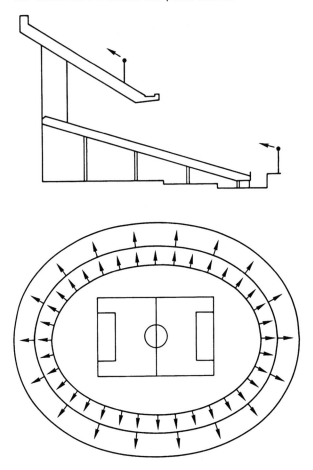

Figure 10.39. Sports stadium distributed loudspeaker system with delayed second tier.

In larger installations, signal delay lines may be employed to keep signals in step with each other. Figure 10.39 shows the possible sound-system layout for a large outdoor (uncovered) stadium using two tiers of distributed loudspeakers with the second-tier loudspeakers delayed to synchronize with the first-tier sound arrivals.

The absence of reverberation, and the potentially long-path echoes associated with enclosed spaces, enable outdoor sound systems to cover large areas and provide good intelligibility at large distances from a loudspeaker source, e.g. up to 200–300 m. Care, however, needs to be taken to ensure that discrete, late reflections are avoided, e.g. from neighbouring structures or buildings. The loudspeakers are generally clustered together to provide the appropriate throw and coverage angles. Positioning the loud-speakers on the roof of a building, or a specially constructed tower, enables the initial effects of the inverse square law to be reduced and gives a more even coverage. For example, the level at 100 m will be only 6 dB down on that at 50 m. By employing long-throw and short-throw devices, and making careful use of off-axis radiation, good coverage can be produced.

For very large events, repeater PA towers are frequently employed with judicious use of signal delay lines to synchronize the sound arrivals. (The main object of their use here is to achieve sound synchronization and intelligibility, to negate the long

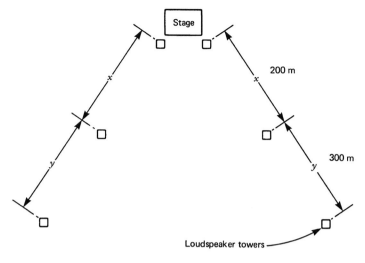

Figure 10.40. Large outdoor PA system employing loudspeaker 'delay' towers.

echoes which would otherwise occur, rather than subtle localization effects.) Although one central repeater tower would provide the best quality/intelligibility, this is not always possible, because of either visual or overall coverage constraints. When using two towers, some loss of quality will inevitably occur within the overlap zone, due to the uneven off-axis radiation patterns likely to be experienced and the phase cancellation effects produced on the centre line of the two widely spaced sources. Figure 10.40 shows a typical layout for such a system. For this particular installation, the towers were separated by approximately 200 and 300 m, necessitating time (signal) delays of around 600 and 1500 ms (1.5 s). Where high quality music reproduction is required, low impedance loudspeaker systems are normally employed but with the power amplifiers sited locally to the loudspeakers. Signal distribution is carried out using balanced line level (0 dBu) feeds.

10.9 Climatic effects

When directing sound over large distances outdoors, a number of climatic effects become significant and must be taken into account.

10.9.1 Relative humidity

Contrary to what most people believe, there is more sound attenuation in dry air than damp air. The effect is a complex one, being highly frequency-dependent. Figure 10.41 illustrates the effect. The excess attenuation is only really significant at frequencies above 2 kHz. This means that the high frequencies (important to intelligibility) will be attenuated more with distance than the low frequencies. Figure 10.41 also shows that the attenuation will be greatest at around 10–20% relative humidity. A loss of 9–10 dB at 4 kHz, for example, over a distance of 100 m, is very significant. Some judicious use of an equalizer may therefore be required – but extreme care must be taken to ensure that the drive unit can handle the additional power applied to it. Furthermore, making up for the loss at 100 m will put a peak in the response at 50 m or nearer. A compromise between the maximum measured loss and the overall coverage response must therefore be chosen.

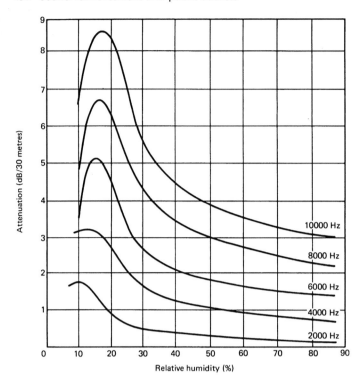

Figure 10.41. Effect of relative humidity on sound absorption/attenuation.

10.9.2 Temperature gradient

Sound propagation over large distances outdoors may often give extremely erratic results (even allowing for the humidity effect). This is because sound refraction often occurs as the velocity of sound changes slightly with variations in temperature. Figure 10.42(a) shows a situation which often occurs at nightfall, when the ground is still warm, while Fig. 10.42(b) shows the corresponding effect which can arise in the morning when the ground is cool but the air temperature is increasing. This 'skipping' characteristic may give rise to local hot spots or dead spots within the listening area.

10.9.3 Wind gradients

Figure 10.43 shows the effect of wind-velocity gradients on sound propagation. The resultant velocity of sound in this case is the velocity of sound in still air (344 m/s at 20°C) plus the velocity of the wind itself.

10.10 Sound-masking systems

The open-plan office, although providing many benefits in terms of improved communication between staff, also by definition produces the disadvantage of the distraction of hearing other people's conversations. It is virtually impossible not to listen to another conversation going on – particularly in small, sparsely occupied offices. In larger open-plan offices a general hubbub of conversation and occupational noise

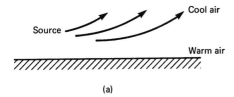

(a)

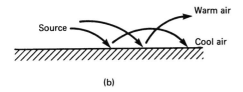

(b)

Figure 10.42. Effect of temperature gradients on sound propagation: (a) ground is warm; (b) ground is cool.

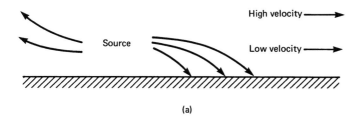

(a)

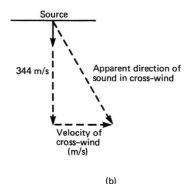

(b)

Figure 10.43. Effect of (a) wind velocity gradient and (b) cross-wind on apparent direction of sound propagation.

occurs, which masks the general intelligibility and reduces the distraction. By electronically introducing into open-plan office areas a bland wash of low-level background sound, with no meaningful or tonal component, the desirable sound masking of the large or busy office can be achieved. In addition, the speech privacy provided by cellular office partitioning (particularly lightweight, de-mountable forms) can be improved by carefully increasing and tailoring the background sound level by means of a sound conditioning system. However, sound masking can work only over

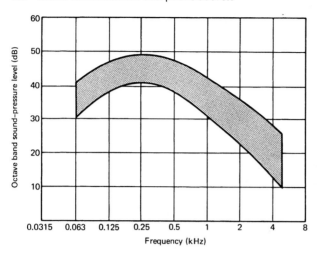

Figure 10.44. Typical noise-masking spectrum.

a limited and well-defined range. If the level is too low, it will not be effective; alternatively, if it is too loud, then it will in itself become distracting and fatiguing. An operating window ranging from approximately 40 dBA minimum to 48 dBA maximum has been found to be satisfactory for a sound-masking/conditioning system.

Many years of experience indicate that an ambient noise having the spectral response shape shown in Fig. 10.44 is the most effective for masking speech. Typically, a nominal level of 40 dB SPL at 1 kHz is set and the spectral response curve referenced to this; 45 dB at 1 kHz should be regarded as the absolute maximum and 35 dB the minimum. The response characteristics shown must be accurately achieved with great uniformity throughout the space (room) for the system to be effective. Pink noise, generated digitally for uniformity, is generally used as the basic sound source and is then filtered to achieve the appropriate spectrum using 1/3 octave filters or equivalent.

It is important that the loudspeakers distributing the masking sound cannot be localized by the room occupants. For this reason, the masking loudspeakers are not set into the ceiling like paging or background music types, but instead are installed in the void above the suspended acoustic ceiling. A variation of no more than ±2 dBA must be achieved for the system to operate effectively in an open-plan area.

Figure 10.45 shows a simplified schematic diagram for a sound-masking system. The masking system can also be used for general paging or emergency PA announcements. A separate signal chain is required, but this increases the potential of the system and effectively reduces the cost. Appropriate equalization of the speech signal is required for it to penetrate the ceiling construction effectively but this is not generally a problem, and an extremely clear and intelligible announcement system can be achieved. Note that although it is possible to use a sound-masking system as an announcement system, the converse is rarely true. The loudspeaker distribution, mounting and zoning are unlikely to meet the design aims described above.

10.11 Reverberation enhancement

Loudspeakers and electroacoustic systems are beginning to play an active and extensive role in concert hall and auditoria design. Unlike its passive acoustic counterpart, an electroacoustic system allows very precise control to be achieved over the four

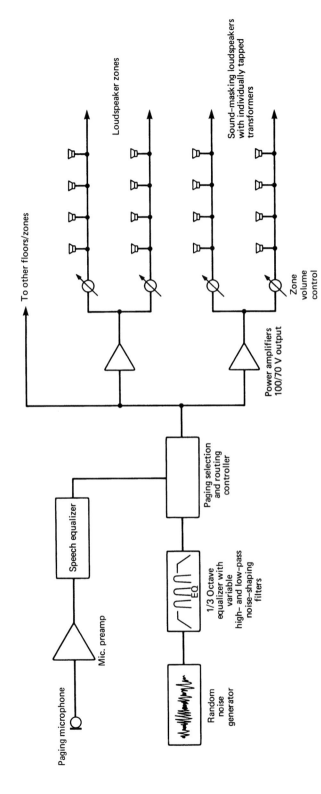

Figure 10.45 Simplified diagram of sound-masking system.

major characteristics of sound in an enclosed space, namely direction, intensity, frequency content and time of arrival. Sound reflectors and canopies have been used in concert hall design for many years, and it would appear to be taking almost as long for electroacoustic 'reflectors' to become accepted. The technique has been with us for over 30 years, since Peter Parkin began his experiments with 'assisted resonance' in London's Royal Festival Hall in the early 1960s[23].

10.11.1 Assisted resonance

When the Festival Hall opened in 1951 it was criticized for having a 'cold' orchestral sound, which was primarily due to the shortness of the low-frequency reverberation time. Although it would have been theoretically possible to correct for this by modifying the structure of the building, e.g. by raising the roof by several feet, the costs of such an approach were clearly prohibitive. Instead, an electronic solution was attempted – the objective being to feed back sound energy into the auditorium to make up for the over-rapid natural absorption of this energy. This is achieved by employing a number of microphone-amplifier-loudspeaker channels to pick up and re-radiate the sound energy.

Each channel is, as far as possible, frequency independent with respect to the others. This is achieved by the use of high Q acoustic filters (Helmholtz resonators) together with the selectivity afforded by the careful location of these units at nodal points within the ceiling. The operating principle of the system is shown in Fig. 10.46, together with a block diagram of a basic acoustic resonance system channel. The system installed in the Festival Hall used 168 of these channels distributed within the ceiling of the hall. The increase in reverberation time is shown in Fig. 10.47. The Festival Hall system operates between 50 Hz and 1200 Hz, and is very much an integral part of the acoustics, being used for every symphonic performance. Since the installation of the first Assisted Resonance system in the 1960s, the system was gradually developed and installed in a number of other auditoria and changing its role from being a corrective system to one which gives an auditorium greater flexibility of use. This is particularly the case with the trend of building multipurpose auditory spaces, supposedly capable of providing the facilities and acoustics to cater for a wide range of uses, e.g. from drama and conferences to flat floor functions; sporting events to jazz; and pop concerts to full-scale symphony concerts.

It has been known for many years that such uses require quite different acoustic conditions. For example, speech requires a reverberation time of around 0.9–1.1 s, with the intelligibility rapidly diminishing with reverberation times exceeding 1.4 s. Symphonic music, however, requires a reverberation time in the range 1.8–2.5 s, in order to achieve appropriate fullness of tone and fluidity. Table 10.6 further illustrates this by giving the preferred reverberation times for a range of presentations. Electronic alteration of an auditorium's acoustic characteristics is an obvious solution to the problem.

10.12 Electronic architecture

Since the 1960s, much research has been carried out in the fields of auditorium acoustics, subjective preference, psychoacoustics and acoustic measurement techniques. This research has shown that, although reverberation time is an important factor in the 'acoustic signature' and subjective assessment of an auditorium, it is far from being the whole story. Therefore, an apparently optimum reverberation time characteristic does not guarantee success and critical acclaim. It has been established that it is the microstructure of the reverberant decay curve which is the important factor. The decay curve can in fact be divided into a number of different constituent parts, each having its own definable influence on the perceived sound quality and characteristic.

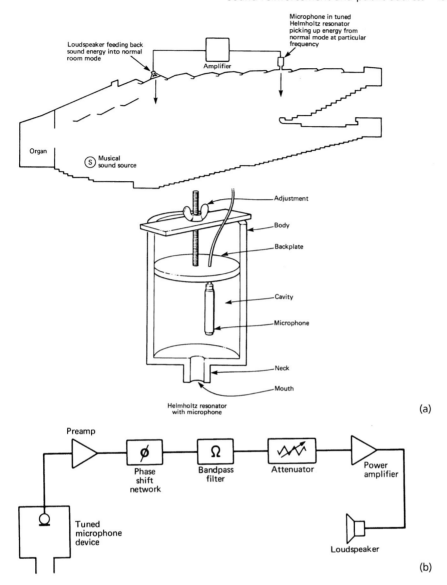

Figure 10.46. (a) Principles of assisted resonance system; (b) single-channel block diagram.

Initial research showed that the sound field in a large room or auditorium could be divided into three main components: the direct sound, the early reflections (and initial time gap) and the reverberant decay (the reverberation time) as shown in Fig.10.48. Later research has shown that the direction from which sound arrives at a listener, its frequency content and time of arrival all have an effect on the subjective impression. (For example, Shultz in 1965 showed that a small change of bass energy in the early portion of the reverberant field produces a much greater change in perceived bass warmth than does an equal change in the bass energy of the direct sound.)

Table 10.6 Preferred reverberation times (at 500 Hz)

Types of presentation	Reverberation time (s)
Cinema	1.0
Pop music	1.0
Speech	0.9–1.2
Drama	1.0–1.4
Comic opera	1.2–1.4
Baroque opera	1.2–1.4
Grand opera	1.4–1.6
Symphonic music (Baroque)	1.2–1.5
Symphonic music (Classical)	1.5–1.8
Symphonic music (Romantic)	1.8–2.5

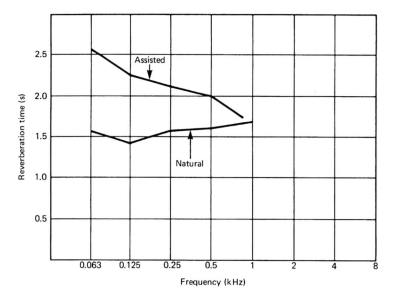

Figure10.47. Assisted resonance system, Royal Festival Hall – effect on reverberation time.

We now divide the sound field into four distinct parts and identify them as follows: direct sound, early reflections, late field and reverberant field. The direct sound provides localization and a foundation to be enhanced by later acoustic interaction. Early reflections, normally from a concert shell or side walls, enhance volume, timbre and articulation. They also provide a sense of liquidity, breadth and immersion. Late reflections, from more distant surfaces and from multiple 'bounces', are important in maintaining the warmth and impact of bass instruments. This is true also of the reverberant field, in addition to its prolonging and enriching effect. Furthermore, lateral (side) reflections are found to create quite different effects and responses from similar overhead reflections, the former being generally more desirable for musical performance.

Based on this knowledge, and current digital audio capabilities, a new breed of electronic acoustic enhancement and creation systems has arisen. Such techniques make full use and control of the four prime factors – direction, intensity, frequency

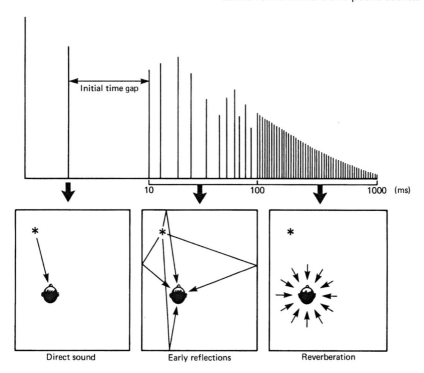

Figure 10.48. Schematic representation of sound field within an auditorium.

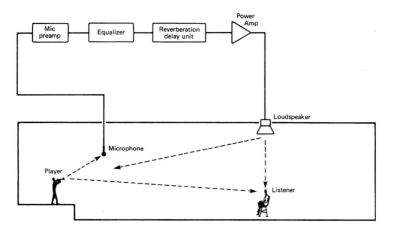

Figure 10.49. Simplified diagram of reverberation enhancement system.

content and time of arrival. Figure 10.49 presents a much-simplified block diagram of such a system. It can be seen that, unlike assisted resonance or multi-channel reverberation, it is the direct or early reflection sound component (and not the reverberant) which needs to be picked up and processed. The processing enables control to be achieved over the above factors as follows:

Table 10.7 ERES channel specifications

Characteristic	Frequency spectrum (Hz)	Time arrival (ms)
Presence/articulation	250–6000	0–20
Liveness	250–2000	300–2500
Reverberation	20–1500	300–3000
Warmth	20–250	60–300

- Direction – via the location and aiming of the enhancement loudspeakers. (These are not merely restricted to the ceiling of the auditorium but should be located in the side walls and under the balconies etc. as well.)
- Intensity – via the gain of each channel or loudspeaker group.
- Frequency content – by filtering and shaping the spectral content of the signal fed out from each loudspeaker group.
- Time of arrival – by electronically delaying the signal before re-radiating it. (Multiple reflection sequences and reverberation may also be added.)

One system which operates on the above principles is the 'Electronic Reflected Energy System' (ERES) conceived by Jaffe Acoustics in the USA[24]. The ERES system consists essentially of four basic channels, each catering for a specific acoustic effect/requirement, as shown in Table 10.7. Figure 10.50 presents a basic block diagram of such a system.

The first two channels are provided through the use of digital delay devices, and strategically located and distributed groups of 50 mm and 100 mm loudspeakers. Delay times are set to provide or augment the envelope of early field reflection patterns for a specific audience area. Each loudspeaker requires 10–20 W of amplifier power.

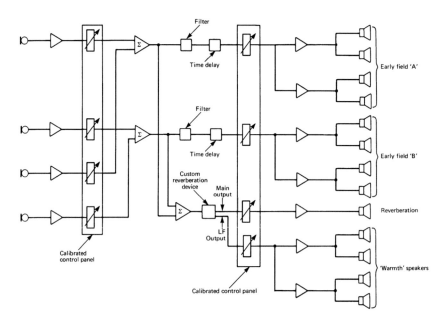

Figure 10.50. Simplified diagram of ERES system.

The second two channels (reverberation and warmth) are provided through the use of a custom, multi-tap, digital-reverberation device and distributed 200 mm coaxial and 300 mm loudspeakers. The custom reverberator provides a diffuse package of late-arriving reflections, gradually decaying but pseudo-randomly varying in amplitude and time spacing, which combines with the natural reverberation decay of the room. The reverberation time or the level of the reverberant field, or both, may be increased. Each transducer requires 20–50 W of amplifier power. Such systems are now enabling the architect to design the shape and size of hall needed to fulfil seating and economic requirements, while the acoustics can be tailored to create a totally different environment to suit the particular presentation required.

A number of other commercial systems have been developed along the above lines. The three main systems currently available are: LARES (USA), SIAP (Netherlands) and ACS (Netherlands). The exact way in which the systems work is commercially very sensitive and few details are available. However, each system consists of microphones, to pick up the direct sound over or near to the stage, signal processors/amplifiers and loudspeakers distributed around the auditorium. Premature feedback and coloration are two of the most difficult aspects to overcome. Clearly any system that relies on picking up sound in the same acoustic space as the radiating loudspeakers will be prone to feedback. SIAP and LARES use time variant processing which effectively can be thought of as continuously moving the loudspeakers around the room before feedback can build up – however, this is done electronically rather than physically. Gains of 6 to 18 dB can be achieved – not only enabling substantial prolongation of the reverberation time but also the running reverberation, the subjectively noticeable part during musical passages. Reverberation time increases of factors of 2–3 times are typically available (i.e. increasing an RT from around 1 second to 3 seconds) but for special effects reverberation times of around 6 seconds are also generally possible. A SIAP system typically employs some 60–70 loudspeakers distributed around the auditorium, whereas LARES generally appears to use rather less[25,26]. ACS employs a rather different approach whereby the microphones over the stage provide an image of the sound to the listener (rather like an acoustic holograph). The system is claimed to increase the reverberation time of a space by up to a factor of 4 (using a reverberant model based on auditoria with good acoustic characteristics). An increase in level of < 3 dB is achieveable[27]. (Further information on the systems can be found in the cited references.)

10.13 Cinema sound systems

Cinema sound systems went through a minor revolution in the 1980s in terms of their potential capabilities and quality of performance. Several factors are responsible for this but the prime one must undoubtedly be the development and introduction of Dolby stereo/surround-sound and Lucas Film THX. Today a cinema may work in mono with just one loudspeaker located behind the screen, or up to five full-range loudspeakers may be used to provide a stereo perspective with further 'surround-sound' loudspeakers located around the auditorium for various effects.

Most present-day soundtracks for commercial films are optical. They are located near to the edge of the film just inside the sprocket holes. The light from a small exciter lamp is focused on to the film, modulated by the film soundtrack and picked up by a photocell, the electrical output of which varies with the amount of light reaching it.

Optical soundtracks date back to the 1930s but in those days, and indeed until comparatively recently, they had a very poor frequency response, typically −25 dB at 9 kHz. Modern stereo optical soundtracks for 35 mm film have a very much improved response, extending to at least 12 kHz, and are often encoded with Dolby noise reduction so that they also offer a presentable signal-to-noise ratio. Matrix encoding is employed to derive from the two soundtracks the stereo left, right and combined centre signals together with the surround sound signal. This format is also made compatible with the old 'academy' mono format (with the mono sound head reading and combining both the left and right channels) so that the film can be shown in old-fashioned one-speaker cinemas.

Magnetic formats for film sound are also available, and there are three principal forms in current use: 35 mm, four-track, 'conventional' 70 mm six-track, and Dolby stereo 6-track. The six tracks of the conventional 70 mm magnetic format are used to provide left, centre and right signals together, and there are two further behind-the-screen channels for half-left and half-right loudspeakers, producing a seamless, stable multichannel sound image spread across a screen perhaps 24 m or more in width. The surround-sound signal is carried on the sixth track. However, when Dolby stereo evolved in the mid-1970s, it was decided that modern recording techniques and the layouts of new 70 mm cinemas rendered the additional full-range loudspeaker channels unnecessary. In the Dolby format, therefore, only the extreme left, centre and extreme right loudspeakers carry full-range information – the two additional channels carry only low frequencies, below 200 Hz. This serves to increase the overall low-frequency power-handling capability of the cinema sound system.

Figure 10.51 shows schematic diagrams for a basic system and the Dolby optical stereo system. Equalization and compensation are employed for the characteristics of both the electronic signal chain and the loudspeaker/auditorium house curve. The house curve equalizers (one per channel) incorporate both high- and low-pass shelving filters together with 27-band 1/3 octave filters. The equalization compensates not only for any loudspeaker or loudspeaker/room response anomalies, but also for high-frequency losses caused by the projection screen through which the sound has to pass. The surround sound channel(s) are routed via an audio signal delay unit, to allow advantage to be taken of the precedence effect, enabling correct localization to be maintained throughout the cinema. Bi-amplification is common, with a typical crossover frequency of 500 or 800 Hz.

Figure 10.52 presents the International Standard Curve for Cinema Sound Systems which, when achieved, should allow the same quality of sound to be produced from a film soundtrack no matter where the film is playing. Sub-woofers are also being used increasingly in cinema systems, operating down to 20 Hz from a synthesized signal. Figure 10.53 shows a typical cinema loudspeaker comprising bass bin and CD horn.

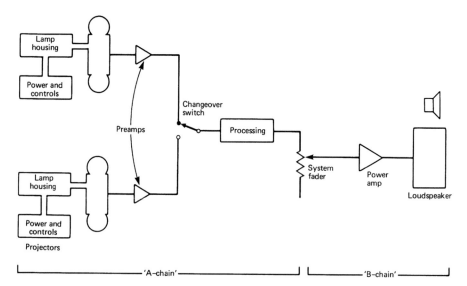

Figure 10.51. (a) Basic block diagram of cinema projection system.

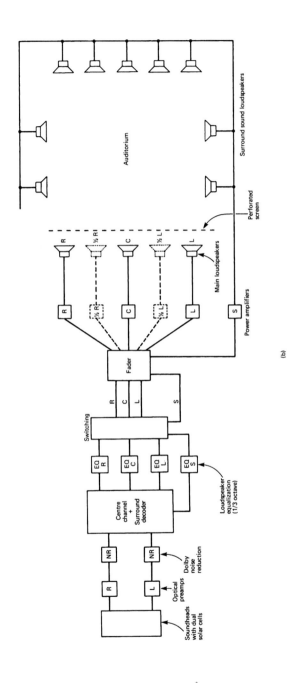

Figure 10.51 (*continued*). (b) Simplified schematic diagram of Dolby stereo cinema sound system.

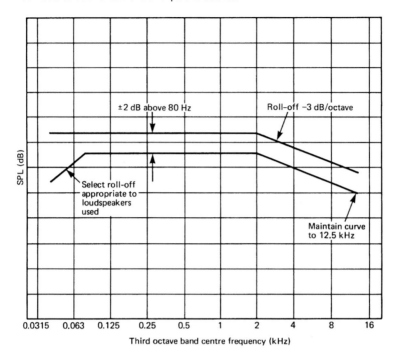

Figure 10.52. International standard response curve for cinema playback systems.

Figure 10.53. Typical commercial cinema loudspeaker (Courtesy JBL).

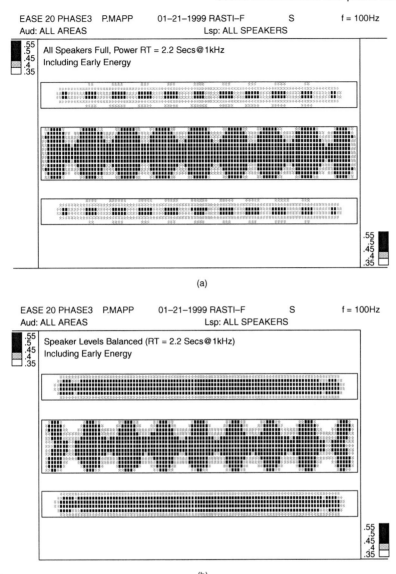

EASE 20 PHASE3 P.MAPP   01–21–1999 RASTI–F   S   f = 100Hz
Aud: ALL AREAS   Lsp: ALL SPEAKERS

.55
.5
.45
.4
.35

All Speakers Full, Power RT = 2.2 Secs@1kHz
Including Early Energy

.55
.5
.45
.4
.35

(a)

EASE 20 PHASE3 P.MAPP   01–21–1999 RASTI–F   S   f = 100Hz
Aud: ALL AREAS   Lsp: ALL SPEAKERS

.55
.5
.45
.4
.35

Speaker Levels Balanced (RT = 2.2 Secs@1kHz)
Including Early Energy

.55
.5
.45
.4
.35

(b)

Figure 10.54. CAD plot of Potential Intelligibility (Rasti) for a two-level shopping mall sound system: (a) result with all loudspeakers operating at full power; (b) improvement in intelligibility when the loudspeaker outputs are appropriately balanced.

10.14 Sound system performance and design prediction

The last 10–20 years has seen sound system performance improve significantly. This is not only in terms of the loudspeakers themselves, but also through a better and more widespread understanding of the acoustic environment in which they operate. The availability of computer aided design (CAD) programs have helped to achieve this, as have the increased demands of the consumer now accustomed to domestic CD, hi-fi and home cinema sound quality.

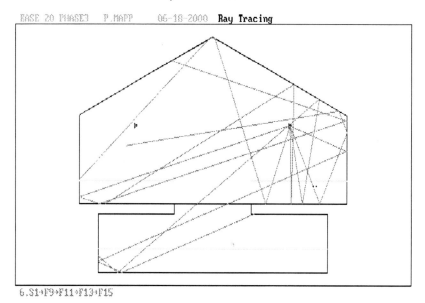

Figure 10.55. CAD model showing ray tracing.

CAD enables those unskilled in acoustics to produce system designs and layouts and then check them before installation. By inputting basic room data and speaker positions, programs are able to carry out an acoustic analysis of a space and predict with good accuracy the likely coverage and potential intelligibility. The accuracy of the results, of course, depends on how well the model is constructed, the skills of the original programmers and the accuracy of the data available. Most programs incorporate an extensive materials and loudspeaker database that allows a wide variety of options to be tried. Apart from carrying out traditional statistically based acoustic analyses, the advantage of the computer is that multiple ray tracing can be performed, allowing a detailed analysis of the sound arrivals at a given position to be made. This enables the effects of the direct sound, early and late reflections and reverberant sound field to be evaluated. Figure 10.54 shows a typical coverage plot and Rasti intelligibility produced by a computer model, while Fig. 10.55 shows a ray tracing analysis. However, as with all computer programs, the old adage of 'garbage in, garbage out' still very much holds true, and it should be remembered the program does not design the system for you, but merely allows the results of a design to be checked.

10.15 Microphones

As with any form of sound reproduction or hi-fi system, the input source is of critical importance to successful operation. If you do not have a good signal to begin with, then there is little or nothing that the system can do to overcome this. A good microphone is therefore an essential prerequisite for a good sound system. The principal factors which influence microphone design and use include frequency response, directivity, impedance matching and physical construction. Of these, directivity has a special relevance since, for example, using a directional (cardioid) microphone instead of a non-directional one may allow the loudspeaker level to be raised several dB before instability arises.

References

1. MAPP, P and COLLOMS, M, 'Improvements in intelligibility through the use of diffuse acoustic radiators in sound distribution', AES 103rd Convention New York (1997).
2. MAPP, P and ELLIS C, 'Distributed Mode Loudspeakers in sound reinforcement design – some case histories', *Proc IOA*, **21** Part 8 (1999).
3. PARKIN, P H and TAYLOR, J H, 'Speech reinforcement in St Paul's Cathedral', *Wireless World*, Feb. (1952).
4. VAN de WERF, J, 'A loudspeaker column with exceptional properties', AES 96th Convention (1994).
5. Glenn Leembruggen JBL Audio Engineering Conference Palm Springs (1999).
6. Duran Audio Intellivox Data sheet and installation guide (1999).
7. KATZ, B, 'The theory and practice of designing with directional beamforming loudspeaker arrays', *Proc IOA*, **21** Part 8 (1999).
8. EAW White Paper, 'Phased Point Source Technology' (1997).
9. HAAS, H, 'The influence of a single echo on the audibility of speech', *Acustica*, **1**, 49 (1951), reprinted *JAES*, **20** (1972).
10. WALLACH, H et al., *Am. J. Psychology*, **62**, 315 (1949).
11. MEYER, E and SCHODDER, C R, *Gottinger Nachrichten Math. Phys. KI.*, **11a** (1962). Also in reference 29.
12. LOCHNER, J P A and BURGER, J F, *Acustica*, **8**, 1 (1958).
13. PEUTZ, U M A, 'Articulation loss of consonants as a criterion for speech transmission in a room', *JAES*, **19** (1971).
14. STEENEKEN, H J and HOUTGAST, T, 'Some applications of Speech Transmission Index (STI) in auditoria', *Acustica*, **51**, 229–234 (1982).
15. STEENEKEN, H J and HOUTGAST, T, 'Rasti: a tool for evaluating auditoria', Bruel & Kjaer Technical Review, No. 3 (1985).
16. IEC 849/BSEN60849.
17. MAPP, P, 'Objective speech intelligibility testing of sound systems', *Proc IOA*, **21**, Part 5 (1999).
18. MAPP, P, 'Practical limitations of objective speech intelligibility measurements'. AES 102nd Convention, Munich, March (1997).
19. FLETCHER, H, *Speech and Hearing*, Van Nostrand, New York (1953).
20. HOCHBERG, I, 'Most comfortable listening for the loudness and intelligibility of speech', *Audiology*, **14**, 27–33 (1975).
21. MAPP, P, 'How loud is loud enough?' *Sound & Video Contractor* **14** No. 1, Jan. (1996).
22. MAPP, P, 'Some effects of equalization and spectral distortion on speech intelligibility'. *Proc IOA*, **19**, Part 6 (1997).
23. PARKIN, P H and MORGAN, K, 'Assisted resonance in the Royal Festival Hall, London: 1965–1969'.
24. JAFFE, C J and LOBB, W, 'Reflected energy designs for auditoria', 64th AES Convention, New York (1979).
25. PRINSSEN, W and KOK, B, 'Technical innovations in the field of electronic modification of acoustic spaces', *Proc IOA*, **16**, Part 4 (1994)
26. GRIESINGER, D, 'Progress in electronically variable acoustics'. *Proc Wallace Sabine Centennial Symposium*, ASA, June (1994).
27. BERKHOUT, A, 'A holographic approach to acoustic control', *JAES*, **36**, Dec. (1988)
28. ISO 2969.
29. CREMER, L and MULLER, H A, *Principles and Applications of Room Acoustics*, Vol. 1. Translated by SCHULTZ, T J, Applied Science Publishers, London (1982).

Bibliography

HARRIS, N and HAWKSFORD M, 'The Distributed Mode Loudspeaker as a broad-band radiator', AES 103rd Convention New York (1997).
DAVIS, D and DAVIS, C, *Sound System Engineering*, Howard Sams, Indianapolis (1987).
KEELE, D B, 'What's so sacred about exponential horns?' AES preprint No. 1038 (1975).
TAYLOR, P H, The line source loudspeaker and its applications', *British Kinematography*, **44**, No. 3, March (1964).
LIM, J S (Ed.), *Speech Enhancement*, Prentice Hall, Englewood Cliffs, NJ (1983).

MAPP, P, and DOANY, P, 'Speech intelligibility analysis and measurement for a distributed sound system in a reverberant environment', AES 87th Convention New York (1989).

MAPP, P, 'Reaching the audience', *Sound & Video Contractor*, **17** No. 11 (1999).

MAPP, P, 'Sound out of doors', *Sound & Video Contractor*, Jan. (2000).

MAPP, P, Audio System Design and Engineering: Part I 'Equalizers and audio equalization', Part II 'Digital audio delay', Klark Teknik (1985).

MAPP, P, 'The loudspeaker–room interface'. *Hi-Fi News & Record Review*, **20**, No. 8 (1981).

PARKIN, P H, HUMPHREYS, H R and COWELL, J R, *Acoustics, Noise and Buildings*, Faber & Faber, London (1979).

BONNER, C R and BONNER, R E, 'The gain of a sound system', *JAES*, **17**, No. 2 (1969).

MAPP, P and ELLIS C, 'Improvements in acoustic feedback margin in sound reinforcement systems', AES 105th Convention San Francisco (1998).

'Sound reinforcement' (Anthology), *JAES*, **1–26**, by various authors.

BARNETT, P W, 'Electro-acoustic reverberation enhancement systems'. *Studio Sound*, June (1984).

DE KONING, S H, 'Multiple channel amplification of reverberation (MCR)', *Proc. IOA*, Sept. (1982).

ALLEN, I, 'The production of wide range, low distortion optical sound tracks utilizing the Dolby noise reduction system', *JSMPTE*, **84**, Sept. (1975).

HODGES, R, 'Sound for the cinema', *dB Magazine*, March (1980).

BLAKE, L, 'Mixing techniques for Dolby stereo film and video releases', *Recording Engineer/Producer*, June (1985).

UZZLE, T, 'Movie picture theatre sound', *Sound and Video Contractor*, **14–20**, June (1985).

MEAD, W, 'A new dimension in cinema sound', *Sound and Video Contractor*, 112–116, June (1985).

11 Loudspeakers for studio monitoring and musical instruments

Mark R. Gander
(with acknowledgement to Philip Newell)

PART 1: STUDIO MONITOR LOUDSPEAKERS

11.1 Introduction

With the possible exception of audiophile hi-fi debates, no areas of loudspeaker design and evaluation are so hotly contested as those concerning studio monitor and musical instrument loudspeakers. These products are used in professional work which must yield saleable commercial results, yet the ultimate judgement is still subjective rather than objective. While sophisticated measurements and objective standards may be used to design and evaluate their characteristics, it is the personal judgements of the musician, recording engineer and ultimately the audience which determine the real value of the units. This chapter will review historical and current practice, using classic or widely accepted designs to illustrate techniques, methods and philosophies.

11.2 Studio monitor performance requirements

Various authors have set down typical criteria for studio monitor loudspeaker performance characteristics[1-4]. These can be ranked in a rough order of importance, though all designs vary in the degree to which they achieve these goals or reorder the priorities.

11.2.1 Smooth wide-range axial frequency response

The accuracy of the on-axis pressure amplitude versus frequency response is the main standard of judgement for a studio monitor loudspeaker. Response is usually quoted as a plus or minus decibel tolerance across the loudspeaker's usable frequency range. Plus or minus 3 dB (6 dB total window) is normal for high-quality designs, with ±2 dB or better being possible with careful control of loudspeaker mounting and other environmental factors. The response roll-off at the low- and high-frequency extremes may be described by the 3 dB down point (half-power), or alternatively the frequency range of the monitor may be specified with a more forgiving 6 dB down (half-pressure) or 10 dB down (half-loudness) tolerance.

The frequency-response range of 50 Hz to 10 kHz is the most important, for it encompasses most of the musical energy; the obvious target of 20 Hz to 20 kHz is possible only in the largest multiway systems, with current technology. Both these specifications for frequency coverage limits conform to the rule-of-thumb recommendation that the product of the low-frequency and high-frequency limits of an audio transmission system should equal approximately 500 000 for subjectively balanced reproduction[5].

Even within a 2 dB tolerance window there is much room for subtle rises or sags in selective frequency ranges, which can allow two monitors with the same specifications to have significantly different sound characteristics. When it is considered that even a good, representative, consumer loudspeaker may exhibit ±6 dB or worse response, differing monitor characteristics and choices based on personal preference are still within acceptable bounds.

11.2.2 High output capability

The studio monitor must be capable of reproducing sound levels which reveal flaws in the recording process, and at least reproduce realistic performance levels. For speech and solo or chamber music, 85–90 dB may be sufficient, for orchestral music 95–100 dB with 105 dB loudest peaks will be adequate, while for rock, pop, or synthesized music, 115–120 dB and above may be required. Assuming that electronic and acoustic noise levels are minimized, these output capabilities will yield realistic dynamic range with which to monitor the recording.

The studio monitor can achieve these sound levels through a combination of high sensitivity/efficiency and high power-handling capability. High power-handling capability has the additional benefit of enhanced ruggedness, protecting the unit from the occasional abuse which is inevitable within the working studio environment. Lower sensitivity and efficiency is often traded-off for increased bandwidth and smaller physical size. Power handling should not be increased at the expense of efficiency without careful note being taken of the actual resultant output. Excessive input requirements will not only decrease reliability, but also decrease dynamic linearity through voice-coil heating and resultant power compression[6].

11.2.3 Low non-linear distortion

The output pressure waveform must be relatively free of harmonic, inharmonic and intermodulation distortion products at the required monitoring levels.

The driving force of the loudspeaker motor (usually a voice-coil and magnet structure) and the restoring force of the suspension elements (the diaphragm surround and/or centring spider) are normally relatively linear over only a very narrow range of motion. Their complex interactions can yield characteristic distortion signatures for different design topologies[7]. Diaphragms are subject to 'break-up' distortions above the piston band, which can create both harmonic and sub-harmonic distortion products. Under severe stress, 'cone cry' will yield non-harmonically related output, i.e. inharmonic distortion[8].

Interaction between multiple drive units and passive network elements can cause distortions, as can enclosure panels, bracing, lining and acoustic loading effects such as porting[6]. Air itself is subject to significant non-linearity past the 150 dB range, which can occur within enclosures and horn throats[9].

11.2.4 Accurate time domain transient response

While virtually all single loudspeaker units exhibit minimum-phase response, virtually all multi-unit systems do not[10]. For the case where the highs arrive today and the lows tomorrow, it is clear that a non-minimum phase response will lead to deteriorated system accuracy[11]. What has been less clear is the extent to which time and phase response alone, or their effect on amplitude accuracy, must be controlled. Experiments have shown that trained observers are able to detect changes of the order of 0.5 ms[12]. Much controversy remains, and there are those who extol both the measured and perceived benefits of approaching perfect zero phase-shift between drive elements.

Transient response is a function of both the amplitude and phase accuracy of a system. Square wave, tone burst and delayed resonance testing should all yield in the frequency domain the same information which relates the amplitude and phase in the time domain from the impulse response (see Chapter 12).

11.2.5 Controlled dispersion

The pressure field from a studio monitor loudspeaker should extend evenly with frequency across some defined horizontal and vertical 'window of consistency'. This is desirable so that, for example, more than one listener can hear the optimum sound field, but also so that the early reflections and reverberant field are excited with substantially the same spectrum. The window of consistency, within which the pressure should drop by no more than 6 dB from the on-axis value, should be a minimum of 45–90° in the horizontal plane and 20–40° in the vertical. Cone diaphragms narrow their directivity with increasing frequency, so that crossing over to progressively smaller drive elements becomes necessary. Horn flares can control the beamwidth angles, within the limits of mouth size constraints.

If coverage angles are held successfully, not only the axial pressure response but also the total power response can be made flat and smooth. If the room acoustics and the reverberation response with frequency are well behaved, then the reverberant field will be consistent with the direct field at the listening positions[13]. In difficult acoustic environments, narrow coverage angles and concentration on the direct field alone, or monitoring in the near-field, will give fewer monitoring problems.

11.2.6 Controlled impedance

The complex load impedance which a typical monitor loudspeaker presents to the driving amplifier can cause variations in the sound character presented. Lead resistance, capacitive and inductive load swings, and back e.m.f. interacting with amplifier feedback all yield measurable and audible changes in speaker and amplifier response. Both the nominal and minimum impedance moduli, as well as the phase swings present, should be kept to within reasonable bounds or clearly noted as potential problems for the driving amplifier[14].

11.3 Significant monitor designs

11.3.1 Coaxial designs

The most long-lived and popular of all the studio monitor design concepts from the 1950s through the 1980s were the coaxial drivers. The concept of a compact, high-efficiency, wide-dispersion point source has great appeal and conforms to the theoretical concept of a perfect loudspeaker[15]. These designs first appeared in the 1940s in both England and the USA and have continued in only slightly modified form to the present day.

The American coaxial, the Altec 604, was developed by James B. Lansing after the Lansing Manufacturing Co. merged with All Technical Services (Altec) theatre maintenance company in 1941. It was first presented as a field-coil energized transducer in 1943, and re-introduced with Alnico permanent magnets in 1945[16,17]. The basic concept was a 380 mm low-frequency driver with a large 75 mm voice-coil diameter, permitting a high-frequency compression driver to be mounted at the rear of the magnet assembly that could radiate through the centre to a small multi-cell horn mounted in the middle of the low-frequency cone. The compression driver had a 44 mm diameter aluminium diaphragm with a tangentially formed surround and circumferential phase plug. A passive crossover in the 1 kHz to 2 kHz range was used, and various sizes of ported enclosure were recommended, depending on the application.

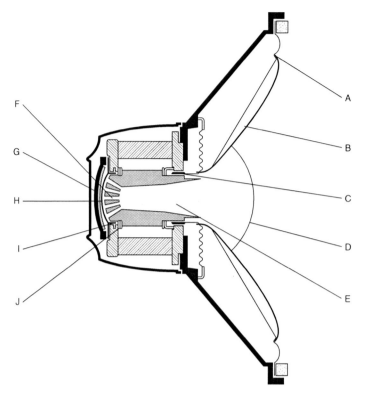

Figure 11.1. Dual concentric loudspeaker with Alnico magnet structure. (Courtesy Tannoy Ltd.) Key: A, rolled surround for stability in low bass response; B, ribbing which virtually eliminates cone break-up, ensuring smooth response and very high power capacity; C, high-temperature voice-coil; D, dustproof, acoustically transparent sealing dome; E, concentric HF horn (completed by curved LF cone); F, phase-compensating multiple throat for extended and smooth HF response; G, acoustic balance cavity for reduced distortion; H, precision contoured high-frequency diaphragm; I, aluminium voice-coil conductor for high power capacity and excellent HF response; J, exclusive magnetic shunt for increased LF flux.

Although it was initially designed for motion-picture monitoring and playback, the Altec 604's quality and performance established it as a standard during the early days of magnetic tape recording in the 1950s. Various improvements and changes were made through the years in the diaphragm, phase plug and network design. Companies other than Altec made successful modifications to the crossover design and augmented the low-frequency response with auxiliary woofers. The Mastering Lab in Los Angeles devised a particularly successful crossover which was marketed as the 'Big Red' series by Audio Techniques of Stamford, Connecticut.

Parallel to the development of the American Altec 604, the British Tannoy Company was working on two loudspeaker concepts. One was to be a wide-band laboratory point source for measurement calibration, the other a high-sensitivity and high-power handling unit for public address and live music reinforcement. Ronnie Rackham of Tannoy designed a coaxial product to do both jobs, the Tannoy Dual Concentric, and moved both the horn and the driver behind the woofer, using the low-frequency cone to complete the horn flare. This solved the problem of diffraction interference from the high-frequency horn blocking the cone, at the expense of

(a)

(b)

Figure 11.2. Time-Align® studio monitors: (a) original 813 using Alnico Altec 604–8G co-axial transducer, with proprietary exponential horn; (b) 813A using ferrite Altec 604–8K co-axial transducer, with improved horn treatments. (Courtesy UREI.)

potentially greater intermodulation distortion. The Tannoy Dual Concentric was first shown in 1947 at the London Audio Show, and it subsequently became very successful with the BBC and, in its 300 mm version, Decca Records. The late 1960s and 1970s brought variations which traded reduced efficiency for increased bandwidth and power handling (Fig. 11.1).

In the late 1970s, the interest in time domain accuracy led to another revitalization of the Altec 604 coaxial design by UREI. Bill Putnam of UREI, working with Ed Long and his patented Time Alignment® process, created the 813 monitor[18, 19]. This made the 604 into an almost perfect point source in the third dimension (effective depth) as well as in height and width[20]. An elaborate passive crossover network incorporated the time-corrective circuitry as well as protective elements for the drivers. An auxiliary woofer and damped-port enclosure completed these Time-Align® monitors. The excellent execution of the concept, together with the already wide acceptance of the Altec coaxial design, ensured success for the UREI 813 as a new dominant studio standard into the 1980s (Fig. 11.2).

The cobalt shortage of the late 1970s caused both Altec and Tannoy to convert their co-axials from aluminium–nickel–cobalt Alnico magnet designs to ferrite structures. The differing magnet shapes, and the consequent necessity for different motor topologies and compression drive throats (and the potential for magnetic circuit distortions), meant that the new drivers were not necessarily exact equivalents in performance[21].

The 1980s saw the availability of co-axial designs from other manufacturers based on simple ferrite low-frequency drivers, some of which could be fitted with any chosen compression high-frequency driver.

11.3.2 Two-way discrete designs

Separate woofer and horn/driver designs predate the co-axials, and as a concept, if not in specific form, have maintained equal status as monitors of popular choice among recording engineers. More flexibility is possible in the design of the individual drive units, and the mid-range crossover and horn coverage can be smoother and much more controlled. The sacrifice of the point source image can be reduced as a major liability if monitoring is done at a greater distance, and if vertical response angle demand is restricted.

Discrete two-way designs were also first developed as compact motion picture playback systems. One of the earliest, dating to the late 1930s, was the Lansing Mfg. 'Iconic'. It consisted of a 44 mm aluminium diaphragm compression driver on a multi-cell horn on top of a 380 mm cone speaker in a vented enclosure (Fig. 11.3).

The Altec-Lansing A7 'Voice of the Theatre' was a compact version of the giant Altec A4 and A2 industry-standard theatre systems[22]. It utilized a combination ported horn low-frequency enclosure, and 'sectoral' exponential-type high-frequency horn rather than the multi-cellular designs used in the larger systems. It also used a 44 mm diaphragm, 25 mm throat driver and a 500 Hz or 800 Hz passive crossover, similar to the original Iconic. Later, Altec adapted these components to more compact direct-radiator low-frequency designs, in both single 380 mm and dual 300 mm configurations. All these variations enjoyed success as studio monitors.

Lansing left the Altec Company and formed James B. Lansing Sound (JBL) in 1946. After Lansing's death in 1949, the designs of Bart Locanthi and the implementations of Ed May dominated JBL monitors through the 1950s and 1960s. Locanthi brought the concept of the acoustic lens to studio monitors for wider dispersion from an exponential horn design[23]. The 4320 Studio Monitor was the professional designation for the S7 system/C50SM enclosure which had established itself from the early 1960s with Capitol Records and others in the growing recording business in Los Angeles. The design still used a 25 mm throat, 44 mm aluminium diaphragm compression driver, but with a larger magnet for more flux, and enhanced phasing plug and diaphragm mounting design. The low-frequency driver used a 100 mm

Figure 11.3. Lansing 'Iconic' two-way system of the late 1930's, with field-coil energized transducers. (Courtesy of JBL Inc.)

diameter, short voice-coil in a deep magnetic gap for long, controlled travel. A subsequent variation, the 4325, became popular in England with Abbey Road and other prestigious studios. In the 1970s the same basic concept evolved into the 4330 and 4331 versions, using a more conventional woofer design with overhanging coil topology.

After the success of the UREI 813, the beginning of the 1980s saw JBL introduce a new discrete two-way concept, the Bi-Radial® Studio Monitor[13]. In place of the acoustic lensed exponential horn, a constant-directivity design with consistent 100° by 100° dispersion was used. The concept was that of smooth off-axis and energy response, so that off-axis listeners and the reverberant room field experienced the same characteristics as the on-axis response. Further woofer and compression drive refinements were added, along with inherent acoustic alignment for phase coherency and minimum group delay (Fig. 11.4).

Custom monitors as recommended by leading studio consultants such as Tom Hidley, founder of Westlake Audio, and George Augspurger have become quite common in large expensive studios. These are typically a single horn or horn/lens with a 100 mm diaphragm large-format compression driver, flush-mounted in the control-room wall above two side-by-side 380 mm low-frequency units (Fig. 11.5). These systems are typically bi-amplified with an electronic crossover at 800 Hz[24].

In 1983 Joseph D'Appolito published a paper on the dispersion and crossover consistency benefits of mounting two low-frequency transducers in an over-under configuration about a central high-frequency device[25]. The practical benefits of this approach spurred the adoption of this format for two-way designs by leading studio designers including Shozo Kinoshita, Philip Newell and others (Fig. 11.6).

(a)

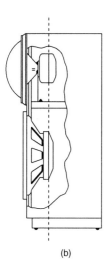

(b)

Figure 11.4. Bi-Radial® studio monitors: (a) 4430 single 380 mm and 4435 dual-380 mm; (b) cutaway side view diagram of driver acoustic alignment. (Courtesy JBL Inc.)

11.3.3 Three-way designs

Whether to increase bandwidth and reduce distortion, or merely to overcome component deficiencies, three-way (or augmented two-way) systems typically add an extra high-frequency tweeter element. The Klipsch 'La Scala', an all-horn-loaded design variation on the well-known Paul Klipsch corner horn designs, and the 'Cornwall', originally with a unique vented low-frequency section, are classic examples which had their share of proponents for monitor use in the 1950s and 1960s[26,27].

Electro-Voice introduced the Sentry III and Sentry IV three-way systems in the early 1970s. Both utilized a larger sectoral (radial) horn with phenolic diaphragm

Figure 11.5. Classic dual-380 mm two-way studio monitor configuration with wide-angle diffraction high-frequency horn. This particular 1970s embodiment also added a 25 mm exit compression driver without a horn for ultra-high-frequency augmentation. (Courtesy Westlake Audio.)

Figure 11.6. Large studio control room with main monitors utilizing round axisymmetric waveguide horn with low-frequency drivers in over/under, D'Appolito configuration. Near-field monitors are visible on stands behind console. (Courtesy Reflexion Arts.)

Figure 11.7. 4333A three-way monitor. (Courtesy JBL Inc.)

driver and a small radial tweeter with an aluminium diaphragm. The Sentry III has a vented 380 mm driver low-frequency section. The Sentry IV mounts two 300 mm drivers in a large folded horn for greater efficiency at the expense of enclosure size and low-frequency response[28].

The JBL 4320 two-way could be upgraded with the addition of an annular-ring diaphragm diffraction slot radiator. The subsequent 4332 and 4333 were configured as three-way systems from conception, with provision for mirror-imaging of the non-symmetrical tweeter position (Fig. 11.7).

Various custom studio monitors have added auxiliary high-frequency devices. Some of the more unusual concepts include mounting two diffraction tweeters with their mouth openings rotated 90° relative to one another, and employing a 25 mm throat compression driver without a horn as a direct diffraction source (Fig. 11.5).

11.3.4 Four-way designs

Part of the concept for the reworking of the JBL monitor line in the early 1970s was that of the four-element design. A single cone driver was added for the two or three lower mid-range octaves, with the result that the monitor could be more efficient and exhibit less intermodulation distortion while maintaining wide bandwidth. The JBL 4350 was the first example of this concept, introduced in 1971 in a bi-amplified configuration (Fig. 11.8). It found acceptance as a main reference monitor in many studios in the British Isles. The 4343, basically a 4333 with an added 250 mm low-mid driver, found similar acceptance throughout Europe and Asia (Fig. 11.9).

By the 1980s, the benefits of active electronics drive and control to handle the complexity of four-way systems was fully realized and accepted. Systems with dedicated signal processing electronics and amplification, in some cases on-board the speaker system itself, were developed by such companies as Westlake Audio in the

Figure 11.8. 4350B four-way monitor system. (Courtesy JBL Inc.)

Figure 11.9. 4343B four-way monitor system. (Courtesy JBL Inc.)

USA, State of the Art Electronik in Canada, Quested in the UK, and others (see Section 11.3.7).

11.3.5 Dome midrange monitors

The dissatisfaction of some users with typical horn characteristics, yet the need for high sound pressure levels, led to specific development of the high output dome midrange. In the late 1970s Bill Woodman of ATC developed a 75 mm diameter soft fabric dome driver. It used a dual suspension system and a short-coil/deep-gap topology in its large magnetic motor system (Fig. 11.10). The dome diameter was small enough to allow wide-dispersion operation up to a 3–5 kHz crossover range. The large 75 mm diameter voice-coil and magnetic motor assembly achieved relatively high output, high power handling for wide dynamic range, low power compression, and low distortion, all from a crossover point as low as 400 Hz. Monitors using the device first achieved prominence in the 1980s from Roger Quested (Fig. 11.11). Ultimately other manufacturers built similar devices of the same diameter to be used in the Quested designs as well as other monitors, and by the late 1980s ATC were successfully marketing monitor systems using their own midrange unit (Fig. 11.12).

A subsequent development was that of a 100 mm dome device designed by Stanley Kelly and employed in Neil Grant's Boxer systems. The larger dome and 100 mm motor system, with a designated bandwidth of 250 Hz to 2.5 kHz, further increases efficiency and power handling, and further decreases excursion requirements and thermal compression. The three-way Boxer system integrates it with a 38 mm dome tweeter and four 300 mm low-frequency drivers, dedicated crossover electronics and amplification (Fig. 11.13)[29].

PMC (The Professional Monitor Company), founded by former BBC engineers, achieved notable success in the 1990s, particularly in British mastering facilities, with dome designs that feature transmission line low-frequency loading.

11.3.6 Compact 'near-field' monitors

In those applications not allowing large physical size and/or not requiring high output levels, smaller two- or three-way systems typically employing all direct-radiator drive elements can be successfully employed. They are usually mounted on the console top or otherwise close to the listener, so that they are referred to as 'close-field' or 'near-field' monitors (Fig 11.6). This monitoring in the near-field reduces the influence of the room acoustics on the perceived sound.

The BBC had a particularly active research department in the 1950s and 1960s, from which such engineers as D. E. L. Shorter and H. D. Harwood produced and published many designs and ideas. The LS series of BBC monitors included innovations in the use of Bextrene cone materials, surround design, equalization and crossover circuitry, driver slot loading, and measurement analysis techniques[30].

The classical music recording community in general, and many classical studios in England and Europe specifically, have tended to avoid horn-loaded monitors. Classical monitoring does not require the acoustic levels necessary in pop recording, and hence the high sensitivity and power handling of horn-loaded systems are not required. More than that, there is a prejudice against the 'horn sound', the inherent choppiness in response and throat distortion of traditional designs. The smoother response, lower distortion and lower efficiency of systems by KEF, B&W and others have often been preferred as the studio reference in such applications (Fig. 11.14). (Fast flare-rate, wide coverage-angle horns, such as those employed in the JBL Bi-Radial® monitors (Fig. 11.4) and the axisymmetric round 'waveguides' such as used in the Reflexion Arts monitors (Fig. 11.6), address this and all but eliminate any of the 'horn sound'.)

Compact monitors are also used for deliberate comparison with the main monitors. This allows the engineer and producer to check how the programme mix will

(a)

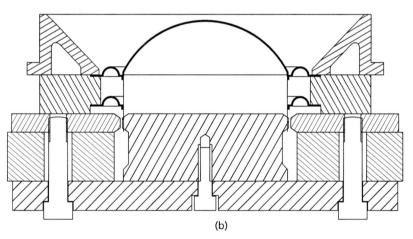

(b)

Figure 11.10. ATC 75 mm soft dome driver. (a) Cutaway view; (b) dynamic section. (Courtesy ATC Loudspeaker Technology Ltd.)

Figure 11.11. Quested Q212 monitor. (Courtesy Quested Monitoring Systems.)

Figure 11.12. ATC SCM 50A system. (Courtesy ATC Loudspeaker Technology Ltd.)

Figure 11.13. Discrete Systems Boxer system. (Courtesy Discrete Systems Ltd.)

(a)   (b)

Figure 11.14. 801F system: (a) front view; (b) cutaway side view. (Courtesy B&W Loudspeakers Ltd.)

sound on typical domestic speakers. For this reason, domestic 'bookshelf' units like the KLH 17, AR-3A and Large Advent first found their way into this application in the late 1960s and early 1970s. One of the first loudspeakers of this size to be specifically designed for professional use was the JBL 4310, later 4311. A three-way 300 mm woofer direct-radiator design, it was originally conceived as a compact monitor for Capitol Records designed to emulate the sound of the large JBL horn two-way units (Fig. 11.15). It was subsequently marketed as the L100 Hi-Fi speaker, thus establishing a position at both ends of the commercial recording chain, and both implementations became extremely successful.

Another type of reference sometimes necessary is a small loudspeaker which simulates the restricted bandwidth and output of a cheap portable or car radio. While studios can employ any inexpensive unit for this type of monitoring, the Auratone cubes have been adopted by many studios as a 'reference' cheap speaker (Fig. 11.16).

Despite their dominance in the hi-fi market, the Japanese have had relatively little success in establishing their designs in the field of studio monitors. The Yamaha NS-1000 has received some acceptance as a monitor as well as a hi-fi speaker[31]. The Nakamichi Reference Monitors, manufactured by Mitsubishi, were found alongside JBL four-way units at NHK Broadcast and large studios in Japan in the 1970s[32,] More recently the TAD/Kinoshita designs have found favour as a home-market reference, as well as appearing in other countries.

In the 1980s one Japanese compact design, the Yamaha NS-10 200 mm two-way hi-fi speaker, did become the next *de-facto* reference standard for console-top 'near-field' monitoring. Its success began when it was adopted by major recording engineers

Figure 11.15. 4310 control monitor and all Alnico-magnet components. (Courtesy JBL Inc.)

searching for the smallest speaker that would still yield a minimally acceptable bandwidth. In this way its success can be seen as the next step in compact monitor evolution, combining the functionality of the JBL 4311 and the Auratone. After sustained industry demand, Yamaha responded with a dedicated pro version, the NS-10M (Fig. 11.17), an interesting reversal of the JBL 4311 to L100, Pro to hi-fi transformation. While widely applied and ubiquitous as a reference in large commercial and home project studios alike, a general agreement as to its harsh and forward high-frequency character led to the practice of adding a layer of tissue in front of the tweeter to attenuate the response in that range[33].

Figure 11.16. Model 5C Super-Sound-Cube. (Courtesy Auratone Corp.)

Figure 11.17. Yamaha NS-10M monitor. (Courtesy Yamaha International Corp.)

The late 1980s and 1990s saw an extraordinarily wide range of compact monitors available in the market, coinciding with the rise in home recording and the project studio. As small loudspeakers are inevitably born of compromises, their acceptance is often highly subjective. Genelec has emerged as the worldwide market leader with their integrated active range, the 200 mm two-way model 1031A being the most popular (Fig. 11.18). In Europe, the most successful in addition to the Genelecs include the Quested F11, and the ATC SCM range. In the USA, models from KRK,

Figure 11.18. Genelec 1031A powered studio monitor. (Courtesy Genelec Inc.)

Alesis, Mackie and Event Electronics have seen significant success. Tannoy are quite well placed on both sides of the Atlantic with numerous models, as are relative newcomers such as Dynaudio from Denmark, and FAR in France and Belgium. The Yamaha NS-10 continues in widespread use, however by the mid-1990s integrally self-powered loudspeakers had become the dominant industry preference.

11.3.7 Integrally powered designs

Early attempts at integrally powered ('active') monitor concepts began with the 'energizer' amplifiers added to JBL designs in the 1960s, and the bi-amp modules available from Altec for their coaxial and other two-way configurations. Also notable are a variety of self-powered designs from Klein + Hummel in Germany, standardized upon by Deutsche Rundfunk (German Broadcast). One of the first integrated active monitors to achieve relatively wide acceptance is the Meyer HD-1 (Fig. 11.19). It is a 200 mm two-way direct radiator design employing bi-amplification and patented amplitude and phase correction circuitry to maintain linear phase and a symmetrical central lobe through the crossover region[34].

The completely integrated multi-way active monitor has perhaps seen its greatest expression in the Genelec designs from Finland. Complete matching of the transducers, active crossover, amplification, and enclosure elements eliminates the unknown factors in the assembled system. Each element is designed to function as a part of the whole, with the net acoustical output the ultimate goal in mind, and the entire product line endeavours to differ only in low-frequency cut-off and acoustic output level. The designs also employ a directivity control guide around the mid and high frequency transducers, designed to match the directivity of the drivers at the crossover point in order to yield a predetermined total directivity and smoother power radiation into the room (Fig. 11.20).

The late 1990s revealed the resurgence of JBL as a significant competitor in the monitor market with the appearance of their LSR line. The concept of Linear Spatial Reference, smooth and even coverage across the listening field, was based on the work of Floyd Toole[35]. The monitors also feature innovative transducer designs using dual coil and dual gap drive, with neodymium magnets and dynamic braking[36] (Fig. 11.21).

Figure 11.19. Meyer Sound HD-1 monitor. (Courtesy Meyer Sound Laboratories, Inc.)

Figure 11.20. Genelec 1035B monitor system. (Courtesy Genelec Inc.)

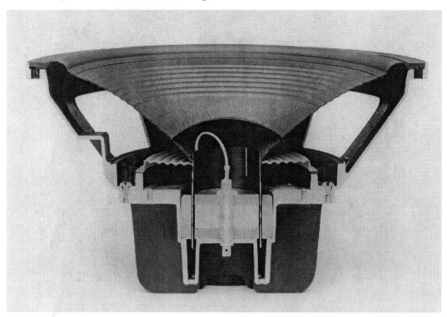

Figure 11.21. 300 mm low-frequency studio monitor transducer utilizing a neodymium magnet energizing dual gaps and driving coils. There is also a third, centrally positioned, shorted coil on the assembly which acts as a dynamic brake. (LSR Series woofer, Courtesy JBL Inc.)

11.3.8 Subwoofers

Low-frequency augmentation is often accomplished through the addition of a supplementary bass bandpass section, or subwoofer. The availability of clean signals down to the lowest frequencies via digital recording methods have spurred interest in reproduction systems that can accurately handle the lowest audible octaves. In some cases, notably hard rock, heavy metal and rap recordings, it may be primarily additional level rather than bass extension that is required. Techno, house and other electronically-derived forms may require both extended bandwidth and additional level, and the purist would argue that this capability is necessary in any case to extend the accuracy of the reproduction.

The most basic methodology is to add one or more 460 mm transducers, or additional matching 380 mm units, to the control room main monitors. Often these are built into the soffit below the main reference monitors, though they can also be utilized to augment console-top near fields. Usually crossed or layered over in the 100 Hz region, this bandpass must be carefully physically positioned and interfaced into the room acoustics to avoid unfortunate interactions with room modes. Psychoacoustic criteria have been developed to determine the required acoustic output, and acceptable levels of distortion[37].

The rise in project studio and workstation environments has led to the availability of powered subwoofer modules designed to complement associated full-range monitors. Many of these include extensive 'bass management' electronics, designed to perform crossover, summation and other signal processing tasks. (Fig. 11.22)

Figure 11.22. Monitor system with small satellite monitors and common subwoofer, which also houses full amplification, signal control and bass management facilities. (Tria system, Courtesy Alesis Corp.)

Figure 11.23. Multi-channel 5.1 surround sound monitoring workstation environment, showing left-centre-right front speakers and subwoofer. Not visible are left and right rear-channel surround loudspeakers. (MPS Series, Courtesy M&K Professional.)

11.3.9 Surround-sound monitoring

The influence of multi-channel motion picture sound combined with the desire to exploit the possibilities of three-dimensional spatial reproduction has given rise to the 5.1 channel surround-sound standard for monitoring. Rather than the aborted quadraphonic attempt in the early 1970s, the 5.1 standard goes back to the original Bell Laboratories auditory perspective experiments from the 1930s which demonstrated that three front channels give the best front spatial positioning. Two rear channels for left and right rear along with the 'point one' subwoofer channel are added (Fig. 11.23). Standard recommendations for speaker positioning and other elements of system implementation have been generated, such as International Telephonic Union publication ITU-R BS.775. Some elements such as the choice of type of surround channel speaker – wide angle direct radiating or controlled directivity; equal size to the three main channels or smaller; single source or dipole – are still subject to aesthetic debates[38].

PART 2: MUSICAL INSTRUMENT LOUDSPEAKERS

11.4 Musical instrument direct-radiator driver construction

High-power full-range cone driver loudspeakers for electronic musical instrument amplification require special variations on the basic components and methods employed for loudspeakers in general[39]. Each element of a loudspeaker will be described in terms of the specific characteristics which apply to typical musical instrument drivers (Fig. 11.24) – though Chapter 2 should also be consulted.

11.4.1 Frame

The loudspeaker frame, chassis or basket is usually constructed of stamped steel or cast aluminium. It must be strong and rigid in order to support the weight of the magnet assembly and absorb the motional forces of the cone/coil assembly under high accelerations and long excursions. It additionally acts as a heat-sinking element for the heat from the voice coil conducted through the top plate. Any warping of the frame due to thermal or mechanical stress can cause mis-centring of the moving assembly, with subsequent coil rubs and failure.

Figure 11.24. Cutaway view of 380 mm musical instrument loudspeaker with ferrite magnet and 63.5 mm diameter voice-coil. (Model EVM-15L, Courtesy Electro-Voice Inc.)

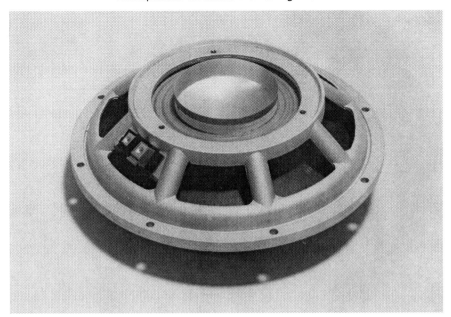

Figure 11.25. Replaceable basket assembly for 300 mm musical instrument loudspeaker. Note terminals, 100 mm diameter voice-coil, and one-piece aluminium dome and coil form. (Courtesy Peavey Electronics Corp.)

The number of support spokes and their structural shape, along with the material chosen, determines the strength of the frame. Flanges on the front edges, and on the multiple support spokes, can increase the rigidity of the normally inferior stamped frame, to approach that of a casting. However, the stamped steel frame will always suffer the problem of shunting magnetic energy away from the air gap. This effect will typically incur a 1 dB gap flux loss when compared with a frame cast from the usually employed non-magnetic aluminium. Cast steel is not normally encountered, since it would rob flux in a similar fashion as stamped steel, although zinc, magnesium and alloys have been employed. Plastic frames have been tried, but their lack of heat-sinking ability, as well as their reduced resistance to thermal and mechanical stress as compared with metals, has hampered their success. The multiple spokes and support members on a metal frame can increase the heat-radiation surface area for improved heat sinking in an analogous fashion to that used for power amplifiers[40]. Field-replaceable frame and cone/coil assemblies have been designed, but these present significant coil-centring tolerance problems in mounting (Fig. 11.25). Standards exist for nominal diameter, mounting flange dimensions and bolt circle, defined by BS 1927 in the UK, EIA RS-278 in the USA, IEC 124 in Europe, and JIS C5501 and C5530 in Japan.

11.4.2 Diaphragm[8]

The most critical element in the response character of a direct-radiator musical instrument loudspeaker is the cone diaphragm. Its material composition, depth, and angle or radius together determine to a major degree the bandwidth and shape of the transducer's amplitude response.

Almost all musical instrument cone drivers still utilize felted paper cones, because of the high stiffness-to-weight ratio. Modern plastic materials popular for hi-fi drivers, such as Bextrene and polypropylene, are too heavy for the efficiencies required in

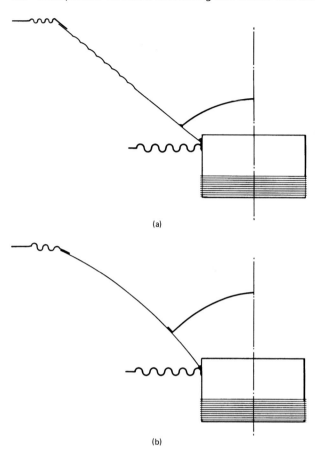

Figure 11.26. Two different types of cone/coil assemblies: (a) straight-sided; (b) curvilinear. (Courtesy Fane Acoustics Ltd.)

musical applications, and lack the necessary stiffness for response extension to the 5–7 kHz range. Even among the traditional paper cone stocks, only the hardest materials are chosen for full-range musical instrument use, with maximum stiffness for minimum mass being the desirable criterion. Cone-forming methods are closely guarded secrets within the loudspeaker cone industry, with the precise formula of wood pulps, water slurry, binders, dyes and other ingredients, as well as the specific screening, suction and drying processes, proprietary to individual manufacturers[41]. Subsequent methods to improve hardness, such as restriking (hard-pressing the cone surface) and various types of lacquer treatments, are also employed in the search for maximum stiffness with minimum mass.

The shape of the cone is also of major importance to the frequency-response character. The two basic shapes employed are the straight-sided and the curvilinear (Fig. 11.26). The straight-sided is a true truncated cone shape, with the apex angle improving the resistance to plate resonances, compared with a simple disc diaphragm. The curvilinear design imparts a changing curve to the cone side walls, which may be a simple radius or may follow an exponential or hyperbolic law. The progressive shape of the curvilinear imparts improved resistance to resonance modes of the diaphragm above the piston-band limit, where the reproduced wavelength is equal to or less than the diaphragm size. A curvilinear design can typically suppress diaphragm break-up modes for an additional octave above that of a similar straight-sided cone. Additionally, its curved shape can allow internal circumferential flexing, eliminating the outer cone moving mass at higher frequencies. Then only the central portion of the cone, the coil and the dome may act together as a more efficient high-frequency reproducer. Curvilinear designs are hence most desirable for maximum efficiency and extended response. Straight-sided designs are employed where the extra mass they require for sufficient thickness and rigidity is desirable for low-frequency to mid-range efficiency balance, and where low-frequency stiffness is needed for long-travel bass designs. If a straight-sided design is used in a high-efficiency application, low mass may be used to allow certain break-up modes for specific response and distortion characteristics.

Cone depth is also a major consideration. A deep cone angle will typically extend high-frequency cone resonances for a smoother, wider bandwidth. A shallow cone will lower the resonant modes, yielding a rising, peaked mid-range in a typical light voice-coil mass design. Deep cone angles, particularly in straight-sided designs, can suffer from cavity-resonance effects, imparting a series of resonant peaks and dips to the response[42].

11.4.3 Surround and centre spider suspension elements

The cone edge surround connects the cone to the frame, provides a mechanical impedance termination to the progressive waves travelling through the cone at mid and high frequencies, and must provide sufficient travel capability to accommodate the loudspeaker motor (coil and magnet) design. Often in musical instrument designs the surround is merely an extension of the cone paper and not a separate attachment. Ridges or folds can be screened into the edges with separate suction and thickness control to determine cone resonance frequency. Plasticizers or other viscous damping compounds will be applied to the edge, either by immersion or surface coating. If an insufficient or improper degree of damping is provided, and if the surround area is too large in comparison with the cone area, a surround anti-resonance can be formed where cone motion is actually out of phase with surround motion. This can cause a characteristic dip in axial frequency response[43]. Similarly, the second resonance of the surround can interact with the travelling waves through the cone at high frequencies, causing a large peak in the amplitude response[44]. These effects can be used to impart a distinctive character to a musical loudspeaker design.

Other common surround constructions which typically try to avoid these aberrant characteristics are separately formed and attached cloth (cambric) edges in accordion pleats, half-rolls, and double half-roll cross-sections. The type of weave, size of thread, and forming treatments combine with the shape in determining stiffness characteristics, and damping treatment is still required to seal the weave and control the aforementioned resonance characteristics. Floppy surrounds of rubber, PVC and foamed plastic common in long-throw hi-fi designs are not normally suitable for musical instrument loudspeakers, because of their lack of damping and poor termination characteristics, as well as reduced ruggedness.

In conventional designs, a centring spider acts with the surround to form a two-point suspension for coil and diaphragm motion. The spider is typically attached to the voice-coil tube just below the cone neck, and is formed as a flat dish with corrugations to allow the desired motion. It is attached at the base of the frame and glued down flat, or in some designs a cup-shaped edge is used to space apart the two connection points and allow for increased travel. Spiders are typically formed of treated cloth, and stiffness is controlled by shape, weave and treatment as with cloth surrounds.

Dual-spider designs have been employed, either to apply a third point of suspension or to allow coil centring independent of cone and surround[45]. The second spider can be added in parallel to the first, attached to the voice-coil tube, attached some-

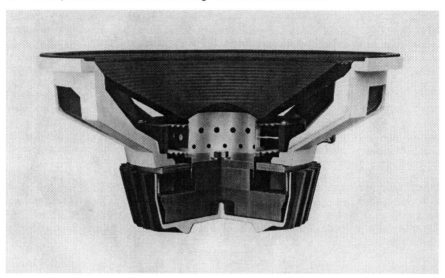

Figure 11.27. Cutaway view of 380 mm loudspeaker with double spider construction. (Courtesy Cetec-Gauss Inc.)

what up from the base of the cone, or mounted inside the voice-coil tube supported by a magnet pole-piece extension (Fig. 11.27). While dual-spider construction can make a more rugged design, the extra mounting mass and demanding control of the varying stiffness elements are often detrimental to response and efficiency.

Under dynamic signal conditions, the suspension elements interact in a complex fashion with the coil motion and magnetic gap. Suspension stiffness is never wholly linear, but is 'progressive' in increasing stiffness with excursion. This changing stiffness can limit coil/diaphragm motion, or in some cases actually increase motional linearity[7].

11.4.4 Voice-coil[46]

The voice-coil is the heart of the loudspeaker, and its size, shape and composition are second only to the cone in dominating performance characteristics. Typical coils are wound of copper or aluminium wire. Copper has higher density and lower intrinsic resistivity; hence, for equal-size coils of equal d.c. resistance the copper coil will have more turns, greater mass and greater inductance. The choice of coil mass as compared to total moving mass of a loudspeaker is critical to the high-frequency bandwidth and response. Theoretical analysis shows that maximum efficiency is obtained when coil mass equals cone and air-load mass, with both as small as possible. At low frequencies, when both are minimized, coil mass is typically much less than that of the cone and air load, depending on the chosen piston size. At high frequencies, however, where the central cone portion effectively decouples from the outer section, the equal ratio can be realized. It then becomes essential to choose as low a voice-coil mass as possible for best high-frequency response; this favours aluminium wire coils. Aluminium coils have the additional benefit of lower inductance for increased high-frequency drive.

The high-frequency requirement of minimum coil mass must be traded off against the piston-band efficiency requirements of, again, minimal mass, but also maximum motor strength $(Bl)^2/Re$. B, the flux density, can be increased independently, but the length of conductor, l, and the d.c. resistance, Re, will favour the characteristics of

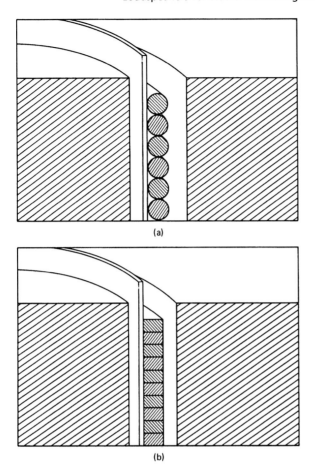

(a)

(b)

Figure 11.28. (a) Round wire and (b) flat wire; cutaways on coil form in magnetic gap showing greater packing density. (Courtesy JBL Inc.)

a copper voice coil. The actual choice will depend not only on whether piston-band efficiency or high-frequency response are of primary importance, but also on low-frequency requirements. The increased motor strength of an equivalent copper coil will also yield increased low-frequency damping, reducing low-frequency output in open-back, infinite-baffle or sealed-box acoustic loading conditions. Conversely, more motor strength yields a better Helmholtz driver, and hence may provide a more favourable vented alignment[47]. The use of rectangular-formed (ribbon) wire rather than round wire will increase the space factor of the coil, allowing the packing of more conductor into the available gap area (Fig. 11.28). This will normally increase the motor strength in much greater proportion than the increase in mass, and will typically yield a 2 dB increase in efficiency. The rectangular wire is typically wound edgewise in a single layer, as opposed to two or four layers for a normal round-wire coil, although layered flat winding of rectangular wire has also been employed.

The handling, soldering and work-hardening liabilities of aluminium wire being well known, copper-clad aluminium wire has been utilized when commercially available. It eliminates most of the manufacturing problems with pure aluminium wire with only a slight mass and resistance penalty, still far removed from pure copper.

Annealing (heat treatment) of the wire can also improve its mechanical strength.

Voice-coil size, both cross-sectional area and diameter, affects both efficiency and thermal capabilities. Even a maximum-efficiency musical instrument cone driver is less than 10% efficient, and the majority of the power must be dissipated as heat in the voice-coil. For a given cross-sectional area and d.c. resistance, a larger voice-coil diameter will use a larger wire gauge with greater current capacity; hence there will be less heat and less efficiency loss through resistance increase. A typical loudspeaker operated at its rated power-handling limits may double its resistance, thus halving its efficiency, and it draws half power from the amplifier for a net output loss of 6 dB over that which would be otherwise expected[6].

Smaller voice-coil designs can minimize their susceptibility by increasing coil and gap size, tightening gap tolerances and maximizing thermal heat-sink masses and areas[40].

11.4.5 Coil form[46]

The form, former, bobbin or tube around which the voice coil is formed connects it to the cone apex and centring spider(s). It must maintain rigidity, transferring the high drive forces to the diaphragm under conditions of both physical and thermal stress.

Early musical instrument speakers, since they were essentially adaptations of hi-fi designs, used simple kraft (wood pulp) or bond (cloth) paper tubes for coil forms. As the power requirements increased, these materials were quickly seen to be inadequate (they burst into flames). The advent of polyamide nylon paper (e.g. Nomex, a Dupont trademark) brought a synthetic material which had increased rigidity and heat resistance. The subsequent development of polyimides (e.g. Kapton, a Dupont trademark) brought even greater thermal and mechanical strength, with only a slight penalty in increased mass and greater adhesive curing complexity. Mica, glass fibres and FRP (Fibreglass Reinforced Plastic) are other materials employed for their stiffness and thermal resistance (Fig. 11.29).

Figure 11.29. Voice-coil on tube former, in this case made of glass-fibre. (Courtesy Fane Acoustics Ltd.)

The other approach to coil forms is to make them heat-conductive rather than heat-resistive. The entire tube can then serve as an additional radiating surface to transmit heat away from the voice-coil, increasing efficiency and power handling. Thin aluminium sheet is normally employed; brass has also been used but is of too high a mass for extended high-frequency musical instrument designs. While aluminium has the required heat resistance (or in this case, conduction) and stiffness, it suffers from a number of other potential liabilities. Its heat conduction will bring high temperatures to the areas of spider, dome and cone neck connection, potentially weakening glue bonds and causing failure. Aluminium is conductive and any breaks in wire insulation can cause a short-circuit to the former. Since the aluminium conductor is part of the moving assembly, localized magnetic fields can be generated, causing an eddy-current brake action. This action will oppose coil and diaphragm motion, increasing damping and potentially decreasing output at both low and high frequencies. It can occur even if the conductive former has a slit to break the continuous loop, due to localized eddy current action, and will increase in proportion to former length and thickness. Because of these liabilities, thinner sections of aluminium may be used in a multiple laminate with other materials.

Perforations in the coil form can improve air circulation around the coil and former area, and are claimed to reduce mass, although usually insignificantly. In addition to the tube itself, speakers often employ a second support layer outside and above the tube for additional stiffening. All materials are subject to thermal failure caused by overpowering or insufficient adhesive curing (Fig. 11.30).

11.4.6 Dust cap

The dust cap or centre dome of the speaker prevents foreign particles from entering the gap, pumps air past the coil to improve cooling, and can improve high-frequency response. A light thin dome, whether of aluminium or stiff paper, connected directly to the top of the coil form, can radiate high frequencies directly from the coil when the outer cone sections decouple at short wavelengths. The dome can also increase the structural rigidity of the coil form. Domes larger than the coil size may be purely for cosmetic purposes or may actually be attached at critical vibration node or peak points for response adjustment[48]. Aluminium domes connected to an aluminium coil form can add extra heat dissipation area (Fig. 11.25). If a non-vented magnet structure is used, a porous dust cap may be required to reduce internal air pressure at the expense of some effective piston area.

11.4.7 Adhesives

As with coil form materials, adhesives have been forced to improve in thermal and mechanical characteristics as increased power demands are placed on musical instrument loudspeakers. Older nitrile rubber and vinyl types, which are heat sensitive and require slow air drying to cure, have largely been replaced by thermosetting and two-part epoxies, which can handle much greater heat and forces. Cyanoacrylate types have rapid cure cycles, but even modern specialty mixes tend to crystallize after sustained periods at high temperatures. In some cases, changing to stronger, high-temperature adhesives can eliminate subtle compliance effects and cause changes in the frequency response.

Curing of the voice-coil bond to the former typically requires special adhesives compatible with both the former material and wire insulation. Complex baking procedures normally include a slow rise to cure temperature and then a sustained period to ensure outgassing of the volatiles for complete cure. Baking mandrel fixtures are normally required to maintain pressure and coil shape during curing.

(a)

(b)

(c)

Figure 11.30. Overdriven voice coil, former, and adhesives: (a) loose turn and distorted former; (b) typical overheated Nylon/paper-based coil; (c) distorted polyimide former causing scraping; (d) distorted aluminium former causing scraping; (e) blistering of former due to insufficient curing of adhesives. (Courtesy Fane Acoustics Ltd.)

(d)

(e)

Figure 11.30 (*continued*)

11.4.8 Lead dress

Coil leads must be brought out to a flexible connection, allowing coil and diaphragm motion, and thence to a set of wiring terminals typically mounted to a frame member. Because of the fragile nature of these connections, they form one of the most vulnerable areas of speaker construction. Leads must be routed with service loops, insulators and reinforcements to prevent shorts or opens. Flexible terminal leads may be made of braid or flexible tinsel, and may be solder-connected to the actual coil wire at the body of the cone, at the cone neck or at the coil tube itself. Lead routing may be through eyelets placed in the cone body, through holes pierced in the cone, or through or under the centring spider. Viscous dampeners or adhesive reinforcements may be added both to insulate and to prevent vibration. Placement of the leadouts can contribute to or suppress asymmetrical rocking modes, and excessive lead mass and stiffness must be avoided as they can also cause the moving assembly to break into these spurious vibrations.

11.4.9 Motor topology[7]

The voice-coil/magnetic gap relationship within a loudspeaker determines the relative efficiency and linearity of the design. There are three possible topologies: the overhanging coil, the underhung coil in a deep gap, and the equal coil and gap (Fig. 11.31). The equal coil and gap topology is normally employed in maximum efficiency wide-range musical instrument designs, as this fills the gap with conductor for maximum motor strength. Since all magnet structures have a significant fringe flux extending above and below the gap, a slightly overhanging voice-coil will actually give the greatest motor strength at the expense of slightly increased mass. More overhang will increase linearity, but at the expense of both mass and motor strength. The underhung coil/deep gap topology is normally used only for high-linearity designs, since it is expensive in terms of magnet and metalwork, and sacrifices efficiency for linear travel, although it does maintain low coil mass.

The use of an undercut pole-piece to focus and linearize the fringe flux both below and above the gap will sacrifice some flux-carrying ability and thermal mass. The

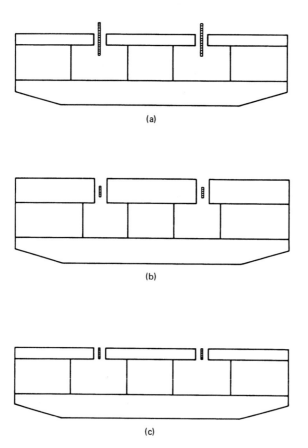

(a)

(b)

(c)

Figure 11.31. Cross-sectional schematic views of three loudspeaker motor voice-coil/magnetic-gap topologies: (a) long coil-shallow gap; (b) short coil-deep gap; (c) equal height coil and gap. (Courtesy JBL Inc.)

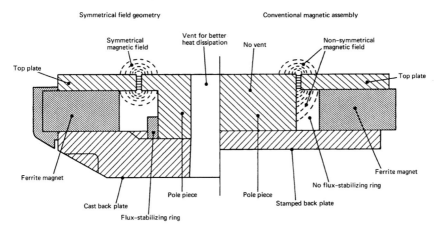

Symmetrical field geometry   Conventional magnetic assembly

Figure 11.32. Half-cutaway views comparing conventional magnetic assembly with symmetrical field geometry utilizing undercut pole piece to equalize fringe flux, and shorting ring to check flux modulation. (Courtesy JBL Inc.)

subjective desirability of its distortion reduction for instrument applications is a topic of debate. Similarly the use of shorting rings to cancel modulation of the magnet by the coil field is a measurable effect which may be selectively kept or removed depending on design judgement (Fig. 11.32). The use of shorted turns within the assembly to reduce coil inductance and increase high-frequency drive is fairly well accepted[46].

Clearance tolerances between voice-coil and pole tips are critical to both flux strength and thermal transfer. Gap space must be as tight as possible to maximize both, but must also be large enough to allow excursion motion and thermal coil expansion. Fluid-magnetic suspensions can improve thermal and magnetic transfer but are subject to stability problems with motion and time[40].

11.4.10 Magnet structure

The most common magnet in use is that made of a slurry of barium or strontium ferrite particles press-formed into shapes typically much larger in diameter than length. While not nearly as efficient as Alnico, a metal alloy of aluminium, nickel and cobalt, it is more economical and also more resistant to demagnetization[49]. Older designs were typically based on centrally placed Alnico slugs but, since the development of ferrite magnets in the 1950s, the majority of designs have been converted, primarily because of economic pressures (Fig. 11.33).

Neodymium is still relatively expensive but is beginning to see some use to create lightweight designs. The 'retro' guitar and amplifier movement of the 1980s and 1990s has furthered an interest in the 'Alnico sound', with re-issues and duplications of classic designs from Jensen, Celestion and others (see Section 11.4.11).

The ferrite magnet will typically extend outwards from the middle of the top plate and back plate, and protective rubber or plastic covers are sometimes affixed to prevent chipping or cracking. Metal rear covers may be placed around the entire magnet structure for cosmetic purposes, or may have fins for improved heat transfer from the structure[36,40].

The top plate and pole-piece must be of mild steel or other low-reluctance metal, with sufficient cross-sectional area to maintain required flux levels. Back plates are typically stamped, cast or cold-forged. Vent holes through the back plate and pole-piece can relieve internal air pressure and increase thermal circulation, but can reduce

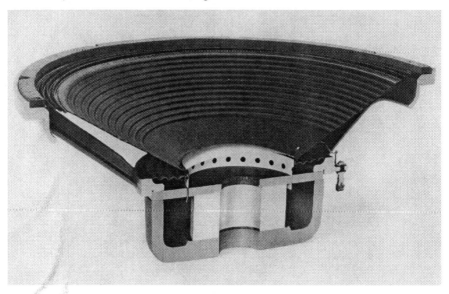

Figure 11.33. Cutaway view of 380 mm bass musical instrument loudspeaker with 100 mm diameter voice-coil and centrally placed Alnico magnet. (D140, Courtesy JBL Inc.)

the flux-carrying capacity because of reduced metal area, and must be blocked with foam or gauze to prevent foreign particles from entering the gap.

11.4.11 Classic and vintage guitar speaker designs

Certain designs found favour with the early amplifier designers and were embraced by the performing guitarists as embodying the right, 'magic' combination of characteristics. The 250 mm and 300 mm cone loudspeakers which were used in the first Fender, Gibson, Vox and other early guitar amplifiers are today considered classics and their sound is highly prized for its authenticity and linkage to the roots of electric music[50]. The early Jensen 'Concert Series' speakers used by Fender and Gibson with their seamed cones, the Celestion Blue 'Bulldog' speaker with its blue magnet cover used in the early Vox designs, the Celestion models supplied to Marshall, are all hallmarks of vintage sound character and performance (Fig. 11.34). All these speakers employed small 25 mm or 37 mm diameter voice-coils and most had lightweight craft paper formers, seamed cones, and were assembled with basic low-temperature adhesives. These factors defined limitations in the power-handling capability of the speakers, but also allowed definite frequency response and distortion characteristics.

The JBL D120 300 mm and D130 380 mm units, with their large 100 mm diameter edge-wound ribbon-wire voice-coil, light and shallow curvilinear cone, high-efficiency Alnico magnet structure, and cast frame, were adopted by Leo Fender and many others in the 1950s and 1960s as a standard or premium upgrade for their amplifiers. The gleaming aluminium centre dome peering out from behind the amplifier grille cloth came to signify the highest standard of sound quality and performance. The low gain of the early amplifiers, and the high capacitive losses inherent in the pickups and cables first used, meant the rising, bright response of these JBLs was ideal for the combined guitar and amp system. Other similar early designs were Altec models, also based on early Jim Lansing designs adapted for musical instrument use.

Figure 11.34. Modern reproductions of classic Alnico magnet guitar speakers. (Courtesy Jensen/Recoton, Inc.)

As amplifier gain increased and signal path quality and flexibility improved during the 1980s, the Electro-Voice EVM-12L 300 mm unit became the standard for high output, desirable frequency balance and extension, and responsive overdrive distortion characteristics. Celestion, in establishing its OEM (original equipment manufacturer) relationship with Marshall amplifiers became the standard speaker for overdrive distortion in 'stack' amplifier/speaker cabinets configurations favoured by heavy-metal aficionados. The Eminence company in Kentucky, USA, deserves mention for its expertise and dominance in the OEM musical instrument raw frame speaker business.

11.5 Musical instrument loudspeaker enclosures and systems

Virtually every hi-fi, cinema or sound-reinforcement enclosure and system design has been adapted or adopted for musical instrument use. While the normal considerations of efficiency and bandwidth are of primary interest, the particular characteristic of each design continues to be its subjective sound character. Distortions are considered only as colorations, good or bad only within the context of the overall subjective judgement of the systems, not as objective characteristics to be measured and analysed.

The individual system and enclosure types will each be described in detail[51].

11.5.1 Open-back 'finite baffle' enclosure

The most common musical instrument system is the open-back cabinet, housing both the amplifier and cone speaker. It is generally referred to as a 'guitar amplifier', although it is employed for virtually every electronically amplified instrument, with varying degrees of success. The front surface comprises the speaker baffle board, covered with grille cloth, with the amplifier and controls typically above or on the

Figure 11.35. Open-back guitar amplifier with single 300 mm or 380 mm loudspeaker. (Courtesy MESA Engineering Inc.)

Figure 11.36. 'Piggy-back', 'stack' type guitar amplifier with sealed speaker enclosures, each mounting four 300 mm units. (Courtesy Marshall Amplification Ltd.)

top panel (Fig. 11.35). The open back of the enclosure allows air circulation to the electronics, and forms an outlet for the rear wave of the loudspeaker(s). This creates a dipole source action, substantially cancelling low-frequency output below approximately 150 Hz depending on placement and positioning. Protective slats over portions of the rear opening can act further to adjust the low-frequency response character, as do the enclosure and baffle size. Speakers with certain infinite-baffle response characteristics can show marked differences when used in open-back configuration.

Virtually every speaker configuration and combination have been successfully utilized in the open-back amplifier or extension cabinet, some of the more popular being single 200 mm, two, three or four 250 mm, single or double 300 mm, and single 380 mm. The front baffle surface can be angled upwards or split, left and right, to improve sound dispersion for the narrow high-frequency beamwidth of full-range cone speakers.

11.5.2 Sealed box

The logical extension, and next most commonly encountered enclosure variation, is the sealed enclosure. It is usually employed to obtain greater bass response from small enclosures, or to restrict cone motion at low frequencies and hence prevent damage to the drive unit. Many small electric-bass amplifiers use a small scale speaker chamber isolated from the amplifier electronics. The most popular of the 'piggy-back' instrument amplifiers have separate amplifier and sealed speaker enclosures. The upper enclosures in multi-unit stacks are often angled for better dispersion (Fig 11.36). Experimentation often leads to operation of the enclosures with the back panels removed for an open-back sound, depending on personal choice and the characteristics of the drive units employed. Some units use a combination of driver types and both open and sealed enclosure sections.

11.5.3 Vented enclosures

For maximum bass output and diaphragm excursion control, and hence improved power handling, the next logical step is to vent the sealed enclosure to create a bass reflex design. This is the usual choice for electric bass guitar and other high-power low-frequency sources. When fitted with, most commonly, 380 mm or 460 mm drive units optimized for ruggedness in both power handling and excursion, the vented box

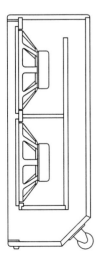

Figure 11.37. Ported direct-radiator bass musical instrument enclosure with two 380 mm drivers. Port extends from baffle up rear of enclosure for labyrinth effect. (Courtesy Gallien-Kruger Inc.)

yields the most versatile bass reproduction system. Port tuning can be varied upwards to give an underdamped resonant bump to accentuate the bass character, or downwards for deep, smooth response extension to the lowest audible notes (Fig. 11.37)[52].

The most popular bass guitar configuration in the late twentieth century was the multiple 200 mm driver configuration, either a quad or octet, although this was often employed in a sealed box rather than vented configuration. The 200 mm units were often supplemented with a tweeter to augment the high-frequency response, and the configuration was about equally used on its own, or with additional 380 mm or 460 mm driver systems to extend the low end.

11.5.4 Front-loaded horn

For efficiency increase and directivity control, the cone musical instrument driver can be mounted driving into a horn flare[53]. The rear of the driver is typically enclosed in a small rear chamber to control excursion and eliminate driver reactance[54]. Flare rate and mouth size of the horn will typically limit the horn response to above 150 Hz for all but the largest mouth sizes, and the loading factor on the cone diaphragm will determine whether the unit operates as a true horn or merely as a 'directional baffle'[55]. In either case, axial sound pressure increases of the order of 6 dB are typical in the lower mid-range, at the expense of low-frequency and high-frequency response reduction outside the active horn pass-band (Fig. 11.38)[56].

11.5.5 Folded horn

In order to extend horn efficiency to a lower frequency, the horn flare may be folded to generate a longer horn within a smaller physical size. The basic design is borrowed

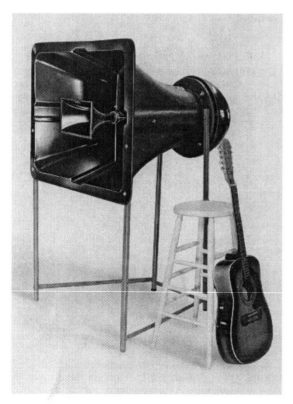

Figure 11.38. Full-range cone-driven horn with centrally mounted high-frequency horn/driver. (Courtesy Community Light and Sound Inc.)

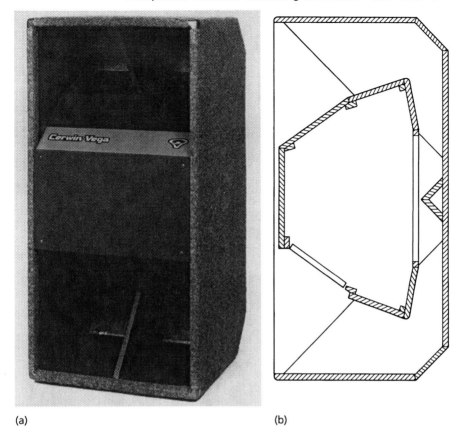

(a)   (b)

Figure 11.39. W-horn enclosure: (a) photo; (b) cutaway side view. (Courtesy Cerwin-Vega Inc.)

from the early cinema systems, and allows response to extend down to the 50 Hz region for a typical 1×0.5 m mouth area (with 0.5 m depth)[57,58]. Since the drivers, typically one or two 380 mm or 460 mm units, are internally mounted and the sound waves must pass around the horn folds, high frequencies are severely attenuated, with response falling off rapidly above 400 Hz. Deep bass response can be somewhat enhanced by venting the rear driver chamber. Response smoothness typically suffers from the flare approximations and insufficient mouth size. Despite its limitations, the high efficiency with which electric bass fundamental tones are reproduced has made this a classic design, and highly successful for those manufacturers who introduced it for musical instrument use. Since the internal construction of the flare in its usual form suggests a W-shape, such loudspeakers are commonly referred to as 'W-bins' or 'W-horns' (Fig. 11.39).

11.5.6 Rear-loaded (folded) horn

In order to allow full-range use, the horn flare can be generated from the rear of the driver, folded to conserve space, and opened out at its mouth to the front baffle for the bass frequencies only. The front of the driver can then operate as a direct radiator for the mid-frequency and high-frequency range[59,60]. It is therefore popular for full-range electric bass guitar use.

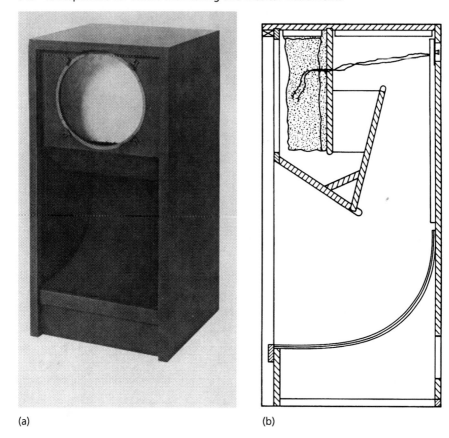

(a)   (b)

Figure 11.40. Rear-loaded horn enclosure: (a) photo; (b) cutaway side view. (Model 4530, Courtesy JBL Inc.)

Successful commercial examples were first designed as cinema cabinets for one or two 380 mm drivers, and have been adapted for 460 mm units. The efficiency is extremely high in the 60–120 Hz octave, but the horn rear wave causes a cancellation with the direct-radiator output in the 125–150 Hz region, and the direct-radiator response is somewhat choppy due to the minimal baffle surface. Their appearance has given them the popular nickname 'sugar scoops' or 'scoop' bins (Fig. 11.40).

11.5.7 Vented (front-loaded) horn

If the requirements for reactance annulling at the rear of the driver are ignored, the rear chamber can be enlarged and vented. The resulting combination enclosure operates as a horn or directional baffle where mouth and flare size are sufficient, typically above 150 Hz, and as a vented-box direct radiator below, typically down to a 40–50 Hz cut-off. The transition character will depend on the driver parameters but, for high-efficiency musical instrument drivers there will typically be a 6 dB axial sensitivity increase from 100 to 200 Hz. Other than this response step, smooth wide-band response is possible. The vented horn is perhaps the best method of obtaining maximum output from a cone drive unit, historically designed around 380 mm drivers (Fig. 11.41)[22].

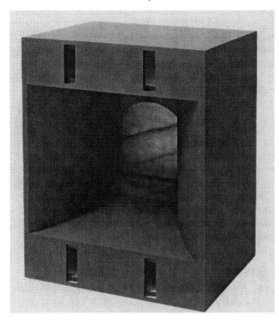

Figure 11.41. Ported horn combination enclosure. (Model 4560, Courtesy JBL Inc.)

Figure 11.42. System using slot loading of cone driver. (Model 380-SD, Courtesy TOA Electronics Inc.)

11.5.8 Slot loading

Restricting the diaphragm area of a direct radiating system provides the loading factor and front cavity volume of a horn design without the subsequent acoustical coupling of the horn flare. The net effect is a slight efficiency increase with an attendant high-frequency bandwidth reduction. An increase in dispersion also results from the smaller radiating aperture (Fig. 11.42)[30].

11.5.9 Karlson enclosure

An original design from the early 1950s Golden Age of hi-fi, the 'Karlson Coupler', was claimed to embody the best characteristics of the sealed box, vented bass reflex and exponential horn. The exponential flare on the front panel was supposed to resonate the enclosed air column progressively and non-selectively at all frequencies, and an internal port vented the driver rear chamber (Fig. 11.43)[61,62]. In practice, the response exhibits a bass efficiency rise and choppy mid-range very similar to a rear-loaded folded horn. The design continues to be popular with electric bass guitarists and is periodically rediscovered by various manufacturers.

11.5.10 Push–pull compound driver system

Placing two identical drivers in front of one another, with a small coupled volume of air between, allows the actual rear enclosure volume to be half of that which would normally be required for a single driver. Olson first described the concept, with the additional detail that, since the compliance of the enclosure is inversely

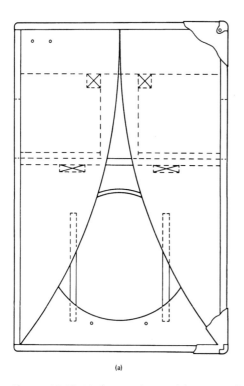
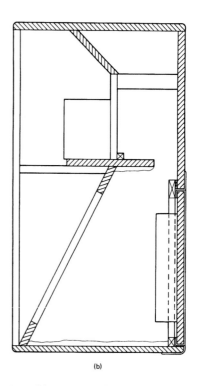

(a)   (b)

Figure 11.43. Karlson enclosure: (a) cutaway front view; (b) cutaway side view. (Courtesy Cetec-Gauss Inc.)

(a)

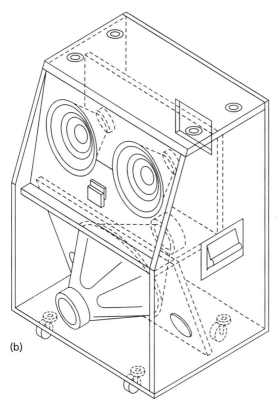

(b)

Figure 11.44. (a) System using push–pull compound woofer configuration; (b) cutaway perspective view. (Model SVT 50 Iso Vent, Courtesy Ampeg division of St Louis Music, Inc.)

proportional to the square of the cone area, the compliance can be further increased by making the driving loudspeaker smaller than the radiating loudspeaker[63]. For practical reasons both drivers are usually identical and the actual effect is to allow two drivers, which would normally need twice the enclosure size, to be used in place of a single driver with the optimum parameters for the available enclosure volume. The system is often referred to as an isobaric design because of the constant pressure nature of the coupled chamber. The trade name 'Isobarik' is used by Linn Products of Scotland, the audiophile manufacturer of the design[64].

Another variation faces the two drivers into one another in a push–pull arrangement, the constant pressure chamber being formed by the volume captured within the cones. The drivers are then driven electrically out of phase, which can result in cancellation of even order non-linearities (Fig. 11.44)[65].

11.5.11 Rotating baffle

The rotating baffle concept was initially created to allow the electric organ to simulate the acoustic interactions of the many pipes of a pipe organ. Beat frequencies, and amplitude and frequency (phase) modulation were created by a motor-driven rotating baffle of cylindrical shape, with an exit at one side, moving in front of the speaker cone. More sophisticated embodiments added counter-rotating horns driven from a compression driver for the high frequencies. Two speeds were provided, slow

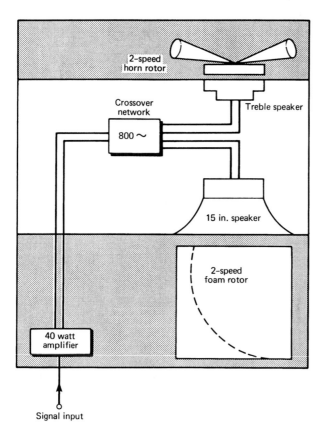

Figure 11.45. Leslie tone cabinet: interior schematic view showing cylindrical rotating bass baffle and rotating high-frequency horns. (Courtesy Hammond Organ.)

for chorus and fast for a tremolo effect. The 'Leslie' speaker, named after the inventor Donald Leslie, has found great popularity with guitarists as well as becoming indispensable to the Hammond Organ sound and to that of electric or electronic organs in general (Fig. 11.45)[66].

11.5.12 Column speakers

The column speaker or line array is technically a multiple-driver variation of the sealed or vented box, but deserves mention on its own. By vertically stacking multiple cone drivers, wide horizontal dispersion can be maintained while reducing vertical beamwidth[67]. The drivers may all be identical or may be of different sizes and characteristics. For proper beam focusing in the vertical plane, the widest dispersion and high-frequency bandwidth devices should be mounted central to the column. The input signal may also be electrically tapered, or the array acoustically tapered, to limit the bandwidth of the outer drivers to control the narrowing of the vertical coverage angle more consistently[68].

11.5.13 Horn-loaded port

This variation of the vented box attempts to simulate a rear-loaded horn by merely giving a flare expansion to an extra-large port area. The mouth size and horn length are insufficient, and the lack of coupling to the driver and the large acoustic capacitance of the major box volume mean that the design operates simply as a bass reflex with too large and wasteful a port (Fig. 11.46).

11.5.14 The 'talk box' vocal actuator

One of the most novel musical instrument speaker applications is a device which allows a musician to speak with the sound of the instrument, typically an electric guitar. The instrument amplifier is connected to a compression driver but, instead of feeding a horn, the driver is connected to a small-diameter flexible tube (Fig. 11.47). The other end of the tube is placed in the musician's mouth. As the musician speaks into the PA system microphone, the sound of the guitar modulates his voice for a wide range of interesting effects. The typical resultant sound is most characteristically heard on the 1970s recordings 'Rocky Mountain Way' by Joe Walsh and 'Do You Feel Like I Do?' by Peter Frampton.

11.5.15 Stereo Field eXpansion

Dual-channel or stereo pickups from guitars, and stereo outputs from keyboards or effects, as well as the ubiquitous drive of the creative musician for new and different sounds, have all created the need for spatial expanded reproduction systems. The most basic and practical solution is to employ twin amplifiers with separate loudspeakers spaced apart to generate the different source and arrival times which create the perception of spaciousness in the sound. Attempts have been made to create a stereo or spatially enhanced sound from a single enclosure system, some as simple as mounting multiple loudspeakers on differently angled baffled surfaces.

One recent development is a system marketed by Fender as SFX, for Stereo Field eXpansion. The basic principle is that of M-S (mid-side) miking in reverse. A stereo signal from a guitar preamp or effects loop feeds two speakers. One channel is routed through a sealed enclosure with its cone facing forward, representing the 'mid' part of the signal. And like the 'side' part of the M-S miking equation, a bottom speaker is mounted on a baffle, perpendicular to the top speaker, within an open-sided enclosure. This too generates a figure-eight dispersion pattern, where one side is out of phase with the other (Fig. 11.48).

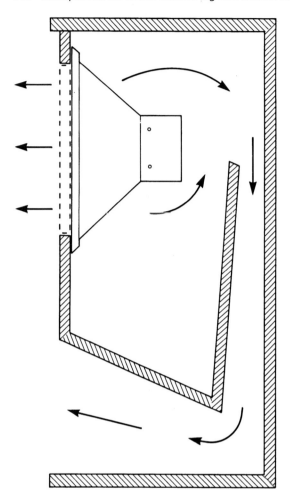

Figure 11.46. Horn-loaded port enclosure. (Courtesy Music Man Inc.)

The SFX electronics process stereo signals into sum and difference signals. The left signals are made into signals that add acoustically in the air around the cabinet, causing the SFX speaker array to steer the combined acoustical output of the two speaker elements toward the left side of the cabinet. Signals originally from the right are made into signals that subtract acoustically, causing the SFX speaker array to steer the acoustical output toward the right side of the cabinet. The effect is a broad and deep spatial impression from a single acoustic source.

11.5.16 Multiway systems

Various combination systems have been devised to extend bandwidth, overcome shortcomings or optimize usage of the various enclosure concepts. For example, 200, 250 or 300 mm drivers have been added in front-mounted sealed sub-enclosures in W-horns. With a high-pass protection capacitor in series, these additional elements restore the high frequencies rolled off in the folds of the horn (Figs 11.49 and 11.50).

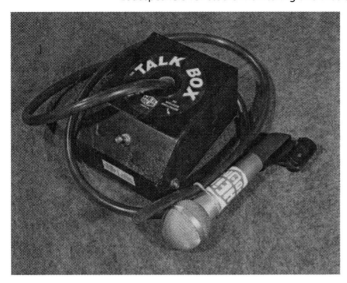

Figure 11.47. 'Talk Box' vocal coupler device. (Courtesy Heil Sound, Ltd.)

Figure 11.48. SFX musical instrument amplifier. Speaker mounting configuration and electronic processing creates spatial effects from a single cabinet enclosure. (Courtesy Fender Musical Instruments, Inc.)

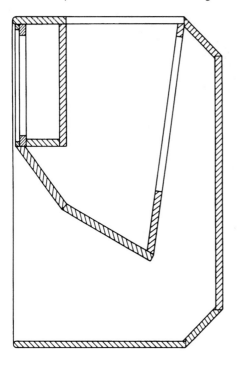

Figure 11.49. Folded horn with rear-firing 460 mm horn driver and front-facing 300 mm direct radiator in sealed sub-enclosure: cutaway side view. (Courtesy Cerwin-Vega Inc.)

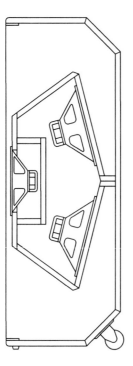

Figure 11.50. Folded horn with four 300 mm horn drivers and two 25 cm direct radiators in sealed sub-enclosure. (Courtesy Gallien-Kruger Inc.)

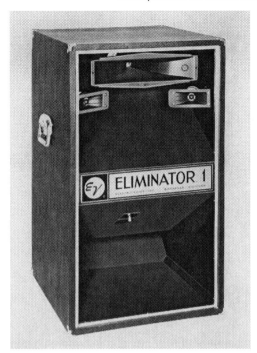

Figure 11.51. W-horn with mid-range horn and tweeters added. (Courtesy Electro-Voice Inc.)

Figure 11.52. Compact ported-horn two-way system. (Model 4115H, Courtesy Yamaha International Corp.)

(a)

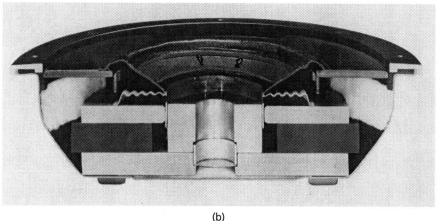

(b)

Figure 11.53. (a) Direct-radiator 460 mm three-way keyboard/ instrument sound-reinforcement system with vented mid-range and horn tweeter. (b) Cutaway side view of vented mid-range driver. (Model S-1803 and VMR, Courtesy Electro-Voice Inc.)

Various horns, lenses, siren-driver tweeters and other high-frequency projection devices have been added to musical instrument amplifier/speaker cabinets for increased projection, as well as for alternative use in PA sound-reinforcement applications. Conversely, PA and theatre loudspeaker sound systems have been used or adapted for electric instruments requiring wide bandwidth and dynamics, such as electric pianos, synthesizers and other electronic keyboards, electric violins, etc. Mid-range horns and tweeters are added to W-horns for full-range response (Fig 11.51).

Figure 11.54. Four-way all-horn loaded loudspeaker system. (Model RS440, Courtesy Community Light and Sound Inc.)

The ubiquitous ported horn enclosure can likewise be made into two- or three-way systems, and miniaturized versions of this design have become extremely popular with musicians (Fig. 11.52). Special designs can combine three-way of four-way systems into compact, portable packages (Figs 11.53 and 11.54). Experiments have also been conducted in an attempt to merge individual instrument amplification into a single system integrated with the overall sound-reinforcement system, such as the Grateful Dead's 'Wall of Sound' in the early 1970s[69].

11.5.17 Fully integrated powered systems

The basic guitar amp or musical instrument amplifier is by definition an integrally powered system. The more sophisticated sound reinforcement system approaches to sound amplification and reproduction have been employed to apply the concepts of electronic crossovers and multiway amplification to the challenges of instrument amplification. Modern high-quality bass guitar 'rigs' are invariably bi-amplified, with larger loudspeakers and higher power employed for the lowest frequency range, electronically crossed over to a smaller amplifier driving a tweeter or smaller cones for the highs.

As the move toward completely integrated loudspeakers and electronics continues to create greater sophistication, these same advances impact and improve on the simple instrument amplifier. JBLs EON® sound systems take the integration a step further. In order to deal with the problem of heat dissipation from both the transducers and amplification, the entire baffle is a single die-cast heat sink assembly. The one-piece integrated baffle includes the high-frequency horn, the woofer frame, with integral mounting for the signal processing and electronic crossover electronics, and bi-amplifier. The electronics are mounted on the rear of the baffle directly above the bass reflex ports, which have moulded heat sink fins within. In this way the pumping action of the vent actually serves to increase air circulation to cool the electronics,

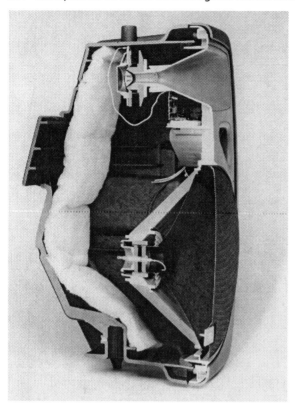

Figure 11.55. Fully-integrated powered loudspeaker system, with input mixing, signal processing and equalization, and bi-amplification. One-piece tooled aluminium baffle integrates mechanical and thermal design of woofer frame, horn, and amplifier heat sinking, with heat sink fins placed within the ports for dynamic cooling. (EON® system, Courtesy JBL Inc.)

and the entire aluminium assembly can radiate excess heat from both the electronics and the loudspeakers[70]. The system also uses a unique low-frequency motor with a neodymium magnet and dual-coil and gap configuration[36]. With both mic and line inputs, mixing facilities, and multi-band equalization on board, the system can be used for keyboard and acoustic guitar amplification as well as for overall sound reinforcement (Fig. 11.55).

PART 3: THE DIGITAL FUTURE

11.6 Electronic speaker modelling, digital input and control, and the possibility of the digital loudspeaker

The digital revolution is clearly taking over from older analogue methods in the audio signal chain. From the microphone preamplifier through to the power amplifier, the economics of DSP – Digitial Signal Processing – have revolutionized recording sound production and reproduction, except for the transducers where the microphone and the loudspeaker at each end of the signal chain are currently still analogue in nature.

While development continues on digital transducers, the power of DSP is pushing outward to add flexibility even to the analogue transducer embodiments.

Musical instrument amplifiers available from Crate, Line 6 and Johnson, among others, include emulation circuitry that attempts to duplicate the tonal voicing characteristics of classic combo and stack amplifiers, especially in the area of tube overdrive distortion qualities. This modelling extends to include the frequency response, distortion and dynamic characteristics of the guitar speakers associated with the classic designs.

While virtually all studio monitors are touted as 'digital ready', we are only on the cusp of true digital monitoring. A few manufacturers, notably Genelec, Marantz and Roland, have monitors available with digital inputs; that is, on-board D-to-A conversion electronics in addition to signal processing and amplification. The Roland system, as well as some other available non-speaker-associated signal processors, can perform speaker-modelling effects. This allows the user to select from a variety of sonic 'models' which attempt to emulate the characteristics of classic and industry-accepted standard monitors. This technology is in its infancy, and primarily addresses frequency response characteristics but, as it develops, other more subtle parameters such as time-domain characteristics and other linear and non-linear distortions will be more closely approximated. The goal is to allow the listener to use one set of loudspeakers to check many different standards, instead of switching between multiple pairs of monitors at mixdown.

In both the musical instrument and studio monitor cases, however, the simulation of other loudspeakers is, of course, limited to signal path modifications. Spatial characteristics such as dispersion and other acoustic phenomena are not capable of being addressed, regardless of the sophistication of the digital processing, these being inherently dependent on the physical radiation characteristics of the transducer. It is ultimately the listeners who must evaluate the utility, usefulness and degree of accuracy of these simulations.

As far as the holy grail of a true digital loudspeaker is concerned, only the most rudimentary of prototype concept models exist as we enter the new millennium, though patents for these and other more theoretical concepts exist. It remains to be seen whether or not the analogue electrodynamic transducer will be superseded in the digital future for both musical instrument sound production and studio monitoring reproduction.

References

1. SHORTER, D E L, 'A survey of performance criteria and design considerations for high-quality monitoring loudspeakers', *Proc. IEE,* **105**, Part B, 607–625, November (1958); also *J. Audio Eng.* Soc., **7**, No. 1, 13–28, 54 (1959).
2. OLSON, H F, 'High-quality monitor loud-speakers', *dB Magazine,* December (1967).
3. EARGLE, J, 'Requirements for studio monitoring', *dB Magazine,* February (1979).
4. COLLOMS, M, *High Performance Loudspeakers,* 5th edition, Halstead Press (a division of John Wiley, New York) (1997).
5. LANGFORD-SMITH, F (ed.), *Radiotron Designer's Handbook,* 4th edition, Amalgamated Wireless Valve Co. Pty Ltd, Sydney, 617 (1952); CD-ROM, Old Colony Sound Labs, Peterborough, NH; reprinted as *Radio Designer's Handbook,* Newnes, Butterworth-Heineman, Oxford (1997).
6. GANDER, M R, 'Dynamic linearity and power compression in moving-coil loudspeakers', *J. Audio Eng. Soc.,* **34**, 627–646 (1986).
7. GANDER, M R, 'Moving-coil loudspeaker, topology as an indicator of linear excursion capability', *J. Audio Eng. Soc.,* **29**, 10–26 (1981).
8. FRANKORT, F J M, 'Vibration and sound radiation of loudspeaker cones', *Phillips Research Reports Supplements,* No. 2, Eindhoven, Netherlands (1975).
9. THURAS, A L, JENKINS, R T and O'NEIL, H T, 'Extraneous frequencies generated in air carrying intense sound waves', *J. Acoust. Soc. Amer.,* **6**, 173–180 (1935).

10. HEYSER, R C, 'Loudspeaker phase characteristics and time delay distortion', *J. Audio Eng. Soc.*, **17**, Part I, 30–40 (1969); Part II, 130–137 (1969).

11. HILLIARD, J K, 'Basic sound recording and reproducing practices between 1927 and 1940', *J. Soc. Mot. Pic. and Tel. Eng.*, **92**, 207 (1983).

12. PRIES, D, 'Phase distortion and phase equalization in audio signal processing – a tutorial review', *J. Audio Eng. Soc.*, **30**, 774–794 (1987).

13. SMITH, D, KEELE, D B, Jr and EARGLE, J M, 'Improvements in monitor loudspeaker systems', *J. Audio Eng. Soc.*, **31**, 408–422 (1983).

14. HEYSER, R C, 'Speaker tests – Impedance', *Audio*, September, 33 (1974).

15. KINSLER, L E and FREY, A R, *Fundamentals of Acoustics*, 247, John Wiley, New York (1962).

16. LANSING, J B, 'The duplex loudspeaker', *Communications*, December (1943); also *J. Soc. Mot. Pict. Eng.*, **43**, 168–173 (1944).

17. LANSING, J B, 'New permanent magnet public address loudspeaker', *J. Soc. Mot. Pict. Eng.*, **46**, 212–219 (1946).

18. LONG, E M, 'A time-align technique for loudspeaker system design', presented at the 54th AES Convention, Los Angeles, preprint 1131, May (1976).

19. AUSTIN, D, 'Time-aligned loudspeaker systems', *dB Magazine*, March, 58–61 (1979).

20. FINK, D G, 'Time offset and crossover design', *J. Audio Eng. Soc.*, **28**, 601–611 (1980).

21. AUSTIN, D, 'Time-aligned loudspeakers revisited', *dB Magazine*, August, 27–29 (1981).

22. LANSING, J and HILLIARD, J K, 'An improved loudspeaker system for theaters', *J. Soc. Mot. Pict. Eng.*, **45**, 339 (1945).

23. FRAYNE, J G and LOCANTHI, B N, 'Theatre loudspeaker systems incorporating an acoustic-lens radiator', *J. Soc. Mot. Pict. and Tel. Eng.*, **63**, 82–85, September (1954).

24. AUGSPURGER, G L, 'Versatile low-level crossover networks', *dB Magazine*, March, 22 (1975).

25. D'APPOLITO, J, 'A geometric approach to eliminating lobing error in multiway systems', presented at the 74th AES convention, New York, October, preprint 2000 (1983).

26. KLIPSCH, P W, 'A high quality loudspeaker of small dimensions', *J. Ac. Soc. Am.*, **17**, 254–258 (1946).

27. KLIPSCH, P W, 'A speaker system with bass back-loading of unusual parameter values', *IRE Trans. Audio*, **AV-8**, No. 4, 120–123 (1960).

28. NEWMAN, R J, 'A high-quality all horn type transducer', presented at the 40th AES Convention, Los Angeles, 27–30 April, preprint (1971).

29. DUNCAN, B, 'Son of Boxer', *Studio Sound*, September, 25 (1992).

30. HARWOOD, H D, 'New B.B.C. monitoring loudspeaker', *Wireless World*, March, April and May (1968).

31. HEYSER, R C, 'Yamaha NS-1000', *Audio*, January, 82 (1979).

32. HEYSER, R C, 'Nakamichi reference monitor', *Audio*, December, 72 (1975).

33. HODAS, B, 'Examining the Yamaha NS-10M "Tissue paper" phenomenon', *Recording Engineer/Producer*, February, 54 (1986).

34. MEYER, J and KOHUT, P, assignors to Meyer Sound Laboratories, Incorporated, U.S. Patent Number 5,185,801, 'Correction circuit and method for improving the transient behavior of a two-way loudspeaker system', issued 9 February 1993, filed 18 July 1991.

35. TOOLE, F T, 'Loudspeaker measurements and their relationship to listener preferences: Parts 1 and 2', *J. Audio Eng. Soc.*, **34**, (227–235, 323–348) 1986.

36. BUTTON, D J, 'Magnetic circuit design methodologies for dual coil transducers', presented at the 103rd AES Convention, New York, preprint 4622, September (1997).

37. FIELDER, L D and BENJAMIN, E M, 'Subwoofer performance for accurate reproduction of music', *J. Audio Eng. Soc.*, **36**, (443–456) 1988.

38. HOLMAN, T, *5.1 Channel Surround Sound: Up and Running*, Focal Press, Oxford (2000).

39. GILLIOM, J R, BOLIVER, P L and BOLIVER, L C, 'Design problems of high-level cone loudspeakers', *J. Audio Eng., Soc.*, **25**, 294–299 (1977).

40. HENRICKSEN, C, 'Heat transfer mechanisms in loudspeakers; analysis, measurement and design', *J. Audio Eng. Soc.*, **35**, 778–791 (1987).

41. JOHNSTON, G C, 'The design and manufacture of loudspeaker cones – an introduction', presented at the 45th Convention of the AES, Los Angeles, 15–18 May, preprint 919 (1973).

42. SAKAMOTO, N, SATOH, K, SATOH, K and ATOH, N, 'Loudspeaker with honeycomb disk diaphragm', *J. Audio Eng. Soc.*, **29**, 711 (1981),

43. COHEN, A B, *Hi-Fi Loudspeakers and Enclosures*, 2nd edition, 84, Hayden Book Co., London (1968).

44. SHINDO, T, YASHIMA, O and SUZIKI, H, 'Effect of voice-coil and surround on vibration and sound pressure response of loudspeaker cones', *J. Audio Eng. Soc.*, **28**, 490 (1980).
45. COHEN, A B, 'The mechanics of good loudspeaker design', *J. Audio Eng. Soc.*, **2**, 176 (1954).
46. KING, J, 'Loudspeaker voice coils', *J. Audio Eng. Soc.*, **18**, 34–43 (1970).
47. AUGSPURGER, G L, 'The importance of speaker efficiency', *Electronics World*, January (1962).
48. SHINDO, T, KYONO, N and YASHIMA, O, 'Role of the dust cap in the cone loudspeaker', presented at the 63rd Convention of the AES, Los Angeles, 15–18 May, preprint 1469 (1979).
49. PARKER, R J and STUDDERS, R J, *Permanent Magnets and Their Applications*, John Wiley, New York (1962).
50. FOX, D, and BUDDINGH, T, 'Alnico Taste Test', *Guitar Player*, **33**, August, 102 (1999).
51. BADMAIEFF, A and DAVIS, D, *How to Build Speaker Enclosures*, Howard Sams, Indianapolis, IN (1966).
52. HOGE, W J J, 'A new set of vented loudspeaker alignments', *J. Audio Eng. Soc.*, **25**, 391 (1977).
53. HILLIARD, J, 'Historical review of horns used for audience type sound reproduction', *J. Acous. Soc. Am.*, **59**, 1 (1976).
54. LEACH, W M, Jr, 'Author's reply', and MCCLAIN, E F., Jr, 'Comments on reactance annulling in horn-loaded loudspeaker systems', *J. Audio Eng. Soc.*, **29**, 532 (1981).
55. OLSON, H F, 'A new high efficiency theatre loudspeaker of the directional baffle type', *J. Acous. Soc. Am.*, **2**, 485 (1955).
56. KEELE, D B, Jr, 'Low-frequency horn design using Thiele/Small driver parameters', presented at the 57th Convention of the AES, Los Angeles, May, preprint 1250 (1977).
57. MASSA, F, 'Horn-type loudspeakers – a quantitative discussion of some fundamental requirements in their design', *Proc. IRE*, **26**, 720 (1938).
58. TREMAINE, H, *Audio Cyclopedia*, 2nd edition, 1105, Howard W. Sams, Indianapolis, IN (1969).
59. OLSON, H F and HACKLEY, R A, 'Combination horn and direct radiator loud-speaker', *Proc. IRE*, **24**, 1557 (1936).
60. CARLTON, D P, 'The CW horn: a constant-width folded exponential loudspeaker horn', *Audio*, November (1955).
61. KARLSON, J E, 'A new approach in loudspeaker enclosures', *Audio Engineering*, September (1952).
62. KARLSON, J E, 'The Karlson speaker enclosure', *Radio and Television News*, January, 58 (1954).
63. OLSON, H F, *Acoustical Engineering*, 157, Van Nostrand Co., Princeton, NJ (1957); reprint, Professional Audio Journals, Philadelphia, PA (1991).
64. COCKCROFT, J, 'An isobarik system', *Speaker Builder*, August, 7 (1985).
65. DICKASON, V, *The Loudspeaker Design Cookbook*, 5th edition, Audio Amateur Press, Old Colony Sound Labs, Peterborough, New Hampshire (1995).
66. HENRICKSEN, C A, 'Unearthing the mysteries of the Leslie cabinet', *Recording Engineer/Producer*, April, 130 (1981).
67. PHELPS, N D, 'Acoustic line loudspeakers', *Electronics*, **13**, No. 3, 30 (1940).
68. KLEPPER, D L and STEELE, D W, 'Constant directional characteristics from a line source array', *J. Audio Eng. Soc.*, **11**, 198 (1963).
69. DAVIS, D and WICKERSHAM, R, 'Experiments in the enhancement of the artist's ability to control his interface with the acoustic environment in large halls', presented at the 51st Convention of the AES, Los Angeles. 13–16 May, preprint 1033 (1975).
70. BUTTON, D J, assignor to JBL, Incorporated, US Patent Number 5,533,132. 'Loudspeaker thermal management structure', issued 2 July 1996, filed 23 January 1995.

Bibliography

General

BERANEK, L L, *Acoustics*, McGraw-Hill, New York (1954); reprint, American Institute of Physics, New York (1986).
BORWICK, J, 'Loudspeakers and headphones', in *Audio Engineer's Reference Book*, 2nd edition (M Talbot-Smith, ed.), Focal Press, Butterworth-Heinemann, Oxford (1998).

DICKASON, V, *The Loudspeaker Design Cookbook*, 5th edition, Audio Amateur Press, Peterborough, New Hampshire (1995).

EARGLE, J M, *Loudspeaker Handbook*, Chapman and Hall, London (1996).

HENRICKSEN, C A, 'Loudspeakers, enclosures and headphones', in *Handbook for Sound Engineers*, 2nd edition (G M Ballou, ed.), Howard Sams, Carmel, IN (1991).

KELLY, S, 'Loudspeakers' and 'Loudspeaker enclosures' in *Audio and Hi-Fi Handbook*, 3rd edition (I R Sinclair, ed.), Newnes, Butterworth-Heineman, Oxford (1998).

KINSLER, L. E, FREY, A R, COPPENS, A B and SANDERS, J V, *Fundamentals of Acoustics*, 3rd edition, John Wiley, New York (1982).

LANGFORD-SMITH, F (ed.), *Radiotron Designer's Handbook*, 4th edition, Amalgamated Wireless Valve Co. Pty Ltd, Sydney, (1952); CD-ROM, Old Colony Sound Labs, Peterborough, NH; reprinted as *Radio Designer's Handbook*, Newnes, Butterworth-Heineman, Oxford (1997). Includes extensive bibliography.

MCLACHLAN, N W, *Loud Speakers: Theory, Performance, Testing, and Design*, Constable, London (1934); corrected edition, Dover Publications, New York (1960).

OLSON, H F, *Acoustical Engineering*, Van Nostrand Co. Inc., Princeton, NJ (1957); reprint, Professional Audio Journals, Philadelphia, PA (1991).

OLSON, H F, *Modern Sound Reproduction*, Van Nostrand Reinhold, New York (1972).

OLSON, H F, *Music, Physics and Engineering*, Dover Publications, New York (1967).

STRAOBIN, B, 'Loudspeaker design', in *Encyclopedia of Acoustics*, John Wiley, New York (1997).

Studio monitor applications

BORWICK, J (ed.), *Sound Recording Practice*, 4th edition, Oxford University Press, Oxford (1994).

EARGLE, J M, *Handbook of Recording Engineering*, 2nd edition, Van Nostrand Reinhold, New York (1992).

EVEREST, A, *Handbook of Multichannel Recording*, Tab Books, Blue Ridge Summit, PA (1975).

MARTIN, G (ed.), *Making Music*, Quill (William Morrow & Co.), New York (1983).

NEWELL, P R, *Studio Monitor Design: A Personal View*, Focal Press, Butterworth-Heineman, Oxford (1995).

HUBER, D M and RUNSTEIN, R E, *Modern Recording Techniques*, 4th edition, Focal Press, Butterworth-Heinemann, Oxford (1995).

WORAM, J, *Sound Recording Handbook*, Howard Sams, Indianapolis, IN (1989).

Musical instrument applications

BACON, T (ed.), *Rock Hardware: The Instruments, Equipment and Technology of Rock*, Harmony Books, Quill, New York (1981).

BROSNAC, D, *The Amp Book: An Introductory Guide to Tube Amplifiers*, d. B. Music Co., Ojai, CA (1983).

DARR, J, *Electric Guitar Amplifier Handbook*, 3rd edition, Howard Sams, Indianapolis, IN (1971).

DOYLE, M and EICHE, J (eds), *The History of Marshall: The Illustrated Story of 'The Sound of Rock'*, Hal Leonard Publishing, Milwaukee, WI (1993).

MORRISH, J, *The Fender Amp Book*, Miller Freeman, San Francisco (1995).

FLIEGLER, R, *AMPS!: The Other Half of Rock 'n' Roll*, Hal Leonard Publishing, Milwaukee, WI (1993).

MACKAY, A, *Electronic Music: The Instruments, the Music, the Musicians*, Control Data Publishing, Harrow House Editions Limited, Minneapolis, MN; Phaidon, Oxford (1981).

TEAGLE, J, and SPRUNG, J, (Contributor), *Fender Amps: The First Fifty Years*, Hal Leonard Publishing, Milwaukee, WI (1995).

TRYNKA, P (ed.), *Rock Hardware*, Balafon/Outline Press: London; Miller Freeman, San Francisco (1996).

WHEELER, T, *The Guitar Book*, 2nd edition, Harper and Row, New York (1982).

12 Loudspeaker measurements

John Borwick
Revised by Julian Wright

12.1 Introduction

Measurements on loudspeakers are made by various sorts of individuals and for a variety of reasons. For example, the loudspeaker designer uses measurements to help him identify the effects on performance of different dimensions, materials or components and to enable him to draw up a specification which his manufacturing colleagues can then use. The manufacturer must carry out quality assurance measurements on drive units and other components – whether constructed in-house or bought in – as well as on completed loudspeaker systems, for instance to confirm that they meet the published specification or to select matched stereo pairs from within the tolerance spread of sensitivity, axial response, etc. The professional user will similarly conduct *in situ* measurements, perhaps to adjust the performance to given room or auditorium requirements, and hi-fi reviewers will often use objective measurements in support of their subjective impressions.

To be of real use, these measurements must be accurate, meaningful and repeatable – though each of these adjectives requires some qualification and explanation, and they are by no means mutually exclusive. To be *accurate,* measurements must employ high-quality test equipment and standard procedures which have been evolved having the twin virtues of maximum reliability and avoidance of error. To be *meaningful,* measurements must correlate as far as possible with audible (subjective) effects and relate to the particular application for which the loudspeaker is designed. Thus it would be possible to conduct tests which are too accurate, say to an accuracy of ± 0.01 dB, for an application where results within ± 1.0 dB are good enough. To be *repeatable,* measurements must conform to strict rules with regard to the instrumentation used, disposition of test microphones, ambient conditions of temperature and humidity, and the specified acoustic environment.

The complex nature of the radiated waves and the even more complex, and as yet imperfectly understood, nature of the human hearing mechanism make it impossible to produce objective measurements which correlate exactly with the subjective qualities as heard by a listener[1]. Instead, the practice is to measure and specify those characteristics of the loudspeaker which seem the best indicators of subjective quality, and continually strive to refine the test procedures so that the objective/subjective correlation is enhanced (see Chapter 13). The most important characteristics include the following[2]:

(a) axial frequency response;
(b) directional characteristics;
(c) sensitivity;
(d) efficiency;
(e) impedance;
(f) linear distortion;

(g) non-linear distortion;
(h) power-handling capacity.

12.2 Measurement standards

For all their inability to provide ideally precise information, loudspeaker measurements remain a mandatory fact of life, and all workers in this field should at least be aware of the published standards, as well as of their limitations. The principal international standard is IEC Publication 60268[3] (see description in Chapter 15), and its recommendations form the basis for most of the guidelines given in this chapter.

Even where local national standards exist, or recommendations are published by individual trade associations or societies – which may have the advantage of more frequent updating – these should be read in conjunction with IEC 60268 Part 5. For example, the Audio Engineering Society published in October 1984 a useful 'AES Recommended Practice Specification of Loudspeaker Components Used in Professional Audio and Sound Reinforcement', AES2–1984 (ANSI S4.26–1984)[4] which has high relevance to the fields specified, and the special sorts of drive units employed therein, but does not replace the wider-ranging IEC document. When consulting such documents, it should be borne in mind that they are all subject to revision (albeit at rather lengthy intervals) and it is essential to get hold of the most recently published version.

12.3 The measurement environment

Except for the category of measurements where purely electrical tests are made (e.g. d.c. voice-coil resistance and modulus of impedance) the basic measurement chain takes the form shown in Fig. 12.1. The loudspeaker is placed in the measurement environment and the test signals are fed to it via a suitable power amplifier. Attention must be paid to the impedance of the cables feeding the loudspeaker: this should be compensated in post-processing or at least minimized. The radiated signals are picked up by a measuring microphone and passed to a data capture device, meter, level recorder or other form of processor for display or storage.

Of great importance is the type of environment in which the measurements are conducted. Ideally, from the point of view of interference-free results, the environment will have zero ambient noise, be free of obstructions and possess no reflecting boundaries (free field) – since the microphone will respond to all sounds reaching it, not just those arriving along the direct path from the loudspeaker. The practical types of measurement environment possess one of the following three basic types of

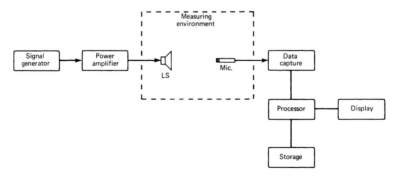

Figure 12.1. Basic measurement schematic.

characteristic: free field (anechoic), diffuse field (reverberant) or typical room (semi-reverberant). These will now be considered, with notes as to their relative advantages and disadvantages.

12.3.1 Free-field

A free sound field is defined as being a region of space whose boundaries have negligible effect on sound waves over the frequency range of interest. Thus a loudspeaker or other source placed in such a sound field is free to radiate in all directions (4π steradians) and the wavefront (in the special case of an omnidirectional or 'point' source) will take the form of a continually expanding sphere. It is often convenient and informative to measure under 'half free-field' or 'half-space' conditions where the source is mounted flush in a very large baffle (approximating to an infinite baffle) and radiates into 2π steradians. This removes the effects of enclosure diffraction from the analysis. The near field technique described by Keele[18] provides a simple method for establishing the low-frequency part of a half free-field measurement, but without the difficulties of providing a reflection-free environment. Two practical implementations for free-field measurements exist: (a) outdoors or (b) in a specially constructed anechoic room.

(a) Outdoors

Since the earliest days, free-field measurements have been conducted in the open air with the loudspeaker and measuring microphone mounted on a tall tower, to minimize the influence of ground reflections, or on a boom projecting from the corner of a tall building. A half-free field is sometimes used (and is recommended in the German standard DIN 45500 Part 7)[5]. This is more suitable for tests on individual drive units than complete systems and involves mounting the unit flush with the ground, facing upwards. Gander[39] has described a 'ground-plane' approach for deriving a free-field measurement.

Advantages: relatively inexpensive; can give near-perfect acoustic performance.
Disadvantages: adversely affected by ambient noise, wind and weather; may create
  a nuisance to neighbours.

(b) Anechoic room

Free-field conditions can be approached except for the lowest audio frequencies by constructing an anechoic (literally 'without echoes') room or chamber (see Fig. 12.2). If sufficiently long wedges of absorbent material are used to cover all six inner surfaces, including the door and floor – with an open-work metal grid over the floor to allow access – surface reflections can be reduced to negligible levels.

Advantages: very low ambient noise, especially if a floating construction is used to
  isolate the chamber from the main building structure; repeatable tests can be
  conducted close to other laboratory facilities; excellent acoustic performance
  down to the cut-off frequency.
Disadvantages: expensive; measuring distance is limited by the room dimensions; inaccurate at low frequencies (roughly below the cut-off frequency for which the
  wedge length is $\lambda/4$) because of incomplete absorption.

Note, however, that for a given system the low-frequency performance can be additionally measured outdoors to derive a correction curve for future measurements in a particular anechoic chamber. This is often used for quality-control purposes.

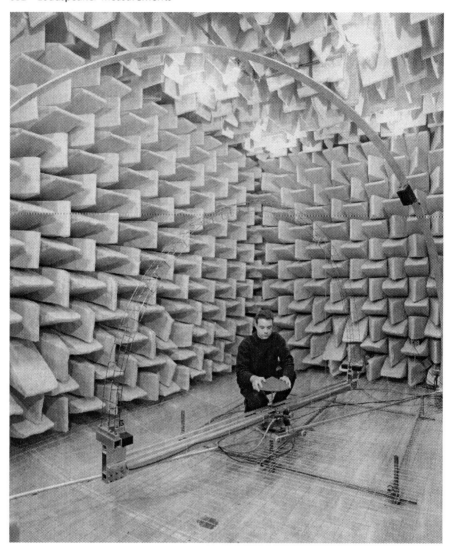

Figure 12.2. Typical anechoic room. (Courtesy Salford University.)

12.3.2 Diffuse field

A diffuse field is defined as being completely homogeneous, with the same average sound-energy density at all points and the same mean acoustic power per unit area in all directions. Such conditions can be approached, for a central space remote from the inner surfaces, by constructing a very reverberant room with hard reflecting walls, floor and ceiling (see Fig. 12.3). Reverberant test rooms are principally used to measure total sound power.

Advantages: effectively integrates the total radiated power regardless of source directivity.

Disadvantages: relatively expensive; field is not perfectly diffuse, particularly close to the boundaries; limited accuracy with directional sound sources.

Figure 12.3. Reverberant room measuring 12 m³. (Courtesy Bang and Olufsen.)

12.3.3 Semi-reverberant field

The sound field in a typical living-room or laboratory is a mixture of direct, reflected and reverberant sounds; the proportions of these various types of sound waves are dependent on the size and shape of the room, the directivity of the source, its position, the microphone location and the distribution of (frequency-selective) sound-absorbing materials and reflecting obstacles within the room.

Advantages: inexpensive; can be used to provide useful information about the performance of a particular loudspeaker in a given listening room.
Disadvantages: results are correct only for the given room and test set-up; ambient noise inhibits low-level accuracy. Note, however, that gating techniques now permit sophisticated testing even in such an environment (see next section).

12.3.4 Simulated free-field

Modern signal-processing techniques make it possible to isolate the direct sound from the reflected sounds and thus permit 'free-field' measurements to be conducted in any physical environment – anechoic, diffuse field or semi-reverberant. As an extension, it may be desirable to select individual reflections for separate examination. The techniques currently used will each be described later.

Advantages: gives repeatable results in any location; provides rapid data on amplitude and phase; microcomputer-based systems are now relatively inexpensive and yet provide powerful and versatile measuring techniques.
Disadvantages: requires skilled analysis; time domain gating sets a limit to the frequency resolution.

12.4 Measuring conditions

As well as employing a specified test environment, it is essential to specify as many as possible of the test conditions which could conceivably have a significant effect on the results. Some of the most basic of these are outlined below; for definitions of the conditions and the parameters to be measured, the reader is referred to IEC 60268–5[3].

12.4.1 Temperature and humidity

Ambient conditions are likely to affect performance, for example the properties of many plastic materials are temperature-dependent and some materials such as paper cones are liable to absorb moisture from the air. In any event, it is important to state temperature and humidity readings as conditions of the measurement.

12.4.2 Test environment

Even in an anechoic chamber the measured response will vary slightly with position, which should therefore be noted. Similarly, outdoor measurements are prone to ground reflections, etc. In addition, the anechoic chamber dimensions should be stated as well as the frequency below which the measurements are invalidated due to insufficient absorption, and the physical details of an outdoor test rig should be specified.

12.4.3 Ambient noise

The average ambient sound-pressure level should be stated. If measuring errors are not to exceed 1 dB, the signal-to-noise ratio must be at least 20 dB for any frequency within the range of interest.

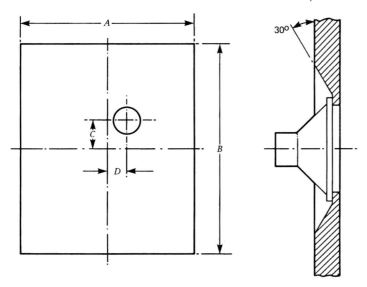

Figure 12.4. Standard baffle recommended in IEC 268–5, where $A = 135$ cm, $B = 165$ cm, $C = 22.5$ cm, $D = 15$ cm. For AES recommended dimensions see Table 12.1.

Table 12.1 Preferred baffle dimensions for lf loudspeaker

Nominal loudspeaker size[a] (mm)	Baffle dimensions[b] (mm)			
	A	B	C	D
200	1350	1650	225	150
250	1690	2065	280	190
315	2025	2475	340	225
400	2530	3090	420	280
500	3040	3715	505	340

[a] The nominal loudspeaker size is defined as the outer diameter of the frame, and the metric equivalent is given in the nearest preferred number according to IEC Standard, Publication 268–14 (1980)
[b] Baffle dimensions have been calculated from the IEC dimensions by multiplying them by the ratio of the nominal size in inches to 8 inches, and rounding the resulting dimensions to the nearest 5 mm.

12.4.4 Loudspeaker mounting

This will affect the acoustic loading and therefore the measured results and should be described clearly. Four possibilities exist:

(a) a standard baffle;
(b) a specified enclosure;
(c) in free air without a baffle or enclosure;
(d) in half-space free-field, flush with the reflecting plane.

In practice, (b) and (d) are the much preferred options, as the others introduce effects which will not exist in the finished product or installation. Complete loudspeaker systems are usually measured without any additional baffle. For drive units, the IEC publication provides details of a standard baffle (see Fig. 12.4) as well as two possible methods of avoiding resonant cavities. The AES recommendation[4] shows essentially the same baffle construction but suggests that the IEC dimensions should be used for drive units up to 200 mm in diameter but must be scaled upwards for larger drivers as shown in Table 12.1.

12.4.5 Measuring apparatus

Where the test equipment used has special characteristics or limitations, these should be specified.

12.4.6 Measuring distance

A distinction is made between near- and far-field measuring conditions[18]. Far-field conditions are said to exist when each doubling of distance from the source results in a 6 dB reduction in sound level (the inverse square law). In general, this situation exists when the microphone is at least two or three times further away than the distance between parts of the source which are radiating energy at the same frequency. Usually this suggests a measuring distance of 2 m or more, and even then some near-field phenomena may exist so that some spatial averaging is advisable.

Note, however, that a near-field measurement may have special usefulness in some circumstances. This is made with a closely placed microphone or contact transducer (see Section 12.5.1(d)).

12.4.7 Reference or measuring axis

Measurements should be made with due regard to the data specified by the manufacturer defining the effective reference plane of the drive unit or system, the reference point (its effective centre) and the reference axis, which should be used as the zero axis for frequency-response and directional-response measurements. The reference axis is usually normal to the front baffle of a loudspeaker but may be skewed in multiple-driver systems due to differences in the effective depth of the individual drivers.

12.5 Characteristics measurements: small signal

It is convenient to group under this heading the category of loudspeaker measurements in which the test signal amplitude and bandwidth are kept to values for which the loudspeaker performance can be assumed to be linear. In reality such amplitudes are unfeasibly small, and the pragmatic approach is to use 'real-world' amplitudes and assume linearity for convenience. This assumption must *always* be borne in mind. Large-signal measurements and non-linearity are discussed in Sections 12.6 and 12.7.

12.5.1 Frequency response (transfer function)

By far the most important and informative test measurement is that of the loudspeaker's axial frequency response. The frequency response has two components, amplitude and phase, which together completely define the linear behaviour of the system. Figure 12.5 shows a symbolic representation of a linear system response in both the time and frequency domains[13].

The time-domain representation takes the form of the response to an ideal impulse, known as the impulse response $h(t)$; the frequency-domain equivalent takes the form of the familiar frequency response or transfer function $H(f)$. The impulse response

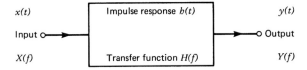

$y(t) = x(t) * b(t)$ convolution of time functions

$Y(f) = X(f) \cdot H(f)$ multiplication of spectra

Figure 12.5. Symbolic representation of a linear system response, expressed in both the time and frequency domains. (After Fincham.)

and frequency response are related mathematically through the Fourier transform[40], and once either of these is known the output may be predicted for any given input signal, steady state or transient. This is calculated in the frequency domain by multiplication of the transfer function with the input expressed as a function of frequency, and in the time domain by convolution of the input waveform with the impulse response as indicated in Fig. 12.5.

Frequency-response measurement techniques have dominated the literature and continue to inspire the development of new high-technology test methods. The methods currently in use may be grouped conveniently in terms of the type of test signal employed. Note that the type of stimulus will to some extent affect the measurement results, such that they may differ for the same test object using any two measurement methods. This is an artefact of the false assumption of linearity.

(a) Continuous sine wave

A sine wave (single-frequency continuous tone) test signal can be used to derive the amplitude response characteristic of any loudspeaker drive unit or system. Usually a constant input level is applied and the test frequency is altered in steps. A sufficient number of such point-to-point measurements of the radiated signal level may

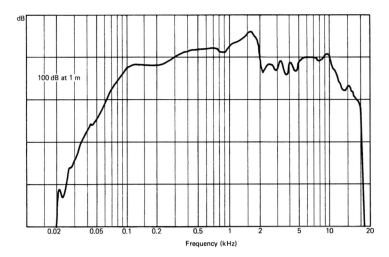

Figure 12.6. Typical swept sine-wave response curve of a medium-quality two-way loudspeaker system. (Courtesy B&K.)

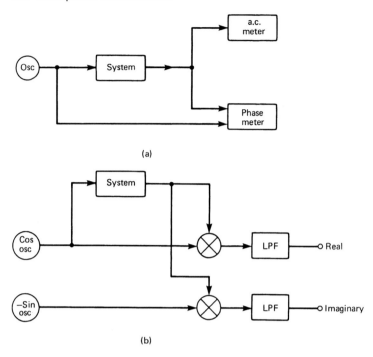

(a)

(b)

Figure 12.7. Two methods of measuring the transfer function (amplitude and phase) of a system: (a) a.c. meter and phase meter; (b) two oscillators in phase quadrature plus multipliers and low-pass filters to extract the real (in-phase) and imaginary (quadrature) components.

be used to derive (or display) the loudspeaker's amplitude response. Such point-to-point measurements suffer from the restriction that the sine wave must be applied for a sufficient time at each spot frequency for the amplitude to reach its steady-state value (and for the previous signal to die away). Thus increasing the number of test frequencies in order to improve the frequency resolution (for example when testing a loudspeaker having high-Q resonances) necessarily increases the duration of the test.

As an alternative, the tuning of the sine-wave generator may be *swept* slowly through the bandwidth of interest (traditionally 20 Hz to 20 kHz) (see Fig. 12.6). Suitable instruments for swept sine wave testing have been available for many years and the method has universal application. Note, however, that the same limitations on sweep rate for a given frequency resolution apply as for point-to-point testing, and errors may be introduced due to noise and the limited number of cycles in the sine wave. Using a synchronized tracking narrow-band filter at the microphone output will of course improve the rejection of noise and other interference.

As has been said, the amplitude response does not totally define the system's performance. The corresponding phase response is required and may be measured, e.g. in the simple way illustrated in Fig. 12.7(a) using a separate phase meter. The measured phase response consists of the phase shift due to the loudspeaker under test plus additional phase rotations due to the time delay between the loudspeaker and the microphone. A delay element in the measuring chain is usually employed to eliminate this additional phase shift so that the phase shift due to the loudspeaker alone is displayed. Alternatively, a more convenient technique known as synchronous detection can be employed.

In Fig. 12.7(b) a pair of oscillators in phase quadrature (90° relative phase) plus a pair of multipliers and low-pass filters[6] are used to extract the real and imaginary components. This arrangement responds to the fundamental frequency and rejects spurious harmonies and noise.

(b) Time-delay spectrometry (TDS)

This method was first described by Heyser[7] in 1967 and has been the subject of several extensions since[41]. The basic principles of TDS can be illustrated as shown in Fig. 12.8. It resembles that of Fig. 12.7(b) with the addition of a variable time delay between the test generator (producing a rapid sine wave sweep or 'chirp') and the two quadrature signals. The delay can, for example, be adjusted to τ, the delay due to the physical distance between the loudspeaker and the test microphone, for correct synchronous demodulation of the direct signal only. Precisely linear sweep of the chirp excitation ensures that a frequency offset in the demodulating sweep accurately corresponds to a particular time delay. The further delay T experienced by a reflected wave will therefore cause it to be rejected by the low-pass filters. The TDS method has advantages for measuring in a noisy environment or normal semi-reverberant room conditions.

The versatility of TDS as a measurement method is due in large part to its ability to represent data in the time domain as easily as in the frequency domain. As with any gating system, it is a feature of the TDS reciprocity that, because of the limited time or frequency range which is available for a measurement to be made, the time and frequency domain descriptions cannot both be of infinite accuracy: one or the other must be limited or 'windowed'. This is known as the uncertainty principle of frequency and time analysis, and can be expressed as follows:

$$f \approx 1/T \quad \text{and} \quad t \approx 1/F$$

where T and F are the chosen time and frequency ranges, and t and f are the resolutions in the respective domains.

The choice or 'segmentation' of the frequency range F to be used will depend on the particular application; the choice of T may need to be deliberately limited, e.g. to remove undesired reflections.

The use of a linear sweep in TDS means that the sweep rate S is time-independent and given by:

$$S = F/T \text{ (Hz/s)}$$

where T = sweep time.

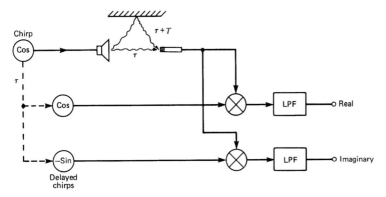

Figure 12.8. TDS method of measuring the transfer function of a delayed signal component. The further delay T of a reflected wave will cause it to be rejected.

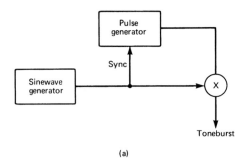

(a)

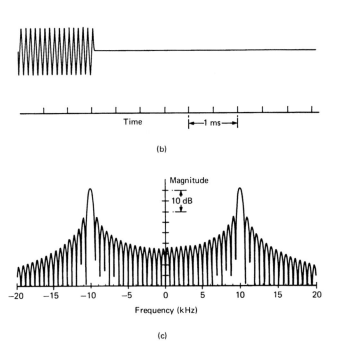

Figure 12.9. Standard method of generating tonebursts: (a) block diagram; (b) 16-cycle 10 kHz toneburst; (c) spectrum of the 16-cycle 10 kHz toneburst.

(c) Triggered sine wave (toneburst)

Toneburst testing has a long history but has been hampered by the difficulty of correctly interpreting the results. A toneburst is generated by switching a sine wave on and off at regular intervals (see Fig. 12.9) and may be regarded as the product of a sine wave and a square pulse. As a rule the toneburst starts and stops at zero crossings of the sine-wave so that the burst and the 'space' contain an integral number of cycles, say 8 cycles on, 16 cycles off.

The usefulness of tonebursts for loudspeaker measurements has been extended by the technique of gating or using a time 'window' to select only that part of the received toneburst which is of interest. For example, selecting only the centre few cycles (see Fig. 12.10) will give the effective steady-state response as from a contin-

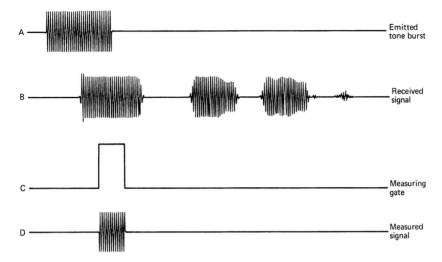

Figure 12.10. Gating technique used with toneburst testing. (Courtesy B&K.)

uous sine wave. On the other hand, as Fig. 12.10 illustrates, such gating may also be used to eliminate reflected sounds and so permit simulated free-field measurements to be made in any room environment.

Useful data can be derived from toneburst responses under a number of headings: the steady-state response at the frequency of the toneburst; the time constants associated with transients (by inspection of the attack and decay waveforms); and the general transient behaviour or 'transient distortion'. However, while enabling comparisons to be made between different systems, the conventional toneburst response provides data which are limited in frequency coverage (unless many measurements are made) and yet affect frequencies remote from that of the burst and can therefore cause confusion. This problem can be reduced by employing band-limited tonebursts[9], or more effectively by modulating the amplitude of the sine wave in a number of steps[10] (approximately 1/3 octave).

In essence, the problem is that a toneburst is essentially a single-frequency signal which is abruptly terminated or truncated by a gating pulse signal – a process of amplitude modulation which will inevitably produce sidebands and excite the test loudspeaker over a wide frequency range (see Fig. 12.9(c)). Bunton and Small[11], and Linkwitz[10] have tackled the problem of these truncation sidebands in toneburst testing, and the related creation of 'spectral leakage' when a data window introduces abrupt signal truncations in the derivation of cumulative spectra. They use the term 'apodize' (literally, remove the unwanted 'feet' of), credited to Heyser[12], to characterize a process of modifying the truncation process to produce a more easily interpreted display.

(d) Impulse

In 'small-signal' measurements, where the loudspeaker can be assumed to be a linear device, a test signal consisting of a short-duration impulse can be used to derive not only the steady-state frequency response, in terms of both amplitude and phase, but also the transient response[13,14]. It is also an extremely rapid method, a measurement being completed in a time equal to the duration of the impulse response.

It could in any case be argued that some form of impulse test signal has a more direct relevance in loudspeaker measurements than simple sine waves, bearing in

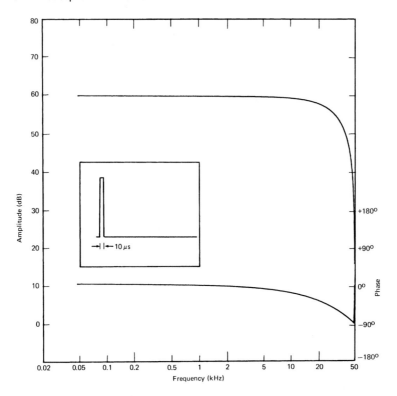

Figure 12.11. Frequency and phase response of a 10 µs rectangular test pulse.

mind the transient nature of most speech and musical sounds. Indeed, this appears to have been recognized by the earliest of researchers, from Rice and Kellogg[15] and McLachlan[16] onwards.

The impulse response (time domain) and frequency response (frequency domain) are related mathematically by the Fourier Transform, as already discussed for tone-burst testing. Since the late 1960s the so-called Fast Fourier Transform (FFT), using a computer, has simplified and speeded up the process and introduced new techniques for investigating system transfer functions. The computer's ability to measure, digitize and process results, and permit a range of test signal types to be used, has meant that the impulse response completely defines the system behaviour for both transient and steady-state signals.

In the digital technique described by Berman and Fincham[13], the test signal is a very narrow square pulse which nevertheless has a wide linear response, as illustrated in Fig. 12.11. The test environment, though not necessarily anechoic, should be large enough for the loudspeaker response to individual impulses to die away before the first reflection arrives at the microphone (to allow gating or truncation without clipping the received impulse tail). It is a feature of impulse testing that the radiated energy is very small, so that the signal-to-noise ratio will be inadequate unless special additive procedures are used. In practice, this problem is solved by a process of signal averaging. The typical measuring chain is shown in Fig. 12.12, with the loudspeaker under test placed close to the centre of the measuring room, remote from all reflecting surfaces.

A repetitive short pulse of the form shown in Fig. 12.13(a) is amplified and fed to the loudspeaker. The microphone detects the radiated acoustic pressure wave, and

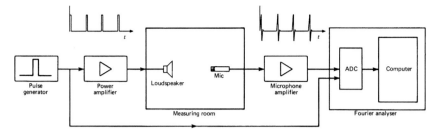

Figure 12.12. Block diagram of impulse response measuring chain. (After Berman and Fincham.)

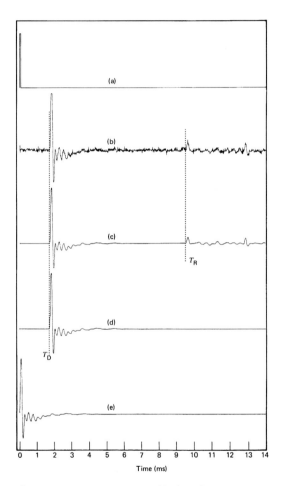

Figure 12.13. Sequence used in impulse response measurement: (a) test pulse; (b) single-pulse response; (c) signal-averaged response; (d) response after removal of room reflections (gating), (e) impulse response, corrected for loudspeaker/microphone time delay.

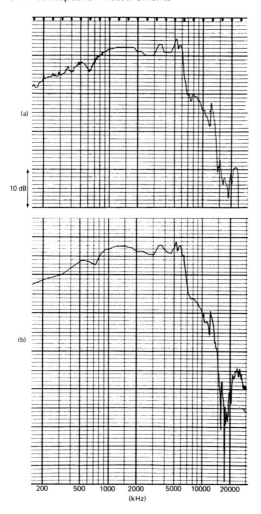

Figure 12.14. Comparison of amplitude responses of 200 mm drive unit in 45 litre enclosure: (a) by swept sine wave outdoors; (b) by impulse digital method.

its output is suitably amplified, digitized by the analogue-to-digital converter and stored in the computer. The response of a typical loudspeaker to a single impulse is shown diagrammatically in Fig. 12.13(b), delayed with respect to the input signal by a time T_D corresponding to the microphone/loudspeaker distance. A smaller peak can be seen at time T_R, caused by the arrival at the microphone of the first room reflection (from the nearest reflecting surface). The interval between the input pulses is made roughly equivalent to the room reverberation time (typically 0.5–1.0 s).

As has been mentioned, the low energy content of the test pulse means that signal averaging must be employed to improve the signal-to-noise ratio of the measurement. The test pulse is repeated at equal time intervals, and the resulting responses are summed. For each doubling of the number of averages, the signal increases by 6 dB, whereas any random noise increases by only 3 dB. Thus the signal-to-noise ratio increases by a factor of $10 \log N$ dB for a total of N averages (see Fig. 12.13(c)). Similarly, by gating the output to set all sampled values beyond T_R to zero, the room

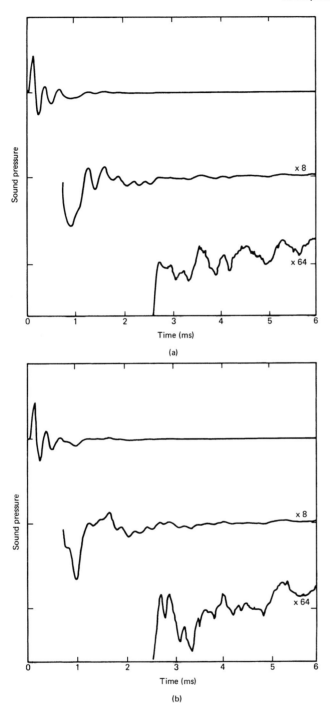

Figure 12.15. Improved resolution obtained by multiplying the tail of the impulse response by 8 and 64 times to reveal significant differences between two nominally identical 110 mm drive units in a 7-litre closed box.

response is eliminated (Fig. 12.13(d)); and the loudspeaker/microphone time delay can be removed by redefining the time origin (Fig. 12.13(e)).

By way of verification, Berman and Fincham show the amplitude responses for a 200 mm drive unit in a 45-litre enclosure measured by the swept sine-wave method outdoors and by the impulse digital method just described (Fig. 12.14). The equivalence of the two methods is clearly demonstrated. Comparisons were also made with analogue toneburst and square wave responses, and the equivalents calculated from the impulse response, again with extremely close correlation. When the loudspeaker behaves as a minimum-phase system, which is the case for most loudspeaker drive units (but not complete systems, due to the effect of the crossover network) over at least part of the spectrum, amplitude and phase are uniquely related; one can be derived from the other by means of the Hilbert Transform. Comparing this calculated phase response with the one measured in the impulse test confirms the frequency region over which minimum-phase behaviour applies. This has direct application to crossover networks, whose design is greatly simplified if the drive units are minimum-phase systems, since only the amplitude response of the crossover filter need be considered.

The impulse response itself can provide a new order of detailed information by magnifying the impulse decay or tail. This is shown in Fig. 12.15, where the tail has been magnified 8 and 64 times, giving a visual dynamic range of over 50 dB for a pair of nominally identical units. Small changes of materials or construction are revealed and can be related to particular components, the surround, etc. The method can also be applied to the measurement of enclosure characteristics by sending short pulses into the cabinet and using a probe microphone.

Fincham[17] has later described refinements to the method providing measurements in a non-anechoic environment to an accuracy better than 1 dB down to 20 Hz. Such accuracy, at previously hard-to-measure but subjectively sensitive low frequencies, has been found necessary if objective testing is to correlate adequately with subjective assessments (see Chapter 13).

This method overcomes some of the errors inherent in the impulse method by replacing the flat-spectrum test pulse with a pre-emphasized pulse following the spectrum of the background noise. A mirror-image de-emphasis circuit is introduced following the microphone amplifier. The improved low-frequency resolution thus achieved is illustrated in Figs 12.16 to 12.18. Spectral shaping of the input signal is also employed to avoid the errors due to the simple time limiting (truncation) necessary in some test rooms to gate the impulse response before it reaches zero because of the imminent arrival of the first room reflection.

(e) Random noise

As an alternative to sine waves or impulses, random noise may be used as the test signal (band-limited to the frequency range of interest). Available noise generators can supply either 'white noise' which contains all frequencies in a random (Gaussian) distribution, i.e. equal energy per Hz, or 'pink noise', which contains equal energy per percentage bandwidth (see Fig. 12.19).

Amplitude response may be measured either (a) by using constant percentage bandwidth (typically 1/3 octave) and pink noise; or (b) by analysing the loudspeaker's output using a constant-bandwidth analyser (heterodyne or Fast Fourier transform spectrum analyser) with white noise. Correlation techniques can also be used with white noise to obtain the cross-spectra or cross-correlation coefficient from which amplitude and phase may be derived.

The signal-to-noise ratio can be improved by using a narrow-band rather than a broad-band signal. This arrangement enables the maximum input level to be concentrated over a narrow band of frequencies. The addition of a synchronized tracking filter in the microphone chain will provide a still greater improvement in S/N ratio.

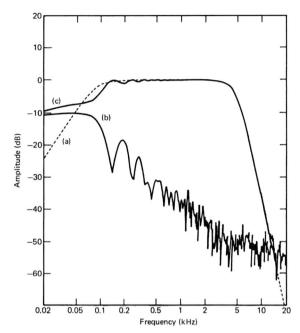

Figure 12.16. Errors in FFT response measurement due to random noise in the measuring chain: (a) true response of a model loudspeaker estimated as if a 10 µs 30 V rectangular pulse (16 repetitions) were used; (b) noise spectrum of chain (16 averages); (c) measured response ((a) + (b)).

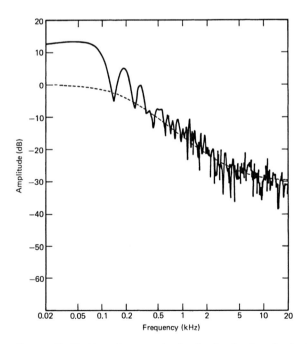

Figure 12.17. Use of pre-emphasized pulse (broken line) which follows that of the background noise.

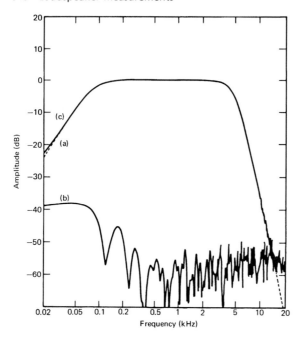

Figure 12.18. Reduced errors through use of pre-emphasis: (a) true response of model loudspeaker; (b) noise spectrum after de-emphasis; (c) measured response ((a) + (b)).

(f) Maximum Length Sequences

In recent years a new stimulus has established itself as a highly effective measurement tool. A Maximum Length Sequence (MLS) is a periodic binary sequence with some attractive properties, including short measurement times and high signal-to-noise performance. The use of the MLS in acoustic measurements is discussed extensively by Rife and Vanderkooy[35].

Computer acquisition offers the rapid availability of three-dimensional cumulative spectra displays, which are obviously of high interest in transducers for speech and music reproduction. By simple computation and the use of the FFT, it is now possible to produce delayed-response curves directly from the impulse response. In the example shown in Fig.12.20, a linear amplitude scale has been chosen to assist examination of the early part of the decay. The logarithmic amplitude scale in Fig. 12.21 makes the low-level information more accessible. The frequency scale is linear in both examples because a log scale has been found to cause confusion in the high-frequency region.

It is interesting to note that Shorter[1], though using tonebursts, anticipated the presentation of three-dimensional displays representing amplitude, frequency and time. Naturally, modern methods utilizing computers have greater speed and versatility both in the generation of suitable test signals and the processing and presentation of results.

12.5.2 Directional response

The on-axis frequency response has been treated above at some length because it is arguably the most important and revealing single criterion for assessing the likely subjective performance in listening tests. Yet in all practical situations the listener

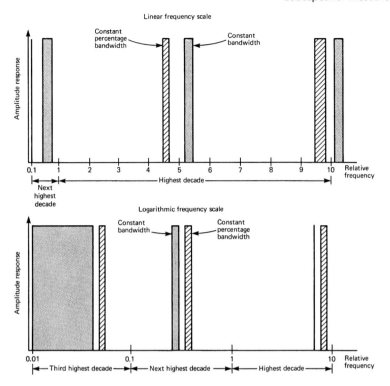

Figure 12.19. Comparison of filtered white noise and pink noise, showing how white noise (constant bandwidth) occupies varying widths on a logarithmic scale whereas pink noise (constant percentage bandwidth) occupies an equal width at all centre frequencies.

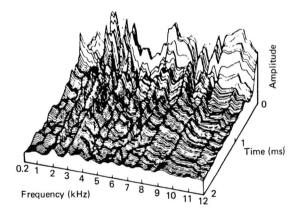

Figure 12.20. Cumulative decay spectra using a linear amplitude scale to assist examination of the early part of the decay.

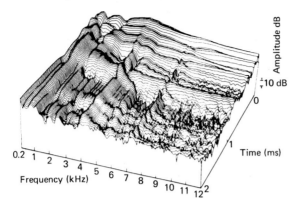

Figure 12.21. Cumulative decay spectra using a logarithmic amplitude scale for examination of the low-level information.

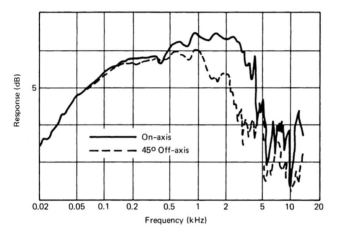

Figure 12.22. Typical plot of directional response, comparing the response on the reference axis with that at 45°.

will receive more than just the direct on-axis sound. He may be located at some other angle, and he will certainly receive a totality of reflected sounds in addition to the direct wave. The way in which a loudspeaker's response varies with direction is therefore an essential element in its overall evaluation.

Directivity can be measured by means of any of the frequency-response test methods already discussed. Instead of being placed on the reference axis, the microphone is located at the desired angle (perhaps 30°, or in incremental steps of 15°) and the response is plotted for direct comparison with the axial result. Figure 12.22 shows a typical presentation. Directivity in the horizontal plane is usually of most interest as it affects the consistency of high-frequency balance with listening position over an arc in front of the loudspeaker, stereo imaging, etc. However, the tests may be repeated in the vertical plane where desirable. In fact, when a loudspeaker drive unit has lateral symmetry, the single-plane measurement is sufficient to produce a three-dimensional directivity pattern by rotation.

When a polar plot at a single frequency is required, the loudspeaker may be mounted on a turntable driven in synchronism with a circular chart recorder.

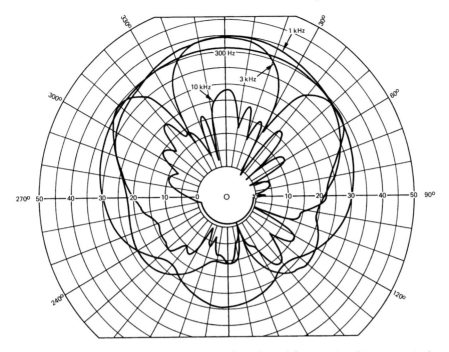

Figure 12.23. Polar plot on circular chart for selected frequencies. (Courtesy B&K.)

Repeating the test at several frequencies of interest produces a polar plot of the form shown in Fig. 12.23. Again, the horizontal plane is generally the most useful, though the loudspeaker may be placed on its side, or at any other angle, if required. If a complete 360° polar plot is not required, the turntable drive can be set to sweep through a restricted angle, say 60°.

For some applications it is desirable to specify the directivity index, which is defined as the ratio of the intensity at a chosen point on the reference axis to that of a point source radiating the same total acoustic power. It is possible, though necessarily rather complicated, to compute the total power radiated into a semi-anechoic field from a series of measurements using different microphone locations on the surface of a hypothetical hemisphere[20] (see Fig. 12.24).

However, two simpler procedures are available. The first relies on an integration of the sound pressure squared, taken from the two-dimensional polar plots described above. The second uses the relationship between the on-axis frequency-response curves measured under both free-field and diffuse-field (reverberation room) conditions. The directivity index D_i is then given by the expression:

$$D_i = L_{ax} - L_p + 10 \log_{10}(T/V) + 25 \text{ dB}$$

where L_{ax} = the SPL measured under free-field conditions, referred to a distance of 1 m

   L_p = the SPL measured in the reverberant room
   T = the reverberation time of the room (s)
   V = the volume of the room (m³)
   25 = a value related to different constant factors.

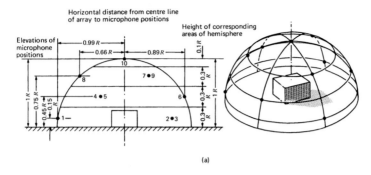

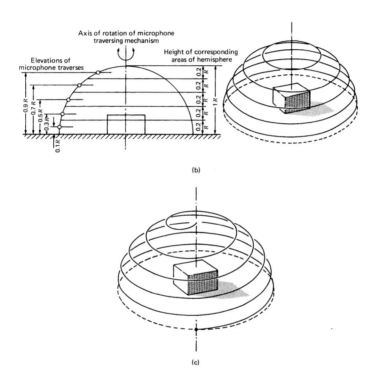

Figure 12.24. Alternative methods for determining sound-power levels in a semi-anechoic field: (a) using fixed microphones at 10 positions; (b) using a circle-traversing microphone at five different heights; (c) using a meridional traversing microphone plus loudspeaker turntable to produce a spiral traverse. (Courtesy B&K.)

12.5.3 Sensitivity

The sensitivity of a loudspeaker is a measure of the acoustic sound-pressure level produced by a stated electrical voltage input – normally V where $V^2/R = 1$ W. For example, an input voltage of 2.83 V corresponds to 1 W into a nominal impedance of 8 Ω.

The test signal is normally pink noise band-limited to the frequency range of interest. The corresponding sound-pressure level is usually measured at a distance of 1 m under free-field conditions.

12.5.4 Efficiency

The efficiency of a loudspeaker is normally expressed as the ratio of the total acoustic power radiated into a free field (4π steradians) – though half-space (2π) is sometimes specified – to the nominal electrical power input ($V_{IN}^2/R_{nominal}$). The total power radiated can be measured in a semi-anechoic field using one of the multiple-microphone arrangements over a hemispherical surface as shown in Fig. 12.24, and integrating the square of the r.m.s. sound pressure. Substitution techniques are available using a proprietary source of known (calibrated) acoustic power output. Alternatively, as in the case of the sensitivity measurements just discussed, a reverberant room (diffuse field) can yield the result less laboriously.

Efficiency is generally quoted as a percentage. Modern high-quality loudspeakers are notoriously inefficient, typically less than 1% for direct radiators and seldom exceeding 5%, the remainder of the input power being expended as heat. Some of the problems which this heat generation can cause are discussed in Section 12.6.3.

12.5.5 Impedance

Loudspeakers are generally designed and assembled with one of a very limited number of nominal impedance values. The most widely used values are $8\,\Omega$ and $4\,\Omega$, conveniently matching transformerless output transistorized amplifiers. Earlier valve (tube) amplifier designs could often be matched to 4, 8 or $16\,\Omega$ by means of a tapped output transformer.

The input impedance of a loudspeaker varies widely with frequency, and in some parts of its working range is largely reactive, with the result that the output current

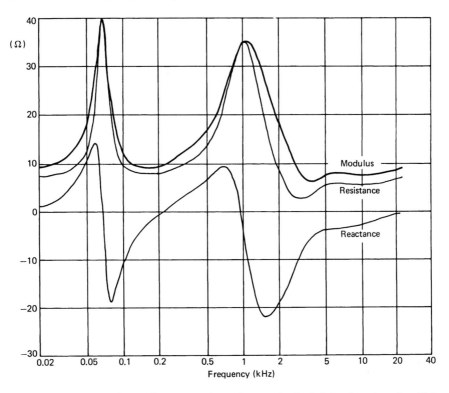

Figure 12.25. Typical impedance variation of a nominally $8\,\Omega$ loudspeaker. It will be seen that the modulus has a minimum of $6.4\,\Omega$ at 4 kHz. (Courtesy KEF Electronics.)

of the amplifier is out of phase with the output voltage. With a predominantly reactive load, the maximum instantaneous current through the output transistors does not occur at the moment when the voltage across them is passing through a minimum, but at some other point in the signal cycle at which the voltage is higher and the power dissipation correspondingly greater. If, as sometimes happens, the resistive component of the loudspeaker impedance is well below the nominal value for which the amplifier is designed, distortion due to current limiting may set in before the rated output level is reached.

Published curves of loudspeaker impedance against frequency usually give only the numerical value or modulus, which does not distinguish between resistance and reactance. It is then impossible to tell whether the resistive component is too low at some part of the frequency range to allow satisfactory operation with certain types of amplifier. Preferably, therefore, the curves should give not only the modulus, but the resistive and reactive components as well. Figure 12.25 shows the impedance of a nominally 8 Ω loudspeaker presented in this way[21]. As can be seen, although the modulus of the impedance is never less than 6.4 Ω, the resistive component in the frequency range around 3 kHz falls to 3 Ω, and this could lead to premature distortion due to waveform clipping by the protective circuit in an amplifier.

Measurement methods employing either constant voltage or constant current are in common use. In the simplest constant-voltage method, the signal is switched between the loudspeaker under test and a variable resistance which is adjusted until the voltage readings agree. The loudspeaker impedance is then equal to the resistor value at the frequency in question. Constant-current methods are speedier in action and lend themselves to sweep measurements of impedance modulus. Constant current can be derived from a 'compressor' generator but a simpler technique is to use a standard voltage generator connected to the loudspeaker through a resistance having a value much higher than that of the loudspeaker under test, typically 3–4 kΩ for an 8 Ω loudspeaker (Fig. 12.26). Alternatively, impulse or other transfer function methods may be used to establish the modulus of impedance and phase.

12.6 Large-signal measurements: distortion-limited

So far only small-signal measurements have been considered, and so it has been assumed that the loudspeaker under test is behaving as a linear system, i.e. the transfer function (ratio of output amplitude to input) is a constant for all input amplitudes. This assumption may be valid within reasonable limits for small signals, yet some departure from linearity and consequent non-linear distortions are inevitable and should be measured and specified. Two categories of large signal operation can be distinguished: distortion-limited (reversible) and damage-limited (irreversible).

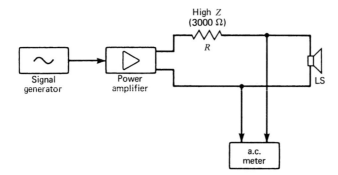

Figure 12.26. Simple constant-current method for measuring loudspeaker impedance.

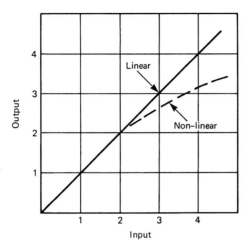

Figure 12.27. Higher input signal amplitudes generally result in a non-linear transfer characteristic.

The distortion-limited case will be discussed here, and damage-limited tests follow in Section 12.7.

12.6.1 Non-linear distortion

Loudspeakers, in common with most audio transducers, amplifiers and signal processors, will progressively depart from the idealized linear characteristic as the signal amplitude is increased (see Fig. 12.27). In the non-linear region the proportionality between input and output no longer applies and the output is characterized by the presence of spurious frequencies not present in the input, compression effects, overheating etc. It is standard practice to carry out distortion measurements at relatively modest levels, usually with an output of 90 dB SPL at a distance of 1 m. Additional tests at high signal levels are discussed below in Section 12.7.

(a) Harmonic distortion

The most common method used to evaluate amplitude-dependent non-linearity is the measurement of harmonic distortion. This particular manifestation of amplitude non-linearity relates to the appearance in the sound output of 'harmonics', whose frequencies are integral multiples of the input frequency, $2f$, $3f$, $4f$ etc.

Total Harmonic Distortion (THD) is the ratio of the r.m.s. output signal due to distortion, to the total r.m.s. output signal. It can be measured in a number of ways but the most simple is to measure the r.m.s. value of the total output, using the same set-up as for frequency-response measurements (Section 12.5.1) and also that obtained when the fundamental tone f is filtered out. THD can then be expressed as the ratio of the second of these measurements to the first, either as a percentage or in dB. Note, however, that system noise such as mains hum will be included in the THD measurement. Accordingly this form of the measurement is often called 'THD + noise' (denoted THD + N) to differentiate it from an FFT-based extraction of individual harmonics followed by summation.

In some circumstances it may be enough to perform such a test at a single frequency, say 1 kHz. However, since THD can change rapidly with signal frequency, it is best to present a continuous plot of THD as a function of frequency using a sweep oscillator and tracking notch filter to remove the fundamental.

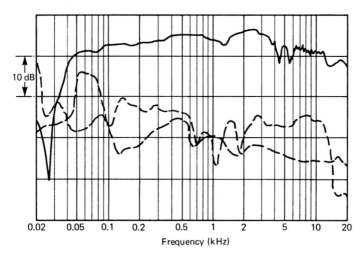

Figure 12.28. Frequency response of a typical loudspeaker with plots of second and third harmonic distortion (raised 20 dB for convenience).

Alternatively, point-to-point or sweep measurements of individual harmonics can be performed using an offset tracking filter. Figure 12.28 shows this type of presentation with the plots of the second and third harmonics raised 20 dB to aid comparison with the frequency response. The very peaky nature of the distortion characteristics does lead to some difficulty in interpretation, and naturally corresponds in part to the uneven response of the loudspeaker itself. Presenting the curves together, as here, does allow a percentage distortion figure to be calculated at any chosen frequency by reference to the continually changing fundamental level.

(b) Modulation distortion

When more than one frequency is present at the input to a non-linear system, amplitude and frequency modulation take place. For example, where the test signal input comprises f_1 and f_2, with f_2 much greater than f_1, the output will contain a plurality of intermodulation products at frequencies taking the form $f_2 \pm f_1$, nf_2, $\pm f_1$, etc. For testing purposes, f_1 and f_2 are chosen well within the rated frequency range (i.e. as with harmonic distortion measurements the upper frequencies are not investigated), with f_1 less than f_2 and only the side-band tones $f_2 \pm f_1$ and $f_2 \pm 2f_1$ are considered. The r.m.s. amplitude of f_2 is generally made higher, say four times that of f_1.

The test arrangement resembles that for THD with a two-tone sweep generator and a tracking harmonic multiplier to allow plots of the modulation products to be recorded. Since narrow bandpass filters can be used in the analyser, noise products are largely eliminated.

(c) Difference frequency distortion

An adaptation of the modulation distortion method may be used in which f_1 and f_2 are set close together and only the difference frequency products which fall within a narrow band (well within the system range) are measured. In a proposal by Thiele[22], for example, $f_1 = 1.5f_2$ so that for values around 12 and 8 kHz the $(f_1 - f_2)$ and $(2f_2 - f_1)$ products both fall on the frequency $0.5f_2$ and can be read through the same narrow bandpass filter – while all other products lie outside the audio spectrum. Similar reasoning allows 1200 and 800 Hz to be used for middle-frequency measurements, with a pair of frequencies in the 40–80 Hz region employed for low-frequency measurements. Instrumentation can be simple, the common source for all tones being a crystal in the 2–4 MHz region, and it is a further advantage that the test signals need not be pure sine waves.

This Thiele proposal is being considered for incorporation in the IEC Publication 268–3, and Small[23] has designed a practical distortion meter giving straightforward and accurate measurements applicable to amplifiers, tape recorders and loudspeakers.

(d) Measurements using tonebursts

Traditionally, all the above distortion measurements have used continuous sinusoidal test signals. However, at high levels high-frequency units may easily be damaged due to overheating, and therefore the maximum continuous rating of such units sets an upper limit to the sine wave signal level which may be used for distortion measurements. Yet it is of interest to know how non-linear a loudspeaker becomes under short-term overloads, since so much programme material contains short peaks in amplitude. Ding Yong-Sheng[24] describes a toneburst method which permits the measurement of harmonic distortion at much higher peak levels, without thermal damage.

The method relies on the reconstruction of the loudspeaker's steady-state response from a digital sample of its toneburst response. The toneburst is made of short enough duration to avoid damage to the loudspeaker, and only a few cycles in the middle of the burst are selected to avoid the effects of transient distortion by the loudspeaker at the beginning and end. By repeating the digital sample, a steady-state response is created. Figure 12.29 shows (a) the input toneburst, (b) the loudspeaker response $P(t)$ consisting of a transient beginning, a steady-state portion, and a transient end, (c) a recreation of $P(t)$ as continuous steady state by repetition, and (d) the spectrum of $P(t)$ obtained by means of a Fast Fourier Transform procedure, with all the harmonics represented.

12.6.2 Excursion-dependent distortions

The most obvious result of applying high peak level signals to a loudspeaker is that the coil/cone mechanism will be constrained to perform large-excursion movements. Gander[25] has analysed the physical topology of loudspeakers and the possibility of predicting from such data the maximum excursion (or 'maximum throw') of particular diaphragm/motor assemblies before severe non-linear or peak clipping distortion or damage will occur. So far as non-linearity is concerned, Gander suggests that a 3% distortion level makes an appropriate choice of linearity limit, compared with the 10% THD level proposed earlier by Small[26], and points out that a 3% (−30 dB) distortion component contributes only about 0.25 dB to the total acoustic output. Direct visual measurements of diaphragm displacements can be made by means of a probe, laser or accelerometer.

12.6.3 Time-dependent distortions

In addition to the distortions related to short-term high-signal amplitudes just discussed, there are distortions and side effects which reveal themselves only when high-level signals are handled over a period of time.

An important factor in long-term testing at high levels is obviously the rise in temperature of the voice coil due to the inefficiency of power transfer, and this will produce a rise in coil resistance. (For copper wire, the resistance rises by 0.4% per °C.) This in turn will lead to reduced efficiency (or thermal compression) and a number of related changes in fundamental parameters and responses. Measurements exist to assess each of these effects, but only a few aspects will be discussed here.

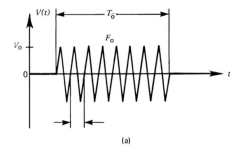

(a)

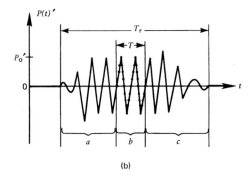

(b)

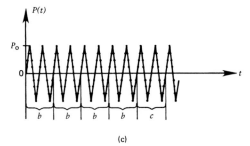

(c)

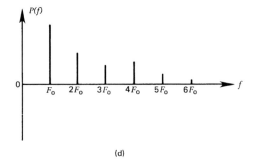

(d)

Figure 12.29. Toneburst measurement of harmonic distortion: (a) the toneburst impulse; (b) the loudspeaker response from which only the central portion is selected; (c) continuous sine wave response obtained by repetition; (d) the resultant harmonic spectrum. (After Ding Yong-Sheng.)

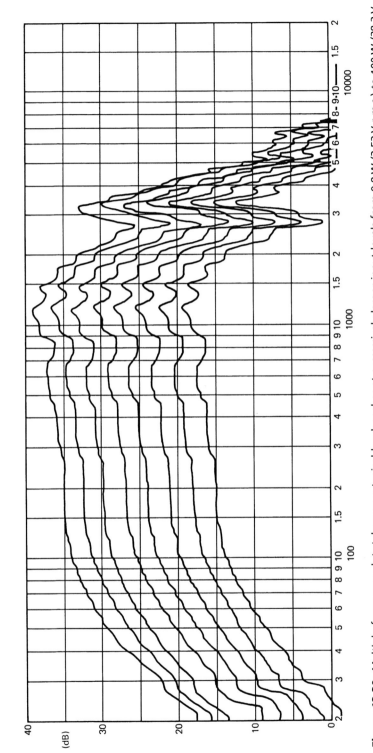

Figure 12.30. Multiple frequency plots taken on a typical loudspeaker at successively larger input levels from 0.8 W (2.53 V r.m.s.) to 100 W (28.3 V r.m.s.) in 3 dB steps. The 0 dB bottom line is 80 dB SPL. (After Gander.)

Changes in a loudspeaker's performance due solely to the heating effect may be regarded as a function of its 'thermal linearity'. Designers of high-power loudspeakers can address this particular parameter by using heat sinks, ferrofluids and other devices to improve the conduction and dissipation of heat.

Another major effect observed when high signal amplitudes are applied to a loudspeaker is the loss of efficiency which takes place due to the aforementioned rise in voice-coil resistance. This compression effect can be considered as affecting the 'dynamic linearity' of the system, and is revealed by non-linearity in the transfer characteristic of the fundamental alone. This can be studied, e.g. by plotting output versus input at a discrete frequency.

Dynamic linearity can also be examined by displaying frequency/amplitude-response curves taken at different input levels on a single graph[27] (see example in Fig. 12.30).

12.7 Large-signal measurements: damage-limited

When the input level to a loudspeaker is sufficient to cause irreversible damage, the latter usually takes the form of either *mechanical damage,* causing fouling between the voice coil and the gap or fracturing of the moving parts, or *thermal damage,* in which the voice coil/former blisters or, in severe cases, the wire itself fuses.

Mechanical overload is usually caused by very short peaks in the input signal, and this will generally occur at low frequencies in normal musical signals. Thermal damage, on the other hand, is due to longer term overloads and is related to the long-term average content of the input signal. With normal music signals, this problem is usually confined to the middle and high frequencies. Under normal conditions of use, where no amplifier clipping is taking place, the peak-to-mean ratio of the programme material is greatest at the low- and high-frequency ends of the signal spectrum, and least at the middle frequencies.

Until recently it has been normal practice to determine loudspeaker power-handling capacity, i.e. the maximum normal continuous programme level which can be accepted without irreversible damage, using a spectrally shaped noise signal which is applied to the loudspeaker for a period of several hundred hours. Changes in the spectral content of programme material, however, due partly to modern recording techniques and partly to the increased use of electronic instruments, have meant that this long-term test can give misleading results. In fact, many loudspeakers which passed this test have later failed in the field.

As a result, this method is now used primarily for assessing the long-term durability of loudspeakers, particularly with regard to fatigue failures in lead-out wires and suspensions. It is no longer used for the purpose of determining the optimum size of power amplifier which may be used with the given loudspeaker system.

In the revised loudspeaker standard IEC 60268–5[3], two new tests have been devised whose primary purpose is to describe with greater accuracy the likely power-handling capacity of a loudspeaker when used with various types of programme, and in this way enable a better match to be made between loudspeaker and amplifier.

These new characteristics are termed 'voltage' and 'power rating of loudspeakers', and two main characteristics are specified: the short-term maximum input voltage and power, and the long-term maximum input voltage and power. The test signal used is a noise signal whose spectrum simulates that of normal programme material (see IEC Publication 60268-IC). The *maximum short-term input voltage* is specified as that which can be handled by the loudspeaker without sustaining permanent damage when the shaped noise signal is applied for 1 s. The *long-term maximum input voltage* is measured in the same way, but the signal is applied for a period of 1 minute. The corresponding short- and long-term input powers may be determined from the expression U^2/R_2 where U is the short- or long-term maximum input voltage, and R_2 is the rated impedance of the loudspeaker.

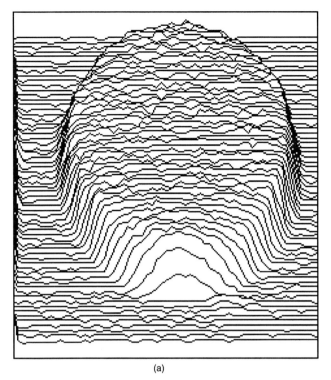

(a)

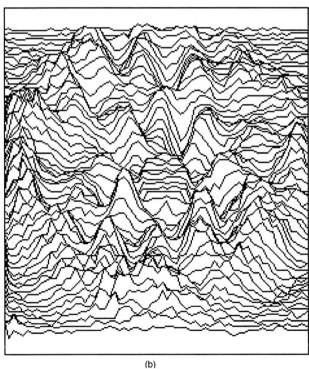

(b)

Figure 12.31. Laser vibrometry images of (a) 'perfect piston' dome tweeter; (b) high-quality guitar cone loudspeaker with designed-in mechanical break-up.

Equivalent ratings are also determined for power amplifiers, and all these characteristics can then be used to establish the optimum matching between amplifier and loudspeaker. The exact figures chosen will, of course, depend on the conditions of use; i.e. whether there is a low probability of incorrect operation leading to clipping of the power amplifier (this condition would apply to most serious high-fidelity situations) or a significant likelihood of severe clipping, e.g. in stage PA systems or discotheques. Matching requirements which will allow optimum matching, taking into account the likely conditions of use, are described in IEC Publication 60268–15.

12.8 Mechanical measurements

Thus far we have concentrated on acoustic measurements. Many anomalies in the acoustic data are actually due to mechanical phenomena, and therefore no analysis of a loudspeaker is complete without an investigation of its mechanical behaviour. In essence this means a requirement to visualize the vibration patterns on any or all moving surfaces, including enclosure walls. Corrington[42] distributed lycopodium powder over diaphragms to show modal behaviour, and stroboscopic methods have been in use for many years, but it was not until the advent of lasers that vibration analysis became truly powerful.

12.8.1 Laser vibrometry

The use of lasers in vibration analysis has become known generically as laser vibrometry. The two main techniques are holography and interferometry. Holography was first used to study loudspeakers by Hladky[43] and Fryer[44]. This technique is limited by the difficulty in extracting phase (directional) information. Interferometry inherently measures complex velocity, which allows computer animations of the scanned surface, effectively producing a 'vibration microscope'[45]. Although large numbers of measurements are required, the system is readily automated. Figure 12.31 shows some examples.

12.9 Integrated measurement systems

Transducer measurement technology has developed over several decades and is now highly sophisticated. The ready availability of computer resources has resulted in the evolution of versatile, integrated systems which are predominantly implemented in software. This has the added advantage of convenient upgrades. Many commercial systems are now available, from the low-cost 'PC card + floppy disk' package for the home enthusiast to expensive, high-precision dedicated DSP systems[46].

Acknowledgements

The author acknowledges with gratitude the help received from L. R. Fincham in the original preparation of this chapter, and from Julian Wright for valuable update material.

References

1. SHORTER, D E L, 'A survey of performance criteria and design considerations for high-quality monitoring loudspeakers', IEE Paper No. 2604, April (1958).
2. BERANEK, L L, *Acoustic Measurements*, John Wiley, New York (1949).
3. IEC 60268–5, 'Specifying and measuring the characteristics of sound system equipment, Part 11, Loudspeakers', 2nd edition (1989).

4. AES2–1984, *JAES*, **32**, Oct. (1984).
5. DIN 45500, Part 5.
6. VANDERKOOY, J, 'Another approach to time delay spectrometry', *JAES*, **34**, July/Aug. (1986).
7. HEYSER, R C, 'Acoustical measurements by time delay spectrometry', *JAES*, **15**, Oct. (1967).
8. HELMHOLZ, H VON, *Sensations of Tone*, Dover Publications Paperback, New York (1954).
9. BAUER, B, 'Speaker tests can be relevant to the listening experience', *High Fidelity*, June (1970).
10. LINKWITZ, S, 'Shaped toneburst testing', *JAES*, **28**, April (1980).
11. BUNTON, J D and SMALL, R H, 'Cumulative spectra, tonebursts and apodization', *JAES*, **30**, June (1982).
12. HEYSER, R C, 'Determination of loudspeaker signal arrival times', JAES, **19**, Oct-Dec. (1971).
13. BERMAN, J M and FINCHAM, L R, 'The application of digital techniques to the measurement of loudspeakers', *JAES*, **25**, June (1977).
14. BROCH, J T, 'On the measurement of frequency response functions', *Bruel & Kjaer Technical Review*, **4** (1975).
15. RICE, C W and KELLOGG, E W, 'Notes on the development of a new type of hornless loudspeaker', *JAIEE*, **9** (1925).
16. MCLACHLAN, N W, *Loudspeakers*, Oxford University Press, Oxford (1934).
17. FINCHAM, L R, 'Refinements in the impulse testing of loudspeakers'', *JAES*, **33**, March (1985).
18. KEELE, D B, 'Low frequency loudspeaker assessment by nearfleld sound-pressure measurement', *JAES*, **22**, April (1974).
19. SMALL, R H, 'Simplified loudspeaker measurements at low frequencies', *JAES*, **20**, Jan./Feb. (1972).
20. LARSEN, H, Bruel & *Kjaer Technical Review*, **4** (1964).
21. KEF Topics, 4, No. 1, KEF Electronics.
22. THIELE, A N, 'Measurement of non-linear distortion in a band-limited system', IREE Convention Digest (Aug. 1975), reprinted in *JAES*, **31**, June (1983).
23. SMALL, R H, 'Total difference-frequency distortion: practical measurements', *JAES*, **34**, June (1986).
24. DING YONG-SHENG, 'A toneburst method for measuring loudspeaker harmonic distortion at high power levels', *JAES*, **33**, March (1985).
25. GANDER, M R, 'Moving coil loudspeaker topology as an indicator of linear excursion capability', *JAES*, **29**, Jan./Feb. (1981).
26. SMALL, R H, 'Closed-box loudspeaker system analysis', *JAES*, **20**, Dec. (1972).
27. GANDER, M R, 'Dynamic linearity and power compression in moving coil loudspeakers', *JAES*, **34**, Sept. (1986).
28. FRANKORT, G J M, 'Vibration patterns and radiation behaviour of loudspeaker cones', *JAES*, **26**, Sept. (1978).
29. FRYER, P A, 'Holographic investigation of speaker vibrations', AES 50th Convention, London (1975).
30. FRYER, P A, 'Inexpensive system for live stroboscopic holographic interferometry', *Optics Technology* (1980).
31. SUZUKI, K and NOMOTO, I, 'Computerised analysis and observation of the vibration modes of a loudspeaker', *JAES*, **30**, March (1982),
32. KYOUNO, N et al., 'Acoustic radiation of a horn loudspeaker by the finite element method', *JAES*, **30**, Dec. (1982).
33. JANSE, C P and KAIZER, A J M, 'Time-frequency distribution of loudspeakers: The application of the Wigner Distribution', *JAES*, **31**, April (1983).
34. BANK, G and HATHAWAY, G T, 'Three-dimensional energy plots in the frequency domain', 67th AES Convention, New York, Preprint no. 1678, October (1980).
35. RIFE, D D and VANDERKOOY, J, 'Transfer-function measurement with Maximum-Length Sequences', *JAES,* **37**, June (1989).
36. DUNN, C and HAWKSFORD, M O, 'Distortion immunity of MLS-derived impulse response measurements', *JAES,* **41**, May (1993).
37. AES, 'ALMA Standard Test Method for audio engineering – Measurement of the lowest resonance frequency of loudspeaker cones'. *JAES*, **49**, April (1992).
38. MORENO, J N, 'Measurement of loudspeaker parameters using a laser velocity transducer and two-channel FFT analyser', *JAES*, **39**, April (1991).

39. GANDER, M R, 'Ground-plane acoustic measurement of loudspeaker systems', *JAES,* **30,** Oct. (1982).
40. RAMIREZ, R W, *The FFT, fundamentals and concepts,* Prentice Hall, Englewood Cliffs, NJ (1985).
41. STRUCK, C J and BIERING, C H, 'A new technique for fast response measurements using linear swept sine excitation', AES Preprint 3038 (1991).
42. CORRINGTON, M S, 'Transient testing of loudspeakers', *Audio Engineering,* August 1950.
43. HLADKY, J, 'The application of holography in the analysis of vibrations of loudspeaker diaphragms', *JAES* , **22,** April (1974).
44. FRYER, P A, 'Holographic investigation of speaker vibrations', AES 50th Convention, London (1975).
45. BANK, G and HATHAWAY, G T, 'Three-dimensional energy plots in the frequency domain', 67th AES Convention, New York, preprint No. 1678, Oct. (1980).
46. www.harmony-central.com/Other/speakerfaq.

13 Subjective evaluation

Floyd E. Toole and Sean E. Olive

13.1 Introduction

A loudspeaker isn't good until it sounds good. The traditional problem has been: how do we know what is good? Whose opinion do we trust? In the early days of our industry, the decision was easy; it was the designer, whose inspiration and skill created the product and, often, whose charisma marketed it. As the industry grew, loudspeaker companies became organizations, with engineering, sales and marketing people who all felt that they deserved a voice, and senior managers who felt obliged to express their views. In some companies it is corporate rank, responsibility for the business, that decides the issue.

Even with the best of intentions, there is no assurance that these people, whoever they may be, are qualified, by virtue of their hearing abilities or listening expertise, to perform the task. Besides, in today's world, with remarkably few exceptions, loudspeakers fall short of our technical performance objectives. Consequently, one finds that the comparative ratings of a group of popularly priced competitive products is not so much an assessment of relative perfection as it is a balance of positive and negative attributes as they are revealed by different kinds of music. These complex decisions can reveal personal biases, requiring that there be an evaluation in which the result is a statistical combination of several individual preferences. A single person's opinion is clearly inadequate.

In practice, the principal difficulty with subjective evaluations is to control what the listeners are responding to. Listeners who, because of room acoustics, hear different sounds from the same loudspeakers cannot possibly agree. Listeners who see the products they are evaluating cannot completely distance themselves from the biasing influences of style, price, size, brand reputation, and so on. Listeners who hear sounds that are unreasonably altered by atypical acoustical surroundings cannot express a view that represents that of a larger population. Music that reflects only a narrow spectrum of what customers play cannot be assured of revealing a balanced perspective of performance. We need to get serious about the conduct of listening tests – so that 'opinions' can take on the attributes of 'facts', to create a system of subjective evaluation that is a proper companion to technical measurements. We need to do 'subjective measurements', not 'listening tests'.

13.2 Turning listening tests into subjective measurements: identifying and controlling the nuisance variables

Typical listening tests are, in scientific terms, fairly loose exercises. Intense concentration, furrowed brows, regular concert attendance, and even professional audio experience, cannot compensate for other factors that conspire to make the results little more than opinions that should be viewed with suspicion. That suspicion is supported, in common experience, when different individuals express quite different opinions about the same products. Does this mean, as audio folklore suggests, that 'we all hear differently'; that it is normal for personal preferences in loudspeakers to be as variable as preferences in many other categories?

No, it does not. It turns out that much, if not most, of the variability in commonplace opinions can be traced to differences in the listening experiences that led to the opinions. These differences have several causes, which will be discussed in some detail in the following sections. When these 'nuisance variables' are controlled, and listeners have opportunities to focus their judgements on the sounds of the loudspeakers under test, it is remarkable how consistent, and how similar, human listeners can be. We can be very good measuring instruments[1-4].

13.2.1 Blind versus sighted tests

It is well known, in subjective evaluations, that humans are susceptible to influences other than the parameter or device under test. We simply find it very difficult to ignore the evidence of brand, size, price, and so on. Amazingly, even at this stage in the development of the audio industry, some people still do not accept the need for blind evaluations.

Complaints seem to centre around the additional 'stress' that such tests impose on listeners, such that they cannot focus in a totally natural way on the task at hand, thereby missing some important nuances. However, the real evidence indicates the opposite. In loudspeaker evaluations, there is no argument that there are audible differences between products, and clearly audible differences traceable to interactions with programme and rooms. In comparisons of evaluations done in both blind and sighted conditions, it has been observed that listeners substantially altered their ratings of products when they were in view, following biases suggested by visual cues. In addition, in blind tests, loudspeakers were rated differently when they interacted with different programme material and different room locations, yet the same listeners in the same circumstances stubbornly refused to alter their opinions of the loudspeakers when they were visible. If there is any stress, it would seem to be in the sighted tests, where listeners clearly had difficulty reconciling the facts that the same loudspeaker can, at different times, elicit different subjective reactions, and that good sound can emit from mundane looking loudspeakers, and vice versa[5].

Judges of wine, and the like, are willing to put their reputations on the line in blind evaluations; it is time for the audio industry to do likewise. The results of listening tests that are not done blind should not be trusted. The evidence is that, if one believes something, one is very likely to 'hear' it.

The step beyond blind is double-blind, meaning that the listener does not know what is being auditioned and, in addition, the experimenter has no real-time control over the sequence of presentations. This does not mean that the experimenter is not in control, it means that the presentation sequence of loudspeaker, programme, position, or whatever other parameters are in the test, is determined (or predetermined) by a randomizing process.

13.2.2 The listening room

The listening room and the loudspeaker form a system, together determining what the listener hears. They cannot completely be separated. Therefore, the choice of room, the arrangement of loudspeakers and listeners within it, and the acoustical treatment and furnishings, are all variables in the evaluations [6-8]. How, then, can we evaluate loudspeakers? Very carefully, is the answer.

The room cannot be eliminated, which is a problem for the entire loudspeaker industry, whether one is a manufacturer or a consumer. What we can do is to start with a room that, in essential dimensions and construction, resembles rooms our customers are likely to have. Within that context, it is necessary to adjust the physical environment to eliminate strong biasing influences, such as powerful bass resonances, attributable to that specific room. Clearly, we are aiming for a 'generic', but good, room.

Standardized listening rooms have been proposed by several standards organizations, indeed the authors of this chapter have contributed to some of them[9, 10]. That said, it is our opinion that such rooms serve little practical purpose unless the specifications are so detailed that builders could source the specific materials and construct rooms from them. The resulting, absolutely identical, rooms could then be used in a systematic manner for comparative subjective evaluations at different locations.

Such a standard, if written, would probably never be used. It also raises the question: do we know enough to be so dogmatic? Consequently, the standards tend to be written as recommendations, and are diluted to the point that they allow for the possibility of rooms of quite widely varying dimensions and acoustical performance. This raises the question: have we been too permissive? Claiming to have an IEC, or other, standard room has become a kind of 'license to practice' for some people in the audio business. The rooms themselves can be quite different and, occasionally, not what was intended in either spirit or reality, but the standards organization appellation gives an appearance of credibility.

Nevertheless, embodied in the standard documents are useful suggestions for the setting up and running of listening tests and guidance on how to perform acoustical measurements. Some of the newer standards even relate to multichannel audio – our future. They are worth reading.

(a) Dimensions and proportions

The notion of an ideal room is attractive, and several investigators have furthered the appeal with analyses of various room dimensional ratios, stating that certain ones are inherently superior. The mathematical analysis of sound in enclosures normally begins with the assumptions that they are rectangular with perfectly flat and reflecting boundaries. From this basis, one can calculate the frequencies of resonant modes arising from the reflection of sound between opposite parallel surfaces (axial modes), among four adjoining surfaces and parallel to the remaining two (tangential modes), and among all surfaces and parallel to none (oblique modes). Analysis reveals that the ratio of the length-to-width-to-height determines the distribution of room resonances in the frequency domain, while the dimensions themselves define the frequencies.

It has long been assumed that a good room should have a uniform distribution of modes in the frequency domain if sounds in the room are not to be coloured[11]. Mathematically based ranking of room ratios lead to very specific recommendations of what is, and is not, a 'good' room [12, 13]. To our knowledge, none of these has ever been subjectively tested in any scientific way. However, several important assumptions distance these theoretical exercises from real-world experience.

1   The assumption that rooms are perfect rectangles, with perfectly flat, perfectly reflecting walls, floor and ceiling, without openings coupling to other spaces, is a formidable restriction. If we had such an idealized room we might wish to change it, so that it would be a pleasant listening space.
2   Most predictions assume that all calculable modes are of equal importance. This is not true. Morse and others have noted that, because they have the longest mean-free path between reflections, the axial modes are usually the most energetic resonances, with the tangential modes next, and the oblique modes the weakest. It is therefore necessary to weight the classes of modes in their order of importance. In normal rooms, experience tells us that audible room problems are almost always traceable to prominent axial modes. In rooms with massive, rigid walls, some low-order tangential modes may be evident.
3   All predictions assume that the loudspeakers and listeners have equal acoustical coupling to all of the modes. This is not true. The location of the loudspeaker

determines the strength with which individual modes are energized and the listener location determines the audibility of the modes that have been energized. Only at the intersection of the three room boundaries (e.g. on the floor in a corner), will all of the modes be energized by the loudspeaker. A subwoofer may be used in such a location, but not full-range systems. Any other location will result in selective coupling to the modes, a condition that also applies to the listeners, who rightfully refuse to sit in corners.

4   Since low bass signals in stereo recordings and film sound tracks are commonly monophonic, the presence of two or more woofers or subwoofers will influence the acoustical coupling to individual modes. There can be constructive and destructive interference, meaning that some modes will be reinforced, and others cancelled, depending on the placement of the multiple low-frequency sources.

5   Finally, in real rooms, the walls, floor and ceiling can behave as frequency-dependent absorbers, as can such surface features as windows, paintings (membrane absorbers), carpets and draperies (resistive absorbers). Such is the scale of these effects that duplicating the basic dimensions of a room will not ensure duplicate acoustical performance.

In other words, the performance of a room can be best evaluated only after it is constructed. Even then, its performance must be assessed through a specific arrangement of loudspeakers and listener(s) within it. Generalizations will not suffice.

Modal calculations can nevertheless be very helpful in identifying the origin of bothersome resonances so that suitable remedies can be brought to bear. With high-resolution (preferably 1/10- to 1/20-octave) measurements as a guide, it is usually possible to arrive at loudspeaker and listener positions that are relatively neutral.

The basic rule to be remembered in this context is that the important room resonances are those that participate in the transfer of energy from the loudspeakers to the listener. The satisfactory acoustical performance of low-frequency sources, and the constraints of stereo and multichannel listening severely reduce the options for room arrangement. Separate subwoofers, operating below about 100 Hz, and satellite loudspeakers operating at higher frequencies are more flexible, since the low-frequency room coupling can be optimized independently of the conventional requirements for loudspeaker positioning. This is a clear advantage for recreational listening, but another complicated variable to be considered in listening tests.

(b) Acoustical treatment

Reflections and reverberation in rooms are major determinants of sound quality, imaging, spatial impression, and envelopment. In addition, it matters greatly whether we are listening to comparisons of single loudspeakers, stereo pairs, or multichannel arrays. Finally, the microphone technique, and the manner in which the recording is mixed, can be as important as almost anything else. We have a complex situation.

Reverberation time calculations and measurements are not difficult to do, and there is good agreement between the predictions and reality in large rooms in which sound absorbing materials are well distributed, with the consequent good diffusion of the sound field. However, domestic listening rooms violate this condition in several ways. It is common for large areas of carpet on the floor to oppose a hard flat ceiling. Further concentrations of absorption appear as drapery that can cover much of one wall and, perhaps, none of the others. These surfaces and upholstered furniture absorb much of the direct sound radiated by the loudspeakers, preventing it from participating in the reverberation. The frequency-dependent directivity of the loudspeakers is another factor separating simple theory from practice.

The result is that listeners seated in typical rooms at moderate distances, say 3 m, from typical forward-radiating loudspeakers are in a sound field that is a mixture of direct sound, early-reflected sounds reflected from the immediately-adjacent bound-

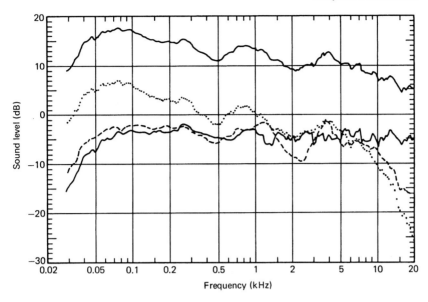

Figure 13.1. Estimates of the direct sound (——), the early-reflected sound (– – –), and reverberant sound (· · · · ·) for a loudspeaker at the stereo-left position in the IEC recommended listening room. The top curve is the energy sum of the three components, shifted vertically for clarity, which is a good prediction of a listening room curve measured at the forward listener location. The prediction is idealized to the extent that it does not include the effects of interference from adjacent boundary reflections and standing waves which would normally dominate the form of the curve below about 300 Hz. (Reproduced from 'Loudspeaker measurements and their relationship to listener preferences' by F. E. Toole in *JAES*, **54**, parts 1 and 2.)

aries, and reverberant sound. At low frequencies the sound field is predominantly reverberant, at mid-frequencies it is a mixture of all three types, and at high frequencies the direct sound dominates.

Figure 13.1 illustrates the situation for a loudspeaker in a typical room, with normally reflective boundaries (carpeted floor, reflective walls and ceiling). For the loudspeaker it is a difficult test, since the on-axis (direct sound), 40–80° off-axis (early-reflected sounds) and sound power (reverberant sounds) radiations all are energetically represented at the listener's ears. Serious misbehavior in any of the categories has an opportunity to be audible. Such demands are fitting, if not compulsory, for a test environment, in that a loudspeaker that sounds good there may also perform well in most other typical rooms.

The direct sound component is influenced only by listening distance, and the orientation of the listener with respect to the loudspeaker. Early-reflected sounds are affected by the acoustical properties of the room boundaries at the points of reflection and the performance of the loudspeaker at the relevant angles off axis. The reverberant sound field is dictated by the overall acoustical properties of the room and by the directivity of the loudspeaker. In typical situations the reverberant sound field is not well diffused. This is neither good nor bad, it is merely a matter of fact. The normal arrangement of carpet and room furnishings results in the reverberation, such as it is, mainly taking place in the upper portion of the room, energized by sounds radiated horizontally and upwards from the loudspeakers. The height and orientation of the mid- and high-frequency radiators of the loudspeakers can therefore be significant factors in the generation of the reflected/reverberant sound field.

The choice of reverberation characteristics for a listening room must be influenced by what is typical in the market area for the products being evaluated. Published data show substantial regional differences due to the influences of climate, culture and fashion in interior decoration. A survey of Swedish rooms[15] indicated middle-frequency reverberation times averaging 0.55 s, and ranging from 0.25 s to 0.9 s. Measurements in living rooms in the Netherlands and the United Kingdom[16] yielded mean values between 0.4 s and 0.5 s, although a more recent survey found that mean values had dropped to 0.3–0.35 s at mid frequencies[17]. Measurements of US and Canadian living rooms have yielded mean values in the range 0.35–0.4 s[18]. Desirable listening conditions for stereophonic or multi-channel reproduction may require a revision downward from the prevailing mean value in some areas. Of course, relatively high levels of acoustic absorption in small rooms, combined with loudspeakers that are relatively directional, begs the question: is there, in any meaningful sense, reverberation at all, or are we simply dealing with a sequence of early reflections? It is our opinion that we are, and that most measures of reverberation time in typical listening spaces are of questionable value.

Listening rooms in professional buildings can require special attention. It is common, for example, for the floor to be hard and the ceiling absorbing: an acoustical inversion. Masonry floors and walls lack low-frequency sound absorption compared to the lower mass, flexible structures used in domestic housing. In such cases, low-frequency absorbers must be added to reduce the reverberation times to typical values.

Rigid, massive walls also increase the Q of room resonances. This results in two unfortunate effects. First, the resonances tend to be very energetic, colouring the sound and exhibiting prolonged ringing that is clearly undesirable. Second, bass performance varies dramatically from place to place in the room. In multichannel systems, where one should be designing for a listening area, not a sweet spot, this is a serious problem. One solution is to build all, or some portion of, the interior surface of the room of material that can act as a membrane (diaphragmatic) absorber, damping the resonances. A single layer of gypsum wallboard on wooden studs is such a surface. Another is to add specifically designed, and appropriately positioned, custom membrane absorbers. In our experience, the rooms with the most stubborn bass problems tend to be those that, by chance or mistaken ideals, have been built too solidly.

A normally furnished domestic room can, with very little effort, be made into an acceptable listening environment. However, rooms in commercial buildings, engineering labs, or retail showrooms often have problems because of sparse furnishings, a lack of typically rich carpeting and drapes, and an abundance of hard surfaces. In these cases it is common to use special acoustical devices: absorbers, diffusers, etc. to tailor the sound field. Purveyors of such products have strong motivations to load a room with more of these devices than may be prudent, if the purpose of the listening evaluation is to assess how a loudspeaker may sound in a typical customer's home. They work very well, if used in moderation.

(c) Background noise

The concern over background noise is based on notions that it will mask other sounds and thus prevent listeners from hearing subtle musical nuances or distortions. Frankly, steady-state background noise would have to be quite high before it could compete with the masking effects of the music itself. Experience with lossy digital perceptual encoders has given us all a new appreciation for just how powerful masking really is.

Nevertheless, a low background noise is worth striving for, especially low-frequency rumbles from traffic or air-handling systems. A simple, inexpensive, way to deal with HVAC noise is to install a motorized damper, allowing the air flow to be shut off during tests.

As loudspeakers are increasingly being equipped with power amplifiers, an evaluation of hiss and hum levels is essential. This cannot be done unless the ambient room noise is as low as the quietest customer dwelling. This sets a requirement that is much more demanding than the typical one of judging sound quality, where one may indeed get away with 30–35 dBA. A better assessment standard may be the popular NC contours, used for evaluating noise in studios and concert halls, and levels of NC 20 or lower should be aimed for.

Structural or other buzzes and rattles are another, insidious, noise problem, because they are frequency and level dependent. The only cure for these is to ensure the integrity of the room structure and surfaces while it is being built, and to test it thoroughly with slowly swept pure tones afterwards.

13.2.3 Loudspeaker and listener positions

In general, loudspeakers should be placed according to the manufacturers' recommendations. This is absolutely critical in those cases where the acoustical design embraces the adjacent wall or floor. For most products, however, there are few restrictions, in which case the locations should be chosen to ensure the neutrality of the acoustical coupling to the listeners. A one metre or more distance from the side walls is commonly regarded as sufficient to reduce adjacent-boundary interactions at mid and high frequencies. To avoid large adjacent-boundary effects at low frequencies, the distances to side and rear walls should be different[19].

The commonly recommended practice of placing the mid- and high-frequency units at the listener's ear height means that some loudspeakers must be elevated. This decides, arbitrarily, the condition of another experimental variable: the distance of the woofer above the floor. Obviously, this does not apply to floor-standing systems.

To avoid reflections close to the ears, listeners should be seated in chairs with backs no higher than the shoulders. All listeners in a group should have a clear acoustical view of the loudspeakers, which may require elevating the chairs behind the front row. The wall behind the listeners should be a metre or more away. In general, this wall should be treated with acoustical absorption if it is close to the listener. Diffusing surfaces close to, and behind, the listener can cause a lack of spatial clarity in the virtual stereo images comprising the soundstage between the loudspeakers, and should be avoided, although they are acceptable at greater distances.

Monophonic listening places no constraints on the physical layout other than the general requirement that listeners should be at a realistic distance from the loudspeakers: not less than 2 m and probably not more than 4 m. Stereo reproduction, on the other hand, imposes a requirement for symmetry, not only of the loudspeakers with respect to the listeners, but also of the loudspeakers with respect to the side walls of the room. The latter requirement places the listeners on the mid-line of the room and at the mid points of the family of transverse axial modes. Since the requirement for symmetry applies to the zone ahead of the listener, the walls beside and behind can be treated with absorbers and large surface irregularities to shift some of the nulls away from the listener locations. Some standards state that stereo loudspeakers should subtend an angle of 55–65° at the nearest listener position, however it is common practice, even among reviewers, to employ smaller angles. Care must be exercised, since this is a variable that influences the impressions of soundstage and spaciousness.

Multichannel listening is another matter. The common recommendation for professional installations places identical loudspeakers at 0°, ± 30°, and ± 110°[20]. Domestic installations rarely manage to achieve the 60° spread of the front loudspeakers, and the surround loudspeakers are often multidirectional designs intended to enhance the reflected sound field in the listening room. Each scenario has its own legitimacy, so the experimenter must make the decision and adequately document why the choice was made.

Even with the most elaborate preparations, the acoustical coupling between all loudspeaker locations and all listener locations will not be identical. Elaborate studies have shown that deviations from carefully optimized conditions, such as listener locations, resulted in increased variations in the ratings of sound quality by listeners, reducing the resolution of the tests[2]. In the evaluation of loudspeakers which are similarly high in performance, location can be as important a determinant of sound quality as the loudspeaker itself [6]. Consequently, spatial averaging must be used to minimize the residual biases associated with position. In the course of repeated listening comparisons, the loudspeakers and listeners should be systematically moved through the optional positions. Positional substitution of the loudspeakers under test is the solution, if one wishes to expedite the process.

(a) Randomization

If there are several loudspeakers in a multiple comparison set-up, it is virtually impossible for all of them to be in acoustically 'neutral' positions, as perceived by the listeners. Likewise, if there are several listeners, it is unlikely that all of them will be unbiased by their locations. If we wish to reach a conclusion about the relative merits of these products using these listeners, there are two options.

The first option is to repeat the test with each loudspeaker in each location, with each listener in each listening position. The number of combinations can become quite large, and the time and energy expended in shuffling loudspeakers and keeping

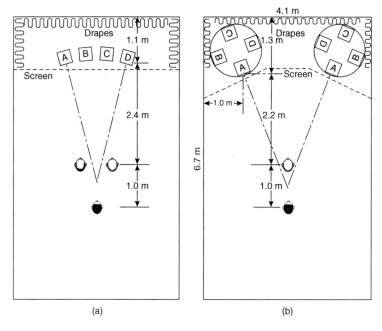

(a)   (b)

Figure 13.2. The listening room arrangements for (a) mono and (b) stereo loudspeaker comparisons. In the stereophonic comparisons both listeners are within 5° of the loudspeaker axes. Front row chairs are slightly lower than the rear chair so that all listeners have an unobstructed acoustical view of the loudspeakers. The floor is carpeted, the ceiling is hard, the sidewalls between the loudspeakers and listeners are hard and flat, and the remaining walls are covered with sound-absorbing and scattering objects. (Reproduced from 'Subjective measurements of loudspeaker sound quality and listener performance' by F. E. Toole in *JAES*, **33**.)

track of listeners becomes a serious burden. But, it must be done if the influences of position are to be removed as a biasing factor.

A variation on this uses only a single listener at a time, reducing the repetitions, and at the same time, eliminating another major problem in jury evaluations – group interactions. One should not underestimate the ability of humans to 'follow the leader'. Even non-verbal cues can be sufficient to eliminate the independence of the individual opinions among the listening panel. A further advantage of single listeners is that the loudspeaker selection is under his/her control, as is the timing of the experiment, which should lead to better data.

(b) Control by substitution

Anyone doing listening tests on a regular basis will find the above-mentioned procedures to be quite tiresome and time consuming. Shortcuts yield less reliable data, compromising the entire exercise, which is a complete waste of time. The real answer is to create a device that can bring different loudspeakers to the same location in a room, so that loudspeaker position is simply eliminated as a variable. Each repetition then becomes a valid data point, not just a means to average out the effects of a nuisance variable.

Figure 13.2 shows how the matter was handled in some early experiments, using turntables. Figures 13.3 and 13.4 show a recent solution, involving computer-controlled pneumatically-driven pallets which permit up to three sets of Left/Centre/Right channel or stereo Left/Right comparisons, up to four single-speaker comparisons.

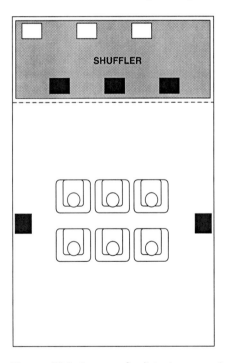

Figure 13.3. Layout of a listening room in which the loudspeakers are on pneumatically driven pallets which can position single loudspeakers, stereo pairs, or multichannel left, centre, right combinations so that active loudspeakers are always in the same forward positions, and inactive loudspeakers are parked against the wall. In this installation, up to four single loudspeakers, three stereo pairs, or three L, C, R sets can be compared in the same experiment. Loudspeakers can be exchanged in 3 to 4 seconds.

Figure 13.4. A photograph of the shuffler end of the room described in Fig. 13.3, showing a stereo pair of loudspeakers in the listening position, and another pair in the rest postition.

In this elaborate scheme, it is possible to eliminate loudspeaker position as a variable, or to make it a variable, by choosing different positions on different occasions[21].

13.2.4 Programme material

In subjective measurements the programme material is the test signal, and the listener is the measuring instrument. As with any measurement, not all problems are revealed by all test signals so it is necessary to select signals that have the potential to reveal the significant defects in loudspeakers. To maximize the efficiency and sensitivity of the tests, care should be taken to avoid redundancy, and music that is merely entertaining; every programme excerpt should be potentially productive.

In addition there is the essential matter of acknowledging what listening tests reliably can and cannot reveal. Some kinds of loudspeaker defects are simply difficult to hear. They may be more reliably detected by technical measurements than by listening. Distortion products, for example, are often masked by the very sounds that generate them[22]. It has long been known that our tolerance for distortions is much higher when listening to complex programme material than with special test signals[23]. To demonstrate the audibility of distortions by using technical sounds, like pure tones, is arguably little different from performing a physical measurement using the same signals. In either case, there is not a clear relationship to the audible consequences of the distortion in music.

Resonances, and defects in spectral balance are both more efficiently revealed in complex orchestrations than in solo instruments with their necessarily limited spectral content[24]. Recordings of single instruments and voices are also critically dependent upon the choice and placement of, perhaps, a single microphone in the highly directional and frequency-dependent sound field of the source. The possibility of bias is thereby increased. Ensembles of instruments provide an enriched sound field in which microphone placement has less of an effect on the sound quality.

Superimposed on all of this is the fact that, in general, the recording engineer selects and locates the microphones, and equalizes the signal, to produce a pleasing effect in the recording control room, while listening through unspecified and unstandardized loudspeakers in an atypical room. These uncontrolled listening experiences at the origin of the programme impose biases on the results of the loudspeaker evaluations.

The expedient method of dealing with the problem of variable recording standards is to employ a variety of different sources and trust that, in the average, a neutral perspective will be achieved. The experimenter should attempt to reject obviously coloured recordings, assembling a collection of recordings from independent sources. A reasonable beginning would be a foundation of classical and 'acoustical' jazz, and some purists argue that nothing else will suffice. Such an elitist approach is inappropriate, however.

To be realistic, tests must include popular music, reproduced at appropriately high levels, so that the products are stressed as they will be in the real world. The wide-band spectrum of this music, with the persistent high sound levels at both frequency extremes, turns out to be useful in its own right, adding a confirmation of the assortment of defects found in other programme selections, while adding new perspectives. Those with open minds may discover that some popular music is every bit as demanding, and revealing of problems, as the classical repertoire.

Audio systems nowadays are increasingly expected to handle the full bandwidth and sound power of dramatic, often exaggerated, pyrotechnical effects in blockbuster movies. For these it is necessary to calibrate the system to the required home-theatre reference sound levels, and to select some appropriately difficult passages. This is not so much a test of fidelity as is it of survival ability.

The human voice is a signal for which there is a special affinity. Both solo and choral selections have been the traditional backbone of listening tests. Broadcasters, especially, like to include passages of spoken word extending, perhaps, to live-versus-reproduced comparisons with the original announcer. Those who pursue this will find out just how dependent vocal sound is on such things as the choice of microphone, its angle and distance, reflections from the script table, and so on. It is by no means a foolproof signal. Because of its limited bandwidth, it is also a signal of limited ability to reveal loudspeaker ills. Unless there is a special reason to use them, spoken word recordings are a poor use of a listener's time.

The listener training programme, described in Section 13.2.6(b) below, yielded an interesting result: a rating of different programme material according to its ability to reveal loudspeaker colorations. The differences were huge, and the results have provided useful guidance in selecting programme that makes good use of listeners and their valuable time[25].

In situations where the special advantages of monophonic listening are utilized, the programme must undergo a further selection process. In particular, it is important to avoid stereo recordings that significantly alter sound quality when the channels are summed, indicating severe interference between left- and right-channel signals. It may be safer to use only a carefully-chosen single channel for such listening.

13.2.5 Loudness level

In evaluations of sound quality, it is essential that perceived loudness of all the comparison sounds be equal. This is not always easy to achieve. If the products all have very similar bandwidth and spectrum shapes, almost any measurement will suffice, even, perhaps, a pure tone. This will work for amplifier comparisons but, for loudspeakers, that is not the case. The more different the comparison sounds are in bandwidth and timbre, the more difficult it is to maintain a constant loudness balance across a range of different programme material.

One investigation of alternative measurement methods concluded that B-weighted measurements of pink noise were about as good as very elaborate loudness computations. This was for loudspeakers of very different spectra – clearly not typical of the middle to high-end products of today[26]. In our experience, we find that no measurement system is flawless. With the best of intentions, there will occasionally be combinations of programmes and loudspeakers that provoke differences in perceived loudness. Since it is the perception of loudness that we are attempting to equalize, the listeners' opinions must prevail. B-weighted measurements of pink noise are a good beginning, though, and if this is not available, A-weighting will have to suffice.

Equal loudness also applies in an absolute sense, the playback level. This is a difficult decision at times; too high, and small loudspeakers are pushed into distress, too low and the dynamic capabilities are not properly evaluated. Whatever levels are selected, for whatever reasons, they should be documented using a recorded reference sound such as noise, and maintained for the rest of the experimental series.

Multichannel listening raises new issues of loudness balancing, this time of the five surround channels with each other. With five identical loudspeakers, at identical distances, in an 'idealized' room environment, it has been found that the choice of calibrating signal is not very critical. When the loudspeakers were not at identical distances, it was observed that the subjective loudness judgements were based principally on distance[27]. In real-world situations, with loudspeakers of different directivities being a common choice for surround channels, and reflected sounds playing a strong, if not dominant role, any such level alignments are likely to be subjects of discussion, if not controversy.

13.2.6 Listeners

(a) Selection

Listeners should be selected on the basis of their ability to evaluate sound quality in a discriminating and repeatable fashion. While it is commonly believed that musicians and regular concert goers are better suited for this task, research has shown that they do not distinguish themselves as being any better than experienced hi-fi listeners [2, 28, 29]. Given proper training, subjects with the right aptitudes, but no prior musical or critical listening experience are generally found to perform similarly to the most seasoned critical listeners.

A listener's performance can be quantitatively measured by statistically analysing their listening test data. By performing analysis of variance (ANOVA) on each individual listener's data, the error variance and F-statistic scores can be computed to establish who are the most discriminating and reliable judges of sound quality[31].

Hearing loss can significantly affect listeners' judgement of loudspeaker sound quality[1, 32]. The effect is measurable at audiometric threshold elevations of 10–20 dB affecting listeners' error variance and bias in judgment. It is interesting to note that the effect seems to be more related to broadband loss, rather than to high frequency loss, such as presbycusis, which occurs naturally with age. It is therefore prudent to perform routine audiometry exams on the listening panel and eliminate those with low- and middle-frequency losses greater than 20 dB.

(b) Training

It should be no surprise that training and familiarity with the listening task can improve a listeners' performance. One trained listener is the equivalent to using seven untrained listeners in terms of achieving the same statistical power[31]. Using fewer listeners without giving up statistical power can save considerable time and money.

Training can take on many different forms. The simplest and most direct one is to run several practice or pilot trials to familiarize the listener with the range of stimuli, their task and listening conditions. Pilot tests are also beneficial for identifying unreliable listeners and revealing any potential flaws in the experimental design

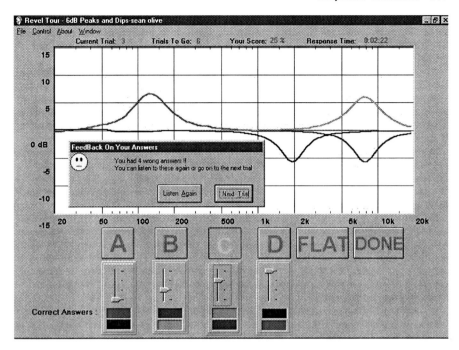

Figure 13.5. The GUI for 'Eartrain', a Harman software program that teaches listeners to identify correctly resonances added to program according to centre frequency, Q and amplitude. The listener must match each resonance A-D to its correct frequency response drawn on the graph. The program can be heard without any resonance ('FLAT'). Feedback is given on listeners' responses.

or test set up. It is better to discover these problems early on, than later after several hundreds of trials have been run.

More specialized forms of training may be necessary for more difficult listening tasks. For example, some tests may require the listener to give detailed analysis of several different attributes of sound quality. To partially address this need, the authors have developed a computer program called 'Eartrain' used to train listeners who participate in product evaluations [25]. The GUI of the software is shown in Fig. 13.5. The software teaches listeners to identify resonances added to various programmes according to their centre frequency, Q and level. A reference of the sound without equalization indicated as 'FLAT' is provided, and feedback is given on the listeners' responses.

The self-administered computer-based training enables listeners to report audible differences among loudspeakers in unambiguous precise terms that product engineers can directly correlate to the loudspeaker's technical performance, and make improvements if necessary. In this way, the listening test becomes a more integral feedback loop in the product development process.

13.2.7 Mono/stereo/multichannel modes

With stereo and multichannel being the dominant modes of listening, it may seem heretical even to suggest monophonic tests. However, there is convincing evidence that listeners are more sensitive to some sound-quality aberrations when listening monophonically[2]. The results from interleaved mono and stereo listening tests indicated that listeners awarded similar scores to highly rated loudspeakers, but in stereo they tended to be less offended by problems that elicited strong criticism in mono tests. The consistency of product ratings was also affected; variations in repeated assessments were as much as a factor of two higher in the stereo tests, making all ratings less certain.

In evaluations of sound-image and spatial quality, there are important similarities between the responses to mono and stereo presentations although, obviously, stereo presentations contain more information. Much of the sense of depth and space seems to be conveyed in single-channel presentations, an observation that confirms the close relationship that has been noted between ratings of sound quality and spatial quality. A good loudspeaker, used in pairs, is likely to be a good stereo loudspeaker[2].

Stereo comparisons of loudspeakers in which the products are placed side-by-side expose listeners to a sound stage that either shrinks and expands, or shifts, laterally. This, combined with the varying distances to the side walls, can mask differences between closely-rated loudspeakers. Positional substitution is the preferred method, as shown in Fig. 13.2.

Multichannel audio presents more challenges. Multichannel audio for movies has been with us for many years and has evolved from matrixed surround algorithms, to various digital discrete forms. Even the number of channels is expanding. However, confusion about how to evaluate such systems is eased by the fact that the film industry has a long tradition of establishing somewhat standardized circumstances, system equalization and sound levels for the creation of the sound tracks, and for their subsequent playback.

Multichannel music is a new and different subject. In all the years we have enjoyed stereo, the industry has not seen fit to establish much in the way of standards. If this continues into the multichannel era of the business, we all have a problem. One can only hope that reason prevails. As an example of something in need of standardization, consider the absolutely enormous differences in bass performance in a room resulting from the use of five full-range loudspeakers, versus five satellites and one or more subwoofers with bass management. With five full-range loudspeakers, the panning of the bass signals will determine the coupling of the low-frequency energy to the room modes, and therefore the quantity and quality of the bass that is heard. It is a situation out of control. Here we offer no solutions, just a very strong caution.

13.3 Electrical requirements

The complications of a system for loudspeaker comparison invite problems from a variety of sources. Relay and switch contacts, lengths of wire in the signal path, and loudspeakers with unusual impedances require careful preparation of the system and monitoring during its use.

The switching system should provide adjustments for equal-loudness comparisons by means of gain adjustments situated ahead of the power amplifiers. The adjustments themselves can be done successfully using pink noise measured at the listener's location using an omnidirectional microphone. For the measurement, B-weighting is preferred, especially if the loudspeakers are quite different in their spectra. If the loudspeakers are similar, the commonly available A-weighting can be used.

Power and signal grounding should avoid hum-producing loops, with special care being paid to the isolation of the power amplifier input and output grounds; some amplifiers can behave abnormally if they are connected. It is useful if the switcher can accommodate equalizers and the outboard signal conditioners used by some loudspeakers. For both of these reasons, it is a great simplification if each test loudspeaker has its own identical power amplifier.

The circuit from the output of the amplifier to the loudspeakers should not represent an abnormal impedance. If there are relay contacts, they must be generously rated, particularly those that switch the inductive loudspeaker loads. Loudspeaker wires should be of sufficient gauge that the total resistance of the loudspeaker circuit, including the relay contacts, does not exceed 0.2 Ω, or 1/40 of the lowest rated imped-ance of the loudspeakers under test[9].

The power amplifier must be capable of driving low-efficiency, low-impedance loud-speakers to the highest sound levels without limiting, instability or protective-circuit activity. A nominal power rating of 200 watts per channel into 8 Ω, with a capability for close to double that into 4 Ω is a reasonable basic requirement. A clipping indi-cator or a peak-power meter is a necessity.

Loudspeakers are designed to be driven by constant-voltage amplifiers. Therefore, power amplifiers with significant output impedances (e.g. most vacuum tube devices and many Class D and 'digital' amplifiers) should not be used. The output imped-ance is in series with that of the loudspeaker wire – it is the combined effect that matters. When such a device is used to drive loudspeakers having frequency-depen-dent impedances (the majority of products) the performances of loudspeakers are altered in ways not intended by the manufacturers. This invalidates the test.

With digital signal sources, it is possible to accomplish all of the signal routing, adjustments, and programme selection, in the digital domain, placing high-quality D/A converters upstream of the power amplifiers.

13.4 Experimental method

13.4.1 Comparison alternatives

A listening test is essentially a comparison between one or more products or sounds. The manner in which the products are presented can seriously bias the performance of the listener and the test results. Nuisance factors to consider include the number of products compared in each trial, the time interval between the comparisons, the order of presentation, and the length of the test itself. Several possible comparison alternatives are available: single stimulus, paired comparisons, multiple comparisons, ABC (with hidden reference), and ABX.

Single stimulus comparisons, by definition, require the listener to make successive judgments of one or more products in isolation one at a time. Common practitioners of this technique include audio reviewers, hi-fi salespersons and some audiophiles who claim that single stimulus comparisons produce more accurate and sensitive judgements. This notion is quite contrary to scientific evidence[31]. Since our acoustic memory is notoriously short and error-prone, single stimulus comparisons are gener-ally not recommended where the audible differences between test objects are small.

Paired comparisons allow immediate switching back and forth between two different products with a time gap of less than a few seconds. Using this technique, very precise and repeatable judgements are possible, even when the audible differ-ences among products are very small. Many researchers believe paired comparisons used in conjunction with the ordinal scale greatly facilitate the listeners' task. On the negative side, if there are more than two products to be tested, paired comparisons require 6 times as many trials to obtain the equivalent data from using a multiple (i.e. A/B/C/D) comparison protocol. Multiple comparisons among more than four loudspeakers at a time may be too difficult a task for the listener, but any number greater than two leads to great improvements in the efficiency of the test.

The ABC (with hidden reference) method is widely used for testing very small audible differences, such as those found in low bit-rate perceptual coders [30]. The listener is given a two-alternative forced choice where the 'reference' signal is always identified as A. The compressed signal and reference are randomly assigned to B or C. The listener must determine whether the 'hidden reference' is B or C. To estab-lish the magnitude of the impairment the listener must rate what they believe to be the compressed signal on a 5 point impairment scale relative to the uncompressed signal. This method has limited application to loudspeakers since, typically, the audible differences are comparatively large.

The ABX protocol is just a variation of ABC except that no 'reference' is identified. The listener switches among three stimuli, two of which are identical. The listener must determine whether X is A or B[33].

No matter what comparison technique is employed, the presentation order of products should always be randomized and balanced to avoid possible bias in the results. A sufficient number of trials and repetitions should be run to establish the reliability of the listeners' responses and achieve the necessary statistical power. The duration of the test should not exceed 30–40 minutes since fatigue can affect listener performance [31, 32].

Ratings for each product should be averaged using several different programme selections since they are known to bias listeners' judgement of the product [6,7,25]. The programmes should be selected on the basis of their ability to reveal the audible strengths and weakness of the test products. As nearly as can be judged, they should be free of spectral or spatial artifacts. Finally, the listeners' task is considerably eased if repeated 20–30 s loops of programme are used for each selection.

13.4.2 Rating scales

Rating scales enable the listeners' opinions to be transformed into numerical data that through statistical analysis can examine possible psychoacoustic relationships between the stimulus and its perception. The scale of choice should be appropriately defined for the attribute(s) being measured. Its meaning should be clearly understood by the listeners so that it is used in a consistent fashion. Rating scales can generally be classified as one of four types: nominal, ratio scale, ordinal scale, and interval scale. Nominal and ratio scales are not commonly used for loudspeaker evaluations, so we will limit our discussion here to the ordinal and interval scales.

Ordinal scales are used to establish the order or ranking of two or more products according to some attribute. A chief disadvantage of the ordinal scale is that it does not tell us the magnitude or distance between objects being ranked. However, some researchers feel the simplicity of ordinal scales makes the listeners' task much easier and less stressful, which will lead to improvements in their performance.

Interval scales give both ranking and magnitude information making their use appealing. Examples include the 0-to-10 Fidelity Scale [1–3] and Preference Scale [6–8]. A variation on the latter is the bi-polar Reference Scale that includes a reference or anchor product. Using a reference or anchor can help establish consistency in the use of the scale, and continuity between tests. Attaching meaningful verbal attributes to the intervals or the end-points will help ensure consistent interpretation and use of the scale.

The major problem with using interval scales is that you may find some listeners will interpret and use the scale differently, which can cause scatter and outliers in the data. Normalizing the data can make visualization of any trends more apparent.

13.4.3 Questionnaires

A test questionnaire or response form should be provided for each listener indicating a clear explanation of the listeners' task and the purpose of the test. Increasingly computers are being used to accomplish this task, and for several good reasons. Computers can perform the tedious, labour intensive task of monitoring, collecting, storing and analysing the data more efficiently than humans and with fewer errors. Computer-automating this process along with the control of all devices used in the listening test can, in the long term, lead to more efficient and cost-effective listening tests. An example of a computer-generated questionnaire used by the authors is shown in Fig. 13.6. The listener can switch between the test loudspeakers by pressing buttons A through D. Ratings of various timbral and spatial attributes for loudspeakers A–D are entered below. Finally a text box is included so that listeners can include any relevant comments.

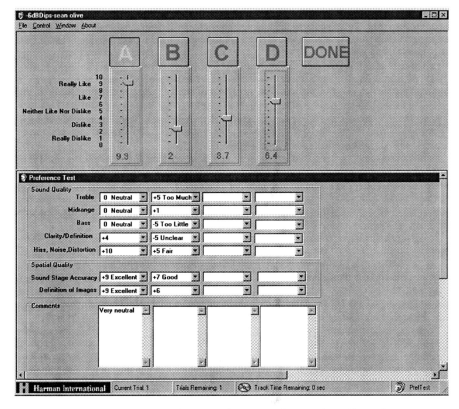

Figure 13.6. The user interface for the software 'Pref. Test' that allows the listener to control switching among loudspeakers in a multiple-comparison test and enter ratings and comments on interval scales for various quality attributes. Comments are optional.

13.5 Statistical analysis of results

Statistical analysis of the listening test data is required to determine what factors, if any, are responsible for the variance in the listeners' ratings. Fortunately, there are several commercial statistical packages available that can perform many kinds of analysis presented in a wide choice of output formats. The relative ease in which these programs can put out statistical results does not preclude the requirement that the users fully understand which statistical method is most appropriate for their application, and that the underlying assumptions of the statistical test are met. The old phrase 'there are lies, damned lies, and then there are statistics' is entirely appropriate here.

Simple descriptive statistics can be used to calculate the individual and group averages, the variability of the data and its distribution. Graphing the individual and group data for each variable and their interactions can help visually clarify the relationship between the data and any possible effects that may exist. The plotted means should include the 95% confidence interval indicators, providing an estimate of the range within which the true mean lies with a 5% chance of error.

It is customary to subject the data to a significance test to determine the probability (expressed as the p-value or significance level) that the experimental results could have occurred from chance, due to errors in the measurement and performance of the listeners. By convention, a p-value less than 0.05 means there is a less than 5% chance that the results occurred by chance, and therefore the results can be considered statistically significant. There is always a danger that the significance test may detect a significant result when none exists (Type I error) and vice versa (Type II errors). Type I errors can be reduced by choosing a more stringent significance level, whereas Type II errors can only be guarded against through proper experimental design. Increasing the number of subjects and/or observations in the test, can increase the statistical power of test which will reduce the likelihood of Type II errors. There are several different types of significance tests including analysis of variance (ANOVA), the F-test, the chi-square test, and the t-test. For details on which one to use, you should consult statistical textbooks [34-36].

Significance testing falls under one of two classifications: (1) parametric or linear tests or (2) non-parametric or non-linear tests. Both tests assume that the individual observations are independent. The parametric tests assume the observations have a normal distribution and they are from distributions of equal variance. Non-parametric tests make no assumption about the distribution of the data, and for that reason, they are sometimes referred to as 'distribution-free' tests.

It is important first to verify that the data are normally distributed before parametric tests are used. This can be done by plotting histograms of the ratings given to each product and checking to see they have a normal distribution. Another quick test is to compare the mean, median and modal values – in a normal distribution, they will be very close. A third method is to use the Kolmogorov–Smirnoff test, a non-parametric test which determines whether the underlying distributions between the data and a hypothetical normally-distributed data set are the same or not.

If the data are normally distributed, an ANOVA can be applied to determine the effect of one or more test variables on the listeners' ratings. A special kind of ANOVA known as Repeated Measures ANOVA is commonly used in subjective measurements when a measurement is being repeated on the same subject at a different time or under different experimental conditions. Since each subject acts as his or her own control, individual effects are removed, thus eliminating the need for a large random sample of listeners. Analysis of variability both within-subjects and between-subjects is included in assessing the effect of each variable.

Non-parametric analysis differs from parametric analysis essentially in that it examines the data on an ordinal scale rather than an interval scale, and performs sequential ranking of the data. It is possible to transform interval data to ordinal data so that both parametric and non-parametric analysis can be performed on the same data set. There are several types of non-parametric tests that perform non-parametric versions of t-tests and ANOVA. For example, the Friedman test performs a non-parametric two-way ANOVA by ranks for matched samples and is useful for tests with two or more treatments.

If the significance test indicates that the main test variable produced a significant result, you will probably want to examine the mean values of each level within the variable to determine which ones are different from each other. To do this, a *post-hoc* or multiple comparison test performs a series of two-tailed null hypothesis comparisons between each level to determine where the statistically significant differences are coming from. There are several parametric *post-hoc* tests to choose from. Among the more popular and robust ones are the Scheffe's and Games–Howell. For non-parametric tests, the Wilcoxon signed rank test performs a non-parametric version of the t-test to determine whether the ranks of each pair are statistically significant.

13.6 Conclusions

Listening evaluations are routine in the audio industry. They remain the final arbiters of sound quality. Sadly, most are done under such relaxed conditions that the results are of questionable validity. If this industry is to be thoroughly professional, we must put as much effort and intellectual investment into subjective measurements as we do into their technical counterparts. We make accurate technical measurements that we have difficulty in correlating with listener evaluations, and then compound the problem by making subjective evaluations that are unreliable. It is very difficult to make progress under such circumstances. Would any serious engineer make a voltage measurement with an uncalibrated voltmeter? Not likely.

References

1. TOOLE, F E, 'Listening tests, turning opinion into fact', *JAES*, **30**, 431–445, June (1982).
2. TOOLE, F E, 'Subjective measurements of loudspeaker sound quality and listener performance', *JAES* , **33** (1985).
3. TOOLE, F E, 'Loudspeaker measurements and their relationship to listener preferences', *JAES*, **34**, pt 1 227–235, April (1986), pt 2, 323–348, May (1986).
4. TOOLE, F E, 'Listening tests – identifying and controlling the variables', *Proceedings of the 8th International Conference*, AES, May (1990).
5. TOOLE, F E and OLIVE, S E, 'Hearing is believing vs. believing is hearing: blind vs. sighted listening tests and other interesting things', 97th Convention, AES, preprint No. 3894 Nov. (1994)
6. OLIVE, S E, SCHUCK, P, SALLY, S and BONNEVILLE, M, 'The effects of loudspeaker placement on listener preference ratings', *JAES*, **42**, 651–669, Sept. (1994).
7. OLIVE, S E, SCHUCK, P, RYNA, J, SALLY S and BONNEVILLE, M, 'The variability of loudspeaker sound quality among four domestic-sized rooms', presented at the 99th AES Convention, preprint 4092 K-1, Oct. (1995).
8. SCHUCK, P L, OLIVE, S E, RYAN, J, TOOLE, F E, SALLY, S, BONNEVILLE, M, VERRAULT, E and MOMTAHAN, K, 'Perception of reproduced sound in rooms: some results from the Athena project', pp.49–73, *Proceedings of the 12th International Conference*, AES, pp. 49–73, June (1993).
9. IEC Publication 268–13: Sound System Equipment, part 13. Listening Tests on Loudspeakers (1985)
10. AES20–1996: Recommended Practice for Professional Audio – Subjective Evaluation of Loudspeakers (1996)
11. GILFORD, C L S, 'The acoustic design of talks studios and listening rooms', 1959, reprinted in *JAES*, **27** (1979).
12. LOUDEN, M M, 'Dimension ratios of rectangular rooms with good distribution of eigentones', *Acoustica*, **24**, 101 (1971).
13. BONELLO, O J, 'A new criterion for the distribution of normal room modes', *JAES*, **29** (1981).
14. MORSE, P M, *Vibration and Sound*, 1964, reprinted by Acoust. Soc. Amer., New York (1976).
15. OLOFSSON, J, Statens Provingsanstalt Rapport Sp-Rapp 1974:13, Stockholm (1974)
16. JACKSON, G M and LEVENTHALL, H G, 'The acoustics of domestic rooms', *Applied Acoustics*, **5** (1972).
17. BURGESS, M A and UTLEY, W A, 'Reverberation times in british living rooms', *Applied Acoustics*, **18** (1985).
18. BRADLEY, J S, 'Acoustical measurements in some Canadian homes', *Canadian Acoustics*, **14**, 19–25 (1986).
19. ALLISON, R F, 'The influence of room boundaries on loudspeaker power output', *JAES*, **22**, 314–319 (1974).
20. ITU-R Recommendation BS 775–1, Multichannel stereophonic sound system with and without accompanying picture, Geneva, 1994.
21. OLIVE, S E, CASTRO, B, and TOOLE, F E, 'A new laboratory for evaluating multichannel audio components and systems', preprint, 107th AES Convention, San Francisco, Sept. (1998).
22. CABOT, R C, 'Perception of nonlinear distortions' presented at the 2nd AES International Conference, Anaheim, preprint No. C1004 (1984).
23. PREIS, D, 'Linear distortion, measurement methods and audible effects – a survey of existing knowledge', presented at the 2nd AES International Conference, Anaheim, preprint No. C1005 (1984).

24. TOOLE, F E and OLIVE, S E, 'The modification of timbre by resonances: perception and measurement', *JAES*, **36**, 122–142, March (1988).
25. OLIVE, S E, 'A method for training of listeners and selecting program material for listening tests', 97th Convention, AES, preprint No. 3893, Nov. (1994).
26. AARTS, R M, 'A comparison of some loudness measures for loudspeaker listening tests', *JAES* **40**, No.3, 142 (1992).
27. ZACHAROV, N, BECH, S and SOUKUISMA, P, 'Multichannel level alignment, Part II: The influence of signals and loudspeaker placement', 105th Convention, AES, preprint No. 4816 (1998).
28. GABRIELSSON, A and SJORGEN, H, 'Perceived sound quality of sound reproducing systems', *J.Acoust. Soc. Am.*, **65**, 1019 (1979).
29. GOULD, G, 'An experiment in listening – who are the most perceptive listeners?' *High Fidelity Magazine*, **25**, 54–59, Aug. (1975).
30. ITU-R Recommendation BS.1116: Methods for Subjective Evaluation of Small Impairments in Audio Systems including Multichannel Sound Systems, 2nd edition (1997).
31. BECH, S, 'Listening tests on loudspeakers: a discussion of experimental procedures and evaluations of the response data' *Proceeding from the AES 8th International Conference*, Washington, DC, May (1990).
32. BECH, S, 'Selection and training of subjects for listening tests on sound-reproducing equipment', *JAES*, **40**, No. 7, 590 (1992).
33. CLARK, D, 'High resolution subjective testing using a double-blind comparator', *JAES*, **30**, 330–338, Aug. (1992)
34. FREEMAN, D, PIASANI, R and PURVES, R, *Statistics*, 3rd edition, W.W. Norton, New York (1998).
35. BOX, G E P, HUNTER, W G and HUNTER, J S, *Statistics For Experimenters: An Introduction to Design, Data Analysis and Model Building*, John Wiley, New York (1978).
36. LEHMANN, E, *Nonparametrics: Statistical Methods Based on Ranks*, Prentice Hall, Englewood Cliffs, NJ (1998).

14 Headphones

C. A. Poldy

14.1 Introduction

The first difference that comes to mind when comparing how loudspeakers and headphones produce a sound signal at the ear is that in one case the ear is immersed in a propagating sound field, and in the other it registers the SPL (sound pressure level) in a leaky pressure chamber. 'Since the ear drum is essentially a pressure detector, pressure gradient, particle velocity and other effects do not influence the final sound image.' This is a daring claim, since much research has been performed dealing precisely with this point, for example such elusive phenomena as 'the missing 6 dB between LS (loudspeaker) and HP (headphone) listening', and whether bone conduction plays a role in hearing external sources under normal conditions, as well as the question whether the ear is sensitive to absolute phase, by which is meant whether it can detect any difference between two sounds with identical amplitude response, but different phase. Research continues irrespective of whether a single investigation comes up with the answer yes or no. This is understandable, since such effects are extremely elusive. For example, a trivial confirmation is that white noise does indeed sound different from a delta impulse. Such statements reveal only that the writer has misunderstood the problem, not realizing that the issue is confined mainly to short-term effects involving phase shifts of less than about 90°. Evidently artificial constraints often have to be introduced. The very existence of research activities on this elusive issue, whether non-pressure phenomena 'influence the final sound image', already provides the answer. In this chapter some results of such work will be reviewed, but we shall proceed assuming the answer is no.

Of all the components in the electroacoustic transmission chain, headphones are the most controversial. High fidelity in its true sense, involving not only timbre but also spatial localization, is associated more with loudspeaker stereophony due to the well-known in-head localization of headphones. And yet binaural recordings with a dummy head, which are the most promising for true-to-life high fidelity, are destined for headphone reproduction. Even in their heyday they found no place in routine recording and broadcasting. At that time the causes were unreliable frontal localization, incompatibility with loudspeaker reproduction, as well as their tendency to be unaesthetic. Since digital signal processing (DSP) can filter routinely using binaural head-related transfer functions, HRTF, dummy heads are no longer needed.

Still the most common application of headphones is to feed them with stereo signals originally intended for loudspeakers. This raises the question of the ideal frequency response. For other devices in the transmission chain (Fig. 14.1), such as microphones, amplifiers and loudspeakers, a flat response is usually the design objective, with easily definable departures from this response in special cases. A loudspeaker is required to produce a flat SPL response at a distance of typically 1 m. The free-field SPL at this point reproduces the SPL at the microphone location in the sound field of, say, a concert being recorded. Listening to the recording in front of a LS, the head of

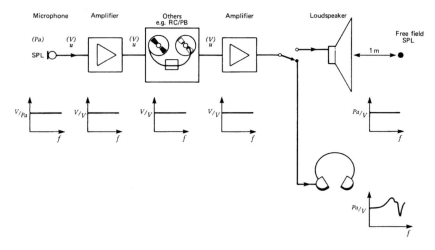

Figure 14.1. Block diagram comparing headphones with other components in the hi-fi chain.

the listener distorts the SPL linearly by diffraction. His ear signals no longer show a flat response. However, this need not concern the loudspeaker manufacturer, since this would also have happened if the listener had been present at the live performance. On the other hand, the headphone manufacturer is directly concerned with producing these ear signals. The requirements laid down in the standards have led to the free field calibrated headphone, whose frequency response replicates the ear signals for a loudspeaker in front, as well as the diffuse field calibration, in which the aim is to replicate the SPL in the ear of a listener for sound impinging from all directions. It is assumed that many loudspeakers have incoherent sources each with a flat voltage response.

The use of headphones in communications and audiometry differs markedly from high-fidelity applications, where the subjective impression in music is the main quality criterion. The desired frequency range is often narrower and other aspects of human hearing, such as the occlusion effect, ambient noise attenuation, bone conduction, interaural attenuation and leak susceptibility can become significant. Technical requirements are more strictly complied with, inasmuch as they are objectively measurable.

Computer simulation, using electrical equivalent circuits with lumped elements has become an essential tool. Since most circuit simulation programs are for electrical circuits, modifications using macros are needed before the circuits can become routinely useful in acoustics. To quantify the effects of tolerances in mechanical dimensions of components, the input data for bottlenecks such as holes and slits should normally be the mechanical dimensions in millimetres, and not the Ohm- and Henry-value representing the bottleneck. There is another reason why labelling bottlenecks with R and L values in 'acoustic' circuits is not always meaningful. Unlike electrical resistances and inductances they are essentially frequency dependent. Despite this, in the simulation circuits of this chapter, for some elements R and L labels are used. How can such elements be of any use to readers, or mean anything to them? The answer is that, where they are used at all, the frequency dependence is not dramatic. Where it is felt that the dimensions will mean more to the reader, and also where the influence of frequency dependence is the issue, macros with millimetre dimensions have been used.

At higher frequencies in tubes, which are distributed elements, cavity resonances occur. It is often essential to take these into account somehow. Genuine lumped elements cannot do this accurately. Where a tube is almost one-dimensional the transmission line can describe it very well. It is a small step to extend the lumped element approach to include such two-ports. The detailed information provided should enable the reader to apply the techniques described here to other systems.

14.2 Acoustics of headphones

14.2.1 The pressure chamber principle – headphones compared with loudspeakers

Unlike loudspeakers, which produce a propagating sound field around the head of the listener, the sound field of headphones is confined to a relatively small volume of up to about 30 cm³. Since the ear is essentially a pressure detector, this difference in mechanism is of little consequence as far as the subjective sound image is concerned.

For a closed back loudspeaker at low frequencies, the significant 'source quantity' for SPL at a distance is the volume acceleration:

$$P \ [\text{Pa}] = \rho \ [\text{kg/m}^3] \times (\text{volume acceleration} \ [\text{m}^3/\text{s}^2])/(4 \times \pi \times \text{Distance} \ [\text{m}])$$

For a flat SPL response, the acceleration should be independent of frequency. To fulfil this, the excursion must drop by 12 dB/octave on raising the frequency. This happens automatically above the main resonance, where compliance is unimportant compared with inertial mass. Here Newton's law, Force = mass × acceleration, guarantees a constant acceleration for a given force Bll.

In a closed headphone, the force pumps on a cavity, giving a sound pressure proportional to excursion. Up to about 2 kHz, where the wavelength of sound is still large compared with the dimensions of the cavity, the sound pressure is distributed uniformly in the volume, even in the presence of mild leaks. The SPL and phase of this uniformly distributed pressure does, of course, depend on the extent of the leaks. In practice, leaks are always present whose influence dominates at low frequencies. For intermediate frequencies (1 kHz), where their influence is very low, the cavity can often be regarded as a pressure chamber. Here the pressure is in phase with the volume displacement of the transducer membrane and its amplitude is proportional to it. This is a convenient point to introduce the equivalent circuit approach with its electroacoustic analogies and the physical background. Air compression due to sound usually takes place too fast for heat exchange with the surroundings or walls to occur – adiabatic rather than isothermal. According to the adiabatic compression law (the isothermal one is without the exponent 1.4) we have:

$$pV^{1.4} = \text{constant} \tag{14.1}$$

The quantity p is the total pressure, including the d.c. component, atmospheric pressure. V is the volume of a fixed quantity of air. The factor 1.4 is the ratio of specific heats at constant pressure and constant volume, called γ or κ, the convention depending on language. The pressure p is not linearly related to volume V. However, sound is only a small a.c. ripple (of a few Pascal), superposed on 1 atmosphere (approx. 100 000 Pa). If p is sinusoidal, V will be so too, with opposite phase. Omitting the derivation, the resultant rise of pressure dp (Pa) due to the membrane pumping dV (m³) into the headphone cavity of volume V (m³) is:

$$\frac{dp}{-dV} = \frac{1.4 \times 100\,000}{V} \tag{14.2}$$

where the factor 100 000 is the normal atmospheric pressure P_{atmos} (see also Fig. 14.2). The pressure chamber is effectively an air cushion, whose stiffness is proportional to the atmospheric pressure. Constant excursion gives constant pressure. Loudspeakers

S.I. Units throughout	Flow velocity profile	Equivalent circuit	R	L	C
Hole 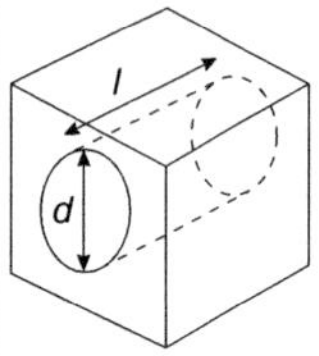 $f \prec f_0$	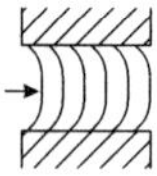	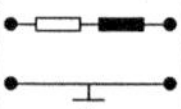	$\dfrac{8\eta l}{\pi\left(\dfrac{d}{2}\right)^4}$ $\left(=\dfrac{7.58\times10^{-4}l}{d^4}\right)$	$\dfrac{4}{3}\times\dfrac{\rho l}{\pi\left(\dfrac{d}{2}\right)^2}$ $\left(=\dfrac{2.04l}{d^2}\right)$	—
$f \succ f_0$ $f_0=\dfrac{64\eta}{\pi\rho d^2}$ $=\dfrac{3.16\times10^{-4}}{d^2}$	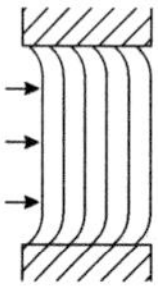		$\dfrac{8\eta l}{\pi\left(\dfrac{d}{2}\right)^4}\sqrt{\dfrac{f}{f_0}}$ $\left(=\dfrac{7.58\times10^{-4}l}{d^4}\sqrt{\dfrac{f}{f_0}}\right)$	$\dfrac{\rho l}{\pi\left(\dfrac{d}{2}\right)^2}$ $\left(=\dfrac{1.53l}{d^2}\right)$	—
Slit $(d \prec b)$ 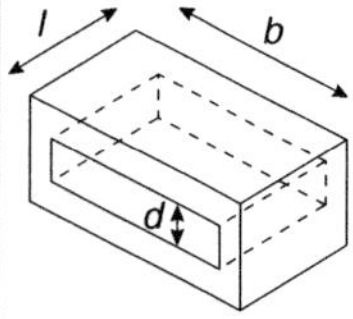 $f \prec f_0$	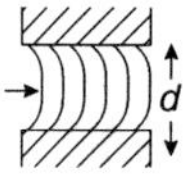	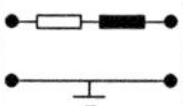	$\dfrac{12\eta l}{bd^3}$ $\left(=\dfrac{2.23\times10^{-4}l}{bd^3}\right)$	$\dfrac{6}{5}\times\dfrac{\rho l}{bd}$ $\left(=\dfrac{1.44l}{bd}\right)$	—
$f \succ f_0$ $f_0=\dfrac{36\eta}{\pi\rho d^2}$ $=\dfrac{1.78\times10^{-4}}{d^2}$	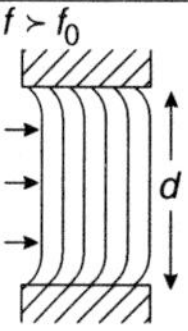		$\dfrac{12\eta l}{bd^3}\sqrt{\dfrac{f}{f_0}}$ $\left(=\dfrac{2.23\times10^{-4}l}{bd^3}\sqrt{\dfrac{f}{f_0}}\right)$	$\dfrac{\rho l}{bd}$ $\left(=\dfrac{1.2l}{bd}\right)$	—
Cavity compliance $V=$ volume		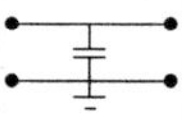	—	—	$\dfrac{V}{\gamma P_{atmos}}$ $\left(=\dfrac{V}{1.4\times10^5}\right)$
Membrane plus coil $S=$ area $m=$ mass $f_R=$ resonance	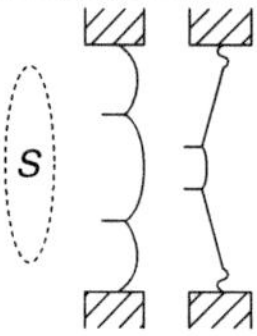	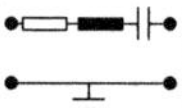	≈ 0	$\dfrac{m}{S^2}$	$\dfrac{S^2}{4\pi^2 m f_R^2}$
Radiating piston (full space) $S=$ area	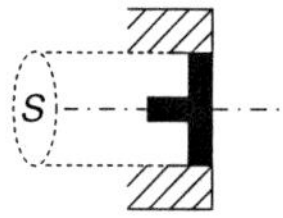	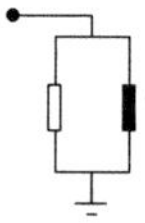	$\dfrac{\rho c_0}{S}$ $\left(=\dfrac{405}{S}\right)$	$\dfrac{0.85\rho}{\sqrt{\pi S}}$ $\left(=\dfrac{0.56}{\sqrt{S}}\right)$	—

Figure 14.2. The acoustic elements, their electrical equivalent circuits, and R, L, C values expressed in terms of mechanical dimensions: l, b, d (m), S (m^2), V (m^3), frequency f (Hz). Effective total length of a constriction in flow direction is l. For a circular hole the total effective length l equals the geometrical length l_0 plus the end correction $0.85d$. The viscosity coefficient under normal conditions is $\eta = 1.86 \times 10^{-5}$ kg/(ms). It is proportional to pressure and roughly proportional to absolute temperature (K). Air density is $\rho = 1.2$ kg/m^3. Speed of sound $c_0 = 340$m/s. Specific heat ratio $\gamma = 1.4$ for air. $P_{atmos} = 10^5$ Pascal. The impedance analogy is used, where sound pressure [Pa] is equivalent to voltage (V), and volume velocity (m^3/s) is equivalent to current (A). The equivalences for L, R, and C are 1H $\equiv$ 1 kg/m^4, 1Ω $\equiv$ 1 kg/(m^{4}s), 1F $\equiv$ 1 m^{4}s^2/kg.

below resonance also have constant excursion, being dictated only by the compliance, but the SPL drops by 12 dB per octave, due to the acceleration drop.

The above equation is reminiscent of the electrical condenser equation:

$$\frac{dU}{dQ} = \frac{1}{C} \tag{14.3}$$

where the voltage change dU is analogous to the pressure change dp above, and the charge increase dQ is analogous to the volume compression $-dV$. For the analogy to be useful, an expression is required relating the volume V with the value of the condenser C in the analogy, C (m³/Pa) or (acoustic Farad SI) called the acoustic compliance of the cavity:

$$C = \frac{V}{\gamma \times P_{atmos}} \tag{14.4}$$

14.2.2 The electrical analogies used in this chapter (Fig. 14.2)

It will pay us to describe first the nomenclature, concepts and analogies used here, before actually using them. They are by no means universally agreed upon, various acoustics specialists using different analogies and units. In the analogy, it is important to decide whether the electrical symbolism is used for expressing acoustical or mechanical quantities. For acoustic quantities, pressure p (N/m²) is equivalent to voltage U, and volume velocity q (m³/s) is equivalent to current I (see Olson[162]). For mechanical quantities, force F (N) is equivalent to voltage U, and velocity v (m) is equivalent to current I. Note here the difference m² corresponds to an area S (for later: Section 14.2.2(e) where the ideal transformer connects the mechanical and acoustical worlds). Expressing acoustic elements in mechanical units can also be done without transformers if the mechanical resistances and inductances are divided by S^2, and mechanical condensers are multiplied by S^2, where S is the moving cross-sectional area perpendicular to flow direction:

$$R = \frac{r}{S^2} \quad L = \frac{m}{S^2} \quad C = c \times S^2 \tag{14.5}$$

Capital and small letters indicate acoustical and mechanical quantities, respectively. It may seem inconsistent to use m rather than small L. However the concept of mass is expressed intuitively better this way, apart from the visual problem of confusing l with the number one (1). In the mechanical domain some acousticians use the admittance analogy: force≡current and velocity≡voltage. They have good reasons for doing so, circuit topology corresponding better to geometry. However, most of us are more familiar with the impedance analogy, which is used here.

The components of headphones comprise mainly the following categories of building blocks:

(a) Cavity: C (acoustical compliance) connected to ground;
(b) Acoustical bottlenecks, porous paper, holes, slits: R (resistance) , L (acoust. mass or inertance), in series;
(c) Compliant membrane: L (mass), C (compliance), R (damping resistance), all in series;
(d) Radiation impedances: (R, L in parallel);
(e) Mechanical–acoustical interface: ideal transformer;
(f) Electrical–mechanical interface: gyrator.

We deal with these groups in turn.

(a) Cavities

From equation (14.4) the acoustic compliance of a cavity can be expressed as:

$$C = 7.14 \times 10^{-6} \times V \text{ (acoustic Farad SI)}\tag{14.6}$$

where V is in m^3. One of the condenser terminals is invariably grounded (Fig. 14.2).

(b) Acoustical bottlenecks or constrictions

Porous elements can be used for damping resonances. One needs large R (resistance) and small L (air mass), to prevent ωL from competing with R. For this, paper and woven resistances are excellent, since the pores, though irregular, are very small. They can be regarded as a collection of holes in parallel. For large holes, viscosity effects only occur within a thin boundary layer near the walls, not in the middle. In the small pores of paper the boundary layer extends over the entire pore cross-section. R decreases with increasing diameter to the 4th power (Fig, 14.2), and L only to the 2nd power. Therefore small diameters give good damping. In slits this is less extreme, powers 3 for R and 1 for L for the slit thickness d. Unlike their electrical counterparts, R and L in acoustical bottlenecks always occur together.

(b1) Frequency dependence of R and L values in holes (Fig. 14.2)

The discussion will be confined to holes, since the main features apply also to slits, except for the factor 6/5 instead of 4/3 in L. Unlike genuine electrical inductances, L, where frequency dependence is only found in the impedance $Z = \omega L$ via ω, not in L itself, acoustical masses have a frequency-dependent L value. Acoustical resistance values R are also frequency dependent. For low frequencies (capillary regime I, $f < f_0$) the boundary layer permeates the whole cross-section, giving a frequency-independent R value with $L = (4/3) \times m/S^2$ for holes, where m can be verified to be the physical mass of air in the air plug. Comparing with L in equation (14.5) makes the mass appear larger than expected. In reality this comes from the effective area S being smaller since the parabolic flow profile constricts it. For long narrow tubes[186] the factor (4/3) should be replaced by:

$$\frac{4}{3} - \frac{\gamma}{3}\left(\frac{8\eta l}{\rho c_0 r^2}\right)^2$$

where r is the tube radius, and l its length. In regime II ($f > f_0$) the R value is proportional to $\sqrt{(f/f_0)}$ and L drops to the value m/S^2 expected for a rigid air plug of mass $m = \rho \times l \times S$.

The need for two different formulae for bottlenecks makes simulation difficult, since the cross-over region (f_0) then requires cosmetic treatment. One would hope that a single formula or set of formulae would deal with all frequencies automatically. This is the case for the lossy tube. The basic equations came from Egolf[185]. In the equivalent circuit for this two-port (Fig. 14.3) the H matrix, otherwise hardly used outside transistor theory, can be conveniently applied. Of course, the concept of impedance of such a two-port is not really meaningful, since the currents at the ends of the tube are different. However, by choosing a very short length 1 mm and grounding the far end, the input impedance can be meaningfully compared with the R and L values of holes for regimes I and II. To demonstrate frequency dependence, a hole diameter 0.55 mm was chosen, aiming for the crossover f_0 to be at 1 kHz (Fig. 14.4). Being really rational, one must admit there is no real need to differentiate between holes and circular cavities, the lossy tube dealing correctly with both, except for the waste of computation time for Bessel functions. The main result here is that a single physical model confirms the two tendencies: L drops by 25% and R rises. To obtain round numbers in L, 66 tubes were chosen in parallel.

The problem of how to treat R and L in a short hole for all frequencies was tackled by Ingard[179] which gives a higher R than the lossy tube model. Better agreement is

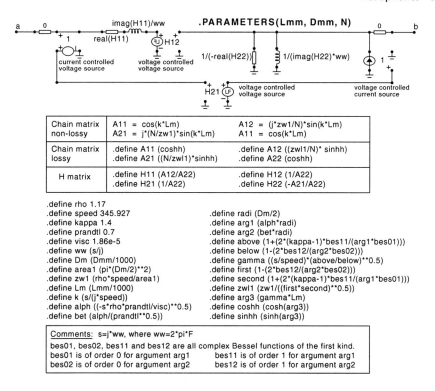

Figure 14.3. The equivalent circuit for the lossy tube two-port.

achieved by Slotte[184]. He used polynomial approximations of the Bessel functions of the first kind of order 0 and 2. The accuracy of the polynomials is ± 2% for the real part and ± 0.6% for the imaginary part, approaching perfection in the lowest and highest frequency limits. His expression for the acoustic resistance and acoustic mass are:

$$R_a = 8f\rho_0 \times \text{length} \times \sqrt[4]{\frac{4096 + 45x^3 + 4x^4}{nd^2x^2}}$$

$$L_a = \frac{332 - 13x + 7x^2 + 2x^3}{249 - 9x + 4x^2 + 2x^3} \times \frac{4\rho_0 \times \text{length}}{n\pi d^2}$$

where $x = \dfrac{d}{\sqrt{2} \times skin}$

The parameter *skin* is the viscous boundary layer thickness equal to $\sqrt{(\eta/\rho_0\pi f)}$, which of course is smaller than the hole cross-section only for high frequencies above the transition frequency f_0. The parameter *length* is the effective length of the holes including end correction, and d is the hole diameter for n holes. The viscosity coefficient $\eta = 1.86 \times 10^{-5}$ kg/(ms) for air at 23°C and 1000 mb.

(b2) End correction for holes Returning now to the simpler treatment, the effective length of the air plug l for a hole of diameter d in an infinite wall of physical thickness l_0 is $l = l_0 + l'$ where:

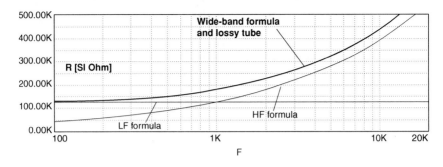

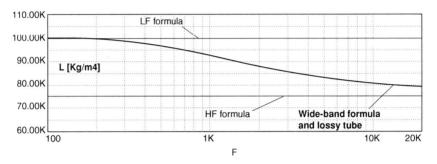

Figure 14.4. The R and L values calculated from the lossy tube two-port and the wide-band formulae of Slotte[184], compared with the low- and high-frequency equations of Fig. 14.2 for holes.

$$l' = 0.85d \qquad (14.7)$$

This so-called end correction arises from the constriction of flow lines. One study[176] has indicated that the same end correction factor 0.85 applies not only to L, but also to R. This is a strange coincidence, since they are physically unrelated. For a constriction d in a tube of finite diameter D the expression[164,165]

$$l' = 0.85d[1 - 1.47(d/D) + 0.47(d/D)^3] \qquad (14.8)$$

is preferable since it vanishes for $D = d$.

(b3) Two useful types of R,L element for simulations: RacoustDC, LacoustDC While simulating it is often difficult to adjust R and L simultaneously. However, if the low-frequency value of either R or L has been measured or is approximately known, rather than making a blind guess for the unknown component it is better to make a realistic estimate of the pore size. The equations in Fig. 14.2 for R and L can then be used with full frequency dependence, without any need to decide on the number of holes or on end-correction factors, since there is reason to believe that they are equal[176]. The input is either R and pore size, or L and pore size.

(c) Membranes

Membranes are represented by R, L and C in series. These acoustic parameters are related to the mechanical ones *Rmech, Mms, Cms* as shown in equation (14.5), using the effective membrane area *Sd*. Here we have used the nomenclature of

(a)

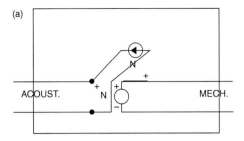

(b)

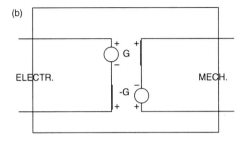

Figure 14.5. The simulation circuit for (a) an ideal transformer and (b) a gyrator, using dependent sources, with multiplication factor N or G.

the Thiele–Small procedure (see Section 14.5.3), which is the most reliable way to obtain these parameters for a moving-coil transducer. The coil must be in place, to ensure correct compliance. $Rmech$ is normally negligible, except where it is deliberately cultivated using sandwich membranes with adhesive between. L and C are readily determined from the total mass m and resonance f_R (Fig. 14.2), if the effective area S is known.

(d) Radiation impedances

Since most headphones are not entirely closed, there are also radiating components (Fig. 14.2) due to openings or leaks. These are represented in the first approximation by a parallel combination of R and L connected to ground, which is rigorously true only for a spherical zero-order source. The real component R predominates at high frequencies, and represents energy loss by radiation. This component can be utilized for damping resonances. For this purpose it is just as effective as a viscous resistance. At low frequencies the radiation impedance becomes reactive and is equivalent to a mass loading. Since radiation impedances in headphones are very low compared with the impedances behind the openings whence they radiate, they can be replaced by a short circuit in most approximations.

(e) Acoustic transformers

As seen above there are two analogies $(U, I) \equiv (F, v)$ and $(U, I) \equiv (p, q)$, where F is force (N), v is velocity (m/s), p is pressure (Pa), and q is volume current (m³/s). Where both systems appear in the same electrical equivalent circuit, an ideal transformer is required as an interface. The turns ratio is 1:S, where S is the area of the moveable mechanical surface upon which the pressure p is acting. Thus the pressure p on the primary side is transformed into the force $F = pS$ on the secondary side, where the (F, v) system is used. Similarly, the velocity v on the secondary side is transformed into volume velocity $q = v \times S$ on the primary side. Figure 14.5(a) shows the simulation circuit of an ideal transformer, using dependent sources.

(f) Transduction two-ports

To complete the electroacoustic equivalent circuit, the genuine electrical part must be included. This is connected using a transduction two-port. Figure 14.7 shows the two-ports for electrostatic and dynamic transduction in the $(U, I) \equiv (F, v)$ analogy. The mechanical analogy is the only physically meaningful one, since the acoustical components are only indirectly moved by the mechanical parts. For an explanation of the gyrator see Leach[163], for the electrostatic two-port see Hunt[167]. Figure 14.7 shows the T two-port of the electrostatic (C-transducer) and the two-port (gyrator) for the dynamic (D-transducer or moving-coil transducer). The former is symmetrical and the latter is regarded as having a form of asymmetry. This so-called antireciprocity property of gyrators is well known. However, Hunt has shown that this is not of physical origin, that it is an artefact, resulting from an inadequacy of the conventional symbolism (see the 2nd edition of this book, p. 498). Reversing the input and output of the gyrator does not influence results. Figure 14.5(b) shows how to simulate a gyrator using current controlled voltage sources. If one prefers to eliminate the gyrator interface, Fig. 14.102(a) shows the equivalent circuit replacing the electrical source by a force source.

For the single-sided C-transducer:

$$c_k = \frac{d_0}{U_{pol}} \qquad (14.9)$$

where d_0 is the electrode-membrane spacing and U_{pol} is the d.c. polarizing voltage between membrane and electrode. For a push–pull system:

$$c_k = \frac{d_0}{2U_{pol}} \qquad (14.10)$$

On the mechanical side there may be reasons to prefer the (p, q) system, rather than (F, v). Though this is unphysical, the results will still be correct. Then the two-port constants of Fig. 14.7 must be modified. For dynamic transducers Bl is replaced by Bl/S. For electrostatic transducers the c_k is replaced by $c_k \times S$. Then $c_k = c_k \times S$, expressing acoustic quantity in terms of a mechanical quantity.

The gyrator contributes to damping in a moving-coil sound source. Like a microphone, the LS registers its own motion irrespective of the cause, itself or a sound wave. The induced voltage opposes the generator voltage, suppressing the current. This is effective only for strong Bl values. Progressively doubling Bl from a small value initially raises the whole response by 6 dB for each step. Then for large values the shape changes. In the resonance, increasing Bl can even weaken the SPL output (Fig. 14.6).

(g) Loss of volume due to the idea that bottlenecks are just R and L in series

Fortunately this rarely causes serious errors in simulations. But it is permanently present, if the formulae of Fig.14.2 are used without further thought. For example, take a wide bottleneck Vb between two cavities V1 and V2. The bottleneck is important for high frequencies. If V1 is connected to the rear of a LS and V2 is not leaky, the low-frequency behaviour will hardly be influenced by the bottleneck since ωL is small. However, the total volume, including that in the bottleneck Vb, contributes to the overall stiffness via the condenser element Cb as in Fig.14.2. Therefore, when using lumped elements, the following routine is recommended: two grounded condensers, each of value Cb/2 are placed, one at each end of the RL bottleneck.

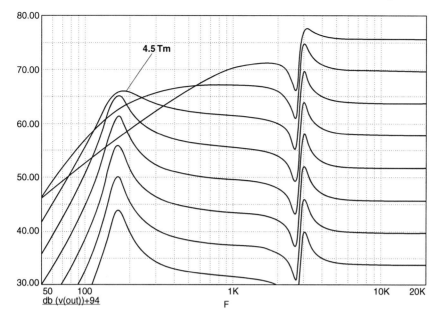

Figure 14.6. Loudspeaker responses for progressively increased magnetic field strength. From curve to curve the *Bl* value is doubled. This is the transducer of Fig. 14.18.

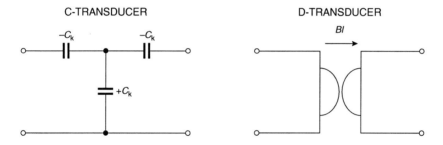

Figure 14.7. The purely transductional electroacoustic two-ports for electrostatic (C-) and dynamic (D-) transducers.

This should be done in such a way as not to cause spurious resonances with L from end corrections EC. Each Cb/2 should merge unobtrusively with V1 and V2. Therefore Cb/2 should be introduced before and after the EC–bottleneck–EC combination. This avoids non-physical resonances occurring at high frequencies. The one and only function of the Cb/2 elements is to conserve physical volume.

14.2.3 Types of headphone design – circumaural, supra-aural, closed, open, intra-aural

Regarding the coupling of the transducer to the ear, we distinguish between circumaural, supra-aural and intra-aural. Circumaural headphones, where the cushion surrounds the ear, tend to have better bass response than the supra-aural type (which lie flat on the ear). By 'better' is not meant 'loud', but rather 'reproducible'. The earpiece can be quite well sealed in the circumaural type – the only irreproducible

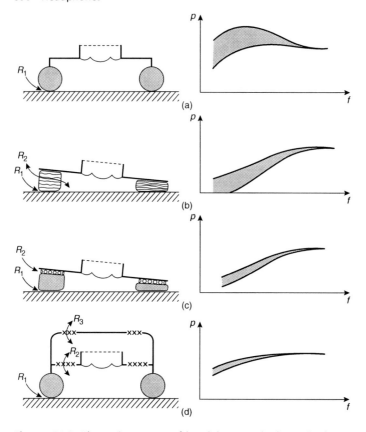

Figure 14.8. The various types of headphone and schematic characteristic responses at low frequencies: (a) closed; (b) open foam cushion; (c) cushion of impermeable foam with fixed resistance; (d) integrated open.

leaks being due to the presence of hair. These can load the large cavity without much SPL loss, due to its low impedance. Reproducible leaks are deliberately incorporated into the earphone towards the back.

Circumaural headphones are characterized by a relatively large coupling volume in excess of 30 cm³, and an inner diameter of the cushions of at least 55 mm. The upper limit for the validity of the lumped-element approach is about 2 kHz, since standing waves in the coupling volume occur above this frequency, associated with a non-uniform sound-pressure distribution. The upper limit for supra-aural headphones is higher, but not as high as one would expect from the diameter of the coupling volume alone, which can be as small as 20 mm, since the length of the ear canal, about 25 mm, now contributes to the largest dimension of the coupling volume. However, improving on the lumped element approach is much easier for the ear canal (Section 14.3.3) than for the three-dimensional coupling volume (Section 14.2.4(b)).

The reproducibility of the given frequency response of a supra-aural headphone is lower than that of its circumaural counterpart, due to the relative ambiguity of the positioning of the earpiece. The advantage of supra-aural headphones is their lightness. The extreme of this is seen in the popular mini-earphones, which can have excellent sound quality, including bass, whose reproducibility is still a problem. Between the two extremes – namely the circumaural and mini-earphones – the supra-aural headphones attempt to cover the whole surface area of the ear, thus reducing the unwanted leaks to a minimum. The remaining leak R_1 (Fig. 14.8), which is variable, influences the frequency response. In order to combat this effect, the cushions are frequently made of a porous material such as foam. The acoustic resistance R_2 of the foam is chosen to be a little lower than that of the leak, so that this combination of two resistances in parallel is mainly determined by the foam, rather than the non-reproducible leak R_1. This situation is depicted schematically in Fig. 14.8(b). Since the porosity of the foam is also influenced by the mechanical pressure against the ear, a further improvement is obtained by dividing the foam into two parts. The first part, which is in contact with the ear, is non-porous but soft, providing the best seal. The second part is rigid but porous, providing the well-defined low acoustic resistance R_2 (Fig. 14.8(c)).

Figure 14.8 depicts schematically the types of headphone construction and the characteristics of their frequency responses. The closed headphone type (Fig. 14.8(a)) shows a frequency response which is determined at low frequencies by the leak. Though a circumaural headphone is characterized by less leak than the supra-aural variety, the above argument about the reproducibility of the frequency response applies here too. The well-defined parallel resistance R_2 need not be in the cushion. It can be integrated into the body of the earpiece. A modified version of this, using two resistances R_2 and R_3, is shown in Fig. 14.8(d), which represents the integrated open headphone[9].

This well-defined acoustic resistance R_2 *is* in reality a combination of R_2 with an acoustic mass L_2. The pure acoustic resistance without any reactive component, unlike

Figure 14.9. Membrane of an isodynamic headphone.

an electrical resistance, cannot be realized. This means that the impedance of the well-defined leak $Z_2 = R_2 + j\omega L_2$ increases with frequency, together with that of the unavoidable leak $Z_1 = R_1 + j\omega L_1$, thus closing the headphones for relatively low frequencies as the frequency is increased.

One can conclude from the above that all headphones with reproducible bass response – which is one of the high-fidelity criteria – are open to some extent, except those with fluid-filled cushions (see Section 14.2.8). There is, however, another type of openness which is not related to the leaks, though they may also be present. This is found in the group of headphones where the transducer membrane itself is transparent to sound, as, for example, in the isodynamic headphones (Fig. 14.9) or other types with a large membrane driven over its whole surface. Such membranes are light and are characterized by a low resonance frequency. Though the coupling volume to the ear may itself have a fairly tight seal, thus providing the pressure chamber effect, extraneous noises are transmitted through the membrane unattenuated. The high dynamic range of digital recordings is not taken advantage of. However, in-head localization is less for open headphones (Section 14.4.4).

14.2.4 Types of transducer

The isodynamic headphone (Fig. 14.9), being acoustically less complicated than the moving-coil system, can be taken as the starting point for discussing the acoustics of headphones.

(a) The isodynamic transducer

A schematic cross-section is shown in Fig. 14.10. On one surface of the membrane a maze of conducting material has been deposited, usually aluminium for lightness, producing a low-mass vibrating system of about 100 mg. On either side of the membrane is a system of magnetic rods magnetized perpendicular to their length, and so organized that opposite neighbours repel each other. In the diagram, the cross-section of each rod is visible. The resultant magnetic field at the membrane is in the plane of the membrane and perpendicular to the current in the conductors. A typical value for B is 0.08 T. The pattern of the conducting layer, which resembles a printed circuit, is arranged so that the whole membrane is driven in phase. In order to increase the total force, each conductor is divided into a number of tracks which are mechanically in parallel but electrically in series, thus achieving a total conductor length of the order of 10 m, with an electrical resistance of 200 Ω. Since the surface S of the membrane is large, it constitutes a significant fraction of the boundary surface of the coupling volume to the ear.

The influence of the membrane on the acoustics of such a headphone is determined primarily by the acoustic mass of the membrane, L_M. The low mechanical tension results in a low membrane resonance of less than 200 Hz which is, however, not seen in the resulting frequency response, since the dominant acoustic compliance C_O of the air volume is much lower than that of the membrane. The result is effectively a series resonance circuit of L_M with C_O, having a resonance at about 4 kHz. Figure 14.11 (curve 1) shows the calculated response for the equivalent circuit (Fig. 14.12) leaving out all acoustic resistances except for the leak impedance R_0, L_0. To produce a good frequency response, this resonance must be almost critically damped. The natural damping effect of the radiation resistance R_R is shown in curve 2. This being as yet insufficient, a porous resistance R_H is also provided to give curve 3. Above 5 kHz simulation with lumped elements is inaccurate, due to the inhomogeneity of the sound field in the cavity. The resonances due to standing waves are useful in raising the response above that expected from the lumped-element calculation using a single condenser for the cavity compliance. Although details of the resonances can only be determined empirically, lumped elements can be used in order to demonstrate the main features, as is shown in Fig. 14.23 for the moving-coil head-

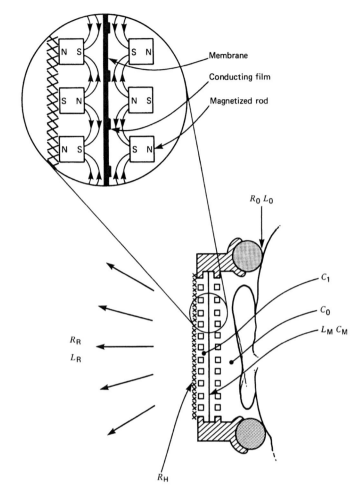

Figure 14.10. An isodynamic headphone.

phone. Careful development can produce an excellent response up to about 5 kHz with or without a higher SPL around 3 kHz.

The flat response down to the lowest frequencies is a natural result of the relatively low mechanical impedance of the membrane, and requires no special effort on the part of the acoustics engineer. Even the effect of leaks under the cushion has no dramatic influence on the response, since practically the whole force Bll from the transducer comes to bear on the coupling volume, very little being wasted in moving the extremely compliant low-mass membrane. The phase of the pressure at low frequencies is, of course, influenced by leaks, since they have a large inductive component. The effect of leaks on the amplitude, however, is seen only by increasing the membrane tension. Curve 4 shows the effect of increasing the stiffness by a factor of 10. The leak elements R_o and L_o, taken into account in the calculation of all these curves, correspond to a typical leak in a circumaural headphone cushion.

The potential efficiency of the isodynamic transducer is compared with that of the moving-coil transducers (described in the next section) in Table 14.1. After comparing the maximum available pressures from the two transducer types, it is evident that

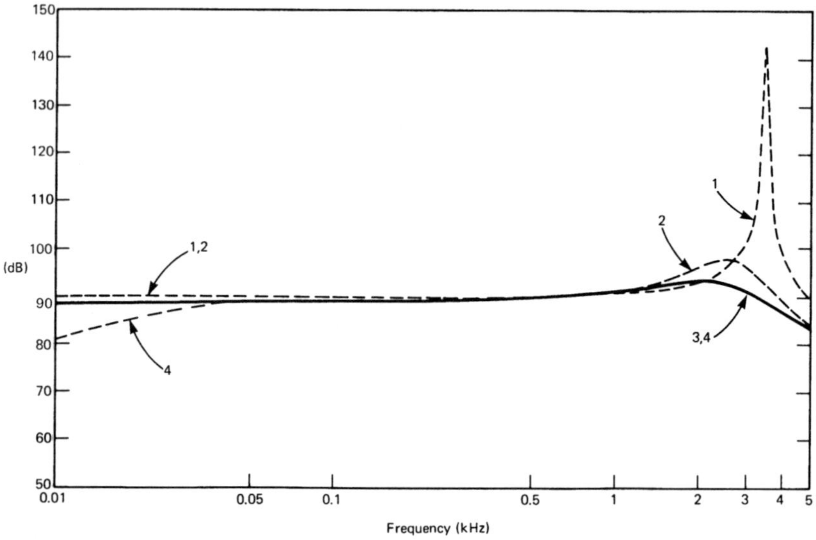

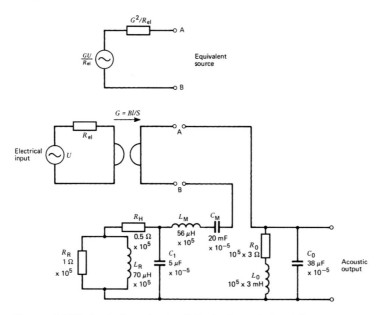

Figure 14.11. Response for isodynamic headphone circuit of Fig. 14.12. Curve 1: resistance $R_R = 0$, $R_H = 0$, leak Z_0 active; curve 2: with radiation impedance Z_R normal (R_H still $= 0$) and leak; curve 3: with all resistances; curve 4: membrane ten times stiffer, the leak starting to have an effect.

Figure 14.12. Equivalent circuit of the isodynamic headphone.

the isodynamic has to give all it can, whereas the moving-coil system has plenty to spare. This will be visible in the pressure chamber results of the next section. One is prepared to waste more than 20 dB of the potentially available pressure at low frequencies.

(b) The moving-coil transducer

A high-end commercially available circumaural dynamic headphone (Fig. 14.13) is described and simulation results presented, which, of course, due to the large cavity, cannot be expected to be accurate above 1 kHz. Nevertheless, the effect of the integrated leak and cushion can be demonstrated. It is also reasonable to ask what happens if the cushion is made impermeable, or, for that matter, if it is left out altogether. Here it is assumed the cup is immovable (for influence of cup motion, see Section 14.2.6(a).

Figure 14.14 shows the Thiele–Small parameters of the driver, in vacuum and in air. To simplify it as far as possible, the paper resistance was removed. This improves the accuracy due to the higher quality factor and prevents air cushion effects from interfering with the series resonance circuit approximation assumed in the Thiele–Small algorithm. Measurements with paper in place (omitted here) were also performed to estimate the paper resistance 8.5e5. The pore size was estimated at 120 microns to get a realistic L value (Section 14.2.2(b3)). The technique of combining vacuum and air measurements to obtain the circuit element values is described in Section 14.5.3.

In the following, the influence of the various components is shown in measurements (Figs 14.15–14.17) with the corresponding simulations (Figs 14.18–14.22). The equivalent circuit is shown in Fig. 14.23. If the driver circuit is left in the simplified form with only one transformer, as was done here (Fig. 14.24), the elements can be taken directly from the Thiele–Small parameters. To avoid unnecessary complications in the model, the damping from each of the two magnet gaps is not treated individually, since this would require two ideal transformers[7,8] (as in Fig. 14.102). Instead, its influence was estimated, and thrown wholesale into the resistance with 1e5. This is reasonable here since the intricacies of the double transformer model are only seen at high frequencies, and these will be masked by the cavity resonances, which cannot be simulated here properly anyway. Dome decoupling effects were also neglected for this reason. Baffle measurements of the driver (Fig. 14.15(a)) as a LS agree so well (Fig. 14.18) with the simple model as to justify the stiff dome assumption. Inspection reveals a relatively stiff dome, compared with the torus. This can be achieved by a two-step deep-drawing technique. The anomaly at 3 kHz comes from the Helmholtz resonator of the back openings 70 Henry with the volume 5.8 cm^3 assumed under the membrane to tune this resonance. Removing the paper does not leave $R = 0$ and $L = 0$, but an acoustic mass of 70 Henry, mainly from end correction, which is always present, also when the paper is in place.

Table 14.1 Typical parameters for the isodynamic and moving-coil transducers

	Isodynamic	Moving-coil
Bl	0.72 Tm	4.5 Tm
Diaphragm area Sd	42×10^{-4} m^2	8.8×10^{-4} m^2
Electrical d.c. resistance	200 Ω	124 Ω
Transducer force for 1 V	36×10^{-4} N	363×10^{-4} N
Available pressure from transducer P_0	0.86 Pa	41 Pa
Normal listening level in headphone	1 Pa	1 Pa

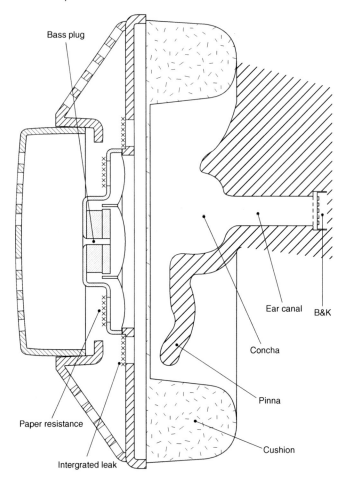

Figure 14.13. Section through a supra-aural moving-coil, i.e. dynamic, headphone.

With the paper replaced, the back was closed completely using putty, taking care not to block the paper pores, and for this a protective layer of normal paper was put between the two. Closing the back shifts the main resonance from 160 Hz up to 1.2 kHz (Fig. 14.15(b), curve a). The simulation (Fig. 14.19, curve a) automatically gave the correct resonance, confirming that 5.8 cm³ under the membrane was correct. Opening only the bass plug gives front and back responses curves b and c. With the paper resistance active again, the effect of the bass plug is small (Fig. 14.15(c), curves ab1 and ab2). Up to this point all responses apply to the free field SPL at 12.5 cm distance. The rest deals with the periphery of the driver on an ear model made of rubber. The cylindrical ear canal of 7.5 mm diameter was 19 mm long with hard walls and the condenser microphone at the end was effectively a hard termination. For simplicity, no attempt was made to reproduce the ear drum impedance. How serious this omission is, can be demonstrated using simulations, since the drum impedance is known (Fig. 14.93). Before evaluating the effects of the integrated leak and cushion, both were deactivated to make the so-called pressure chamber situation (Fig. 14.16). This was done by replacing the cushion by a Perspex ring of the same dimensions, and blocking the integrated leak. Curve a is for the back of the driver closed, curve

Mass Extra	361.0E-6 kg

(Group 4)	Acoustical parameters	
Vas	270.00E-6 m3	1
AREA	877.3E-6 m2	2
DIAMETER	33.4E-3 m	3
R(Air)	19.8E+3 Pa/(m3/s)	
L(Air)	94.6 kg/m4	

(Group 5)	more electro-mech. parameters
Mass(Air)	72.8E-6 kg
Cms(mean)	2.51E-3 m/N
Bl(mean)	4.56 Tm
Bl/sqrt(RdcLS)	0.41 Tm/sqrt(Ohm)

(Group 6)	DC resistances
RdcLS	124.3 Ohm
RdcTotal	126.0 Ohm
RdcAmmeter	1.7 Ohm

(Group 1)	Main input	
	IN VACUUM	IN AIR
FsExtra	124.5 Hz	114.2 Hz
Fs	182.4 Hz	158.7 Hz
Us	0.1947 V	0.1947 V
Is	110.00E-6 A	230.00E-6 A
F1	171.0 Hz	144.0 Hz
F2	195.5 Hz	174.5 Hz
Ir0	412.28E-6 A	596.16E-6 A

(Group 2)	Main electro-mechanical Thiele–Small parameters	
Mms	314.9E-6 kg	387.7E-6 kg
Cms	2.42E-3 m/N	2.59E-3 m/N
Bl	4.61 Tm	4.51 Tm
Rms	12.9E-3 N/(m/s)	28.2E-3 N/(m/s)

(Group 3)	More electro-mechanical Thiele–Small parameters	
SQRT(F1F2)	182.8 Hz	158.3 Hz
Accurancy	0.24 %	-0.26 %
Qms	27.9	13.7
Qes	2.1	2.4
Qts	2.0	2.0
Zres	1770.0 Ohm	846.5 Ohm
r0	14.0	6.7
SQWRT(r0)	3.7	2.6
Zr0	472.2 Ohm	326.6 Ohm

Figure 14.14. Thiele–Small measurements of the large 45 mm driver in the supra-aural dynamic headphone.

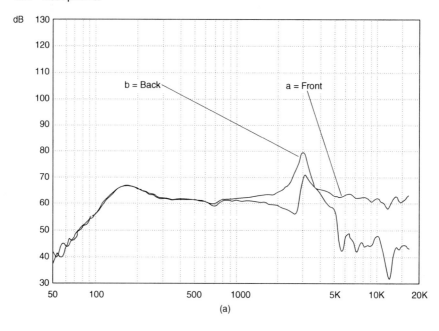

(a)

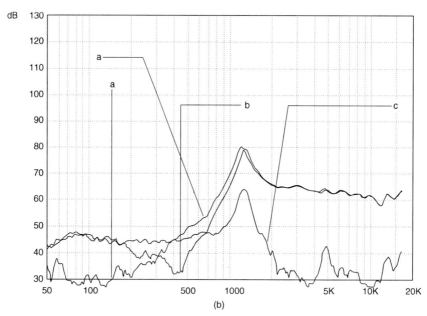

(b)

Figure 14.15. Free-field measurements of SPL of the 45 mm driver at 12.5 cm in a large baffle, in front and behind, for 200 mV. (a) Back opened by removing porous paper resistance. (curve a) = front: (curve b) = back. (b) Back completely closed (curve a); only bass plug opened, curves b (front) and c (back). (c) Porous paper resistance active, with bass plug blocked (curves ab1), and open (curves ab2), curve a = front, curve b = back.

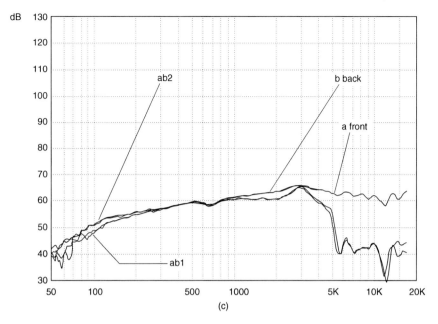

Figure 14.15 (*continued*)

b with only the bass plug open, and curve c with both paper resistance in the driver and bass plug open. To simulate the low-frequency behaviour in these curves an overall cavity volume of 65 cm^3 was needed. Then, with the driver in its normal state, the headphone was measured (Fig. 14.17), with cushion blocked (curve a), with cushion functioning normally (curve b), and finally the cushion was omitted (curve c), with thin posts instead to keep the correct distance. According to the pressure chamber idea described above, one might expect the result of leaving out the cushion to be catastrophic.

Making the cushion porous smooths the response around 1 kHz, without much loss of low-frequency SPL. This it does by virtue of its highly resistive character $R = 0.8e5$ SI Ω, a very low R value with hardly any L. Omitting it results in only 10 dB loss at 100 Hz. This shows how ear loudspeakers with acceptable bass, such as the K1000 of AKG, are possible at all, with freely suspended drivers near and over the ears, but very close. In the K1000 the membrane area is even larger than in the headphone studied here, making a cushion that much less necessary. Another approach[141] for small earphones suggests suspending them near the ear but, due to the small size, there is loss in bass.

The anomaly at 1.8 kHz is only visible when the cushion is impermeable, and becomes more pronounced in the pressure chamber mode (Fig. 14.17). Treating the volume 65 cm^3 as a simple condenser (Fig. 14.25) is inadequate in two ways: (1) The dip of Fig. 14.17 at 1.8 kHz is absent, and (2) the high frequency SPL is much weakened. Figure 14.26, the equivalent of Fig. 14.22, but with a single undivided condenser for 65 cm^3, shows the high frequency SPL loss. This SPL is regained by progressively decoupling the outer part of the cavity from the inner part as the frequency rises. This occurs at the edge of the pinna, which almost touches the inner surface of the headphone, leaving a gap of 2 mm over a distance of several cm. This slit is mainly L, whose impedance ωL rises with frequency. The cushion communicates directly with the outer cavity. Of course, the inner cavity is not isolated from the cushion, since the front of the ear has no pinna, but neither is its connection via a short circuit. Instead, one

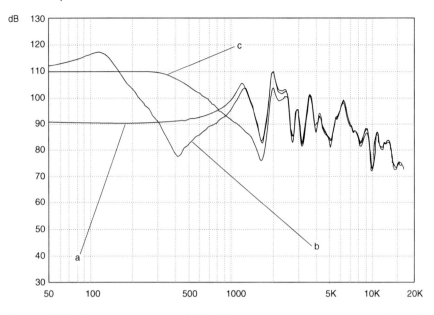

Figure 14.16. Ear measurements of SPL of the supra-aural headphone. In the rubber ear model the ear canal had a diameter of 7.5 mm and length 19 mm, with a 12.7 mm B&K pressure microphone at the ear drum end. It had no eardrum impedance. The cushion was replaced by a hard impermeable Perspex ring of same dimensions, and integrated leaks were carefully blocked with putty; (curve a) no openings whatsoever; (curve b) only bass plug open; (curve c) both bass plug and paper resistance in driver active.

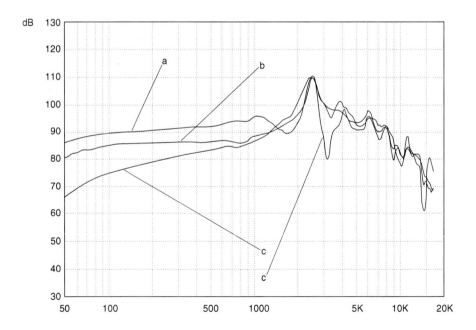

Figure 14.17. To show contribution of cushion on Fig. 14.16, with integrated leak active throughout: (curve a) cushion impermeable, i.e. replaced by Perspex ring; (curve b) cushion normal; (curve c) cushion removed.

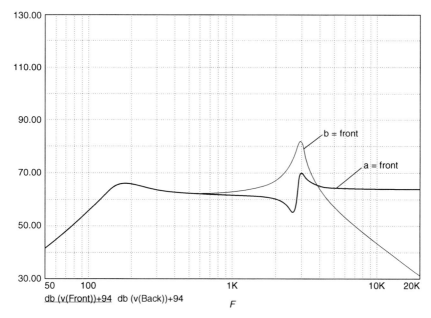

Figure 14.18. Simulation corresponding to Fig. 14.15(a).

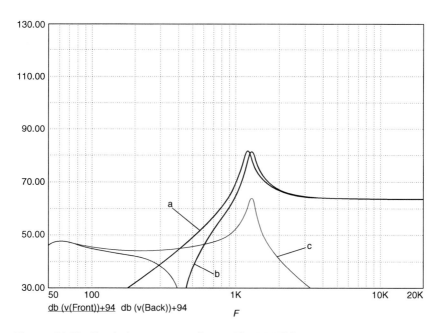

Figure 14.19. Simulation corresponding to Fig. 14.15(b).

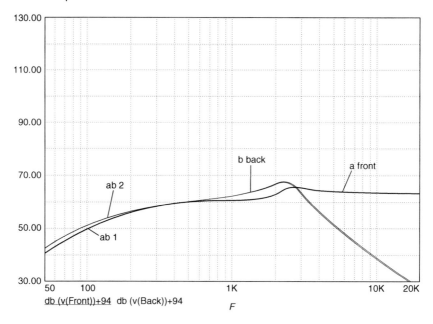

Figure 14.20. Simulation corresponding to Fig. 14.15(c).

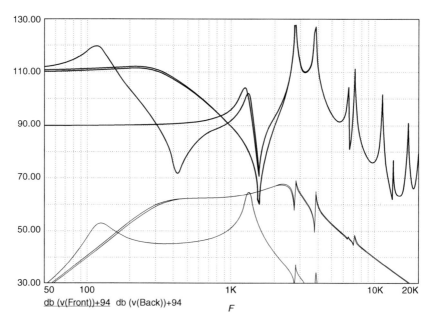

Figure 14.21. Simulation corresponding to Fig. 14.16. The thin curves are the free-field outputs of the back of the driver, a relict from the free-field circuits. Since they occupy otherwise empty space on the graphs, and since they may be of some interest, they were not removed. Of course, to be really informative, the cushion and integrated leak outputs would need to be added as well for the total SPL nearby.

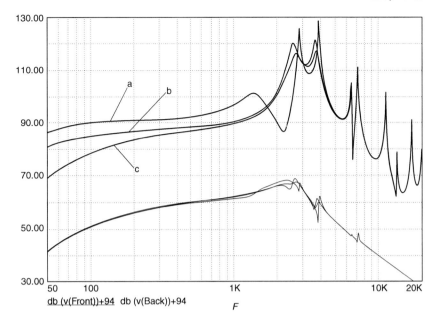

Figure 14.22. Simulation corresponding to Fig. 14.17, without curve c.

needs a transmission line of length 26 mm, whose volume XYZ must be recognized as part of the total volume 65 cm³. Apart from introducing the dip at 1.8 kHz, dividing the cavity in this way prevents the outer condenser, for 55 cm³, from short-circuiting the high frequencies before they can enter the concha and ear canal.

(c) The electrostatic transducer

The electrostatic transducer (see also Chapter 3) consists of a charged membrane, usually supported at its edges, and driven in an electric field modulated by the sound signal. Figure 14.27 shows a schematic cross-section of the push–pull construction, where the membrane is supported between two conducting perforated electrodes. Single-sided transducers with only one electrode are known, but the superiority of the push–pull construction lies in its low harmonic distortion, which is particularly important at low frequencies. This is the case even in headphones, since the membrane excursions, unlike in a pure pressure chamber system, increase in amplitude due to the inductive load arising from leaks.

An external high-voltage d.c. supply is required to maintain the membrane polarization. In some models this voltage is obtained from the audio signal itself by rectification. Since a linear deflection characteristic is given only with a constant charge, any modulation of charge by the audio signal must be suppressed, by introducing a large resistance in series with the polarizing voltage. The charging time constant is set to be considerably larger than the period of the lowest frequency of interest. At high signal levels, however, if the membrane touches an electrode it will discharge, thus interrupting the signal until the recharging process has taken place.

Insulating the surface of the membrane is no solution, since this results in the membrane remaining attached to the electrode. In order to avoid this, the large resistance is shifted into the membrane itself by making it weakly conducting. The result is then a partial discharge at the point of contact, which has a milder effect. Many electrostatic headphones are based on this principle.

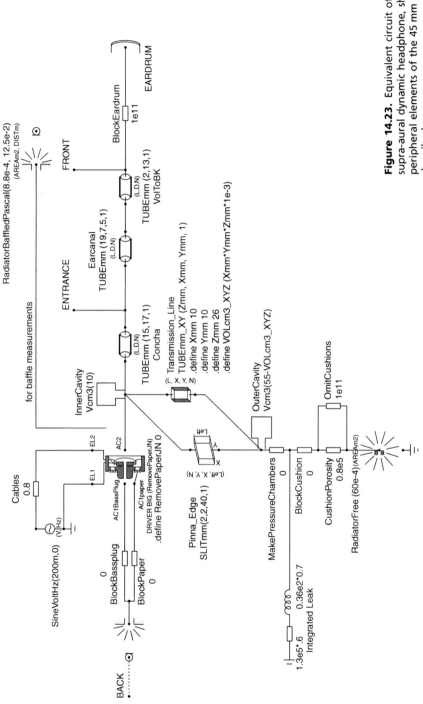

Figure 14.23. Equivalent circuit of the supra-aural dynamic headphone, showing peripheral elements of the 45 mm driver, described as a macro.

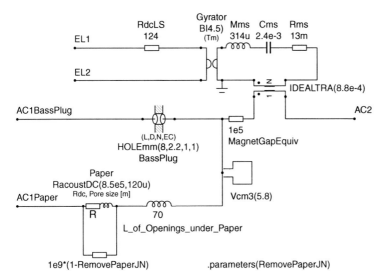

Figure 14.24. Equivalent circuit of the 45 mm driver. Many of the elements come directly from the Thiele–Small parameters.

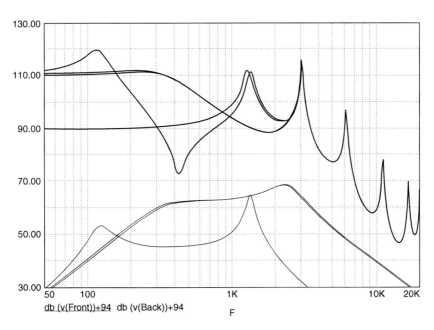

db (v(Front))+94   db (v(Back))+94

F

Figure 14.25. As for the simulation of Fig. 14.21, but with the 65 cm³ cavity treated as a single condenser.

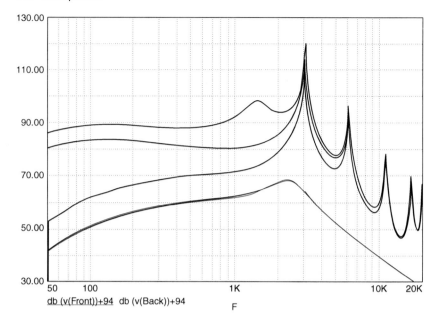

Figure 14.26a. As for the simulation of Fig. 14.22, but with the 65 cm³ cavity treated as a single condenser.

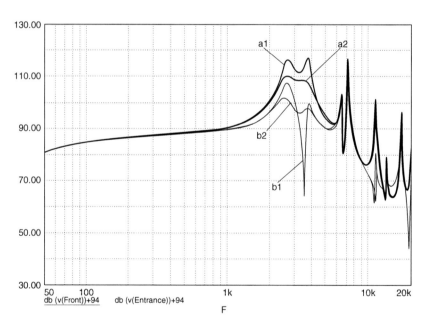

Figure 14.26b. Influence of eardrum impedance on response at (a) eardrum, (b) entrance of ear canal: 1 = hard, 2 = soft.

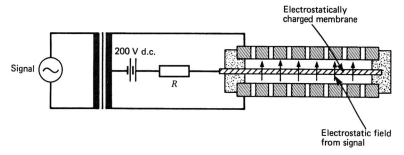

Figure 14.27. Electrostatic push–pull transducer (not drawn to scale).

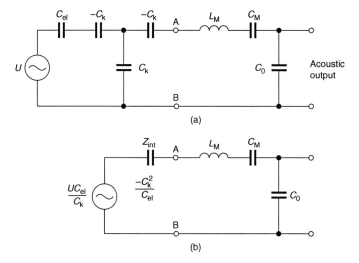

Figure 14.28. Equivalent circuit of an electrostatic transducer demonstrating the negative compliance effect (see text). (a) Complete electroacoustic circuit; (b) equivalent with pressure source only p, q system).

Figure 14.28(a) shows the equivalent circuit of an electrostatic headphone. The number of elements has been reduced to a minimum in order to demonstrate a curiosity of electrostatic transducers, namely the negative compliance. The transducer two-port, already introduced in Section 14.2.2(f), consists of condensers C_k connecting the electrical and acoustic portions. The output terminals AB of the two-port (Fig. 14.28(a)) are loaded with the acoustic impedance of the headphone. The circuit portion to the left of AB can be replaced by the equivalent pressure source UC_{el}/C_k, where C_{el} is the electrical capacitance of the electrodes. The internal impedance of this source, Z_{int}, is that of a negative capacitance $-C_k^2/C_{el}$. This has the desirable property of lowering the resonance of the system as a result of the increased effective compliance. What was said above about isodynamic transducers applies even more here; namely, the low membrane mass allows a high system resonance and relative ease in producing a flat response over the whole audio range. An outstanding property of electrostatic headphones is the transparency of the sound impression. The faithful impulse response is a result of the low mass of the membrane.

(d) The electret transducer

This is identical to a normal electrostatic transducer, as regards its transduction properties and acoustics. The membrane, however, possesses a permanent polarization. This can be a volume polarization, where charges are trapped within the foil material, or a dipole polarization, where the two surfaces are oppositely charged. The membrane is normally supported at its edges. The additional polarizing unit is not required. However, one still cannot dispense with an accessory, since a step-up transformer for the signal voltage is required. A typical transformer ratio is 1 : 60. When used in the push-pull mode, the harmonic distortion components can be made low, better than −40 dB. This distortion, however, often arises not from the transducer but from the transformer.

(e) The electromagnetic transducer

This type of transducer finds no application in high-fidelity headphones, due to the large mass of the vibrating component. However, its efficiency is high and it was used in communications headphones (Section 14.4.3) until recently. Figure 14.29(a) is a schematic diagram to show the working principle. The d.c. flux in the air gap is modulated by the signal current in the coil. A characteristic feature is the force which

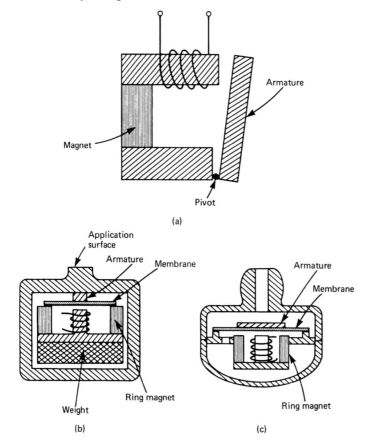

Figure 14.29. Electromagnetic transducer. (a) Schematic of principle[161]; (b) bone conduction vibrator; (c) insert earphone. The armature or membrane consists of magnetic material, e.g. vanadium permendur[167] ((b) and (c) taken from reference 175).

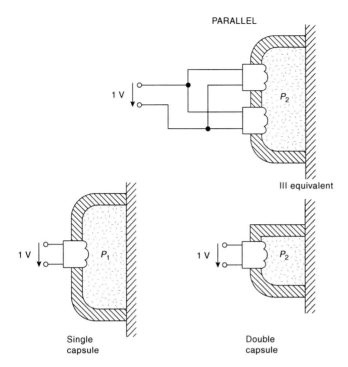

PARALLEL

1 V

P_2

III equivalent

1 V

P_1

1 V

P_2

Single
capsule

Double
capsule

Figure 14.30. Parallel connection of two dynamic capsules in a schematized closed headphone. The effective volume per capsule is halved. The double-capsule analysis here is confined to a half-headphone.

continually acts against the pivot compliance to close the air gap. The permanent magnetic polarization can be replaced by a d.c. current in the coil. As in the moving-coil system, the transduction two-port consists of a gyrator, but with an additional negative compliance component whose presence is related to the stability condition, requiring a permanent elastic mechanical load to prevent the air gap from closing. Figures 14.29(b) and (c) show a bone conduction vibrator and an EM insert earphone.

14.2.5 Two-way systems

When speaking of two-way systems for headphones, a combination of different types of transducer mechanism is usually implied, e.g. dynamic-electrostatic, generally aimed at covering a wider frequency range. However, there are also systems using identical transducers. The increase in total membrane area reduces non-linear distortion.

Such distortion is potentially low in the following types: (1) electrostatic (push–pull), (2) isodynamic (this is essentially also push–pull), (3) double-transducer dynamic. Some examples of low total harmonic distortion (THD) in these three categories are (1) Stax SR-Lambda, (2) the isodynamic closed headphone, YH100 of Yamaha, (3) AKG K280, with two identical dynamic capsules. In dynamic transducers, for ideally low THD, either the coil should be much longer than the magnetic field gap or vice versa. Neither condition is ever the design objective in practice. Therefore reducing the moving-coil excursion required for a given Pa/V-sensitivity is desirable, as is possible with two capsules. This turns out to be highly beneficial.

Connecting the dynamic transducers electrically in parallel is preferable, resulting in a higher sound pressure for a given low-impedance generator. Figures 14.30–14.33 illustrate this. A highly schematic closed headphone at a low frequency is assumed

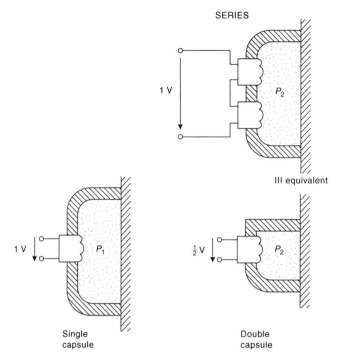

Figure 14.31. Series connection of the two dynamic capsules of Fig. 14.30.

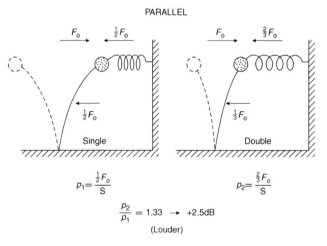

$$p_1 = \frac{\frac{1}{2}F_o}{S}$$

$$p_2 = \frac{\frac{2}{3}F_o}{S}$$

$$\frac{p_2}{p_1} = 1.33 \rightarrow +2.5\text{dB}$$

(Louder)

Figure 14.32. The mechanical system of Fig. 14.30 (parallel connection) reduced to the bare essentials. 1 V applied to a single capsule is assumed here to give $F_o(= Bll)$. The membrane is represented as a mass on a spring (drawn here like a metronome). The other spring (drawn coiled) is the volume compliance. The force in this latter spring is a measure of the sound pressure in the cavity: p_1 and p_2 for single- and double-capsule headphones, respectively. S is the membrane area. Arrows pointing left represent the spring forces. Those pointing right represent the source force. Only 1/3 of the F_o force source per capsule is wasted in the membrane, compared with 1/2 for a single-capsule headphone.

SERIES

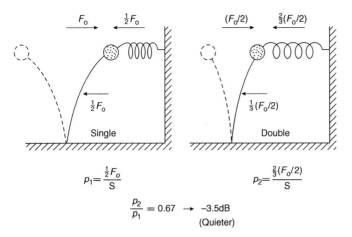

$$p_1 = \frac{\frac{1}{2}F_0}{S} \qquad\qquad p_2 = \frac{\frac{2}{3}(F_0/2)}{S}$$

$$\frac{p_2}{p_1} = 0.67 \quad\rightarrow\quad -3.5\text{dB}$$
(Quieter)

Figure 14.33. The mechanical system of Fig. 14.31 (series connection) simplified as in Fig. 14.32. Here the force source per capsule is reduced to $F_0/2$.

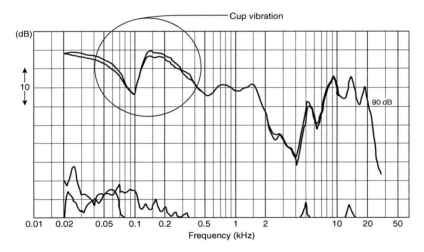

Figure 14.34. The frequency response of a circumaural dynamic communications headphone, showing influence of main cup resonance at 150 Hz. Also visible are the cavity resonances above 3 kHz.

here. The spring compliance of a single transducer is assumed equal to that of the cavity. One capsule of the doubled system effectively sees half the original volume (doubly stiff spring, equation (14.4)). Hence we are justified in imagining a symmetrical separating wall. Of course, at most frequencies in reality we are dealing with damping resistances as well. But here pure reactances have been chosen to simplify the discussion, at the same time doing justice to the fact that a dynamic transducer must overcome considerable internal impedance (Section 14.2.4b)), thus wasting much of its *Bll*-force in some manner, be it resistive or reactive. The two capsules are geometrically so mounted as to converge the high-frequency radiation on the ear canal entrance.

The above argument applies for fixed voltage. On the other hand, if the electrical power is constant[178] the parallel and series combinations are identical, the lower distortion being the only advantage remaining. However, for loudspeakers, where the electromechanical damping is significant and desired, parallel is still preferable.

14.2.6 Vibration anomalies

(a) Influence of cup vibration

Up to now it has been assumed that the light membrane mounted on the relatively heavy ear-piece cup is the only source of motion. However, practically all circumaural dynamic headphones show anomalies below 200 Hz due to cup vibration (e.g. Fig. 14.34). One main resonance at 150 Hz arises from the rigid mass m_{cup} vibrating on a spring, comprising cushion c_c and flesh c_f (Fig. 14.36). The sound pressure p is due to both the membrane and cup contributing volume currents $v_{mem}S_{mem}$ and $v_{cup}S_{cup}$ (m³/s). To minimize cup vibration influence, the following have to be maximized: m_{cup}/m_{mem}, S_{mem}/S_{cup}. This condition is well fulfilled by isodynamic and electrostatic headphones. Anomalies at middle frequencies (Fig. 14.34) are due to decoupling of various components in the cup. Therefore, a robust heavy cup is desirable with a resilient cushion (c_c minimal). The flesh resilience gives an upper limit of about 200 Hz for the main resonance.

(b) Rocking motion

In addition to the main vibration mode of a dynamic transducer membrane there usually appears an undesired one at less than twice the main resonance. Figure 14.35 should make this behaviour plausible. As is well known, a one-dimensional string shows modes with harmonic integer frequency ratios 1, 2, 3, etc.. A circular drum skin or thin membrane also shows vibration modes above the fundamental F1, but they are derived from Bessel functions, and have nothing to do with the integer series. F2 = 1.59 × F1. Next comes F3 = 2.14 × F1, which is a quadrupole mode, occurring before the next rotationally symmetrical mode F4 = 2.3 × F1. That being the case, there is no reason why the moving-coil transducer should be limited to F1.

Only rotationally symmetrical modes, such as F1, F4, F8 (=3.6 × F1), can inherently move air, and will normally be induced even in an ideal transducer, for both dynamic and electrostatic types. In all the other modes plus and minus cancel out. Therefore one may well expect the rocking mode F2 to be undetectable and harmless, even if present. Slight asymmetries in the magnetic field, which are inevitable, can induce it. Even then, it should still be undetectable. It is only when a second asymmetry in the loading occurs, for example in the compliance, that plus and minus no longer cancel, giving the characteristic notch in the response. The wires to the coil, which are usually led out on one side, are an obvious cause of asymmetrical loading.

A second consequence of rocking is non-linear distortion, even if no second asymmetry is present. The rocking mode is easily induced, since any porous resistances, which can only dampen modes which result in net air motion, are ineffective here. There is a high chance of the coil scraping the walls of the magnet gap, causing distortion.

14.2.7 Influence of cushion compliance

The influence of cushion porosity was treated in Section 14.2.4(b). Mechanical compliance in a cushion has two effects: first, the mechanical spring c_c effect of Section 14.2.6(a), and second, an acoustic breathing compliance C_b (Fig. 14.36). In a particular commercially available headset, $C_b = 21 \times 10^{-10}$ F (equivalent air volume 300 cm³), and $C_{vol} = 5 \times 10^{-10}$ F of the 70 cm³ cavity. This cushion causes a 12 dB loss in sensitivity below 1 kHz. To a first approximation a cushion can behave as a lossy passive

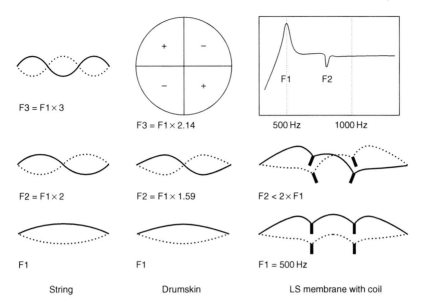

Figure 14.35. Vibration modes.

membrane, with R, L and C in series. Due to the reactance ωL_b, the cushion influence is small above 1 kHz.

14.2.8 Sound insulation

Especially for communications headphones, the sound insulation can be at least as important as the transducer acoustics. The insulation is invariably weakest at low frequencies, rarely exceeding 15 dB below 200 Hz. This is not due to leaks, which can be made negligible using fluid-filled cushions, but rather to vibration[96] of the cup or ear-piece as a rigid entity (Fig. 14.37(b)). The rocking motion[97] (Fig. 14.37(a)), which does not contribute to air compression, is of no consequence. Figure 14.38(a) illustrates the main elements governing vibration and Fig. 14.38(b) is the electrical equivalent circuit in the (F, v) system[98]. The cup mass vibrates on a lossy spring due to the cushion and flesh. These two compliant elements, which are in series mechanically (a), appear in parallel in the analogy (b). Figure 14.39 shows the calculations of Schröter[98], where the cup mass and flesh impedance are varied. The leaks are assumed to be negligible.

The fluid-filled cushions[97] provide excellent sealing. They must fulfil the requirements of adapting to head contours as well as behaving as a stiff spring. The cushion impedance depends primarily on the sheath material, which is only partly dilated with fluid. The limit of low-frequency attenuation is due to flesh impedance. Measurements of Shaw and Thiessen[97] demonstrate this (Fig. 14.40). The cushion was sufficiently stiff to yield a 28 dB transmission ratio on a rigid surface (curve I), but only 10 dB on a real head (curve II). Another transmission mechanism, the so-called flanking path, was postulated by Shaw[86] after observing, among other things, that curve II above resonance does not show the usual 12 dB/octave slope (dashed line, Fig. 14.40). In this mechanism the flesh transmits vibration waves along the surface under the cushion. At frequencies above 1 kHz, the attenuation can be made so good that the limit for hearing is governed by bone conduction. Cavity resonances are damped using fibrous materials[118], and cup resonances can be suppressed by using semiplastics with high internal viscosity[119]. See also Shaw[87].

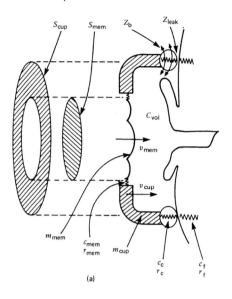

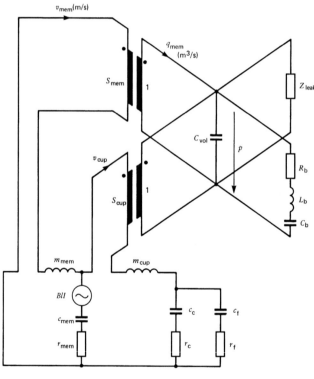

Figure 14.36. The essential elements for influence of cup vibration (a), and the equivalent circuit (b). For values of r_c, r_f, c_c, c_f, and m_{cup} see Fig. 14.38. The acoustic cushion elements R_b, L_b, C_b are also included. $R_b = 6 \times 10^5$W, $L_b = 600$ H, $C_b = 21 \times 10^{-10}$F, $C_{vol} = 5 \times 10^{-10}$F, $m_{cup} = 0.06$ kg.

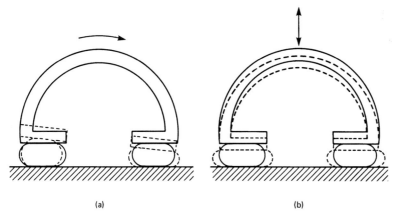

Figure 14.37. Modes of vibration of a hearing protector. (a) Rocking motion without air compression. (b) Translational vibration detrimental to attenuation. (From Shaw and Thiessen[97].)

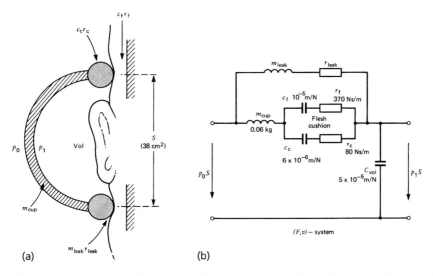

Figure 14.38. The significant acoustic elements for attenuation of a 60 g hearing protector: (a) schematic section; (b) equivalent circuit. The (F, v)-system is used. The ambient sound pressure p_0 is attenuated to the level p_1. (Values from Schröter[98].)

14.2.9 Small in-ear headphones

The acoustics of small in-ear headphones, used frequently with portable tape machines and CD players, is described here. The various constructions deal differently with the inevitable and desirable leak at the main sound outlet, which limits the low-frequency SPL. Without a leak, which is mainly inductive, the low frequency SPL in the pressure chamber, whose load is capacitive, is too high. In the following, RL and V represent bottlenecks and cavities respectively. The design shown in Fig. 14.41 illustrates one of the most common ways of dealing with the leak. RL8 is between the front surface of the earphone and the ear surface, described as a slit, which is, of course, not very reproducible. RL9 is the additional mainly resistive

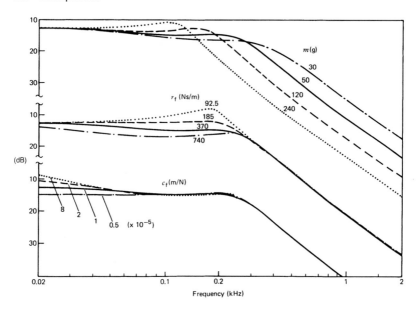

Figure 14.39. Influence of cup mass m and flesh resistance r_f and flesh compliance c_f on attenuation of hearing protector in Fig. 12.34. (From Schröter[98].)

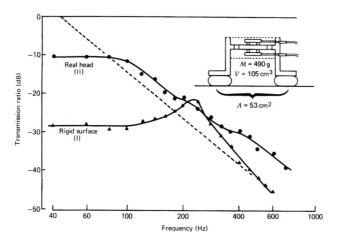

Figure 14.40. Rigid surface (I) and real head (II) transmission curves for an aluminium cup sealed by a water-filled PVC annulus. Inset shows the cup with built-in microphones. The broken line is the calculated mass line. (From Shaw and Thiessen[97].)

component due to the foam cover, usually supplied with the earphone, for hygiene. Another, more reproducible, way to deal with the leak is to funnel the sound into the ear canal with a rubber adapter, and make a leak using a paper resistance (inset of Fig. 14.41). This is indeed more reproducible but tends to be too closed, giving higher SPL below 1 kHz, since the inevitable impedance of RL8 is still present. Also, the bottleneck in the sound output causes a roll-off above 5 kHz. The sample described here is not claimed to be ideal. Rather, the purpose is to illustrate how

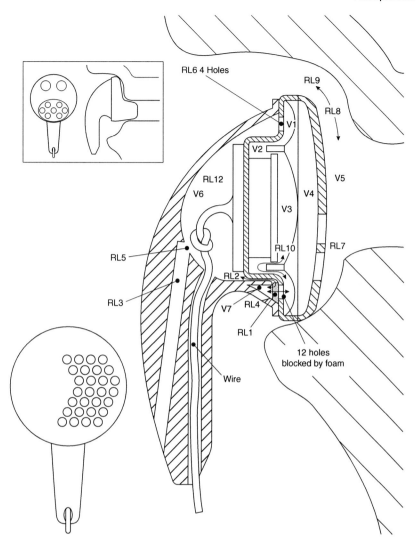

Figure 14.41. Representation of a small earphone applied to an ear. The upper inset shows another way of coupling the earphone to the ear. The two circles represent openings containing paper resistances. The lower small drawing depicts the front openings RL7 of the earphone investigated. The large drawing is not a rigorous construction drawing, for the purpose of demonstrating the functional elements.

the various elements contribute to the response. In the housing there are two main rear openings. One is a 3/4 circle of foam resistance, RL1, glued into the housing. The driver, whose rear has 16 holes, presses against RL1, 12 of these holes communicating with RL1, which leads to RL4, comprising a row of small openings of two types: holes and slits. The other is an inductive bass pipe, tube RL3. The simulation circuit is shown in Fig. 14.42 (main circuit) and Fig. 14.43 (driver). The wire leak is omitted here. Its effect is usually negligible, especially if there is a large bass pipe parallel to it. The knot can potentially block the bass pipe, thus making it less effective, as can the bottleneck RL5. This burr is presumably an unintentional result from

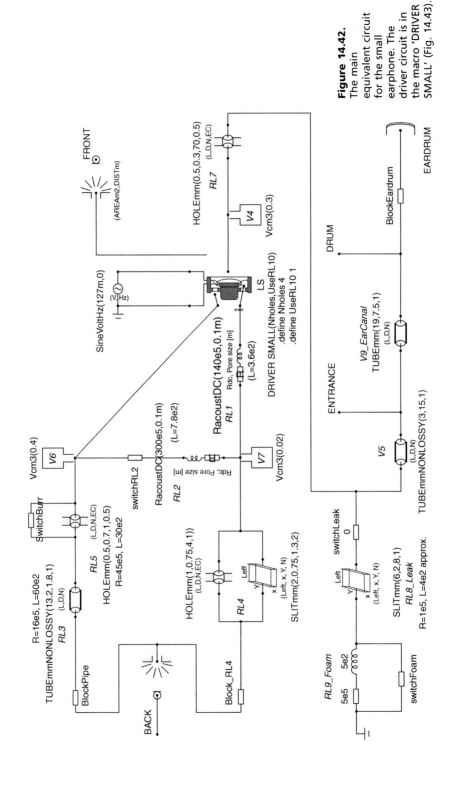

Figure 14.42.
The main equivalent circuit for the small earphone. The driver circuit is in the macro 'DRIVER SMALL' (Fig. 14.43).

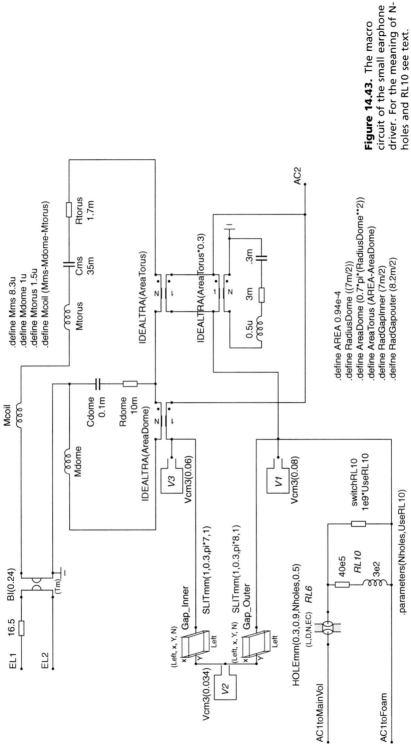

Figure 14.43. The macro circuit of the small earphone driver. For the meaning of N-holes and RL10 see text.

Main input

(Group 1)	IN VACUUM	IN AIR
FsExtra	195.7 Hz	179.4 Hz
Fs	294.2 Hz	252.8 Hz
Us	0.0608 V	0.0608 V
Is	1.51E-3 A	1.81E-3 A
F1	257.1 Hz	215.4 Hz
F2	334.0 Hz	295.0 Hz
Ir0	2.25E-3 A	2.46E-3 A

(Group 2) Main electro-mechanical Thiele–Small parameters		
Mms	8.3E-6 kg	10.6E-6 kg
Cms	35.14E-3 m/N	37.23E-3 m/N
Bl	0.24 Tm	0.25 Tm
Rms	2.7E-3 N/(m/s)	3.9E-3 N/(m/s)

(Group 3) More electro-mechanical Thiele–Small parameters		
SQRT(F1F2)	293.0 Hz	252.1 Hz
Accuranry	−0.39 %	−0.29 %
Qms	5.7	4.3
Qes	4.7	5.1
Qts	2.6	2.3
Zres	40.3 Ohm	33.6 Ohm
r0	2.2	1.8
SQWRT(r0)	1.5	1.4
Zr0	27.1 Ohm	24.7 Ohm

Mass Extra	10.5E-6 kg	

(Group 4) Acoustical parameters		
Vas	44.50E-6 m3	1
AREA	93.7E-6 m2	2
DIAMETER	10.9E-3 m	3
R(Air)	138.1E+3 Pa/(m3/s)	
L(Air)	263.7 kg/m4	

(Group 5) more electro-mech. parameters	
Mass(Air)	2.3E-6 kg
Cms(mean)	36.18E-3 m/N
Bl(mean)	0.24 Tm
Bl/sqrt(RdcLS)	0.06 Tm/sqrt(Ohm)

(Group 6) DC resistances	
RdcLS	16.5 Ohm
RdcTotal	18.2 Ohm
RdcAmmeter	1.7 Ohm

Figure 14.44. Thiele–Small measurements of the driver in the small earphone.

the injection moulding tool. The procedure to obtain the elements (described in Section 14.5.3) was that used for the large circumaural headphone of Section 14.2.4(b), theoretical responses adjusted to agree with measured responses. To save space, only the simulated responses are shown. Therefore the coupling volume V8 to the microphone, which was included in the circumaural headphone simulation (Section 14.2.4(b)), becomes of no interest and is omitted here.

Figure 14.44 shows the Thiele–Small parameters of the driver. Due to the low movable mass of 8 mg (hardly achievable without an aluminium coil) the low resonance Fs requires a very soft membrane: 35 mm/N. The wires emerging from the coil are glued to the membrane, a common technique to minimize rattling and dampen unwanted resonances, which would shorten the life of the transducer. Apart from lowering either of the Qms values, from which the mechanical resistance Rms was derived, this causes an asymmetry which can result in anomalies due to rocking (Section 14.2.6(b)), which, however, were not seen here. The air mass (difference between air and vacuum measurements) comes partly from the 16 rear holes and magnet gap.

Starting with the driver alone, Fig. 14.45 shows the responses at 12.5 cm (with 127 mV, for 1 mW) in front and behind a large baffle. The peak at 9 kHz in the rear output through the 16 holes shows the characteristic Helmholtz resonance. The front output would also show an anomaly at 9 kHz, if the dome were stiff. Closing the back (Fig. 14.46(a)) allowed the total volume (V1 + V2 + V3) under the membrane to be estimated. With the housing mounted, but with the rear openings RL3 and RL4 blocked (Fig. 14.46(b)), gave the main housing volume V6. One peculiarity of the driver macro is the resistive component R10, which turned out to be needed when the housing was mounted. The constriction RL6 is then asymmetrically placed, causing resistive losses for air flowing within the cavity V1 + V2 + V3, which did not occur when all 16 holes were open., and which also could not be accounted for by the inherent R of four holes, due to their large diameter. This effect, which is similar to the losses in a condenser microphone between membrane and electrode, could not have been isolated if the openings in the housing had also been active. RL10, which is approximated here with frequency independent R and L values, is switched on and off as appropriate.

The foam resistance RL1 does not purely lead the air into V7. There is also a slight leak sideways into the main cavity V6 via RL2. Both RL1 and RL2 were measured using an acoustic multimeter (Fig. 14.96 shows an example of such equipment). The element RacoustDC is explained in Section 14.2.2(b3). Opening both RL3 and RL4 gave front and back responses shown in Figs 14.47(a) and (b). The responses with these opened individually, not shown here, contributed to the circuit values used.

Now we can make the step from loudspeaker response (a) to the final headphone response (d). Evidently, they are not very different in shape. The response (d) was with the leak RL8 (without RL9 as yet). Letting the earphone pump on a pressure chamber gives the curve (c). The knee at 100 Hz is due to the bass pipe. The effect of blocking it is shown in Fig. 14.48. Blocking the openings RL4, as shown in Fig. 14.49, reveals the quality factor of the bass pipe dip resonance, with and without the burr, i.e. RL5. The burr, which is a narrow resistive bottleneck, makes the bass pipe less effective. The shift in dip resonance from 1 kHz to 800 Hz (RL4 assumed blocked) is due to the additional air mass in RL5. The length and diameter of the pipe are easily measured: 13.2 mm and 1.8 mm. Not even with the help of end corrections (Section 14.2.2(b2)) could the low value 800 Hz be accounted for, assuming an ideal pipe. This led to the discovery of the burr at the inner end. The same effect can come from the cable knot blocking the bass pipe.

Figure 14.50 shows the effect of neglecting the eardrum impedance. Figure 14.51 shows how using the foam cover element RL9, typically supplied with an earphone, can raise the bass SPL. Reduction in high frequency SPL due to part of the foam covering RL7, though detectable, is small by comparison. This is because the air path

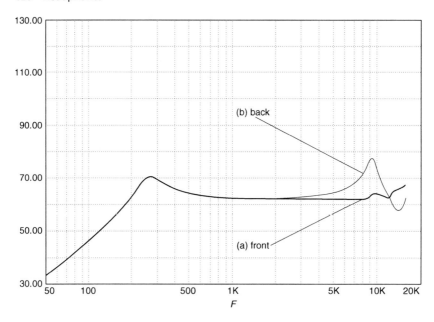

Figure 14.45. Baffle response of driver (a) front, (b) back.

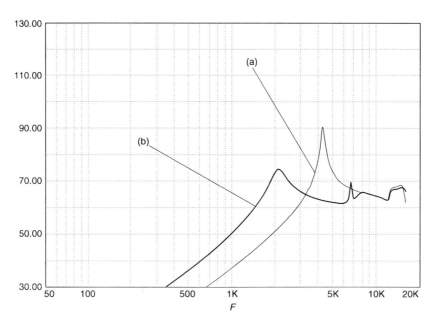

Figure 14.46. Baffle response of driver (a) with closed back, (b) with housing mounted but closed.

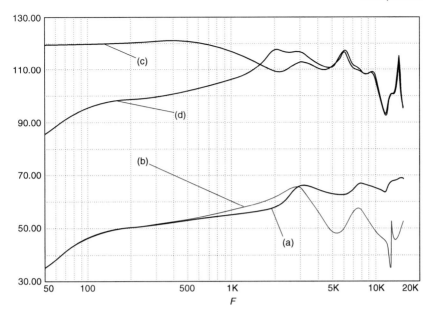

Figure 14.47. Baffle responses (a) front, (b) back with housing mounted and back openings active; the same on the ear (c,d). Leak RL8 blocked (c), active (d).

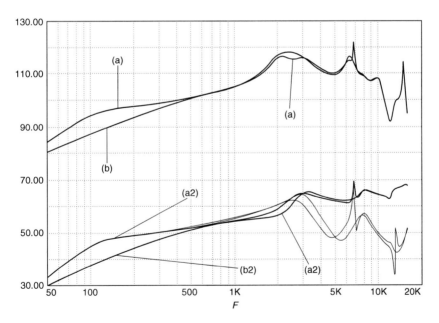

Figure 14.48. Ear response of earphone: (a) bass pipe open, (b) closed. The lower curves (2) show the same combination (without RL7) radiating from a large baffle (heavy curves: front, light curves: back).

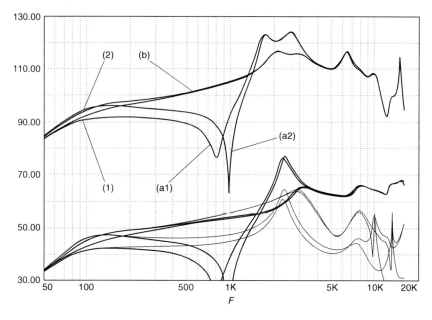

Figure 14.49. Ear response (a) RL4 blocked, (b) RL4 active. (1) means with burr, (2) means without. The equivalent curves on the baffle (without RL7) are shown also, below, thick curves: front, thin curves: back.

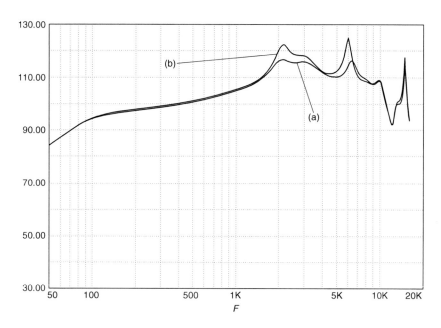

Figure 14.50. Ear response with eardrum impedance (a), and without (b).

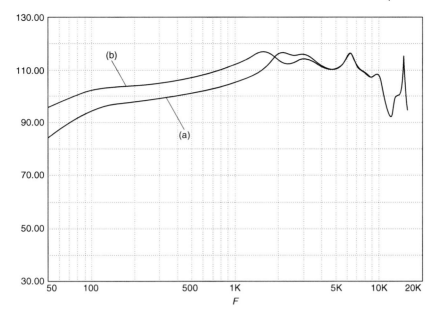

Figure 14.51. Ear response without the foam cover (a) and with it (b).

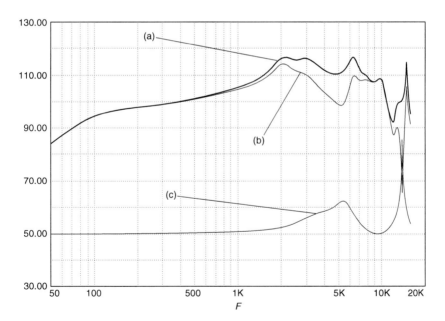

Figure 14.52. Ear response at drum (a) and at ear canal entrance (b). The transfer function from entrance to canal, ratio of pressures, is shown below (c) shifted up by 50 dB.

in RL9, influencing the LF range, is along the compressed foam, whereas the HF path out of RL7 is only across the foam thickness.

Finally, since transfer functions and responses are often referred to the ear canal entrance, Fig. 14.52 compares the response at the entrance with that at the drum. Also shown is the transfer function within the ear canal, shifted up by 50 dB to make it visible. The peak at 5 kHz corresponds to the ear canal length being a quarter wavelength. The double hump comes from using Pösselt's Circuit Fig. 14.92(a) rather than Fig. 14.92(b). The height of this peak 10 dB is governed by the eardrum impedance, to be compared with Fig. 14.64.

14.3 The hearing mechanism

14.3.1 The head and outer ear characteristics

We begin with the transfer function from free field to the ear canal entrance, which is relevant to binaural localization. That portion of the path between this point and the eardrum, which has no bearing on localization, is dealt with in Section 14.3.3.

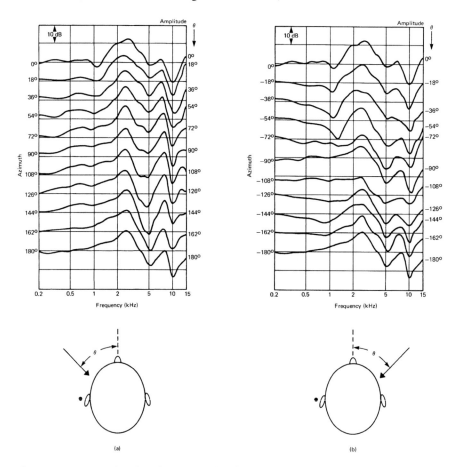

Figure 14.53. Amplitude of the transfer function from free-field to the ear canal entrance, sound incidence from the horizontal plane: (a) ear facing source; (b) ear in shadow zone. Frontal incidence curve at top. (From Mehrgardt and Mellert[151].)

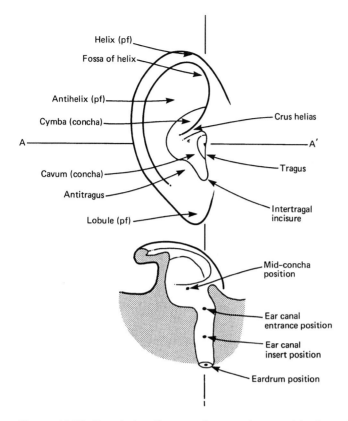

Figure 14.54. Descriptive diagram of external ear and horizontal cross-section at AA' (From Shaw[51].)

(a) Amplitude distortions

Sound is diffracted by the head, body and ears before reaching the eardrum[6]. Figures 14.53(a) and (b) show the SPL transformations from free field to the ear canal entrance for various incident directions in the horizontal plane[151]. These results were averaged by a special technique involving shifts along the frequency axis in order to preserve fine structure.

The minimum at 8–10 kHz depends on details of pinna geometry[80], in particular the cavum (concha), cymba and crus helias (Fig. 14.54). The minimum at 1200 Hz for frontal incidence comes from shoulder reflections[135]. It is absent for rear incidence. The maximum at 2–3 kHz is a property of the entire pinna geometry. At high frequencies for $\theta = 90°$, the average SPL is 6 dB above free field, consistent with pressure doubling due to an obstacle[1]. In the centre of the shadow zone ($\theta \sim -90°$) response is flat up to 2 kHz. For a more regular shape, the response would be flat to higher frequencies (white spot effect).

(b) Interaural time delay (ITD)

Phase delay   Interaural phase is of far greater importance than monaural phase effects (Section 14.4.2(b)). Phase-shift phenomena are expressed here as time delay according to:

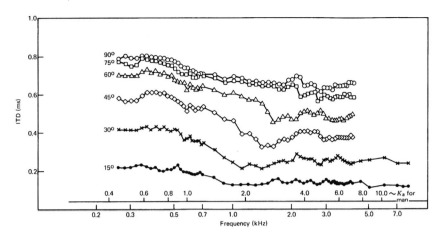

Figure 14.55. Interaural time delay ITD measured on a manikin's head surface as a function of frequency and angle of incidence with no torso. $K_a = 2\pi a/\lambda$ for effective head radius a and wavelength λ. (From Kuhn[52].)

$$\tau_{ph} = \frac{\phi}{2\pi f} \tag{14.11}$$

where positive and negative results here mean delay and anticipation respectively. Since we are not dealing at the moment with energy or signal envelope arrival times, negative values do not violate causality.

The results of Kuhn[52] are shown in Fig. 14.55 for various frontal angles of incidence. As expected, 90° incidence gives the largest interaural delay, of 0.8 ms at low frequencies. Low-frequency values are larger than high-frequency values by approximately 3/2. Kuhn defined the non-dimensional parameter:

$$\Pi = \frac{\text{ITD}}{a \, (\sin \theta)/c_0} \tag{14.12}$$

where a is the radius of the equivalent sphere and c_0 is the sound velocity. The data of Fig. 14.55, when plotted in terms of this normalized parameter, were reduced to a single curve, resulting in:

$$\Pi = \begin{cases} 3 \text{ for low frequencies} \\ 2 \text{ for high frequencies} \end{cases} \tag{14.13}$$

The two values 3 and 2 were accounted for by diffraction theory and the theory of creeping waves respectively (Kuhn[52]).

This distortion in the interaural delay does not result entirely from an increased delay around the head for low frequencies at the ear in the shadow zone. The ear facing the source shows an anticipating phase response – the phase arrives earlier due to the presence of the head in the sound field, as depicted schematically in Fig. 14.56 for a sphere. The incident and reflected components p_0 and p_1 are superposed, giving pressure doubling (Fig. 14.56(d)) at high frequencies for normal incidence. The lines of constant phase are drawn in towards the obstacle (Fig. 14.56(a)), giving the time-shift responses shown in Fig. 14.56(c). For completeness the group delay τ_{gr} (see next section) is also included. The equivalent circuit of Fig. 14.56(b) with $R\bar{C} = 0.56a/c_0$, accounts quantitatively for p, τ_{ph} and τ_{gr} for low and high frequencies, and qualitatively at mid-frequencies. For a somewhat different treatment of this problem see Bauer[92]. The drawing in of lines of constant phase towards the obstacle happens also in the shadow zone, causing a genuine delay at the far ear.

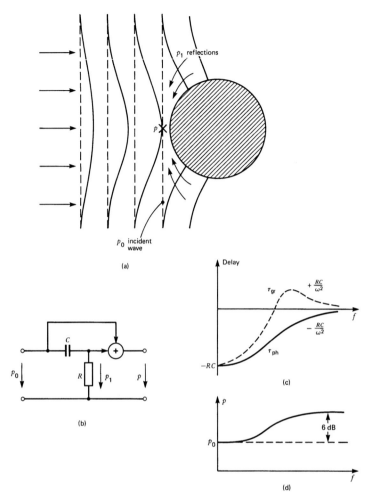

Figure 14.56. (a) Descriptive diagram demonstrating phase anticipation at the ear facing the sound source. (b) Equivalent two-port circuit with $RC = 0.56a/c_0$, a = radius of equivalent sphere, c_0 = sound velocity. (c) Time 'delay' for phase and group delay. (d) Amplitude function. R, C can be replaced by L, R. Then $L/R = 0.56a/c_0$.

Envelope delay As regards the arrival times of the ear signal envelopes, whose ITD is especially important at high frequencies (see next section), the group delay τ_{gr} is generally used, as defined by:

$$\tau_{gr} = \frac{1}{2\pi} \frac{d\phi}{df} \tag{14.14}$$

Heyser [44] has shown that the group delay as defined in equation (14.14) is a meaningful measure of signal arrival time only for an all-pass transmission system, i.e. with a flat amplitude response. For non-all-pass systems, negative group delays are frequently obtained which are non-causal and obviously not a measure of signal arrival time. Calculations can also yield positive values, which are often identified with signal delay. However, causality alone is not a reliable indication of the correctness of the value obtained. A preferred method is to measure signal arrival times empirically by envelope detection using band-limited impulses.

14.3.2 Sound localization

Out-of-head localization of headphone sound images is not in itself a problem as long as the simulated source direction is far left or right. The main problem lies in the out-of-head localization of sources near the median plane where front–back discrimination is not reliable. For a comprehensive review of localization see Blauert[11,45,46].

(a) Left–right hearing

The discrimination of sidedness in localization is otherwise known as lateralization[16,17,46]. However, this also includes the left–right perception of acoustic images in the head. The projection of the image out of the head normally requires that both the interaural amplitude and arrival time differences approximate those of a real source.

Low frequencies Low-frequency sources result in ear signals of practically equal amplitude. Even these sources can be localized. This applies both to pure sine waves under anechoic conditions and to more complex signals. The main cue available for sine waves is the phase-derived interaural time delay (ITD). This is unambiguous only at low frequencies (Fig. 14.57(a)). The electrical impulses sent to the brain from the basilar membrane of the cochlea (Section 14.3.3(b)) are triggered every second half-cycle during rarefaction. For higher frequencies such as 900 Hz, ambiguity sets in. It is then not clear which ear signal is leading, as far as ITD is concerned. At 1250 Hz (phase shift 180°) the ambiguity is complete. Interaural amplitude difference (IAD) then becomes more significant. Indeed, investigations with sinusoidal signals have revealed double images, known as time image and intensity image. Time images are relatively stable with respect to IAD changes, whereas intensity images can be offset by IAD[10]. The latter effect is known as time-intensity trading[47], i.e. the sidedness due to ITD can be neutralized to a large extent by an increased intensity of the delayed signal (5 dB is sufficient to bring the lateralized image due to an ITD of 0.5 ms back into the median plane[45,46]). Complex signals in natural hearing provide other cues (see next section) which can override the phase-derived interaural time delay. If this were not the case, then additional phase distortion due to room resonances would disturb the localization of the source (law of the first wave front, Section 14.3.2(d)). However, for artificial out-of-head localization of headphone sound images, simulation of phase delay (Section 14.4.4) for low frequencies is indispensable. Comparing intensity stereophony with time-delay stereophony at low frequencies under non-reverberant conditions reveals the following curious effect: an amplitude difference in the loudspeaker signals produces a phase shift in the ear signals and vice versa. Vector addition as shown in Fig. 14.58 taken from Blauert[45,46] demonstrates this. Intensity stereophony is therefore more compatible with natural hearing than time-delay stereophony, since the latter gives in-phase ear signals of unequal amplitude at low frequencies. Moreover, the amplitude difference is the inverse of that expected from shadowing in natural hearing (Fig. 14.58(b)).

High frequencies and complex signals At frequencies above about 1600 Hz where phase delay information is ambiguous, the fine structure of a waveform is ignored and attention is focused on the envelope delay. There are three kinds of ITD[12]: (a) the onset flank ITD is important for brief impulses. For sound longer than a few milliseconds there are ongoing ITDs: (b) for sinusoidal components of a signal ITD equals the interaural delay of the fine structure (e.g. zero crossings) which is only audible below 1600 Hz: (c) for complex waveforms there are also ongoing envelope delays which are detectable in the whole audio range. For details refer to the

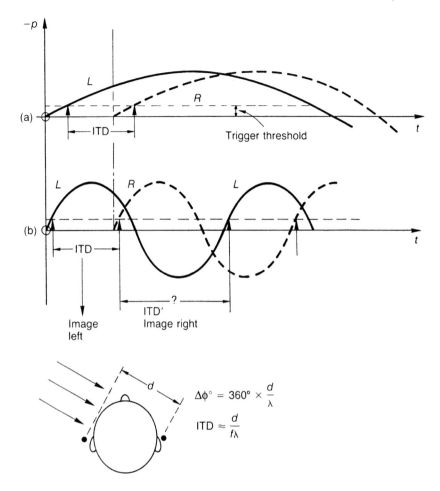

Figure 14.57. Phase-derived ITD for (a) low frequencies, (b) high frequencies giving ambiguous left–right information. The auditory neurons fire during rarefaction half-cycles -p.

literature[13,14,48,50,94]. Figure 14.59 shows the frequency ranges for the main interaural lateralization cues.

Threshold for a binaural image Headphone experiments have been reported[15] aimed at determining the critical threshold for the appearance of a binaural image. It was shown that the decisive factor is not the interaural amplitude difference (IAD), but rather the absolute amplitude of the weaker of the two ear signals, whose threshold is 1–6 dB above the monaural audibility threshold. This is as in stereoscopy, where the weaker of the two images is still able to contribute to 3D, even when it is extremely dark.

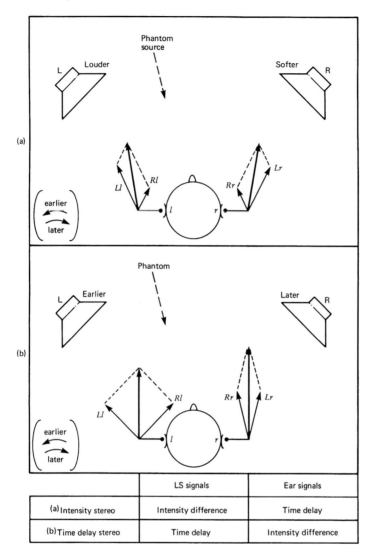

	LS signals	Ear signals
(a) Intensity stereo	Intensity difference	Time delay
(b) Time delay stereo	Time delay	Intensity difference

Figure 14.58. Vectorial superposition of the two loudspeaker contributions at each ear for (a) intensity- and (b) time delay-stereophony. This applies to low frequencies where the head shadow gives phase delays but no attenuation. (After Blauert[45,46].)

(b) Front-back hearing

Front–back discrimination is a special case of localization in the median plane, including elevation perception. Unlike left–right hearing, which rests on a good theoretical foundation involving only physical parameters, front–back hearing is only partly understood. The many relevant parameters belong both to the realms of physical acoustics and psychoacoustics. The various parameters can take turns in dominating the localization, depending on the listening conditions. Otherwise well-reproduced physical localization parameters can be overridden by other factors, such as visual cues and familiarity with the acoustics of the listening room.

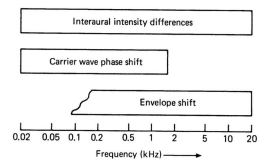

Figure 14.59. Frequency ranges for influence of various interaural cues on left–right hearing. (After Blauert[45,46].)

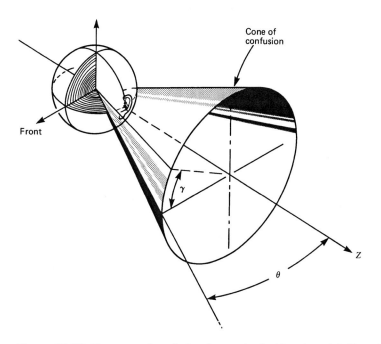

Figure 14.60. The cone of confusion for a spherical head model. (From Searle *et al.*[18].)

Cone of confusion A sound source produces the interaural time and amplitude cues ITD and IAD. The direction is perceived primarily using ITD, which is relatively stable. However, the information is ambiguous since a given ITD could arise from a source at any point on a cone[18,54] whose axis coincides with the line joining the ears (Fig. 14.60). In the localization task the probable source direction is narrowed down further using IADs. However, IADs are less stable than ITDs, since IADs are more susceptible to influences arising from posture and reflecting obstacles. The errors in localization are thus confined to the cone. A special case is the median plane (cone angle $\theta = 90°$): in front is frequently confused with behind (front-back inversion) or above (elevation error). Another special case ($\theta = 0°$ or 180°) is given by a source far left or right. This is the only case where the ITD information alone is sufficient for a unique localization.

Binaural and monaural cues ITD and IAD as used in lateralization are binaural cues, since their definition must involve both ears. For different angles in the median-plane, the spectral content of the ear signals varies. This quantity, assuming identical ears, can be meaningfully defined for each ear independently and is therefore primarily a monaural cue.

Directional bands Blauert[54] developed the directional-band model of median-plane localization. In this monaural cue model, the spectral content of a signal is responsible for the perceived direction. Each frequency band is associated with a given direction. Boosting one band at the expense of the others gives an increased tendency for localization in the direction of the boosted band. The frequency ranges of the directional bands are as follows:

Front: 260 Hz-550 Hz, 2.5 kHz-6 kHz
Rear: 700 Hz-1.8 kHz, 10 kHz-13 kHz
Above: 7 kHz-10 kHz.

Psychoacoustic experiments with narrow-band sound sources in the median plane without head movements have shown[54] that the direction of the sound sensation depends only on frequency according to the above scheme, and does not depend on the actual direction of the source. Figure 14.61 implies that this model has more than a purely empirical foundation. The SPL is plotted at the ear with loudspeaker in front, minus SPL at the ear with loudspeaker behind. The positive and negative values correlate with the frontal and rear bands fairly well. One may infer that daily experience with real sources, coupled with a subconscious learning process, has led to the directional bands in any individual which may be used for localizing familiar sounds.

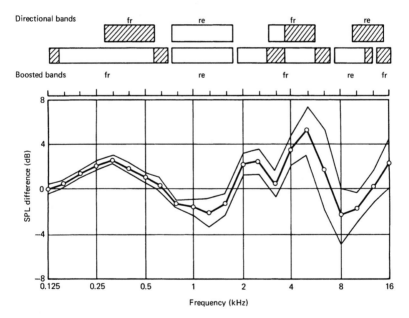

Figure 14.61. SPL at the eardrum with loudspeaker in front minus that for loudspeaker at rear. (Mean of 10 subjects and confidence interval of the mean at a 95% level of significance.) For the boosted bands (fr: front, re: rear) the majority of subjects have a higher SPL at the eardrum for sound incidence from the respective direction. Border areas: 95% level of significance; shaded areas: most likely. The subjective directional bands are also included. (After Blauert[54].)

That the directional-band model is not the only mechanism for frontal localization is shown by the fact that boosting the frontal bands in headphone listening with a wide-band mono signal does not give a headphone sound image in front. Exaggerating the IAD spectrum of the frontal IIRTFs of an individual can improve frontal localization and has also been replaced[26] to reduce front–back confusion.

Head movement In natural hearing, head movements provide a powerful though not essential dynamic localization mechanism. Both binaural ITDs and IADs as well as monaural spectra are varied synchronously, giving a pattern of variation which uniquely defines the source direction. Sound source motion also provides an effective cue.

Elevation effect The distinct minimum at 8–10 kHz (Fig. 14.53) has been shown to determine the apparent elevation of the acoustic image, which rises with increasing frequency. Measurements[59,60] show a shift of the minimum from 8 kHz for $-15°$ source elevation up to 11.5 kHz for 45°. The elevation effect is not confined to the median plane. This is a dynamic monaural cue – a motionless minimum has no effect.

Monaural pinna cues Models of localization assuming purely monaural cues have a fundamental drawback. They require the listener to be familiar with the sound source for successful localization, unless there is motion. Otherwise, he has no means of knowing whether the perceived spectral colouring comes from his own ear filter or is an inherent property of the source.

Binaural pinna disparity This has been extensively investigated[53,55,56,90]. A binaural cue in which spectral differences of the ear signals are processed to determine source position was reported by Searle *et al.*[18,57] to be as important as the monaural cue. Genuit[116] also found evidence for the binaural cue.

The significance of the binaural cue is demonstrated by the following experiment. A binaural recording is made using the listener's own head instead of an artificial one (see Section 14.4.4(b) for binaural technique). Small electret microphones in the listener's ear canals detect the ear signals for a wide-band source in front. The procedure is repeated with other persons. On playback via symmetrical headphones, the listener easily recognizes his own recording, since he hears it in front. Other recordings are heard elevated or in-head localized (IHL). However, listening to his own recording with the left and right channels reversed gives the worst IHL. It appears that any statistical choice of interaural spectral cues has more in common with his own ears than his own reversed. The headphone response is of secondary importance. Reversing channels has no influence on localization of sources behind.

Bone conduction Although bone conduction (BC) has often been put forward as the missing link to explain frontal localization, it is not generally believed today to have any influence on unoccluded free-field hearing. The argument is simple – due to its attenuation 50 dB below air conduction (AC), its influence must be masked by the latter. One of the proponents of the bone conduction influence was Scherer[4,36,37.]. In his model, which does not require AC and BC to compete additively, the AC component is FM-modulated by BC shaking the cochlea. Sone *et al.*[102] report that frontal out-of-head localization with hearing protector headphones (Section 14.4.3) can be achieved by irradiating the head with a loudspeaker in front of the listener, but not fixed to him. The headphone signal (AC) was obtained from two microphones next to the listener, with SPL adjusted to be equal to that outside the headphones. The arrival time of the BC component was critical. It could be misadjusted by leaning back. However, they were unable to achieve out-of-head localization using BC vibrators on the forehead.

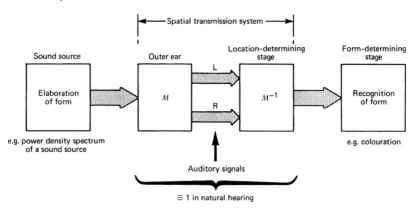

Figure 14.62. Association model for spatial transmission of hearing. (From Theile[20].)

(c) Association model

In the association model of Theile[20] attention is focused on the effects of the linear distortions arising from the head diffraction in a sound field. It is stated that these spectral distortions are perceived as an unnatural colouring of the acoustic image if the correct localization process does not take place. The model is depicted in Fig. 14.62. The spectral distortions arising from the source direction are symbolized by a filtering operation M which gives the linearly distorted ear signals. If these distortions are recognized for what they are, and associated with the correct source position, then the brain contributes an inverse filtering M^{-1}. The source is then perceived uncoloured. If the localization mechanism is somehow suppressed, then the inverse filtering stage is omitted. The spectral distortions M are then perceived as an inherent property of the source, thus resulting in coloration of the image.

(d) The precedence effect

The precedence effect[24], otherwise known as the law of the first wavefront[22] or the Haas effect (see also Chapter 10), has made a great impact on audio engineering and room acoustics. The original paper of Haas[23] in German has been translated into English[58]. The effect was also discovered independently by Wallach et al.[24]. If a primary sound is followed within a limited time by a similar sound from another direction, e.g. a reflection, then only one sound is perceived. For a delay in excess of 0.6 ms, the localization is determined purely by the primary sound. Thus room reflections contribute only to a sensation of spaciousness without interfering with the localization. For less than 0.6 ms delay, the apparent direction is determined by both primary and secondary sounds. For complex sounds such as speech and for single clicks, the echo thresholds are reached for delays of 40 ms (Fig. 14.63) and 5 ms respectively.

The precedence effect is most effective for sounds with discontinuous or transient properties[24], such as piano music or speech, not for steady tones or continuous noise. The start of a signal is clearly of importance in the precedence effect. Abel and Kunov[13], however, report (see also Section 14.3.2a) that not only are onset envelope and intensity information used in the precedence effect but also phase in the initial portion of the signal. They conclude that 'for signals without strong envelope information, an initial portion, as opposed to the exact moment in time when the signal starts, is used'. In the model of Zurek[81] both the precedence effect and the ability of the auditory system to choose the right interaural phase delay (Section 14.3.2a) are aspects of a more fundamental process involving time windowing (window duration 0.7 ms, sampling every 10 ms).

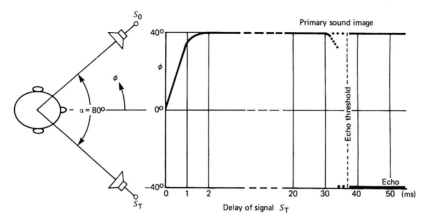

Figure 14.63. Direction heard as function of delay of signal S_T, schematic, for speech $S_O = S_T = 50$ dB. (From Blauert[45,46].)

Although the influence of reverberation on localization is essentially suppressed by the precedence effect, out-of-head localization for headphone listening was reported by Sakamoto *et al.*[21] to be improved by adding reverberation, and depends on the acoustic energy density ratio of reflected sound to direct sound.

14.3.3 Behind the ear canal entrance

The ear canal plays no part in localization[138,151]. It serves essentially as an acoustic transmission line from the ear canal entrance to the eardrum. The same transfer function (Fig. 14.64) applies irrespective of the original direction of incidence up to at least 10 kHz, the same polar diagram being obtained in front of a blocked ear canal and at the eardrum of an unoccluded ear canal[51]. A systematic investigation[112] on humans and on an artificial head has confirmed this. Thus the ear canal input can be treated as a two-pole in the electrical analogy. At about 3 kHz the quarter-wave resonance occurs, giving a 10 dB amplification. This, together with 12 dB from the transfer function of the external ear (Fig. 14.53), results in a maximal sound pressure at the eardrum about 20 dB[87] above the free-field value. The standing-wave ratio rises from 10 dB at about 3 kHz to 20 dB at 6 kHz and above[28]. The impedance of a typical eardrum is shown in Fig. 14.65. The low-frequency behaviour[60] is that of a compliant membrane up to 800 Hz. Above 2 kHz the eardrum divides spontaneously into zones vibrating independently[51,61,182]. The impedance at high frequencies was measured by Hudde[25], taking into account the non-uniform ear canal cross-section. The result cannot be interpreted theoretically, first, because of the complicated membrane motion, and second, due to conceptual difficulties in defining impedance Z at a point near the inclined eardrum where the sound waves are not plane; a shift of reference point by 1 mm at 10 kHz can change a value from primarily real to imaginary. Therefore, the reflectance is a preferable measure of eardrum behaviour at high frequencies (Fig. 14.66) as defined[25] by:

$$r = \frac{(Z/Z_W) - 1}{(Z/Z_W) + 1} \tag{14.15}$$

where Z_w is the wave impedance $\rho c/S$ at a plane of area S near the eardrum. For $S = 0.55$ cm^2, $Z_w = 9.2$ MΩ. Only the phase of r is influenced by the position of the reference plane. The linear phase (Fig. 14.66) at high frequencies is interpreted as arising from a reflection behind the eardrum.

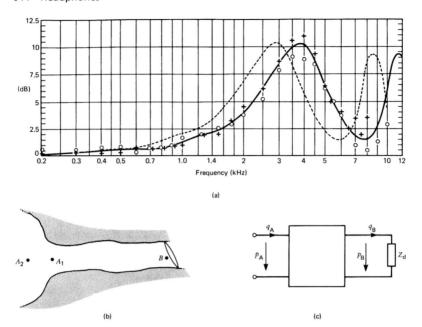

Figure 14.64. (a) Amplitude of transfer function from ear canal entrance A_1, (solid line) or mid-concha position A_2 (broken line) to the eardrum B. (From Shaw[29].) The ear canal (b) is treated as a two-port transmission line terminated by the eardrum impedance Z_d.

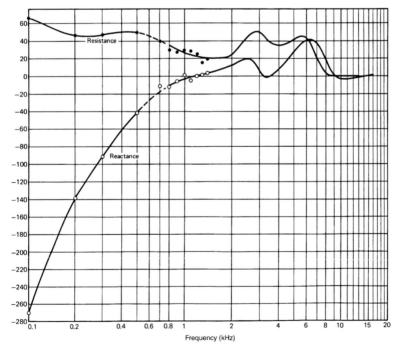

Figure 14.65. The eardrum impedance. Results of Zwislocki[60] up to 1 kHz and Mehrgardt and Mellert[151] above 1 kHz.

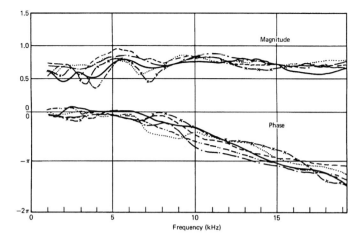

Figure 14.66. Eardrum reflectances of six ears. (From Hudde.[25])

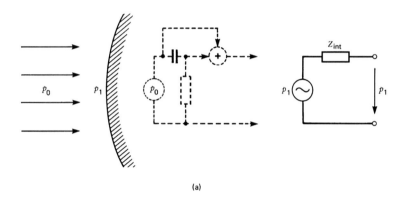

(a)

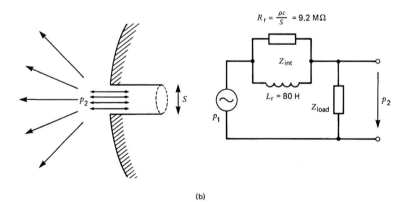

(b)

Figure 14.67. (a) The free-field source p_1, at the surface of a rigid obstacle. In the absence of the obstacle pressure is p_0. (Fig. 14.56). Surface fragment S is occluded. (b) Removal of the occlusion reveals the internal impedance of the source p_1. This is the radiation impedance Z_r (Fig. 14.2(e), Section 14.2.2).

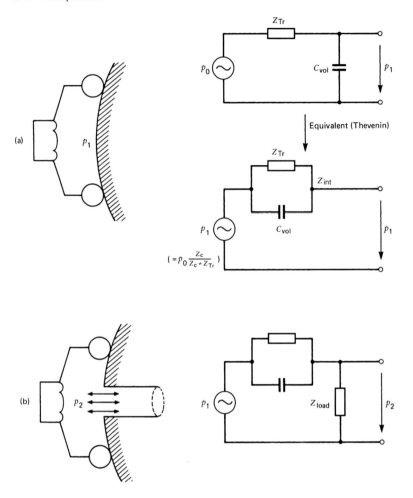

Figure 14.68. Coupling of a headphone to the ear: (a) occluded; (b) unoccluded. For f < 2 kHz $p_2 = p_1$.

(a) Acoustic loading by the eardrum

The following are two extreme cases of acoustic loading: (a) free-field sources, and (b) hearing aids. In the former, the exact value of the eardrum impedance has little influence[51] on the sound pressure in the ear canal below 2 kHz (Fig. 14.67). In the latter, the influence is significant for all frequencies due to the relatively high internal impedance of the transducer. Headphones represent a loading situation between these two extremes (Fig. 14.68). Although the internal transducer impedance Z_{Tr} can also be high, the coupling volume of about 30 cm³ is much larger than the equivalent volume of the eardrum compliance (0.8 cm³) . Thus Z_{drum} has little influence below 2 kHz, since the impedance of the parallel combination of C_{vol} and Z_{Tr} of the Thevenin source is low (Fig. 14.68(b))

At about 3 kHz, where the quarter-wavelength resonance occurs, the transmission line properties of the ear canal are important. The input impedance of the canal is given by:

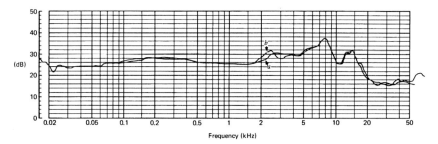

Figure 14.69. Frequency response of a headphone measured at the ear canal entrance: (a) normal; (b) eardrum stiffened on inside by excess pressure via the eustachian tube.

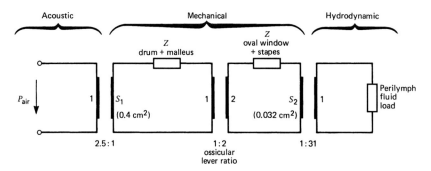

Figure 14.70. Approximate equivalent circuit for the middle ear (CGS units).

$$Z_{\text{load}} = \frac{Z_w^2}{Z_{\text{drum}}} \tag{14.16}$$

Thus the harder the termination, the lower the input impedance of the canal entrance. At this frequency, the pressure in the ear canal is more sensitive to the exact value of Z_{drum} than at any other frequency. This applies also to ¾ wavelength, ¾, etc. With $Z_w = 9.2\ M\Omega$, and taking typical values of $Z_{\text{drum}} = 20$ and $40\ M\Omega$, one obtains 4.2 and 2.1 $M\Omega$ respectively for the load impedance. Figure 14.69 qualitatively shows this effect. The frequency response of a headphone was measured with a probe microphone at the ear canal input under normal conditions (curve a) and with the eardrum stiffened (curve b) by an excess internal pressure produced by the lungs through the Eustachian tube with closed nostrils. The resulting abnormal eardrum impedance influences the frequency response only around 2–3 kHz.

(b) The middle and inner ear

The chain of elements (Fig. 14.70) from the ear canal through the middle ear to the basilar membrane of the inner ear (cochlea) constitutes a series of transformers for acoustic, mechanical and hydrodynamic vibrations. In the cochlea the transduction process into electrical impulses takes place.

The middle ear as a transformer The sound pressure in front of the eardrum (effective area $S_1 = 0.4\ cm^2$) is transformed by the ossicular chain into the pressure of the peri-lymphatic fluid behind the oval window ($S_2 = 0.032\ cm^2$) of the cochlea. In addition to the step-up transformation $S_2 : S_1 \simeq 1 : 31$, there is a mechanical transformation of

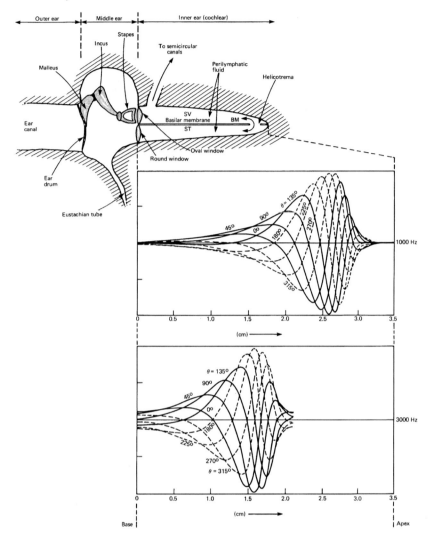

Figure 14.71. The inner ear (cochlea), showing vibrational profile of the basilar membrane for two frequencies (curves of Bogert[65]). The position of the envelope maximum depends on frequency.

1:2 due to the lever action of the ossicles, giving a resultant ratio[67] of 1:25. An approximate equivalent circuit for this process is shown in Fig. 14.70 (the acousto-mechanical transformers are not dimensionless; see Section 14.2.2).

Static pressure equalization Variations in ambient air pressure cause a bulging of the eardrum which lowers its sensitivity especially for low frequencies. This is equalized from time to time through the Eustachian tube (Fig. 14.71) as a result of jaw movements. In its normal state the Eustachian tube is closed.

The acoustic reflex The middle ear also functions as a limiter to protect the cochlea from overstimulation[68]. In this feedback mechanism two muscles in the middle ear are activated by the brain to fix the malleus and stapes (Fig. 14.71). This reflex begins at an SPL of 80–90 dB and reaches its full strength above 100 dB. Unfortunately, the reaction time of 10 ms is too long to offer protection against short impulses.

The inner ear (cochlea) The cochlea consists essentially of two fluid-filled ducts separated by the basilar membrane (Fig. 14.71). The structure is coiled (almost three turns) and is embedded in bone, but is shown here unwound to demonstrate its function. The stapes is connected to the oval window, which is the normal signal input. The more compliant round window is for pressure relief. At the apex, the two ducts (scala vestibuli, (SV) and scala tympani, (ST)) are connected through an opening (helicotrema). The SV is connected to the semicircular canals of the vestibular apparatus.

Distributed along the SV/ST partition are hair cells which detect displacement of the basilar membrane (BM). The electrical impulses are transmitted along 30 000 auditory nerve fibres[69]. Different fibres have different stimulation thresholds. The number of discharges of one fibre increases with loudness up to a maximum of 100 per second. At about 40 dB above loudness threshold, some nerve fibres saturate while others are just coming into action. In the theory of Howes[69], considering all fibres, the resultant waveform for a tone is approximately half-wave rectified, making the auditory system a squaring device. In the absence of a signal, the firing rate is low but not zero. Downward BM excursions suppress the firing.

Spectral-analysis in the cochlea The BM, which becomes wider towards the apex, represents a tapered dispersive transmission line. Travelling waves (Békésy[70]) propagate from the base to the apex. For a given frequency, these have an envelope maximum at some position on the BM. Increasing frequency shifts this maximum towards the base. This is demonstrated in Fig. 14.71 for two frequencies. Thus a spatial position is allocated to each frequency. The response is further sharpened[71] by a feedback mechanism from the brain which inhibits the neural firing on either side of the envelope maximum. However, this form of filtering is still not sharp enough to account fully for pitch perception, which is to a large extent a time-domain phenomenon.

(c) Bone conduction

In the normal mechanism of hearing, air conduction (AC), the foundation to which the eardrum is attached, as well as the cochlea casing and skull, are essentially stationary. The small vibrations have not been shown to influence the sound image. Where they do have an influence, one speaks of bone conduction (BC), which is normally regarded as an unwanted effect[67]. Noises originating in an animal's own body, such as chewing and breathing, are heard via BC. It serves only to decrease the signal-to-noise ratio for a creature which might be anxious to detect the approach of a predator. A damping factor BC/AC of 45–55 dB has been found. This is analogous to the problem in microphone technology of suppressing the sensitivity to handling noise. The BC/AC ratio was obtained from free-field threshold measurements[62,95,129]. The skull vibrates in its first dipole mode as a rigid mass at frequencies below 500 Hz. At higher frequencies, flexural distortions occur, giving resonances of a quadrupole nature which are well established in the range 1–1.6 kHZ[63,66] (Fig. 14.72).

The following three mechanisms lead to cochlea stimulation by BC:

(a) cochlea compression;
(b) inertia;
(c) air conduction detour.

Cochlea compression Figure 14.73, taken from Békésy[66], demonstrates this mechanism schematically. If the two cochlea ducts and windows were identical (a), a

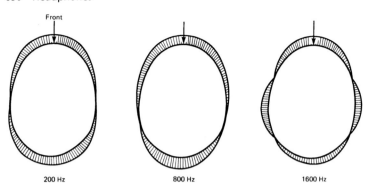

Figure 14.72. Vibrational modes of the skull for vibration applied to the forehead. (From Békésy[70].)

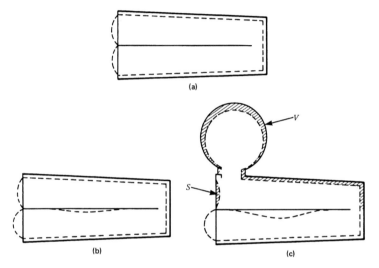

Figure 14.73. Mechanism of cochlea compression: (a) symmetrical; (b) and (c) progressive introduction of asymmetry gives basilar membrane deflection. (From Békésy[66].) S= oval window, V= vestibular apparatus.

compression would serve only to deflect the windows without moving the basilar membrane. Making the oval window stiffer allows the BM to respond to compression (b). Introducing an additional asymmetry by connecting the vestibular apparatus, which also contributes compression, enhances the response of the BM (c).

The signal input is no longer the oval window as in air conduction. However, travelling waves are still generated, propagating in the same direction irrespective of the location of the input[65,67,75,76]. This strange situation, where the waves propagate towards their own source, has been theoretically explained[67,75] and shown to be in line with Hamilton's principle[67]. Thus BC gives rise to the same type of response as AC. This explains the possibility of AC/BC cancellation experiments[32].

Audibility of ultrasound   Ultrasound up to at least 100 kHz can be heard by BC[30,31]. The pitch is constant from 15 to 100 kHz. The normal limit of hearing at around 18 kHz comes from the transmission limitation of the middle ear due to inertia of the ossicles.

Inertia mechanisms Vibration of the temporal bone to which the ossicles are attached results in a force from the stapes against the oval window. This inertia component of BC is negligible below 800 Hz[74] and above 2 kHz[32]. Inertia of the perilymphatic fluid in the cochlea has also been put forward as a mechanism for BC[77].

Air conduction detour (ACD) An important mechanism involves skull vibrations first being transformed in the ear canal into airborne sound due to relative motion between the jaw bone and skull[73]. This is then heard by the normal mechanism via the eardrum and ossicles (Fig. 14.74(a)). From threshold and cancellation experiments, Khanna *et al.*[32] conclude that this component is equal to the compressional component below 900 Hz. At isolated frequencies from 600 to 700 Hz, ACD can be larger. Above 900 Hz the inner ear component dominates the total BC response.

(d) The occlusion effect

The ACD component is the foundation of the occlusion effect. In the unoccluded state it can only just compete with cochlea compression. With ear occlusion (e.g. by an ear muff or headphone), however, it dominates. This is because the SPL in the ear canal arising from the vibrations of its own walls (current source[95]) is no longer determined by the low radiation impedance but by the high reactance of the pressure chamber (Section 14.2.1). This is illustrated in Fig. 14.74. Amplification of BC by ear occlusion occurs up to 2 kHz[66] (Fig. 14.75). The effect can easily be demonstrated by closing the ear canal while humming a low tone. Clenching the jaws weakens this effect as expected. The occlusion effect is negligible for volumes in

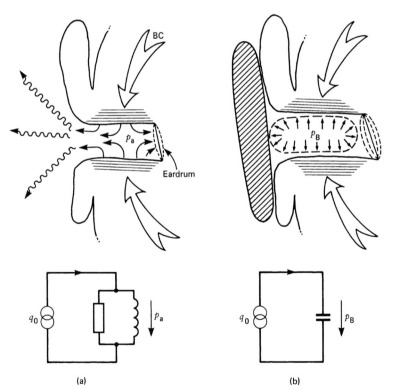

Figure 14.74. The occlusion effect: (a) pressure in canal partially lost by radiation; (b) pressure build-up due to occlusion.

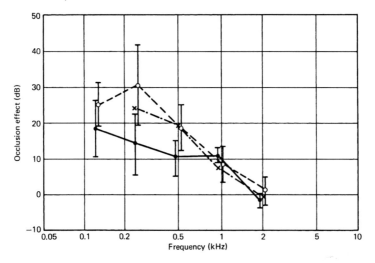

Figure 14.75. The occlusion effect as a function of frequency for supra-aural head-phones, measured by threshold shifts. Bone conduction vibrator on (a) mastoid, open circles[35], (b) forehead, filled circles[35], open circles[168]. (From Brinkmann and Richter[35].)

excess of 2 litres[32,66]. The occlusion effect measured by threshold shifts is at most 25 dB. Objective ear canal SPL measurements[95], however, reveal SPL shifts due to occlusion of up to 40 dB at 150 Hz.

14.4 Headphone applications

14.4.1 High-fidelity headphones (normal application)

In normal application, headphones are fed with stereophonic signals originally intended for loudspeaker reproduction. The rigorous definition of high fidelity implies that the listening experience should be indistinguishable from the original. This is a legitimate aim in dummy head stereophony (Section 14.4.4(b)), but not here (mainly due to in-head localization). Despite this, the concept of hi-fi is widely used. It appears to have undergone a change in its definition to accommodate high-quality headphones with misused loudspeaker signals. In this adapted definition, it is sufficient that the listening experience be brilliant and pleasing. Exaggerated brilliance, however, results in unnatural colouring of the acoustic image. Thus a hinting reference to the original concept of hi-fi is still maintained, but only as regards the sound quality, not its localization. As mentioned in Section 14.1, a flat frequency response of sound-pressure level (SPL) is rarely the design objective.

(a) Free-field response

For want of any better reference, the various international and other standards[114,115,169–172] have set up the following requirement for high-fidelity head-phones: the frequency response and perceived loudness for a constant voltage mono signal input is to approximate that of a flat response loudspeaker in front of the listener under anechoic conditions. The free-field (FF) transfer function of a head-phone at a given frequency (1000 Hz chosen as 0 dB reference) is equal to the amount in dB by which the headphone signal is to be amplified to give equal loudness. Averaging over a minimum number of subjects (typically eight) is required. The

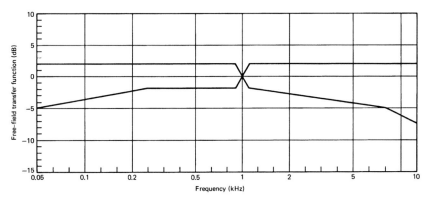

Figure 14.76. Typical tolerance envelope for free-field transfer function of a hi-fi head-phone, (From DIN 45500[115].)

procedure is described in Section 14.5.1. Figure 14.76 shows a typical tolerance field. A free-field headphone has a flat FF transfer function response. The frequency response of SPL at the ear canal entrance then has a 10 dB peak at 2 kHz and minima near 1.2 kHz and 8–10 kHz (Fig. 14.53, 0°). A headphone with flat SPL frequency response (lacking in brilliance) has a saddle in the FF transfer function at 2 kHz (i.e. the mirror image of the 0° curve of Fig. 14.53).

(b) Diffuse-field response

During the 1980s there began a movement[20] to replace the free-field standard require-ments with another, where the diffuse field (DF) is the reference. As it turned out, it has made its way into the standards, but without replacing the old one. The two now stand side by side[183]. The dissatisfaction with the FF reference arose principally from the magnitude of the 2 kHz peak. It was held responsible for coloration of the image, since frontal localization is not achieved even for a mono signal. The way in which the hearing mechanism perceives coloration is described by the association model of Theile[20] (Fig. 14.62). A comparison of the ear responses for diffuse field and free field is shown in Fig. 14.77[38]. Experiments[39] with 1/3 octave random noise in an anechoic chamber showed that the loudness sensation depends only on the SPL in the ear and not on the type of field FF or DF. The concept of coloration in music, however, requires signals with larger bandwidths.

Since the subjective listening test is the one that counts, FF headphones have so far been more the exception than the rule. A palate of different frequency responses is available to cater for individual preferences, and each manufacturer has its own headphone philosophy with frequency responses ranging from flat to free field and beyond[88–90].

14.4.2 Headphones compared with loudspeakers

(a) The apparently missing 6 dB

A discrepancy between headphones and loudspeakers at low frequencies has puzzled acousticians for the last 60 years; it concerns the SPL in the ear required for a given loudness sensation. Up to 10 dB more SPL is required for headphones. The effect begins at 300 Hz and increases with decreasing frequency[93,100,101]. This so-called missing 6 dB problem has triggered off much speculation about an apparent funda-mental difference between listening by headphones and loudspeakers, and has even thrown doubt on the assumption that the ear is a pure pressure detector. There are

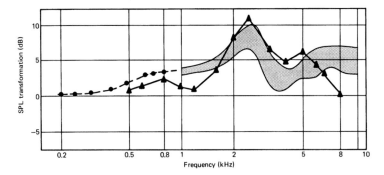

Figure 14.77. Comparison of diffuse-field and frontal-incidence SPIL transformations from free-field to the ear canal entrance. DF: filled circles and shading[38], FF: triangles[29]. (From Kuhn and Guernsey[38].)

two aspects to the problem: one concerning threshold measurements (comparing minimum audible fields, MAF, with loudspeakers and minimum audible pressures, MAP, with headphones) and the other concerning loudness comparisons well above threshold.

Both aspects have been reported[41,91] to arise from various artefacts and subtle effects which had previously been neglected. The discrepancy in threshold measurements arises from masking of the low-frequency tones by physiological noise[41,91,100] amplified by the occlusion effect (Fig. 14.74). Physiological noise under an earphone is a low-frequency phenomenon due to respiration, blood flow and muscle tremor with SPL up to 30 dB and spectrum dropping by 5–15 dB/octave[84] from 32 to 250 Hz.

The above-threshold aspect is more difficult to explain away, as physiological noise cannot mask the signals so easily. Rudmose[41] reports a number of effects such as: (a) mechanical chair vibrations via the floor[41] when using loudspeakers; (b) a psycho-acoustic phenomenon whereby more SPL is required for a nearby source than a distant loudspeaker for a given subjective loudness. In the words of Rudmose[41], the distant source has a 'larger acoustic size' and consequently appears louder. He reports that 'once a subject discovers this phenomenon he can be trained to eliminate it', as originally observed by von Békésy. Other effects involve (c) distortions, (d) procedural details for loudness balances requiring more time for the headphone judgement, and (e) in monaural tests the problem of effectively excluding the non-test ear (Section 14.4.3). Whether or not the missing 6 dB problem has really been put to rest will be revealed in years to come.

The above involved loudness. However, a persistent difference between headphones and loudspeakers is one of quality, whereby 'the sources don't sound alike even when judged to be equally loud'[41]. Whether or not bone conduction plays a role (Section 14.3.2(b)) has yet to be confirmed.

(b) Audibility of phase distortions

Unlike harmonic distortions, which are more audible with loudspeakers, monaural phase distortion is more audible on headphones. In the former case, where standing waves are present, the head can be moved to a position where the fundamental is weak, but not the harmonics[42]. In the latter case, phase distortion is presumed to be smeared by room reverberations[85], making it practically inaudible for music signals over loudspeakers. Therefore, investigations have been confined mainly to headphones[85], where the audibility is far greater, though admittedly subtle.

This section is not concerned with interaural phase, which was discussed in Section 14.3.1(b), but rather differences in sound between signals of identical amplitude spectrum but different phase spectrum. Neither are we concerned here with trivial cases of such phase distortions, such as time-reversed speech, or comparison of continuous wide-band noise and short clicks, which also can have identical energy spectra, but are obviously distinguishable. As Schroeder[71] pointed out, when discussing 'phase deafness' one speaks only of 'short-time spectra'.

Ohm's law of acoustics In 1843 Ohm[103] formulated the acoustic phase law, which states that the ear only perceives the frequency response of amplitude but not phase. This law is also associated with Helmholtz[104], who postulated that the inner ear consists of a set of tuned resonators whose intensity outputs are sent to the brain. The failure of Helmholtz to hear phase effects has been attributed[105] to his unwieldy instrumentation, with expiry of large time intervals from one setting to the next. More recent work by Hansen and Madsen[106,107] and Lipschitz *et al.*[85], among others, has led to a revision of the phase law, especially for headphones below 1.6 kHz but also for loudspeakers under anechoic conditions and for synthesized signals.

Genuit[117] reported that the cause of the superiority of electrostatic compared with dynamic headphones is the monaural phase distortion in the latter below 300 Hz, reaching a maximum of about 70° at 20 Hz. Since dynamic headphones are more sensitive to leaks (Section 14.2.4) than headphones with large light membranes, the phase distortion is to be expected. However, Genuit speaks of the all-pass component, which shows that a dynamic headphone is not a minimum phase system at low frequencies. Ohm's law of acoustics seems to apply, except for some exotic special cases.

Phase audibility in tone combinations The types of signal where the ear is most sensitive to phase distortion have pronounced asymmetry in waveform. The concept of relative phase of two tones is only meaningful for harmonic combinations, such as musical tones where the harmonics f, $2f$, $3f$, etc. are phase locked. Monaural phase is audible only for f, $2f$ combinations and of these only when $2f$ is below 1000 Hz, as for interaural phase. A 180° phase shift of the $2f$-component is equivalent to a polarity reversal. Although monaural phase distortion is audible, the effects are so subtle that suggestions for polarity standardization in audio systems, (e.g. Stodolsky[109]) have rarely been implemented. (Of course, left and right must agree in polarity, but that is another issue.)

All-pass phase distortions and minimum-phase systems The type of phase distortion causing genuine signal degradation is of the all-pass type. Minimum-phase systems give natural phase distortions which go hand in hand with amplitude distortions: a maximum in the amplitude response corresponds roughly to a point of inversion in phase response (Fig. 14.78). They are related by the Hilbert transform[44]. In nature such behaviour is more the rule than the exception. Since equalizers are also minimum phase systems, ironing out the amplitude distortions automatically flattens the phase response. Fortunately, the majority of audio components are minimum-phase systems. Even acoustic diffraction effects, e.g. around the human head and ear (Section 14.3.1), show minimum-phase character, with exceptions in the shadow zone and for some pinna shapes around 10 kHz (see Fig. 14.79). Thus simple electrical filtering can simulate most head diffraction effects of amplitude and phase in one step. Single transducer systems also show predominantly minimum-phase character.

On the other hand, most crossover networks and analogue tape recorders are not minimum-phase systems, and cause distortions which can be corrected only using all-pass filters[110]. In addition, wrong acoustic positioning of a transducer in multiple transducer systems can cause irreversible phase distortions. To minimize this component caused by sound path delay, the phase must be determined outside the normal pass-band of the transducer. The correct time delay will have been subtracted, and consequently the acoustic position found, if the phase is constant for indefinitely high frequencies[108].

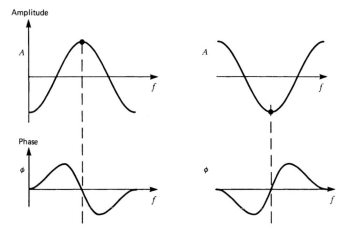

Figure 14.78. Schematic diagram of amplitude and phase relation (Hilbert transform) for a minimum-phase system.

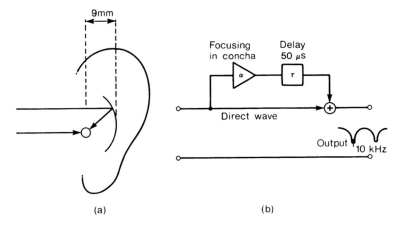

Figure 14.79. Descriptive diagram demonstrating minimum-phase condition for cavum concha reflection in pinna diffraction. (a) Geometrical components considered. (b) Equivalent circuit. As long as the later component is weaker than the direct wave ($\alpha < 1$) the result has minimum-phase character. If $\alpha > 1$, non-minimum-phase results.

(c) Advantages and disadvantages

Some of the more obvious aspects are listed in the following:

(a) Isolated listening for headphones – one neither disturbs nor is disturbed;
(b) Mobility in loudspeaker listening – no cable. However, movement away from the ideal listening spot deteriorates the stereo image. The headphone image is more stable in this respect;
(c) Excellent quality of sound reproduction with headphones for relatively little expenditure compared with loudspeakers of similar quality;
(d) In-head localization for headphones is a disadvantage;
(e) Headphone advantage in that the listening room has no effect on the sound image.

Apart from these self-explanatory factors, some other aspects deserve mention.

Hearing risk for headphone users, particularly Walkman type

Headphones are commonly used in public places. Point (a) above is clearly the implicit expectation, otherwise loudspeakers would be chosen instead. Point (a) is not generally fulfilled, for two reasons:

(i) Effect on the surroundings It is not possible to produce mini-earphones with a hermetically closed housing. The back of the membrane always radiates practically unhindered into the surroundings. Otherwise the membrane motion would be hindered by the stiffness of the rear cavity. The high-frequency radiated energy increases with the area of the radiating surface (see Fig. 14.2(e)), because the relative short-circuiting effect of L on R is reduced. A large membrane size also contributes to the volume current radiated. Thus the smaller the earphone, the less the disturbance. The small type (Section 14.2.9) which rests inside the cavum concha (Fig. 14.41) is best in this respect. On the other hand, supra-aural headphones are a public nuisance. Even small earphones are frequently a public disturbance, due to the excessive loudness commonly desired (see next section). When transistor radios appeared for the first time they were used even on beaches, but it was not long before their use in public was banned. In the case of supra-aural headphones no such effective action has been taken, presumably because of the implicit belief that point (a) above applies to all headphones, by definition.

(ii) Effect on the listener Usually the loudness chosen is tantamount to a misuse of the headphones. The level is gradually raised, since the listener desires increasingly large doses. The incurred hearing loss, advanced by years, is then irreversible. The usual behavioural feedback mechanism of pain, beneficial as a warning, which operates in other parts of the body, is not present in this case. There have been a few attempts (initiated, for example, by the German Ministry of Health in June 1988) at introducing obligatory qualitative warnings such as: 'Headphone listening at high volume levels – particularly over extended periods of time – may damage your hearing', but no effective implementations have yet materialized. Part of the problem is the difficulty in quantifying the risk. Suggestions have been made for the wording of the warnings, for example 'Adjust the level for a *pleasing* effect without damaging your hearing'. Such wordings are inadequate. This is because modern man's sensory appreciation is unreliable (Alexander[177]), and pleasure is experienced particularly within the danger zone. Since a long time will certainly elapse before legislative measures become effective it may be worthwhile quoting the following table for the daily dosage.

Total listening time (min)	Danger threshold dB_{SPL} (A-weighted)
120	99
60	102
30	105
15	108

As regards the difference between headphones and loudspeakers, much scientific work shows that there is no difference in principle. However, practice shows that headphone users choose to listen at levels about 8 dB louder than they would with loudspeakers. Table 14.2 shows the results of Mathers and Landsdowne[40]. The subjects were asked to adjust loudness levels to 'normal' and 'loud'. The criteria were a comfortable level for critical listening for the former, and the highest level tolerable without discomfort or fatigue for the latter.

Table 14.2 Loudness levels chosen for five music excerpts with means for 18 subjects: A, heavy rock; B, calypso; C, D, classical; E, Latin-American. (From Mathers and Lansdowne[40].)

Sound source	Level	Musical excerpt					
		A	B	C	D	E	Mean
Headphones	Normal	91.2	90.8	87.8	89.4	91.9	90.2
	Loud	100.4	98.5	95.4	97.1	98.9	98.1
Loudspeakers	Normal	81.4	84.9	83.4	85.2	82.7	83.5
	Loud	90.1	91.7	89.8	91.5	89.4	90.5

This is reminiscent of the elusive missing 6 dB problem (Section 14.4.2(a)), which has apparently been put to rest by closer scientific scrutiny. Psychophysical measurements of loudness discomfort levels have also been reported[111] to be identical for headphones and loudspeakers. And yet observation of daily practical listening does reveal a difference.

Audibility of FM-vibrato   Musicians modulate the pitch of tones where possible, e.g. for the violin, left-hand vibration, or for the electric organ, by means of FM. The FM itself is not what is usually heard, but rather a secondary AM effect resulting from steep flanks in frequency response. In musical instruments such as those of the violin family, fine structure is present due to natural body resonances. For the electric organ, this fine structure is provided by the loudspeaker and room resonances. Practising with headphones is associated with less vibrato, since the frequency response is smoother.

14.4.3 Special-purpose headphones

(a) Communications headphones

In this category (e.g. headphones for speech under noisy conditions), frequency response requirements are modest, a bandwidth from 300 Hz to 3 kHz being sufficient for speech intelligibility. Electromagnetic transducers (Section 14.2.4(e)) have been replaced by moving-coil systems (Section 14.2.4(b)). Low acoustic damping is applied to achieve high sensitivity with an upper cut-off at 3 kHz. Although good bass response is not a problem if leaks are minimized (Fig. 14.34), the low frequencies are often deliberately removed, e.g. by puncturing the membrane or some other acoustic short circuit. For a well-balanced sound, the centre of gravity, as it were, of the response should be at 1 kHz on the logarithmic scale[123]. Hence the bass frequencies are detrimental if treble is missing.

Low-frequency cut-off has the added advantage that intelligibility of speech in noise is improved. The reasons are that: (a) noise in reverberant surroundings is predominantly low frequency, and (b) low frequencies mask high frequencies effectively, but not vice versa. The noisy surroundings referred to here are not necessarily those of the listener, but rather of the speaker at the other end of the line, which may be identical, e.g. when the transducer is fed from a microphone mounted on the outside of the earpiece. Non-linear transmission for communication in intermittent noise is then advantageous[121].

For suppression of ambient noise from the listener's surroundings, the transducer is built into a hearing protector. The limit of attenuation is that of bone conduction, 40–60 dB, which can be reached for good constructions only above 2 kHz. Noise attenuation at low frequencies is limited by motion of the rigid ear-piece as a whole (Section 14.2.6(a)) with a spring-mass resonance around 200 Hz. The best attenuation here is about 15 dB for a 300 g cup. This could in principle be improved by increasing the mass. However, since comfort is of great importance when the ear-piece is worn for long periods, this is not acceptable. Similarly, larger clamping forces than 6 N are not practicable, and 20 N can cause severe headache[97].

The requirement of good attenuation is incompatible not only with comfort but also with transducer sensitivity. A larger coupling volume improves attenuation but decreases sensitivity[118]. This is demonstrated in Fig. 14.80(a) and (b). The double cavity design as in Fig. 14.80(c) partially overcomes this, since the impedance $R_p + j\omega L_p$ can be made large in the pass-band of the transducer while acting as a short circuit for low frequencies where cup vibration plays a role. More complicated structures involving double cushions[118] have also been proposed (Fig. 14.80(d)).

Headsets for switchboard or office use are required to be light with a small un-obtrusive microphone. Sound attenuation is minimal. The headband is frequently replaced by other constructions, e.g. under the chin, or one-sided mini headsets, hanging on the ear or for attachment to spectacle frames[122].

Leak sensitivity of telephone receivers For applications in light-weight mobile tele-phone systems, small flat dynamic receiver capsules are used. Leak sensitivity is one of the quality criteria. Since the sound outlet is never pressed hermetically tight against the ear there is always a natural sound leakage, which is largest at low frequen-cies. Good leak sensitivity is associated with the concept 'low impedance', referring to the effective internal impedance of the pressure generator.

(The use of the word 'receiver', which has historic origins, can cause confusion: it means sound 'source' or earphone. Thus from the point of view of the acoustical world it is a sender, not a receiver. However, from the electrical point of view it receives electrical signals, hence the common usage.)

Measurements of frequency response and overall sensitivity RLR (receiver loud-ness rating) have normally been performed with an air-tight connection to the coupler

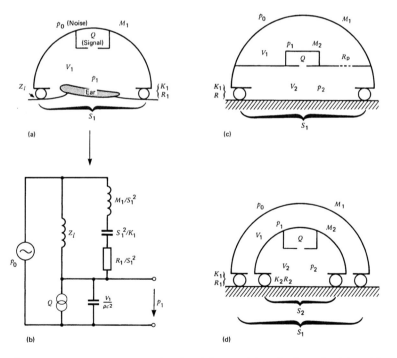

Figure 14.80. Hearing protection headphone with a single cavity (a) and its equivalent circuit (b). Double-cavity designs (c), (d). (Taken from Shaw and Thiessen [118].)

(IEC 318 Standard). Now an additional measurement with a leak simulation is required. Leak rings are used in combination with couplers for simulating the ear leak, mainly R and L, but the simulation can also include an effective increase in cavity volume. A curiosity is that in narrow frequency ranges near 1 kHz the sensitivity can even increase due to the leak simulation, since the parallel combination C (coupler volume) and L (acoustic mass of the leak) has a higher impedance at the resonance than C alone would have (as for a band-stop filter). Changes in RLR of 2 to 3 dB with the IEC-coupler (Section 14.5.1(c)) are considered acceptable, with a standard supra-aural leak simulation ($R = 2.4\,M\Omega$, $L = 700\,H$, $1.2\,cm^3$ volume increase). An additional quality criterion is a minimal change in frequency response shape.

RLR is a single number arrived at by accumulating the sensitivities for 20 frequencies from 100 Hz to 8 kHz, with frequency-dependent weighting factors to take into account the ear's sensitivity at different frequencies. It is a damping rating – thus the more negative it is, the stronger the output.

For low leak susceptibility the effective internal impedance Z_I of the system must be made small compared with the load Z_L. Z_I consists not only of the obvious inner transducer impedances, but also those parts of the coupler which are always present (see Fig. 14.81). Thus one can rearrange the circuit (Fig. 14.81(b)) according to the Thevenin rule such that the coupler volume C (including any side branches for eardrum simulation (Section 14.5.1(c)) no longer appears to be in the load, adopting a new position parallel to the inner-transducer impedances as part of Z_I (Fig. 14.81(c)). The leak can then be regarded as the load Z_L. To make the voltage drop (pressure p_1) across Z_L less sensitive to the value of Z_L, an additional permanent leak Z_R (mainly resistive) is introduced. Due to Z_R there is an inherent low-frequency deficit in p_1, even before introducing the actual leak Z_L. To offset this the membrane mobility can be enhanced by opening the back of the capsule through a porous acoustic resistance R_B.

Underwater headphones   Whereas free-field underwater hearing is via bone conduction (Section 14.3.3(c)), underwater headphones make use of the normal tympanic route by irradiating the eardrums. Figure 14.82 shows an underwater earphone[159] with an electromagnetic transducer. The silicone fluid has a selected viscosity for mechanical damping. The back is sealed by a limp neoprene bladder to alleviate static pressure variations with depth. Such a device is a dipole radiator[3]. Calibration at various frequencies is done by loudness balancing using a 1 kHz reference tone[159] (comparison with a loudspeaker as in Section 14.4.1(a) not being feasible). Underwater hearing sensitivity decreases with increasing frequency, thresholds[2] ranging from 58 to 74 dB SPL. To offset this, the SPL has to rise by 12 dB/octave[160], in order to give a subjectively flat response[3]. It is assumed throughout that the ear canal is filled with water. A bubble trapped at the eardrum increases the sensitivity by about 4 dB[160].

(b) Audiometry headphones

In audiometry, both supra-aural and circumaural headphones are used for air conduction testing. New headphones are usually calibrated by loudness comparison at 20–30 dB above threshold with standard earphones[128]. Frequency-response shape is of secondary importance as long as the standard frequency points are calibrated. Circumaural headphones give more consistent results than the supra-aural type[83,126]. Circumaural systems are also preferred for their superior attenuation of background noise[128] (Fig. 14.83). In addition, supra-aural headphones pressing against the pinna cause a partial collapse of the ear canal[78,127] which can lead to spurious audiograms in about 4% of patients. Errors, typically 10–15 dB where collapse occurs, are greatest at 2 kHz and need not necessarily cause inconsistent results[127]. For a supra-aural earphone the standard centre hole size in the cushion is 3/4 inch[27].

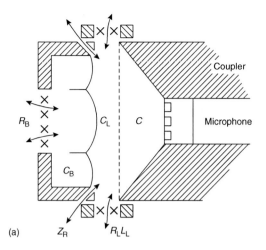

(a)

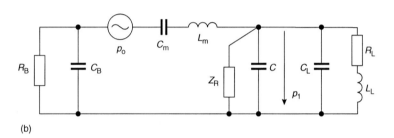

(b)

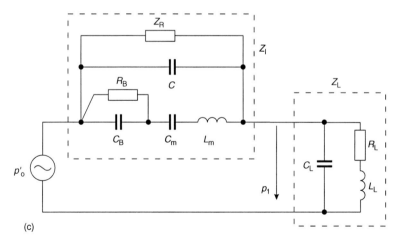

(c)

Figure 14.81. (a). Schematic representation of a simplified telephone receiver on a simplified coupler. C_L, indicates the volume increase due to the leak ring with $R_L L_L$. (b) The conventional equivalent circuit, all in the acoustic regime (no transformers or gyrators). (c) The Thevenin circuit, for analysing the effect of the leak, which is now the only load. The Thevenin pressure generator $p' = p_0 \times Z_{CR}/(Z_{Bm} + Z_{CR})$, where Z_{CR} is the impedance of C and Z_R in parallel, and Z_{Bm} is the inner-transducer impedance consisting of R_B, C_B, C_m and L_m.

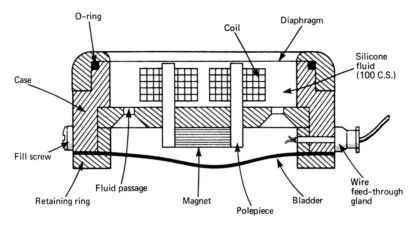

Figure 14.82. Cross-section of an underwater headphone. Diaphragm vanadium permendur, (From Schumann et al.[159].)

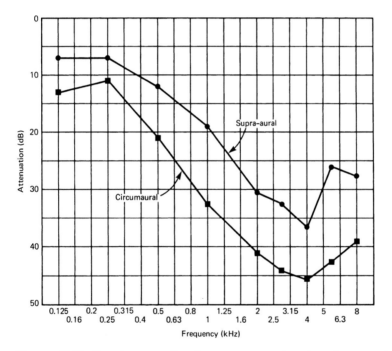

Figure 14.83. Sound attenuation for a supra-aural and a circumaural audiometry headset. Values taken from Michael and Bienvenue[128], Table VI.

Physiological noise   With noise-attenuating headphones (Section 14.2.8), physiological noise becomes a problem because of the occlusion effect. Low-frequency thresholds are raised as a result of masking. The amplitude at 100 Hz is typically 30–40 dB SPL[84] but can reach 60 dB if the ear-piece is held by hand, due to tonic arm muscle contractions[124]. Thus the not-uncommon practice of holding the earphone by hand or thoughtlessly touching it should be avoided.

Bone-conduction audiometry By combining bone-conduction (BC) and air-conduction (AC) audiometry, one can differentiate between conduction impairment (i.e. middle ear) and sensorineural (inner ear) hearing loss. The tone from the bone conduction vibrator is received equally by both inner ears, although the propagation velocity within the head is close to that in air[33,72]. Therefore, masking of the non-test ear is required. Narrow-band masking noise (300 Hz bandwidth) with centre frequency equal to the test tone is more efficient than broad-band noise, since only frequencies near the tone contribute to suppression of the tone sensation. The masking noise is delivered by an earphone covering the non-test ear, but not so loud as to be heard in the test ear. Therefore, there is a restricted range of permissible masking intensity, which varies with the degree and type of hearing impairment. Thus, with a patient whose hearing loss is not known at the outset, the correct masking intensity can only be determined iteratively by the method of successive approximations. According to Studebaker[125], the mid-masking level M_0 can be determined from the expression:

$$M_o = M_{oN} + B_T + AB_M/2 \tag{14.17}$$

where M_{oN} is for normal hearing, B_T is the BC loss in the test ear, and AB_M is the air-bone gap of the non-test ear, all in dB.

Because of the occlusion effect (Section 14.3.3), sensitivity to bone conduction below 2 kHz is raised by 10–25 dB when an earphone is worn over the test ear. Therefore, BC audiometry should be performed in a soundproof room with the test ear free[131].

Interaural attenuation If a patient is totally deaf in one ear, then a sound delivered to this ear by an earphone will be just heard by the contra-lateral ear (assumed to be normal) when the intensity is 40–50 dB above threshold[130]. This acoustic leakage, which determines the value of M_{oN} (equation (14.17)), has been shown by Zwislocki[72] to be due to bone conduction. He also showed that the leakage increases with increasing area under the earphone cushion. The coupling to the skull, bringing it into vibration, increases with area. Insert earphones show large interaural attenuation[34], reaching 100 dB[72]. Their use in audiometry would allow the masking noise to be confined entirely to the non-test ear[132,133].

Central masking In the above section relating to interaural masking, peripheral masking was implicitly assumed; i.e. involving sound conduction from one side of the head to the other. For completeness, however, central masking[72] should be mentioned; this occurs even when the physical masking signal is below threshold. It is attributed to a physiological interaction in the central nervous system[125]. Its influence is small but consistently present, amounting to a threshold shift of about 5 dB.

14.4.4 Out-of-head localization with headphones

(a) Open and closed headphones

In open headphones, either the transducer itself is acoustically transparent (e.g. isodynamic) or the surroundings of the transducer consist of porous material. Reflecting surfaces are kept to a minimum in order to allow the pinna to diffract the sound undisturbed, i.e. without additional reflections as in a closed ear-piece. The resulting frequency response at the ear canal entrance contains the linear spectral distortions above 2 kHz characteristic of the individual ear. This is partially responsible for the natural sound of open headphones. The in-head localization (IHL) for closed headphones is frequently attributed to the lack of natural pinna reflections.

However, from the point of view of the telecommunications engineer, this argument is unsatisfactory: a given frequency response can be produced by electrical filtering just as well. Thus a closed headphone can be corrected to have the same response as a given open headphone. They should therefore sound the same, and yet listening tests with normal stereophonic recordings confirm that IHL is worse with closed headphones.

Of course, genuine out-of-head localization (OHL) is not possible with intensity stereophony when the loudspeaker signals are fed unblended to the headphones. A signal component appearing only in one channel, e.g. cello on far right, is fed only to the right ear. Why is it then, that this cello sounds further away with open headphones and more definitely in the head with closed ones?

For open headphones of the large-membrane type, the airborne interaural leakage increases by 6 dB/octave up to a maximum at 2 kHz, above which it decreases due to shadowing. The attenuation at 2 kHz is 30 dB, with an interaural delay of 1 ms improving out-of-head localization for a single channel when the other is silent. At a listening level of 80 dB SPL, the signal at the opposite ear is therefore 50 dB above threshold, which is well above the critical level (1–6 dB above threshold) for a binaural image[15] (Section 14.3.2). Since a good record player pickup has a 30 dB channel separation, crosstalk without delay may anyway be present, competing with the acoustic leakage. Masking data[134,146] for self-masking of wide-band noise show that a signal is detectable even in the presence of a coherent masker 10 dB louder. Because of the band-pass nature of the leakage signal, coherence is not expected. This, as well as the different delay and transient structure, will enhance detection of the headphone leakage signal.

Electrical simulation of the acoustic crosstalk of an open isodynamic headphone, using closed headphones, shows this effect to be distinctly audible. However, for stereo signals, while both channels are active, this effect is of little significance. It has been shown[166] that for headphones with stereophonic programme material, crosstalk with channel separation of greater than 15 dB is inaudible. Thus the 30 dB channel separation of a good record player pickup (and more so with a CD player) is more than sufficient. In conclusion it can be said that under special conditions acoustic crosstalk can account for the OHL of open headphones, but psychological factors, such as audibility of ambient noise, override any physical mechanisms under normal conditions.

(b) Binaural recording and reproduction

In binaural reproduction, the illusion of being present in an acoustic event is imparted to a listener by means of headphones[155]. His ears receive signals which have been recorded by two pressure microphones, one in each ear of an artificial head. Figure 14.84 depicts the simplest set-up. It is implied that all acoustic parameters, including ear canal shape and eardrum impedance, have been meticulously simulated to produce accurate ear signals for a realistic (including spatial) impression. For this conceptually straightforward case, the frequency response of the headphones, say at the ear canal entrance 4 mm inwards (if that was the microphone position in the artificial head), should be flat[143]. However, insisting on conceptual simplicity involves considerable practical difficulties, such as eardrum simulation, placement of a necessarily small microphone in an artificial ear canal without acoustic interference, and the fact that most available headphones do not have a flat response.

Hudde and Schröter[138] demonstrated that a faithful reproduction of the ear canal and eardrum is unnecessary. The ear canal can be treated as a two-port whose output B is terminated by the eardrum impedance Z_d (Fig. 14.64), and whose input A is a point 4 mm behind the entrance. The directional characteristics of the head depend only on the outer geometry[138, 151], including shoulders and torso. All coloration effects arising from improper eardrum or canal simulation, or even their omission, can be corrected electrically or otherwise. As pointed out by Schöne[137], this gives the designer additional headroom in tackling the problems of noise level and compatibility with normal loudspeaker stereophony. An investigation comparing eight well-known artificial heads and human subjects showed that binaural recordings from microphones in real ears gave much better localization[5].

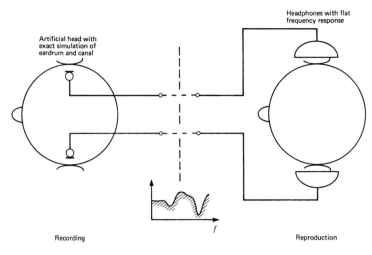

Figure 14.84. Binaural recording and reproduction, diagram demonstrating the fundamental idea.

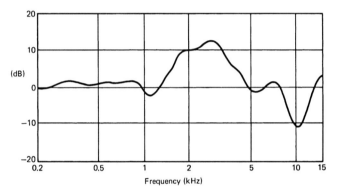

Figure 14.85. Frontal-incidence transfer function from free-field to the ear canal entrance (0° curve from Fig. 14.49)[151].

The free-field accommodated artificial head   The configuration of Fig. 14.84 should supply the headphone user with the spectral distortions shown in Fig. 14.85 for a frontal source, assuming the artificial head simulates a typical individual. However, the standardized headphone (free-field headphone, Section 14.4.1) does this also. Therefore, the artificial head signals must be equalized with the inverse spectrum of Fig. 14.85. A non-conforming headphone can be electrically equalized. Figure 14.86 depicts recording and reproduction for free-field accommodated artificial head signals. The filter $1/K'$ performs not only the inverse of K (Fig. 14.85) but also corrects for any lack of eardrum and ear canal simulation. Thus the output of such an artificial head for frontal incidence corresponds to that of a stereo microphone with flat frequency response[137,140].

Compatibility with loudspeaker stereophony   Artificial head stereophony, a binaural technique, was for a long time regarded more as a curiosity than as a recording technique for routine use in broadcasting. Part of the problem lies in the need for compatibility with loudspeaker stereophony. Since binaural is the newer technique,

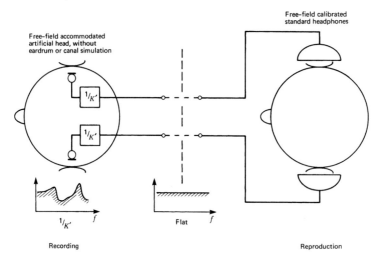

Figure 14.86. Binaural recording and reproduction using a free-field accommodated artificial head.

it must accommodate itself to the established recording technique, and not vice versa[136]. It is necessary, therefore, for the binaurally recorded material also to be reproducible directly over loudspeakers without accessories.

Regarding spectral aspects, the free-field accommodated binaural signals are compatible with loudspeaker stereophony. As regards spatial compatibility, the compression of the stereo panorama[145] is not as bad as one might expect: ± 90° becomes ± 30°, and ± 30° becomes ± 20°. Each of the binaural signals, originally intended for one ear, reaches both ears. However, the situation is not much different from the technique of mixed intensity and time-delay stereophony intended for loudspeakers (e.g. the technique[147] using a felt-covered 200 mm disc). The latter is essentially a squashed artificial head, with approximately the same inter-aural delay as a head. Since this is intended for LS stereophony, it is difficult to see why the inter-aural delay in the microphone set-up should be considered significant at all. The mechanism for a phantom source with loudspeakers is entirely removed from the recording situation, and benefits more by intensity differences than delays (Fig. 14.58).

Mono compatibility Switching from stereo to mono mixes the left and right channels additively. For pure intensity stereophony, the individual components of a stereo recording are added in phase. If time delays are present, then comb filtering will cause coloration: 0.25 ms delay gives a comb spectrum with minima at 2 kHz, 6 kHz, 10 kHz, etc. The frequencies of the minima are inversely proportional to the delay. The depths of the minima depend on the intensity difference of the signals being added. Strictly speaking, not even intensity stereophony is mono-compatible with loudspeakers, only with headphones. Binaural and other recording techniques involving time delay are not mono-compatible.

Noise suppression Artificial heads are generally further from the sound sources than microphones in multi-microphone recordings. Minimizing microphone noise therefore becomes important[19,139,140]. Since the ear canal need not be faithfully simulated, more space is available for acoustic resonators and microphones than would otherwise be the case. This gives added scope for minimizing noise[136,137]. The aim is to keep the sound pressure in front of the microphone membrane as high as possible at all frequencies by means of Helmholtz resonators[148] or, failing that, to compensate for sound-pressure minima by using the mechanical resonance of the membrane with less damping. Genuit[140] used a 12.7 mm condenser microphone with increased polarization voltage, thus lowering the noise floor. The minimum at 10 kHz (Fig. 14.85) was compensated by the mechanical membrane resonance. Thus part of the free-field equalization of an artificial head can be performed mechanoacoustically[144] before the transduction.

Although a noise level comparable to that of the human ear can be reached, an additional noise suppresser is required since the localization of the incoherent microphone noise sources is different from that of the signal[140]. It may be added that an acute ear has an effective noise level equivalent to a microphone with 20 dB SPL (A-weighted) noise level[142]. Killion has shown[142] that hearing aids can be constructed with aided thresholds better than those of an acute ear if the gain is turned up. Therefore, natural hearing does not represent the limit of possible signal-to-noise ratios in binaurally recorded signals.

(c) Blending circuits and 3D audio

Blending circuits have the function of processing signals, usually originally intended for loudspeakers, in such a way as to give the headphone listener a sensation similar to free-field listening. Apart from the obvious and essential aim to avoid in-head localization, frontal localization is desired, but hardly ever achieved reliably. This is because frontal localization depends on the HRTF (head-related transfer function) coinciding with that of the listener. The HRTFs are transfer functions, occurring in pairs, one for each ear, equal to the ratio of complex sound pressure at the ear divided by the free-field sound pressure which would prevail at the point in space where the middle of the head is, in the absence of the listener. Slight differences in the geometry of the two ears can become the deciding factor for frontal localization, since there are no other differences for sound sources, arrival times being identical. In principle this applies to all directions in the median plane. However, years of experience in subconsciously comparing the sound sensations for visible sources makes the listener particularly critical for frontal localization. For invisible sources, those behind or above, he has not been given a chance to learn to associate a certain direction with a certain sound sensation. Thus all binaural sound sensations in the median plane which do not match with any frontal ones are automatically heard behind or above.

In the case of two-channel stereo destined for loudspeakers, left and right have to be cross-coupled, L to L, R to R, as well as L to R, and R to L, as for LS listening. For example, the left channel arrives also at the right ear with a delay of some 250 ms, corresponding to the path-length difference of 90 mm in the free field for a loudspeaker 30° left. Attempts to simulate this have in the past used analog filtering with electrical or acoustic elements, such as hollow tubes to transfer sound with the correct delay and damping to simulate head diffraction and shadow effects. This does not work at low frequencies, since the tube, which is never matched with its characteristic impedance, becomes essentially a pressure chamber, with or without leaks which shift the phase erratically. This shortcoming is serious, since the frequency range where delay is most effective for out-of-head-localization (OHL) is below 1 kHz[149] and this is the phase delay[150]. Electrically the delays and diffraction effects are now realized digitally.

A state-of-the-art product is the HEARO family from AKG-Acoustics which has a digital signal processor with software from Dolby. It processes a multi-channel digital Dolby Surround signal, intended originally for loudspeakers. The Dolby decoder at the input produces the five digital LS signals, each of which is then filtered by a selectable set of HRTFs for the various LS directions to be simulated. FIR filters are used (finite impulse response). The two-channel stereo signal is then transmitted to the wireless headphones. A database of different HRTFs from many individuals is available, through which the listener can surf, till a matching one is found. For a listener with asymmetrical ears, a hopelessly unsuitable HRTF may well become optimal by reversing left and right channels. Indeed, according to Section 14.3.2(b), as far as frontal localization of the centre component is concerned, the more unsuitable the original sensation was, the greater the likelihood of reversal having a beneficial effect. A technique of suspending small earphones close to and in front of the tragus of the pinna was reported[141] to cultivate individual HRTF cues for frontal localization, without explicitly having to reproduce all of them in the electical signals. All high frequency pinna cues, the uncritical as well as the critical ones, one of which is the binaural pinna disparity, are present.

Limitations and compromises   To allow the listener to judge the effectiveness of a blending circuit, equipment is generally equipped with a switch for putting the cross-feed out of action. This is frequently where the disappointment begins, especially if the programme is not chosen carefully. A spectacular effect can be expected only for pure intensity stereophony – any time delays between the left and right channel in the recording serve only to dilute the effect to be heard when switching. The second disappointment is the lack of clear frontal localization, which is the cause of the third drawback – that the cross-feed almost resembles mono. Since frontal localization is not normally achieved, the image for a given angle lies somewhere on the cone of confusion (Fig. 14.60), usually with a diffuse distribution over the cone (Fig. 14.87). The intended line panorama (a) therefore tends to become a curved surface (b). To overcome this, it is advisable to exaggerate the panorama width, and simulate a listener between two loudspeakers. An instrument confined to one channel will then be localized 90° left or right, i.e. a unique direction, not a cone. Now at least fragments of the stereo image are projected out of the head, and the question of frontal localization is discreetly put aside. At this point a psychoacoustic effect is worth mentioning: switching to cross-feed gives the impression of a headphone with weak bass, even if the response remains unchanged. Bass is more strongly felt when the left-right intensity in the ears is unequal, because this is an unnatural condition. Cross-feed with delay balances the intensity and increases the perceived source distance; hence the subjective decrease of sensation.

The serious drawbacks of passive cross-feed systems are no longer an issue. Since digital signal processing has become routine, development work is no longer hampered by secondary difficulties in realizing the cross-feed. As a result, 3D audio techniques for headphones which used to be confined to research institutions, have made their mark commercially.

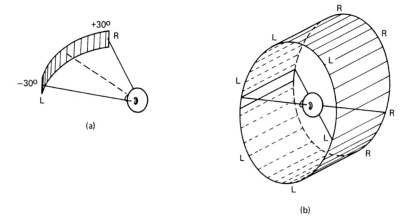

Figure 14.87. Acoustic image localization for cross-feed LS simulation with headphones: (a) design objective; (b) usual case when frontal localization is lacking.

For underwater localization The sound velocity under water is 4.6 times that in air[158]. Therefore, the interaural time delay (ITD) is too short for directional information. Free-field underwater hearing also involves bone conduction (BC) (Section 14.3.3), which is effectively mono. To bypass BC, underwater headphones (Fig. 14.82) feed signals directly to the ears, thus reinstating reception via the eardrum. If each ear-piece is equipped with a hydrophone to receive underwater sounds for that ear, the ITD is still too low for directional hearing (0.17 ms for extreme left or right, as opposed to 0.8 ms in air). Bauer and Torick[158] developed a blending circuit to simulate the ITDs (and IADs) that a diver would receive if he were in air. The two-hydrophone array was combined to form two cardioid polar patterns back to back (Fig. 14.88). Thus an off-centre underwater source gave an intensity difference in the two channels. These were then processed by a cross-feed with delay as discussed above to produce the ITDs. This is an application of intensity stereophonic principles at low frequencies where intensity differences in the signal channels give pure time delays at the ears.

14.5 Practical matters

14.5.1 Testing of frequency response

The three main procedures are: (a) subjective loudness balancing with a loudspeaker; (b) real-ear SPL measurements; (c) coupler measurements.

Procedures (a) and (b) are suitable only in the development laboratory. For routine production line testing and control, couplers (c) are used[152,173], since the absolute response is of less interest than reproducibility and keeping within tolerances.

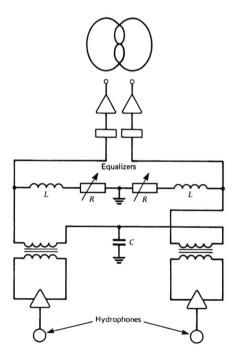

Figure 14.88. Cardioid phase shift network for a pair of spaced hydrophones at low frequencies where λ/4 > spacing. (From Bauer and Torick[158].)

(a) Loudness balancing with a loudspeaker

The equipment required is:

(a) noise generator (pink noise);
(b) filter for stepwise third-octave band-pass filtering at generator output;
(c) sine-wave generator for use below 250 Hz;
(d) loudspeaker without sharp resonances;
(e) third-octave equalizer for the loudspeaker;
(f) click-free foot-switch;
(g) attenuator with at most 2 dB steps;
(h) amplifier for the loudspeaker;
(i) r.m.s. voltmeter for headphone input (mono);
(j) measuring microphone with SPL-meter in dB (unweighted).

The procedure In the absence of the listener, a free-field SPL of about 70 dB is produced at the spatial point otherwise to be occupied by the centre of his head. For third-octave bands up to 4 kHz, a realistic tolerance[114] for the SPL is ± 2 dB for various points in a spherical volume of diameter 150 mm centred about this reference point. For the ear positions, the tolerance is stricter, ± 1 dB. Above 4 kHz these tolerances are ± 3 dB and ± 1.5 dB respectively. A quiet anechoic chamber is required for a propagating free-field (FF) wave. The loudspeaker (LS) response at the reference point is made flat using the equalizer. A block diagram of the set-up is shown in Fig. 14.89.

The listener takes his place, seated with the head centred on the reference point and facing the loudspeaker. For each of the 27 third-octave bands, the listener compares the loudness of the free-field and headphone signals alternately, removing and replacing the headphones in a quick simple movement. The time period for each comparison is 2.5 s (headphones require more[41]), with pauses of similar length. The attenuator is adjusted, whereby an 'equal loudness' judgement is forbidden[79]. The attenuation value for equal loudness is taken from the mean of equally probable judgements 'headphone louder and LS louder', to preclude errors from weakening of concentration. The attenuation, taking 1 kHz as 0 dB reference and averaging over eight listeners for each frequency, is the FF transfer function of the headphone (no

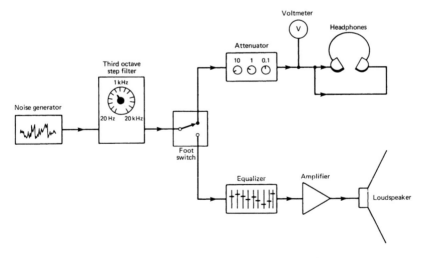

Figure 14.89. Experimental set-up for loudness balancing of headphones with a loudspeaker.

change of sign required). Below 250 Hz, because of the stochastic nature of narrow-band noise, it is advisable to replace the noise source by a sine wave generator. Small anechoic chambers with dimensions less than 5 m are unreliable below 300 Hz. A diffuse field can be simulated in an anechoic chamber using eight loudspeakers[82] with incoherent sources.

(b) Real-ear tests

The frequency response of SPL at a chosen point near the ear canal entrance is measured. A probe microphone (e.g. 3 mm condenser microphone with flat pressure response) is used. The microphone is inserted under the cushion at the inter-tragal incisure (Fig. 14.54), a convenient notch[153] for minimizing any additional leak from the probe. Real-ear tests are a simple means of approximately determining the FF transfer function without going through the tedious loudness-balancing procedure. A free-field headphone should show a 10 dB peak at 2 kHz and a minimum near 10 kHz (Fig. 14.53, 0° curve). Departures from this curve are an approximate measure of the FF transfer function. However, the loudness balancing may produce different results[49].

(c) Couplers and artificial ears

Without eardrum simulation   Modelling an individual's ear raises the problem of description with engineering drawings as well as involving redundant fine structure. A practical artificial ear for routine laboratory and factory use has the following properties.

(a) It is robust;
(b) It simulates the main aspects of ear pinna shape;
(c) It is easily replicable with simplest possible geometry consistent with (b);
(d) It approximates the air volume between headphone and ear drum;
(e) It simulates a typical acoustic leakage;
(f) It accommodates both supra-aural and circumaural headphones.

Figure 14.90 shows a section and plan view of such an artificial ear. The SPL is registered by a 12.7 mm condenser microphone. An artificial ear of rubber for laboratory use is shown in Fig. 14.91. It was cast from an individual. The gauze simulates leakage due to hair. A 3 mm microphone registers SPL at the ear canal entrance.

With eardrum simulation   Artificial ears with eardrum and pinna integrated into an artificial head with[55] or without[98] torso are in common use. Such systems have a multitude of functions: measurement of attenuation of hearing protectors and plug-type ear defenders, frequency response of headphones and insert hearing aids[174]. For attenuation measurements, the bone conduction is significant but cannot be directly simulated mechanically. From a knowledge of BC thresholds, and making reasonable assumptions about phase of the BC component, Schröter[43,98] corrected for BC mathematically.

Various eardrum simulations have been performed, e.g. with acoustic lossy delay lines[98] and acoustic ladder networks[155], Zwislocki coupler[154] with four side branches.

For computer simulation the circuit in Fig. 14.92 can be used. It was derived from the circuit of Pösselt[182] taking into account the inevitable decoupling of the drum from the malleus at high frequencies. The impedances from this circuit are shown in Fig. 14.93 If the simplified circuit (b) is used the resistive part becomes frequency independent.

Another coupler, still commonly used though already on its way out, is the IEC-318. The circuit elements supplied with this coupler are the R,L,C values of the two resonators and the C value of the main 2.5 cm^3 cavity. Improved agreement between

1. Artificial ear
2. Simulation of sound leakage
3. Ear resonances
4. Probe microphone

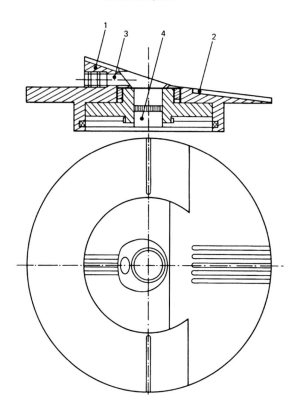

Figure 14.90. Artificial ear assembly of brass for supra-aural and circumaural headphones.

simulations and measured responses is achieved by recognizing that this cavity, of length 8 mm, has delay-line properties, which show up above 3 kHz. Although it is not cylindrical, but blunt conical, it is still better to treat it as an 8 mm cylinder (Fig. 14.94), than to use a capacitor. Of course the volume must be 2.5 cm³, which can be adjusted by choosing the diameter accordingly. Such an approximation can only improve, never worsen, matters. Including the microphone impedance is a luxury.

The BK 4195 artificial ear   A more realistic artificial ear than the IEC-318, suitable for various types of earphones and for wide-band telephone applications, is the P.57.3.2 recommendation of CCITT. At the core of this is the IEC-711 occluded-ear simulator, comprising a simplified eardrum simulation, with an 18-mm tube for that fraction of the ear canal from the ear drum to the tip of an ear insert. For other earphones the rest of the ear canal is simulated by an 8.8 mm extension tube leading either to a concha bottom simulator or to one of two types of pinna simulator. The first type (a), suitable for supra-aural earphones, has a simplified geometry similar to that in (Fig. 14.90). The second (b) has a realistic shape and shore-A hardness defined to simulate a real pinna. It is suitable for both supra- and circumaural headphones. In type (a) two leakage grades simulated by slits are included: high leak (approx. 60 kΩ 90 H) and low leak (approximately 900 kΩ 500 H). Type (b) is not in common use. Although it is described by unambiguous pinna cross-section drawings it can hardly compete with type (a) in simplicity of construction. The eardrum side-branches for drum impedance

Figure 14.91. Artificial ear model of rubber for headphone response measurements in the laboratory.

are distributed along the tube, rather than being at the end. To compare the response with other data whose reference point is at the ear-canal entrance the transfer function of the artificial ear-canal is required as a correction.

The simulation circuit supplied with this system assumes, of course, frequency independent R and L values. Though the agreement with measured responses, as far as overall shape is concerned, is quite good the frequency dependence makes itself felt particularly when the high leak is active, since the large diameter of the holes means the R value falls to a low value towards low frequencies (Fig. 14.2), where ωL loses its influence. In addition the tubes in this coupler need a transition line treatment which influences the frequencies particularly above 2 kHz. The circuit, including frequency dependence and transmission lines, is shown in Fig. 14.95. It was derived using a 28 mm transducer, whose inner elements were known from Thiele–Small measurements and baffle response measurements. The response on the coupler under both leak conditions was measured and the acoustic circuit elements adjusted for agreement. The transducer size was deliberately chosen to match the coupler. Thus the conditions are as one-dimensionally piston-like as can ever be expected here. The aim was to minimize discrepancies due to transverse modes in the concha as far as possible. The diverging lines of flow from a much smaller transducer would have provoked transverse modes. This would have unnecessarily masked the benefits of improving on the one-dimensional treatment, without bringing other benefits. There was considered to be enough room for improvement in the one-dimensional treatment, without having to provoke the worst case, needing a finite-element approach, and even then, not uniquely definable.

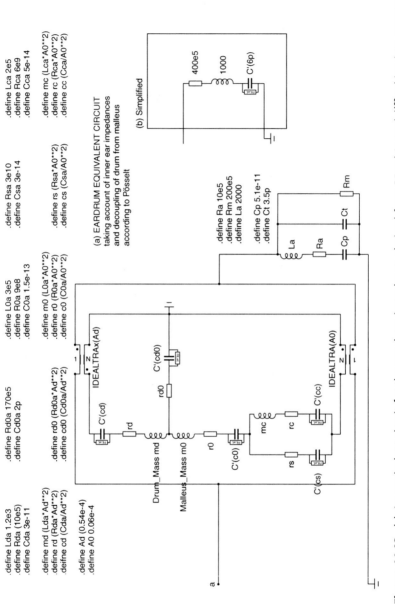

.define Lda 1.2e3 .define L0a 3e5 .define Rda (10e5) .define R0a 9e8 .define Cda 3e-11 .define C0a 1.5e-13

.define Rd0a 170e5 .define Lca 2e5 .define Cd0a 2p .define Rca 6e9 .define Cca 5e-14

.define Lsa 3e5 .define Csa 3e-14

.define md (Lda*Ad**2) .define m0 (L0a*A0**2) .define rd (Rda*Ad**2) .define r0 (R0a*A0**2) .defne cd (Cda/Ad**2) .define c0 (C0a/A0**2)

.define rd0 (Rd0a*Ad**2) .define mc (Lca*A0**2) .define cd0 (Cd0a/Ad**2) .define rc (Rca*A0**2) .define cc (Cca/A0**2)

.define rs (Rsa*A0**2) .define cs (Csa/A0**2)

.define Ad (0.54e-4) .define A0 0.06e-4

(a) EARDRUM EQUIVALENT CIRCUIT
taking account of inner ear impedances
and decoupling of drum from malleus
according to Pösselt

.define Ra 10e5
.define Rm 200e5
.define La 2000

.define Cp 5.1e-11
.define Ct 3.5p

(b) Simplified

Figure 14.92. (a) An equivalent circuit for the ear drum impedance, derived from Pösselt's model[182] taking into account the HF decoupling of the drum from the malleus. The simplified model (b), though giving impedances of the right order of magnitude, cannot reproduce the frequency dependence of *R*. The condensers *C'* are otherwise normal. They have been shunted by a large resistance for computer-technical reasons, to provide a d.c. path to ground.

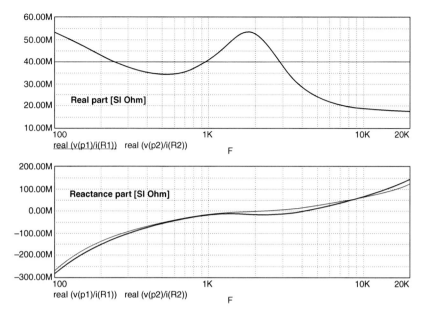

Figure 14.93. The resistive and reactive components of the eardrum impedance from the circuit in Fig. 14.92: heavy curves: Pösselt model; light curves: simplified model.

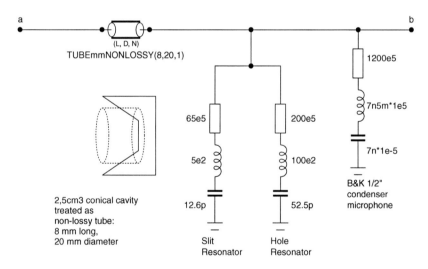

Figure 14.94. IEC-318 coupler.

14.5.2 Control of acoustic components

High-quality acoustic resistances are made of woven metal thread. Other materials such as felt and paper are also used, but have higher tolerances. Resistances can be made using small holes or narrow slits, but they have a relatively large acoustic mass L, unless the holes are smaller than 150 μm. To achieve resistances of low L, comparable to woven metal or felt ($\omega L < R$ below 6 kHz), the holes need to be smaller than 90 μm (Fig. 14.2).

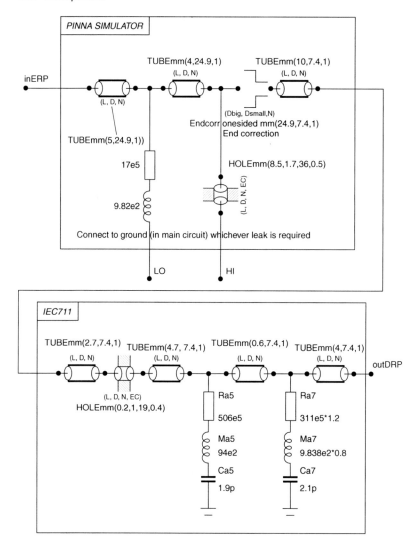

Figure 14.95. Circuit used with B&K coupler with high and low leak. See equation 14.8 for input data to the end correction element.

(a) Membranes

Unlike electrostatic transducer membranes, dynamic membrane systems are relatively heavy. The membrane has two functions: to provide the high compliance needed for a low resonance, and to support the coil. High compliance Cms is achieved in a thin membrane, but the second function, requiring strength, sets a limit to this. The compliance can be raised by pleating[156] the torus part of the membrane. For controlling Cms, it is sufficient to adjust the deep-drawing process parameters to influence the thickness, assuming a given membrane form. The internal mechanical resistance is hardly ever important compared with other acoustic resistances. Likewise the membrane mass is usually of secondary importance compared with coil mass.

(b) Porous resistances

For quick routine control of acoustic resistances, simplicity of operation similar to using a multimeter for electrical resistances is desirable. Then one must forfeit the chance of evaluating acoustic mass L, since this requires phase measurements. Fortunately, in practice R and L go hand in hand for a given material, and a wrong value of L does not go unnoticed since R then also lies outside the tolerance envelope. Low frequencies (below 30 Hz) are chosen, to avoid influence from ωL, though very porous samples may require an even lower frequency.

Using a loudspeaker with a microphone   The principle is that of a potential divider. A loudspeaker (LS) provides a pressure source p_0 which drives a current q through a series combination of a reference resistance R_0, and the sample R to be measured (Fig. 14.96). The microphone detects the pressure p_m, which is a function of R. The rectified and amplified signal is registered by an ampere-meter calibrated in acoustic ohms. The ohm scale is distorted as in an analog multimeter, and 'open circuit' ($R = \infty$) is adjusted by a potentiometer with the sample hole blocked. For kΩ (cgs) ranges, i.e. 100 MΩ (S1), account must be taken of the reactance of the coupling volumes (about 1 cm^3) and microphone load. The reference resistance R_0 is first calibrated (Fig. 14.97) against an acoustic volume compliance $1/\omega C$ (equation (14.4)), whereby the atmospheric pressure P is required. A convenient method is to use an adjustable volume, to keep the compliance constant for different pressure readings. Temperature T is not important since the compliance depends only on pressure. In the system of (Fig. 14.96), once R_0 is calibrated, P and T are no longer required, to first order (except where coupling volume reactance is important). R_0 depends only on viscosity, which is only weakly dependent on T. Moreover, viscosity change affects both R_0 and R, so that the potential divider ratio remains almost constant. The method is suitable for low and medium resistance ranges, the limit being set by the cavity reactance or microphone impedance, whichever is lower.

Using a loudspeaker without microphone   The back e.m.f. of a loudspeaker driven at its natural resonance is a fairly sensitive measure of its acoustic output load for low load values (0–40 Ω cgs or 0–4 MΩ S1). A low resonance (50 Hz or less) is essential to ensure that only the real component R is measured ($\omega L \ll R$), unless the driver

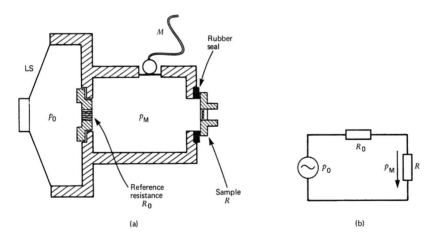

Figure 14.96. (a) Acoustic resistance measuring equipment using a loudspeaker and microphone. (b) Simplified equivalent circuit. In parallel with R are the cavity reactance and microphone impedance, which can be neglected only for low resistance ranges.

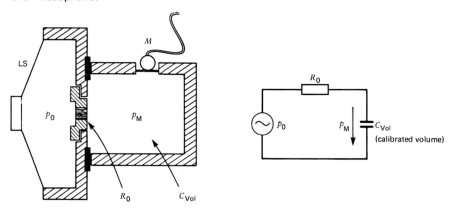

Figure 14.97. Set-up for calibration of the reference resistance R_0 of Fig. 14.96 against a volume reactance of compliance C_{vol}. The cavity is much larger than that shown in Fig. 14.96. Using a calibrated plunger to adjust the volume one can correct for air pressure departures from 1000 mB.

frequency is tuneable (in which case the cavity reactance $1/\omega C_{vol}$ should be kept much greater than R). The method of Fig. 14.98 is suitable for porous foam either uncut (a) or shaped (b), e.g. headphone cushion ring for radial resistance component. A grille (a) is required to prevent mechanical vibrations of the foam. The circuits (c) and (d) are equivalent. The current in the LS leads is detected by a small resistance ($1\,\Omega$) and the amplified rectified signal displayed on an ampere meter, calibrated in acoustic Ω (Fig. 14.99). The zero Ω offset comes from the loudspeaker membrane resistance R_m (typically $0.5\,\Omega$ cgs or $0.05\,M\Omega$ S1). A characteristic value of Bl/S is 10^3 SI. Such equipment is robust and, under normal conditions, requires practically no maintenance.

14.5.3 Obtaining the R, L and C values of elements for simulations

For equivalent circuits to be helpful, the element values must be trustworthy. To ensure this, one begins in the core, i.e. the driver or loudspeaker LS, simplifying as far as possible, till only the LS remains. From Sections 14.5.3(b) onwards the routine is to compare measured and simulated SPL frequency responses at every stage of increased complexity, the LS being mounted in a large baffle. Sections 14.5.3(a)–(d) are devoted to the core. Then the peripheral elements are added step by step (Section 14.5.3(e)). In this final stage, if any discrepancies occur there should, under normal circumstances, be no need to doubt the core parameters, such as the movable mass Mms, mechanical compliance Cms of the membrane, or transduction factor Bl etc. Thus the number of parameters to be questioned at any time is kept to a manageable minimum. Attempting to adjust 20 parameters simultaneously is not only a tedious task but also hopeless. To ensure linearity, low signal levels are chosen throughout.

(a) First step: Thiele–Small procedure

Figure 14.100 shows the checklist for this procedure, and can be seen as two parts: the first aimed at Mms and Cms, the second finally yielding Bl (field strength $\times$ wire length in magnet gap) the transduction factor. Only the integral product of B and l is of interest. The value of B itself is of no significance, since its spatial distribution along the coil is far from homogeneous. To obtain Mms, there is no need to cut out the membrane and weigh it with the coil. Thiele [180] and Small [181] invented a practical

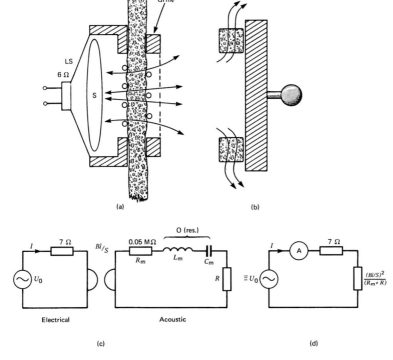

Figure 14.98. Acoustic resistance measuring equipment for (a) porous foam sheets, (b) headphone cushions, (c), (d) equivalent circuits. In parallel with R is the cavity reactance, which might be large enough to neglect if the cavity is small.

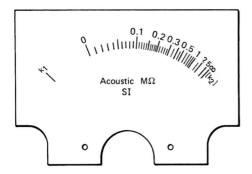

Figure 14.99. Calibrated scale for measuring instrument of Fig. 14.98. The calibration points k_1 and k_2 are for current $I = 0A$ and full scale deflection ($R = \infty$) respectively.

non-invasive procedure, for all main linear parameters of a LS (see also Clark[64] for an extension to non-linear LS parameters). With practice and routine (Excel file) it need take no longer than 20 minutes. Even unavoidable acoustic parameters such as *Lacoust* of air mass in the magnet gap and in rear openings (usually important) can be quantified, as can the air resistance *Racoust* of these elements (usually less important) which may throttle coil motion. The LS driver is stripped to the bare minimum, opening any very closed rear vents if possible, since the inevitable air cushion between

LS Type: _____   Date: _____

Rdc		Ω	Total DC resistance in circuit
MX		Kg	Extra mass of putty (less than expected Mms)
FresX		Hz	Resonance frequency with MX
Fres		Hz	Resonance frequency without MX
Mms		Kg	$\dfrac{MX}{[\,(Fres/FresX)^2 - 1\,]}$ total movable mass (coil, membrane, air etc.)
Cms		m/N	$\dfrac{1}{[\,4\pi^2 \times Fres^2 \times Mms\,]}$ Mechanical compliance of membrane
Vas		m^3	Acoustic compliance of membrane as equivalent air volume at 1000 mBar *Measurable* : with acoustic multimeter.*Otherwise* : If effective membrane area Sd known, Vas = Cms×Sd^2×140000.
Sd		m^2	$\sqrt{\left(\dfrac{Vas}{Cms \times 140000}\right)}$ effective membrane area
Dd		m	$2 \times \sqrt{(Sd/\pi)}$ effective membrane diameter
Us		V	Measured voltage at Fres
Is		A	Measured electrical current at Fres
Zres		Ω	Us/Is Electrical impedance at Fres
r0			Zres/Rdc
√r0			
Zr0		Ω	Rdc×√r0
Ir0		A	Us/Zr0, current where impedance equals Zr0, to look for F1 and F2
F1		Hz	Frequency below Fres, where current Ir0 is found
F2		Hz	Frequency above Fres, where current Ir0 is found
Fcheck		Hz	√(F1×F2) should be close to Fres
Qms			Fres×√r0 / (F2-F1) Mechanical Q factor
Qes			Qms / (r0-1) Electrical Q factor
Qts			1 / ((1/Qms) + (1/Qes)) Total Q factor
Rmech		N/(m/s)	(1/Qms) ×√(Mms/Cms) Damping resistance in mechanical units
B1		Tm	√ (2π×Fres×Mms×Rdc / Qes) Transduction constant

Directly measured parmeters are indicated in **bold** lettering, others are deduced.

Figure 14.100. Checklist for deducing Thiele–Small parameters.

large air masses in these vents and the mechanical mass of the coil plus membrane renders inaccurate the series RLC description upon which this procedure is based. The electrical inductance should be negligible in the frequency range used here, i.e. around the main resonance. The main features are:

1   The resonance of the loudspeaker (LS) is measured by detecting the current minimum for constant applied voltage (Fig. 14.101). The resonance is dictated by mass Mms and compliance Cms, neither of which is known. The mass is increased by a known amount MX using removable putty. It should be intimately in contact with that part of the membrane glued to the coil, without

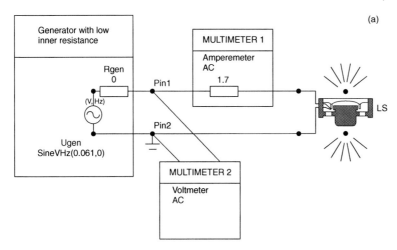

(a)

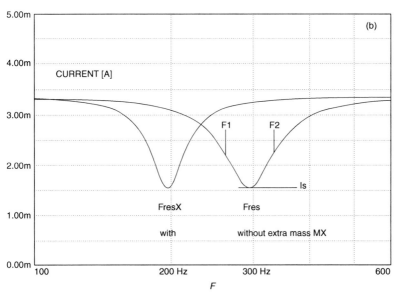

Figure 14.101. (a) Schematic of set-up for Thiele–Small measurements. (b) Simulation of Thiele-Small measurement to illustrate what the parameters mean. In measurement, usually with single frequency points, there is no overview of the curve shapes.

changing the compliance. The two resonances yield two equations and two unknowns:

$$\text{Fres} = \frac{1}{2\pi\sqrt{(\text{Mms}\times\text{Cms})}} \quad \text{and} \quad \text{FresX} = \frac{1}{2\pi\sqrt{((\text{Mms}+\text{MX})\times\text{Cms})}}$$

The 'm' in Mms and Cms means 'mechanical', the s means 'system'. The author sees no pressing reason to depart from this original nomenclature. Note that Cms in vacuum and air should agree.

2 Acoustical compliance Cas (m³/Pa) is equal to Cms × Sd², where Sd is the effective membrane area. Cas can be deduced from Cms and Sd if the effective

diameter Dd is measured with callipers. For torus membranes it may be more reliable to measure Cas directly using an acoustic multimeter of the type described in Fig. 14.96. Then Sd can be deduced combining Cas with Cms. In the absence of such equipment Sd can be adjusted later to make simulated free field SPL agree with baffle measurements.

3   Vas, a well-known acoustic concept among LS engineers, is the acoustic compliance, usually expressed in terms of equivalent air volume in m^3. To be unambiguous one should also state the atmospheric pressure. Assuming air at 1000 mb ($= 10^5$ Pa) Vas $=$ Cas $\times 1.4 \times 10^5$ (see Fig. 14.2).

4   The width of the current minimum is a measure of Bl. The wider the valley, the stronger the field. The Thiele–Small procedure uses the information Us and Is in the middle of the checklist to make a suggestion for further measurements, namely to look for two frequencies below and above Fres, where the current is Ir0. F2–F1 is the width of the valley. The formulae in the checklist then lead directly to Bl.

5   On the way to Bl one obtains the three quality factors: mechanical, electrical and total. Qms gives the mechanical resistance Rmech (N/(m/s)).

6   How do we incorporate Mms in the equivalent circuit? Combining measurements in vacuum and air is a powerful way of obtaining the acoustic elements. The difference in Mms is a direct measure of effective air mass, expressed in mechanical units. Expressed, in keeping with its physical origin, in acoustic units, we get Lacoust $=$ (MmsAir-MmsVacuum)/Sd^2. This can be put into a single element of the gaseous part of the equivalent circuit, assuming the accuracy of the simplified circuit Fig. 14.102(a) suffices. Lacoust represents the total influence of all bottleneck masses, magnet gaps and rear vents etc., thrown into one pot. In the refined circuit (b) this is not so simple. Only inspection of the dimensions within the LS can lead to a reasonable decision as to how much of Lacoust belongs to the magnet gap and how much to the rear vents. Unfortunately we cannot write Lacoust $=$ LrearVents $+$ LinnerGap $+$ LouterGap. Lacoust, coming from Thiele–Small measurements is merely the effective final result from three components not in series.

7   How do we incorporate Rmech in the circuit? Its physical origin can be both mechanical and acoustical. The vacuum value of Rmech is a genuine mechanical damping, representing energy losses within the membrane foil. Unlike viscosity damping there is no reason to expect this to be linear. Fortunately Rmech is almost always so small as to make this limitation imperceptible, which explains why this phenomenon is hardly mentioned in LS literature. The difference of Rmech in air and vacuum, gives the Racoust. As we did for Lacoust we can write Racoust $=$ (RmechAir-RmechVacuum)/Sd^2, which appears explicitly in the simplified circuit. The final remarks about Lacoust above apply also to Racoust.

For reliable measurements note the following:

1   The quantity $\sqrt{(F1 \times F2)}$ should be near Fres, that is, Fres is their average on the logarithmic scale. This is a necessary, but not sufficient indication of a reliable measurement.

2   It is advisable to measure FresX first. Apart from being less demanding (no need to note Us and Is) one might damage the membrane slightly when sticking on the putty. If Cms is thus irreversibly changed, this new (admittedly wrong) value will apply to both measurements, and a reliable measurement of this slightly wrong value will be obtained, (together with a reliable measurement of Mms and Bl). Otherwise there is the double drawback of a wrong value of Cms coupled with internally inconsistent data giving unreliable measurements of everything else. A similar consideration applies to MX: The value of MX should be measured *after* removing it. This ensures that any foreign bodies which were

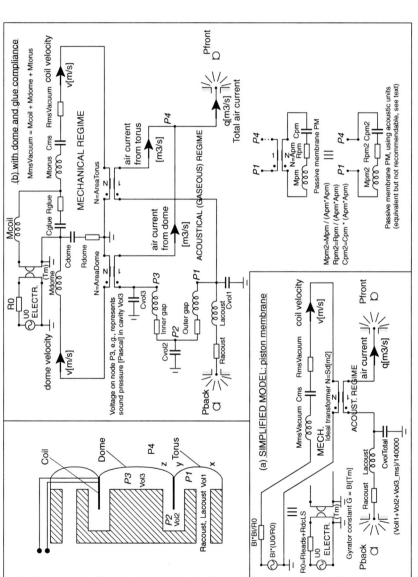

Figure 14.102. The basic equivalent circuits for small loudspeaker drivers with rear opening under the torus. (a) Simplified, (b) more refined.

sticking on the membrane before, and which are removed with the putty, are included. Since the procedure of attaching the ring of putty is somewhat delicate and arduous, one avoids having to do this twice, doing both vacuum and air measurements of FresX first, before removing the putty.

3   The generator should be of negligible internal impedance, otherwise the change in voltage on going from Fres (at which Ir0 was deduced) to F1 or F2 demands a new estimation of Ir0. This iterative procedure, though usually achievable in two steps makes the measurement tiresome, and raises the chances of error.

4   Rdc is the total d.c. resistance in the circuit, not just the LS. Therefore Us should be measured directly at the output of the zero-ohm generator, not at the LS. Although one tends to regard the LS as the sample under test, the ampere-meter with its internal resistance is seen here as an integral part of the sample. Its resistance can be subtracted later if RdcLS is desired. Note that Rdc is highly temperature dependent. It should be measured immediately before doing the measurements. Vacuum and air measurements should be done without too much delay between, to ensure the same conditions. The aim of measuring Rdc is not primarily to obtain the nominal d.c. resistance of the LS, but to know Rdc for the conditions of the Thiele–Small measurements. When connecting the LS terminals, especially for low-impedance loudspeakers, crocodile clips should be avoided. It is worth the trouble of soldering the connections.

5   For torus membranes, the coil may be so fine that the putty cannot be confined to the coil region. It is then acceptable to allow it to touch the dome of the membrane, never the torus, which governs Cms. The quantity of putty should be sufficient to give a generous frequency shift, yet not so much as to load the spring Cms too much. A rule of thumb is to make it clearly less than Mms. This can be decided by trial and error before the actual measurement, making sure that FresX is greater than $Fres/\sqrt{2}$.

6   For high accuracy, if the current minimum is not very sharp, the frequency of the minimum itself should not be directly noted, but rather two frequencies left and right of the minimum where the currents are slightly higher and equal. The average of these two frequencies is that of the current minimum.

7   No generators have zero internal impedance, even those designed especially for LS measurements may have an internal resistance of the order of 0.3Ω, which must be taken into account for a LS of 8Ω or less. When looking for F1 and F2 where the current is Ir0 (higher than the resonance current Is) as in the check list, it is assumed that the applied voltage Us remains unchanged. What we are really looking for is not so much the current Ir0, but the frequency at which the impedance Zr0 applies. Ir0 is just a means of finding this frequency, assuming Us to be still valid. Since the LS impedance at F1 is lower, the voltage at the generator output will decrease at F1 and F2. When near F1 or F2 (let us assume F1) this new voltage Usnew can be noted and the value of Ir0 adjusted to Ir0×Usnew/Us. Then F1 is looked for again, and if necessary Usnew noted again if different from the last value. This is in principle a recursive procedure, but usually the final result is reached in one step. The final Usnew must apply also at F2, so this procedure need only be performed once. If the generator impedance is frequency independent it may be more convenient to include it as one of the parameters in the check list and adjust the formulae accordingly.

8   Here the voltage Us is assumed almost constant. The practice of using a constant current source is not recommended. To maintain the imposed current a strong LS must perform an enormous excursion at resonance, which is likely to make it non-linear.

(b) Simulating sound pressure level (SPL) at a given distance

This will be needed in the following steps. In the radiating piston circuit of Fig. 14.2, R and L are usually so small that a short circuit suffices. In the simulation (Fig.

14.102) the volume current q (m³/s) at the main opening is evaluated. For a large baffle (half space) the resultant sound pressure is:

$$P \text{ (Pa)} = \frac{2 \times \rho \,(\text{kg/m}^3) \times \text{volume acceleration (m}^3\text{/s)}}{4 \times \pi \times \text{distance (m)}}$$

For a LS with closed back in full space (no baffle) the factor 2 is left out. Assuming the air density ρ is 1.2 kg/m³ we can write a more practically useful expression, for a given peak excursion (often of more interest than RMS due to space limitations). For a LS in a large baffle:

$$P \text{ (Pa)} = \frac{5.4 \times \text{area (m}^2) \times F^2 \times \text{peak excursion (m)}}{\text{distance (m)}}$$

For full instead of space (closed back LS) replace 5.4 by 2.7. Directional effects, which start to occur at high frequencies, are ignored here throughout. For an LS smaller than 50 mm on the IEC baffle, a small distance of about 100 mm should be chosen. This ensures that components from the back of the baffle and edge diffraction effects cannot compete with the main output. Measuring not only the SPL Pfront out of the front output but also Pback out of the back yields useful information. The LS need only be reversed.

(c) Evaluating the total cavity volume under the membrane (simple membrane circuit of Fig. 14.102(a))

Usually the geometry is too complicated to calculate the volume from the dimensions. The back vents of the LS should be well closed with putty. Pback should now almost disappear. The main resonance in Pfront shifts far up due to the air cushion. Taking the Thiele–Small parameters as a foundation, we adjust the cavity volume in the simulation until the resonance agrees with measurement. When the back vents are re-opened the values Racoust and Lacoust obtained above should be valid, unless the magnet gap is very narrow. Agreement should be obtained here for mid and low frequencies. Also Pfront and Pback should be equal (figure-of-eight response), unless the back vents are so narrow as to cause air cushion compression. At any rate a Helmholtz resonance always occurs at high frequencies, due to Lacoust and the air cushion under the membrane, giving a doublet in Pfront and a peak in Pback.

(d) Finer points (circuit of Fig. 14.102(b))

Dome compliance At high frequencies the dome compliance Cdome can become important, making the dome permeable to sound, irrespective of coil motion. This compliance can be adjusted in the equivalent circuit to give the measured responses, for both front and back output (which differ at high frequencies due to the air cushion behind the membrane). If the above Helmholtz resonance seems to be absent in Pfront this is typically due to the dome compliance having a critical optimal value. However, the Helmholtz resonance is always visible in Pback, invariably as a peak in the response. The dome compliance represents a spring in the region z (Fig. 14.102(b)). The dome mass Mdome is often so small as to be unimportant. Weighing a piece of foil of the same area gives a rough estimate.

Glue compliance and air gap bottlenecks For loudspeakers with stiff membranes it may be necessary to take into account glue compliance between coil and membrane. Typically this raises the SPL around 10 kHz initially, due to an ultra-sound resonance coming down into the audio range. Now Mtorus and Mcoil can no longer be lumped together. Mtorus is estimated by weighing, as for Mdome. This is also the opportunity to adjust R and L values in the magnet gap, which is usually necessary, since the frequencies at which glue compliance plays a role are high enough to make the magnet gap separate the cavity behind the membrane into three parts, Vol1, Vol2, Vol3. This is often the hardest part of the whole procedure, since a rather large number of parameters must be adjusted simultaneously. However, if not attempted at this stage of simplicity it will be hopeless later.

Torus compliance We have granted that the dome can be compliant enough to be permeable to sound, even for a blocked coil. It would then be inconsistent to ignore this effect in the torus. One might claim already to have allowed for this effect via Cms, but not in such a way as to make the torus itself permeable to sound. Until now the compliance Cms had the one and only function to allow the coil to move. This is insufficient for large flat drivers where the outer torus diameter is much larger than the coil diameter. To illustrate this let us assume a stiff dome. The circuit, as discussed thus far, would predict complete impermeability to sound for a blocked coil, which is unrealistic. The most logical approach to this problem would be to divide the torus into two springs, at x and y (Fig. 14.102(b)), with a torus mass between, extending the circuit accordingly. This does not work. Due to the dynamics of these inherently non-discrete elements the effect of Cms cannot be regarded as resulting from two discrete springs at x and y, in such a way as to account for the permeability to sound realistically at the same time. A practical approach is to add a passive membrane PM with mechanical parameters such as mass Mpm, compliance Cpm, area Apm expressed as estimated percentages of the corresponding torus values Mtorus, Cms, AreaTorus. Unlike the (unsuccessful) logical approach mentioned above, the torus parameters seem to occur twice in the circuit, which is unavoidable. It is an advantage to work consciously with mechanical percentage values. This keeps the estimated numbers from wandering off into totally unrealistic orders of magnitude: clearly the area Apm cannot be ten times the actual torus area. Similarly Mpm and Cpm must bear some relation to Mtorus and Cms. To profit from this advantage the ideal transformer is indispensable. If one leaves it out, using purely acoustic quantities, Mpm2, Rpm2, Cpm2, though this is formally equivalent, one will be working in the dark.

(e) The rest of the system

For systems with non-ideal (e.g. non-cylindrical) bottlenecks and irregular cavities it is not reliable to attempt to deduce R, L, C from first principles using the expressions in Fig. 14.2. A better approach is, again, to simplify the system, as demonstrated below. The acoustic topology must be decided upon, i.e. where the cavities and bottlenecks are likely to be. To illustrate, we assume a not too complicated system with one undivided cavity VOL behind the rear vents of the LS. Communicating with the outside are three groups of different types of openings RL1, RL2, RL3, e.g. holes, slits, porous resistance (or large holes, small holes, porous resistance). All openings are closed, until only VOL behind the LS remains. The response is measured, and the volume adjusted in the simulation for agreement. Next, each of the three groups of openings is opened in turn closing the other two. R and L are adjusted in each case. When all are opened (three impedances in parallel) measurement and simulation should automatically agree. If not, the topology had not been correctly perceived. Another possible source of error is turbulences, which is less likely for low signal levels.

References

1. MÜLLER, G C, BLACK, R and DAVIS, T E, 'The diffraction produced by cylindrical and cubical obstacles and by circular and square plates', *J. Acoust. Soc. Am.*, **10**, 6 (1938).
2. BRANDT, J F and HOLLIEN, H, 'Underwater hearing thresholds in man', *J. Acoust. Soc. Am.*, **42**, 966 (1967).

3. BAUER, B B and TORICK, E L, 'Calibration and analysis of underwater earphones by loudness balance method', *J. Acoust. Soc. Am.*, **39**, 35 (1966).

4. SCHERER, P, 'Inversionsversuch zur Vorn-Hinten-Ortung mit Sinustönen', in *Fortschritte der Akustik, DAGA '84*, DPG-Kongress GmbH, Bad Honnef, 743 (1984).

5. MØLLER, H, HAMMERSHØI, D, JENSEN, C B and SØRENSEN, M F, 'Evaluation of artificial heads in listening tests', *J. Audio Eng. Soc.*, **47**, 83 (1999).

6. SHAW, E A G, 'Ear canal pressure generated by a free sound field', *J. Acoust. Soc. Am.*, **39**, 465 (1966).

7. POLDY, C A, 'Electrical analogs for membranes with application to earphones', *J. Audio Eng. Soc.*, **31**, 817 (1983).

8. BAUER, B B, 'Equivalent circuit analysis of mechano-acoustic structures', *J. Audio Eng. Soc.*, **24**, 643 (1976).

9. FIDI, W and PESCHEL, K, 'Ein integriert-offener Kopfhörer', *Funkschau*, **18**, 693 (1974).

10. BILSEN, F A, CORNELIS, J G and RAATGEVER, J, 'On the time image in lateralization', *J. Acoust. Soc. Am.*, **64**, Suppl. No. 1, 36 (1978).

11. BLAUERT, J, *Räumliches Hören* (Nachschrift, Neue Ergebnisse und Trends seit 1972), S. Hirzel Verlag, Stuttgart, 172 (1974).

12. McFADDEN, D and PASANEN, E G, 'Lateralization at high frequencies based on interaural time differences', *J. Acoust. Soc. Am.*, **59**, 634 (1976).

13. ABEL, S M and KUNOV, H, 'Lateralization based on interaural phase differences: effects of frequency, amplitude, duration, and shape of rise/decay', *J. Acoust. Soc. Am.*, **73**, 955 (1983).

14. KUNOV, H and ABEL, S M, 'Effects of rise/decay time on the lateralization of interaurally delayed 1-kHz tones', *J. Acoust. Soc. Am.*, **69**, 769 (1981).

15. RUDNO-RUDZINSKI, K and RENOWSKI, J, 'Latéralisation en présence de signaux dichotiques à grande différence de niveau sonore', *Acustica*, **41**, 194 (1978).

16. PLENGE, G, 'Über das Problem der Im-Kopf-Lokalisation', *Acustica*, **26**, 241 (1972).

17. PLENGE, G, 'On the differences between localization and lateralization', *J. Acoust. Soc. Am.*, **56**, 944 (1974).

18. SEARLE, C L, BRAIDA, L D, DAVIS, M F and COLBURN, H S, 'Model for auditory localization', *J. Acoust. Soc. Am.*, **60**, 1164 (1976).

19. SCHÖNE, P, 'Der Signalstörabstand bei Kunstköpfen', in *Fortschritte der Akustik, DAGA '80*, VDE-Verlag, Berlin, 835 (1980).

20. THIELE, G, 'Study on the standardisation of studio headphones', Reprinted from EBU Review – Technical, No. 197 (Feb. 1983). (Published by the Technical Centre of the European Broadcasting Union, Avenue Albert Lancaster 32, B-1180 Brussels, Belgium.)

21. SAKAMOTO, N, GOTOH, T and KIMURA, Y, 'On out-of-head localization in headphone listening', *J. Acoust. Soc. Am.*, **24**, 710 (1976).

22. BLAUERT, J, 'Localization and the law of the first wavefront in the median plane', *J. Acoust. Soc. Am.*, **50**, 466 (1971).

23. HAAS, H, 'Einfluß eines Einfachechos auf die Hörsamkeit von Sprache', *Acustica*, **1**, 49 (1951).

24. WALLACH, H, NEWMAN, E B and ROSENZWEIG, M R, 'The precedence effect in sound localization', *(The American J. Psychol.*, **12**, 315 (1949)), *J. Audio Eng. Soc.*, **21**, 817 (1973).

25. HUDDE, H, 'Measurement of the eardrum impedance of human ears', *J. Acoust. Soc. Am.*, **73**, 242 (1983).

26. ZHANG, M, TAN, K C and ER, M H, 'Three-dimensional sound synthesis based on head-related transfer functions', *J. Audio Eng. Soc.*, **46**, 836 (1998).

27. RICHARDS, W D, FRANK T A and PROUT, J H, 'Influence of earphone-cushion center-hole diameter on the acoustic output of audiometric earphones', *J. Acoust. Soc. Am.*, **65**, 257 (1979),

28. STINSON, M R, SHAW, E A G and LAWTON, B W, 'Estimation of acoustical energy reflectance at the eardrum from measurements of pressure distribution in the human ear canal', *J. Acoust. Soc. Am.*, **72**, 766 (1982).

29. SHAW, E A G, 'Transformation of sound pressure level from the free field to the eardrum in the horizontal plane', *J, Acoust. Soc. Am.*, **56**, 1848 (1974).

30. CORSO, J F and LEVINE, M, 'Pitch discrimination at high frequencies by air and bone conduction', *Am. J. Psychol.*, **78**, 557 (1963).

31. PUMPHREY, R J, 'Upper limit of frequency for human hearing', *Nature*, **166**, 571 (1950).

32. KHANNA, S M, TONNDORF, J and QUELLER, J E, 'Mechanical parameters of hearing by bone conduction', *J. Acoust. Soc. Am.*, **60**, 139 (1976).

33. TONNDORF, J and JAHN, A F, 'Velocity of propagation of bone-conducted sound in a human head', *J. Acoust. Soc. Am.*, **70**, 1294 (1981).

34. YUND, E W, EFRON, R and DIVENYI, P L, 'The effect of bone conduction on the intensity independence of dichotic chords', *J. Acoust. Soc. Am.*, **65**, 259 (1979).

35. BRINKMANN, K and RICHTER, U, 'Kopfhörer DT48: Schalldämmung und Ohrverschluß-Effekt', *Acustica*, **47**, 53 (1980).

36. SCHERER, P, 'Inversionsversuch zur Vorne-Hinten-Ortung', in Fortschritte der *Akustik, DAGA '82*, DPG Kongress GmbH, Bad Honnef, 1203 (1982).

37. SCHERER, P, 'Verbesserte Wiedergabe von Phantomquellen', *Proceedings 11th International Conference on Acoustics*, ICA, Paris, 147 (1983).

38. KUHN, G F and GUERNSEY, R M, 'Sound pressure distribution about the human head and torso', *J. Acoust. Soc. Am.*, **73**, 95 (1983).

39. ROBINSON, D W, WHITTLE L S and BOWSHER, J M, 'The loudness of diffuse sound fields', *Acustica*, **11**, 397 (1961).

40. MATHERS, C D and LANSDOWNE, K F, 'Hearing risk to wearers of circumaural headphones: an investigation', BBC Report RD 1979/3.

41. RUDMOSE, W, 'The case of the missing 6 dB', *J. Acoust. Soc. Am.*, **71**, 650 (1982).

42. LIPSCHITZ, S P and VANDERKOOY, J, 'The great debate: subjective evaluation', *J. Audio Eng. Soc.*, **29**, 482 (1981).

43. SCHRÖTER, J and ELS, H, 'On basic research towards an improved artificial head for the measurement of hearing protectors', *Acustica*, **50**, 250 (1982).

44. HEYSER, R C, 'Loudspeaker phase characteristics and time delay distortion: Part 1 and 2, *J. Audio Eng. Soc.*, **17**, 30 and 130 (1969)

45. BLAUERT, J, *Spatial Hearing*, MIT Press, Cambridge, MA (1983).

46. BLAUERT, J, *Räumliches Hören*, S. Hirzel Verlag, Stuttgart (1974).

47. HARRIS, G G, 'Binaural interaction of impulsive stimuli and pure tones', *J. Acoust. Soc. Am.*, **32**, 685 (1980).

48. BLAUERT, J, 'Binaural localization: multiple images and applications in room- and electroacoustics', in *Localization of Sound: Theory and Applications*, GATEHOUSE, R W, Ed., Amphora Press, Connecticut, 65 (1982).

49. RHENIUS, V, 'Entwicklung eines diffusfeldentzerrten Kopfhörers', *Fernseh- und Kino-Technik*, **40**, 201 (1986).

50. SCHARF, B, FLORENTINE, M and MEISELMANN, C H, 'Critical band in auditory lateralization', *Sensory Proc.*, **1**, 109 (1976).

51. SHAW, E A G, '1979 Rayleigh medal lecture: the elusive connection', in *Localization of Sound: Theory and Applications*, GATEHOUSE, R W, Ed., Amphora Press, Connecticut, 13 (1982).

52. KUHN, G F, 'Model for the interaural time differences in the azimuthal plane', *J. Acoust. Soc. Am.*, **62**, 157 (1977).

53. HEBRANK, J H, 'Pinna disparity: a case of mistaken identity?' *J. Acoust. Soc. Am.*, **59**, 220 (1976).

54. BLAUERT, J, 'Sound localization in the median plane', *Acustica*, **22**, 205 (1969/70).

55. BURKHARD, M D and SACHS, R M, 'Anthropometric manikin for acoustic research', *J. Acoust. Soc. Am.*, **58**, 214 (1975).

56. HEBRANK, J and WRIGHT, D, 'Are two ears necessary for localization of sound sources in the median plane?' *J. Acoust. Soc. Am.*, **56**, 935 (1974).

57. SEARLE, C L, BRAIDA, L D, CUDDY, D R and DAVIS, M F, 'Binaural pinna disparity: another auditory localization cue', *J. Acoust. Soc. Am.*, **57**, 448 (1975).

58. HAAS, H, 'The influence of a single echo on the audibility of speech', *J, Audio Eng. Soc.*, **20**, 146 (1972).

59. SHAW, E A G, 'External ear response and sound localization', *in Localization of Sound: Theory and Applications*, GATEHOUSE, R W, Ed., Amphora Press, Connecticut, 30 (1982).

60. ZWISLOCKI, J, 'Some impedance measurements on normal and pathological ears', *J. Acoust. Soc. Am.*, **29**, 1312 (1957).

61. TONNDORF, J and KHANNA, S M, 'Tympanic-membrane vibrations in human cadaver ears studied by time-averaged holography', *J. Acoust. Soc. Am.*, **52**, 1221 (1972).

62. ZWISLOCKI, J, 'In search of the bone-conduction threshold in a free sound field', *J. Acoust. Soc. Am.*, **29**, 795 (1957).

63. FRANKE, E K, 'Response of the human skull to mechanical vibrations', *J. Acoust. Soc. Am.*, **28**, 1277 (1956).

64. CLARK, D, 'Precision measurement of loudspeaker parameters', *J. Audio Eng. Soc.*, **45**, 129 (1997).

65. BOGERT, B P, 'Response of an electrical model of the cochlea partition with different positions of excitation', *J. Acoust. Soc. Am.*, **29**, 789 (1957).

66. BÉKÉSY, G VON, , 'Zur Theorie des Hörens bei der Schallaufnahme durch Knochenleitung', *Ann. Physik*, **13**, 111 (1932).

67. SILVER, S L, 'Acoustic energy transformation in the human auditory system', *J. Audio Eng. Soc.*, **23**, 33 (1975).

68. FELDMAN, A S and ZWISLOCKI, J, 'Effect of acoustic reflex on the impedance at the eardrum', *J. Speech and Hearing Res.*, **8**, 213 (1965).

69. HOWES, W L, 'Loudness of steady sounds', *Acustica*, **41**, 277 (1979).

70. BÉKÉSY, G VON, *Experiments in Hearing*, McGraw-Hill, New York (1960).

71. SCHROEDER, M R, 'Models of hearing', *Proc. IEEE*, **63**, 1332 (1975).

72. ZWISLOCKI, J, 'Acoustic attenuation between the ears', *J. Acoust. Soc. Am.*, **25**, 752 (1953).

73. FRANKE, E K, VON GIERKE, H E, GROSSMAN, F M and VON WITTERN, W W, 'The jaw motions relative to the skull and their influence on hearing by bone conduction', *J. Acoust. Soc. Am.*, **24**, 142 (1952).

74. BÁRÁNY, E, 'A contribution to the physiology of bone conduction', *Acta Oto-laryngol* (Stockholm), Suppl. 26 (1938).

75. TONNDORF, J, 'Compressional bone conduction in cochlear models', *J. Acoust. Soc. Am.*, **34**, 1127 (1962).

76. BÉKÉSY, G VON, *J. Acoust. Soc. Am.*, **27**, 155 (1955).

77. WEVER, E G, 'Recent investigations of sound conduction, Part II: The ear with conductive impairment', *Ann. Otol. Rhinol. and Laryngol.*, **59**, 1037 (1950).

78. VILLCHUR, E, 'Free field calibration of earphones', *J. Acoust. Soc. Am.*, **46**, 1527 (1969).

79. BOCKER, P and MRASS, H, 'Zur Bestimmung des Freifeldübertragungsmasses von Kopfhörern', *Acustica*, **9**, 340 (1959).

80. PUDDIE RODGERS, C A, 'Pinna transformations and sound reproduction', *J. Audio Eng. Soc.*, **29**, 226 (1981).

81. ZUREK, P M, 'The precedence effect and its possible role in the avoidance of interaural ambiguities', *J. Acoust. Soc. Am.*, **67**, 952 (1980).

82. VETT, I and SANDER, H, 'Production of spatially limited 'diffuse' sound field in an anechoic room', *J. Audio Eng. Soc.*, **35**, 138 (1987).

83. LIPPMANN, R P, 'MX41/AR earphone cushions versus a new circumaural mounting', *J. Acoust. Soc. Am.*, **69**, 589 (1981).

84. BERGER, E H and KERIVAN, J E, 'Influence of physiological noise and the occlusion effect on the measurement of real-ear attenuation at threshold', *J. Acoust. Soc. Am.*, **74**, 81 (1983).

85. LIPSCHITZ, S P, POCOCK, M and VANDERKOOY, J, 'On the audibility of midrange phase distortion in audio systems', *J. Audio Eng. Soc.*, **30**, 580 (1982).

86. SHAW, E A G, 'Hearing protector attenuation: a perspective view', *Applied Acoustics*, **12**, 139 (1979).

87. MOORE, B C J, GLASBERG, B R and BAER, T, 'A model for the prediction of thresholds, loudness and partial loudness', *J. Audio Eng. Soc.*, **45**, 224 (1997).

88. RHENIUS, V, 'Kopfhörer: Messverfahren im Vergleich, wer hat das richtige Maß?' *Funkschau*, **17**, 44 (1983).

89. POLDY, C A, 'Philosophy of modern headphone design', *Australian Sound and Recording*, **3**, 31 (1981).

90. WISCHGOLF, K J, 'Der halboffene Stereokopfhörer', *Funkschau*, **4**, 73 (1981).

91. KILLION, M C, 'Revised estimates of minimum audible pressure: where is the missing 6 dB?' *J. Acoust. Soc. Am.*, **63**, 1501 (1978).

92. BAUER, B B, 'On the equivalent circuit of a plane wave confronting an acoustical device', *J. Audio Eng. Soc.*, **24**, 653 (1976).

93. MUNSON, W A and WIENER, F M, 'In search of the missing 6 dB', *J. Acoust. Soc. Am.*, **24**, 498 (1952).

94. BLAUERT, J, LINDEMANN, W and GRUBER, K, 'Model of a central lateralization processor – quantitative evaluation', *J. Acoust. Soc. Am.*, **71** (Suppl.), SB7 (1982).

95. PÖSSELT, C, 'Der Verschlusseffekt als Funktion der Verschlussimpedanz', in *Fortschritte der Akustik, DAGA '82*, DPG Kongress GmbH, Bad Honnef, 1137 (1982).

96. ZWISLOCKI, J, 'Factors determining the sound attenuation produced by earphone sockets', *J. Acoust. Soc. Am.*, **27**, 146 (1955).

97. SHAW, E A G and THIESSEN, G J, 'Improved cushion for ear defenders', *J. Acoust. Soc. Am.*, **30**, 24 (1958).

98. SCHRÖTER, J, 'Messung der Schalldämmung von Gehörschützern mit einem physikalischen Verfahren (Kunstkopfmethode)', Doctoral thesis, Ruhr-Universität Bochum (1983).

99. KNUDSEN, E I, 'The hearing of the barn owl', *Scient. American*, **83** (Dec. 1981).

100. ANDERSON, C M B and WHITTLE, L S, 'Physiological noise and the missing 6 dB', *J. Acoust. Soc. Am.*, **24**, 261 (1971).

101. SIVIAN, L J and WHITE, S D, 'On minimum audible sound fields', *J. Acoust. Soc. Am.*, **4**, 288 (1933).

102. SONE, T, EBATA, M and NIMURA, T, 'On the difference between localization and lateralization', 6th ICA Congress (Tokyo), A-29 (1968).

103. OHM, G S, 'Noch ein Paar Worte über die Definition des Tones', *Ann. Physik, Chem.*, **62**, 1 (1844).

104. HELMHOLTZ, H VON, *On the Sensations of Tone*, translated by ELLIS, A J, 2nd edn, Longmans, Green and Co., London (1885), Dover, New York (1954).

105. CRAIG, J H and JEFFRESS, L A, 'Why Helmholtz couldn't hear monaural phase effects', *J. Acoust. Soc. Am.*, **32**, 884 (1960).

106. HANSEN, V and MADSEN, E R, 'On aural phase detection', *J. Audio Eng. Soc.*, **22**, 10 (1974).

107. HANSEN, V and MADSEN, E R, 'On aural phase detection: Part II', *J. Audio Eng. Soc.*, **22**, 783 (1974).

108. HEYSER, R C, 'Determining the acoustic position for proper phase response of transducers', *J. Audio Eng. Soc.*, **32**, 23 (1984).

109. STODOLSKY, D S, 'The standardization of monaural phase', *IEEE Trans. on Audio and Electroacoustics*, **AU-18**, 288 (1970).

110. PREIS, D, 'Phase distortion and phase equalization in audio signal processing – a tutorial review', *J. Audio Eng. Soc.*, **30**, 774 (1982).

111. MORGAN, D E and DIRKS, D D, 'Loudness discomfort level under earphone and in the free field: the effects of calibration methods', *J. Acoust. Soc. Am.*, **56**, 172 (1974).

112. ALGAZI, V R, AVENDANO, C and THOMPSON, D, 'Dependence of subject and measurement position in binaural signal acquistion', *J. Audio Eng. Soc.*, **47**, 937 (1999).

113. HASEGAWA, I, OCHI, M and MATSUZAWA, K, 'Acoustic radiation pressure on a rigid sphere in a spherical wave field', *J. Acoust. Soc. Am.*, **67**, 770 (1980).

114. DIN 45619/1, 'Headphones: determination of the free-field sensitivity level by loudness comparison with a progressive sound wave' (1975).

115. DIN 45500/1, 'Hi-Fi technics; requirements for electrodynamic headphones' (1975).

116. GENUIT, K, 'Untersuchungen zur Bedeutung von einzelnen Strukturen der Aussenohrübertragungsfunktion auf das räumliche Hören', in *Fortschritte der Akustik, DAGA '86*, DPG-Bad Honnef, 485 (1986).

117. GENUIT, K, 'Untersuchungen der psychoakustischen Eigenschaften von Höreignissen bei der Kopfhörerwidergabe', in *Fortschritte der Akustik, DAGA '86*, DPG-GmbH, Bad Honnef, 489 (1986).

118. SHAW, E A G and THIESSEN, G J, 'Acoustics of circumaural earphones', *J. Acoust. Soc. Am.*, **34**, 1233 (1962).

119. ZWISLOCKI, J, 'Development of a semiplastic earphone socket', *J. Acoust. Soc. Am.*, **27**, 155 (1955).

120. THOMPSON, W Jr, 'Collection of equivalent circuit representations of electric-field coupled transducers', *J. Acoust. Soc. Am.*, **72**, 1062 (1982).

121. WILLIAMS, P and WILKINS, P, 'A hearing protector for improved communications in intermittent noise', 11th International Congress on Acoustics, ICA, Paris, **3**, 281 (1983).

122. GÄNSLER W, 'Sprechzeuge – Grundsätzliches und Entwicklungstendenzen, *Z. Post. u. Fernmeldewes.*, **21**, 562 (1969).

123. ROTH, W, 'Für Hörbehinderte: Musikhören mit Genuss', *Funkschau*, **3**, 67 (1982).

124. BROGDEN, W J and MILLER, G A, 'Physiological noise generated under earphone cushions', *J. Acoust. Soc. Am.*, **19**, 620 (1947).

125. STUDEBAKER, G A, 'On masking in bone-conduction audiometry', *J. Speech and Hearing Res.*, **5**, 215 (1962).

126. VOGEN, C S and COPELAND, A B, 'Comparison of the reliability of an Auraldome headset and a standard headset in obtaining pure-tone audiometric threshold measures', *J. Acoust. Soc. Am.*, **60**, 256 (1976).

127. HILDYARD, V H, DENVER, M D and VALENTINE, M A, 'Collapse of the ear canal during audiometry', *Arch. Otolaryngol.*, **75**, 422 (1962).

128. MICHAEL, P L and BIENVENUE, G R, 'Calibration data for a circumaural headset designed for hearing testing', *J. Acoust. Soc. Am.*, **60**, 944 (1976).

129. WATSON, N A and GALES, R S, 'Bone conduction threshold measurements: effects of occlusion, enclosures, and masking devices', *J. Acoust. Soc. Am.*, **14**, 207 (1943).

130. HOOD, J D, 'Bone conduction: a review of the present position with especial reference to the contributions of Dr Georg von Békésy', *J. Acoust. Soc. Am.*, **34**, 1325 (1962).

131. HUIZING, E H, 'Bone conduction – the influence of the middle ear', *Acta Oto-Laryngologica*, Suppl. 155, 1–99 (1960).

132. LITTLER, T S, KNIGHT, J J and STRANGE, P H, 'Hearing by bone conduction and the use of bone-conduction hearing aids', *Proc. Roy. Soc. Med.*, **45**, 783 (1952).
133. HOOD, J D, 'The principles and practice of bone conduction audiometry', *Proc. Roy. Soc. Med.*, **50**, 689 (1957), and *Laryngoscope*, **70**, 1211 (1960).
134. ZWISLOCKI, J J, 'Masking: experimental and theoretical aspects of simultaneous, forward, backward and central masking', in *Handbook of Perception*, Vol. 4, CARTERETTE, E C and FRIEDMAN, M P, Eds (1978).
135. GENUIT, K, 'Strukturbestimmende Merkmale von Aussenohrübertragungseigenschaften und deren Abhängigkeit von der Schalleinfallsrichtung', in *Fortschritte der Akustik, DAGA '82*, DPG Kongress GmbH, Bad Honnef, 1195 (1982).
136. SCHÖNE, P, 'Ein Beitrag zur Kompatibilität raumbezogener und kopfbezogener Stereofonie', *Acustica*, **47**, 170 (1981).
137. SCHÖNE, P, 'Zur Nutzung des Realisierungsspielraums in der Kopfbezogenen Stereophonie', *Rundfunktechn. Mitteilungen*, **28**, 1 (1980).
138. HUDDE, H and SCHRÖTER, J, 'The equalization of artificial heads without exact replication of the eardrum impedance', *Acustica*, **44**, 301 (1980).
139. KLEINER, M, 'Problems in the design and use of dummy-heads', *Acustica*, **41**, 183 (1978).
140. GENUIT, K, 'Ein Beitrag zur Optimierung eines Kunstkopfaufnahmesystems', *Studio*, **10**, Dec. (1981).
141. CHONG-JIN TAN, WOON-SENG GAN, 'Direct concha excitation for the introduction of individual hearing cues', *J. Audio Eng. Soc.*, **48**, 642–653 (2000).
142. KILLION, M C, 'Noise of ears and microphones', *J. Acoust. Soc. Am.*, **59**, 424 (1976).
143. BLAUERT, J, 'Vergleich unterschiedlicher Systeme zur originalgetreuen elektroakustischen Übertragung', *Rundfunktechn. Mitteilungen*, **18**, 222 (1974).
144. HUDDE, H and SCHRÖTER, J, 'Verbesserung am Neumann-Kunstkopf-system', *Rundfunktechn. Mitteilungen*, **25**, 5 (1981).
145. THIELE, G, 'Zur Kompatibilität von Kunstkopfsignalen mit intensitätsstereofonen Signalen bei Lautsprecherwiedergabe: Die Richtungsabbildung', *Rundfunktechn. Mitteilungen*, **25**, 11 (1981).
146. MILLER, G A, 'Sensitivity to changes in the intensity of white noise and its relation to masking and loudness', *J. Acoust. Soc. Am.*, **19**, 609 (1947).
147. JECKLIN, J, *Musikaufnahmen*, Franzis Verlag, Munich (1980).
148. WOLLHERR, H, 'Mikrofonankopplung an das Aussenohr eines neuen Kunstkopfes', *Rundfunktechn. Mitteilungen*, **25**, 27 (1981).
149. BAUER, B, 'Stereophonic earphones and binaural loudspeakers', *J. Audio Eng. Soc.*, **9**, 148 (1961).
150. BLAUERT, J and LAWS, P, 'Verfahren zur orts- und klanggetreuen Simulation von Lautsprecherbeschallungen mit Hilfe von Kopfhörem', *Acustica*, **29**, 273 (1973).
151. MEHRGARDT, S and MELLERT, V, 'Transformation characteristics of the external human ear', *J. Acoust. Soc. Am.*, **61**, 1567 (1977).
152. DIN 45581, 'Headphones: conditions and procedures for type tests' (1975).
153. SANK, J R, 'Improved real-ear tests for stereophones', *J. Audio Eng. Soc.*, **28**, 206 (1980).
154. ZWISLOCKI, J J, 'An acoustic coupler for earphone calibration', Special Report LSC-S-7, Syracuse University, New York (1970).
155. BLAUERT, J, MELLERT, V, PLATTE, H J, LAWS, P, HUDDE, H, SCHERER, P, POULSEN, T, GOTTLOB, D and PLENGE, G, 'Wissenschaftliche Grundlagen der kopfbezogenen Stereofonie', *Rundfunktechn. Mitteilungen*, **22**, 195 (1978).
156. DE-OS 30 26 291 (Offenlegungsschrift), (1980).
157. BUTLER, R A and BELENDIUK, K, 'Spectral cues utilized in the localization of sound in the median sagittal plane', *J. Acoust. Soc. Am.*, **61**, 1264 (1977).
158. BAUER, B B and TORICK, E L, 'Experimental studies in underwater directional communication', *J. Acoust. Soc. Am.*, **39**, 25 (1966).
159. SCHUMANN, J, ABBAGNARO, L A and BAUER, B B, 'An improved underwater earphone', *J. Acoust. Soc. Am.*, **50**, 1217 (1971).
160. BAUER, B B, 'Comments on 'Effect of air bubbles in the external auditory meatus on underwater hearing thresholds'', *J. Acoust. Soc. Am.*, **47**, 1465 (1970).
161. MERHAUT, J, *Theory of Electroacoustics*, McGraw-Hill, New York (1981).
162. OLSON, H F, *Acoustical Engineering*, Van Nostrand, Princeton, NJ (1957).
163. LEACH, W M, Jr, 'On the specification of moving-coil drivers for low-frequency horn-loaded loudspeakers', *J. Audio Eng. Soc.*, **27**, 950 (1979).
164. RSCHEVKIN, S N, *Hochfr. u. Elektroak.*, **67**, 128 (1959).
165. CREMER, L and MÜLLER, H A, *Die wissenschaftlichen Grundlagen der Raumakustik*, Vol. 2, Hirzel, Stuttgart (1976).

166. ADKINS, J M and SORKIN, R D, 'Effect of channel separation on earphone-presented tones, noise, and stereophonic material', *J, Audio Eng. Soc.*, **33**, 234 (1985).
167. HUNT, F V, *Electroacoustics*, Harvard University Press/John Wiley, New York (1954).
168. DIRKS, D and SWINDEMAN, J G, 'The variability of occluded and unoccluded bone-conduction thresholds', *J. Speech Hear. Res.*, **10**, 232 (1967).
169. DIN 45500/10, 'Hi-Fi technics: requirements for electrodynamic headphones' (1975).
170. DIN 45619/2, 'Headphones: determination of the free-field sensitivity level by loudness comparison with a reference headphone' (1975).
171. IEC 581, 'High fidelity audio equipment and systems: Minimum performance requirements', 581–10 (Part 10): Headphones (1986).
172. IEC 267-7, 'Sound system equipment', Part 7: Headphones and headsets (1984).
173. ANSI S3.7:1973 (R1979), 'American National Standard Method for Coupler Calibration of Earphones'.
174. ANSI S3.25–1979, 'American National Standard for an Occluded Ear Simulator'.
175. *Reallexikon der Akustik*. Erwin Bochinsky, Frankfurt am Main, pp. 168 and 207 (1982).
176. STINSON, M R and SHAW, E A G, 'Acoustic impedance of small, circular orifices in thin plates', *J. Acoust. Soc. Am.*, **77**(6), 2039, June (1985).
177. ALEXANDER, F M, *Constructive Conscious Control of the Individual*, Victor Gollancz, London (1992).
178. THEILE, A N, 'Force conversion factors of a loudspeaker driver', *J. Audio Eng. Soc.*, **41**, 701 (1993).
179. INGARD, K U, *Notes on Sound Absorption Technology*, Noise Control Foundation, p.26 of Chapter 2 (1994).
180. THEILE, A N, 'Loudspeaker in vented boxes: Part 2', *Proceedings of the IRE Australia*, **22**, pp.487–508 (Aug. 1961). For part 1 see *JAES*, **19**, 382–392 (May 1971).
181. SMALL, R H, *IEEE Transactions on Audio and Electroacoustics*, vol. AU-19, 269–281 (Dec. 1971).
182. POSSELT, C, Dissertation 'Einfluss der Verschluss-spezifischen Knochenschallhörschwelle des Menschen auf die objektive Messung der Schalldämmung von Gehörschützen', Bochum (1984).
183. *Sound system equipment – Part 7: Headphones and earphones*. International Standard CEI/IEC 268–7 (1996).
184. SLOTTE, B, 'Acoustics simulation in mobile phone audio design' Master of Science thesis. Espoo, Finland, Helsinki University of Technology (1999).
185. EGOLF, D P, 'Experimental scheme for analyzing the dynamic behaviour of electroacoustic transducers', *J.Acoust. Soc.Am.*, **62**, No. 4, October (1977).
186. RASMUSSEN, K, 'A note on the acoustic impedance of narrow tubes', *Acustica*, **51**, 72 (1982).

15 International standards

J. M. Woodgate

15.1 Introduction

In terms of the practical use of standards (and standards are pointless if they are not used), conventional dictionary definitions of the word 'standard' are inappropriate. Nor is the official definition given in ISO Guide 2[1] a very satisfactory statement. It reads: 'A technical specification or other document available to the public, drawn up with the cooperation and consensus or general approval of all interests affected by it, based on the consolidated results of science, technology and experience, aimed at the promotion of optimum community benefits and approved by a body recognized on the national, regional or international level.' This definition is rather verbose and contains both implicit and explicit value judgements. The definition given in BS 0[2] is not much better.

A simpler, and therefore perhaps a better definition is: 'A standard is an agreement by a group of people to adopt a defined practice.' Presumably the participants to the agreement expect some benefit from their participation, but that is not a property of the agreement itself, which may in the event prove to bring quite unexpected results, both desirable and, occasionally, otherwise. It is important to recognize that if the practice is not, in fact, defined, or is inadequately defined, the resulting document will give rise to contention rather than agreement, the consequences of which may be costly.

The content of a standard may be a definition, an abbreviation or symbol, a method of measurement or a performance requirement (or, of course, a related series of any of these). However, it is necessary to distinguish carefully between these concepts themselves in order to maintain a clear understanding of the purpose and scope of individual standards.

A standard draws only initial authority from the status of the issuing body. Its ongoing authority derives from the degree of acceptance that it receives in the field and the use that is made of it in practice. Standards-making authorities do well to bear this principle in mind, because it has a major influence on the number of copies of the standard that will be sold, and thus on the economic viability of drawing up the standard in the first place.

A powerful standards-making body may issue a standard which does not in fact embody a consensus agreement of interested parties. Such a standard is basically flawed, and its effect may range from simply being a nuisance to being a serious barrier to innovation, communication or trade. These are all situations whose main adverse effects fall on the relevant sector of industry. It is essential, therefore, that industry should clearly recognize that its interests lie in adequate participation in the standards-making process, at both national and international level. The existence of a regional level, referred to in the ISO definition above, tends to exacerbate the difficulty of avoiding standards which serve purely regional interests; and, in the fullness of time, bodies at this level may be seen to serve no useful purpose and may be eliminated.

The most important standards in the present context ought to be those produced by the International Electrotechnical Commission (IEC), because they represent the results of consensus among the major manufacturing and using countries. However, these standards are not as well known in many countries as they should be. This is partly because, in the past, international standards-making was a very time-consuming process, with the result that published standards were often not available when wanted, and were out of date when they eventually appeared. Considerable improvements have been made in recent years, and most standards are now relevant and up to date when published. Indeed, some are sufficiently forward-looking to be still satisfactory at the end of the normal five-year planned lifetime. This ideal will never be universally achieved because commercial considerations can in some cases influence the speed of change, both by delaying the introduction of new material into standards under revision (for fear of making current products obsolescent overnight or by causing confusion in the marketplace because of changes in methods of specification) and by forcing innovation at an unforeseen rate in order to sustain demand.

National standards can often be revised on a shorter timescale, but this possibility is not without dangers, since it provides temptation to recreate national differences subsequent to the achievement of international agreement, perhaps with the purest of motives such as a desire to include the results of recent research. The dangers can be minimized if national standards bodies bring their new material at an early stage to the appropriate international forum for speedy consideration. By exposure to a wider group, improvement of the new material may be a further beneficial consequence of this process.

15.2 Standards-making bodies

One of the most confusing aspects of the standards business is the proliferation of acronyms for the lengthy titles of standards-making bodies. The following list is not by any means exhaustive; references to national committees are confined to those which participate in international standardization work on loudspeakers through the IEC.

15.2.1 International bodies

ISO (International Organization for Standardization). Concerned with standardization in fields other than electrotechnology, including mechanical and purely acoustic technology.

IEC (International Electrotechnical Commission). Concerned with international standardization in electrotechnology. Electroacoustics is within the terms of reference of the IEC and close liaison is maintained with ISO on matters of joint interest.

ITU Radio Communications Bureau, formerly CCIR (Comité Consultatif International de Radio). The bureau would deny that it is a standards-making body, but the Recommendations and Reports that it produces in the field of broadcasting technology become *de facto* standards in many countries.

15.2.2 Regional or supranational bodies

CENELEC (European Committee for Electrotechnical Standardization). Comprises 18 member countries from the former EEC and EFTA.

ARSO (African Regional Organization for Standardization). Comprises members of the UN Economic Commission for Africa, and of the OAU.

ASMO (Arab Organization for Standardization and Metrology).

15.2.3 National bodies

GOSTR (Committee of the Russian Federation for Standardization, Metrology and Certification).

ENIU (Italian National Board of Standardization).

PKNMJ (Polish Committee for Standardization).

SAA (Standards Association of Australia).

ANSI (American National Standards Institute).

ANSI delegates some of its activities to other bodies, and there are other autonomous standards-making bodies in the USA. Consequently, the standards scene appears complex. The Institute of Electrical and Electronic Engineers (IEEE), the Audio Engineering Society (AES) (qv.), and the Electronic Industry Association (EIA) issue industry standards or recommendations which may become *de facto* national or even international standards. The same is to some extent true of the former Institute for High Fidelity (IHF), although IHF standards which have not been endorsed by IEE or EIA are less widely recognized outside the USA. Such standards may be endorsed by ANSI but be published by the originating organization(s).

BSI (British Standards Institution).

NEC (Netherlands Standardization Institute).

JIS (Japanese Industrial Standards Committee).

Also JSA (Japanese Standards Association). The Electronic Industries Association of Japan (EIAJ) issues industry standards which are eligible for adoption as national standards after two years, and may also become *de facto* international standards.

DIN (German Institute for Standardization).

Because the range of DIN standards is very comprehensive, especially in the fields of electronics and mechanical engineering, these standards may in some cases become *de facto* international standards.

AFNOR (French Association for Standardization),

CSA (Canadian Standards Association).

15.2.4 The Audio Engineering Society

The Audio Engineering Society (AES) is a special case. Truly an international body, its members are individual engineers and scientists, mostly from industry and academic institutions, whereas the actual members of other international bodies (ISO, IEC) are national standards committees, which are governmental or quasi-autonomous national bodies. Because of the greater degree of concentration and control which can be exercised, the AES is often the audio industry's forum of choice for the discussion of important matters, especially concerning new technology. However, because of the complexity of US law concerning any procedures which might result in a restraint of trade, the Society has adopted Standards Committee procedures that ensure wide representation of all interests in the drafting processes, and the Society has now published a considerable number of standards.

By the terms of its constitution, the IEC was, until recently, unable to receive delegates other than from national standards committees; thus it could not formally recognize the AES. (Certain other bodies were in a similar position.) The IEC has now introduced a 'D-liaison' status, which allows appropriate representative industry bodies to participate directly in IEC work.

15.3 Standards for loudspeakers

15.3.1 International standards

IEC standards have been prepared on: (a) methods of measurement for loudspeakers, (b) dimensions, and (c) performance requirements for household high fidelity loudspeakers.

(a) Methods of measurement

This standard is known as IEC 60268–5[3]; the First Edition was published in 1972, and the Second Edition in 1989. A revision is under way (2000).

The standard is written in such a way as to encourage uniformity in the preparation and presentation of specifications. In the first section, the conditions for measurements are discussed and stipulated. It is clearly necessary to define these 'rated conditions' (see Section 16.2.2) first of all. The second section is devoted to describing the important characteristics of loudspeakers for purposes of specification, giving the agreed methods of measurement for these characteristics and indicating how the results should be presented. The latter is most important in countering 'specmanship', the deliberate manipulation of specification data to give a false or misleading impression of the performance of a product. The final section of the document lists the characteristics which the manufacturer should include in his specification, indicating data to be marked on the loudspeaker itself, data to be included in the published specification, and additional data which may be provided at the manufacturer's option.

Considerable increases in understanding of the functioning of loudspeakers and enclosures have occurred in the last two decades[4–6] and the results of much of this research have been included in the Second Edition. The most significant changes are in the determination of the maximum tolerable energy input (where the true nature of the long-term life test described in the First Edition[3] is recognized) and methods are given for measuring the maximum tolerable energy input over shorter, and at least equally important, time periods, and in the determination of the drive unit characteristics which form the basis of closed and vented enclosure designs, the so-called Thiele–Small parameters[5].

(b) Dimensions

The relevant standard is IEC 60268–14, the First Edition[7] of which was published in 1971. Plans to revise this standard disclosed an example of the non-technical (in the sense, here, of electroacoustic technology) considerations which affect standards-making. Many countries, especially in eastern Europe, adopted the First Edition and equipped all loudspeaker factories with press-tools, cone tools and machines to make loudspeakers conform to the 1971 standard. Any significant change to the dimensions given in that edition would thus result in the need to replace a large amount of expensive equipment, and possibly significant disruption of production, both of the loudspeakers and of the equipment in which they are used.

Nevertheless, a sufficient number of countries were in favour of a revision, which proved to be possible without making unacceptable changes to the requirements, and this was published as the Second Edition of IEC 60268–14[8] in 1980. The majority of currently mass-produced loudspeakers either comply precisely with this standard or are close to compliance. Sixteen countries voted explicitly in favour of publication of the Second Edition.

(c) Performance requirements for household high fidelity

Work on this subject began in the IEC in 1968, and it soon became clear that the treatment of loudspeakers would be difficult and contentious, not least because the *de facto* standard at the time was the German national standard *DIN 45500* Part 7[10], which was the subject of considerable criticism by experts in other countries. A particular difficulty with the German standard was its attempt to obtain correlation between subjective assessment, in representative real rooms, of the balance between low-frequency and high-frequency sensitivity on the one hand, and measured frequency response, on the other, by measuring the frequency response when the loudspeaker was mounted with its front surface flush with that of a very large supplementary flat baffle. This results in the loudspeaker radiating into half-space (a solid angle of 2π steradians), which could be considered reasonably representative, at low frequencies, of the situation where a loudspeaker is mounted on a bookshelf or wall unit; a situation which was very common in Germany at that time (and still is) but which is not so common in other countries, and is far from ideal for achieving the best overall results. Where space is available in the listening room, free-standing loudspeakers are usually preferrred, although there are some special designs which turn the proximity of the room boundaries to advantage, at low frequencies at least.

The objective measurement suffers in any case from a significant defect (see also Chapter 12). The presence of the supplementary baffle modifies the axial frequency response not only at low frequencies but also, by altering the diffraction which would normally occur at the edges of the enclosure, at higher frequencies. The important off-axis responses are often even more profoundly altered. This problem cannot be overcome by mounting the loudspeaker with its *back* surface flush with the baffle, because reflections from the baffle cancel the forward radiation when the enclosure depth is an odd multiple of half a wavelength, resulting in the measured response being a multiplication of the true frequency response of the loudspeaker by that of a classical comb filter.

The alternative of requiring the measurement to be carried out invariably under free-field conditions (radiation into a sphere, 4π steradians) was equally unacceptable, both because of the status of the German standard and because loudspeakers, particularly the popular, smaller enclosures, balanced for a flat response under these conditions, sound unacceptably bass-heavy when wall- or shelf-mounted.

Ingenious techniques are now incorporated into some designs to allow the user to adjust the loudspeaker to compensate for such room-related effects, but these are not sufficiently simple or cheaply realized to permit their universal adoption in products complying with the standard, whose expressed intention is to define the borderline between 'high fidelity' and 'non-high fidelity', rather than (uselessly) to attempt to define excellence.

After a great deal of discussion and negotiation, agreement was finally reached. In the standard IEC 60581-7[9], the manufacturer is required to state how the loudspeaker is to be mounted and, consequently, whether it is to be measured in free field or with a supplementary baffle. A decision on this subject has to be made, in any case, during development of the loudspeaker, in order to determine the correct balance between low- and high-frequency response. There remains the problem of the directional response and diffraction effects, and this has been dealt with by considering the half-space frequency response to be valid up to a critical frequency, and the free-field response to be valid at higher frequencies. The critical frequency is inversely proportional to the smaller linear dimension of the front face of the enclosure, considered as a rectangle:

$$f = 360/a \tag{15.1}$$

where a = dimension (m)

This artifice, while difficult to defend in purist terms, provides a practical solution, and any error in the results is likely to be small and restricted to a narrow frequency range; therefore, it is unlikely to have critical consequences.

15.3.2 National standards

While many countries have adopted the IEC standards mentioned above, there are some national standards that are of significant importance.

Australia

The SAA standard AS 1127, Part 5[11],which is otherwise equivalent to IEC 60268–5, contains supplementary clauses on the practical measurement of Thiele-Small parameters.

France

While France has adopted IEC 60268–5, it is a matter of some regret that the French National Committee was unable to accept IEC60581–7 and has produced a national standard, NFC97–405[12], which differs fundamentally in some respects from the international standard.

United States of America

Underwriters' Laboratory standard UL1270[15] contains flammability tests for grille coverings.

EIA standards RS 276A, 278B, 299A and 438 give measurement methods for various characteristics, and mounting dimensions.

AES S2–1984 (ANSI S4.26–1984) recommends methods of specifying the performance of loudspeakers used mainly for sound reinforcement and signalling purposes. Because this was considered to be a new area of standardization, the drafting committee have not followed the terminology and concepts used in other related standards, and the manufacturer is required to specify a large number of characteristics of the product. It remains to be seen whether this standard becomes widely used, and whether it provokes an international standard on the same subject.

References

1. ISO Guide 2: 1996, 'General terms and their definitions concerning standardization and certification', International Organization for Standardization, Geneva (1980).
2. British Standard BS 0: Part 1: 1997, 'A standard for standards', British Standards Institution, London (1991).
3. IEC 60268, *Sound system equipment*, Part 5, 'Loudspeakers', 2nd edition, International Electrotechnical Commission, Geneva (1989).
4. THIELE, A N, 'Loudspeakers in vented boxes', *Proc. IREE (Australia)*, **22**, 487 (1961): republished in *JAES*, **19**, 382 and 471 (1971),
5. SMALL, R H, 'Direct radiator loudspeaker system analysis', *IEEE Trans. Audio Electroacoust.*, **AU-19**, 269 (1971): republished in *JAES*, **20**, 383 (1972).
6. BERMAN, J M and FINCHAM, L R, 'The application of digital techniques to the measurement of loudspeakers', *JAES*, **25**, 370 (1977).
7. IEC 60268, *Sound system equipment*, Part 14, 'Mechanical design features', 1st edition, International Electrotechnical Commission, Geneva (1971).
8. IEC 60268, *Sound system equipment*, Part 14, 'Circular and elliptical loudspeakers; outer frame diameters and mounting dimensions', 2nd edition, International Electrotechnical Commission, Geneva (1980).

9. IEC 60581, *High fidelity audio equipment and systems: minimum performance requirements,* Part 7, 'Loudspeakers', International Electrotechnical Commission, Geneva (1986).

10. DIN 45500, 'Heimstudio-Technik (Hi-Fi) Teil 7 Mindestanforderung an Lautsprecher', Beuth Verlag, Berlin, Cologne, Feb. (1971). Also, draft revision, Entwurf, June (1980).

11. AS 1127, Part 5, 1977: Standards Association of Australia, Sydney (1977).

12. Registered French Standard NFC 97–405, AFNOR, Paris (1977).

13. DIN 45572 Teil 2 (Draft): 'Loudspeakers, systematical classification, nomenclature', Beuth Verlag, Berlin, Cologne (1983).

14. DIN 45578 Teil 1 (Draft): 'Magnet systems for moving coil loudspeakers, definitions and methods of measurement, dimensions', Beuth Verlag, Berlin, Cologne (1983).

15. UL1270 Standard for Radio Receivers, Audio Systems and Accessories, Underwriters' Laboratories Inc., New York (looseleaf publication).

16. AES-2–1984 r 1997 (ANSI S4.26–1984), AES Recommended Practice, 'Specification of loudspeaker components used in professional audio and sound reinforcement', Audio Engineering Society, New York (1984) (under revision).

16 Terminology

J. M. Woodgate

16.1 Introduction

This terminology is not intended to be an exhaustive list of all possible terms that may be used in connection with loudspeakers, partly because such a list would be very long, and partly because there are differences in the terminology used in specialized sub-fields; some of this terminology can only be justified by usage.

This chapter is in two parts. The first part deals with definitions of the types that appear in standard vocabularies such as the International Electrotechnical Vocabulary (IEC 60050-801[1]) and BS 4727 Part 3 Group 08[2], while the second part deals with quantities used in calculations and the symbols usually assigned to them. It has proved possible to submit some of the definitions prepared for this book to the IEC for inclusion in the International Electrotechnical Vocabulary Chapter 801[3].

16.2 Definitions

16.2.1 General terms

acoustic (*adj*)   Pertaining to the sense of hearing or the theory of sounds: operated by sound: (of musical instruments) producing sound directly, without ancillary equipment.
Note   Some authors draw a distinction in meaning between acoustic and acoustical. Such a distinction is not supported by most lexicographers and appears to be of little value.

acoustics (*noun*)   (1) (plural in form but treated as singular) The study of sound and similar vibrations. (2) (treated as plural) Acoustic properties or characteristics (for example, of an enclosure).

acoustic oscillation   Movement of particles in an elastic medium about an equilibrium position.

sound   Acoustic oscillation capable of exciting the sensation of hearing.

noise
(1)   Sound having no clearly-defined frequency components.
(2)   Electrical signal producing such a sound when applied to an electroacoustic transducer.
(3)   Any unwanted sound, or the electrical signal producing it.

white noise Stationary, random noise having a Gaussian probability distribution of amplitude.

Notes

(1) 'Stationary' means that the average properties of the noise do not change with time. A continuous sinusoidal signal is also a stationary signal.

(2) 'Random' means that the amplitude of the signal at any instant cannot be predicted from a knowledge of previous values. A continuous sinusoidal signal clearly does not satisfy this requirement because it is completely predictable.

(3) 'Gaussian' means that the amplitude probability distribution (P) obeys the equation

$$P = \frac{1}{\sqrt{2\pi}} \exp\left(\frac{-Z^2}{2}\right)$$
(16.1)

where

$$Z = X - \frac{\mu}{\sigma}$$
(16.2)

X is the instantaneous amplitude, μ is the mean value and σ is the standard deviation (usually normalized to unity in equation (16.1)).

(4) White noise has the property of containing equal energy in each unit bandwidth. The definition does not preclude the existence of a d.c. component ($\mu \neq 0$), but in practice there must be an upper limit to the spectrum, because otherwise the energy would be infinite. So actually we are invariably dealing with *band-limited white noise* and the bandwidth may be limited at the lower end as well as at the upper.

pink noise Stationary, random noise having equal energy in each unit *fractional* bandwidth.

Note Pink noise is used as a test signal in electroacoustics because its spectrum more closely resembles that of real audio signals (averaged over a long period) than does that of white noise, which contains more high-frequency energy. Pink noise also, by definition, has a level spectrum when measured with octave or third-octave filters.

sound field Region of space containing sound energy.

free (sound) field Sound field in which the effect of the boundary is negligible.

diffuse (sound) field Sound field in which the sound energy density is the same at all points and the mean acoustic power per unit area is the same in all directions.

level The logarithm of the ratio of a value of a quantity analogous to power to a stated or standard reference value.

Notes

(1) Quantities analogous to power include the squares of voltages, currents and pressures.

(2) The definition applies to levels expressed in bel (base of logarithm = 10) or neper (base of logarithm = e). Levels expressed in decibel (base of logarithm = $10^{0.1}$) are naturally 10 times larger numerically than the same levels expressed in bel.

(3) The reference value for sound pressure in air is 20 (Pa), or -94 dB (Pa). The reference value in water is usually 1 Pa.

volume velocity The product of the area of a surface element and the component of the particle velocity normal (perpendicular) to it.
Note The surface may be notional. For instance, a surface may be imagined within a homogeneous body of fluid.

16.2.2 Systems and their elements

acoustic system System designed to generate, transmit, process or receive acoustic energy.

excitation (syn. stimulus) Energy presented to an input port of a system.

response Energy emitted by a system due to an excitation.

transfer function Relationship between response and excitation in which excitation is the independent variable.

transducer Device designed to receive one form of energy and emit a different form, in such a manner that desired characteristics of the received energy (such as those conveying information) are transmitted.

passive transducer A transducer in which the emitted energy is derived entirely from the received energy.

active transducer A transducer in which the emitted energy is derived in part from sources other than the received energy.

reversible transducer A transducer which will function with net energy flow in either direction through it.
Notes
(1) Most transducers are reversible but many appear to be very inefficient when used in the opposite direction to that for which they are designed. In fact, the efficiencies in each direction are not independent.
(2) Net energy flow is the long-term average of the difference between the energy entering a specified port and the energy emitted by it.

linear transducer A transducer in which the emitted energy is strictly (linearly) proportional to the received energy.

reciprocal transducer A linear, reversible transducer.
Note This is the IEV definition. Some authors use this term instead of 'reversible transducer' (q.v.), without implying strict linearity.

16.2.3 Concepts for calculating, measuring and specifying behaviour and performance

sensitivity Quotient of a specified quantity describing the response of a system by another specified quantity describing the excitation.
Note For a loudspeaker, the response is usually expressed in terms of sound pressure, and the excitation is then preferably expressed in terms of the input voltage, since this is a quantity of the same kind as sound pressure and there is a quasi-linear relationship between them. The sensitivity expressed in this way is substantially constant with respect to amplitude, and it is usually a design goal to make it substantially independent of signal frequency as well. Other logically satisfactory pairs of quantities may also be used to express sensitivity, such as sound power and the corresponding electrical power in the rated impedance:

$$P = V^2/R \qquad\qquad (16.3)$$

where V is the applied voltage and R is the *rated* impedance (q.v.). Because the *actual* impedance of most types of loudspeaker varies considerably with frequency, both in magnitude and phase, the *actual* power input is not appropriate as a factor in determining sensitivity. Similarly, since a loudspeaker is normally supplied from a signal source whose internal impedance is low compared with the rated impedance of the loudspeaker, the current flowing through the loudspeaker is also not appropriate for the specification of sensitivity.

mechanical impedance   Quotient of the force at a point in a mechanical system by the velocity in the direction of the force at its point of application.
Notes
(1) In the frequency domain the impedance is in general a complex quantity (one having real and imaginary parts).
(2) Mechanical systems exist in which the application of a force results in motion normal to the applied force. Such mechanical systems do not often occur in electroacoustics, but electromagnetic transducers exhibit analogous behaviour, which is termed 'gyroscopic'.

driving-point impedance   The impedance at the point of application of an excitation.

load   A port of a system or an energy sink intended to receive energy from another system.
Note   The concept includes loads whose impedances are zero or infinite, which can therefore receive only infinitesimal energy from a finite system.

loaded impedance   The driving-point impedance of a system when its output is presented with a load of specified impedance.

blocked impedance   The driving-point impedance of a system when its output is presented with a load of infinite impedance.
Note   A load of infinite impedance is one which permits no flow; thus current, velocity or volume velocity is zero. Great care is necessary with this and allied concepts if mobility analogies are used.

free impedance   The driving-point impedance of a system when its output is presented with a load of zero impedance.
Note   A load of zero impedance is one which supports no tension; thus potential difference, force or sound pressure is zero. Great care is necessary with this and allied concepts if mobility analogies are used.

motional impedance   The difference between the loaded impedance and the blocked impedance of a transducer.

mechanical admittance   Reciprocal of the mechanical impedance.

mechanical immitance   Pantechnicon expression for impedance or admittance.

mechanical resistance   Component of the mechanical impedance in which energy can be dissipated.
Note   With sinusoidal excitation the force applied to a mechanical resistance is in phase with the resulting velocity, thus resistance is represented by a real quantity in the frequency domain.

mechanical reactance Component of the mechanical impedance in which energy can be stored in kinetic or potential form.
Note With sinusoidal excitation the force applied to a mechanical reactance is in quadrature with the resulting velocity, thus reactance is represented by an imaginary quantity in the frequency domain.

stiffness (mechanical) Quotient of change of force by change of displacement of an elastic element.
Note Energy is stored in an elastic element in the form of potential energy. Such an element therefore possesses mechanical reactance. With sinusoidal excitation the force and velocity are in quadrature, with the phase of the velocity leading.

compliance (mechanical) The reciprocal of stiffness.

mass (mechanical) The quantity of matter present in a body.
Note Energy is stored in a moving mass in the form of kinetic energy. A mass therefore possesses mechanical reactance. With sinusoidal excitation the force and velocity are in quadrature, with the phase of the velocity lagging.

electromechanical transducer Transducer designed to receive electrical excitation and provide a mechanical response.
Note The term may also be used for a transducer which receives a mechanical excitation and provides an electrical response. However, some transducers which provide a mechanical response to an electrical excitation are not reciprocal, and it is therefore preferable to use the term *mechanoelectrical* for transducers having mechanical input and electrical output.

electromechanical force factor Quotient of the output force of the blocked mechanical system by the input current of the electrical system.
Note For a passive reciprocal transducer, the quotient of the open-circuit output voltage of the electrical system by the input velocity of the mechanical system is numerically equal to the electromechanical force factor: if the coupling is gyroscopic (as in an electrodynamic transducer), the sign is reversed.

acoustic impedance Quotient of the sound pressure at a surface (which may be notional) by the volume velocity through the surface.
Notes
(1) In the frequency domain the impedance is in general a complex quantity (one having real and imaginary parts).
(2) Acoustic systems which exhibit gyroscopic behaviour are rare, but some electroacoustic transducers behave in an analogous manner. (See 'Mechanical impedance'.)

acoustic admittance Reciprocal of the acoustic impedance.

acoustic resistance Component of the acoustic impedance in which energy can be dissipated.
Note With sinusoidal excitation the pressure applied to an acoustic resistance is in phase with the resulting volume velocity, thus resistance is represented by a real quantity in the frequency domain.

acoustic reactance Component of the acoustic impedance in which energy can be stored in kinetic or potential form.
Note With sinusoidal excitation the pressure applied to an acoustic reactance is in quadrature with the resulting volume velocity, thus reactance is represented by an imaginary quantity in the frequency domain.

specific acoustic impedance Quotient of the sound pressure at a point by the particle velocity at that point.
Note In the frequency domain the impedance is in general a complex quantity (one having real and imaginary parts).

specific acoustic resistance Component of the specific acoustic impedance in which energy can be dissipated.

specific acoustic reactance Component of the specific acoustic impedance in which energy can be stored.

characteristic impedance of a medium Quotient of the instantaneous sound pressure by the instantaneous particle velocity in a free plane-progressive wave.
Note The characteristic impedance is equal to the product of the density of the medium and the speed of sound in it.

acoustic mass (1) An element in which acoustic energy may be stored in kinetic form. (2) Quotient of the magnitude of the impedance of such an element at a given frequency by the angular frequency.
Note This implies that for sinusoidal excitation the volume velocity through the element is in lagging quadrature with the sound pressure across it.

acoustic stiffness (1) An element in which acoustic energy may be stored in potential form. (2) Product of the magnitude of the impedance of such an element at a given frequency and the angular frequency.
Note This implies that for sinusoidal excitation the volume velocity through the element is in leading quadrature with the sound pressure across it.

acoustic compliance (1) Synonym of 'acoustic stiffness (1)'. (2) Reciprocal of 'acoustic stiffness (2)' (q.v.).

electroacoustic transducer Transducer designed to receive electrical excitation and provide an acoustic response.
Note The term may also be used for a transducer which receives an acoustic excitation and provides an electrical response. However, some common transducers which provide an acoustic response to an electrical excitation are not reciprocal, and it is therefore preferable to use the term *acoustoelectrical* for transducers having acoustic input and electrical output. Normally, of course, the term 'microphone' or 'hydrophone' would be used.

electroacoustic force factor Quotient of the output sound pressure of the blocked acoustic system by the input current of the electrical system.
Note For a passive reciprocal transducer, the quotient of the open-circuit output voltage of the electrical system by the input volume velocity of the acoustic system is numerically equal to the electroacoustic force factor: if the coupling is gyroscopic the sign is reversed. Note that the electroacoustic coupling in an electrodynamic transducer is gyroscopic.

16.2.4 Loudspeakers

loudspeaker An electroacoustic transducer intended to produce sound to be heard at a distance from the transducer.
Note The term is also used as a synonym for 'drive unit' (q.v.), which can be very confusing. In an attempt to avoid this problem, the second edition of IEC 60268-5[4] uses the term 'loudspeaker system' to emphasize the distinction.

drive unit   A basic electroacoustic transducer forming an article of commerce and used in conjunction with other parts (such as baffles, enclosures, horns and electronic elements) to construct a loudspeaker or loudspeaker system.

loudspeaker system   An assembly of one or more drive units with other parts (such as baffles, enclosures, horns and electronic elements), forming a loudspeaker.

multi-unit loudspeaker   A loudspeaker assembly including two or more drive units.

dividing network (syn. crossover network)   An electrical filter, having more than one pair of output terminals, designed to distribute signals of different frequencies to the appropriate drive units of a multi-unit loudspeaker.

reference plane   A plane specified by the manufacturer, with respect to some physical feature of the loudspeaker or drive unit, in order to define the reference point and the direction of the reference axis.
Notes
(1) For symmetrical structures, the reference plane should preferably be specified to be parallel or tangential to the radiating surface or to a plane defining the front of the drive unit or loudspeaker system. For asymmetrical structures, the position of the plane may be indicated by means of a diagram.
(2) In strict terminology, this term should be 'Rated reference plane', as it is to be specified by the manufacturer and cannot be independently measured. (See 'rated value' (1).)

reference point   A point in the reference plane, specified by the manufacturer. It is the point of intersection of the reference axis with the reference plane.
Note   Strictly, this term should be 'rated reference point', as it is to be specified by the manufacturer and cannot be independently measured. (See 'rated value' (1).)

reference axis   A line passing through the reference plane at the reference point in a direction specified by the manufacturer. It is used as the zero reference axis for directional and frequency response measurements.
Notes   (1) For symmetrical structures, the reference axis is usually normal to the radiating surface or to the reference plane. (2) Strictly, this term should be 'rated reference axis', as it is specified by the manufacturer and cannot be independently measured. (See 'rated value' (1).)

rated value (1) (syn. rated condition)   A value specified by the manufacturer to establish a specified operating condition.
Note   Rated values in this sense are not subject to verification by measurement.

rated value (2) (syn. rated value of a characteristic)   The value, specified by the manufacturer, of a characteristic, based on measurements by standard methods under rated conditions.

rated impedance   The value, specified by the manufacturer, of pure resistance to be substituted for the loudspeaker to define the available electric power of the source.
Note   For almost all types of loudspeaker, the *actual* impedance varies considerably with frequency, and the sound output is proportional, not to the input current but to the input voltage. For a loudspeaker assembly supplied with signals from a source whose impedance is low compared with the lowest value of the actual impedance of the loudspeaker (at any frequency), the *actual* impedance is only of consequence for determining the necessary maximum output current capability of the signal source. For a drive unit, the actual impedance is also important in the design of any associated dividing network. For a multi-unit loudspeaker with a passive dividing network,

the impedance may fall to very low values at some frequencies, because parts of the network behave as a series-tuned circuit or an impedance-transforming filter. Under conditions of successive sharp reversals in signal polarity at critically unfavourable time intervals, the current demanded by the loudspeaker may be much greater than would be expected from measurements with sinusoidal signals[5], but opinions differ as to the incidence of such signals in real programme material. There can be no standard method of measuring this effect, since the characteristics of the necessary (most critical) test signal depend on the precise characteristics of the drive units and dividing network. Approximations to the most critical signal can be deduced theoretically from the circuit values and the electrical equivalent circuits of the drive units, but the process may be complex and tedious. It is possible to design for the avoidance of significant effects of this nature, and to provide very adequate output current capability in amplifier design, and these measures can result in satisfactory performance.

16.2.5 Constructional features of drive units

chassis (syn. bucket)   The supporting structure of a drive unit.

diaphragm   The vibrating, radiating element of a drive unit, especially of the electrodynamic type. The diaphragm may be described as a *cone* or a *dome,* or *planar,* according to its shape.

surround   An annular structure with radial stiffness but very little axial stiffness, connecting the outer edge of the diaphragm to the surrounding chassis.

suspension   An annular structure with radial stiffness but little axial stiffness, connecting the inner edge of the diaphragm of an electrodynamic drive unit, at or near its junction with the former of the voice-coil, to the chassis.
Notes
(1) In modern designs the suspension usually lies outside the diameter of the voice-coil former, but in early designs it was located inside the former at its junction with the diaphragm, and attached to the pole-piece at its centre. Due to the shape, such a suspension is termed a *spider.*
(2) Usually the stiffness of the suspension is considerably greater than that of the surround and, for an unmounted drive unit, provides the major part of the force restoring the diaphragm to its rest position. In a loudspeaker assembly, the stiffness of the air in an enclosure may provide a major part of the restoring force, under dynamic conditions.

voice-coil   The coil through which the signal current flows.

pole-piece   The cylinder of magnetic material lying inside the voice-coil and concentric with it.

top-plate   The sheet of magnetic material through which the voice-coil passes, and which forms the magnetic pole opposite to the pole-piece.

16.2.6 Constructional features of loudspeakers

baffle   A structure intended to separate the forward and rearward radiations of a drive unit.

infinite baffle (syn. closed box)   An enclosure which has rigid walls and is substantially airtight.
Note   A slow air-leak (anti-aneroid) is desirable, to allow equalization across the enclosure boundary of atmospheric pressure changes. However, the effective acoustic mass of such a leak must be very large.

bass-reflex enclosure (syn. vented box) An enclosure provided with openings for acoustic radiation both directly from a drive unit, and for the rear radiation of the unit in a certain frequency range, over part of which the radiation originating from the rear of the drive unit tends to reinforce the forward radiation.

labyrinth An enclosure in which the rear radiation of a drive unit is absorbed in an acoustic resistance. The term refers to the construction of the acoustic resistance as a folded pipe lined with absorbing material.

double-chamber enclosure (from German 'doppelkammer') An enclosure divided into two parts by a partition having either an opening acting as an acoustic mass or a passive diaphragm, providing acoustic coupling between the two parts.

passive diaphragm A diaphragm with a surround and, perhaps, a suspension, acting as an acoustic component having mass and compliance.

passive radiator (syn. ABR, auxiliary bass radiator) A passive diaphragm providing useful acoustic radiation.

vent An opening in the wall of an enclosure for the emission of sound other than directly from a drive unit.

port (1) syn. of vent. (2) Combination of vent and tunnel.

tunnel Structure extending the depth of a vent, normally towards the interior of the enclosure. This is usually intended to increase the acoustic mass.

distributed vent A vent formed by a number of adjacent, relatively small, holes. It offers an acoustic impedance whose ratio of mass to resistance can be adjusted by changing the size and number of the holes.

16.3 Quantities and symbols

The inclusion of a table of symbols is characteristic of many papers on loudspeaker technology, and has been usual from a very early date. It is a particularly convenient practice, because the number of symbols used is often large, and numerous super- and subscripts are used, in addition to many Greek letters and boldface type.

16.3.1 Basic symbols

The following symbology is in common use[6-8] but naturally each paper requires particular usage or variant forms. Nevertheless, the use of *de facto* standard symbols is a considerable aid to the rapid comprehension of a paper.

These basic symbols are extended by the addition of subscripts and primes. Some of the subscript notation is used consistently by different authors, and is detailed in Section 16.3.2 below.

a	radius, especially of a diaphragm.
A	area.
b	radius; dimension.
B	magnetic flux density; susceptance.
c	speed of sound.
C	capacitance.
C_m	mechanical compliance.
C_a	acoustic compliance.

d	distance.
E	e.m.f.; modulus of elasticity.
f	frequency.
F	force.
G	response function; conductance.
h	tuning ratio: in a vented box system, the ratio of box resonant frequency to driver resonant frequency.
H	function; magnetizing force.
i	instantaneous current.
I	current; modified Bessel function.
j	imaginary operator.
J	Bessel function.
k	constant.
K	Bessel function.
l	dimension.
L	inductance.
M_a	acoustic mass.
M_m	mechanical mass.
p	pressure; complex frequency $(s + j\omega)$ in the Laplace transform.
P	power.
Q	(1) quality factor of a resonant system (ratio of energy stored to energy dissipated per cycle).
	(2) directivity factor.
	(3) electric charge.
	Note. The use of the symbol Q for all of these quantities is sometimes impossible and may often cause confusion. In case of conflict, the symbol γ has been used for directivity factor. There is no widely used alternative symbol for quality factor.
r	radius.
R	resistance.
s	complex frequency $(\sigma + j\omega)$ in the Laplace transform.
S	area, especially S_D, area of diaphragm.
t	time.
T	time constant.
U	volume velocity; potential difference.
V	volume.
x	displacement, especially of a diaphragm.
X	reactance; displacement function.
Y	admittance; Bessel function.
Z	impedance.
α	compliance ratio: ratio of drive unit compliance to enclosure volume compliance.
β	flaring index of exponential horn.
γ	ratio of adiabatic to isothermal specific heat of a gas; directivity factor.
η	efficiency.
λ	wavelength of radiation.
μ	magnetic permeability.
ζ	displacement.
ρ	density.
ω	angular frequency.

16.3.2 Subscripts

Dual subscripts are commonly used because of the large number of quantities of the same type involved in some calculations. While the assignment of meanings to these subscripts is not completely consistent, the following represents a guide to common practice.

Where an asterisk is shown in the list, no usage is common.

Symbol	1st sub.	2nd sub.
a, A	acoustic; air	*
b, B	*	box
d, D	diaphragm	diaphragm
e, E	electric(al)	electric(al)
m, M	mechanical	*
p, P	port; vent	port; vent
s, S	drive unit	suspension
t, T	total	test conditions

16.3.3 Representation of instantaneous, peak and r.m.s. values

It is conventional, especially in electrical engineering, to use a lower-case symbol to represent the instantaneous value of a time-dependent quantity, an upper-case symbol with a circumflex accent ($\hat{}$) to represent the peak value, and the plain upper-case symbol to represent the r.m.s. value. While this is a valuable convention in some circumstances, great care must be taken to avoid ambiguity, because upper- and lower-case letters are also used as symbols for *different* quantities. An important case is the symbol p, where the lower-case represents sound pressure and the upper-case represents sound power.

References

1. International Electrotechnical Vocabulary, (IEC 60050-801) *Electroacoustics,* International Electrotechnical Commission, Geneva (1960).
2. British Standard BS 4727 Part 3 Group 08:1995, *Glossary of electrotechnical, power, telecommunications, electronics, lighting and colour terms,* 'Acoustics and electroacoustics terminology', British Standards Institution, London (1995).
3. International Electrotechnical Vocabulary, Chapter 801 'Acoustics and electroacoustics', International Electrotechnical Commission, Geneva (1984).
4. IEC 60268-5 *Sound System Equipment,* Part 5 'Loudspeakers', 2nd edition. International Electrotechnical Commission, Geneva.
5. MARTIKAINEN, I, VARLA, A, and OTALA, M, 'Input current requirements of high quality loudspeaker systems', AES 73rd Convention, NL, 1983. Preprint 1987.
6. *Loudspeakers: an Anthology,* Audio Engineering Society, New York (1978).
7. MCLACHLAN, N W, *Loudspeakers,* Clarendon Press, Oxford (1934).
8. BERANEK, L L, *Acoustics,* McGraw-Hill, New York (1954).
9. OTALA, M and HUTTUNEN, P, 'Peak current requirement of commercial loudspeaker systems', *JAES,* **35**, 455 (1987).

Index

Printed in the United Kingdom
by Lightning Source UK Ltd.
125053UK00001B/13-18/A